Joseph M. Moran

OCEAN STUDIES

Introduction to Oceanography, Third Edition

Education Program

American Meteorological Society
Boston, MA

The American Meteorological Society Education Program

The American Meteorological Society (AMS), founded in 1919, is a **scientific and professional** society. Interdisciplinary in its scope, the Society actively promotes the development and dissemination of information on the atmospheric and related oceanic and hydrologic sciences. AMS has more than 13,000 professional members from more than 100 countries and over 175 corporate and institutional members representing 40 countries.

The Education Program is the initiative of the American Meteorological Society fostering the teaching of the atmospheric and related oceanic and hydrologic sciences at the precollege level and in community college, college and university programs. It is a unique partnership between scientists and educators at all levels with the ultimate goals of (1) attracting young people to further studies in science, mathematics and technology, and (2) promoting public scientific literacy. This is done via the development and dissemination of scientifically authentic, up-to-date, and instructionally sound learning and resource materials for teachers and students.

AMS Ocean Studies is an introductory undergraduate oceanography course offered partially via the Internet in partnership with college and university faculty. **AMS Ocean Studies** provides students with a comprehensive study of the principles of oceanography while simultaneously providing pedagogically appropriate investigations and applications focusing on web-delivered real-world current data. It provides real experiences demonstrating the value of computers and electronic access to time-sensitive data and information.

Ocean Studies: Introduction to Oceanography / Joseph M. Moran — 3rd Edition
ISBN-10 1-878220-48-9 ISBN-13 978-1-878220-48-6

Published by the American Meteorological Society
45 Beacon Street, Boston, MA 02108

Printed in the United States of America

The first edition of this book was published under the title Online Ocean Studies.

Cover: Large waves characterize the Southern Ocean surrounding Antarctica. Photo courtesy of Maria Stenzel/National Geographic Stock.

BRIEF CONTENTS

CONTENTS

CHAPTER 1 OCEAN IN THE EARTH SYSTEM 1

CHAPTER 2 OCEAN BASINS AND PLATE TECTONICS 33

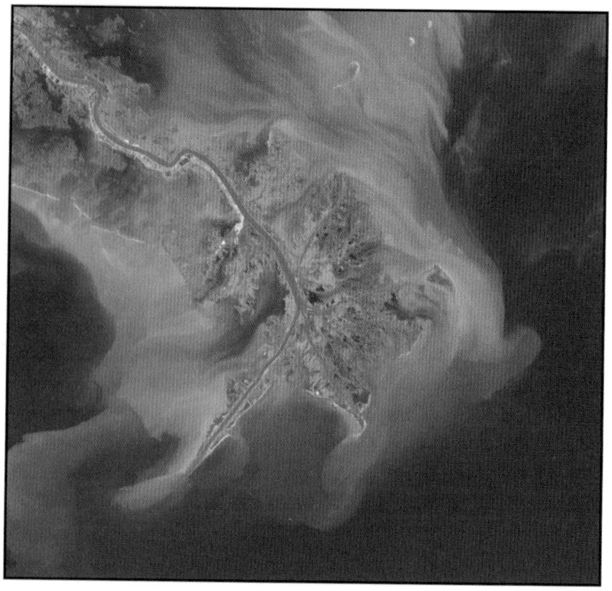

CHAPTER 5 THE ATMOSPHERE AND OCEAN 121

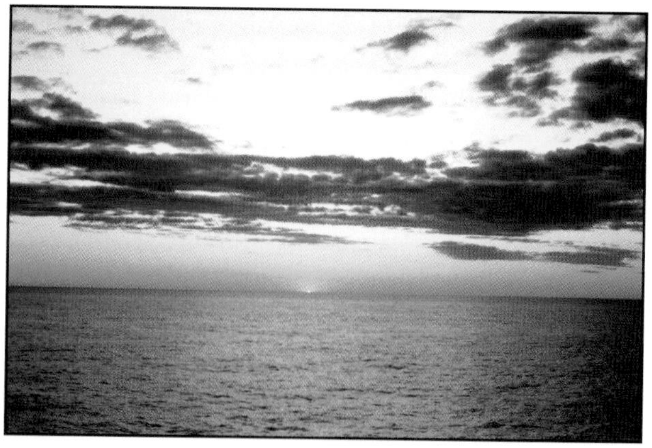

CHAPTER 8 THE DYNAMIC COAST 215

CHAPTER 9 MARINE ECOSYSTEMS 253

CHAPTER 10 LIFE IN THE OCEAN 283

CHAPTER 11 THE OCEAN, ATMOSPHERE, AND CLIMATE VARIABILITY 319

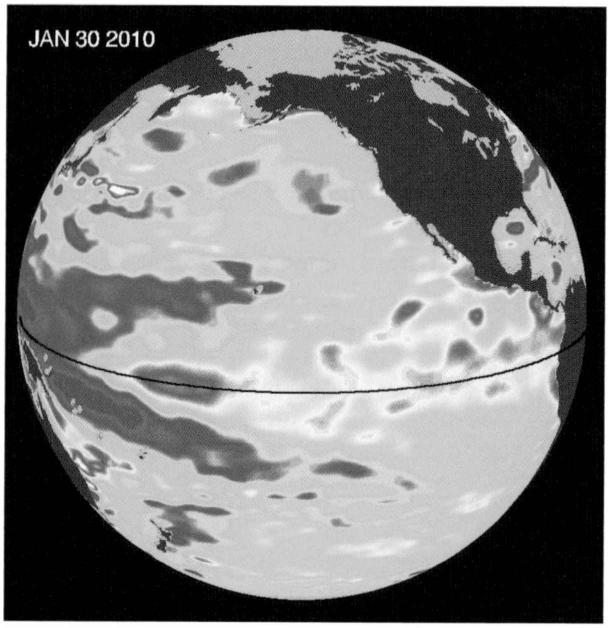

CHAPTER 12 THE OCEAN AND CLIMATE CHANGE 345

CHAPTER 13 THE FUTURE OF OCEAN SCIENCE 391

CHAPTER 14 OCEAN STEWARDSHIP 417

PREFACE

Welcome *to Ocean Studies*! You are about to embark on an analysis of the world ocean and the role of the ocean in the Earth system. *Ocean Studies* is a major initiative of the American Meteorological Society (AMS) and the National Oceanic and Atmospheric Administration (NOAA) Cooperative Program for Earth System Education (AMS/NOAA CPESE). *Ocean Studies* explores the ocean in the Earth system with special emphasis on the flow and transformations of water and energy into and out of the ocean, physical and chemical properties of seawater, ocean circulation, marine life, its habitats and adaptations, interactions between the ocean and the other components of the Earth system (i.e., hydrosphere, atmosphere, geosphere, cryosphere, and biosphere), and the human/societal impacts on and response to those interactions. The purpose of this book is to provide you with background information on the physical, chemical, geological, and biological foundations of oceanography.

Ocean Studies may serve as a stand-alone textbook in an undergraduate college course on oceanography. It is the reference book for *AMS Ocean Studies*, a turnkey course package developed, licensed, and nationally implemented by AMS. Each chapter corresponds to one week of the *AMS Ocean Studies* course. Companion *Investigations Manual* and *Ocean Studies Website* provide students with twice-weekly investigations on ocean science partially delivered via the Internet. The course can be offered in face-to-face, blended, or totally online instructional environments.

AMS Ocean Studies is guided by new findings of learning science that redefine what it means to be proficient in science. According to the National Research Council of the National Academies, Board on Science Education (2007), students who are proficient in science "(1) know, use, and interpret scientific explanations of the natural world, (2) generate and evaluate scientific evidence and explanations, (3) understand the nature and development of scientific knowledge, and (4) participate productively in scientific practices and discourse." These strands of proficiency are learning goals in *AMS Ocean Studies* that address the knowledge and reasoning skills essential for students to become proficient in oceanography and participate as ocean science literate citizens.

The course package, *AMS Ocean Studies*, follows learning science in providing strategically designed "student encounters with science that take place in real time and over a period of months and years (e.g., *learning progressions*)." *AMS Ocean Studies* seeks to engage learners in exploring their world by investigating meaningful questions. Investigations, tied directly to each chapter, have printed and electronic components that make use of oceanographic information/data available via the *Ocean Studies Website*. Investigations engage participants in observation, prediction, data analysis, inference, and critical thinking. The course presents opportunities for students to collaborate with their instructor and fellow students as together they negotiate understanding. Application of information-age technology provides the student with experience in retrieving and analyzing real-world data (some in real-time) and sharing interpretations. Throughout the course, students assemble learning materials for assessment purposes.

Students of *AMS Ocean Studies* explore twelve principal themes plus three optional themes arranged by chapter, each corresponding to one week of the course. Themes are organized so that concepts build logically upon one another as the ocean and its role in the Earth system are demonstrated to follow patterns described by physical laws. The opening chapter introduces the Earth system and examines the ocean's place in the Earth system. Chapter 2 focuses on the characteristics of the ocean basin that are largely the products of plate tectonics. Chapter 3 deals with the unique physical and chemical properties of seawater, and Chapter 4 covers the origin and distribution of marine sediments. Chapter 5 investigates air-sea interactions and the flow of heat energy in the Earth-atmosphere-ocean system. The next three chapters cover the dynamic ocean: surface and deep-ocean currents (Chapter 6), waves and tides (Chapter 7), and shoreline processes (Chapter 8). Chapters 9 and 10 examine marine ecosystems and life in the ocean. The next two chapters explore the ocean's role in short-term climate variability (Chapter 11) and long-term climate change (Chapter 12). The remaining chapters consider the future of ocean science (Chapter 13), ocean stewardship (Chapter 14), and public policy and the ocean (Chapter 15).

Each chapter opens with a *Case-in-Point*, an authentic event or issue that highlights or applies one or more of the main concepts introduced in the chapter. In essence, the Case-in-Point previews the chapter and engages reader interest in the topic early on. Chapter 9, for example, opens with a discussion of harmful algal blooms in coastal waters. The *Case-in-Point* is followed by a *Driving Question*, a broad-based query that links chapter concepts and provides a central focus from the beginning of the chapter. Content is science-rich and informs additional driving questions. Each chapter closes with a list of *Basic Understandings*, *Enduring Ideas* (new to this edition), review questions, and critical thinking questions. *Essays* at the end of each chapter address in some depth a specific topic that builds on a concept introduced in the narrative. Examples include *Ocean Acidification*, *Burgess Shale: A Glimpse into Ancient Marine Life*, *The State of the Sea, Moving the Cape Hatteras Lighthouse,* and *Monitoring the Ocean with Animal-Borne Instruments*. All bold-faced terms are defined in the *Glossary* at the back of the book.

This third edition continues our emphasis on the ocean in the Earth system, the ocean in Earth's climate system, plus methods of in situ and remote sensing of the ocean. New, expanded, or significantly revised topics include the origin of seawater, hydrothermal vents, seawater pH and ocean acidification, marine sediments, meridional overturning circulation (MOC), marine life and the ozone shield, global radiation budget, large-scale ocean circulation, water wave generation, rogue waves, the tsunami hazard, monitoring sea level from space, methane hydrates, harmful algal blooms (HABs), coastal dead zones, dams and marine habitats, marine protected areas, threats to coral reefs, threatened and endangered species, the ocean as a reservoir in the global carbon cycle, lessons of the climate record, climate change and the climate future, global warming and rising sea level, trends in Arctic sea ice cover, El Niño and La Niña, beaches and barrier islands, coastal storms, restoration of Chesapeake Bay, remote sensing of the ocean, responding to marine oil spills, and scientific ocean drilling. This edition includes new Case-in-Points on the 2010 Deepwater Horizon oil spill, the Japanese earthquake and tsunami of 2011, polar bears as a threatened species, and lost wetlands. New Essays examine the future of the Nile Delta and the Atacama hyper-arid coastal desert. This edition also features many new full-color photographs and line art.

AMS Ocean Studies learning materials are the products of collaboration among many individuals with extensive scientific backgrounds and teaching experience. The first edition of this textbook was derived from a manuscript co-authored by M. Grant Gross of Washington College (MD) and Elizabeth Gross of the Scientific Committee on Oceanic Research, International Council for Science. Joseph M. Moran, Associate Director of the AMS Education Program and Professor Emeritus at the University of Wisconsin-Green Bay was managing editor for editions one and two and is principal author for this third edition. Elizabeth W. Mills, Associate Director of the AMS Education Program, had a major editorial role in locating reviewers, interpreting and synthesizing their suggestions, finding appropriate figures, and assembling the final manuscript for internal consistency and publication. Bernard Blair, Education Publications & Instructional Technology Manager of the AMS Education Program, met the numerous technical challenges in formatting the final manuscript into this book with his exceptional skill, attention to detail, dedication, and perseverance. Katie O'Neill, Content Specialist with the AMS Education Program, provided editorial skills that were particularly important to the effort.

Textbook development was a team effort that benefited greatly from suggestions and constructive criticisms provided by the professional staff of the AMS Education Program: James A. Brey, Director, Ira W. Geer, Senior Education Fellow, Robert S. Weinbeck, Associate Director (also of SUNY College at Brockport), Edward J. Hopkins of the University of Wisconsin-Madison, Heather Hyre, Kira Nugnes, and Maureen Moses. Each individual brought special expertise and perspective to the task. For their contributions and encouragement, thanks are due David R. Smith and Andrew C. Muller of the U.S. Naval Academy, and Ronald D. Stieglitz of the University of Wisconsin-Green Bay. A special thank you is extended to Wolfgang H. Berger, Professor of Oceanography Emeritus, Scripps Institution of Oceanography, UCSD, for his thorough, insightful, and most helpful review of the text content prior to the revision.

We are grateful for the interest, encouragement, and assistance provided by David Kennedy, Assistant Administrator for NOAA's National Ocean Service, John Oliver, NOAA Fisheries Deputy Assistant Administrator for Operations, and Peg Steffen, Educational Coordinator, NOAA's National Ocean Service. Many thanks to Stephen Gill, Bruce Moravchik, and Carol Kavanagh of NOAA's National Ocean Service, and Becky Allee and Heather Stirratt of NOAA's National Marine Fisheries Service. We extend a special thanks to the following persons for their thorough chapter reviews: from the NOAA Fisheries Office of Science and Technology: Kristan Blackhart, Fisheries Biologist, Laura Oremland, Marine Biologist, and

Kenric Osgood, Chief, Marine Ecosystems Division; and from the NOAA Fisheries Office of Sustainable Fisheries: Wendy Morrison, Fishery Policy Analyst. We greatly appreciate the continued support from Louis Uccellini, Director of NOAA's National Centers for Environment Prediction (NCEP), Ben Kyger, Head of NCEP Central Operations, and their colleagues in providing current Earth science data and visualizations used throughout the AMS courses.

We are most grateful for the assistance with this project that was enthusiastically provided by Leslie Peart, Consortium for Ocean Leadership/Deep Earth Academy, Bill Crawford, Integrated Ocean Drilling Program-U.S. Implementing Organization/Texas A&M University, Kim Fulton-Bennet, Monterey Bay Aquarium Research Institute, Kevin Schabow, NOAA Chesapeake Bay Office, Howard Diamond, Ethan Gibney, Paula Hennon, Kenneth Knapp, Michael Kruk, and Carl Schreck, NOAA National Climatic Data Center, Paula Dunbar, NOAA National Geophysical Data Center, Eric Blake, James Franklin, and Chris Landsea, NOAA National Hurricane Center, Molly Harrison, NOAA National Marine Fisheries Service, Kristen Crossett, Karen Kavanaugh, and Darren Wright, NOAA National Ocean Service, Tomohiko Fukushima, The University of Toyko, Shelley Martin, U.S. Department of Energy/National Energy Technology Laboratory, Jonathan Cogan and Amy Sweeney, U.S. Energy Information Administration, Lisa Ramirez Rukstales and Lisa Wald, U.S. Geological Survey, John Clarke, U.S. Geological Survey/Georgia Water Science Center, Joellen Russell, University of Arizona, Michael Ledbetter, University of Arkansas at Little Rock, Andy Coburn, Western Carolina University, Stace Beaulieu, Porter Hoagland, and Mary Zawoysky, Woods Hole Oceanographic Institution, the Woods Hole Oceanographic Institution Information Office, the Food and Agriculture Organization of the United Nations, NOAA National Data Buoy Center, and U.S. Geological Survey Gas Hydrates Project. Norman J. Frisch of Brockport, NY, did an excellent job of turning line drawings into final art. Special thanks are extended to the group of outstanding K-12 teachers who serve as AMS Educational Resource Associates (AERAs) for providing valuable advice and encouragement during the development, national implementation, and revision of *AMS Ocean Studies* learning materials.

A note concerns the use of units in *Ocean Studies*. Generally, the International System of Units (abbreviated SI, for Systèm Internationale d'Unitès) is employed with equivalent English or other units following in parentheses. Exceptions are units used by convention or convenience in oceanography (e.g., pressure is given in units of decibars). Also, the equivalence between units is given in context; that is, where general estimates are given, approximate values are shown for all units. Conversion factors are in Appendix I and an ocean timeline is in Appendix II.

Ocean Studies is a major initiative of AMS/NOAA CPESE, a program designed to enhance public understanding of the fluid Earth system emphasizing the atmospheric, oceanic, and hydrologic sciences and to promote activity that will contribute to greater human resource diversity in the nation's scientific workforce. Through CPESE, the AMS assists NOAA in the advancement of its goals directed toward environmental assessment and prediction, protection of life and property, and the fostering of global environmental stewardship. NOAA's success in meeting its mission objectives is highly dependent upon synergistic relationships between it and the users of its products and services. CPESE nurtures this synergy through precollege teacher, introductory undergraduate and general educational activity. Fundamental to CPESE are (1) *breadth*, demonstrating the comprehensive need for describing and predicting changes in the Earth's environment and conserving and wisely managing the nation's coastal and marine resources; (2) *visibility*, increasing public awareness of the ways environmental assessment, prediction, and stewardship touch the lives of all Americans every day; and (3) *diversity*, promoting educational activity and outreach to attract members of groups underrepresented in science, technology, engineering and mathematics to study and consider careers in those fields, including those for which NOAA has employment needs and opportunities. For additional information on CPESE and the AMS Education Program, go to http://www.ametsoc.org/amsedu.

James A. Brey, Ph.D.
Director, AMS Education Program

CHAPTER 1

OCEAN IN THE EARTH SYSTEM

The ocean continuously interacts with the other components of the Earth system. [Courtesy of NOAA]

Case in Point

On 11 March 2011, a succession of tsunami waves struck the northeast coast of Japan's main land mass and largest island, Honshu, decimating the city of Sendai and numerous other coastal communities. The tsunami was caused by the largest earthquake to affect Japan since instrumental recording began 130 years ago, and the fourth largest earthquake in the world since 1900. At the time of this writing, there were 15,093 deaths and 9,093 missing in Japan. Another very powerful tsunami devastated coastal areas along the Indian Ocean on 26 December 2004. That death toll of 227,900 came from 14 countries, and was, as noted in the *Tsunami Evaluation Coalition Synthesis Report*, the greatest ever recorded due to a tsunami. The difference in mortality rates between these tsunamis reflects, in part, the benefits of understanding the ocean's role and educating citizens to make scientifically sound and potentially life-saving decisions.

A tsunami is a series of rapidly propagating, shallow-water ocean waves that develops when a submarine earthquake, landslide, or volcanic eruption perturbs large volumes of water (Chapter 7). Both the 2004 and 2011 tsunamis were caused by powerful earthquakes, with magnitudes of 9 or greater, when one tectonic plate plowed beneath another. The focus of the 11 March 2011 earthquake was at a depth of 32 km (20 mi) about 130 km (81 mi) east of the city of Sendai. This location is on the boundary between the Pacific plate to the east and North American plate to the west, where the northern portion of Honshu is located. The boundary fractured, releasing energy, and elevated portions of the seafloor. This drastic movement generated tsunami waves that radiated outward, severely impacting the nearby coast and was felt around the globe within hours (Figure 1.1). Residents of Japan's coastal communities only had minutes to evacuate to higher ground.

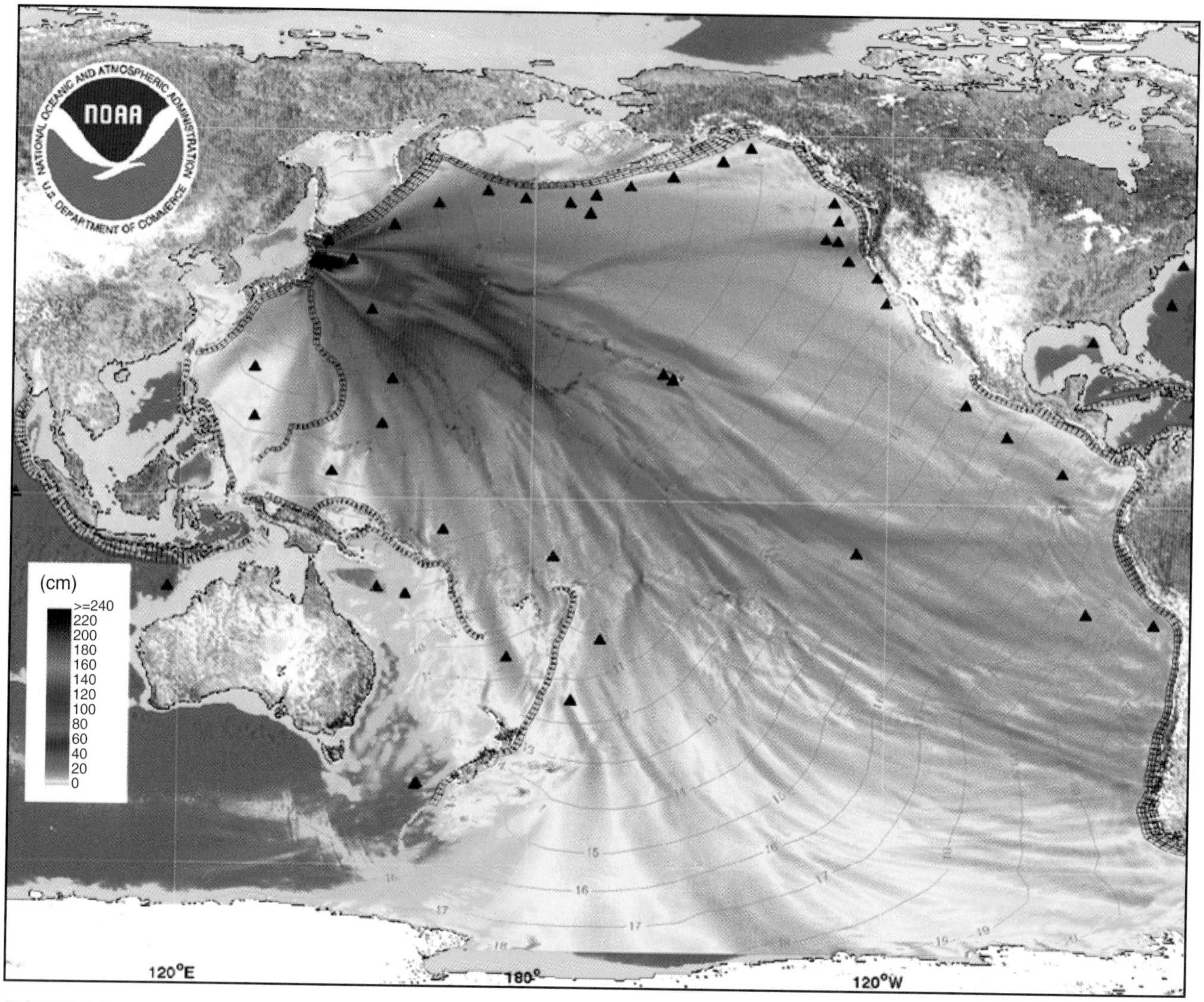

FIGURE 1.1
The MOST (Method of Splitting Tsunami) model simulation of maximum 11 March 2011 tsunami amplitude in cm during 24 hours of wave propagation. [Courtesy of NOAA Center for Tsunami Research]

Quickly moving, these tsunami waves traveled at 500 to 1000 km per hr (300 to 600 mph) with a height of only 0.5 m (1.5 ft). However, as they reached shallow coastal waters, they slowed and their momentum built them to devastating heights as they pushed inland. Waves of heights up to 10 m (33 ft) struck Japan on 11 March 2011, with some traveling up to 10 km (6 mi) inland. In Sumatra's Aceh Province, the maximum height of the 2004 tsunami exceeded 30 m (100 ft), almost the height of a 10-story building.

In 2004, there was no tsunami warning system operating in the Indian Ocean and there was neither advanced warning nor preparation for a tsunami of that magnitude. In the Pacific, work had been underway since 1968 through the *Intergovernmental Coordination Group for the Pacific Tsunami Warning System (ICG/PTWS)*. Additionally, earthquake, volcano, and tsunami-prone Japan trains its people on proper responses to a disaster, communicates risks, and has installed warning systems throughout the region. These precautionary measures saved many lives during the 2011 tsunami.

Understanding tsunamis and undersea earthquakes, as well as volcanoes, is the first step for effective planning, but even a country as well prepared as Japan still suffers major disasters. Only by studying the processes and consequences, and the interactions across the entire ocean system can future tragedies be mediated. Achieving this goal calls for a perspective enveloping not only oceanographic and geological processes, but also humans and their lifestyles, an *Earth system perspective*.

Driving Question:
What is the relationship between the ocean and Earth system?

This book examines Earth's ocean and the energy and processes responsible for formation of its basins, the supply of seawater and dissolved salts, the circulation of its waters, and the abundant life it supports. We also consider the ocean's interactions with the continents, atmosphere, the rocks and sediments that form the ocean floor, and living organisms. By studying the ocean in this way, we are adopting an Earth system perspective, viewing the planet as a set of interacting subsystems. Thus, we examine the hydrosphere (Earth's water and ice, including the ocean), atmosphere (its gaseous envelope), geosphere (Earth's solid portion), and biosphere (all living organisms). In this opening chapter we introduce the Earth system and briefly describe Earth's various interacting subsystems with special emphasis on the ocean.

In short, the Earth system perspective guides our investigation of the ocean. Space explorers use this same systems approach in studying the other planets of the solar system. We also examine the flow and transformations of matter and energy within and among Earth's subsystems. We stress the transport and cycling of water, salts, carbon, and oxygen as we seek to understand how the ocean functions and how it influences our lives. Furthermore, the systems approach is valuable in studying how the planet responds to large-scale environmental change. An example is presented in the first Essay at the end of this chapter.

In this opening chapter, we survey some of the advances in direct (in situ) and remote sensing of the ocean made possible by modern technology (e.g., Argo floats, Earth-orbiting satellites). We introduce the contributions that specialized electronic sensors, computers and numerical models are making in the collection, analysis, and interpretation of environmental data. Finally, we briefly examine how human activity influences the ocean and how the ocean influences humanity.

Earth as a System

What is the Earth system and more fundamentally, what is a system? A **system** is an interacting set of components that behave in an orderly way according to the laws of nature. One familiar example of a system is the human body, which consists of various subsystems including the nervous, respiratory, and reproductive systems plus energy and matter input/output. In a healthy person, these subsystems function internally and interact with one another in regular and predictable ways. Based on extensive observations and understanding of a system, scientists can predict how the system and its components are likely to respond to changing conditions. This predictive ability is important, for example, in dealing with the complexities of global climate change and its potential impacts on Earth's subsystems (Chapter 12).

The Earth system consists of four major interacting subsystems: hydrosphere, atmosphere, geosphere, and biosphere. Here we briefly examine each subsystem, its composition, basic properties, and some of its interactions with other components of the Earth system. The view of Planet Earth in Figure 1.2, resembling a "blue marble," shows all the major subsystems of the Earth system. The ocean, the most prominent feature, appears blue; clouds mostly obscure the ice sheets that cover much of Greenland and Antarctica; and the atmosphere is made visible by swirling storm clouds over the Pacific near Mexico and the middle of the Atlantic Ocean. Viewed edgewise, the atmosphere appears as a thin, bluish layer. Land (part of the geosphere) is mostly green because of vegetative cover (biosphere). The dominant color of Earth is blue because the ocean covers more than two-thirds of its surface; in fact, often Earth is referred to as the "blue planet" or "water planet."

HYDROSPHERE

The **hydrosphere** encompasses water in all three phases (i.e., ice, liquid, and vapor) that continually cycles from one reservoir to another within the Earth system. (We discuss the global water cycle in more detail later in this chapter.) Water is unique among the chemical components of the Earth system in that it is the only naturally occurring substance that co-exists in all three phases at normal temperatures and pressures near Earth's surface. The ocean, by far the largest reservoir of water in the hydrosphere, covers about 70.8% of the planet's surface but accounts for only 0.02% of Earth's mass. About 97.2% of the hydrosphere is ocean salt water; other

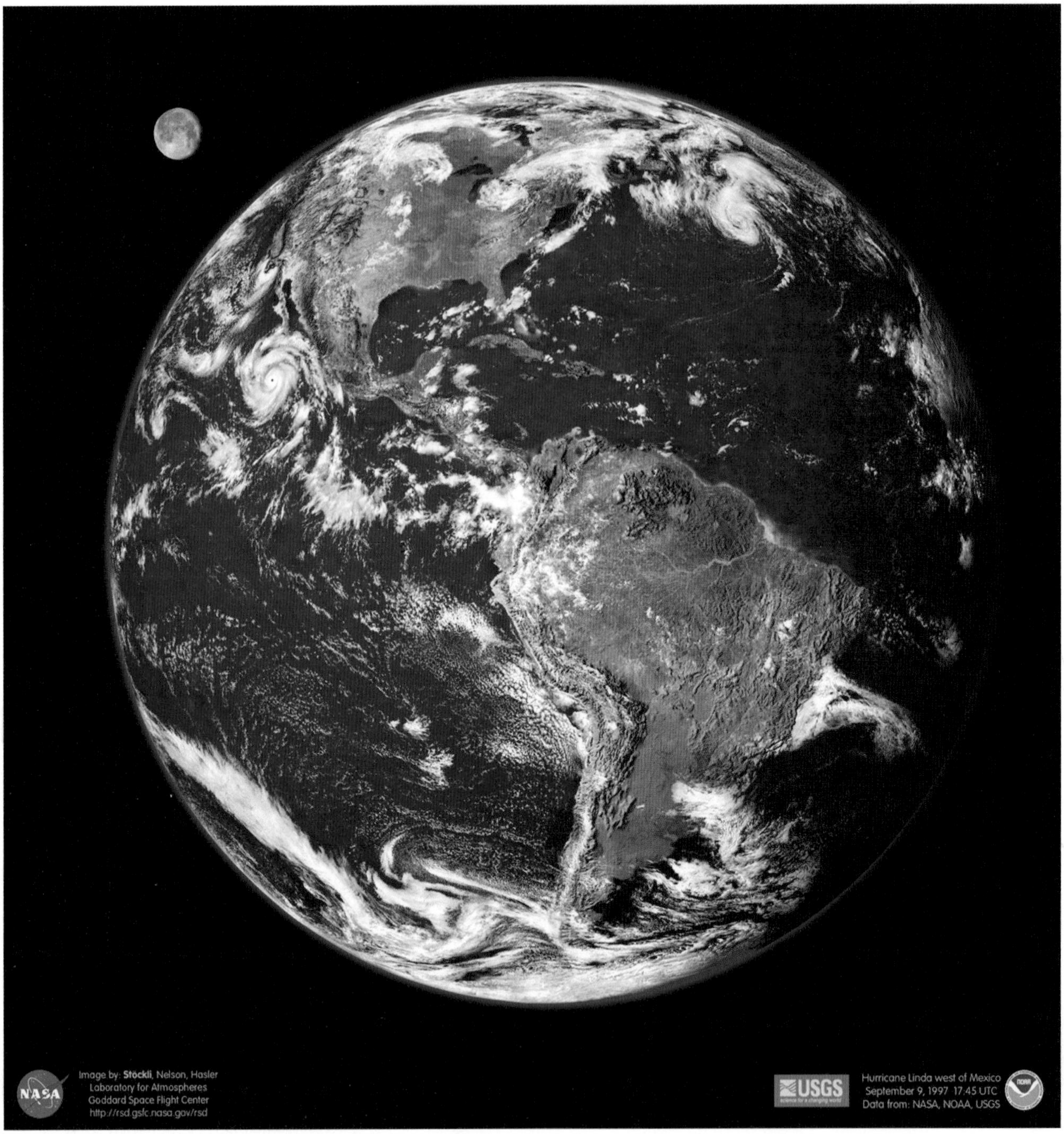

FIGURE 1.2
Planet Earth, viewed from space, appears as a "blue marble," with its surface mostly ocean water and partially obscured by swirling cloud masses. The Moon, appearing in the upper left limb of the Earth, is the planet's only natural satellite and exerts a major influence on the ocean through tides. [Courtesy of NASA, Goddard Space Flight Center]

saline bodies of water account for 0.6%. Using an Earth-orbiting satellite altimeter, M. Charette of Woods Hole (MA) Oceanographic Institution (WHOI) and W.H.F. Smith of NOAA's National Environmental Satellite, Data and Information Service in 2010 determined that the ocean has a volume of 1.332 billion cubic km and a mean depth of 3682.2 m (12,081 ft). The next largest reservoir in the hydrosphere is glacial ice, most of which covers much of Antarctica and Greenland. Ice and snow make up 2.1% of water in the hydrosphere. Considerably smaller quantities of water occur on the land surface (lakes, rivers, streams), in the subsurface (soil moisture, groundwater), the atmosphere (water vapor, clouds, precipitation), and biosphere (plants, animals). The possible origins of water on Earth are described in this chapter's second Essay.

The ocean and atmosphere are coupled such that winds drive surface ocean currents. Wind-driven currents are restricted to a surface ocean layer typically about 100 m (300 ft) deep and can take as much as a few months to years to cross an ocean basin (Chapter 6). Deep-ocean currents, at depths greater than 100 m (300 ft), are much more sluggish and more challenging to study than surface currents because of the considerable difficulties in making measurements at great depths. Movements of deep-ocean waters are caused primarily by small differences in water density (mass per unit volume) arising from small differences in water temperature and salinity (a measure of dissolved salt content). Cold seawater, being denser than warm seawater, tends to sink whereas warm seawater, being less dense, is buoyed upward by (or floats on) colder seawater. Likewise, saltier water is denser than less salty water and tends to sink whereas less salty water is buoyed upward. The combination of temperature and salinity determines whether an ocean water mass remains at its original depth, moves vertically to a level where surrounding water has the same density, or sinks to the bottom. Even though deep currents are relatively slow, they keep ocean waters well mixed so that the ocean has a nearly uniform chemical composition (Chapter 3).

The densest ocean waters form in polar or nearby subpolar regions. Salty waters become even saltier where sea ice forms at high latitudes because growing ice crystals exclude dissolved salts (Chapter 3). Chilling of this salty water near Greenland and Iceland and in the Norwegian and Labrador Seas further increases its density so that it sinks and forms a bottom current that flows southward under equatorial surface waters into the South Atlantic as far south as Antarctica. Here, deep water from the North Atlantic mixes with deep water around Antarctica (Chapter 6). Branches of that cold bottom current then spread northward into the Atlantic, Indian, and Pacific basins. Eventually, the water slowly moves to the surface, mainly in the Pacific, and starts its journey through the islands of Indonesia, across the Indian Ocean, around South Africa, and into the tropical Atlantic. There, intense heating and evaporation make the water hot and salty. This surface water is then transported northward in the Gulf Stream thereby completing the cycle. The transport of heat energy and salt in this ocean-basin scale circulation is an important agent of climate change (Chapter 6).

The frozen portion of the hydrosphere, known as the **cryosphere**, encompasses massive glacial ice sheets, mountain glaciers, ice in permanently frozen ground (*permafrost*), and the pack ice and bergs floating at sea. All of these ice types except pack ice (frozen seawater) consist of fresh water. A **glacier** is a mass of ice that flows internally under the influence of gravity. In places, the Greenland and Antarctic ice sheets are up to 3 km (1.8 mi) thick. The Antarctic ice sheet contains 90% of the ice in the Earth system. Much smaller glaciers (tens to hundreds of meters thick) primarily occupy the highest mountain valleys on all continents. At present, glacial ice covers about 10% of the planet's land area but at times during the past 1.8 million years, glacial ice expanded over as much as 30% of the land surface, primarily in the Northern Hemisphere. At the peak of the last major glacial advance, about 20,000 to 18,000 years ago, the Laurentide ice sheet covered much of what is now Canada and the northern tier states of the United States. At the same time, a smaller ice sheet buried the British Isles and portions of northwest Europe. Meanwhile, mountain glaciers worldwide thickened and expanded.

Glaciers form where annual snowfall exceeds annual snowmelt. As snow accumulates, the pressure of the new snow transforms underlying snow to ice. As the ice forms, it preserves traces of the original seasonal layers of snow and traps gas bubbles. Chemical analysis of the ice layers and air bubbles in the ice provides clues to climatic conditions at the time the original snow fell (Chapter 12). Ice cores extracted from the Greenland and Antarctic ice sheets yield information on changes in Earth's climate and atmospheric composition extending back in time hundreds of thousands of years—to about 650,000 years ago in Antarctica.

Under the influence of gravity, glacial ice flows slowly from sources at higher latitudes or higher elevations (where some winter snow survives the summer) to lower latitudes or lower elevations, where the ice either melts or flows into the nearby ocean. Around Antarctica, streams of glacial ice flow out to the ocean. Ice, being less dense than seawater, floats, forming ice shelves (typically about 500 m or 1600 ft thick). Thick masses of ice break off the shelf edge, forming flat-topped icebergs that are carried by surface ocean currents around Antarctica. Likewise, irregularly shaped icebergs break off the glacial ice streams of Greenland and flow out into the North Atlantic Ocean (Figure 1.3). In 1912, the newly launched luxury

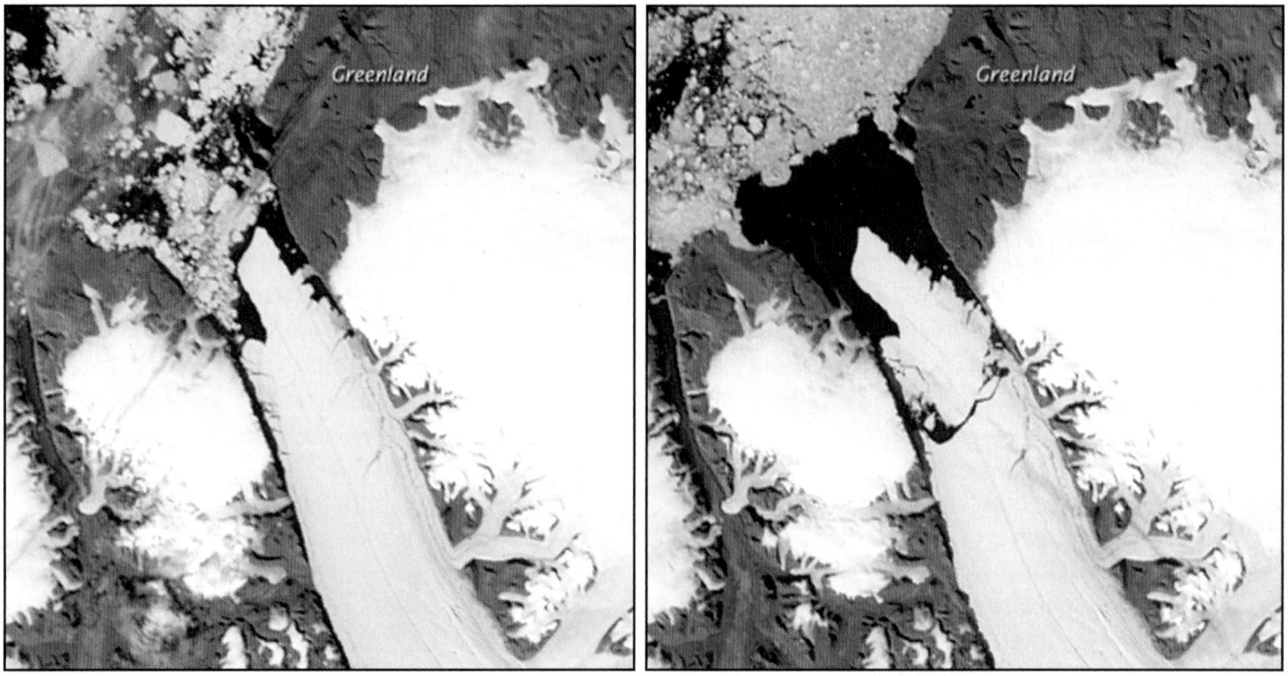

FIGURE 1.3
A massive iceberg (roughly 251 square km or 97 square mi) is shown breaking off the Petermann Glacier along the northwestern coast of Greenland. The left image is from 28 July 2010 and the right image is from 5 August 2010. Images of the glacier were obtained by the Moderate Resolution Imaging Spectroradiometer (MODIS) instrument onboard NASA's Terra spacecraft. [Courtesy of NASA]

liner, *RMS Titanic*, struck a Greenland iceberg southeast of Newfoundland and sank with the loss of more than 1500 lives. Most sea ice surrounding Antarctica forms each winter through freezing of surface seawater. During summer most of the sea ice around Antarctica melts whereas in the Arctic Ocean sea ice can persist for several years before flowing out through Fram Strait into the Greenland Sea and eventually melting. This "multi-year" ice loses salt content with age, so that Eskimos can harvest this older, less salty ice as a source of drinking water.

The hydrosphere is dynamic; water moves continually although at different rates through different parts of the Earth system. The ocean is the ultimate destination of all water moving on or beneath the land surface. Water flowing in river or stream channels may take a few weeks to reach the ocean. Groundwater typically moves at a very slow pace through fractures and tiny openings in subsurface rock and sediment and feeds into rivers, lakes, or directly into the ocean. Water in large, deep lakes also moves slowly, in some cases taking centuries to reach the ocean.

How long is water frozen into glaciers? Glaciers normally expand (thicken and advance) and shrink (thin and retreat) slowly in response to changes in climate (long-term average temperature and snowfall). Mountain glaciers respond to climate change on time scales of a decade. Until recently, scientists had assumed that the response time for the Greenland and Antarctic ice sheets is measured in millennia; however, in 2007 scientists reported that two outlet glaciers that drain the Greenland ice sheet exhibited significant changes in discharge in only a few years. This finding was confirmed by changes in ice surface elevation detected by sensors onboard NASA's Ice, Cloud, and Land Elevation Satellite (ICESat). This unexpectedly rapid discharge is likely due to the flow of large ice streams over subglacial water. Hence, outlet glaciers behave more like mountain glaciers, raising questions regarding the long-term stability of polar ice sheets and their response to global climate change (Chapter 12).

ATMOSPHERE

Earth's **atmosphere** is a relatively thin envelope of gases and tiny suspended particles surrounding the planet. Compared to Earth's diameter, the atmosphere is like the thin skin of an apple accounting for only about 0.07% of the mass of the Earth system. But the thin atmospheric skin is essential for life and the orderly functioning of physical and biological processes on Earth. Unlike the nearly incompressible ocean water, air is compressible, so that air density decreases with increasing altitude above Earth's surface. About half of the atmosphere's mass is concentrated within about 5.5 km (3.4 mi) of Earth's surface and 99% of its mass occurs below an altitude of 32 km (20 mi). At an altitude of about 1000 km (620 mi), Earth's atmosphere merges with the highly rarefied interplanetary gases, hydrogen and helium.

Based on the average vertical temperature profile, the atmosphere is divided into four layers (Figure 1.4). The **troposphere** (averaging about 10 km or 6 mi thick) is where the atmosphere interfaces with the hydrosphere, geosphere, and biosphere and where most weather takes place. In the troposphere, the average air temperature drops with increasing altitude so that it is usually colder on mountaintops than in lowlands. The troposphere contains 75% of the atmosphere's mass and 99% of its water. The *stratosphere* (10 to 50 km, 6 to 30 mi above Earth's surface) contains the *ozone shield*, which protects organisms from exposure to potentially lethal levels of solar ultraviolet radiation. Above the stratosphere is the *mesosphere* where the average temperature generally decreases with altitude and above that is the *thermosphere* where the average temperature increases with altitude but is particularly sensitive to variations in incoming solar radiation.

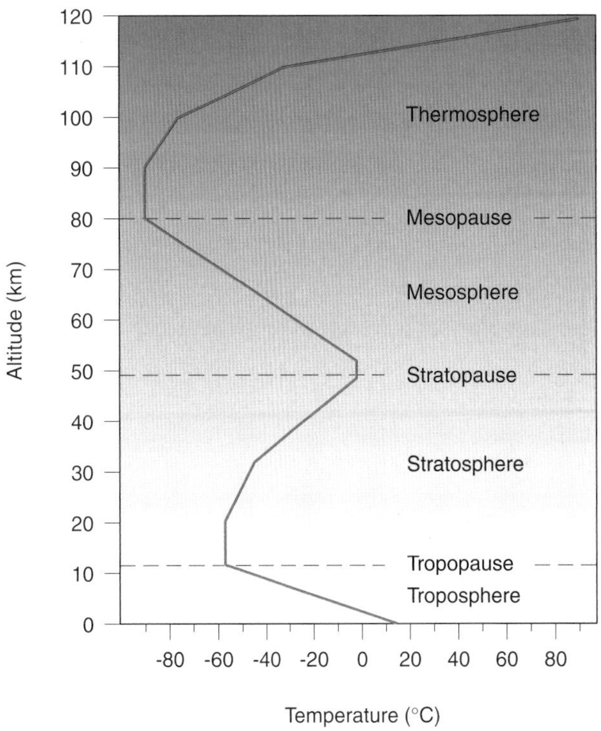

FIGURE 1.4
Based on variations in average temperature with altitude, the atmosphere is divided into the troposphere, stratosphere, mesosphere, and thermosphere.

Nitrogen (N_2) and oxygen (O_2), the chief atmospheric gases, are mixed in uniform proportions up to an altitude of about 80 km (50 mi). Not counting water vapor (which has a highly variable concentration), nitrogen occupies 78.08% by volume of the lower atmosphere (below 80 km), and oxygen is 20.95% by volume. The next most abundant gases are argon (0.93%) and carbon dioxide (0.039%). Many other gases occur in the atmosphere in trace concentrations, including ozone (O_3) and methane (CH_4) (Table 1.1). Unlike nitrogen and oxygen, the percent volume of some of these trace gases varies with time and location.

In addition to gases, minute solid or liquid particles of various compositions, collectively called **aerosols**, are suspended in the atmosphere. A flashlight beam in a darkened room reveals an abundance of tiny dust particles floating in the air. Individually, most atmospheric aerosols are too small to be visible but in aggregates, such as water droplets and ice crystals composing clouds, they may be visible. Most aerosols occur in the lower atmosphere, near their sources on Earth's surface; they are derived from wind erosion of soil, ocean spray, forest fires, volcanic eruptions, industrial chimneys, and

TABLE 1.1
Gases Composing Dry Air in the Lower Atmosphere (below 80 km)

Gas	% by Volume	Parts per Million
Nitrogen (N_2)	78.08	780,840.0
Oxygen (O_2)	20.95	209,460.0
Argon (A)	0.93	9,340.0
Carbon dioxide (CO_2)	0.03886	388.6
Neon (Ne)	0.0018	18.0
Helium (He)	0.00052	5.2
Methane (CH_4)	0.00014	1.4
Krypton (Kr)	0.00010	1.0
Nitrous oxide (N_2O)	0.00005	0.5
Hydrogen (H)	0.00005	0.5
Xenon (Xe)	0.000009	0.09
Ozone (O_3)	0.000007	0.07

the exhaust of motor vehicles. Although the concentration of aerosols in the atmosphere is relatively small, they participate in some important processes. Aerosols function as nuclei that promote the formation of clouds, essential for the global water cycle. And some aerosols (e.g., volcanic dust, sulfurous particles) influence air temperatures by interacting with incoming solar radiation.

The significance of an atmospheric gas is not necessarily related to its concentration. Some atmospheric components that are essential for life occur in very low concentrations. For example, most water vapor is confined to the lowest kilometer or so of the atmosphere and is never more than about 4% by volume even in the most humid places on Earth (e.g., over tropical rainforests and seas). But without water vapor, the planet would have no water cycle, no rain or snow, no ocean, and no fresh water. Also, without water vapor, Earth would be much too cold for most forms of life to exist. Although comprising only 0.039% of the lower atmosphere, carbon dioxide (CO_2) is essential for photosynthesis. Without carbon dioxide, green plants and the food webs they support could not exist. Although the atmospheric concentration of ozone (O_3) is minute, the chemical reactions responsible for its formation (from oxygen) and dissociation (to oxygen) in the stratosphere (mostly at altitudes between 30 and 50 km) shield organisms on Earth's surface from potentially lethal intensities of solar ultraviolet radiation.

The atmosphere is dynamic; that is, the atmosphere is always circulating in response to differences in rates of heating and cooling within the Earth system. On an average annual basis, Earth's surface experiences net radiational heating (more heating than cooling mainly due to the Sun) and the atmosphere undergoes net radiational cooling (to space). Also, net radiational heating occurs in the tropics, while net radiational cooling characterizes higher latitudes. Variations in heating and cooling give rise to *temperature gradients*, that is, differences in temperature from one location to another. In response to temperature gradients, the atmosphere (and ocean) circulates and redistributes heat within the Earth system. Heat is conveyed from where it is warmer to where it is colder: from the Earth's surface to atmosphere and from the tropics to higher latitudes. As discussed in detail in Chapter 5, the global water cycle and accompanying phase changes of water play an important role in this planetary-scale transport of heat energy.

GEOSPHERE

The **geosphere** is the solid portion of the planet consisting of rocks, minerals, and sediments. Most of Earth's interior cannot be observed directly—the deepest mines and oil wells do not penetrate the solid Earth to any great depth. Most of what is known about the composition and physical properties of Earth's interior comes from studying seismic waves generated by earthquakes and explosions. In addition, meteorites provide valuable clues regarding the chemical composition of Earth's interior. From study of the behavior of vibrations that penetrate the planet, geologists have determined that Earth's interior consists of four spherical shells: crust, mantle, and outer and inner core (Figure 1.5). Earth's interior is mostly solid and accounts for much of the mass of the Earth system. The outermost solid skin of the planet, called the *crust*, ranges in thickness from only about 8 km (5 mi) under the ocean to about 70 km (45 mi) in some mountain belts. We live on the crust and it is the source of almost all rock, mineral, and fuel (e.g., coal, oil, and natural gas) resources that are essential for our industrial-based economy. The rigid

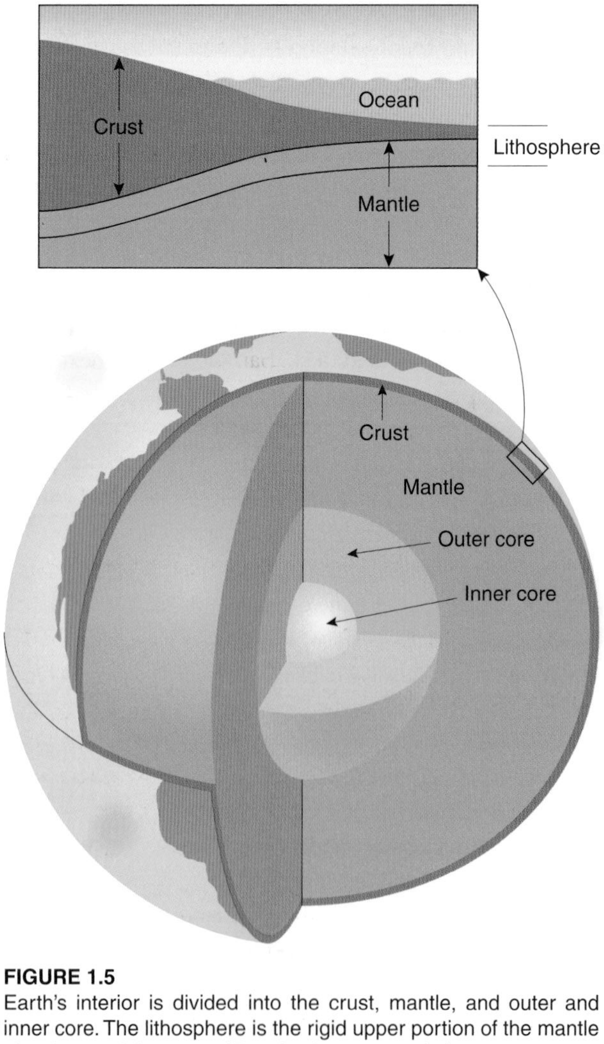

FIGURE 1.5
Earth's interior is divided into the crust, mantle, and outer and inner core. The lithosphere is the rigid upper portion of the mantle plus the overlying crust. (Drawing is not to scale.)

uppermost portion of the *mantle* plus the overlying crust constitutes Earth's **lithosphere**, averaging about 100 km (62 mi) thick. Two sets of geological processes continually modify the lithosphere: surface geological processes and internal geological processes.

Surface geological processes encompass weathering and erosion taking place at the interface between the lithosphere (mainly the crust) and the other Earth subsystems. **Weathering** entails the physical disintegration, chemical decomposition, or solution of exposed rock. Fragments of rock, including particles of organic (e.g., shells of tiny ocean animals) and inorganic origin, produced by weathering are known as *sediments*. Water plays an important role in weathering by dissolving soluble rock and minerals and participating in chemical reactions that decompose rock. Water's unusual physical property of expanding (by 9%) when freezing can produce pressure that fragments rock if the water saturates the tiny cracks and pore spaces in the rock and the temperature drops below 0°C (32°F). More likely, however, the water is not so confined and fragmentation is due to stress caused by the growth of ice lenses within the rock.

The ultimate weathering product is *soil*, a mixture of organic (humus) and inorganic matter (sediment) on Earth's land surface that supports rooted plants and supplies mineral nutrients and water for plants. Soils derive from the weathering of bedrock or sediment and vary widely in texture (particle size). A typical soil is 50% open space (pores) that is occupied by air and water in roughly equal proportions. Plants also participate in weathering through the physical action of their growing roots and the carbon dioxide they release to the soil.

Erosion is the removal and transport of sediments by gravity, moving water, glaciers, and wind. Running water and glaciers are pathways in the global water cycle. Erosive agents transport sediments from source regions (usually highlands) to low-lying depositional environments (e.g., ocean, lakes). Weathering aids erosion by reducing massive rock to particles that are sufficiently small to be transported by agents of erosion. Erosion aids weathering by removing sediment and exposing fresh surfaces of rock to the atmosphere and weathering processes. Together, weathering and erosion reduce the elevation of the land.

Internal geological processes counter surface geological processes by uplifting the land through tectonic activity, including volcanism and mountain building. Most tectonic activity occurs at the boundaries between lithospheric plates. The lithosphere is broken into a dozen massive plates (and many smaller ones) that are slowly driven (typically less than 20 cm per year) across the face of the globe by huge convection currents in Earth's mantle. Continents are carried on the moving plates and, as we will see in Chapter 2, ocean basins are formed by seafloor spreading.

Plate tectonics probably has operated on the planet for at least 3 billion years, with continents periodically assembling into supercontinents and then splitting into smaller segments (Chapter 2). The most recent supercontinent, called *Pangaea* (Greek for "all land"), broke apart about 200 million years ago and its constituent landmasses, the continents of today, slowly moved to their present locations. Plate tectonics explains such seemingly strange discoveries as glacial sediments in the Sahara Desert and fossil coral reefs indicative of tropical climates in Wisconsin. Such discoveries reflect climatic conditions hundreds of millions of years ago when the continents were at different latitudes than they are today.

Geological processes occurring at boundaries between plates produce large-scale landscape and ocean bottom features, including mountain ranges, volcanoes, deep-sea trenches, and the ocean basins themselves. Enormous stresses develop at plate boundaries, bending and fracturing bedrock over broad areas. Hot molten rock material, known as **magma**, wells up from deep in the crust or upper mantle and migrates along rock fractures. Some magma pushes into the upper portion of the crust where it cools and solidifies into massive bodies of rock that form the core of mountain ranges (e.g., Sierra Nevada). Some magma feeds volcanoes or flows through fractures and spreads over Earth's surface as lava flows that cool and solidify (e.g., Columbia Plateau in the Pacific Northwest). At spreading plate boundaries on the sea floor, upward flowing magma solidifies into new oceanic crust (Chapter 2). Plate tectonics and associated volcanism are important in geochemical cycling and release water vapor, carbon dioxide, and other gases to the atmosphere and ocean.

Volcanic activity is not confined to plate boundaries. Some volcanic activity occurs at great distances from plate boundaries and is due to hot spots in the mantle (Chapter 2). A **hot spot** is a long-lived source of magma caused by rising plumes of hot material originating in the mantle (*mantle plumes*). Where a plate is situated over a hot spot, magma may break through the crust and form a volcano. The "Big Island" of Hawaii is volcanically active because it sits over a hot spot located in the mantle under the Pacific plate. A hot spot underlying Yellowstone National Park is the source of heat for geyser eruptions (including Old Faithful). Further complicating matters, however, both lithospheric plates and hot spots move.

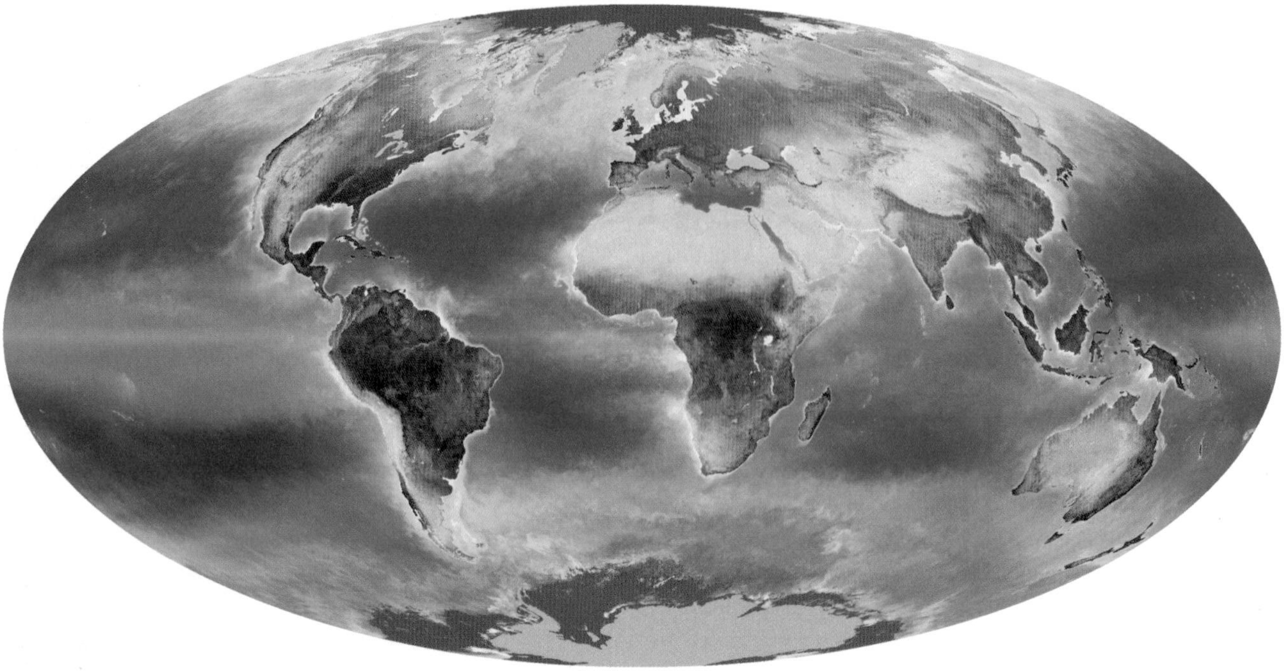

FIGURE 1.6
A view of Earth's biosphere using ocean chlorophyll data provided by NASA's SeaWiFS (Sea-viewing Wide Field-of-view Sensor) project and vegetation data courtesy of the Global Mapping and Modelling & Mapping Studies (GIMMS) project. Chlorophyll concentration is highest where the ocean displays light greenish-yellow shadings. The concentration of plant growth on land is greatest where it is dark green. Medium gray indicates snow and ice covered areas. [Courtesy of NASA.]

BIOSPHERE

All living organisms on Earth are components of the **biosphere** (Figure 1.6). They range in size from microscopic single-celled bacteria to the largest plants and animals that ever lived (e.g., redwood trees and blue whales). Bacteria and other one-celled organisms dominate the biosphere, both on land and in the ocean. The average animal in the ocean is the size of a mosquito. Large, multi-cellular organisms (including humans) are relatively rare on Earth.

Organisms on land or in the atmosphere live close to Earth's surface. However, marine organisms occur throughout the ocean depths and even inhabit rock fractures, volcanic vents, and mud on the ocean floor. Certain organisms live in so-called *extreme environments* at temperatures and pressures previously thought impossible for life. In fact, some scientists estimate that the mass of organisms living in fractured rocks on and below the ocean floor may vastly exceed the mass of organisms living on or above it. Most of these deep-dwelling organisms remain a mystery.

Photosynthesis and cellular respiration are essential for life near the surface of the Earth and exemplify the interaction of the biosphere with the other subsystems of the Earth system. **Photosynthesis** is the process whereby green plants use light energy from the Sun to combine carbon dioxide from the atmosphere with water to produce sugars, a form of carbohydrate containing a relatively large amount of energy. Oxygen (O_2) is an important byproduct of photosynthesis. Animals that consume plants harvest some of that energy. Other animals consume those animals and energy progresses up the food chain. Through **cellular respiration**, an organism processes food and liberates energy in a form that can be used for maintenance, growth, and reproduction. Thereby, carbon dioxide, water, and heat energy are released (cycled) within the Earth system.

Sunlight is the principal source of energy for most organisms living on land or in the ocean's surface waters. However, for specialized organisms inhabiting the darkness of the ocean depths and rock fractures on the ocean floor, chemical energy rather than sunlight drives biological processes. Through a process known as **chemosynthesis**, these marine organisms derive energy from substances such as hydrogen sulfide (H_2S) or methane (CH_4) that originate in Earth's interior (Chapter 9).

Dependency of organisms on one another (e.g., as a source of food) and on their physical and chemical environment (e.g., for water, oxygen, carbon dioxide, and habitat) is embodied in the concept of ecosystem. The biosphere is composed of **ecosystems**, communities of plants and animals that interact with one another, together with the physical conditions and chemical substances in a specific geographical area. Deserts, tropical rain forests, tundra, estuaries, marshes, lakes, streams, and coral reefs are examples of natural ecosystems. Most people live in highly modified terrestrial ecosystems such as cities, towns, farms, or ranches.

An ecosystem is home to producers (plants), consumers (animals), and decomposers (bacteria, fungi). *Producers* (also called *autotrophs* for "self-nourishing") form the base of most ecosystems. Using solar energy, their photosynthetic pigments form energy-rich carbohydrates. Consumers that depend directly or indirectly on plants for their food are called *heterotrophs*. Animals feeding directly on plants are called *herbivores*; those that prey on other animals are called *carnivores* and animals that consume both plants and animals are *omnivores*. After death, the remains of organisms are broken down by decomposers, usually bacteria and fungi, which cycle nutrients back to the environment.

Feeding relationships, called a **food chain**, can be quite simple. In a hypothetical ocean-based (or marine) food chain, microscopic plants called phytoplankton (primary producers) are eaten by tiny drifting ocean animals called zooplankton (herbivores), which in turn are eaten by juvenile fish (carnivores). In a food chain, each stage is called a *trophic level* (or feeding level). On average, only about 10% of the energy available at one trophic level is transferred to the next higher trophic level. *Biomass*, the total weight or mass of organisms, is more readily measured than energy so that scientists describe the transfer of energy in food chains in terms of so many grams or kilograms of biomass. Thus 100 g of phytoplankton are required to produce 10 g of zooplankton, which in turn produces 1 g of juvenile fish. Terrestrial and marine food chains are often more complex than our phytoplankton-zooplankton-juvenile fish example. With some notable exceptions, marine and terrestrial organisms usually eat many different kinds of food, and in turn, are eaten by a host of other consumers. These more complicated feeding relationships constitute a *food web* (Figure 1.7).

An example of an ecosystem that is particularly important in the coastal zone is an estuary, where the hydrosphere, lithosphere, and biosphere interact in special ways. An estuary forms where fresh and salt water mix usually at the mouths of rivers and in tidal marshes and bays. Water in an estuary undergoes daily tidal oscillations (i.e., periodic rise and fall of sea level in response to the gravitational attraction of the Moon and Sun). Organisms living in an *estuary* are adapted to frequent fluctuations in currents, temperature, salinity, and the concentration of suspended sediment. Among the nation's best-known estuaries are Chesapeake Bay (largest in the U.S.), San Francisco Bay, and Puget Sound (Washington).

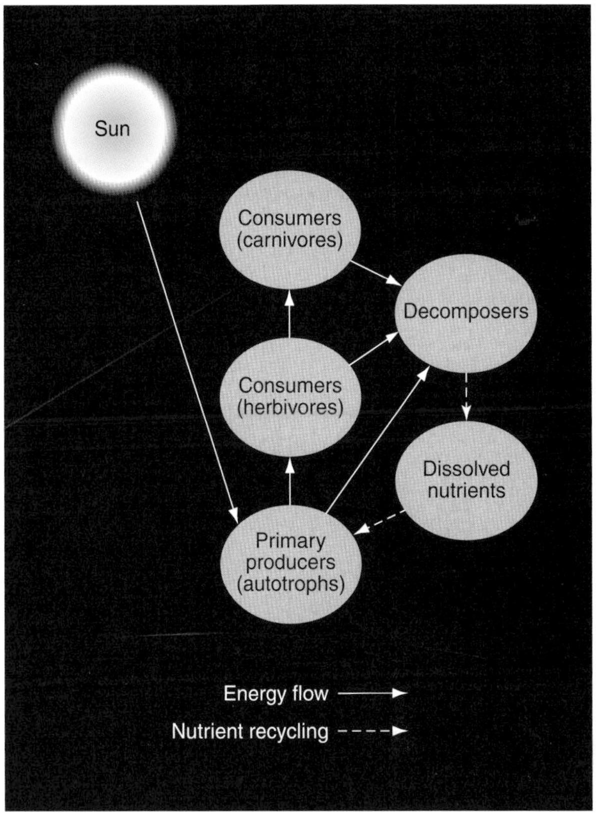

FIGURE 1.7
Schematic diagram of a marine food web in which solar radiation powers photosynthesis and energy is conveyed to successively higher trophic (feeding) levels.

Estuaries are among the most productive ecosystems on the planet because of a special combination of biological and physical characteristics. They receive a continual supply of nutrients and organic matter delivered by both rivers and ocean tidal currents. Water circulates in an estuary in a way that traps and re-circulates nutrients and detritus (dead or partially decomposed remains of plants and animals). These conditions support luxuriant plant growth (e.g., phytoplankton, sea grass) as well as large populations of animals that feed on detritus. The abundant food supply coupled with the unique environmental conditions of an estuary favor development and protection of juvenile fish. Low salinity and shallow water deter many ocean predators that would feed on juvenile fish. Estuaries provide habitat favorable for oysters, mussels, clams, lobsters, shrimp, and snails. These organisms, in turn, are an abundant food source for fish, birds, and humans. Estuaries are also home to juvenile *anadromous fish* (fish that swim from the sea to fresh water upriver to spawn) such as striped bass and salmon. We have more about estuaries in Chapter 8.

Through the vast expanse of geologic time, the biosphere co-evolved with the ocean and atmosphere. For more on this topic, refer to this chapter's third Essay.

Biogeochemical Cycles

A **biogeochemical cycle** consists of pathways along which solid, liquid, and gaseous materials move among the various reservoirs of the Earth system. Accompanying this flow of materials may be chemical reactions that effectively remove materials from a reservoir and the transfer and transformation of energy. Examples of biogeochemical cycles are the water cycle, carbon cycle, oxygen cycle, and rock cycle. The Earth system is an open (or flow-through) system for energy where *energy* is defined as the capacity for doing work. The Earth system receives energy from the Sun and from Earth's interior while emitting energy in the form of infrared radiation to space. Within the Earth system, energy is neither created nor destroyed although it is converted from one form to another. This is the *law of energy conservation* (also known as the *first law of thermodynamics*).

The Earth system is essentially closed for matter; that is, it neither gains nor loses matter over time (except for meteorites and asteroids). All biogeochemical cycles obey the *law of conservation of matter*, which states that matter can be neither created nor destroyed, but can change in chemical or physical form. When a log burns in a fireplace, a portion of the log is converted to ash (and heat energy), and the rest goes up the chimney as carbon dioxide, water vapor, creosote and heat. In terms of accountability, all losses from one reservoir in a cycle can be accounted for as gains in other reservoirs of the cycle. Stated succinctly, for any reservoir

Input = Output + Storage

The amount of some substance stored within a reservoir of a biogeochemical cycle depends on the rates at which the material is cycled into and out of the reservoir. *Cycling rate* is defined as the amount of material that moves from one reservoir to another within a specified period of time. If the input rate exceeds the output rate, the amount in the reservoir (storage) increases. If the input rate is less than the output rate, the amount decreases. Over the long term, the cycling rates of materials among the various global reservoirs have been relatively stable; that is, equilibrium tends to prevail between the rates of input and output.

Closely related to cycling rate is residence time. **Residence time** is the average length of time for a substance in a reservoir to be replaced completely, that is,

$$\textbf{Residence time} = \frac{\textbf{(amount in reservoir)}}{\textbf{(rate of addition or removal)}}$$

The residence time of a water molecule in the various reservoirs of the hydrosphere varies from only about 10 days in the atmosphere to tens of thousands of years or longer in glacial ice sheets. Residence time of dissolved constituents of seawater ranges from 100 years for aluminum to 260 million years for sodium.

Consider the global cycling of carbon as an illustration of a biogeochemical cycle (Figure 1.8). Through photosynthesis, carbon dioxide (CO_2) cycles from the atmosphere into green plants where carbon is incorporated into sugar ($C_6H_{12}O_6$). Plants use sugar to manufacture other organic compounds including fats, proteins, and other carbohydrates. As a byproduct of cellular respiration, plants and animals transform a portion of the carbon in these organic compounds into carbon dioxide that is released to the atmosphere. The same cycle occurs in the ocean when carbon dioxide is cycled into and out of marine organisms through photosynthesis and respiration. In addition to the uptake of CO_2 via photosynthesis, marine organisms also use carbon for calcium carbonate ($CaCO_3$) to make hard, protective shells. Furthermore, decomposer organisms (e.g., bacteria) both on land and in the ocean act on the remains of dead plants and animals, releasing CO_2 to the atmosphere and ocean through respiration.

When marine organisms die their shells and skeletons settle to the ocean bottom. Over long periods of time, these organic remains accumulate, are compressed by their own weight and the weight of other sediments, and gradually transform into solid, carbonate rock. Common carbonate rocks are limestone ($CaCO_3$) and dolostone ($CaMg(CO_3)_2$). Subsequently, tectonic processes uplift these rocks and expose them to physical and chemical weathering processes. For example, atmospheric CO_2 dissolves in rainwater producing carbonic acid (H_2CO_3) that, in turn, dissolves carbonate rock (e.g., limestone) releasing CO_2. As part of the global water cycle, rivers and streams transport these weathering products to the sea where they settle out of suspension or precipitate as sediments that accumulate on the ocean floor. Over the hundreds of millions of years that

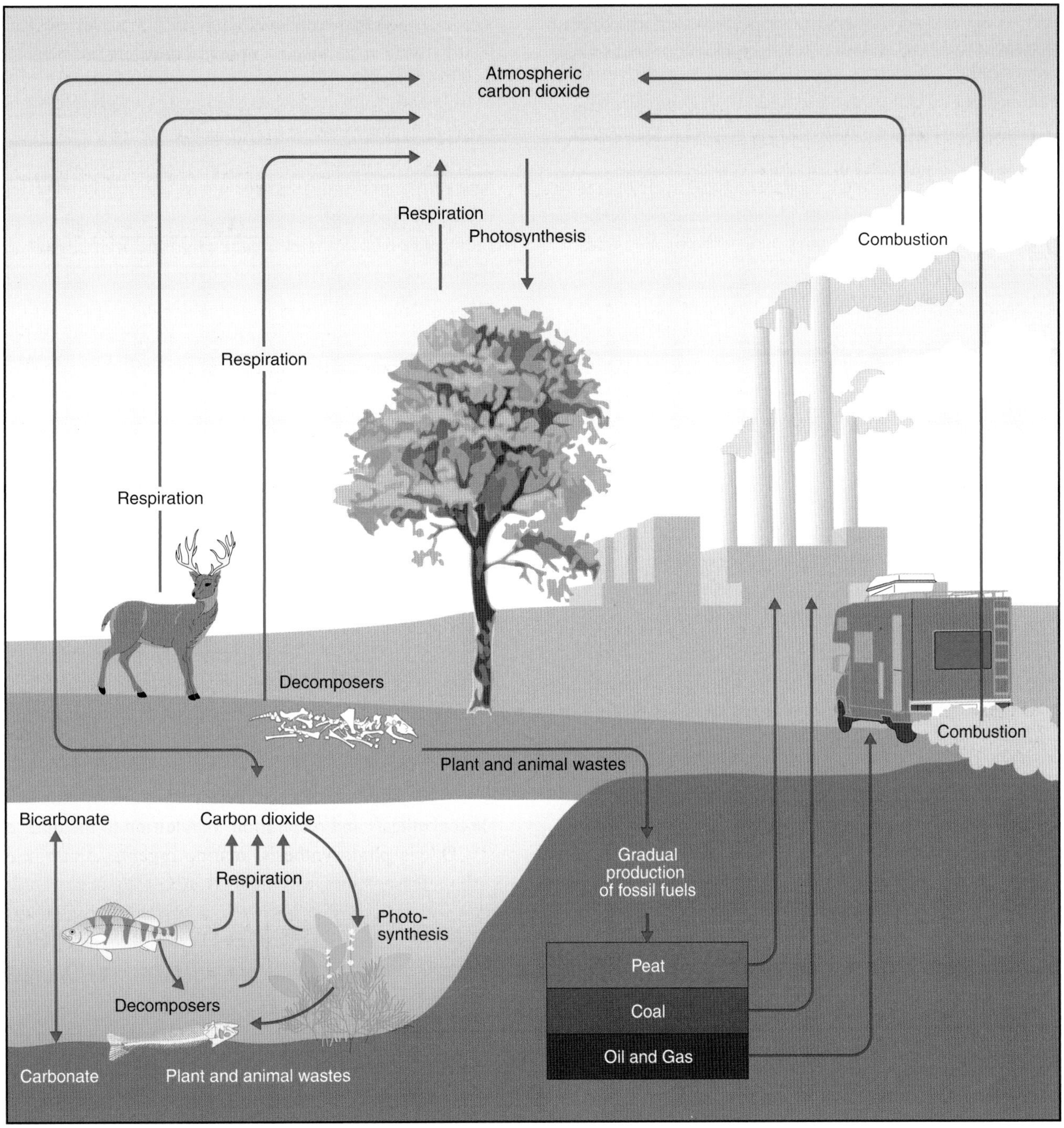

FIGURE 1.8
Schematic representation of the global carbon cycle.

constitute geologic time, weathering and erosion of carbon-containing rocks have significantly altered the concentration of carbon dioxide in the atmosphere.

About 280 to 345 million years ago, during the geologic time interval known as the Carboniferous period, trillions of metric tons of organic remains (detritus) accumulated on the ocean bottom and in low-lying swampy terrain on land. The supply of detritus was so great that decomposer organisms could not keep pace. In some marine environments, plant and animal remains were converted to oil and natural gas. In swampy terrain, heat and pressure from accumulating organic debris concentrated carbon, converting the remains of luxuriant swamp forests into thick layers of coal. Today, when we burn coal, oil, and natural gas, collectively called *fossil fuels*, we are tapping energy that was originally locked in vegetation through photosynthesis hundreds of millions of years ago. During combustion, carbon from these fossil fuels combines with oxygen in the air to form carbon dioxide which escapes to the atmosphere.

Another important biogeochemical cycle operating in the Earth system is the global water cycle, which is closely linked to all other biogeochemical cycles. Reservoirs in the water cycle (ocean, atmosphere, lithosphere, living organisms) are also reservoirs in the other cycles. Furthermore, water is an important mode of transport in biogeochemical cycles. In the nitrogen cycle, for example, intense heating of air associated with lightning combines atmospheric nitrogen (N_2), oxygen (O_2), and moisture to form droplets of extremely dilute nitric acid (HNO_3) that are washed by rain to the soil. In the process, nitric acid converts to nitrate (NO_3^-), an important plant nutrient that can be taken up by plants through their root system. Plants convert nitrate to ammonia (NH_3), which is incorporated into a variety of compounds including amino acids, proteins, and DNA. On the other hand, both nitrate and ammonia readily dissolve in water so that heavy rains can deplete soil of these important nutrients and wash them into waterways.

The Ocean in the Global Water Cycle

We can reasonably assume that the total amount of water in the Earth system is neither increasing nor decreasing although natural processes continually generate and breakdown water. Volcanic activity is more or less continuous on Earth and adds to the supply of water. Water vapor accounts for perhaps half of all gases emitted during a volcanic eruption; at least some of this water originally was sequestered in magma and solid rock. A minute amount of water is contributed to Earth by meteorites and other extraterrestrial debris continually bombarding the upper atmosphere. At the same time, intense solar radiation entering the upper atmosphere converts (*photodissociates*) a small amount of water vapor into its constituent hydrogen and oxygen atoms, which may escape to space. Also, water reacts chemically with other substances and thereby is locked up in various compounds. Annually, the addition of water from volcanic eruptions roughly equals the loss of water through photodissociation of water vapor and chemical reactions. This balance of give and take has prevailed on Earth for perhaps hundreds of millions of years.

The fixed quantity of water in the Earth's system is distributed in all three phases among oceanic, terrestrial (land-based), atmospheric, and biospheric reservoirs. The ceaseless movement of water among its various reservoirs at the planetary scale is known as the **global water cycle** (Figure 1.9). In brief, water vaporizes from ocean and land to the atmosphere where winds can transport water vapor thousands of kilometers. Clouds form and rain, snow and other forms of precipitation fall from clouds to Earth's surface and recharge the ocean and the terrestrial reservoirs of water. From terrestrial reservoirs, water seeps and flows back to the ocean. The continuity of the global water cycle is captured in a verse from Ecclesiastes: "Every river flows into the sea, but the sea is not yet full. The waters return to where the rivers began and starts all over again."

Transfer of water between Earth's surface and the atmosphere is the key pathway in the global water cycle. Water from Earth's surface cycles into the atmosphere via evaporation, sublimation, and transpiration. Ultimately, solar radiation supplies the heat energy for these phase changes of water and powers the global water cycle. **Evaporation** is the process whereby water changes from a liquid to a vapor. Water evaporates from the surface of bodies of water including the ocean, lakes and rivers as well as from soil and the damp surfaces of plant leaves and stems. About 85% of the total annual evaporation in the Earth system takes place at the ocean surface so that the ocean is the principal source of water in the atmosphere. During **sublimation** water changes directly from a solid to a vapor without first becoming liquid. Through sublimation, snow banks shrink and patches of ice on sidewalks and roads disappear even while the air temperature remains below freezing.

Transpiration is the process whereby water that is taken up from the soil by plant roots eventually escapes as vapor through the tiny pores (called stomates) on the underside of green leaves. On land during the growing season, transpiration is considerable and is often more important than direct evaporation of water in delivering water vapor to the atmosphere. A single hectare (2.5 acres) of corn typically transpires 28,000 to 38,000 liters (7400 to 10,000 gallons) of water per day. Transpiration accounts for about 10% of the water vapor component of the atmosphere.

Water moves from the atmosphere to land and ocean via condensation, deposition, and precipitation. **Condensation** is the process whereby water changes phase from vapor to liquid (in the form of small droplets). Water droplets forming on the outside surface of a cold can

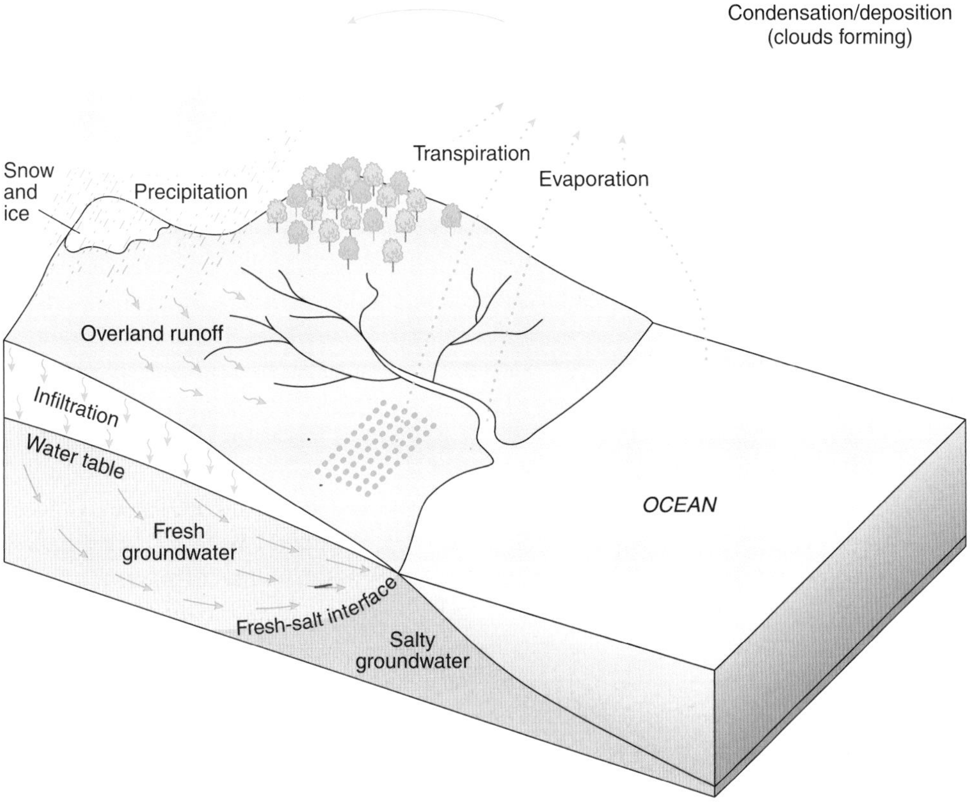

FIGURE 1.9
Schematic representation of the global water cycle.

of soda on a warm, humid day is an example of condensation of water vapor. **Deposition** is the process whereby water changes directly from vapor to solid (ice crystals) without first becoming a liquid. Appearance of frost on an automobile windshield is an example of deposition of water vapor. Condensation or deposition within the atmosphere produces clouds, only a small fraction of which produce precipitation. **Precipitation** is water in liquid or frozen form (i.e., rain, drizzle, snow, ice pellets, or hail) that falls from clouds and reaches Earth's surface.

Significantly, evaporation (or sublimation) followed by condensation (or deposition) purifies water. As water vaporizes (evaporation, transpiration, or sublimation), all suspended and dissolved substances such as sea salts are left behind. Through this natural cleansing mechanism, ocean water that originally was too salty to drink eventually falls as freshwater precipitation, replenishing reservoirs on Earth's surface. Purification of water through phase changes is known as **distillation**. Although precipitation is fresh water, it is not free of all dissolved or suspended materials. Condensation and deposition take place on tiny aerosols suspended in the atmosphere, slightly altering the chemical composition of cloud and precipitation particles. Also, as precipitation falls through the atmosphere, it dissolves or captures gases or suspended particles that also change its composition. (This is the origin of *acid rain*.)

Return of water from the atmosphere to Earth's surface via condensation, deposition, and precipitation completes an essential sub-cycle in the global water cycle. To learn more about the atmosphere-Earth's surface sub-cycle of the global water cycle, compare the movement of water between the continents and the atmosphere with that between the ocean and atmosphere. The balance sheet for inputs and outputs of water to and from the various global reservoirs is the *global water budget* (Table 1.2).

Over the course of a year, the total mass of water that falls as precipitation (rain plus melted snow) on land exceeds the total mass of water that vaporizes (via evaporation, transpiration, and sublimation) from land by about one-third. (This imbalance occurs because landmass characteristics force certain precipitation-forming mechanisms.) Over the same period, the total mass of water falling as precipitation on the ocean is less than the total mass of water that evaporates from the ocean. (Evaporation is more important because the ocean surface is a nearly limitless source of water vapor.) The global water budget thus indicates an annual net gain of water mass on the continents and an annual net loss of water mass from the ocean. The annual excess of water on the continents about equals the annual deficit from the ocean. But year after year the continents do not get any wetter and the ocean basins are not drying up because the excess water on land seeps and flows back to the sea, thus completing the global water cycle. The net flow of water

TABLE 1.2

Global Water Budget

Source	*Cubic meters per yr*	*Gallons per yr*
Precipitation on the ocean	$+3.24 \times 10^{14}$	$+85.5 \times 10^{15}$
Evaporation from the ocean	-3.60×10^{14}	-95.2×10^{15}
Net loss from the ocean	-0.36×10^{14}	-09.7×10^{15}
Precipitation on land	$+0.98 \times 10^{14}$	$+26.1 \times 10^{15}$
Evapotranspiration from land	-0.62×10^{14}	-16.4×10^{15}
Net gain on land	$+0.36 \times 10^{14}$	$+09.7 \times 10^{15}$

from land to sea also implies an equivalent net flow of water in the atmosphere from sea to land.

The flow of excess surface and subsurface water from the continents to the ocean has important implications for the composition of seawater and the global distribution of water pollutants. The ocean is the ultimate destination for all substances dissolved or suspended in surface or subsurface waters. This is one (but not the only) source of salt in seawater. The ocean is also the ultimate destination of contaminants discharged into waterways (e.g., rivers, streams, canals). These contaminants primarily end up in estuaries and other coastal waters, which are also the most productive portion of the ocean. Not surprisingly, considerable debate centers on the capacity of the ocean to assimilate waste—especially toxic and hazardous industrial waste.

Precipitation reaching the land surface can follow one of several pathways. Some precipitated water vaporizes (evaporates or sublimates) directly back into the atmosphere and some is temporarily stored in lakes, snow and ice fields, or glaciers. Some either flows on the surface as rivers or streams (*runoff component*) or seeps into the ground as soil moisture or groundwater (*infiltration component*). The amount of water that infiltrates the ground compared to the amount that runs off depends on rainfall intensity, vegetation, topography, and physical properties of the land surface. For example, rain falling on frozen ground or city streets mostly runs off whereas rain falling on unfrozen sandy soil readily soaks into the ground.

During the growing season, plant roots take up some soil moisture and almost all of that water (98%) transpires back to the atmosphere. Water that seeps to greater depths may completely fill the spaces between particles of soil or sediment or the fractures within rock, constituting *groundwater*. In most environments, groundwater flows very slowly in the subsurface from recharge areas at Earth's surface toward discharge areas including wells, springs, rivers, lakes, and the ocean. Where the groundwater reservoir is located well below surface water reservoirs, water may seep out of rivers and lakes and recharge the groundwater reservoir.

Rivers and streams plus their tributaries drain a fixed geographical area known as a *drainage basin* (or watershed). The quantity and quality of water flowing in a river depends on the climate, topography, geology, and land use in the drainage basin. A drainage basin may also include lakes or glaciers, temporary impoundments of surface water.

Observing the Ocean

Scientists observe the ocean not only out of curiosity but also to better understand the ocean's role in the Earth system. Scientists want to know, for example, the physical and biological factors governing the distribution of marine life and how the ocean influences seasonal and long-term variability in climate. Answers to these and other questions require observational data on the ocean's properties and processes. Today, scientists observe the ocean using both in situ and remote sensing techniques. With an *in situ* measurement, the sensor is immersed in the medium that is being monitored. *Remote sensing* refers to acquisition of data on the properties of some object without the sensor being in direct physical contact with the object.

IN SITU MONITORING OF THE OCEAN'S DEPTHS

While data collected remotely via sensors onboard Earth-orbiting satellites are extremely valuable, oceanographic ships, instrumented buoys, floats, gliders, piloted submersibles, autonomous instrumented platforms and vehicles, and undersea observatories are in situ sources of essential ocean data for surveys and field experiments (Chapter 13). A great deal of ocean research has been and is still done from ships (Figure 1.10). Getting out and staying at sea for an extended period is still a reliable way to investigate what cannot be detected remotely. Furthermore, satellite-borne monitoring instruments, such as radar, can directly observe only the ocean's surface because water is nearly opaque to electromagnetic radiation. Therefore, other techniques are necessary to examine the ocean below the surface. Instruments lowered from ships are used to sample ocean water at depth. Instruments moored on the sea floor monitor properties of the ocean beneath the surface waters. As we will see in subsequent chapters, a variety of in situ instruments and techniques have contributed substantially to our understanding of the ocean. Consider some examples.

One in situ method uses the speed of sound waves in water to measure ocean-water temperature and its variation through time. Ocean water is highly transparent to sound, just as the atmosphere is nearly transparent to visible light. A sound pulse from a transmitter on one side of the ocean can be detected by a sensitive receiver thousands of kilometers away on the other side of the ocean. Whales, for example, take advantage of this property of seawater to communicate long distances across an ocean basin, to attract mates and to locate food.

Sound travels faster in warm water than in cold water. Thus measuring the speed of sound in seawater can be used to determine the average water temperature between the transmitter and the receiver. Furthermore, variations in sound speed between many transmitters and receivers can be combined to obtain a three-dimensional representation of seawater temperature. This technique, called *acoustic thermometry*, is similar in concept to the X-ray scans used to examine the internal organs of humans. We have more to say about acoustic thermometry in Chapter 3.

FIGURE 1.10
Launch of a CTD over the side of the R/V Thompson. The CTD measures conductivity (salinity), temperature, and depth of the water. [Courtesy of NOAA Ocean Explorer]

A **profiling float** is an instrument package that measures vertical profiles of ocean water temperature and salinity (Figure 1.11). In 1998, scientists from more than 30 nations plus the European Union began deploying Argo floats worldwide. (*Argo* is the name of the ship whose captain was the Greek mythological figure Jason.) The initial goal was to deploy 3000 free-drifting floats providing coverage of one sensor per 3 degrees latitude and longitude. That goal was realized on 1 November 2007. Typically, an Argo float is programmed for a 10-day cycle during which it sinks to a depth of 1000 m (3300 ft) where it drifts for approximately 9 days before sinking to a maximum depth of 2000 m (6600 ft). From that depth, sensors measure temperature and conductivity (a measure of salinity) through ocean layers as the float ascends to the surface. At the surface, data and the float's location are transmitted to a communications satellite and made available to the scientific community and the public via the Internet within 24 hours.

Already, the Argo float has proven its value by increasing the accuracy of estimates of the heat storage in the ocean, an important factor in predicting the climate

FIGURE 1.11
A Canadian APEX float prior to launch. This instrument is designed to take profiles of temperature and salinity to a depth of 2000 m. [Photo by Howard Freeland. Courtesy of the Argo Project.]

future and changes in sea level (Chapter 12). These data have also made possible development of more realistic coupled ocean/atmosphere models used for seasonal climate forecasts. Finally, Argo float data provide insights on the dynamics of air-sea interaction during hurricanes and tropical storms. We have more to say about Argo floats in Chapter 6.

Recent years have seen the emergence of *deep-sea cabled observatories*, which are placed on the sea floor and linked to a mainland facility by fiber-optic and power cables. Instruments on the sea floor and in the water column, supported by a power grid, provide real-time data. We have more to say about this topic in Chapter 13.

REMOTE SENSING BY SATELLITE

An important source of remotely sensed ocean data is Earth-orbiting satellites, which routinely monitor the ocean's surface waters, the atmosphere, and other components of the Earth system. Powerful computers collect, process, and analyze enormous quantities of satellite-acquired environmental data. In only minutes, sensors onboard an ocean-observing satellite collect as much data as an ocean-research vessel operating at sea continuously for a decade or longer. Instruments observing Earth from orbiting spacecraft measure selected wavelengths (or frequencies) of electromagnetic radiation reflected or emitted by the various components of the Earth system. **Electromagnetic radiation** (or simply radiation) describes both a form of energy and a means of energy transfer. Forms of radiation include gamma rays, X-rays, ultraviolet (UV) radiation, visible light, infrared (IR) radiation, microwaves, and radio waves. Together, these forms of radiation make up the **electromagnetic spectrum** (Figure 1.12). All types of radiation travel as waves at the speed of light and the different segments of the electromagnetic spectrum are differentiated by wavelength or frequency. *Wavelength* is the distance between successive wave crests or troughs whereas *frequency* is the number of wave crests or troughs passing a point during a period of time, usually a second (Figure 1.13).

The clear atmosphere is essentially transparent to visible light and is selectively transparent by wavelength to other types of radiation such as infrared. Satellite-borne sensors monitor these forms of radiation to gather information on atmospheric processes and properties. Ocean water is much less transparent to electromagnetic radiation than is the atmosphere so that remote sensing by satellite is essentially limited to obtaining data on surface or near-surface ocean processes and properties.

The earliest ocean observations from space came from sensors on meteorological satellites orbited in the 1960s. Beginning in 1978, satellites were launched specifically to monitor the ocean. Now instruments flown on Earth-orbiting satellites routinely provide global images of ocean conditions, which are typically summarized and updated every few days.

Satellite-borne sensors are either passive or active. Passive sensors measure radiation coming from the ocean surface, that is, visible solar radiation reflected by the ocean surface and invisible infrared radiation emitted by the ocean surface. Among the ocean properties measured by passive radiation sensors are sea-surface temperatures (SST) and water color, a measure of marine productivity or sediment concentration. Radar instruments mounted on satellites are active sensors; they emit pulses of microwave radiation and then record the reflected signal to measure surface roughness (an indicator of surface wind speeds and wave heights) as well as ocean-surface elevations, which are used to map bottom topography and surface currents. Techniques to measure salinity and other ocean properties

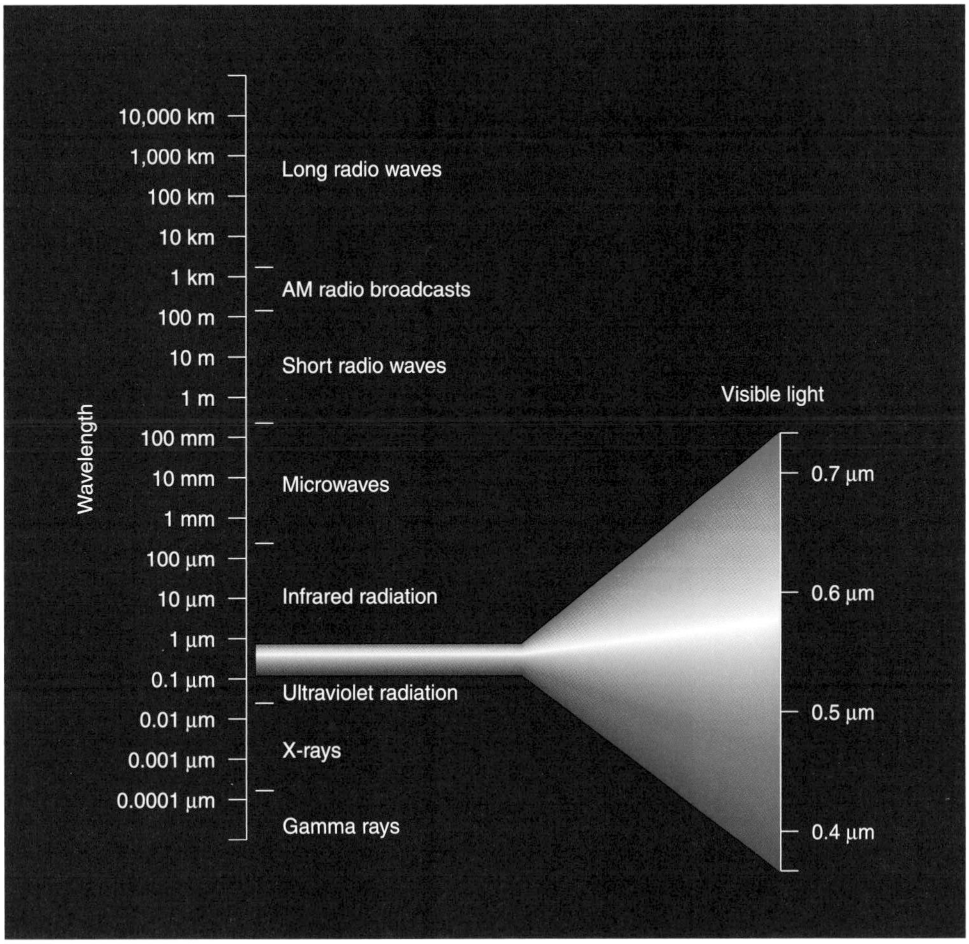

FIGURE 1.12
The electromagnetic spectrum. The various forms of electromagnetic radiation are distinguished by wavelength in micrometers (μm), millimeters (mm), meters (m), and kilometers (km).

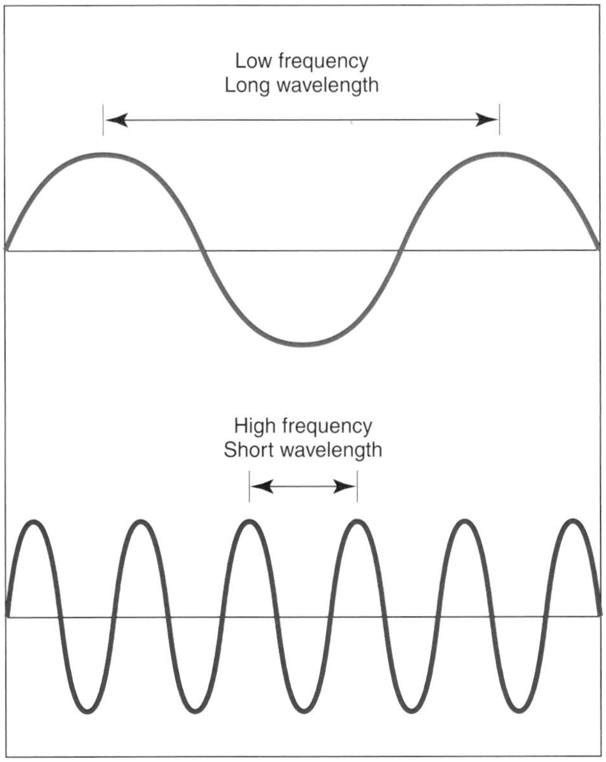

FIGURE 1.13
Wavelength is the distance between two successive crests or, equivalently, the distance between two successive troughs. Wavelength is inversely related to wave frequency.

remotely are under development. Such measurements permit studies of ocean surface "weather" (week-to-week variations) as well as ocean "climate" (variability over decades to centuries).

Satellites that monitor the ocean are in either geostationary or polar orbits. Most of us are familiar with images provided by geostationary satellites, which are launched into relatively high orbits (36,000 km or 22,300 mi) (Figure 1.14). A **geostationary satellite** revolves around Earth at the same rate and in the same direction as the planet rotates so that the satellite always remains over the same point on the equator and its sensors monitor the same portion of Earth's surface. For this reason, these satellites are sometimes described as *geosynchronous*. Each geostationary satellite views about one-third of Earth's surface and five satellites are needed to provide complete and overlapping coverage of the globe between about 60 degrees N and 60 degrees S. Considerable distortion sets in poleward of about 60 degrees latitude.

A **polar-orbiting satellite** is in a relatively low-altitude (800 to 1000 km, 500 to 600 mi) orbit that passes near the north and south poles (Figure 1.15). Earth rotates eastward under the satellite whose orbit remains fixed in space so that sensors view successive strips of Earth. A polar-orbiting satellite that follows the Sun (Sun-synchronous) passes over the same area twice each 24-hr day. Other polar-orbiting satellites are positioned so that they require several days before passing over the same point on Earth's surface.

The flood of observational data from satellite-based sensors is analyzed and stored by computers. Internet websites now make available details of ocean surface properties that were impossible to obtain only a decade ago.

An example of the value of remotely-sensed data is the determination of the distribution and abundance of chlorophyll (plant pigments). Satellite-derived images of "ocean color" locate much of the food used in marine ecosystems and indicate how its concentration varies with time (Chapter 9). (These same sensors also are used to map the distribution of vegetation on land.) Thus large areas of Earth's marine ecosystems can be routinely studied. (Refer back to Figure 1.6 for a sample image.)

Another example of remotely sensed data used in ocean studies is sea-surface temperature (SST) patterns obtained by infrared (IR) sensors onboard satellites. The intensity of radiation emitted by an object increases rapidly as the surface temperature of the object rises. By calibrating temperature against IR-emission, a satellite sensor measures SST and can distinguish between warm and cold ocean currents. For example, Figure 1.16 is an infrared satellite image of the world ocean SST averaged over three days. SST is color coded so that oranges and reds represent the highest temperatures. In this Northern Hemisphere winter depiction, the highest SST values were primarily in the Southern Hemisphere.

Studies of the ocean in the Earth system also require satellite-based communications systems. With communications satellites, scientists working onboard research

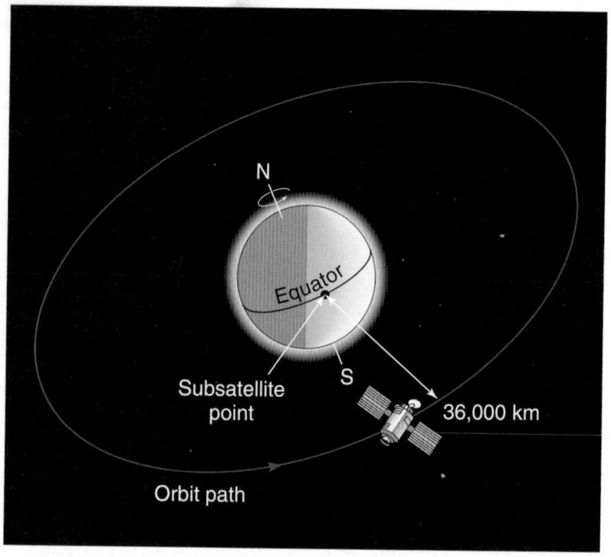

FIGURE 1.14
A satellite in geostationary orbit about the Earth.

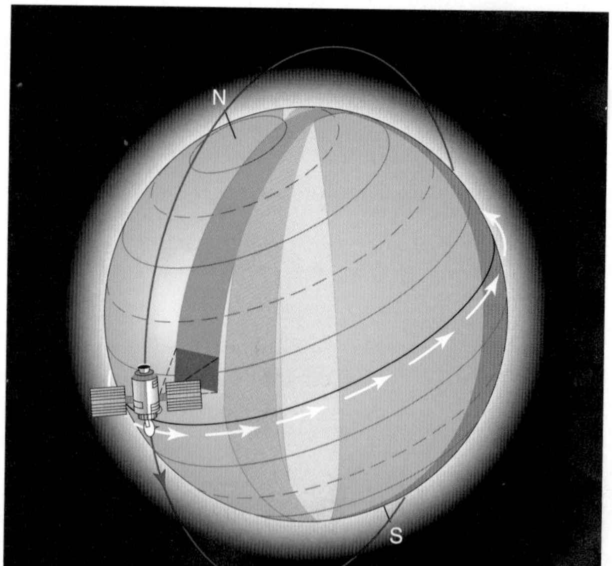

FIGURE 1.15
A satellite in polar orbit about the Earth.

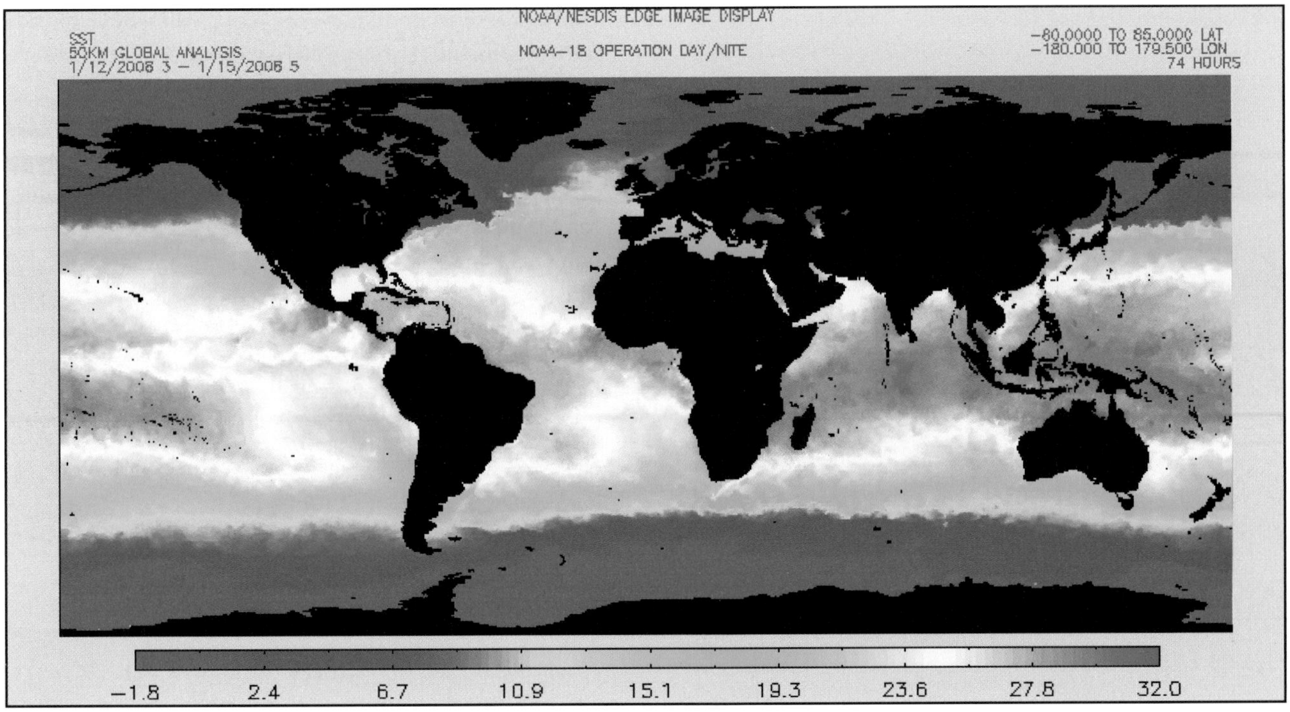

FIGURE 1.16
Satellite-derived sea-surface temperatures (SST) color coded in °C averaged over the period 12-15 January 2008; highest temperatures are shown in orange and red. [Courtesy of NOAA/NESDIS]

ships at sea can communicate nearly instantaneously with labs, computers, or colleagues anywhere in the world. They can also work with schools to provide a virtual shipboard experience for students in their classroom. Remotely operated instruments on drifting buoys, or on unmanned autonomous underwater vehicles, can send the data they collect to scientists on shore via communications satellites. These communications capabilities are now routinely combined as one system for observing both the ocean and atmosphere.

Modeling the Ocean

A **model** is an approximate representation or simulation of a real system. It incorporates only the essential features (or variables) of a system while omitting details considered non-essential. Models are widely used to investigate the Earth system and its components, including the ocean. Models can be conceptual, graphical, physical, or numerical.

A *conceptual model* is a statement of a fundamental law or relationship. Such a model is used to organize data or describe the interactions among components of a system. For example, conceptual models are used to explore linkages among physical and biological subsystems in the ocean and they enable us to understand why and how ocean water circulates. We use conceptual models throughout this book. A *graphical model* compiles and displays data in a format that readily conveys meaning. For example, ocean scientists make extensive use of bathymetric charts, maps of seafloor elevations and the variations in water depth (Figure 1.17). A *physical model* is a small-scale (miniaturized) representation of a system. For example, before powerful computers were widely available, physical models were used to study the flow of water in harbors and estuaries, such as Puget Sound in Washington. Physical models were used to predict changes in currents that would result from deepening the shipping channels leading into New York Harbor. Although the results of physical modeling were generally useful and easily understood, these models were expensive to build, maintain, and operate.

The physical evolution of some systems (e.g., the orbit of the Moon) can be precisely determined by solving a specific mathematical equation. Other systems such as the fluid ocean or atmosphere are too complex to be described by a simple mathematical formula. In these cases, mathematics approximates the behavior of the system on a computer. A *numerical model* consists

FIGURE 1.17
A color-coded bathymetric chart. [Courtesy of National Geophysical Data Center, NOAA Satellite and Information Services]

of mathematical equations that simulate the processes under study. Observational data are used as initial and boundary conditions as well as to guide and verify model predictions. Usually, a numerical model is initialized using current observational data. With the current state of the ocean and/or atmosphere as a starting point, a prediction is made for some subsequent time interval, say 10 minutes hence. Repetition of this process eventually generates a prediction of conditions for the subsequent 12, 24, 36 hrs or longer. Through this iterative process, a numerical model predicts the future state of a complex system.

Today, because computers are readily available, numerical models have essentially replaced physical models in investigating the Earth system. Weather forecasts are probably the most familiar products of numerical models; in fact, numerical models of the Earth-atmosphere system have been used to forecast weather since the 1950s. Very powerful computers are needed to handle the complex mathematics and huge data sets required for ocean, atmosphere, or coupled ocean-atmosphere numerical models. A single set of complex calculations may require hours to a few days to run. Equations may be altered or new data provided to simulate different situations. For example, a numerical model of New York Harbor is used to predict movements of water and sediment in response to changing wind patterns. For many years, Japanese and American scientists have been developing numerical models that simulate the propagation of a tsunami in the open ocean. Numerical models are powerful tools with which to study the response of Earth's subsystems to long-term environmental change (e.g., global climate change).

As noted, a model is an approximation of a real system and there may be advantages in comparing different models and model runs. Climate change studies, for example, routinely compare the output of multiple numerical models. Interacting atmospheric and oceanic processes are simulated using coupled models of the ocean, atmosphere, and biosphere for predicting responses to global climate change. Numerical models can also be used to simulate currents in ancient ocean basins based on reconstructions of global climate and the shapes of ocean basins at the time. Time scales for these models of ancient oceans can be millions of years.

Humans and the Ocean

Humans impact the ocean and the ocean impacts humans. Humans have altered the Earth system to such an extent that we must now consider human activity to be equal in magnitude to many geological processes. Humans impact the orderly functioning of the ocean (and other components of the Earth system) by altering cycling rates and disrupting the equilibrium of biogeochemical cycles. Agriculture, mining, and electrical power generation are examples of human activities that affect the ocean and other components of the Earth system. Most of the waste byproducts of these activities are widely distributed by winds and/or rivers that discharge into the ocean. The ocean can be responsible for coastal flooding and extensive beach erosion, and supply latent heat for hurricane development. As a preview of topics covered in greater detail later in this book, this section summarizes some of the interactions between society and the ocean.

The ocean influences weather and climate. Ocean waters have a relatively great thermal inertia (resistance to temperature change) so that localities immediately downwind from the ocean have a maritime climate with less summer-to-winter air temperature contrast than inland localities. Among other things, this affects the energy demand for space heating and cooling. The global-scale ocean circulation transports heat energy poleward thereby helping to maintain the air temperature gradient between the equator and poles. As a key player in Earth's climate system, the ocean has a stabilizing effect on climate fluctuations. The ocean is an important factor in both short-term (e.g., El Niño) climate variability and long-term (e.g., global warming) climate change.

By burning fossil fuels, deforestation, and other activities, humans are elevating the level of atmospheric carbon dioxide and enhancing Earth's natural greenhouse effect. As a component of the global carbon cycle, the ocean is the principal sink for atmospheric CO_2, including anthropogenic CO_2, and helps regulate its concentration in the atmosphere. The consequences of global warming include melting glaciers and ice sheets, and expanding seawater that raises sea level. Shrinking sea ice cover in the Arctic is likely to exert a positive feedback that will accelerate warming and melting of sea ice and permafrost in the Arctic region. The impact of global warming on marine organisms and ecosystems is likely to be far-reaching and disruptive.

Humans depend on the ocean for water, food, and other resources. Through the global water cycle, seawater eventually falls to Earth's surface as freshwater precipitation that replenishes domestic water supplies. Marine ecosystems are an important source of food, but some human activities threaten that food supply. Overfishing, bycatch, and introduction of exotic species are among the most challenging problems. In addition, runoff of excess nutrients (i.e., nitrogen and phosphorus compounds) from land to sea contributes to harmful algal blooms and the development of dissolved-oxygen-depleted dead zones. In some locations, humans extract seafloor resources including oil and natural gas, sand and gravel, and mineral resources (e.g., certain metals and diamonds). These activities are governed to a large extent by demand, economic conditions, and environmental regulations. To help conserve marine resources (e.g., coral reefs), some areas of the ocean have been set aside as sanctuaries and reserves.

With continued rapid population growth in the coastal zone, humans are becoming more vulnerable to natural hazards including storm surges (generated by tropical cyclones or other coastal storms), tsunamis (most triggered by submarine earthquakes), and seiches (caused by atmospheric conditions or earthquakes). Humans alter the coast for many reasons including control of flooding from storm surges and sea level rise, preservation of recreational beaches, maintenance of harbors and navigation channels, and protection of homes, roads and other structures from wave erosion or slumping of coastal bluffs. (In some cases human activity actually exacerbates the hazard by developing barrier islands or removing protective mangrove swamps.) The traditional approach has been to armor the coast by constructing breakwaters, jetties, groins and seawalls or to artificially nourish storm-eroded beaches with imported sand. These strategies often are ineffective in the long-term, prompting other approaches that seek to accommodate rather than confront the forces of nature (e.g., strategic retreat).

Conclusions

The Earth system consists of several subsystems: hydrosphere (including the ocean and cryosphere), atmosphere, geosphere, and biosphere. These subsystems adhere to natural laws as they interact globally over time scales ranging from seconds to hundreds of millions of years. Matter and energy cycle through the Earth system via global biogeochemical cycles. Matter cycles in closed systems that obey the law of conservation of matter whereas energy cycles in open systems, as the Earth system continually receives energy from the Sun and from Earth's interior while emitting infrared radiation to space.

Application of natural laws combined with monitoring of the Earth system permits subsystem behavior to be understood, modeled, and future conditions to be predicted. Each subsystem is studied using a variety of observation techniques. In situ methods include profilers and ships. Remote sensing by satellite of Earth's subsystems has greatly improved our ability to predict their behavior as a whole and on global scales over long periods of time. The huge quantity of environmental data that is collected via Earth-orbiting satellites is fed into computer models that generate forecasts of the future state of components of the Earth system. In the next chapter, we focus on ocean-geosphere interactions with special emphasis on plate tectonics and its implications for the evolution and characteristics of ocean basins.

Basic Understandings

- A system is an interacting set of components that behaves in an orderly way according to natural laws. The Earth system consists of four major interacting subsystems, the hydrosphere (including the ocean and cryosphere), atmosphere, geosphere, and biosphere.
- The hydrosphere encompasses water in all three phases (solid, liquid, and vapor), which continually cycles from one reservoir to another in the Earth system. The ocean is the largest reservoir, containing 97.2% of all water on the planet and covering 70.8% of Earth's surface. The next largest reservoirs are the glacial

ice sheets that cover much of Antarctica and Greenland.

- In addition to the Antarctic and Greenland ice sheets, the cryosphere encompasses all of the frozen portion of Earth's hydrosphere, including mountain glaciers, permafrost, sea ice (frozen seawater), and ice bergs.
- Surface currents are wind-driven while deep ocean currents are driven by density changes caused by slight differences in temperature and salinity.
- The hydrosphere is dynamic, with water flowing at different rates through and between reservoirs within the Earth system. It can take days to weeks for water to reach the ocean in river channels and millennia for water locked in ice sheets.
- Earth's atmosphere is a relatively thin envelope of gases and tiny suspended solid and liquid particles (aerosols) that surrounds the planet. Based on the vertical temperature profile, it can be divided into the troposphere, stratosphere, mesosphere, and thermosphere. The atmosphere interfaces with the other subsystems of the Earth system and most weather takes place within the troposphere, the lowest layer.
- Nitrogen (N_2) and oxygen (O_2), the principal atmospheric gases, are mixed in uniform proportions to an altitude of 80 km (50 mi). Not counting water vapor, which has a highly variable concentration, nitrogen occupies 78.08% of the lower atmosphere and oxygen 20.95%.
- The significance of atmospheric gases and aerosols is not determined by concentration. In fact, some of the essentials for life occur in very low concentrations: Water vapor (for the water cycle), carbon dioxide (for photosynthesis), and stratospheric ozone (for protection from ultraviolet radiation).
- The atmosphere is dynamic and circulates in response to temperature gradients that arise from radiational heating and cooling within the Earth system.
- The geosphere is the solid portion of the planet composed of rocks, minerals, and sediments. The lithosphere is the section of the geosphere containing the rigid uppermost portion of Earth's mantle and the overlying crust. The lithosphere is continually modified by surface geological processes (weathering and erosion) and internal geological processes (mountain building, volcanic eruptions).
- Weathering refers to the physical and chemical breakdown of rock into sediments. Agents of erosion (rivers, glaciers, wind, and gravity) remove, transport, and subsequently deposit sediments.
- The biosphere encompasses all life on Earth and is dominated by bacteria and single-celled organisms. Cellular respiration and photosynthesis, where sunlight is available, are processes that are essential for life and exemplify the interaction of the biosphere with the other subsystems of the Earth system. The biosphere is composed of ecosystems, which are communities of organisms interacting with each other and the physical and chemical conditions around them. These organisms include producers (plants), consumers (animals), and decomposers (bacteria, fungi) and occupy trophic levels in food webs.
- The paths of solids, liquids, and gases among the reservoirs of Earth's subsystems are the biogeochemical cycles. These cycles follow the law of conservation of matter, which states that matter can neither be created nor destroyed, but can change chemical or physical form.
- Powered by solar radiation, the movement of water through the reservoirs in the hydrosphere is the global water cycle. Key to the global water cycle is the transfer of water between the Earth's surface and atmosphere. Water enters the atmosphere from the ocean and continents via evaporation, sublimation, and transpiration and leaves via condensation, deposition, and precipitation. The net flow of water is from ocean to atmosphere, atmosphere to land, and land to ocean.
- Scientists observe the ocean using techniques both *in situ*, when the sensor is immersed in the medium being monitored, and *remote sensing*, when the sensor does not directly contact the monitored medium.
- In situ techniques, such as acoustic thermometry, instrumented floats, ships, and buoys are used to monitor ocean properties at depth while remote sensing techniques, such as satellite sensors, can only directly observe the ocean's surface.
- Satellites orbiting the planet are platforms for sensors that monitor the Earth system, including reflected or emitted electromagnetic radiation. Electromagnetic radiation is both a form

of energy and a means of energy transfer, distinguishable based on wavelength and frequency.

- A geostationary satellite orbits the planet matching Earth's rotation so it is always positioned above the same spot on Earth's equator. A polar-orbiting satellite travels along relatively low north-south trajectories that take the satellite across the equator and over polar areas.
- A model is a simulation of a real system that includes only variables considered essential to the system. Conceptual, graphical, physical, and numerical models are used to simulate the Earth system and its component subsystems. Numerical models, which consist of one or more mathematical equations describing the relationship among variables, are increasingly important in studying the ocean.
- The ocean impacts humans and humans impact the ocean. The ocean affects humans through its role in weather and climate, natural hazards, and as a source of resources. Human activities affect the ocean (and other components of the Earth system) by altering cycling rates and disrupting the equilibrium of biogeochemical cycles.

Enduring Ideas

- Earth is a system composed of subsystems (hydrosphere, atmosphere, geosphere, and biosphere) linked by biogeochemical cycles with interactions that are governed by natural laws.
- Water in the Earth system is distributed in all three phases among oceanic (the largest), terrestrial (land-based), atmospheric, and biospheric reservoirs. The ceaseless movement of water and energy among these reservoirs constitutes the global water cycle, with the net flow of water at Earth's surface directed from land to sea.
- Scientists gather data on the ocean via in situ methods and remote sensing. In situ methods include instrumented buoys, floats, ships, piloted submersibles, and undersea observatories. Remote sensing by Earth-orbiting satellites can monitor only the ocean's surface waters because water is nearly opaque to electromagnetic radiation.
- A scientific model is designed to simulate a real system. Conceptual, graphical, physical, and numerical models are widely used in ocean studies, incorporating essential features (or variables) while omitting non-essential details.

Review

1. What is the second largest reservoir in the hydrosphere, and where is it located?
2. Identify and describe the layer of the atmosphere that interfaces directly with the ocean.
3. Distinguish between weathering and erosion.
4. What is the significance of a food web?
5. Define *residence time* and compare the residence time of a water molecule in the atmosphere with that of a water molecule in a glacial ice sheet.
6. What role does the ocean play in the global carbon cycle?
7. Identify the various sources of atmospheric water vapor. Which one of these is the principal source?
8. Compare the rates of precipitation and evaporation for the ocean versus the continents over the course of a year. What does this imply about the direction of the net horizontal flow of water on Earth's surface and in the atmosphere?
9. Distinguish between geostationary and polar-orbiting satellites.
10. Remote sensing by sensors onboard Earth-orbiting satellites enables scientists to monitor ocean properties no deeper than surface waters. Explain this limitation.

Critical Thinking

1. Describe how an understanding of the workings of the Earth system relates to our ability to predict how that system might respond to a large-scale disturbance.
2. Provide some examples of how the atmosphere is coupled with the ocean.
3. Identify and describe some of the major interactions between the hydrosphere and geosphere.
4. Speculate on the source of the salt that is dissolved in seawater.
5. About 99% of the water in the atmosphere occurs in the troposphere. Explain why.
6. Satellite-borne sensors that monitor the ocean are either active or passive. Provide examples of each type of sensor.
7. Explain how the ocean stabilizes Earth's climate system.
8. Give an example of how a human activity has altered the rate of biogeochemical cycling within the Earth system.
9. Give some examples of how the various subsystems (e.g., geosphere, hydrosphere) overlap within the Earth system.
10. List several means whereby a human activity affects the orderly operation of the marine environment.

ESSAY: Asteroids, Climate Change, and Mass Extinctions[a]

Geologists and other scientists have gathered evidence from the fossil record of five major mass extinctions that occurred over the past 550 million years (Table 1). At those times, 50% or more of all species died, indicating a drastic change in Earth's environment because it exceeded the tolerance limits of such a vast number of species. But what caused these mass extinctions?

TABLE 1
Major Mass Extinctions of Plant and Animal Species over the past 550 Million Years

End of Ordovician period	444 million years ago
End of Devonian period	359 million years ago
End of Permian period	251 million years ago
End of Triassic period	200 million years ago
Cretaceous-Tertiary boundary	66 million years ago

Prior to 1980, the most popular explanation for mass extinctions was a gradual decrease in species number (perhaps over millions of years) due to long-term climate change coupled with ecological forces. The father-son team of scientists Luis (1911-1988) and Walter Alvarez of the University of California, Berkeley, proposed a much more dramatic explanation for the *K-T mass extinction* about 66 million years ago at the boundary between the Cretaceous and Tertiary periods. The Alvarez team presented convincing evidence of an asteroid impact, including discovery of iridium (Ir) in sedimentary layers from around the world dating from 66 million years ago. Iridium is a silver-gray metallic element that is extremely rare in Earth's crust but found in high concentrations in asteroids.

The Alvarez hypothesis was further bolstered by features found within and near the impact site. The K-T asteroid, at least 10 km (6 mi) in diameter, produced the Chicxulub crater, a 180-km (112-mi) wide crater on the floor of the ancient Caribbean Sea (Figure 1). Marine sediments gradually filled the crater and geological forces elevated a portion above sea level. Radar images obtained by the Space Shuttle *Endeavour* in 2000 revealed a 5 m (16 ft) deep, 5 km (3 mi) wide trough on the Yucatán Peninsula, Mexico, that may mark the outer rim of the crater. Drilling through the layers of sediment on the floor of the nearby Gulf of Mexico recovered cores of fractured and melted rock from the impact zone.

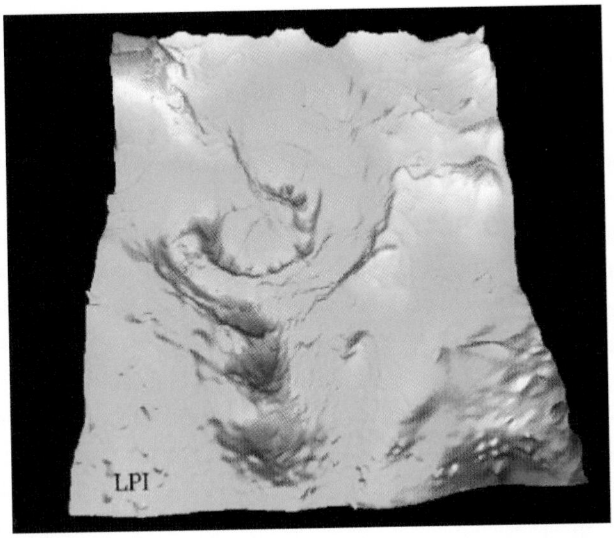

FIGURE 1
The Chicxulub Crater, centered near the town of Chicxulub on Mexico's Yucatán Peninsula, is about 180 km (112 mi) in diameter, represented here as gravity and magnetic field data. It formed about 65 million years ago when a mountain-size asteroid (at least 10 km or 6 mi across) struck Earth's surface. The effects of the impact were thought to be responsible for the extinction of the dinosaurs and about 70% of all species then living on the planet. [Courtesy of NASA, Lunar Planetary Institute, V.L. Sharpton]

Bits of tiny bead-like spherules of glassy rock, which originated as droplets of molten rock blasted into the atmosphere by the impact, are also evidence of the asteroid impact. These droplets, recovered from nearby deep-ocean sediments, cooled as they fell through the atmosphere. Many rocks on land contain mineral grains deformed by the extreme heat and pressure produced by the impact (e.g., shocked quartz) and unusual sediment deposits were produced by enormous waves (tsunamis) generated when the asteroid struck the ocean surface. In addition, a layer of soot indicates burning of considerable vegetation on land.

The K-T asteroid impact had a catastrophic effect on life. Best known is the extinction of the dinosaurs, which had dominated life on Earth for more than 250 million years; but they were not the only victims. The asteroid impact destroyed more than 50% of the other life forms that existed on the planet and caused major extinctions among many groups of marine organisms, including species of plankton.

What caused this ecological disaster? One widely accepted theory is that the asteroid impact vaporized large amounts of sulfur-containing deep-sea sediments, which were blown into the atmosphere. Enormous clouds of tiny sulfate particles resulted, likely augmented by meteoric and Earth materials also thrown into the atmosphere by the impact, greatly reducing the sunlight reaching Earth's surface for 8 to 13 years. Furthermore, precipitation decreased by up to 90%. In this dark, cold and dry environment, most plants died, which starved the herbivores and then the carnivores. Only small animals (such as mammals) could survive by eating dead plants and animals. Eventually, the sulfurous aerosols settled out of the atmosphere, dormant seeds sprouted, and small mammals evolved rapidly to take the place of the dinosaurs. Another possibility is that red-hot, impact-generated particles rained down through the atmosphere making it so hot that most plants and animals were killed directly.

In the 1980s and 1990s, the Alvarez theory of asteroid impact was widely accepted as explaining the cause of all but one of the five major mass extinctions (Table 1). However, a vocal minority of scientists took exception to the preeminent role of asteroid impact, arguing that many of the major mass extinctions were linked to volcanic activity and increased levels of atmospheric CO_2. The largest eruptions of flood basalts closely correspond in age to most of the major mass extinctions. *Flood basalts* consist of many successive lava flows erupting from fissures in Earth's crust, and accompanied by toxic gases released into the atmosphere, including hydrogen sulfide (H_2S), and the greenhouse gases carbon dioxide (CO_2) and methane (CH_4).

Flood basalt eruptions can be enormous. The world's largest flood basalt eruptions, which produced the Siberian Traps, delivered about 4.2 million cubic km (1 million cubic mi) of lava over an area of nearly 7.8 million square km (3 million square mi), approximately 252 to 248 million years ago. This eruption overlaps the time of the great Permian mass extinction (around 250 million years ago), when 90% of all ocean species and 70% of terrestrial vertebrates on Earth were wiped out. No evidence of an asteroid impact has been found to explain the Permian extinction.

According to research conducted by Lee Kump and his colleagues at Pennsylvania State University, the late Permian ocean was stratified. The bottom water had little or no dissolved oxygen while the shallow surface layer was oxygenated. (Most of today's ocean is oxygenated from top to bottom.) With the release of greenhouse gases to the atmosphere during the eruptions that produced the Siberian Traps, the global temperature rose dramatically. This warmed the surface ocean waters, reducing the amount of oxygen absorbed from the atmosphere and reducing the equator to pole temperature gradient, which caused weaker winds and wind-driven ocean surface currents. Consequently, the ocean circulation shifted until the ocean bottom was filled with warm, nearly oxygen-free water. In this environment, anaerobic bacteria dominated, consuming huge quantities of sulfur and producing hydrogen sulfide.

Biomarkers, chemical residue of organisms extracted from ancient strata, of green sulfur bacteria and photosynthetic purple sulfur bacteria were extracted from strata of this age. In time the layer of oxygen-poor, H_2S-rich water became thicker and reached the ocean surface where it escaped to the atmosphere. Highly toxic, especially at high temperatures, H_2S reacts with and destroys stratospheric ozone, allowing lethal levels of solar ultraviolet radiation to reach Earth's surface, which ended the Permian period.

By 2005, the new hypothesis attributing most major mass extinctions to a combination of chemical and circulation changes in the ocean, coupled with global warming due to an enhanced greenhouse effect, was firmly in place.

[a]For much more on this topic, see Ward, Peter D., 2007. *Under A Green Sky*. Washington, DC: Smithsonian Books, 242 p.

ESSAY: Where Did the Water Come From?

Earth is known as the water planet because ocean waters cover almost 71% of its surface. Yet, given how the solar system likely formed, it is surprising that there is any present on Earth at all. Where did the water come from? While there are several hypotheses, there isn't a complete explanation and, in recent years, scientists have offered new alternatives for the origin of water on Earth.

The components of the Earth system (atmosphere, hydrosphere, cryosphere, geosphere, and biosphere) co-evolved through the vast expanse of Earth history. According to astronomers, over 4.5 billion years ago, the Earth, Sun, and entire solar system evolved from an immense rotating cloud of cosmic dust, ice and gases, called a *nebula* (Figure 1). The temperature, density, and pressure were highest at the center of the nebula, and gradually decreased outward. With temperatures exceeding 400 °C (750 °F) at the nebula's center, ice and the lighter elements were vaporized, driving them toward the nebula's outer reaches. Consequently, residual dry rocky masses formed the inner planets, including Earth, suggesting that our water came from elsewhere.

FIGURE 1
The leftmost "pillar" of interstellar hydrogen gas and dust in M16, the Eagle Nebula. [Courtesy of NASA/NSSDC Photo Gallery]

Most of the vaporized and dispersed water in the nebula condensed within comets beyond Jupiter and Saturn. Composed of approximately equal amounts of meteoritic dust and ice, a *comet* is a relatively small mass that moves in a highly elliptical orbit around the Sun. As Jupiter's gravitational attraction strengthened, it may have drawn ice-rich comets from the outer to the inner reaches of the solar system. This would have put them on a collision course with Earth during the latter stages of its formation, producing a veneer of water on the surface.

Accepted until the past decade, the comet hypothesis came into question with the discovery that the ice in comets is chemically distinct from that on Earth. Spectral analyses of three comets that recently approached Earth revealed that their water contains twice as much deuterium as water on Earth. The nucleus of a hydrogen atom consists of a single proton, but *deuterium*, an isotope of hydrogen, has both a proton and a neutron in its nucleus. Deuterium is very rare on Earth. Based on this finding, comets might have accounted for less than half the water on Earth.

While the ice of an *asteroid*, a large rocky body a few kilometers across, contains less deuterium, they are only 10% ice by mass. Some scientists propose that asteroids delivered Earth's water, coming from just inside Jupiter's orbit. At that point in time, the asteroid belt consisted of rocks ranging in size from dust to small planets, and these were scattered inward, including towards Earth. However, such material would have also struck Mars and greatly increased its mass beyond what it is today. Also, on Earth, the ratio of chemicals, such as certain noble gases, is not comparable to that of asteroids. The asteroid hypothesis, therefore, is often dismissed.

A more recent hypothesis centers on a traveling Jupiter. Alessandro Morbidelli of Côte d'Azur Observatory in Nice, France, and colleagues proposed that while the inner planets were forming, a swirl of dust and gases pulled the already formed Jupiter (Figure 2) through the asteroid belt and into the inner solar system. Trailing behind, Saturn's gravitational interaction caused Jupiter to stop its inward motion towards the Sun near Mars' present location, and then reverse direction. In the process, Jupiter forced ice-rich bodies from the outer solar system into new orbits, some of which headed toward Earth and delivered water.

Alternatively, our water may be indigenous to Earth. Though the super hot temperatures would normally vaporize water, on the surface of grains of olivine (a ferromagnesian silicate mineral common on Earth) magnesium holds and stabilizes the oxygen in water molecules. Proposed by Michael Drake of the University of Arizona at Tucson, and colleagues, they suggest that, as the molten Earth cooled and the magma crystallized (to minerals), the water was released into the atmosphere, eventually condensing into clouds that produced rain that filled the ocean basins.

FIGURE 2
Current images of the eight planets in our solar system, placed together for size comparison. [Courtesy of International Astronomical Union]

ESSAY: Co-Evolution of the Ocean, Atmosphere, and Life

Earth's ocean and atmosphere co-evolved through the vast expanse of geologic time. In the beginning, Earth's atmosphere was mostly hydrogen (H) and helium (He) with a few hydrogen compounds including methane (CH_4) and ammonia (NH_3). Because these molecules are relatively light and have high molecular speeds, Earth's weak gravitational field and high temperatures allowed the early atmosphere to escape into space. By about 4.4 billion years ago, the planet's growing gravitational field, however, was sufficiently strong to retain a thin gaseous envelope of volcanic origin, Earth's primeval atmosphere.

The principal source of Earth's atmosphere is *outgassing*, the release of gases from rock through volcanic eruptions and the planet's rocky surface when meteorites struck. Perhaps as much as 85% of all outgassing took place within a million years of the planet's formation, though it continues to this day at a much slower pace. The primeval atmosphere was mostly carbon dioxide (CO_2), with nitrogen (N_2) and water vapor (H_2O), and trace amounts of methane, ammonia, sulfur dioxide (SO_2), hydrogen sulfide (H_2S), and hydrochloric acid (HCl). Radioactive decay of the potassium-40 isotope (^{40}K) in the planet's bedrock added argon (Ar), an inert (chemically non-reactive) gas, to the atmospheric mix. Dissociation of water vapor into its constituent atoms, hydrogen and oxygen, by high-energy solar ultraviolet (UV) radiation contributed a small amount of free oxygen as well while the lighter, faster hydrogen escaped to space. Oxygen also combined with other elements in chemical compounds.

Between 4.5 and 2.5 billion years ago, the Sun was 30% fainter than today. The planet was still warm because of the abundance of CO_2. (Earth's CO_2-rich atmosphere was 10 to 20 times denser than today.) Carbon dioxide slows the escape of Earth's heat to space, and increased the average surface temperature as high as 85 °C to 110 °C (185 °F to 230 °F), significantly elevated from the current average surface temperature of 15 °C (59 °F).

By 4 billion years ago, the planet began cooling and the Earth system changed. Cooling caused atmospheric water vapor to condense into clouds that produced rain. Precipitation and runoff from landmasses gave rise to the ocean that eventually covered 95% of the planet's surface. The global water cycle, which helped cool Earth's surface even more through evaporation, and its largest reservoir, the ocean, were in place. Rain also brought about a substantial decline in the concentration of atmospheric CO_2, which dissolves in rainwater. The weak carbonic acid produced reacts chemically with bedrock, which locked carbon chemically in rocks and minerals. The diminished atmospheric CO_2 lowered surface temperatures further. On land, the physical and chemical breakdown of rock (*weathering*) and erosion delivered carbon containing sediment to the ocean. Runoff washed dissolved CO_2 directly into the sea and some atmospheric CO_2 dissolved in ocean surface waters as their temperatures fell. (CO_2 is more soluble in cold water.)

Although carbon dioxide has been only a minor component of the atmosphere for at least 3.5 billion years, its concentration fluctuated greatly during the geologic past and altered the global climate and life on Earth. All other factors being equal, more CO_2 in the atmosphere means higher temperatures at the Earth's surface. Since peaking at 5000 ppmv about 550 million years ago, the concentration of atmospheric CO_2 generally has declined. However, this interval has been punctuated by many episodes of large-scale volcanic activity, which create temporary upturns in CO_2 concentration and a considerably warmer global climate. An example is the *Middle Eocene Climatic Optimum (MECO)*, about 40 million years ago, when atmospheric CO_2 may have peaked at 4000 ppmv and the deep-sea warmed by an estimated 4 Celsius degrees (7.2 Fahrenheit degrees). Peaks in atmospheric CO_2 correspond to most major mass extinctions of plants and animals on land and in the ocean.

Atmospheric CO_2 levels fluctuated during the Pleistocene Ice Age, 1.8 million to 10,500 years ago, decreasing during episodes of glacial expansion and increasing during episodes of glacial recession. It is not always clear, however, whether variations in atmospheric CO_2 were the cause or effect of these global scale climate changes.

Living organisms also played an important role in Earth's evolving atmosphere, primarily through *photosynthesis*, in which green plants use sunlight, water, and carbon dioxide to produce sugars and oxygen (O_2). Although vegetation is a *sink* for carbon dioxide, geochemical processes were more important than photosynthesis in removing CO_2 from the atmosphere. Based on the fossil record, photosynthesis dates to 2.7 billion years ago when cyanobacteria first appeared in the ocean. However, it was not until 2.3 billion years ago that significant amounts of oxygen began accumulating in the atmosphere. Hundreds of millions of years passed before atmospheric oxygen levels spiked upward in the *Great Oxidation Event*. Why the lengthy delay?

Initially, aqueous oxygen (*dissolved oxygen*) combined with marine sediments before ever escaping to the atmosphere. Once the marine sediments were oxidized, the oxygen dissolved into ocean water and, once that was saturated, into the atmosphere. Also, according to findings reported in 2007 by researchers Lee R. Kump of Pennsylvania State University and M. Barley of the University of Western Australia, the geologic record indicates a shift 2.5 billion years ago from underwater volcanism to terrestrial volcanism, which changed the composition of the eruptive gases from molecules that reacted with oxygen to unreactive molecules. With the subsequent buildup of atmospheric oxygen and the concurrent decline in atmospheric CO_2, oxygen became the second most abundant atmospheric gas after nitrogen within 500 million years.

During the Cambrian period, 542 to 488 million years ago, oxygen vanished from the atmosphere and declined dramatically from ocean waters. Though it is not known why oxygen diminished, anoxic (oxygen-free) conditions spread widely and coincided with a mass extinction. Then, suddenly, oxygen in the ocean and atmosphere surged to levels higher than before with O_2 making up 30% of gases in the lower atmosphere. Research reported in February 2011 by Ohio State University scientist Matthew Saltzman and colleagues attributed this burst of oxygen to giant plankton blooms. Huge amounts of organic matter buried in ocean sediments from that time period would have removed CO_2 from the atmosphere and released O_2. With greatly elevated levels of oxygen, life flourished. New species appeared that were more complex, diverse, and larger than the organisms that had evolved before.

As oxygen emerged as a major component of Earth's atmosphere, the ozone shield formed. Within the *stratosphere*, incoming solar UV radiation drives reactions that convert oxygen (O_2) to ozone (O_3) and back (Chapter 5). Absorption of UV radiation in these reactions prevents potentially lethal intensities of UV radiation from reaching Earth's surface. UV radiation can penetrate ocean water only to shallow depths, so that marine life was able to exist only deep in the ocean but, by 440 million years ago, formation of the *stratospheric ozone shield* made it possible for organisms to thrive in surface waters, and eventually on land.

The concentration of atmospheric oxygen has fluctuated significantly over the past 550 million years, varying between 13% and 31%. These fluctuations were linked to imbalances in the rates of weathering of organic carbon and pyrite (FeS_2), which decreases atmospheric oxygen, and the deposition of these materials with other sediment which increases atmospheric oxygen. Oxygen today composes 21% of the air we breathe.

Nitrogen (N_2), a product of outgassing, is the most abundant atmospheric gas because it is relatively inert and the molecular speeds of N_2 are too slow to escape Earth's gravitational pull. Furthermore, nitrogen is less soluble in water than other atmospheric gases. These factors greatly limit the rate at which nitrogen cycles out of the atmosphere. While nitrogen continues to be generated as a minor component of volcanic eruptions, today the principal source of free nitrogen entering the atmosphere is denitrification, which accompanies the bacterial decay of plants and animals. Countering this input is nitrogen removed from the atmosphere by *biological fixation*, direct nitrogen uptake by leguminous plants such as clover and soybeans, and *atmospheric fixation*, when the heat of lightning causes nitrogen to combine with oxygen to form nitrates (Figure 1).

During the more than 4.5 billion years since Earth's formation, the composition of the ocean and atmosphere co-evolved. Interactions with the geosphere (outgassing, weathering, erosion), hydrosphere (global water cycle), and biosphere (photosynthesis) were key in shaping the chemistry of the modern atmosphere and ocean.

FIGURE 1
High temperatures of lightning causes nitrogen to combine with oxygen to form nitrates that are then washed to the Earth's surface. [Courtesy of Los Alamos National Laboratory]

CHAPTER 2

OCEAN BASINS AND PLATE TECTONICS

A skylight looking into an active lava tube on Kilauea Volcano, Hawaii. Lava flow into the ocean is indicated near the top of the image by the plume of steam. [Photo by Christina Heliker]

Case in Point

Tectonic activity (deformation of Earth's lithosphere) plays a central role in the evolution of the basins occupied by the ocean. In addition, through the years volcanic eruptions and earthquakes produced by tectonic activity both at sea and on land have claimed many lives and caused considerable property damage. (Recall, for example, the role of submarine earthquakes in the generation of tsunamis as discussed in the Case-in-Point of Chapter 1.) Study of the Earth system has greatly improved scientific understanding of tectonic processes and the conditions that contribute to volcanic eruptions, earthquakes, and their after-effects. This basic understanding is enabling scientists to develop techniques to provide the public with some advance warning of volcanic eruptions and, to a lesser extent, earthquakes, thereby promising to save many lives.

The seventh greatest volcanic explosion of the past 10,000 years took place about 3650 years ago on the Greek island of Santorini in the eastern Mediterranean Sea. This violent eruption and the gigantic ocean waves it generated nearly destroyed the island, possibly giving rise to the legend of the "lost city" of Atlantis. About 30 cubic km (7 cubic mi) of ash drifted downwind over the eastern Mediterranean and Turkey. Widespread destruction throughout the region apparently hastened the decline of the Minoan civilization on the nearby island of Crete and the subsequent spread of Greek colonies.

One of the greatest natural disasters of the past few centuries was the eruption of Krakatoa, a volcano on a small island in the Sunda strait between Java and Sumatra in Indonesia. In 1883, after 200 years of inactivity, Krakatoa came to life and was moderately active for several weeks. Then, on 26-27 August, the volcano erupted and violent explosions were heard as far away as Rodriguez Island, 4653 km (2885 mi) across the Indian Ocean, and the atmospheric pressure waves generated were recorded around the world. Two days of gigantic explosions obliterated two-thirds of the volcanic island and triggered enormous tsunami waves which reached estimated heights of 40 m (130 ft). Tsunamis swept inland 16 km (10 mi) on the densely populated islands of Java and Sumatra, killing at least 36,000 people and destroying 165 coastal villages.

Thick clouds of ash carried by winds settled out of the atmosphere producing deep deposits over an area of about 780,000 square km (300,000 square mi), roughly equivalent to the size of Turkey or Pakistan. Thinner deposits covered 3.9 million square km (1.5 million square mi), equal to nearly half the area of the United States. Thick beds of ash also accumulated on the surrounding ocean floor. In all, an estimated 21 cubic km (5 cubic mi) of ash and other larger ejecta were blown into the air, with significant amounts of ash rising 50 km (30 mi) into the stratosphere. Ash remained suspended in the upper atmosphere for several years and was transported by winds around the world, causing brilliantly colored sunsets.

In 1815, another Indonesian volcano, Tambora, erupted violently and impacted the climate in many parts of the world. The eruption blasted a considerable amount of sulfur dioxide (SO_2) into the stratosphere where the gas combined with moisture to produce plumes of tiny sulfurous acid droplets and sulfate particles (collectively called *sulfurous aerosols*). Winds transported sulfurous aerosol plumes around the globe. In the stratosphere, sulfurous aerosols reflect some incoming solar radiation back to space and absorb some solar radiation. The net effect was less solar radiation reaching Earth's surface and cooling of the lower troposphere. Cooling was particularly notable during 1816, referred to in New England as the "year without a summer." Late spring snows and midsummer freezes destroyed crops in the northeast United States and across eastern Canada, bringing considerable hardship to many people.

More recently, in June 1991, Mount Pinatubo, located on Luzon Island in the Philippines, erupted violently, ejecting huge amounts of ash and sulfurous aerosols into the stratosphere (Figure 2.1). Among all 20th century volcanic eruptions, Pinatubo ranked second in size to the 1912 eruption of Katmai in Alaska. The eruption of Mount Pinatubo reduced the summit elevation from 1745 m (5725 ft) to approximately 1485 m (4870 ft) above sea level. Sulfurous aerosols in the stratosphere decreased the amount of solar radiation reaching Earth's surface. The consequent cooling temporarily interrupted the post-1970s global warming trend and was likely responsible for the relatively cool summer of 1992 over continental

FIGURE 2.1
The June 1991 explosive eruption of Mount Pinatubo in the Philippines was rich in sulfur dioxide (SO_2). The resulting sulfurous aerosol veil in the stratosphere caused cooling at the Earth's surface, interrupting the post-1970s global warming trend for a few years. [Courtesy of U.S. Geological Survey photo]

areas of the Northern Hemisphere. The death toll of approximately 350 was relatively low in this eruption in large part due to advance warning and better communications and transportation systems.

Several major volcanic eruptions have occurred in North America, but fortunately none since the continent became heavily populated. Even the spectacular eruption of Mount St. Helens in Washington on 18 May 1980 was relatively small compared to the many massive eruptions that shaped Earth and impacted its inhabitants in the past. Processes that take place in Earth's interior, discussed in this chapter, are responsible for volcanic eruptions.

Driving Question:
What processes shape the lithosphere and how do these processes affect ocean basins?

Many forces working together shape and reshape Earth's lithosphere, creating and destroying both ocean basins and continents. (Recall from Chapter 1 that the *lithosphere* is the rigid outer part of the planet, consisting of the upper mantle and crust.) Forces arise primarily from movement of tectonic plates and geological processes that occur mostly at plate boundaries. In this chapter we learn about these forces, their origins, and the features they produce on the ocean floor. We compare the properties of oceanic crust with those of continental crust in terms of rock type, density, and thickness. The ocean bottom profile shows the transition from continental crust to oceanic crust and is divided into the continental shelf, slope, and rise. Although the shelf, slope, and rise are submerged, they are part of the continental crust whereas the abyssal plains, trenches, and ridges beyond are portions of the oceanic crust. Once thought to be flat and featureless, modern technology has revealed many features on the ocean bottom that provide clues as to the origin of the ocean basins; features include prominent volcanic mountain ranges, deep trenches, and mid-ocean ridges. These and other characteristics of the ocean bottom are associated with processes that take place at divergent, convergent, and transform plate boundaries. We begin by examining the geographical distribution of the world ocean.

Distribution of the World Ocean

Ocean basins and continents are unevenly distributed over Earth's surface (Figure 2.2). The ocean covers about 71% of the planet with the remaining 29% land. However, ocean is less dominant in the Northern Hemisphere than in the Southern Hemisphere. The Northern Hemisphere is 39.3% land and 60.7% ocean whereas the Southern Hemisphere is 19.1% land and 80.9% ocean. The world ocean is composed primarily of the separate, but connected ocean basins, the Pacific, Atlantic, and Indian Oceans as the largest, with the smaller Arctic Ocean basin connected to the far north Atlantic. In 2000, an international agreement defined the Southern Ocean as the waters between 60 degrees S and the Antarctic continent. In the Southern Ocean, the Antarctic Circumpolar Current flows eastward around the Antarctic continent virtually unimpeded by landmasses. The only exception is the northward deflection of currents caused by a combination of the relatively narrow Drake Passage (about 650 km or 400 mi across between the southern tip of South America and Antarctica), the wind field, and bottom topography. A comparison of polar areas shows the Arctic as an ocean surrounded by continents whereas the Antarctic is a continent surrounded by ocean.

Where no convenient continental boundaries exist, arbitrary lines delineate ocean basins. For example, the 150 degrees E longitude line from Australia extending southward to the 60 degrees S latitude line (boundary of the Southern Ocean) separates the Pacific and Indian Ocean basins. North of Australia, the islands of Indonesia divide the Pacific and Indian Ocean basins. Far to the north, the Bering Strait (between Alaska and Siberia) separates the Pacific and the Arctic Oceans. The 70 degrees W longitude line between Cape Horn (the southern tip of South America) and 60 degrees S is considered the boundary between the Pacific and Atlantic Oceans. Still another north-south meridian (20 degrees E) from the Cape of Good Hope (southern tip of Africa) to 60 degrees S separates the Atlantic and Indian Oceans. In addition, numerous seas and gulfs occur adjacent to and connected to the main ocean basins.

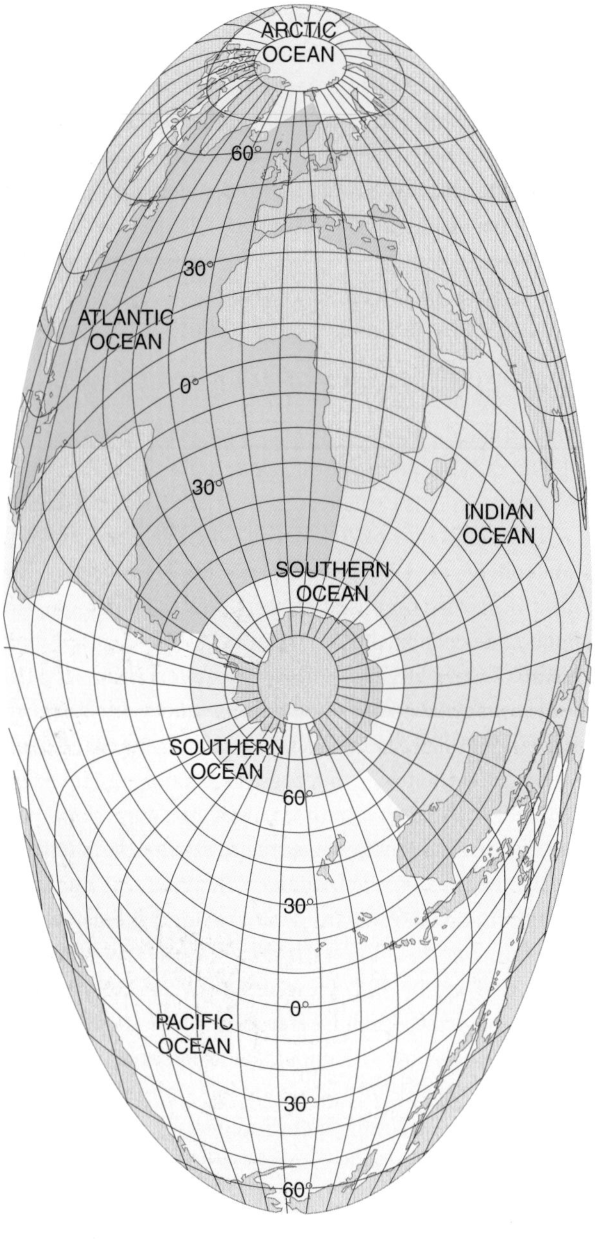

FIGURE 2.2
Ocean basins and continents are unevenly distributed on Earth's surface. Ocean dominates both hemispheres whereas most land is in the Northern Hemisphere.

Oceanic Crust and Continental Crust

Earth's relatively thin solid crust is the portion of the lithosphere that interfaces with the ocean, atmosphere, cryosphere, and biosphere. The crust beneath the ocean differs from the crust of the continents in composition, density, and thickness. In this section, we compare the characteristics of oceanic and continental crust beginning with the basic distinction among rock types.

EARTH MATERIALS

As noted in Chapter 1, internal and surface geological processes continually shape and reshape the lithosphere. These same processes produce a multitude of rock types that compose the crust. Rocks, in turn, are made up of one or more minerals. A *mineral* is a naturally occurring inorganic solid characterized by an orderly internal arrangement of atoms (the basic structural units of all matter) and fixed physical and chemical properties. Some rocks and minerals on land and on the sea floor are economically important resources (Chapter 4).

Rocks composing the crust are classified as igneous, sedimentary, or metamorphic based on the general environmental conditions in which the rock formed. Cooling and crystallization of hot molten *magma* produces **igneous rock** (Figure 2.3). Magma originates in the lower portion of the lithosphere or upper mantle and migrates upward towards Earth's surface. Magma may remain within the crust and cool slowly forming coarse-grained igneous rock such as granite or it may spew onto Earth's surface as *lava* through vents or fractures in bedrock and solidify rapidly forming fine-grained igneous rock such as basalt or glassy material such as obsidian.

Sedimentary rock may be composed of any one or a combination of compacted and cemented fragments of rock and mineral grains, partially decomposed remains of dead plants and animals (e.g. shells, skeletons), and minerals precipitated from solution. Sediments form as rocks undergo physical disintegration and chemical

FIGURE 2.3
ROV (Remotely Operated Vehicle) *Jason* image of magma and igneous rock formed by the eruption of the West Mata volcano nearly 4,000 feet below the surface of the Pacific Ocean, in an area bounded by Fiji, Tonga and Samoa. The orange glow of magma is visible to the left of the sulfur-laden plume. [Courtesy of NOAA and NSF]

decomposition when exposed to rain, atmospheric gases, and fluctuating temperatures at or near Earth's surface. These are weathering processes. Sediments are washed into rivers that transport them to the sea and other standing bodies of water where they settle out of suspension, accumulate on the bottom, and eventually compact into layers of solid sedimentary rock. Sediments are also transported and deposited by wind, glaciers, and icebergs at sea. Most sedimentary rocks have a granular texture; that is, they are composed of individual grains that are compressed or cemented together (e.g., sandstone, shale), although some consist of precipitated minerals and are crystalline (e.g., limestone, rock salt).

Like many sedimentary rocks, **metamorphic rock** is derived from other rocks. A rock is metamorphosed (changed in form) when exposed to high pressure, intense heat, and chemically active fluids—conditions that exist in geologically active mountain belts. Like igneous rocks, metamorphic rocks are crystalline; that is, they are composed of crystals that interlock like the pieces of a jigsaw puzzle. Marble is a common metamorphic rock formed by the metamorphism of limestone ($CaCO_3$) and quartzite is a very durable metamorphic rock formed by metamorphism of sandstone (mostly SiO_2).

Most of the bedrock composing Earth's crust is igneous with some metamorphic rock locally. In many places, thick layers of sedimentary rock and unconsolidated sediments overlie crystalline igneous and metamorphic rocks. Unconsolidated sediments include soil (on land) and clay, silt, sand, or gravel (on land and ocean bottom). Deposits of sediment vary widely in thickness, from a thin veneer to thousands of meters.

Continental crust is mostly granitic, a coarse-grained rock rich in minerals containing silica and aluminum. **Oceanic crust**, on the other hand, is mostly basalt, a fine-grained rock rich in minerals containing iron and magnesium. Continental crust is thicker (20 to 90 km or 12 to 56 mi) and less dense than oceanic crust (only 5 to 10 km or 3 to 6 mi thick). As noted in Chapter 1, both continental and oceanic lithosphere includes the crust and the rigid upper portion of the mantle (Figure 1.5). Oceanic lithosphere has a maximum thickness of about 100 km (62 mi) whereas continental lithosphere ranges in thickness from 100 to 150 km (62 to 93 mi). The lithosphere floats on the underlying **asthenosphere**, a deformable portion of the upper mantle. With temperatures at the base of the lithosphere ranging from about 450 °C to 750 °C (840 °F to 1380 °F), the asthenosphere exhibits plastic-like behavior; that is, it readily deforms in response to stress. As we will see later in this chapter, the lithosphere is continually generated in the oceanic ridge system and drawn down into the mantle at subduction zones.

Granite is less dense than basalt so that continental lithosphere floats higher and extends deeper into the mantle than oceanic lithosphere. In fact, most of the continental lithosphere is below sea level. The continents float in the upper mantle similar to icebergs floating in the ocean. (This form of buoyancy is known as *isostacy*.) Figure 2.4 is a comparison between land elevation and ocean depth. Where ocean waters are less than 1000 m (3300 ft) deep, the ocean bottom is usually a submerged portion of the continental crust (the *continental margin*). Where the ocean is deeper than 4000 m (13,000 ft), the bottom is usually oceanic crust. Mountains higher than 6000 m (20,000 ft) are rare on land, as are portions of ocean basins deeper than 6000 m (20,000 ft). These extreme heights and depths occupy less than 1% of Earth's surface.

Traditionally, *mean sea level (msl)* has served as the fixed reference point for specifying topography on land and bathymetry at sea. Topography is a measure of the elevation of the land surface above msl whereas bathymetry is a measure of ocean depth below msl. In coastal areas, a tide gauge is used to determine mean sea level. However, significant real or apparent changes occur in mean sea level due to global climate change, postglacial rebound, tectonic processes, or coastal subsidence. In response, NOAA's National Geodetic Survey has changed the official reference point from mean sea level to the North American Vertical Datum (a vertical point of reference established in 1988) for the U.S., Canada, and Mexico.

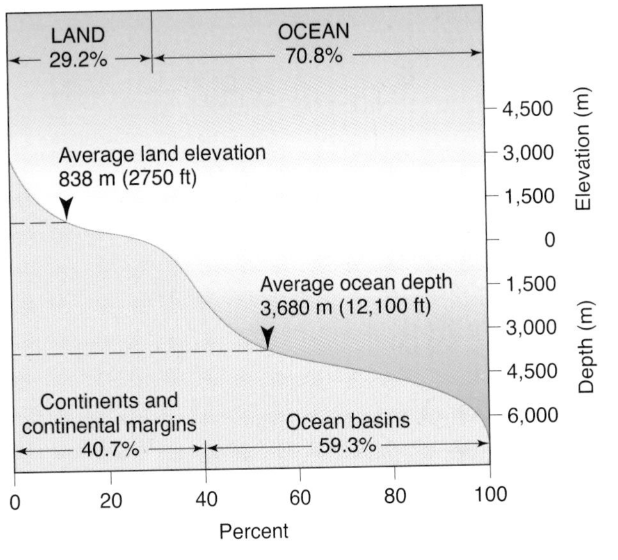

FIGURE 2.4
Hypsographic curve of Earth's surface, showing the relative distribution of ocean and land.

A major factor contributing to the elevation of land above sea level is the temperature of continental crust. Like most other solids, the continental crust expands when heated. Heat originating in Earth's interior and from the decay of radioactive elements in the crust causes the continental crust to expand and rise. Researchers D. Hasterok and D.S. Chapman of the University of Utah argue that heating from below accounts for about half of the elevation of the North American continent. In fact, without this heating, practically all of our continent would be below sea level. Hasterok and Chapman also point out that in addition to crustal temperature, land elevation varies with differences in the composition, density, and thickness (due to tectonic stresses) of continental crust from one place to another.

ROCK CYCLE

Surface and internal geological processes transform rock from one type to another in the **rock cycle** (Figure 2.5). Through the rock cycle, rocks and their component minerals are continually regenerated. Consider an example. Physical and chemical weathering processes fragment an igneous rock mass that is exposed to the atmosphere into sediments which are subsequently transported by running water and deposited in a low-lying basin. In time, the accumulated sediments compact and are cemented together as they gradually convert to sedimentary rock. The mounting weight of the continually accumulating sediments forces the sedimentary rock to greater depths within the crust. (As we will see later in this chapter, subduction also transports crustal material to great depths in the mantle.) Temperature and confining pressure increase with depth so that the rock is eventually metamorphosed, that is, recrystallized into metamorphic rock. At some depth in the subsurface, the temperature is so high that the metamorphic rock melts into magma. The magma may subsequently migrate upward along fractures within the crust, and then cool and crystallize into igneous rock, thereby completing the rock cycle.

Rock transformations involved in the rock cycle are extremely slow; regeneration of rocks and minerals may take many millions of years. Hence, in the time frame of a human lifetime or even of civilization, the rate of regeneration of rock, mineral, and fuel resources (e.g., coal, natural gas) is so slow that for all practical purposes the supply is fixed and finite. For this reason, the resources found in the Earth's crust are essentially *nonrenewable*. We have more to say about resources from the ocean bottom in Chapter 4.

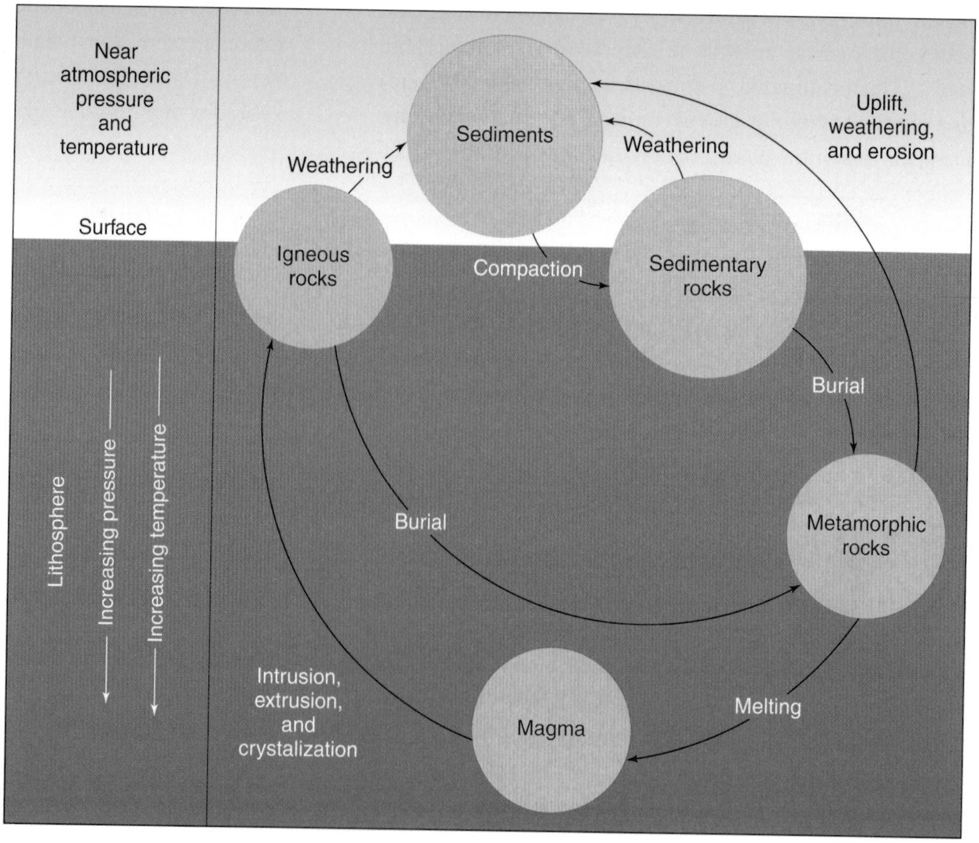

FIGURE 2.5
Through the rock cycle, geological processes (black arrows) convert rock from one type to another.

Ocean Bottom Profile

Ocean depth varies markedly from one place to another. Over large areas water depth is less than 200 m (650 ft); in other areas the water is as deep as 11,000 m (36,100 ft). Average ocean depth is about 3700 m (12,100 ft). Here, we examine the vertical cross-sectional profile of the ocean bottom, including the continental margin and ocean basin (Figure 2.6). In places the ocean bottom is nearly flat and essentially featureless whereas in other places the ocean floor exhibits considerable bathymetric relief. Parts of the ocean bottom are volcanically active with lava interacting chemically with seawater. But over vast areas of the ocean floor, the only significant geological process operating is a very gentle rain of particles that resembles dust settling in a classroom.

CONTINENTAL MARGINS

Based on measurements of water depth, ocean scientists delineate three distinct zones seaward from the coastline. The zone closest to the beach features a very gentle slope extending out to a water depth that averages about 130 m (430 ft). Seaward from there to a depth of about 3000 m (9800 ft) the water depth increases much more rapidly with distance offshore. Then a relatively narrow zone is transitional from the steep slope of the previous zone to the more-or-less flat ocean basin. The initial, gently sloping zone is the **continental shelf**; the second, more steeply sloping zone is the **continental slope**; and the third transitional region is the **continental rise**. The continental shelf, slope, and rise together comprise the *continental margin*. This is not a misnomer because the bedrock of the continental margin is the same as the continental crust. From a geological perspective, the continents do not end at the beach, but at the end of the continental margin. At its outer edge, the continental margin merges with the deep-sea floor or descends into an oceanic trench (Figure 2.6).

The continental shelf is nearly flat, sloping downward less than 1 degree seaward. Although water depth generally increases with distance offshore, the rate of increase is quite small—averaging about 2 m per km (10 ft per mi). Shelf width generally ranges from a few tens of meters to 1000 km (620 mi). As a general rule, the shelf is narrowest where the continental margin is tectonically active (e.g., subduction zones) and is widest where the continental margin is passive (i.e., no plate boundary nearby). Hence, the North American shelf is wider along the passive East Coast (as much as several hundred kilometers) than the tectonically active West Coast (a few kilometers). About 7.5% of the total surface area of the ocean overlies the continental shelves.

The inclination of the continental slope averages about 4 degrees but ranges between 1 and 25 degrees; water depth typically increases by about 50 m per km (265 ft per mi) in a seaward direction, a significantly greater rate of increase than in the shelf zone. In many places such as around the margin of the Atlantic Ocean, the continental slope merges with the more gently sloping continental rise. (The term "rise" here means sloping less steeply to the flat

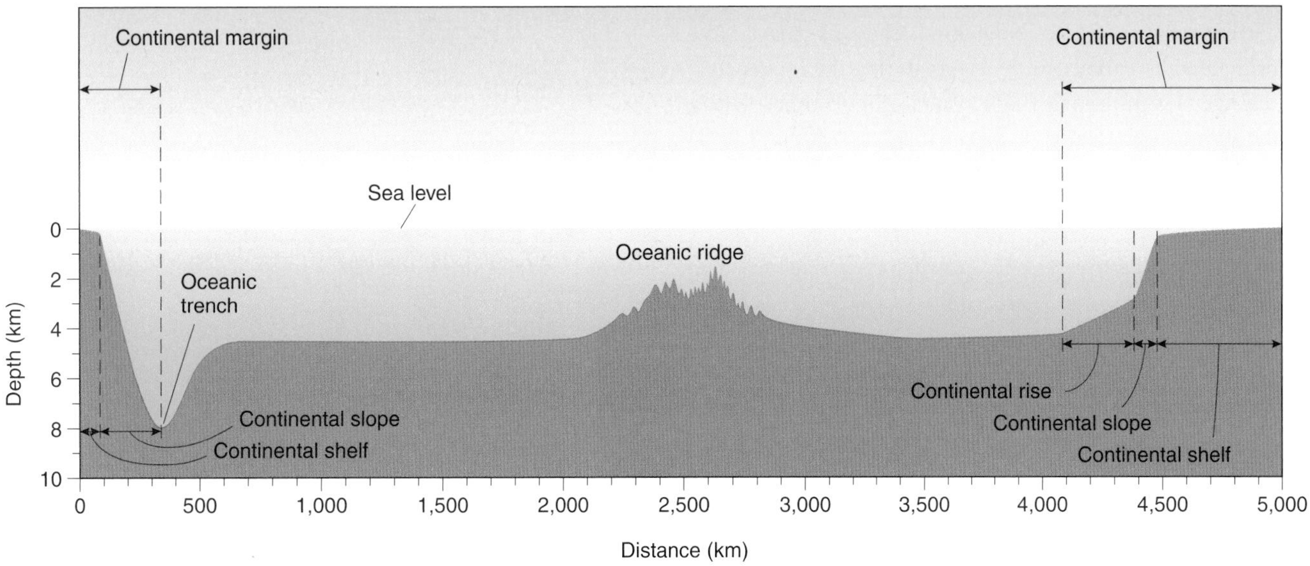

FIGURE 2.6
Cross-sectional profile of the continental margin and ocean bottom with the vertical scale greatly exaggerated.

ocean bottom.) In this passive continental margin, sediment spreads over the ocean floor forming vast, flat **abyssal plains** seaward of the continental rise. Along tectonically active continental margins, such as that surrounding nearly the entire Pacific Ocean basin, the slope descends directly into deep ocean trenches. Land-derived sediment flows into trenches, is incorporated into downward moving rock materials, and is not available to form either a continental rise or abyssal plains. About 15% of the total surface area of the ocean is situated over the continental slopes and rises.

Large numbers of deep, steep-sided **submarine canyons** slice into the continental slope and some run up onto the continental shelf; many of these canyons cut into solid rock (Figure 2.7). A variety of mechanisms are responsible for creating submarine canyons including erosion by turbidity currents or ancient rivers. A **turbidity current** is a gravity-driven flow of water heavily laden with suspended sediment (silt, sand, and gravel) making it denser than normal seawater. In many respects, turbidity currents resemble underwater avalanches. Sediment delivered to the ocean by rivers accumulates on the continental shelf. Eventually the pile of sediments builds to an unstable height and suddenly (perhaps triggered by an earthquake) slides downhill as a unit, scouring the ocean bottom in the process. These sediments accumulate at the base of the continental rise as a series of overlapping **submarine fans**. Turbidity currents have been clocked at speeds as great as 100 km per hr (60 mph), so it is easy to imagine these flows gouging the ocean bottom and clearing submarine canyons of accumulated sediments.

Many submarine canyons appear to be natural extensions of existing rivers lending support to a turbidity current origin. Some submarine canyons may have originated as valleys cut by rivers during the Pleistocene Ice Age (1.8 million to 10,500 years ago). At times during the Pleistocene, sea level was as much as 130 m (425 ft) lower than today (because so much water was locked up in glaciers) and much of the continental shelf was above sea level. With a steeper gradient, rivers flowed across exposed shelves and eroded deep canyons that were subsequently flooded at the close of the Pleistocene when glaciers melted and sea level rose. For example, the Hudson Canyon off the New York Bight likely formed during the last glacial maximum about 18,000 to 20,000 years ago when sea level was lowest (Figure 2.8). At the time, the Hudson River delivered sediment directly to the head of the canyon and turbidity currents were active. Today, the head of Hudson Canyon is located about 250 km (155 mi) offshore in 90 m (300 ft) of water. Extending about 500 km (310 mi) offshore, the Hudson Canyon is one of the largest canyons

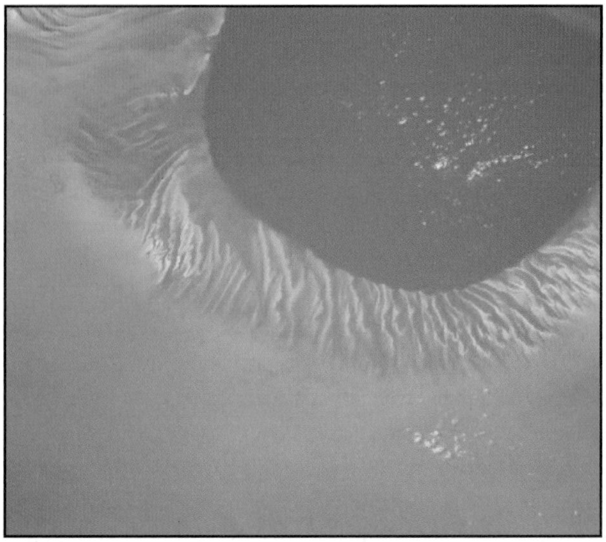

FIGURE 2.7
The darker water in the upper right of this NASA Space Shuttle image is the southern portion of the Tongue of the Ocean in the Bahamas, one of two main branches of the Grand Bahama Canyon. This submarine canyon's nearly vertical walls rise 4285 m (14,000 ft) above the canyon floor. The Grand Bahama Canyon has a length of more than 225 km (140 mi) and a width of 37 km (23 mi) at its deepest point. The lighter blue water covers the shallow Bahama shelf, much of which was above sea level during the last glacial maximum. At the time water draining into the canyon eroded the gullies that are visible at the shelf edge. [Courtesy of NASA]

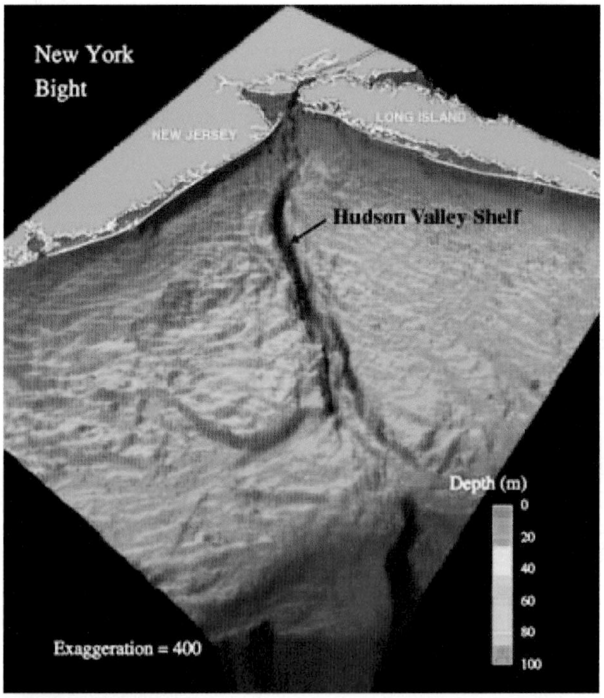

FIGURE 2.8
Bathymetry of the coastal ocean showing New York's Hudson Canyon. This submarine canyon appears to be the natural extension of the Hudson River. [Courtesy of NOAA Ocean Explorer]

on the Atlantic continental shelf. For submarine canyons having no obvious association with existing rivers, sediment sources may be located upstream of longshore currents. As discussed in Chapter 8, longshore currents transport sand and other sediment along the shoreline.

OCEAN BASINS

Fringed by continental margins, **ocean basins** encompass the remaining portion of the oceanic area. Ocean basins have a varied topography featuring deep trenches, seamounts, and submarine mountain ranges. Indeed, undersea terrain is just as diverse as terrestrial terrain and exhibits even greater relief. Nevertheless, much of the ocean bottom (about 42%) is comprised of plains and low hills, most rising no more than about 100 m (330 ft) above the plain. Blanketed with sediments that tend to smooth out any irregularities in the bedrock below, abyssal plains and abyssal hills are typical of ocean basin bathymetry—apart from the 23% of the ocean bottom that is covered by ridge systems.

Ocean scientists use many different instruments and techniques to investigate the properties and composition of the ocean bottom. For more on this topic, refer to this chapter's first Essay.

Plate Tectonics and Ocean Basin Features

The most prominent features of the ocean basins, including trenches and mid-ocean ridge systems, are products of tectonic stresses. Earth's lithosphere is divided into many plates that move very slowly over the face of the globe (Figure 2.9). All plates, except the Pacific, include parts of continents as well as ocean basins. Mountain building and most volcanic activity and earthquakes take place at boundaries between plates, many of which occur either within the ocean basins or along their margins.

EVIDENCE FOR PLATE TECTONICS

Plate tectonics is a unifying concept that combines continental drift plus sea floor spreading to describe the generation, movement, and destruction of Earth's lithosphere and the formation of ocean basins. The

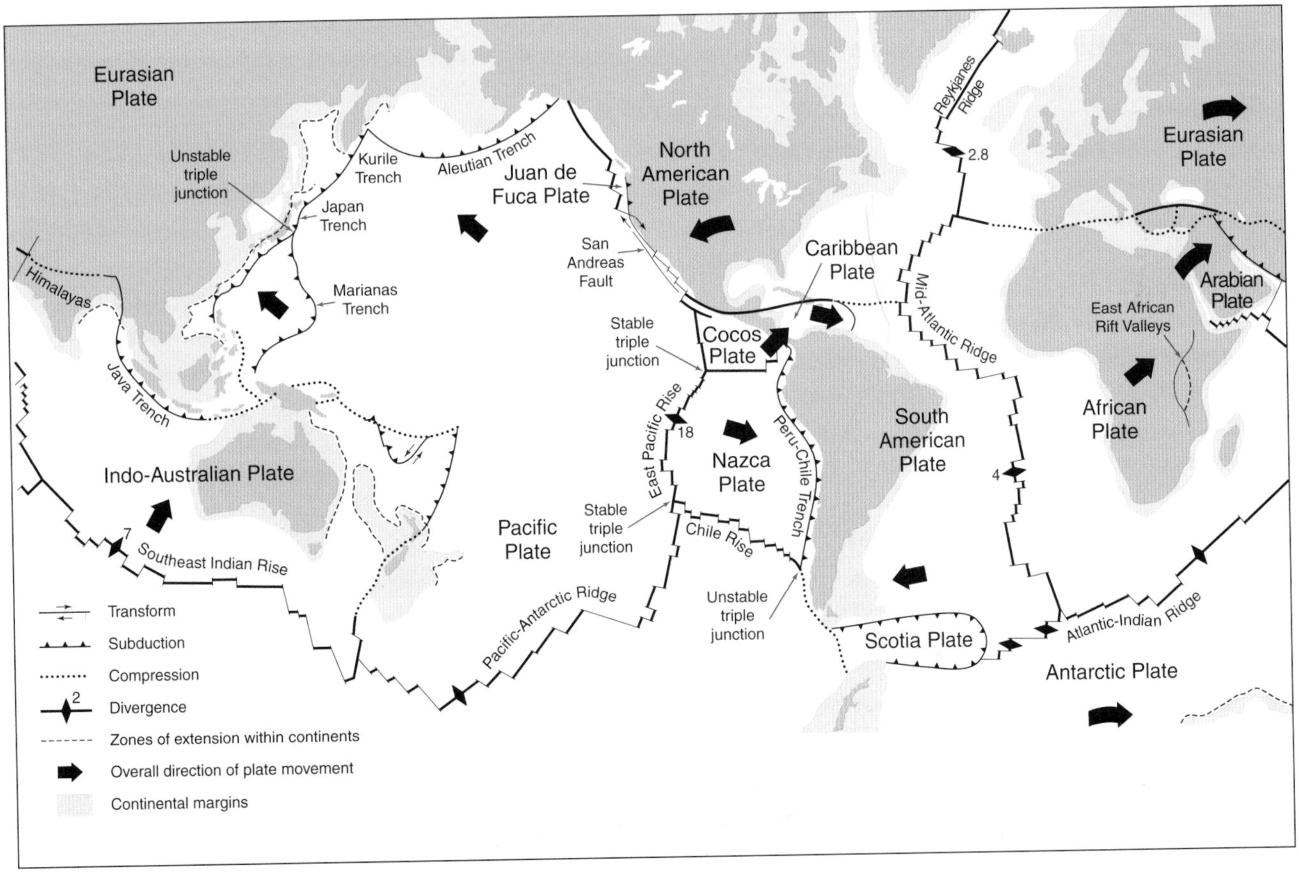

FIGURE 2.9
Divergent, convergent, and transform boundaries of major lithospheric plates. [Adapted from U.S. Geological Survey]

historical evolution of this concept illustrates a systematic form of inquiry involving observation, interpretation, speculation, and reasoning—all components of critical thinking.

Evidence for the continental drift part of plate tectonics dates back at least to the time of the British Philosopher Francis Bacon (1561-1626) who noticed how the eastward bulge of South America closely fits the configuration of a portion of the west coast of Africa. (Comparing the edge of the continental shelves reveals an even better fit than coastlines.) By the middle of the 19th century scientists and others proposed that the Atlantic Ocean formed when landmasses separated during some catastrophic event. Early in the 20th century, the Austrian geologist Eduard Suess (1831-1914) hypothesized that at one time the individual continents of the Southern Hemisphere were together as one huge continent, which he called Gondwanaland. Around 1910, the German meteorologist Alfred Wegener (1880-1930) and the American geologist Frank B. Taylor (1860-1939) independently proposed **continental drift**, the hypothesis that the continents move over the surface of the planet. Wegener proposed that the supercontinent *Pangaea* split into *Laurasia* (encompassing present day North America and Eurasia) and *Gondwanaland* (encompassing South America, Africa, India, Australia, and Antarctica). He based his idea on the observed close fit between continental margins, similarities in fossil plants and animals, and the continuity of rock formations and mountain ranges between continents on either side of the Atlantic Ocean.

Wegener's hypothesis was ridiculed and mostly ignored by geologists on the grounds that he did not provide a scientifically sound mechanism for driving the continents. At the time, most scientists believed that continents and ocean basins were fixed in place; it was difficult to understand how Wegener's continents could plow through solid rock on the bottom of the ocean. At the time, little was known about the ocean floor or Earth's interior. All this began to change in the 1950s and 1960s when investigations of the deep sea floor revealed prominent mountain ranges and deep trenches. In the early 1960s, these findings inspired Harry Hess (1906-1969) of Princeton University and the geological oceanographer Robert S. Dietz (1914-1995) to resurrect an idea first presented in 1929 by the British geologist Arthur Holmes (1890-1965). Hess and Dietz proposed that convection in the Earth's mantle was the driving force behind continental drift.

According to Hess and Dietz, rock in the lower mantle is heated by the decay of certain radioactive elements. This warmer less dense and buoyant material gradually rises within the mantle and then flows along the base of the cool, rigid lithosphere. This process causes the lithosphere above to rise up. Cooling of the mantle material increases its density, causing it to slide and sink back toward the lower mantle where it is reheated, thus completing the convective circulation. These convection currents and the downsloping seafloor drag the lithospheric plates, and are responsible for sea-floor spreading, the divergence of adjacent plates on the ocean bottom. They supply heat and magma to the ocean ridges, and cause subduction, the descent of lithospheric plates into the mantle. Today, some scientists argue that two zones of convection operate in the mantle: deep convection (below about 700 km or 435 mi) and shallow convection (above 700 km) that interfaces with the lithosphere. As described later in this chapter, the convection currents in the mantle can stretch and eventually form a small break in the crust, known as a *rift*.

In subsequent years, evidence accumulated in support of the Hess-Dietz model and sea-floor spreading. Much of this evidence came from deep-sea drilling from specially outfitted ships (discussed in this chapter's first Essay) and includes: (1) the concentration of earthquake activity along plate boundaries, with the deepest earthquakes generally associated with subduction zones (Figure 2.10), (2) enhanced heat flow from Earth's interior along mid-ocean ridges, (3) increasing age of the oceanic crust with increasing distance on either side of the mid-ocean ridges, and (4) increasing thickness of sea-floor sediment with increasing distance from the mid-ocean ridges.

Perhaps the most convincing argument for plate tectonics was the discovery of a distinctive pattern of anomalies in the Earth's ancient magnetic field on the sea floor. We can visualize Earth's magnetic field as emanating from a huge bar magnet centered in the Earth's core with a north and south magnetic pole. Earth's magnetic poles, however, do not coincide with the planet's geographical poles and precise locations vary over time. Currently, the north magnetic pole is located in northwest Canada while the south magnetic pole is in the Southern Ocean. Magnetic lines of force converge toward the two magnetic poles so that the needle on a compass aligns with the magnetic lines of force and points toward the magnetic north.

Oceanic crust contains the mineral magnetite, an iron oxide, which acts like a compass needle in aligning with the planet's magnetic lines of force. This alignment is frozen into the rock when the temperature of cooling lava drops below the *Curie temperature* of 580 °C or 1076 °F, named for its discoverer Pierre Curie

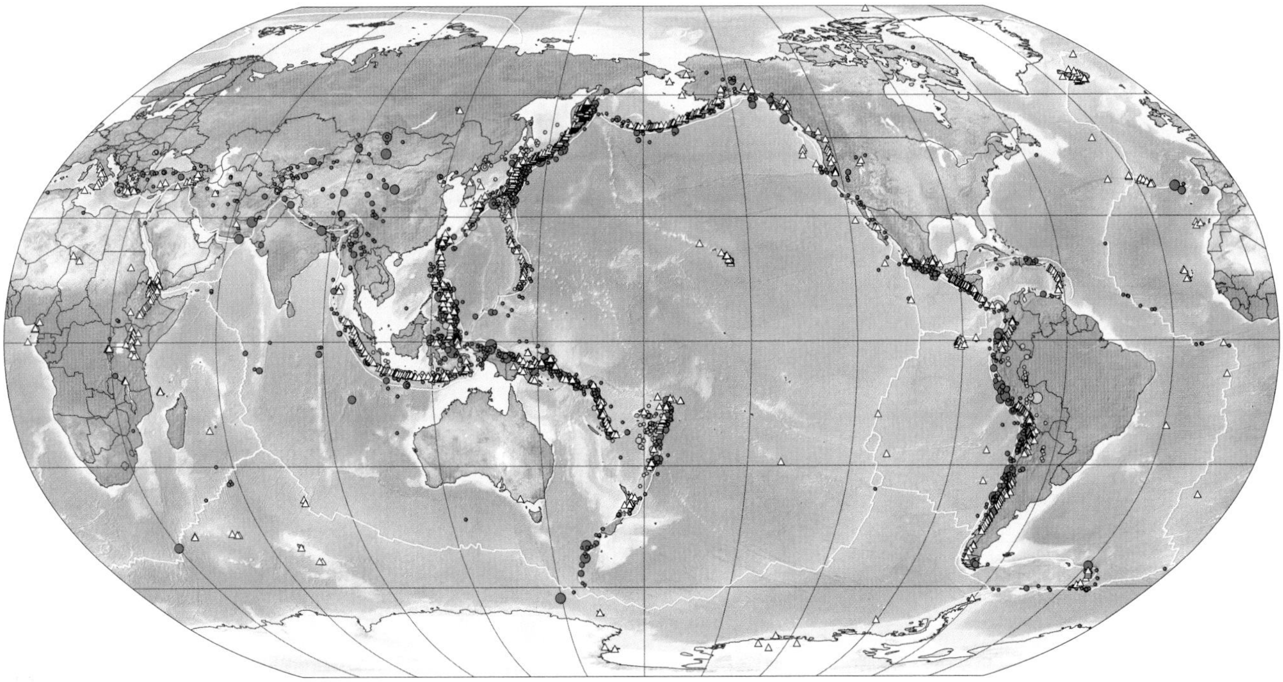

FIGURE 2.10
Most earthquakes (and volcanoes) occur along or near the boundaries of major lithospheric plates. In the figure above, yellow jagged lines represent plate boundaries. Earthquakes are represented by circles; the color red indicating that the earthquake was centered at 0-69 km depth, green indicating 70-299 km depth, and blue indicating 300-700 km depth. The largest circles denote the strongest earthquakes. Active volcanoes are shown by yellow triangles. [Image courtesy of Tarr, A.C., Villaseñor, Antonio, Furlong, K.P., Rhea, Susan, and Benz, H.M., 2010, Seismicity of the Earth 1900–2007: U.S. Geological Survey Scientific Investigations Map 3064, 1 sheet, scale 1:25,000,000.].

(1859-1906). Hence, information on the direction of the planet's magnetic field at the time lava cooled and solidified is locked in the rocks.

Studies of the same magnetite-bearing rock on land reveal times in the past when the polarity of the Earth's magnetic field reversed; that is, the north magnetic pole became the south magnetic pole and vice versa. This rock record indicates 170 reversals over the past 76 million years. Beginning in the 1950s, research vessels and aircraft mapped magnetic polarity reversals over much of the ocean floor. Instruments towed behind ships measured the intensity of the magnetic field and revealed ribbon-like patterns of magnetic anomalies (Figure 2.11). Each anomaly strip was a few kilometers to tens of kilometers wide and thousands of kilometers long and oriented roughly parallel to the mid-ocean ridges. Application of the magnetic land-based chronology to the sea floor enabled scientists to date remotely large areas of the sea floor. Most intriguing was the discovery that the magnetic anomaly pattern on one side of an ocean ridge was the mirror image of the pattern on the other side of the ridge. Furthermore, dating indicated that the oceanic crust became progressively older with increasing distance away from a mid-ocean ridge. In 1963, F.J. Vine and D.H. Matthews of Cambridge University interpreted these anomaly patterns as a record of past magnetic reversals indicating that newly formed crust moved away from the mid-ocean ridges. That is, magnetic anomaly patterns confirmed sea-floor spreading and continental drift and

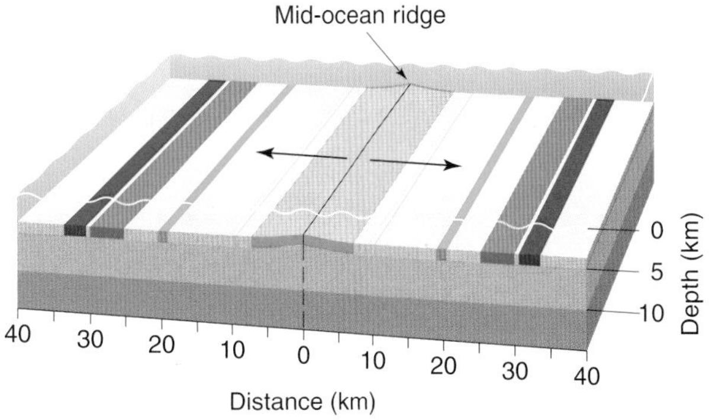

FIGURE 2.11
This block diagram shows the regular pattern of magnetic anomalies on either side of the mid-ocean ridge system. The age of the ocean crust increases with increasing distance from the ridge axis. Light blue stripes represent normal polarity; other colored stripes indicate reversed polarity.

the unifying theory of plate tectonics soon became widely accepted.

Boundaries between the rigid lithospheric plates are principal locales of geological activity. Depending on relative movement, plate boundaries are designated divergent, convergent, or transform. Movements along plate boundaries give rise to a variety of geographic and geological features on the ocean floor and on land.

DIVERGENT PLATE BOUNDARIES

When adjacent plates move apart, they create a **divergent plate boundary**. Diverging lithospheric plates produce rifts (fractures) in the crust through which magma wells up from below (Figure 2.12). Magma (called *lava* when it reaches Earth's surface) at a temperature of about 1200 °C (2200 °F) cools and solidifies rapidly when it comes in contact with much colder seawater (at 2 °C to 5 °C or 36 °F to 41 °F). Through this process, new oceanic lithosphere is generated at divergent plate boundaries. In fact, more than 18 cubic km (4 cubic mi) of new lithosphere is produced in this way every year.

On the seafloor, the outer surfaces of lava flows exhibit characteristic tube-like and pillow-shaped structures (Figure 2.13). These volcanic eruptions at spreading centers are generally tranquil events, unlike the violent explosions that characterize volcanoes associated with subduction zones (discussed below). Also, shallow earthquakes (centered a few kilometers to a few tens of kilometers below the ocean floor) are often associated with these volcanic eruptions.

Newly formed oceanic crust stands higher than nearby older oceanic crust because recently formed warm rock is less dense than older and colder rock. As oceanic crust ages and moves away from plate boundaries, it cools and becomes denser. Also, as the crust is slowly loaded with sediments, it sinks deeper into the underlying asthenosphere (not unlike a ship as it is loaded). Consequently, the older the crust, the deeper is the water above it (Figure 2.14). At divergent plate boundaries (mid-ocean ridge), the ocean bottom might be about 2500 m (8200 ft) deep, but away from the boundary where the crust is 160 million years old, the oceanic crust might be as much as 6000 m (19,500 ft) below sea level. Thus, the ocean floor not only contains information on magnetic reversals and plate movements, but functions as conveyor belts on either side of the mid-ocean ridge system, moving newly formed crust away from divergent plate boundaries.

The mid-ocean ridge is a volcanic mountain range that winds its way from the Arctic Ocean down the middle of the Atlantic Ocean, curves around South Africa and into the Indian Ocean (*Mid-Indian Ridge*), and then into the Pacific Ocean (*East Pacific Rise*). Its total

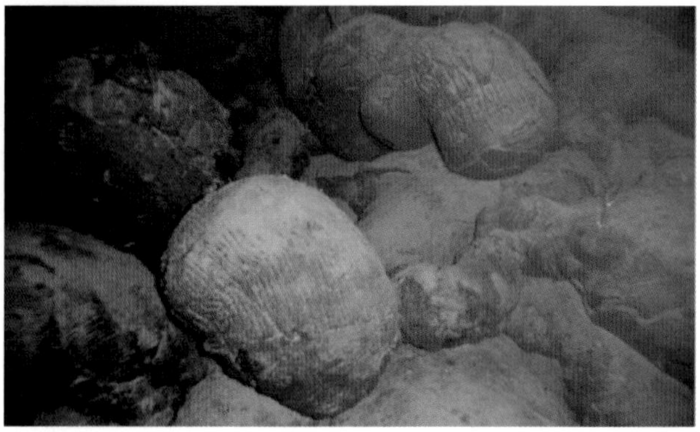

FIGURE 2.13
The pillow shape exhibited by these basaltic flows on the slope off Hawaii is characteristic of lava that cools rapidly on the seafloor. [OAR/National Undersea Research Program (NURP); NOAA]

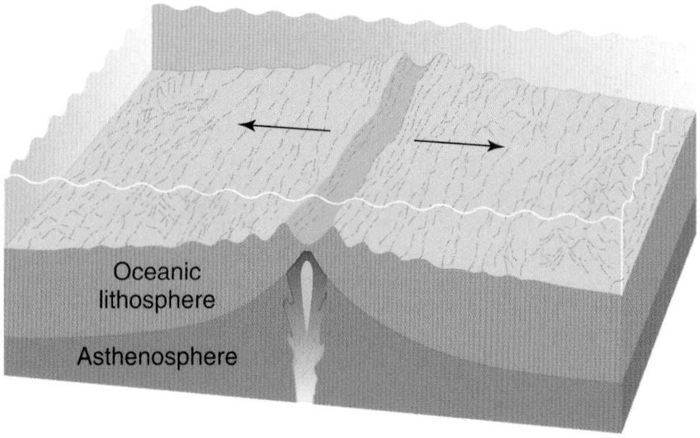

FIGURE 2.12
Divergent plate boundary at a mid-ocean ridge.

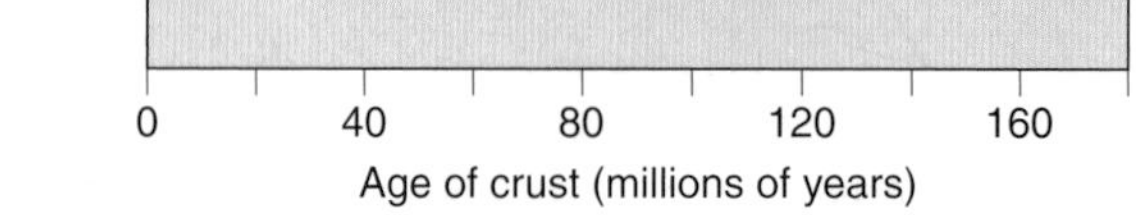

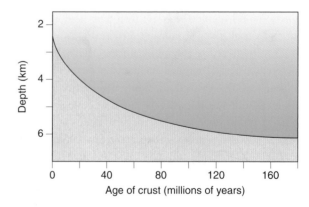

FIGURE 2.14
When formed at a mid-ocean ridge, oceanic crust is about 2.5 km (1.6 mi) below the sea surface. As oceanic crust ages, it cools and becomes denser and moves away from the mid-ocean ridge, accumulating sediment and sinking deeper into the upper mantle. At an age of 160 million years, the oceanic crust is at a depth greater than about 6 km (3.7 mi).

length is about 55,000 km (34,000 mi), much longer than any mountain range on land. Along the mid-ocean ridge a distinction is made between *fast-spreading* and *slow-spreading* boundaries. At fast-spreading boundaries, plates move apart at 100 to 200 mm per yr. With rapid spreading, hot magma is abundant and lava flows as sheets from a central peak, giving the ridge a narrow tent-like profile. An example is the East Pacific Rise, located in the east Pacific Ocean near South America (Figure 2.9). At slow-spreading boundaries, plates move apart at less than 55 mm per yr. With a slower magma supply, the bathymetry is broader and rougher. Rift valleys, 1 to 2 km (0.6 to 1.2 mi) deep and a few tens of kilometers across, occur along the ridge axis. The Mid-Atlantic Ridge is an example of a slow-spreading boundary.

The Mid-Atlantic Ridge divides the Atlantic Ocean bottom into two (western and eastern) roughly symmetrical halves. Each side can be as much as 1600 km (1000 mi) wide with the rim rising about 2500 m (8200 ft) above the adjacent sea floor (Figure 2.15). The Mid-Atlantic Ridge marks a boundary between oceanic plates that have been diverging for perhaps 200 million years. That divergence split apart the ancient supercontinent *Pangaea*, creating the Atlantic Ocean basin. Rifting is evident in the gaping fractures that dissect solidified lava flows in Iceland, a volcanic island formed over a localized magma source (called a *hot spot* and described in detail in this chapter's second Essay) along the ridge (Figure 2.16).

Lengthy, nearly straight, parallel fractures are oriented perpendicular to the mid-ocean ridge and offset segments of the ridge. These *transform faults* allow the spreading motion to adjust to the curvature of the Earth. Rugged sea bottom topography as well as lines of submarine volcanoes or islands mark some fracture zones. The Cape Mendocino Fracture Zone intersects the California coast at Cape Mendocino and extends westward thousands of kilometers on the North Pacific Ocean floor.

Based on spreading rates and geological features, a third category of divergent plate boundary was discovered recently. At *ultraslow-spreading* boundaries, plates move apart at less than 20 mm per yr. Volcanoes occur at more widely spaced intervals and there are no transform faults and few earthquakes. Between volcanoes, the crust fractures and huge slabs of mantle rock rise to the seafloor, providing a glimpse into Earth's interior. Two known sites of ultraslow-spreading are the 1800-km

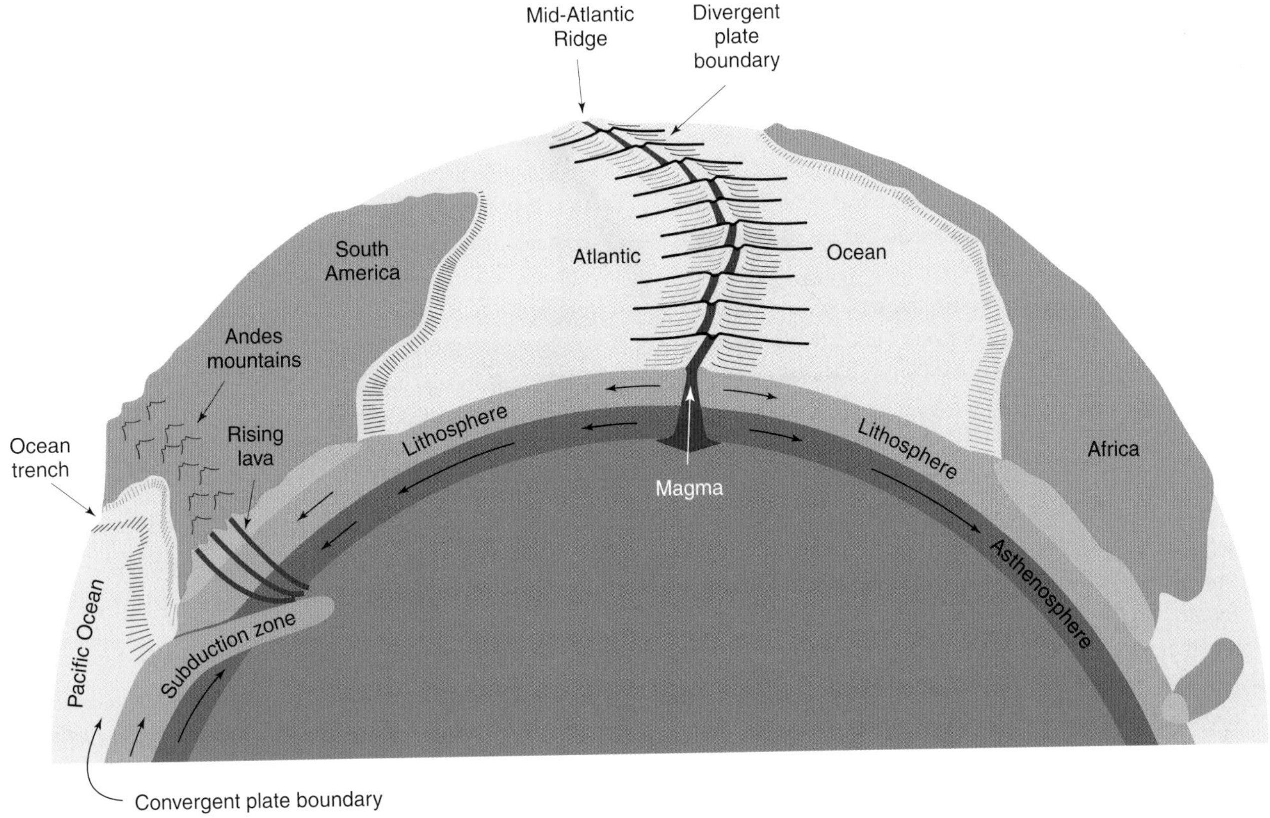

FIGURE 2.15
The Mid-Atlantic Ridge occurs where adjacent plates diverge and is part of a lengthy ridge system that winds through the ocean basins.

FIGURE 2.16
Iceland is a volcanic island formed along the Mid-Atlantic Ridge. The volcanic rocks exposed in this scene fractured as lava flows cooled, congealed and contracted. [Photo by J.M. Moran]

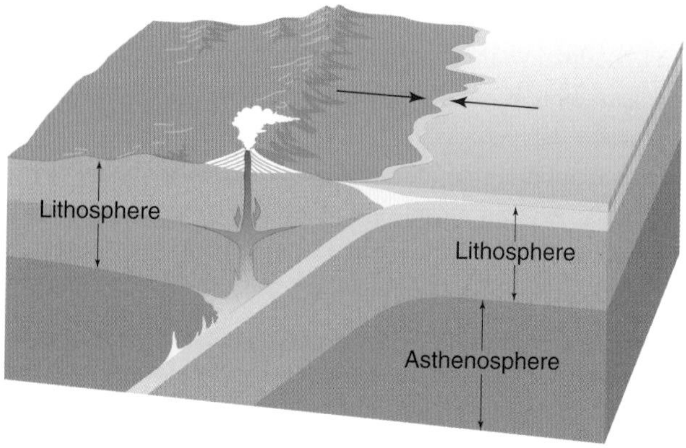

FIGURE 2.17
A convergent plate boundary where an oceanic plate subducts under a continental plate.

(1100-mi) long Gakkel Ridge in 2000 m (6500 ft) of water under the Arctic ice off the northeast coast of Greenland and the 8800-km (5400-mi) long Southwest India Ridge in the South Atlantic. Some marine geologists estimate that ultraslow-spreading boundaries characterize perhaps one-third of the mid-ocean ridge system.

CONVERGENT PLATE BOUNDARIES

When adjacent plates move toward one another, they create a **convergent plate boundary**. There are three possibilities: (1) two oceanic plates collide (e.g., the Pacific and Philippine plates), (2) an oceanic plate and continental plate collide (e.g., the Nazca and South American plates), and (3) two continental plates collide (e.g., the Indian and Eurasian plates). For locations, refer to the map in Figure 2.9.

A **subduction zone** forms where two oceanic plates converge or where an oceanic plate converges with a continental plate (Figure 2.17). In a subduction zone, the denser oceanic plate slips under the other plate and descends into the mantle. With downward transport of rock and sediment, a deep elongated depression (trench) forms on the ocean floor. Temperature and confining pressure increase with depth within Earth's interior so that a plate is heated and compressed as it descends into the mantle. Also, the friction of plates grinding past one another generates heat as well as stresses, which create earthquakes. (Sudden release of the energy created by these stresses is the origin of almost all tsunami-generating earthquakes.) At depths approaching 80 km (50 mi), the subducting plate begins to melt into magma, which migrates upward toward Earth's surface and contributes to mountain building and explosive volcanic eruptions. (The eruptions of Mount St. Helens and Mount Pinatubo are examples.) Whereas new lithosphere forms at divergent plate boundaries, lithosphere is destroyed by being incorporated into the mantle in subduction zones at convergent plate boundaries.

In oceanic-to-oceanic plate collisions, volcanic activity associated with subduction forms **island arcs**, curved chains of volcanic islands. Trenches usually lie on the seaward sides of island arcs with relatively shallow seas near continents. For example, the islands of Japan are part of a volcanic island arc with the Sea of Japan on the continental side and a trench on the Pacific Ocean side. An **ocean trench** is relatively narrow, typically 50 to 100 km (30 to 60 mi) wide, and may be thousands of kilometers long. In cross-section, a trench is slightly asymmetrical with the steeper slope on the continental side. Depths greater than 8000 m (26,250 ft) are not uncommon and some trenches have deep holes called *deeps* that exceed 10,000 m (32,800 ft) in depth. The deepest place on the ocean floor is probably the Challenger Deep, in the Mariana Trench in the Western Pacific, with a maximum depth of about 11,000 m (36,100 ft). For all their importance in plate tectonics, trenches make up only 1 to 2% of the ocean basin floor. Trenches surround most of the Pacific Ocean basin but volcanic island arcs are mostly on the western and northern sides of the Pacific. Besides Japan, Alaska's Aleutian Islands/Aleutian Trench is another example of a volcanic island arc/trench system.

At the convergent plate boundary just off the west coast of South America, the Nazca plate subducts under the South American plate, producing the offshore Peru-Chile Trench, which is 5900 km (3660 mi) long, 100 km (62 mi) wide, and more than 8000 m (26,300 ft) deep (Figure 2.18). Associated with subduction of the Nazca plate is the Andes, a prominent mountain range

FIGURE 2.18
Color-coded bathymetry/topography of the South Pacific Ocean and adjacent South America. The narrow dark blue band paralleling the coast represents the Peru-Chile Trench produced by subduction of the Nazca plate under the South American plate. [Courtesy of National Geophysical Data Center (NGDC), NOAA Satellite and Information Services]

including many volcanoes forming the backbone of South America.

The only gaps in the **circum-Pacific subduction/trench system**, also known as the *Pacific Ring of Fire*, are in Antarctica and in western North America. The Pacific Ring of Fire includes the U.S. West Coast, an active continental margin, whereas the East Coast is not a plate boundary and is passive. The action in the Atlantic Ocean is down the middle (along the Mid-Atlantic Ridge) whereas the action in the Pacific is around the rim. This difference is related to the fact that the Atlantic is spreading with little subduction, while the Pacific is shrinking at the subduction zones.

Deep earthquakes (originating at depths greater than 100 km or 60 mi) are usually associated with subduction zones. Earthquakes occur where adjacent crustal blocks move past each other, releasing energy stored in rocks deformed during plate movements when descending into the mantle at subduction zones. This energy travels as waves both along Earth's surface as well as through its interior. The depth to earthquake energy release generally increases with distance inland with the deepest earthquakes originating at depths of about 700 km (430 mi).

Plate movements also cause continents to collide, but such convergence does not form trenches or island arcs. Continents are much thicker and less dense than oceanic crust so they do not subduct into the mantle. A complex continental convergence is taking place in the Mediterranean region, where Africa and Europe are colliding as evidenced by active volcanism in Italy, dormant volcanoes in France, and earthquakes in the eastern Mediterranean. (In this case plate collision combines with some subduction.) The Himalayan Mountains are the result of the continuing collision of the Asian and Indian plates. Over the past 40 million years, the plate containing the Indian subcontinent has moved thousands of kilometers northward through the Indian Ocean and is now colliding with Asia. The Himalayan Mountains occur where the leading edge of the Indian plate thrusts under the Asian plate. The Rocky Mountains of North America were formed in a similar manner when the Pacific plate thrust beneath the North American plate. Their present height results from the buoyant, relatively young Pacific plate material lying beneath the North American continental crust.

TRANSFORM PLATE BOUNDARIES

Adjacent plates that slide laterally past one another produce a **transform plate boundary** (Figure 2.19). Although crust is neither created nor destroyed and these boundaries are generally free of volcanic activity, slippage can deform rock and trigger earthquakes. The San Andreas Fault of California, site of frequent earthquakes, occurs along a transform plate boundary where the Pacific plate (carrying a piece of California) slides toward the northwest, past the North American plate (Figure 2.20). The Anatolian fault in Turkey is another transform plate boundary and has been the site of some particularly disastrous earthquakes.

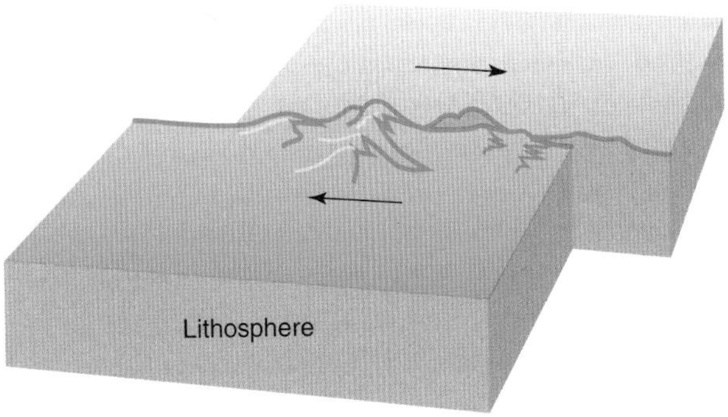

FIGURE 2.19
A transform plate boundary is often the site of shallow earthquakes.

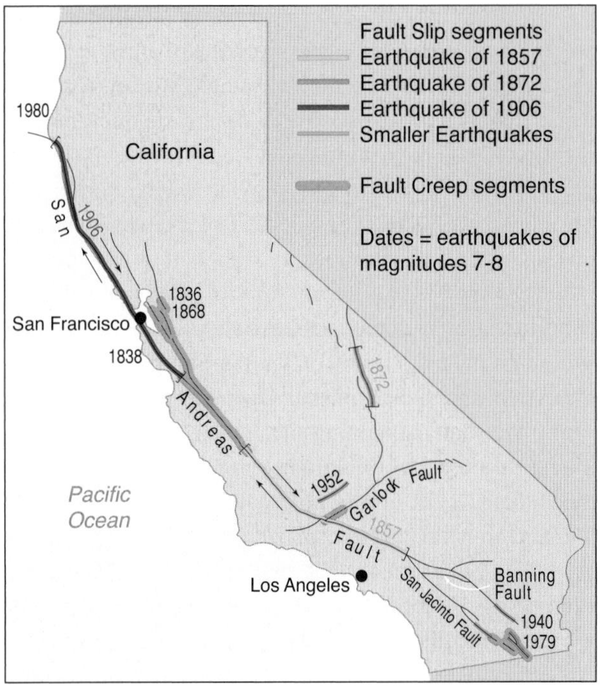

FIGURE 2.20
Map showing the position of the San Andreas Fault in California. [Based on image provided by the US Geological Survey]

MANTLE CONVECTION AND SEA LEVEL

As noted above, convection in Earth's mantle is responsible for the horizontal movement of tectonic plates leading to sea floor spreading, subduction, and transform faults. In addition, mantle convection causes vertical (up and down) movements of the continents accompanied by apparent changes in sea level. For example, in southern Africa, an expansive plateau about 1600 km (1000 mi) across and almost 1600 m (5200 ft) high has been slowly rising over the past 100 million years. Geologists do not attribute this vertical motion to tectonic activity—the region has been tectonically inactive for 400 million years or so. Rather, a hot mushroom-shaped mass several kilometers across (a *superplume*) ascending within the underlying mantle is pushing the plateau upward. And as the plateau rises, sea level falls.

While hot and buoyant superplumes in Earth's mantle cause upward motion of landmasses, other continental areas are sinking due to mantle convection. In some subduction zones, cold dense slabs of sea floor plunge into the mantle, dragging the overlying landmasses downward, accompanied by an apparent rise in sea level. Geologists have found evidence of downward and then upward movement of portions of North America and Australia during the geologic past, far from the possible influence of tectonic plate collisions.

MARINE VOLCANISM

Marine volcanic activity occurs where magma erupts through vents or fissures in Earth's solid surface, that is, along divergent and convergent plate boundaries and over hot spots. Products of volcanic eruptions include materials that are molten (e.g., lava), solid (e.g., ash, cinders, rock blocks), and gaseous (e.g., water vapor, carbon dioxide). Viscosity (the internal resistance of a substance to flow), gas content, and rate of extrusion of molten material govern volcanic activity and the shape of a volcano built by successive eruptions. Volcanoes range from small cinder cones to massive shield-like accumulations of solidified lava. Whereas about 20 major volcanic eruptions occur annually on the continents, many more eruptions occur on the ocean bottom and almost always are unobserved. Only shallow submarine eruptions and those occurring on islands have been extensively studied, although this is changing because of the development of new underwater observing techniques.

As noted earlier in this chapter, volcanic activity on the ocean floor consists of mostly quiescent lava flows. However, continental volcanoes often erupt violently. This difference in volcanic activity stems from the silica (SiO_2) composition and gas content (volatiles) of the magma that feeds a volcano. Magmas that are silica-rich are relatively viscous and may block volcanic vents. Pressure produced by volatile components (mostly water vapor) beneath the block increases and eventually the volcano explodes. Continental crust is mostly silica-rich granite so that magma formed from melting this crust is also silica-rich and viscous with a relatively high volatile content.

A high volatile content magma develops when water-rich oceanic crust subducts under continental crust and melts at depth. This magma migrates upward carrying volatiles, melting and mixing with continental crust, and forming andesitic to granitic magmas that are 60-70% silica and 3-6% volatiles. Such magmas are responsible for violent eruptions characteristic of volcanoes of the Cascade and Andes mountain ranges. On the other hand, oceanic crust is mostly silica-poor basalt derived from the upper mantle. It is also relatively water-poor with only 1-2% volatile content. Hence, the magma that composes the oceanic crust is volatile-poor, silica-poor, has low viscosity, and flows readily as lava. This explains why mid-ocean spreading centers exhibit quiescent volcanic activity (e.g., Iceland). On the other hand, eruptive behavior is usually violent in volcanic island arcs (e.g., the Aleutian Islands) where one oceanic plate subducts under another oceanic plate. Although the subducting plate is intermediate in silica content, intense heating of the wet oceanic crust causes explosive volcanic activity.

A series of volcanic eruptions, usually persisting over a long period of time, builds **seamounts**, underwater volcanic peaks rising more than 1000 m (3300 ft) above the ocean floor. Examples include the Emperor Seamounts in the North Pacific Ocean basin. Massive eruptions may cause the top of the volcano to break the water surface, rising above sea level and forming a volcanic island. The Hawaiian Islands, Iceland, and most other oceanic islands are products of such submarine volcanic activity. With the passage of time and the decline and eventual end of volcanic activity, weathering and erosion rapidly dissect and greatly reduce the elevation of volcanic islands. Meanwhile, the subsurface void produced by the discharge of lava during island formation may start to collapse, causing the island to sink back into the sea. Movement of the oceanic crust away from divergent plate boundaries also transports volcanoes into deeper waters. Such a deep-water seamount featuring a wave-eroded flat top is called a **guyot** (Figure 2.21).

Another example of an ocean basin feature produced when subsidence follows volcanic activity is the coral atoll. A **coral atoll** is a ring-shaped island surrounding a seawater lagoon (Figure 2.22). While serving as a naturalist on the British research ship *HMS Beagle*, Charles Darwin (1809-1882) first proposed that atolls were produced by coral growing upward around the margin of a volcanic island at the same rate that the island gradually sank into the sea. Modern drilling into atolls finds coral remains continuously down to the underlying volcanic rock thus confirming Darwin's hypothesis. Most corals live in warm near-surface waters, so that drilling indicates that the volcanic material was once at the surface, just as Darwin had proposed.

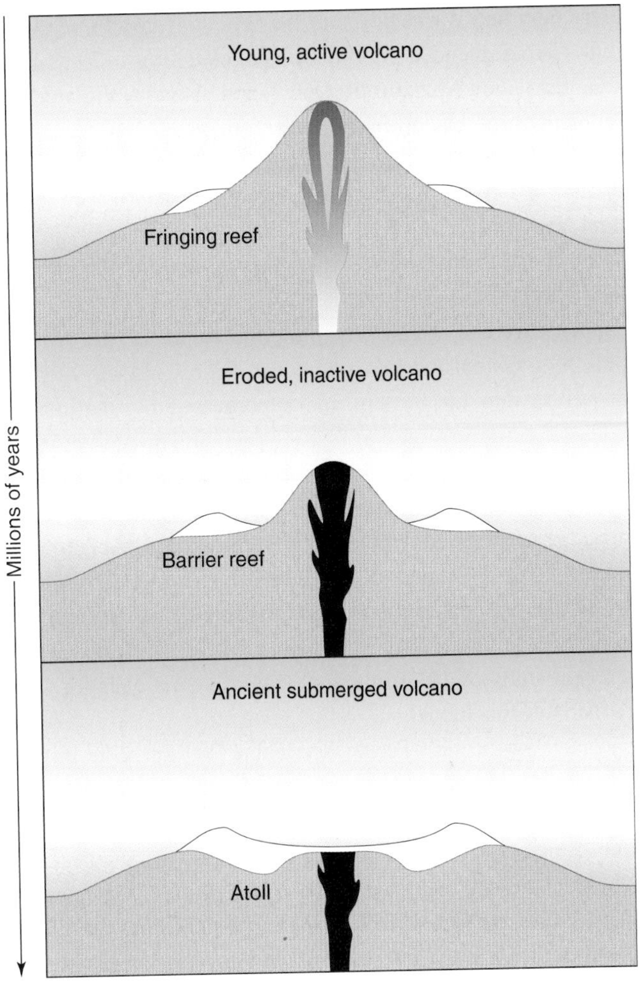

FIGURE 2.22
Stages in the development of a coral atoll.

FIGURE 2.21
This image of Rodriquez Seamount (a guyot) was created by the Monterey Bay Aquarium Research Institute (MBARI) using multibeam sonar data. In 2004, MBARI researchers used the ROV Tiburon to explore this guyot, and discovered ancient beach sands and tide pools around its crest (now 600 m below the sea surface). [Courtesy of MBARI]

Lithospheric plates are rigid and little volcanic activity takes place away from plate boundaries. An important exception is hotspot volcanism, which is responsible for the Hawaiian Islands, as described in the second Essay of this chapter.

HYDROTHERMAL VENTS

At locations near divergent plate boundaries, superheated water (with temperatures as high as 400 °C or 750 °F) circulates through rock fractures in the seafloor and discharges through spectacular **hydrothermal vents**. While a few locations in the ocean were known to be extremely hot, it was not until the Woods Hole submersible *Alvin* descended into the Galápagos Rift in 1979 that scientists viewed hydrothermal vents. Now associated with all three categories of divergent plate boundaries (*fast-spreading*, *slow-spreading*, and *ultraslow spreading*), hydrothermal vents are even more common than previously believed. Evidence suggests they are along most of the mid-ocean ridge system at intervals of less than 200 km (125 mi).

We can think of hydrothermal vents as part of a plumbing system under the mid-ocean ridge. Cold seawater is pulled into the system and superheated by shallow volcanic activity. As it comes in contact with crustal rock, the hot water dissolves minerals and is then ejected out of the vents. Initially, it was assumed that the high confining pressure of the water above forced seawater into the large faults along the flanks of the mid-ocean ridge, through which the water ascended toward the middle of the ridge and exited into the ocean via the cluster of vents. However, a recent study of a hydrothermal vent system argues for a revision of this model.

On the East Pacific Rise about 900 km (565 mi) southwest of Acapulco, Mexico, scientists placed seismometers in 2500 m (8200 ft) of water around the ridge to record microseisms (small earthquakes). Microseisms are believed to occur when cold seawater comes into contact with hot rock. The chilling causes the rock to contract and fracture, which releases seismic energy. Over a seven month period in 2003 and 2004, about 7000 shallow microseisms occurred and, by analyzing the microseisms, scientists were able to reconstruct the flow of water through the hydrothermal vent system. In early 2008, participating scientists proposed a new model for the hydrothermal vent system. The cold seawater descends through a 200-m (650-ft) wide chimney atop the ridge and then flows through a tunnel-like structure along the ridge axis just above a magma chamber. Once superheated, it bubbles up through a series of vents located further along the ridge.

Hydrothermal vent waters are mineral-rich. Sulfur compounds, iron and other metals, precipitate out of solution as the hot oxygen-deprived water mixes with the surrounding cold, oxygenated water. Hydrothermal activity on the ocean bottom is a source of some dissolved constituents of seawater although the principal source is suspended and dissolved substances delivered to the ocean by rivers and streams (Chapter 4). Furthermore, seawater circulating through fractures in newly formed oceanic crust (to depths of 600 m or 1950 ft) removes about two-thirds of the heat escaping from recently erupted lavas near ocean ridges with the balance of the heat conducted into the ocean floor.

FIGURE 2.23
Black smoker emitted at a submarine hydrothermal vent on the Mid-Atlantic Ridge. [Courtesy of OAR/National Undersea Research Program (NURP); NOAA]

The escaping precipitates may appear as dense clouds, called **black smokers** or **white smokers**, depending on the minerals, and accumulate in the form of conical chimneys that may exceed 10 m (33 ft) in height (Figure 2.23). One of the largest hydrothermal chimneys is near the top of a seamount in the Atlantic Ocean about 2500 km (1500 mi) east of Bermuda; it rises to the height of an 18-story building. Whereas most hydrothermal vents occur along the mid-ocean ridge, this one is located about 15 km (9 mi) from the Mid-Atlantic Ridge.

Hydrothermal vents are home to diverse communities of exotic marine organisms, including bacteria, crabs, shrimp, mussels, and tubeworms. One type of tubeworm found in the Galápagos area of the Pacific Ocean floor may grow to a length of 2.4 m (8 ft). In most ecosystems, photosynthesizing organisms form the base of food chains but, in the deep water where hydrothermal vents occur, there is no sunlight. The base of the food chain (and source of energy) consists of chemosynthetic bacteria (Chapters 1 and 9). Chemosynthesis is the process of deriving energy from sulfur compounds, common within the hot vent waters. We have more on hydrothermal vent ecosystems in Chapter 10.

The October 2010 report of previously unknown deep-sea hot springs in the middle of the Atlantic Ocean suggests that hydrothermal activity along the Mid-Atlantic Ridge may be even more extensive than previously believed. While on board the German research vessel *Meteor*, scientists from the MARUM Center for Marine Environmental Sciences and the Max Planck Institute for Marine Microbiology in Bremen used a new precision multibeam *echosounder* to obtain detailed images of the water column above the ocean floor. (For information on echosounders, refer to this chapter's first Essay.) At a site about 500 km (310 mi) southwest of the Azores and a depth of 1000 m (3280 ft), they identified the first of five plumes containing methane (CH_4) and hydrogen sulfide (H_2S) emanating from the ocean floor. This vent is located about 5 km (3.1 mi) from the large Menez Gwen vent field. Using a remote-controlled submarine, scientists observed a smoker with a chimney up to 1 m (3 ft) in height, fluids at temperatures up to 300 °C (572 °F), and animals typically associated with hydrothermal vents.

Researchers speculate that the presence of many more active sites than expected may call for an upward revision of the contribution of hydrothermal vents to the ocean's heat budget. This discovery may solve the mystery of how chemosynthetic organisms are able to migrate the hundreds of kilometers across 'empty' ocean, between large hydrothermal vents. These smaller hydrothermal sites may provide stepping stones for dispersal.

Spreading and Closing Cycles

By about 2.5 billion years ago, Earth's lithosphere had much the same make up as it does today, and cycles of ocean basin formation and closing were underway. Since then, the planet probably has undergone three or more cycles of supercontinent formation and break-up. That is, during at least three long-term cycles, all the continental pieces were brought together as one huge continent and subsequently split apart into smaller continents. Associated with supercontinents are super-oceans so that when a supercontinent breaks up, one or more new ocean basins form. For example, when the supercontinent Pangaea split into Laurasia (to the north) and Gondwanaland (to the south), *Tethys*, a narrow seaway formed between the two. New ocean basins first expand and later contract as a new supercontinent forms. The entire cycle from start to finish may take 500 to 600 million years. Cycles of ocean basin spreading and closing are called **Wilson cycles** after the Canadian geologist J. Tuzo Wilson (1908-1993) who first recognized and described the stages in the life span of an ocean basin (Figure 2.24).

We know the most about ocean basin changes over the past 500 to 600 million years, the most recent cycle. During this time, marine animals formed durable shells, which as fossils can be used to date the rocks containing them fairly precisely. Dating older rock devoid of fossils requires analysis of radioactive elements. These techniques have become more accurate and now permit reconstruction of earlier Wilson cycles. Here we describe the stages in the most recent Wilson cycle.

A Wilson cycle consists of six stages: embryonic, juvenile, mature, declining, terminal, and suturing. The cycle begins because thick continental crust does not conduct heat as readily as thinner oceanic crust. A supercontinent that remains in one location for hundreds of millions of years acts like a blanket, retarding heat flow from Earth's interior. This causes the mantle beneath the supercontinent to warm. As the underlying mantle warms, it expands, elevating the overlying continent and stretching the continental crust. Convection currents in the mantle also contribute to this stretching and eventually the crust fractures, forming a rift valley. This is Wilson's *embryonic stage* (Figure 2.24A). As noted earlier in this chapter, a *rift valley* is an elongated topographic depression bordered by fractures (faults) in the bedrock. The African continent has been in its present location for about 200 million years and now stands about 400 m (1300 ft) higher than other continents as a consequence of the heating and expansion of the underlying mantle. Fracturing and some spreading of the crust produced the rift valleys of East Africa.

With rifting of the continental crust, the broken sides rise about a kilometer enclosing a valley that often fills with fresh water. In East Africa, long, deep lakes now occupy narrow rift valleys. Rift valleys gradually widen and eventually connect to the ocean and the freshwater lakes become narrow saline gulfs. This is happening now in the Red Sea and Gulf of California and marks the beginning of Wilson's *juvenile stage* (Figure 2.24B). With continued lateral spreading of the rift valley, the divergent plate boundary widens and additional oceanic crust is generated signaling the *mature stage* of the Wilson cycle

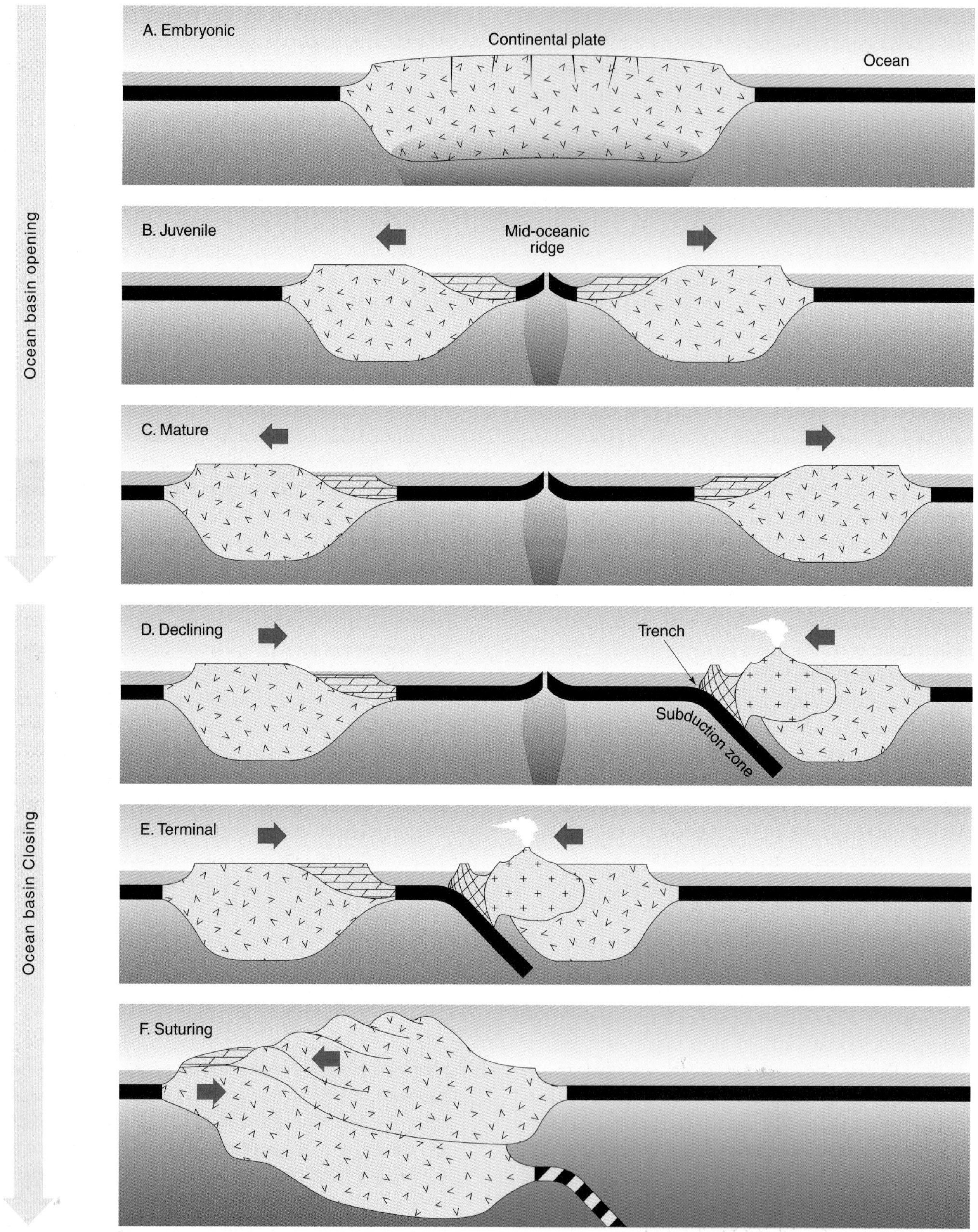

FIGURE 2.24

A simplified model of the Wilson cycle, showing the formation and closing of an ocean basin. [Modified from a drawing by Lynn S. Fichter]

(Figure 2.24C). Today, the Atlantic is a mature ocean with geologically passive margins.

Subduction becomes more widespread around the border of the ocean basin during the *declining stage* of the Wilson cycle (Figure 2.24D). In time, cooling and the loss of volatiles increase the density of oceanic plates. Under the influence of gravity, these plates slide down and away from the bathymetric high of the mid-ocean ridge and sink into the asthenosphere at subduction zones. Today, the Pacific is the best example of a declining ocean with subduction zones forming the Pacific Ring of Fire. This is also now happening on a smaller scale in the South Atlantic (Scotia Arc near Antarctica) and in the West Indies (near Barbados). Typically an ocean basin widens for about 200 million years before subduction begins. As evidence of this, no oceanic crust has been found that is more than 200 million years old. Eventually, the basin begins to close as subduction rates (at trenches) exceed spreading rates (at mid-ocean ridges). For Earth as a whole, spreading must equal subduction—otherwise the planet would be shrinking or expanding. Over the subsequent 200 million years, the ocean basin continues to close and its sediment deposits are deformed and uplifted, creating mountain ranges on the newly assembled supercontintent at the site of the former ocean.

Following the declining stage of the Wilson cycle, the ocean basin closes through subduction as continents from opposite sides of the ocean basin bear down on one another and eventually collide. These events signal the final two stages of the Wilson cycle: the terminal and suturing stages. In the *terminal stage*, the continents are not yet touching but subduction of the intervening oceanic crust causes a narrowing of the sea separating the continents (Figure 2.24E). Volcanic eruptions, earthquakes, uplift, and mountain building accompany subduction of the oceanic crust. An example of this today is the African continent converging with the European continent producing the intervening Mediterranean Sea with nearby volcanism, earthquakes, and young rugged mountains (e.g., the Alps).

During the *suturing stage*, collision of the continents is complete and the intervening sea is gone (Figure 2.24F). The two colliding continental crusts, being less dense than the oceanic crust, do not subduct but rather override one another causing uplift and mountain building. Collision of the continents squeezes out the intervening ocean and causes subduction of oceanic crust. Today, the suturing stage is illustrated by the collision of the Indian and Eurasian plates generating the Himalayan Mountains. The Appalachian Mountains of eastern North America also formed in this way at the end of the previous Wilson cycle, about 450 million years ago.

The Pacific began as a vast super-ocean, known as *Panthalassa*, surrounding the supercontinent Pangaea. When other ocean basins such as the Atlantic started to form, the Pacific Ocean entered the declining stage of the Wilson cycle with concomitant changes in its size and shape. As an ocean basin narrows, subduction accommodates the lithospheric plates that are colliding at or along its margins. Except in a few locations, subduction zones border the Pacific basin. When the Atlantic basin begins closing by expansion of subduction along its margins, subduction in the Pacific basin will diminish or cease.

Wilson cycles also influence sea level relative to the continents over periods of hundreds of millions of years. When supercontinents dominate, continents stand high as the mantle under them warms. When spreading begins, the continental fragments move off the heated mantle and thus stand lower with respect to the sea surface. Furthermore, the newly formed ocean basin floor is relatively shallow reducing the capacity for seawater, and thus low-lying continental areas are flooded.

As the newly formed basin widens and its crust ages, cools, and deepens (under the weight of accumulating sediments), sea level falls. Sea level has apparently varied by as much as 250 m (820 ft) over the past 100 million years. During times of higher sea level, shallow seas, similar to Canada's Hudson Bay, covered large areas of continents. At such times, about 80% of Earth's surface was ocean water. During episodes of lower sea level, shorelines migrated seaward to the continental margins; the ocean then covered about 65% of Earth's surface. Note that sea level also rises and falls in response to changes in Earth's glacial ice cover plus expansion and contraction of seawater as global temperatures fluctuate.

Conclusions

In a geological sense, the present ocean floor is relatively young—the oldest oceanic crust is only about 200 million years old whereas some bedrock on the continents is billions of years old. Oceanic crust is continually generated at divergent plate boundaries (at the mid-ocean ridge system) and incorporated into the mantle as it is drawn down at convergent plate boundaries (**subduction zone**). Over periods of perhaps 500 million years, ocean basins have opened and closed (the Wilson cycle). In this way, plate tectonics explains the origin of the large-scale features of the deep ocean bottom (e.g., trenches, ocean ridges) and the geological processes (e.g., volcanism, earthquakes) operating mostly at plate boundaries. Furthermore, as we will see in Chapter 12, plate tectonics helps explain some long-term changes in climate.

Processes operating at the interface between the ocean and lithosphere (e.g., hydrothermal vents) affect the composition of seawater. Earth materials dissolve in seawater and influence its salinity. The salinity of seawater coupled with its temperature control its density, which in turn affects its vertical motion (thermohaline circulation). Water is a substance with unique physical and chemical properties that not only influence the salinity of seawater but also the exchange of heat energy and moisture between the ocean and atmosphere. In the next chapter, we examine the uniqueness of water and its implications for the ocean in the Earth system

Basic Understandings

- Ocean basins and continents are unevenly distributed over Earth's surface; ocean covers about 71% of the Earth while the remaining 29% is land. The Northern Hemisphere is 39.3% land and 60.7% ocean whereas the Southern Hemisphere is 19.1% land and 80.9% ocean.
- Earth's relatively thin solid crust is the part of the geosphere, which interacts with the atmosphere, hydrosphere, and biosphere. Rocks composing the crust are classified as igneous, sedimentary or metamorphic based on the environmental conditions that formed them. The rock cycle is the transformation of one rock type to another via internal and surface geological processes.
- Earth's crust is composed of mostly igneous rock with metamorphic rock in some localities. In most areas, sediments and sedimentary rock of varying thickness overlie the igneous and metamorphic rock of the crust.
- Oceanic crust is chiefly basaltic (volcanic) rock that is denser and thinner than the mostly granitic rock of continental crust. The lithosphere includes the crust and the rigid upper portion of the mantle and is thicker under land than ocean. The lithosphere floats on the underlying asthenosphere, a deformable (plastic-behaving) region of the mantle.
- The continental margin, the submerged portion of the continent, is a shallow, gently sloping continental shelf that is bordered seaward by the continental slope until it deepens into the continental rise.
- At geologically passive continental margins where plate boundaries are absent, such as surrounding the Atlantic basin, sediment spreads over the ocean floor forming flat abyssal plains adjacent to the continental rise. At geologically active continental margins with convergent plate boundaries, such as surrounding much of the Pacific basin, the continental slope descends into deep ocean trenches.
- A turbidity current is a gravity-driven, sloping flow of water heavily laden with suspended sediments. Fed by river-borne sediment, turbidity currents will periodically scour submarine canyons cut into the continental shelf.
- The deep ocean basin has a varied topography featuring mountain ranges, seamounts, plains, and trenches.
- Plate tectonics describes the fragmentation and movement of Earth's lithospheric plates, and the opening and closing of ocean basins, with continental drift and sea floor spreading. Evidence for plate tectonics consists of cross-Atlantic similarities in continental margins, fossils, rock formations, and mountain ranges. Sea floor spreading, a key aspect of plate tectonics, occurs at divergent plate boundaries and is confirmed by many lines of evidence including the systematic pattern of magnetic anomalies on the sea floor.
- Convection currents in the mantle are proposed as the driving force for plate movement, sea floor spreading, and subduction.
- New oceanic crust forms within divergent plate boundaries, at the mid-ocean ridge system. This volcanic mountain range winds through the world ocean, forming one of the deep-ocean's most conspicuous features.
- Along the mid-ocean ridge there are *fast-spreading*, *slow-spreading*, and *ultraslow-spreading* boundaries. At fast-spreading boundaries, plates move apart at 100 to 200 mm per yr. With rapid spreading, hot magma is abundant and lava flows as sheets from a central peak, giving the ridge a narrow tent like profile, such as at the East Pacific Rise. At slow-spreading boundaries, plates move apart at less than 55 mm per yr and the topography is broader, rougher, and features rift valleys, such as in the Mid-Atlantic Ridge. At ultraslow-spreading boundaries, plates

move apart at less than 20 mm per yr and great slabs of mantle rock rise to the sea floor.

- Where two oceanic plates converge or where an oceanic plate converges with a continental plate, subduction takes place. With subduction, the denser oceanic plate slips under the other plate and descends into the mantle. Subduction zones produce trenches on the deep ocean floor and are associated with shallow to deep earthquakes and violent volcanic eruptions.
- A transform plate boundary occurs where adjacent plates slide laterally past one another and often is the site of earthquake activity but not volcanoes.
- Hot, newly formed oceanic crust is chilled by contact with cold seawater circulating through fractures on the sea floor. This hydrothermal circulation also extracts minerals and salts from rock. Minerals precipitate out of the hot waters and build spectacular vents, tens of meters high, on mid-ocean ridges.
- With a period of about 500 million years, the Wilson cycle describes the successive opening and closing of ocean basins through plate tectonics.

Enduring Ideas

- The lithosphere, the portion of Earth's geosphere consisting of the crust and the rigid upper portion of the mantle, floats on the underlying asthenosphere, a deformable portion of the mantle. The thicker and less dense continental lithosphere floats higher and extends deeper into the mantle than the thinner, denser oceanic lithosphere.
- The many large plates of the lithosphere move slowly over the Earth's surface. Most geological activity, including volcanic eruptions, earthquakes, and mountain building, occur primarily along plate boundaries.
- Plate tectonics is a unifying theory that explains continental drift, sea floor spreading, and the origin of many ocean basin features including the oceanic ridge system, hydrothermal vents, and deep-sea trenches.
- Where adjacent plates diverge, crust fractures, and magma flows from Earth's interior to the surface where it crystallizes into new crust. Where adjacent plates converge, subduction gives rise to explosive volcanic eruptions and high magnitude earthquakes. Along a transform plate boundary, one plate moves laterally past the other, triggering shallow earthquakes.
- Geologically, continental margins are either active or passive depending on the distance to a plate boundary.

Review

1. What percentage of Earth's surface is covered by ocean water? In which hemisphere is most land located?
2. Name the five ocean basins. Which ocean basin is largest?
3. Describe the general origins of igneous, sedimentary, and metamorphic rock. Which type of rock composes most of Earth's crust?
4. Compare and contrast oceanic and continental crust. Which type of crust typically comprises the continental margin?
5. Describe the general relationship between continental shelf width and tectonic activity in the continental margin.
6. Describe the origins of most submarine canyons.
7. Define sea-floor spreading and its significance for oceanic crust.
8. Describe two different types of tectonic plate interactions that form a subduction zone.
9. Describe a transform plate boundary and give an example of this type of boundary.
10. What are Wilson cycles? What is the approximate period of each cycle?

Critical Thinking

1. How does the lithosphere of the deep-ocean basins differ from the lithosphere of the continents?
2. What are some of the implications of the slow pace of the geological processes involved in the rock cycle?
3. Compare and contrast the geological characteristics of the continental margins of the west and east coast of North America.
4. Explain why oceanic crust becomes older with increasing distance from the mid-ocean ridge system.
5. What conditions are required for violent volcanic eruptions in the continental margins?
6. Summarize the key evidence in support of the theory of plate tectonics.
7. How does land subsidence influence mean sea level in coastal areas?
8. How might investigating ultraslow divergent plate boundaries benefit our understanding of Earth history?
9. How do turbidity currents help maintain submarine canyons?
10. Explain why a hydrothermal vent community may have a limited life expectancy (10 to 100 years or so).

ESSAY: Investigating the Ocean Floor

Oceanographers gather information about the ocean floor using a variety of techniques; some relatively simple, such as devices that scoop up rock and sediment samples from the ocean floor, whereas others, such as acoustic and gravimetric instruments, are more sophisticated. Some measurements are made *in situ* (immersed in the medium being measured) whereas others depend on remote sensing methods.

The likely oldest measurement of the ocean floor is depth of the water. Historically, this was done using a *sounding line*, a weighted rope marked off in fathoms (a *fathom* being 1.8 m or 6 ft), that was thrown over the side until the weight touched the ocean floor. As early as 1888, using lead weights affixed to rope, the oceanographer Sir John Murray was able to calculate the volume of the ocean (mean depth × ocean area). Even with such a simple method, his result was only 1.2% greater than the volume of the ocean determined by Earth-orbiting satellite altimeters over 120 years later.

Beginning in the early 1920s, *acoustic sounders* (at first called *fathometers*) significantly improved the accuracy of water depth determination. With this instrument, a sound pulse is directed toward the sea floor and its travel time to the bottom and back is recorded from which the depth is calculated using the speed of sound through water. Acoustic sounders first came into common use on U.S. Navy ships during World War II, and depth measurements led to the discovery that ocean bottom bathymetry is much more variable than previously believed.

Depth data are needed to prepare navigation charts, but acquiring these data via acoustic sounders can be time consuming because oceanographic ships are relatively slow, averaging only 19-33 km per hr (10-18 knots), and very expensive. The U.S. Navy estimates that using an acoustic sounder to measure the depth of the entire ocean floor would require one ship operating for 200 years or 10 ships for 20 years at a cost of $2 billion. An alternative is airborne *laser bathymetry*. A downward directed laser gun is mounted on the underside of an aircraft timing the return of light pulses aimed at the ocean bottom. An aircraft can travel nearly ten times faster than a ship, so close to 1000 square km (386 square mi) can be surveyed in a single day. This is limited, however, by the depth the laser can penetrate within the water column, confining the technique to relatively clean and shallow waters. For depths seaward of the continental shelf, ocean scientists depend on acoustic sounding or satellite-borne laser altimeters. (For more on these remote sensing techniques, see the second Essay of Chapter 7.)

The most straightforward way to sample the surface of the ocean floor is with a *grab sampler* (Figure 1). This mechanical device operates much as a human hand grasping an object. More often, scientists require a vertical section within the ocean bottom. For that, a weighted coring tube is lowered a few meters above the ocean floor and allowed to fall, the weight of the system driving it into the bottom sediments. The sediment core, retrieved when the tube is brought to the surface, reveals a record of local sedimentation over time. Where the ocean bottom is bare rock or covered by thick accumulations of sediment, deep-sea drilling is used to obtain rock cores.

Once a core is brought to the surface, it is labeled with the location on the seafloor, the top and bottom are coded, and the length is measured. The core is then cut into smaller sections and sliced in half lengthwise. One half is stored

FIGURE 1
A Young Modified Van Veen grab sampler ("Young Grab") is being lowered over the side of a ship. When it reaches the ocean bottom, the grab sampler will scoop sediment from a 0.4 square m area. The Young Grab digs down about 10 cm (4 in.) and scoops up the top layers of sediment along with the organisms living in the sediment (infauna). Most benthic infauna live in the top 5 cm (2 in.) of sediment. [NOAA Ocean Explorer]

while the other is analyzed to reconstruct a chapter in Earth's history. Paleontologists examine microfossils to determine the age of the rock or sediment and paleomagnetists use state-of-the-art instruments to read the record of Earth's magnetic field changes, information that helps determine when and where specific rocks were formed. Other scientists measure physical properties such as density, strength, and heat conductivity. After the expedition, cores are transported to one of four repositories for storage and future research. Scientists are able to access these repositories much as the general public uses a library. Also, data from most cores are available in digital form.

One of the first research vessels capable of retrieving cores from the ocean floor was the oil drillship *CUSS I*, which operated in the early 1960s as part of *Project Mohole*, an attempt to drill through the oceanic crust into the mantle. The *Glomar Challenger*, operated from 1968 until 1983 as part of the *Deep Sea Drilling Program (DSDP)*. The *Glomar Challenger* drilled 1092 holes at 624 deep ocean sites. The extracted cores, if laid end-to-end, would be 96 km (60 mi) long. The success of the DSDP inspired the formation of the *Ocean Drilling Program (ODP)*, an international partnership of scientists and research institutions that explore the evolution and structure of Earth by studying deep-ocean cores. ODP was sponsored by the National Science Foundation (NSF) and agencies in 20 other nations under the management of the *Joint Oceanographic Institutions for Deep Earth Sampling (JOIDES)*.

The 143-m (469-ft) long drill ship *JOIDES* (pronounced *joy-deez*) *Resolution* (Figure 2), is the centerpiece of the Integrated Ocean Drilling Program. (It was named after the *HMS Resolution*, commanded by Captain James Cook during his exploration of the Pacific Ocean over 200 years ago.) Built in 1978 in Halifax, Nova Scotia, the ship originally was a conventional oil-exploration vessel. It was converted for scientific research in 1984, equipped with the world's finest

FIGURE 2
The *JOIDES Resolution* is a drill ship that was the centerpiece of the Ocean Drilling Project (ODP) from 1985 to 2003, extracting sediment and rock cores from the ocean bottom around the world. It is presently performing the same service for the Integrated Ocean Drilling Program (IODP). [Courtesy of William Crawford, IODP/TAMU]

shipboard laboratories, and went into service for ODP in January 1985. Annually the *JOIDES Resolution* embarks on four scientific expeditions, each lasting about two months, allowing ODP scientists to extract sediment and rock cores in water as deep as 8235 m (27,018 ft). It has drilled in all the world's ocean basins, including north of the Arctic Circle (to 85.5 degrees N) and south of the Antarctic Circle in the Weddell Sea (to 70.8 degrees S). As of October 2003, when the *Integrated Ocean Drilling Program (IODP)* succeeded the ODP, the cores extracted, if laid end-to-end, totaled more than 226 km (140 mi) in length.

Sponsored initially by the U.S. and Japan, the IODP currently involves 16 additional nations. The IODP plan called for two deep-sea drill ships, a renovated *JOIDES Resolution* and a second, new drilling ship capable of investigating larger portions of the ocean floor, as well as special drilling platforms. After about 20 years of service as a scientific drilling ship, the *JOIDES Resolution* temporarily ceased operations in late 2005 for a 3-year $130 million refurbishment. Lab space was increased by one-third, lab equipment was updated, and accommodations for the scientific crew were upgraded. After work was completed in 2009, the ship joined the *Chikyu*, a new ship specifically designed for ocean science research (Figure 3).

Built by Japan and managed by the *Japan Agency for Marine-Earth Science and Technology (JAMSTEC)*, the *Chikyu*, which means Earth in Japanese, became fully operational in the autumn of 2007. At 210 m (690 ft) in length, it is much larger than earlier drill ships, technologically more advanced, and equipped with specialized equipment designed to minimize environmental hazards. It can accommodate up to 10,000 m (32,800 ft) of drill strings (pipes) enabling the ship to drill up to 7000 m (23,000 ft) into the ocean floor in water as deep as 2500 m (8200 ft), twice as deep as previous drill ships. *Chikyu's* first project was to investigate earthquake mechanisms by boring into the Nankai Trough subduction zone offshore of Honshu, Japan, a location inaccessible to the *JOIDES Resolution* because of the potential for a petroleum blowout. Filling in the gaps of knowledge regarding subduction zones near Japan has occupied the Chikyu ever since. It is ironic that the vessel suffered sufficient damage during the 2011 Tohoku (Honshu) earthquake and tsunami to require the cancellation of an expedition and a trip to dry dock for repairs. Another option with IODP is the use of the specialized drilling platforms for areas where drill ships could not safely or efficiently operate, such as in the Arctic multi-year sea ice.

In addition to deep-sea drilling, other methods are available for study of the materials composing the ocean bottom. One of these techniques uses a sound source, such as an underwater explosion, coupled with a number of strategically positioned acoustic receiving devices. Knowing the speed of sound varies with changes in the physical properties of the medium, much can be learned about the thickness, extent, and composition of oceanic crust. When combined with gravimetric data (obtained from a *gravimeter* that measures the acceleration of gravity), information about local crustal mass and composition, can be inferred. Other ways of directly investigating the ocean bottom include piloted deep-sea research submersibles (e.g. Woods Hole's *Alvin*), sensors tethered to a ship that are dropped overboard, and autonomous underwater vehicles (AUVs). We have more to say about these techniques in Chapter 13.

FIGURE 3
The Japanese drill ship Chikyu (Earth). [JAMSTEC (Independent Administrative Institution, Japan Agency for Marine-Earth Science and Technology), photo taken at Yokosuka New Port, Kanagawa, Japan, by Gleam/License: Creative Commons Sharelike 3.0]

ESSAY: Hotspot Volcanism and the Hawaiian Islands

Most volcanic activity on Earth takes place along convergent and divergent plate boundaries and is readily explained by plate tectonics theory. However, volcanic activity also occurs on the ocean floor and continental interiors at great distances from plate boundaries. These sites of anomalously high volcanic activity are known as *hot spots*. A prominent product of *hotspot volcanism* is the Hawaiian Islands (Figure 1), a segment of a 6000-km (3700-mi) long chain of volcanoes and seamounts on the Pacific Ocean floor, thousands of kilometers from the nearest plate boundary.

FIGURE 1
True-color Terra MODIS satellite image of the Hawaiian Islands. View is toward the north with the "Big Island" (Hawaii) to the lower right. From the youngest (the "Big Island") located over the Hawaiian hot spot, progressively older volcanic islands stretch towards the northwest. [Courtesy of NASA]

In 1971, only a few years after widespread acceptance of the theory of plate tectonics, Jason Morgan of Princeton University proposed that narrow plumes of hot material originating at the mantle/core boundary and ascending within the mantle delivers copious amounts of melt to the lithosphere where it feeds volcanic eruptions. A hot spot is at the top of a mantle plume. Morgan and other scientists assumed that plumes, rooted deep in the mantle, anchored hot spots at fixed locations. Over many millions of years, as an oceanic plate moved over a hot spot, submarine volcanic eruptions built massive accumulations of basaltic lava on the ocean floor that eventually emerged above sea level as a volcanic island. Eventually, the plate carried the island beyond its lava source (the hot spot) and volcanic activity died out. With a hot spot stationary and the plate moving, an age-progressive straight-line chain of volcanoes developed.

Today, active volcanism is confined to the "Big Island" at the southeast end of the Hawaiian Chain near 19 degrees N. The "Big Island" is composed of five overlapping volcanoes (Figure 2). Kilauea, the Island's youngest volcano, has been erupting continuously since the early 1980s. The rate of lava emission per unit area at this hot spot is greater than at any other place on the planet. A new, still submerged volcano called Loihi is forming on the Pacific Ocean floor southeast of Hawaii. The other Hawaiian Islands are eroded remnants of now extinct volcanoes. The oldest and most extensively eroded volcanoes, marked by coral atolls, lie at the northwestern end of the Hawaiian chain (i.e., Kauai). This southeast-to-northwest trending chain of progressively older volcanic islands and seamounts was assumed to mark the track of the Pacific plate toward the northwest at about 10 cm per yr passing over a hot spot fixed at 19 degrees N. Beyond the Hawaiian

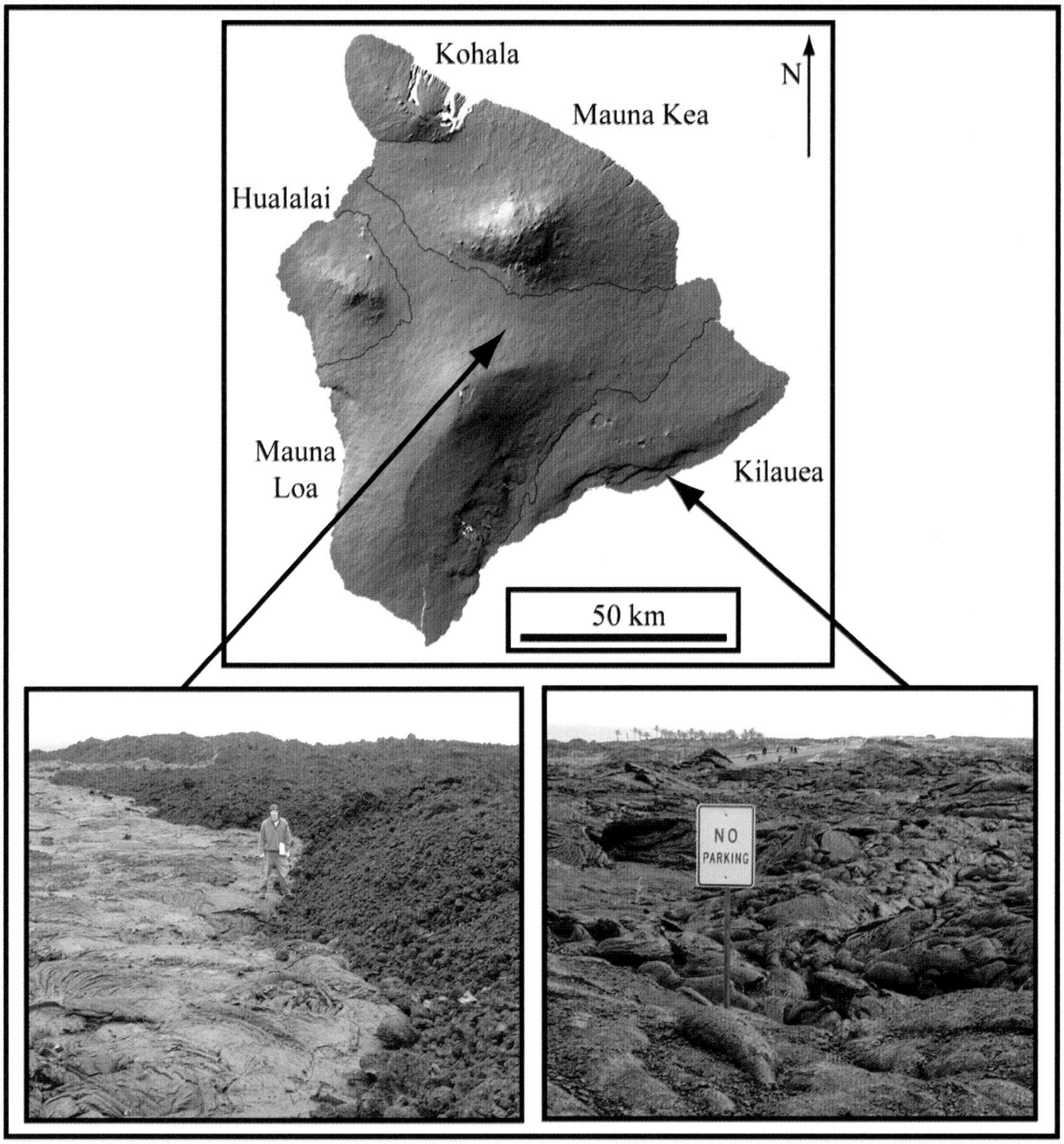

FIGURE 2
Shuttle Radar Topography Mission (SRTM) shaded relief image showing the five shield volcanoes of Hawaii, which increase in age to the northwest. [NASA and Jacob Bleacher]

Islands is a chain of deeply submerged seamounts, the Hawaiian-Emperor Seamount Chain, trending to the northwest on the Pacific Ocean floor (Figure 3). The 60° bend (near 170 degrees E) toward a more northerly heading was attributed to a change in direction of the Pacific plate about 47 million years ago. At the northern end of the Emperor Seamount Chain, extinct submarine volcanoes subduct into the Aleutian Trench.

Based on analysis of deep-sea rock cores, evidence is mounting that hot spots are not stationary. In fact, migration of a hot spot rather than movement of the Pacific plate over a fixed hot spot likely explains the origins of the Hawaiian-Emperor Seamount Chain. This conclusion follows from reconstruction of the latitude of volcanic islands at the time of their formation. If a hot spot were stationary, all currently extinct volcanoes and seamounts would have been active when their latitude was 19 degrees N, the same as the now-active "Big Island."

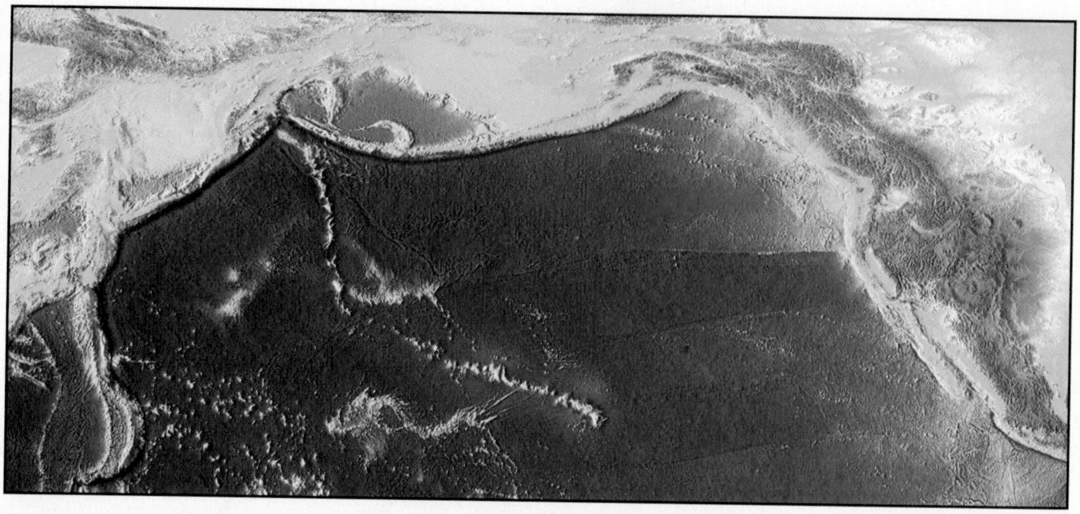

FIGURE 3
The Hawaiian Islands and Emperor Seamount chain stretch more than 6000 km in the Pacific Ocean. [National Geophysical Data Center, NOAA]

Analysis of rock magnetization enables scientists to reconstruct latitude. As noted elsewhere in this chapter, when magma or lava cools and solidifies, certain minerals (e.g., magnetite) act like tiny bar magnets and align with Earth's magnetic field. Analysis of a volcanic rock sample provides information not only on the direction of Earth's magnetic lines of force at the time of crystallization, but also their inclination, a direct measure of latitude. This technique was applied in 2001 by J.A. Tarduno of the University of Rochester and colleagues to rock samples cored from the Emperor seamounts Detroit, Nintoku, and Koko. Based on the reconstructed latitude at which the volcanic rock formed, Tarduno and colleagues concluded that from 81 to 47 million years ago, the hot spot moved southward at more than 4 cm per yr. Other scientists determined that the 61-million year old Suiko seamount had formed at 27 degrees N.

Some scientists argue that the Hawaiian-Emperor Seamount Chain was the product of the combined southward motion of the Pacific plate and southward movement of the hot spot (now under Kilauea). In addition, evidence is not convincing that the Pacific plate actually changed direction 47 million years ago. Movement of the Hawaiian hot spot relative to the Pacific plate may be a more plausible explanation for the abrupt bend in the trend of the Hawaiian-Emperor Seamount Chain.

What might explain the mobility of a hot spot? One possibility is that the base of a hot spot in the deep mantle may move. Another possibility is that a rising plume may be swayed by convection currents in the mantle. Furthermore, many scientists question whether deep mantle plumes are associated with all hot spots. In 2003, scientists at Princeton University reported on the results of a new technique to investigate Earth's mantle using seismic (earthquake) waves that penetrate Earth's interior. Similar to tomography used to examine the internal organs of humans, this technique enables scientists to search for and locate mantle plumes. They found that plumes originating near the mantle/core boundary feed some hot spots (e.g., Hawaii) but other hot spots (e.g., Iceland, Galápagos) have relatively shallow plumes (originating at depths near 660 km or 410 mi), and some hot spots (e.g., Yellowstone) have no plumes at all.

CHAPTER 3

PROPERTIES OF OCEAN WATER

Sea ice on the Beaufort Sea, a marginal sea of the Arctic Ocean bordering the northern coast of Alaska. [Courtesy of NOAA]

Case in Point

Since antiquity, people have used salt from seawater to meet their nutritional needs and, prior to the age of refrigeration, to preserve food for storage and transport. Sodium chloride (NaCl), the chief salt dissolved in seawater, is hygroscopic, that is, it has a chemical attraction for water. By drawing water out of bacteria and also molds and yeasts via osmotic pressure, salt keeps these microbes from reproducing thereby inhibiting food from spoiling. In moderate amounts, NaCl is also safe for human consumption.

Jericho, an oasis at the northern end of the Dead Sea, was founded almost 10,000 years ago as a salt trading center. The Dead Sea, located on the Israeli-Jordanian border, has long been a source of salt (Figure 3.1). The lowest spot on Earth's land surface at 400 m (1300 ft) below sea level, the Dead Sea occupies a depression in the transform fault complex that separates the African and Indian plates (Chapter 2). Massive salt deposits (mostly calcium sulfate rather than sodium chloride) formed when tectonic movements cut off the Mediterranean Sea from the Atlantic Ocean about 6 million years ago. As the isolated seawater evaporated, salt crystallized and accumulated as sedimentary layers on the sea floor. Today, some nearby mountains are almost pure salt but have persisted because of the arid climate and meager runoff.

Originally, demand for salt was great but the supply was limited, making it a valued trade commodity. Salt was so highly valued that it was used as money in Tibet and in many parts of Africa. In his *Natural History*, the Roman historian Pliny the Elder (CE 23-79) noted that in Rome ". . . the soldier's pay was originally salt and the word salary derives from it . . ." Others claim that soldiers were given an allowance to purchase salt. In the Middle Ages, salt cod was an important commodity for members of the Hanseatic League, a trading group of merchants in northern Germany and the Baltic Sea coast, and later an important food source for American colonists. In northern Germany, close to the coast, peat was burned to evaporate seawater and accumulate salt. Consequently, a large part of former peat lands are now below sea level.

FIGURE 3.1
Landsat satellite image showing the Dead Sea. Deep waters on the northern side are shaded dark blue, while lighter blue shadings show Israeli and Jordanian salt evaporation ponds at the south end of the Dead Sea. [Courtesy of NASA]

Traditionally, most commercial salt was derived by solar evaporation of seawater or natural brines (in wells, salt lakes, and enclosed seas) in hot dry climates. Today, salt is mass produced, abundant, and relatively inexpensive. Much of the salt supply in the U.S. comes from Great Salt Lake, a remnant of the much larger Lake Bonneville that dates from the late Pleistocene Ice Age. With solar evaporation, saltwater is let into shallow ponds separated by dikes where suspended materials such as sand or clay gravitationally settle to the bottom. Water then flows on to crystallizing pans where water evaporates and salt precipitates, accumulating as layers on the bottom of the pan. Many different salts precipitate from saltwater in addition to sodium chloride, including calcium carbonate, calcium sulfate, and magnesium sulfate. As saline water evaporates, different salts precipitate at different times because each salt has a different solubility in water. Salt is also mined from subsurface deposits of rock salt by direct extraction or solution mining, whereby water is pumped into underground salt beds to create brine that is recovered and then evaporated at the surface.

Salt is still used as a seasoning or preservative in meatpacking, fish curing, and food processing as well as for curing hides. About half of the world's output of salt is spread on the ground for de-icing roads and walkways in winter in cold climates. (In this case, calcium chloride ($CaCl_2$) is preferred over NaCl.) In the chemical industry, salt is an ingredient in the manufacture of baking soda, hydrochloric acid, and chlorine, and a reagent in metallurgical processes. In water softeners, salt removes calcium and magnesium compounds from tap water.

Driving Question:

How do the properties of water and dissolved salts affect the physical and chemical properties of seawater?

Although water is a very common component of the Earth system, it has some very uncommon properties compared to other substances of similar molecular size or chemical composition. Water's unusually high freezing and boiling temperatures, coupled with the temperature range at Earth's surface, mean that water exists naturally in the Earth system in all three phases, as solid (ice), liquid, and gas (water vapor). In fact, the three phases of water can coexist in equilibrium with each other such as at the edge of pack ice. Water frequently changes phase and during these transitions, unusually large amounts of heat energy are either absorbed from or released to the environment. Furthermore, large quantities of heat are required to change the temperature of water. Water dissolves a wide variety of solids, liquids, and gases and is aptly described as the *universal solvent*. Salinity and temperature primarily affect the density of seawater, which in turn influences the circulation of the ocean.

In this chapter we examine water's unique properties, the fundamental reasons for those properties, and some of the implications for the functioning of the ocean in the Earth system. We begin by describing the structure of the water molecule and explaining how this structure is the principal reason for water's unique physical and chemical properties. We then explore the chemical properties of seawater, emphasizing the types, sources, and cycling of dissolved salts and gases. The closing discussion of the physical properties of seawater covers the effect of temperature and salinity on water density, pressure exerted by seawater, the formation of sea ice, and the use of sound transmission in the ocean to determine water depth and temperature.

The Water Molecule and Hydrogen Bonding

Compared to other naturally occurring substances, water's thermal properties are unique. For example, based on water's molecular weight as well as the freezing and boiling temperatures of chemically related substances, fresh water should freeze at about –90 °C (–130 °F) and boil at about –70 °C (–94 °F). Actually, fresh water's freezing point is 0 °C (32 °F) and its boiling point is 100 °C (212 °F) at average sea level air pressure. Water's unusual properties arise from the physical structure of the water molecule (H_2O) and the bonding that occurs between water molecules. Without this intermolecular force of attraction, known as **hydrogen bonding**, water would exist only as a gas within the range of surface temperature and pressure on Earth. Furthermore, the planet would have no water cycle, no ocean, no ice caps, and probably no life as we know it.

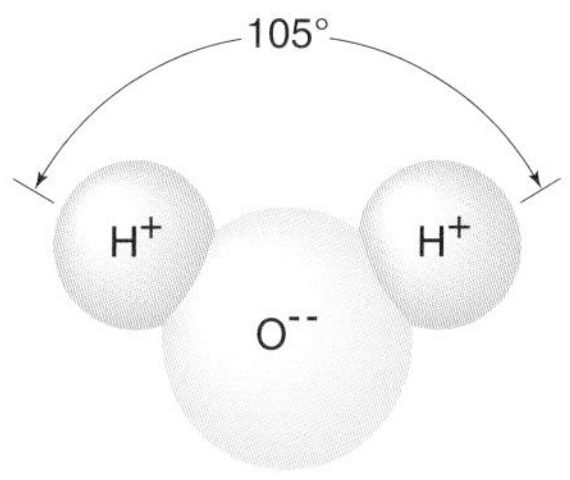

FIGURE 3.2
The water molecule consists of two hydrogen atoms and one oxygen atom.

A water molecule consists of two hydrogen (H) atoms bonded to an oxygen (O) atom (Figure 3.2). Within the water molecule, bonding between hydrogen and oxygen atoms involves sharing of electrons, one from each hydrogen atom and two from the oxygen atom. (An *electron* is a negatively charged subatomic particle.) In this bonding, the electrons spend more time near the oxygen so that the oxygen acquires a small negative charge and the hydrogen is left with a small positive charge. This *covalent bonding* is relatively strong so that the water molecule resists dissociation into its constituent hydrogen and oxygen atoms. The 105-degree angle formed by the arrangement of the hydrogen-oxygen-hydrogen atoms produces a charge separation in the water molecule. Molecules having a separation of positive and negative charges are described as *polar*.

Opposite electrical charges attract so that, like tiny magnets, neighboring water molecules link together. The positively charged (hydrogen) pole of one water molecule attracts the negatively charged (oxygen) pole of another water molecule; this attractive force constitutes hydrogen bonding. Each water molecule can form as many as four hydrogen bonds with surrounding water molecules. A hydrogen bond has about 5% to 10% of the strength of a covalent bond between hydrogen and oxygen atoms in individual water molecules. Nonetheless, hydrogen bonding is sufficiently strong to significantly influence the physical and chemical properties of water. Hydrogen bonding inhibits changes in water's internal energy so that it absorbs or releases unusually great quantities of heat energy when changing phase. Because of hydrogen bonding, greater additions or losses of heat are required to change water temperature as compared to other chemically related substances.

WATER AS ICE, LIQUID, AND VAPOR

Matter exists in three basic phases: solid, liquid, and gas. Transformations from one phase to another always involve a large amount of energy to alter the molecular configuration. Water is one of very few substances that can occur naturally in all three phases within the temperature and pressure ranges found at and near Earth's surface. In this section, we examine how water's molecular structure influences its properties in each phase.

Like all crystalline solids, ice has a regular internal three-dimensional framework consisting of a repeated pattern of molecules. A physical model of the *crystal lattice* of ice is shown in Figure 3.3. Each water molecule is bound tightly to its neighbors but intermolecular bonds are elastic (acting like springs) so that molecules vibrate about fixed locations in the lattice. For this reason, an ice cube retains its shape. Hydrogen bonding is responsible for the ordered arrangement of water molecules in the crystal lattice and the hexagonal (six-sided) structure of ice crystals. (Many of us are familiar with the six-sided symmetry of snow flakes.) Because ice's internal framework is an open network of water molecules, the molecules in ice crystals are not as closely packed as a similar number of molecules in liquid water. At 0 °C (32 °F), ice has a density of about 0.92 g per cubic cm whereas pure liquid water at the same temperature has a density of nearly 1.0 g per cubic cm. This density difference explains why ice floats on the surface of liquid water. Because of their greater density, most common solids would sink if placed in their liquid phase.

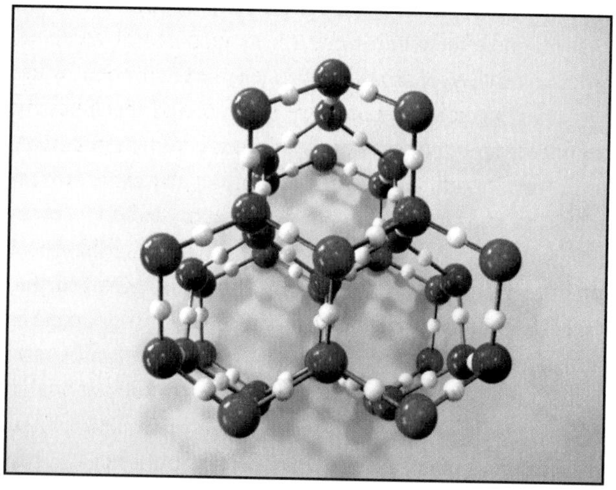

FIGURE 3.3
A physical model of the crystal lattice of ice. Each water molecule is bound tightly to its neighbors but intermolecular bonds are elastic so that molecules vibrate about fixed locations in the lattice. For this reason, an ice cube retains its shape. The ordered arrangement of water molecules in the crystal lattice is responsible for the hexagonal structure of ice crystals. The internal framework of ice is an open network of water molecules, so that the molecules in ice crystals are not as closely packed as a similar number of molecules in liquid water. This property causes ice to be less dense than liquid water.

Despite the openness of the ice lattice, sea ice (formed when seawater freezes) contains less salt than does seawater. Salt ions are too large to fit into the empty spaces within the crystal lattice of ice and cannot substitute for water molecules. Hence, most impurities (salts and gases) are excluded from ice that forms when water freezes.

When ice melts, it becomes liquid water. After observing ice crystals disappearing during melting, we might expect that hydrogen bonds between water molecules are absent when water is in its liquid phase. This is not the case. Instead, many water molecules are linked by hydrogen bonding as ice-like clusters of molecules surrounded by non-bonded (free) water molecules (Figure 3.4). These clusters are constantly forming and breaking up. When water temperatures are near freezing, the transient ice-like clusters are largest and most numerous. Although clusters are present in the liquid phase, water molecules exhibit more activity in the liquid than solid phase. In the liquid phase, water molecules undergo vibrational, rotational and translational (straight-line) motion thus inhibiting formation of ice-like clusters. Also, this greater freedom of movement in the liquid phase explains why liquid water takes the shape of its container.

When liquid water changes to vapor, essentially all hydrogen bonds are broken. Individual molecules move about with even greater freedom than in the liquid phase, diffusing rapidly to fill the entire volume of its container. Gas molecules exhibit vibrational, rotational, and translational motion and exert a force as they bombard a solid or liquid surface. Force per unit area is defined as *pressure*. Heating a closed rigid container of gas accelerates the activity of gas molecules so that they collide more frequently with the inside surfaces of the walls of the container and exert more pressure. Hence, if the volume is held constant, the pressure exerted by a gas increases as the temperature rises. The situation is more complicated for air (a mixture of gases) because the atmosphere is only bounded below (by continents and ocean) and a unit mass of air is free to change volume. When heated, air in the open atmosphere expands and its density decreases.

Water readily changes phase, contributing to the dynamic nature of the Earth system. In changing phase from ice to liquid to vapor, the energy state (i.e., molecular activity) of water increases. Hence, when water changes phase, heat is either absorbed from the environment or released to the environment (Figure 3.5). Melting, evaporation, and sublimation are phase changes that absorb heat. Phase changes that release heat to the surroundings are

FIGURE 3.4
In liquid water, ice-like clusters of molecules linked by hydrogen bonding are constantly forming and breaking up.

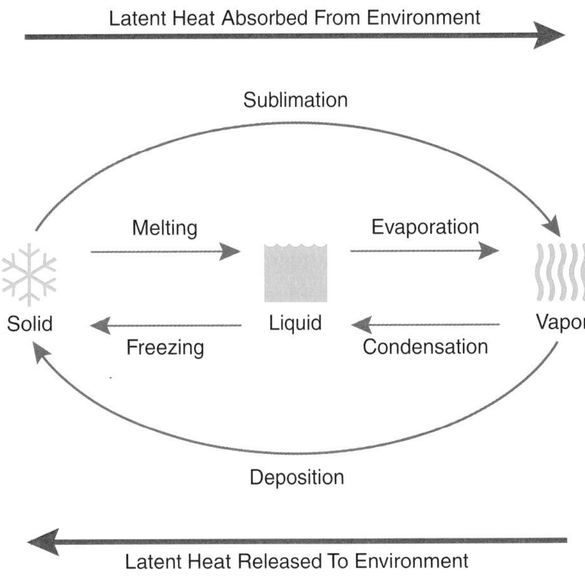

FIGURE 3.5
When water changes phase, heat energy is either absorbed from the environment (melting, evaporation, sublimation) or released to the environment (freezing, condensation, deposition).

freezing, condensation, and deposition. Before examining these phase changes of water in detail, we need to review the distinction between temperature and heat.

TEMPERATURE AND HEAT

From everyday experience, we know that temperature and heat are closely related. Heating a pan of soup on the stove raises the temperature of the soup whereas dropping an ice cube into a warm beverage lowers the temperature of the beverage. Although sometimes used interchangeably, temperature and heat are distinctly different concepts.

All matter is composed of atoms or molecules that are in continual vibrational, rotational, and/or translational motion. The energy represented by this motion is referred to as *kinetic molecular energy* or just *kinetic energy*, the energy of motion. In any substance, atoms or molecules actually exhibit a range of kinetic energies. **Temperature** is directly proportional to the average kinetic energy of atoms or molecules composing a substance. At the same temperature, the average kinetic molecular energy of water is the same regardless of the volume of water.

Internal energy encompasses all the energy in a substance, that is, the kinetic energy of atoms and molecules plus the potential energy arising from forces between atoms or molecules. If two objects have different temperatures (different average kinetic molecular energies) and are brought into contact, energy will be transferred between objects; we call this energy in transit, **heat**. Heat transferred from an object reduces the internal energy of that object whereas heat absorbed by an object increases its internal energy. Heat transferred to or from water brings about a change in temperature or a change in phase. An addition or loss of heat can also do work by bringing about a change in volume of a substance. For example, ice expands when heated and contracts when cooled.

Differences in temperature rather than differences in internal energy govern the direction of heat transfer. Heat energy is always transferred from a warmer object to a colder object. Heat is not necessarily transferred from an object having greater internal energy to an object with less internal energy. Consider, for example, a hot marble (at 40 °C) that is dropped into 5 liters of cold water (at 4 °C). The water has much more internal energy than the marble; nonetheless, heat is transferred from the warmer marble to the cooler water.

The following illustration makes clearer the distinction between temperature and heat. A cup of water at 60 °C (140 °F) is much hotter than bathtub water at 30 °C (86 °F); that is, the average kinetic molecular energy of water molecules is greater at 60 °C than at 30 °C. Although lower in temperature, the greater volume of water in the bathtub means that it contains more total kinetic molecular energy than does the cup of water. If in both cases the water is warmer than its environment, heat is transferred from water to its surroundings. However, much more heat energy must be removed from the bathtub water than from the cup of water for both to cool to the same temperature.

Through the years, scientists have devised various scales that express the temperature of an object by a number representing the degree of warmth. Temperature scales were originally derived in the 18th century using water. Pure water's fixed points of an ice-water mixture and its boiling point at sea level air pressure, are unique. In this book we use the Celsius temperature scale primarily with the Fahrenheit equivalent in parentheses. The Celsius scale, devised by the Swedish astronomer Anders Celsius in 1742, has the numerical convenience of a 100-degree increment between the freezing point (0 °C) and boiling point (100 °C) of fresh water at average sea level air pressure. The United States is one of only a few nations still commonly using the Fahrenheit temperature scale, introduced by the German physicist Gabriel Daniel Fahrenheit in 1714. The Fahrenheit temperature scale features a 180-degree increment between the freezing point (32 °F) and boiling point (212 °F) of fresh water at average sea level air pressure. Formulas for converting between the Celsius and Fahrenheit temperature scales are presented in Table 3.1.

A convenient unit of heat energy is the *calorie*, defined as the amount of heat needed to raise the temperature of one gram of water by one Celsius degree (technically, from 14.5 °C to 15.5 °C). (The Calorie used to measure the energy content of food is actually 1000 heat calories or 1 kilocalorie.) Although the preferred unit of energy in any form, including heat, is the *joule (J)*, we generally use the calorie in this book because of its numerical convenience. Furthermore, thermal characteristics of water were used to define the calorie. For conversion purposes, one calorie equals 4.1868 J and one joule equals 0.239 calorie.

CHANGES IN PHASE OF WATER

A mixture of fresh water and ice has an equilibrium temperature of 0 °C (32 °F). Adding heat to the mixture causes ice to melt, whereas removing heat causes water to freeze. For that reason, 0 °C is called the freezing point of fresh water. Substances dissolved in water suppress the equilibrium temperature of a mixture of ice and water to temperatures below the freezing point of fresh water. The temperature at which seawater begins freezing varies with *salinity*, a measure of the mass (grams) of salts dissolved in a kilogram of seawater. We have more on the effect of salinity on the freezing point of seawater later in this chapter.

TABLE 3.1
Temperature Conversion Formulas

$F = 9/5\ C + 32$
$C = 5/9\ (F - 32)$

When water freezes, **latent heat** is released to the environment and for ice to melt an equivalent amount of latent heat is absorbed from the environment. The word *latent* means "hidden" and refers to the fact that this heat energy is used only to change the phase of water and not the temperature of the water. If heat is added to a mixture of ice and water at 0 °C (32 °F), the temperature of the ice-water mixture remains constant until all the ice melts. All available heat is used to bring about the phase change by breaking the hydrogen bonds that maintain water in the solid (crystalline) phase. Whether freezing or melting is taking place, the latent heat involved is commonly called the **latent heat of fusion**.

At the interface between liquid water and air (e.g., the sea surface), water molecules continually change phase: some crossing the interface from water to air and others from air to water. If more water molecules enter the atmosphere as vapor than return as liquid, a net loss occurs in liquid water mass. This process is known as **evaporation**. Evaporation explains the disappearance of puddles following a rain shower. On the other hand, if more water molecules return to the water surface as a liquid than escape as vapor, net gain of liquid water mass results. This process is called **condensation**. Invisible water vapor condenses on the cold surface of an aluminum beverage can (as tiny visible droplets) on a humid summer day. Heat absorbed from the environment during evaporation and heat released to the environment during condensation are known as the **latent heat of vaporization** and the **latent heat of condensation** respectively.

All of us have experienced **evaporative cooling**. We are chilled upon stepping out of a shower or swimming pool. Water droplets evaporating from the skin absorb heat, lowering the skin's temperature. On a global scale, evaporative cooling is the most important process whereby heat is transferred from the ocean to the atmosphere (Chapter 5). When water evaporates at Earth's surface, water vapor moves into the atmosphere where it may subsequently condense into clouds. Heat absorbed as water evaporates at the surface is later released to the atmosphere during condensation. This latent heat transfer mechanism is also important in powering storms, especially thunderstorms and tropical cyclones (e.g., hurricanes).

Water's latent heat of fusion is 80 calories per g at 0 °C. Considerably greater amounts of heat are required for water to vaporize because essentially all hydrogen bonds must be broken during the phase change. (Recall that clusters of bonded water molecules are present in the liquid phase.) In fact, the magnitude of water's latent heat of vaporization is about seven times that of its latent heat of fusion and the highest of all common substances. The latent heat of vaporization varies with temperature from 597 calories per g at 0 °C (32 °F) to 540 calories per g at 100 °C (212 °F). More energy is needed to break the tighter and more numerous hydrogen bonds at lower temperatures.

At the interface between ice and air (e.g., the surface of a snow cover), water molecules are also continually changing phase: directly from ice to vapor and from vapor to ice. If more water molecules enter the atmosphere as vapor than become ice, a net loss of ice mass occurs. **Sublimation** is the process whereby ice or snow becomes vapor without first becoming a liquid. Sublimation explains the gradual disappearance of a snow cover even while the air temperature remains well below freezing. On the other hand, if more atmospheric water molecules become ice than move from ice to vapor, a net gain of ice mass results. **Deposition** is the process whereby water vapor becomes ice without first becoming a liquid. During a cold winter night, the formation of frost on automobile windows is an example of deposition.

Heat is absorbed from the environment during sublimation and heat is released to the environment during deposition. Latent heat involved in sublimation or deposition must equal the total amount of heat absorbed or released during the combined solid-liquid plus liquid-vapor phase changes. Sublimation requires the latent heats of fusion plus vaporization, known as the **latent heat of sublimation**. Deposition releases to the environment an equivalent amount of latent heat, that is, the **latent heat of deposition**. The magnitudes of the latent heats of sublimation and deposition are remarkably uniform, varying only from 677 calories per g at 0 °C to 678 calories per g at –30 °C.

SPECIFIC HEAT OF WATER

The temperature change associated with an input (or output) of a specified quantity of heat varies from one substance to another. The amount of heat that is needed to raise the temperature of 1 gram of a substance by 1 Celsius degree is defined as the **specific heat** of that substance. Joseph Black (1728-1799), a Scottish chemist, first proposed the concept of specific heat in 1760. The specific heat of liquid water is 1 calorie per g per Celsius degree (at 15 °C). The specific heat of ice is about 0.5 calorie per g per Celsius degree (near 0 °C). Specific heats of other familiar substances are listed in Table 3.2. The variation in specific heat from one substance to another implies that different materials have different capacities for storing internal energy.

Suppose that the same quantity of heat energy is supplied to two different substances having different specific heats. The substance with the higher specific heat undergoes a smaller rise in temperature (warms less) than the substance having the lower specific heat. Because of hydrogen bonding, water has an unusually high specific heat, in fact the highest specific heat of any naturally occurring liquid or solid. From Table 3.2, it can be seen that water's specific heat is about 5 times that of dry sand. Whereas one calorie of heat will raise the temperature of one gram of water by 1 Celsius degree, the same quantity of heat will raise the temperature of one gram of dry sand by slightly more than 5 Celsius degrees. This contrast in specific heat helps explain why at the beach in summer the sand feels considerably hotter to bare feet than does the water. This also largely explains why wet sand feels cooler than dry sand.

A simple experiment conducted at sea level summarizes water's unusually high latent heats and specific heat (Figure 3.6). A one-gram ice cube initially at –20 °C is heated to 0 °C. Every one Celsius-degree rise in ice cube temperature requires an addition of 0.5 calorie of heat energy (the *specific heat* of ice). A total of 10 calories

TABLE 3.2
Specific Heat[a] of Some Familiar Substances

Water	1.000
Wet mud	0.600
Ice (at 0 °C)	0.478
Wood	0.420
Aluminum	0.214
Brick	0.200
Granite	0.192
Sand	0.188
Dry air[b]	0.171
Copper	0.093
Silver	0.056
Gold	0.031

[a]Calories per gram per Celsius degree
[b]At constant volume

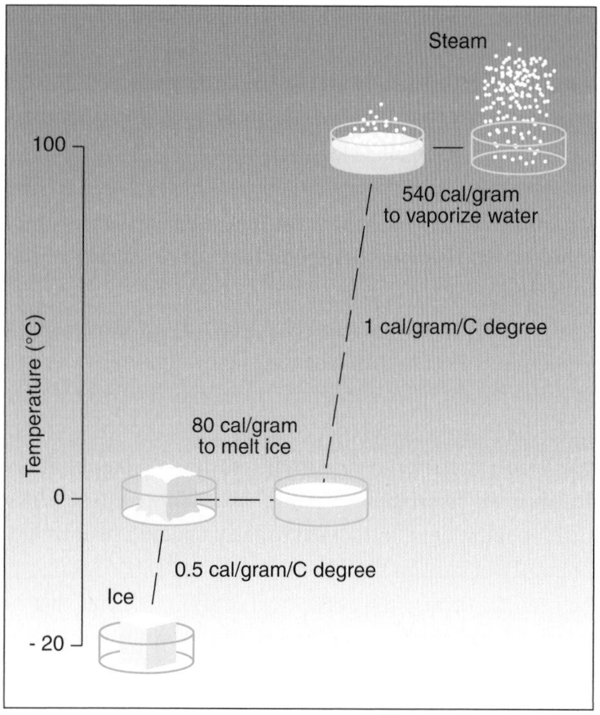

FIGURE 3.6
Heating a 1-gram ice cube causes a rise in temperature plus phase changes, initially to liquid and then to vapor.

of heat (20 C° × 0.5 cal per C°) is required to warm the one-gram ice cube to 0 °C. The temperature then remains constant while 80 calories of heat (*latent heat of fusion*) are added to melt the one-gram ice cube. The liquid water is then heated to 100 °C. Every one-Celsius degree rise in temperature of the one gram of liquid water requires an addition of one calorie of heat energy (*specific heat* of water). If none of the water evaporates, a total of 100 calories of heat (100 C° × 1 calorie per C°) are required to warm the 1 gram of liquid water to 100 °C. At 100 °C, the water vaporizes requiring an input of 540 calories of heat (*latent heat of vaporization*).

MARITIME INFLUENCE ON CLIMATE

Water's exceptional capacity to store heat has important implications for weather and climate. A large body of water (such as the ocean or Great Lakes) can significantly influence the climate of downwind localities. The most persistent influence is on air temperature. Compared to an adjacent landmass, a body of water does not warm as much during the day (or in summer) and does not cool as much at night (or in winter). In other words, a large body of water exhibits a greater resistance to temperature change, called **thermal inertia**, than does a landmass. Whereas the higher specific heat of water versus land is the major reason for the contrast in thermal inertia, differences in heat transport also contribute. Sunlight penetrates water to some depth and is absorbed (converted to heat) through a significant volume of water. But sunlight cannot penetrate the opaque land surface and is therefore absorbed only at the surface. Furthermore, ocean and lake-waters circulate and transport heat through movements of great volumes of water, whereas heat is conducted only very slowly into the soil. The input (or output) of equal amounts of heat energy causes a land surface to warm (or cool) more than the equivalent surface area of a body of water.

Air temperature is regulated to a considerable extent by the temperature of the surface over which air resides or travels. Air over a large body of water tends to take on similar temperature characteristics as the surface water. Consequently, places immediately downwind of the ocean experience moderate contrast between average winter and summer temperatures. Such locales have a **maritime climate**. Places at the same latitude but well inland experience a much greater temperature contrast between winter and summer. Such locales have a **continental climate**. However, in winter the climatic influence of ice-covered bodies of water (e.g., the Bering and Greenland Seas) is more like a landmass than an ocean.

Consider an example of the contrast in maritime versus continental climates. The latitude of San Francisco, CA (37.8 degrees N) is almost the same as that of St. Louis, MO (38.8 degrees N) so that the seasonal variation in the amount of solar radiation striking Earth's atmosphere (due to astronomical factors) is about the same at both places. St. Louis is situated far from the moderating influence of the ocean and its climate is continental. St. Louis' average summer (June, July, August) temperature is 25.6 °C (78.0 °F) and its average winter (December, January, February) temperature is 0.6 °C (33.0 °F), giving an average summer-to-winter seasonal temperature contrast of 25 Celsius degrees (45 Fahrenheit degrees). San Francisco, on the other hand, is located on the West Coast, immediately downwind of the Pacific Ocean; its climate is maritime. The average summer temperature at San Francisco is 17 °C (62.6 °F) and the average winter temperature is 10.2 °C (50.4 °F), giving an average seasonal temperature contrast of only 6.8 Celsius degrees (12.2 Fahrenheit degrees).

The moderating influence of the ocean is also evident in the contrast in climate between Western Europe and Eastern North America. At middle latitudes, prevailing winds blow from west to east so that the maritime influence of the North Atlantic Ocean is much more apparent in

Western Europe than Eastern North America. In the same latitude belt, winters are considerably milder in Western Europe than in Eastern North America. Consider, for example, the contrast in average January temperatures for Montreal, Quebec (45.5 degrees N) versus London, England (51.5 degrees N). In January, the average daily high temperature is –6.1 °C (21 °F) at Montreal and 6.7 °C (44 °F) in London. The January average daily low temperature is –14.4 °C (6 °F) at Montreal and 1.7 °C (35 °F) in London. Although London is considerably farther north than Montreal, its January temperatures are significantly milder.

Chemical Properties of Seawater

Pure water is unknown in the Earth system; that is, water free of all dissolved and suspended material does not occur naturally. This is because water is an excellent solvent for a wide range of materials—in fact, that is why water is sometimes referred to as the *universal solvent.* Water dissolves solids, liquids, and gases to varying extents. The ocean, by far the largest reservoir of water in the Earth system, contains so much salt in solution that it cannot be used for most domestic, agricultural or industrial purposes. In this chapter's first Essay, we describe desalination techniques designed to remove salts from seawater for the purposes of augmenting the supply of fresh water. In this section, we consider water as a solvent and then describe the types and sources of dissolved salts and gases in seawater.

WATER AS A SOLVENT

The polar nature of the water molecule favors the solution (i.e., dissolving) of both ionic and non-ionic substances. Many inorganic materials (primarily salts) are bonded ionically, whereas many organic chemicals have non-ionic bonds. River water, groundwater, and ocean water dissolve some of the rock or sediment (both organic and inorganic) that water contacts and some Earth materials dissolve in water more readily than other Earth materials. In Chapter 4, we consider the types and sources of sediments that dissolve or are suspended in ocean water.

Consider what happens when a pinch of common household table salt is added to water. Table salt is sodium chloride (NaCl), the mineral known as *halite.* In salt's cubic crystalline form, ionic bonds hold the positively charged sodium ions (Na^{+1}) and the negatively charged chloride (Cl^{-1}) ions together (Figure 3.7A). (An *ion* is an electrically charged atom.) Once salt enters the water, however, the hydrogen-bonded complexes of water molecules greatly reduce the force of attraction between oppositely charged sodium and chloride ions. That is, the strength of ionic bonds between sodium and chloride diminishes so that the compound readily dissociates into sodium ions and chloride ions (Figure 3.7B). Sodium ions are attracted to the negatively charged pole of the water molecule while chloride ions are attracted to the positively charged pole of the water molecule. In this way, salt dissolves in water.

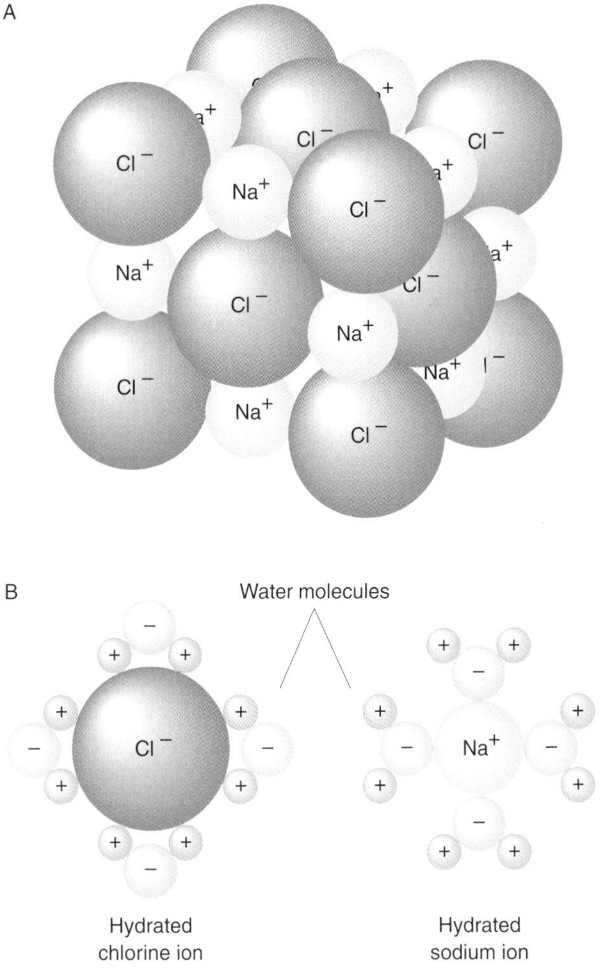

FIGURE 3.7
Common household table salt, sodium chloride (NaCl), dissolves in water. (A) In its cubic crystalline form, ionic bonds hold together the positively charged sodium ions (Na^+) and the negatively charged chloride (Cl^-) ions. (B) Once salt enters the water, however, hydrogen-bonded complexes of water molecules greatly reduce the force of attraction between oppositely charged sodium and chloride ions. The compound readily dissociates into sodium and chloride ions with the sodium ions attracted to the negatively charged pole of the water molecule and chloride ions attracted to the positively charged pole of the water molecule.

SEA SALTS

Seawater is a salt solution of nearly uniform composition, with only slight variation in the relative amount of water. **Salinity** is a measure of the amount of salt dissolved in seawater. On average, seawater is 96.5% water and 3.5% dissolved salts. If all ocean water evaporated and the precipitated salts were spread evenly over the surface of the Earth, the salts would form a layer about 45.5 m (150 ft) thick—about the height of a fifteen-story building.

In the 19th century, chemists began examining the composition of seawater in some detail, and one of them, William Dittmar, verified an important observation made by earlier scientists. In 1884, Dittmar began analyzing the constituents of 77 water samples obtained at various depths and locations in the ocean during the worldwide voyage of *HMS Challenger* (from December 1872 to May 1876). He found that although the total amount of dissolved solids varied among water samples, the ratio of the concentrations of the major constituents of seawater was the same in all samples. That is, the major constituents of seawater occur in the same relative concentrations throughout the ocean, a characteristic of seawater described as the **principle of constant proportions**.

The concentrations of the major dissolved constituents of ocean water, such as chloride (Cl^{-1}) and sodium (Na^{+1}) ions, are *conservative properties* of seawater; that is, these ions occur in constant proportions and change concentration very slowly by mixing or diffusion. They have no sources or sinks within the ocean. Seawater constituents that participate in biogeochemical or seasonal cycles have variable concentrations and are described as *non-conservative properties*. Recall from Chapter 1 that the ocean is a reservoir in all biogeochemical cycles operating in the Earth system. Examples of non-conservative properties of seawater are concentrations of silica and calcium compounds, nitrates, phosphates, and aluminum.

The principle of constant proportions implies that measurement of the concentration of one of the major conservative constituents of seawater is all that is needed to determine the concentrations of any other conservative constituent (as well as the sum of all other conservative constituents). This principle made it possible for ocean scientists to define and measure salinity. Prior to World War II, salinity was defined as the total amount of solid materials in grams dissolved in one kilogram of seawater. The standard method for measuring salinity was chemical titration to determine the concentration of the chloride ion (Cl^{-1}), which was then substituted into a formula to calculate salinity. Salinity was expressed as grams of dissolved material per kilogram of seawater or parts per thousand (ppt). The accuracy of this method was about ± 0.02 ppt.

After World War II, scientists discovered that salinity could be determined from measurements of the electrical conductivity of seawater with ten times the accuracy of the old titration method. In 1978 the international oceanographic community accepted a new definition of salinity based on conductivity measurements; that is, the salinity of a water sample is defined as the ratio of the conductivity of the sample to the conductivity of *standard seawater*. Samples of standard seawater, prepared by Ocean Scientific International in southern England, are supplied to laboratories around the world for use in calibrating salinity-measuring instruments. As a ratio, salinity has no units and may be written as a pure number, but for convenience some ocean scientists prefer to express salinity in *practical salinity units (psu)*. In this book, we present salinity values as a number without units. The conductivity-based definition of salinity was designed to retain the validity of measurements made the old way; that is, a salinity of 35 is essentially the same as 35.0 parts per thousand, or 3.5%. A seawater sample having a salinity of 34.82 contains about 34.82 grams of dissolved materials per kilogram.

Although more than 70 chemical elements are dissolved in seawater, six make up more than 99% of all sea salts: chloride (Cl^{-1}), sodium (Na^{+1}), sulfate (SO_4^{-2}), magnesium (Mg^{+2}), calcium (Ca^{+2}), and potassium (K^{+1}) (Figure 3.8). Common table salt (NaCl) alone accounts for nearly 86% of the total. Numerous trace components include aluminum, chromium, gold, lead, nickel, and zinc.

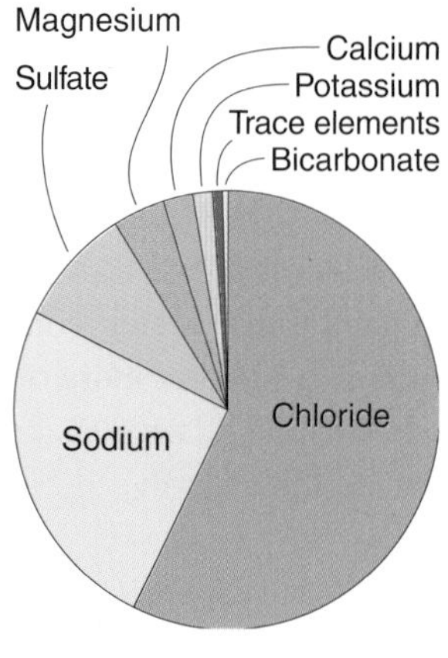

FIGURE 3.8
Although more than 70 different ions are dissolved in seawater, six make up more than 99% of all sea salts: chloride (Cl^{-1}), sodium (Na^{+2}), sulfate (SO_4^{-2}), magnesium (Mg^{+2}), calcium (Ca^{+2}), and potassium (K^{+1}).

What is the origin of the salts dissolved in seawater? The principal source is weathering and erosion of rock and sediment on land and transport by rivers and streams to the ocean (Chapter 4). As the largest reservoir in the global water cycle, the ocean receives most of its water from rivers. Although seawater is considerably more saline than river water, we might expect the ratios of the different chemical constituents to be essentially the same in river and ocean waters. But this is not the case (Table 3.3). Whereas sodium and chloride account for 86% of solids dissolved in seawater, they typically constitute less than 16% of dissolved solids in river water. Calcium (Ca^{+2}) and bicarbonate (HCO_3^{-1}) ions are minor ingredients of seawater (less than 2% of dissolved solids) but major components of river water (almost 50% of dissolved solids). On average, silica (SiO_2) accounts for about 14.5% of the dissolved components of river water but is a minor component of seawater.

What explains the difference in the chemical makeup of dissolved solids in river water versus seawater? For one, marine organisms extract calcium and silica from seawater to build their shells and skeletons. Also some forms of marine life concentrate, secrete, or excrete certain chemical elements. Differences in solubility and rates of physical-chemical reactions among ions also play a role by limiting the concentration of certain substances in seawater (e.g., calcium) or causing some chemicals to precipitate from solution (e.g., manganese). Furthermore, there are other sources of salts dissolved in seawater besides rivers.

Hydrothermal vents and chemical reactions between seawater and recently formed oceanic crust contribute to the salinity of ocean water (Chapter 2). The entire volume of the world ocean cycles through fractures in new oceanic crust about every 10 million years. Suspended particles settle out of or are washed from the atmosphere by rain or snow. For example, gases emitted during volcanic eruptions dissolve in rainwater and enter the ocean contributing chloride and sulfate ions.

The proportionality of the principal sea salts has remained nearly constant for the past 1.5 billion years. This is evident from the composition of salt strata (sedimentary rock) formed from the evaporation of seawater in the geologic past, which has changed little over that period of time. This implies that the rate of addition of new salts to the ocean balances the rate of removal. Processes that remove salt ions from the ocean include sea spray that is blown ashore and isolation of arms of the sea from the ocean followed by evaporation to produce salt deposits. For example, about 6 million years ago, northward movement of the African tectonic plate cut off the Mediterranean Sea from the Atlantic Ocean at Gibraltar. In roughly 1000 years, the Mediterranean almost completely dried up precipitating thick layers of salt.

Other processes that remove salt ions (mostly calcium sulfate) from seawater are the following: certain ions chemically react with one another to form insoluble precipitates and some ions adsorb onto suspended sediments. In both cases, ions settle to the seafloor and mix with other sediments. Compounds used by marine organisms to build shells or skeletons, such as calcium and silica, are removed from seawater and upon death of the organism cycle to the ocean floor as shells, bones, or teeth (Chapter 4). Also, chemical reactions between seawater and newly formed oceanic crust remove some constituents of seawater such as magnesium ions (Mg^{+2}). Residence times of salts in seawater range from hundreds of years (e.g., aluminum, iron) to millions of years (e.g., sodium, potassium).

TABLE 3.3
Comparison of Composition of Ocean Water with River Water[a]

	Percentage of Total Salt Content	
Chemical Constituent	*Ocean Water*	*River Water*
Silica (SiO_2)	-	14.51
Iron (Fe)	-	0.74
Calcium (Ca)	1.19	16.62
Magnesium (Mg)	3.72	4.54
Sodium (Na)	30.53	6.98
Potassium (K)	1.11	2.55
Bicarbonate (HCO_3)	0.42	31.90
Sulfate (SO_4)	7.67	12.41
Chloride (Cl)	55.16	8.64
Nitrate (NO_3)	-	1.11
Bromide (Br)	0.20	-
Total	100.00	100.00

[a]Source: U.S. Geological Survey

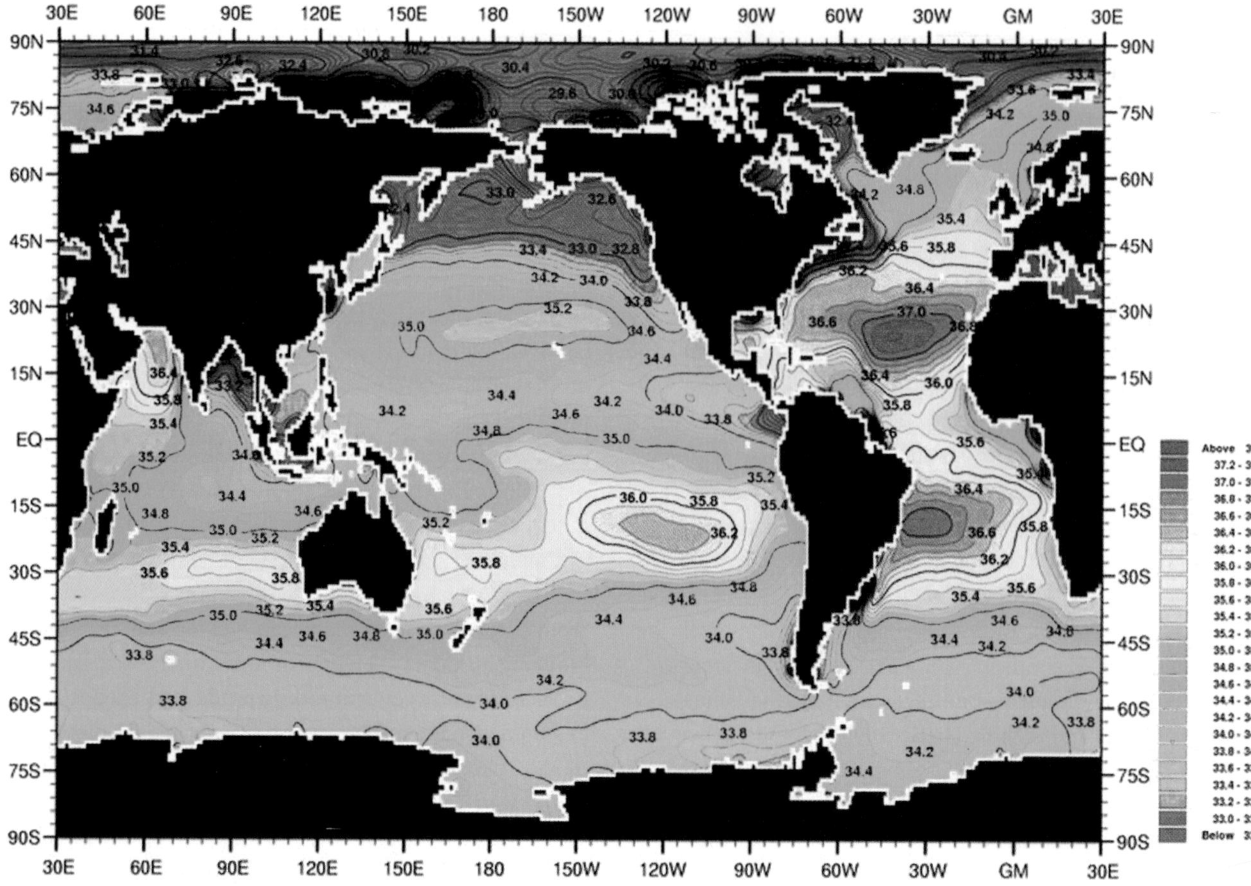

FIGURE 3.9
Global pattern of average annual sea-surface salinity shows the highest values in the subtropics of the Northern and Southern Hemispheres. Salinity values are in parts per thousand. [Courtesy of NOAA, National Oceanographic Data Center]

VARIATIONS IN SALINITY

Processes operating at the atmosphere/ocean interface add or remove water molecules from seawater and largely explain spatial variations in sea-surface salinity. Most dissolved substances are left behind when seawater evaporates or freezes, increasing the surface salinity locally. On the other hand, precipitation, runoff from rivers, and melting ice add fresh water and decrease the local surface salinity. On an average annual basis, freezing of water and melting of ice have a negligible net effect on sea-surface salinity because the increase in salinity produced by ice formation in winter is offset by the decrease in salinity produced by melting ice during summer. Where large rivers enter the ocean, salinity is reduced as fresh water and seawater mix; salinity is also reduced where rainfall is heavy. Runoff is only important in coastal regions, so evaporation and precipitation are the principal processes governing surface salinity over most of the ocean.

Seawater salinity tends to be more variable in coastal areas than the open ocean. The salinity of surface waters of the open ocean averages about 35 and rarely falls below 33 or rises above 38. The surface salinity of the open ocean peaks near the center of each ocean basin, that is, in the subtropics (between 20 and 30 degrees N and between 15 and 25 degrees S) where the annual evaporation rate exceeds precipitation (Figure 3.9). Average sea-surface salinity is generally lowest near the equator and in polar areas where annual precipitation is greater than annual evaporation. Salinity peaks at 40 in the Red Sea and Persian Gulf because of high rates of evaporation brought on by high air temperatures. On the other hand, the discharge of freshwater rivers and streams is responsible for the relatively low salinities of the Baltic Sea (5 to 15), Black Sea (less than 20), and Puget Sound, WA (21 to 27).

DISSOLVED GASES

We are well aware from personal experience that solids such as salt and sugar dissolve in water. What may be less obvious is that gases also dissolve in water. For example, carbon dioxide (CO_2) dissolved under pressure produces a carbonated beverage. If you open a can of cola, tiny CO_2 bubbles escape, giving the drink its fizz. Dissolved gases are also present in ocean water; these include carbon dioxide, nitrogen (N_2), and oxygen (O_2). (Note that dissolved oxygen is not represented by the "O" in H_2O.)

Gases are exchanged between the atmosphere and ocean at the ocean surface. The saturation (maximum) concentration of a gas in water depends primarily on temperature in freshwater bodies such as lakes and rivers, and in seawater. Seawater salinity also impacts solubility to some extent. Almost all gases are more soluble in cold water than in warm water (Figure 3.10). For example, all other factors being equal, we would expect less dissolved oxygen in lakes in summer than in the cooler seasons. As the temperature or salinity of seawater increases, water holds less gas at saturation; of the two factors, temperature is much more important. If seawater is not saturated with a specific gas, the gas is transferred from air to ocean. If seawater is supersaturated with the gas (containing levels of gas above the normal saturation concentration), the gas transfer is in the opposite direction. When water is saturated with a gas, the rate at which the gas dissolves in water equals the rate at which the gas simultaneously escapes to the atmosphere.

Waves on the ocean surface facilitate the transfer of gases between the atmosphere and ocean. Waves roughen the surface and increase the surface area of ocean water exposed to the atmosphere for gas exchange. Another important mechanism in the transfer of gases at the air/sea interface is **bubble injection** whereby breaking waves introduce a foam composed of small bubbles below the ocean surface greatly enhancing exchange rates. On the other hand, in cold climates, sea-ice cover is a barrier to gas transfer at the ocean surface.

Below the ocean-atmosphere interface, biochemical processes play important roles in controlling the proportions of certain dissolved gases. Within the **photic zone**, the sunlit upper layer of the ocean where photosynthesis takes place, dissolved oxygen is enhanced relative to carbon dioxide. As noted in Chapter 1, *photosynthesis* is the process whereby green plants (e.g., phytoplankton) use sunlight, water, and carbon dioxide to manufacture their food and generate oxygen as a byproduct. In surface ocean waters, the principal dissolved gases are nitrogen (48%), oxygen (36%), and carbon dioxide (15%). At ocean depths greater than the photic zone, photosynthesis is absent and decomposer activity and cellular respiration are the most important biochemical processes. Recall from Chapter 1 that *cellular respiration* is the process whereby organisms break down food and release energy in a usable form. Through cellular respiration, marine organisms use dissolved oxygen and release carbon dioxide as a byproduct. For the ocean as a whole, CO_2 is the most abundant dissolved gas, accounting for 83% of the total.

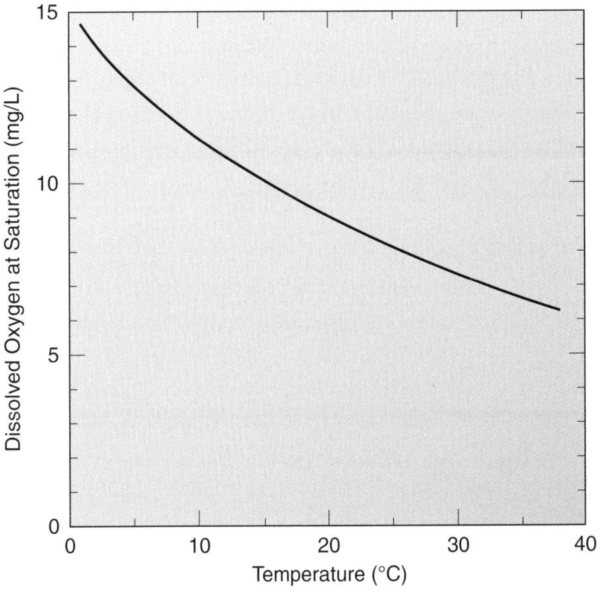

FIGURE 3.10
The saturation value of dissolved oxygen in fresh water decreases with rising temperature.

Carbon dioxide is much more abundant in the ocean than the atmosphere. As noted in Chapter 1, CO_2 is only 0.039% by volume of the gases composing the atmosphere below an altitude of 80 km (50 mi). About fifty-one times as much CO_2 is sequestered in the ocean. Carbon dioxide is highly soluble in water because it reacts with water to produce carbonate (CO_3^{-2}) and bicarbonate (HCO_3^{-1}) ions that readily dissolve in water. At a temperature of 0 °C (32 °F) and sea level air pressure, one liter of water that is saturated with carbon dioxide has dissolved about 1.7 liters of CO_2. The flux of atmospheric carbon dioxide into the ocean has important implications for global climate change (Chapter 12) and ocean acidification (discussed later in this chapter).

If a dissolved gas does not participate in any biochemical process such as photosynthesis or cellular respiration, its concentration in a parcel of seawater remains unchanged except by the relatively slow movements of gas molecules (*diffusion*) through the water or the mixing of water masses containing different amounts of dissolved gas. Generally, nitrogen and inert (chemically nonreactive) gases such as argon and neon behave in this way and their concentrations are described as *conservative properties*. As noted above, biochemical processes influence the concentration of some gases dissolved in seawater, primarily oxygen and carbon dioxide; the concentrations of these dissolved gases are examples of *non-conservative properties*.

SEAWATER pH

Water distributed in the many reservoirs of the Earth system can vary in acidity and alkalinity. An **acid** is a hydrogen-containing compound that releases positively charged hydrogen ions (H^{+1}) when dissolved in water. Strong acids more readily release hydrogen ions than weak acids. An **alkaline substance** (or *base*) releases negatively charged hydroxyl ions (OH^{-1}) when dissolved in water and may also be weak or strong. Pure water has properties of both acids and alkaline materials as water molecules continually break up (into hydrogen ions and hydroxyl ions) and re-form. That is,

$$\mathbf{H_2O \leftrightarrows H^{+1} + OH^{-1}}$$

The acidity of a solution is expressed as **pH**, a measure of its hydrogen ion concentration. On the **pH scale**, the hydrogen ion concentration decreases as the pH increases from 0 to 14 (Figure 3.11). A pH of 7 is neutral with an equal number of $H^{+1}$and OH^{-1} ions; a pH above 7 is increasingly alkaline whereas a pH below 7 is increasingly acidic. The pH scale is logarithmic; that is, each unit increment corresponds to a tenfold change in acidity or alkalinity. Hence, for example, a two-unit drop in pH (e.g., from 5.6 to 3.6) represents a hundred-fold (10×10) increase in acidity. *Acidification* refers to a decline in pH anywhere along the scale.

Surface ocean waters are slightly alkaline with the global mean pH ranging between 8.0 and 8.3. This relatively narrow variation in the pH of seawater is important for those marine organisms whose shells and skeletons are composed of calcium carbonate ($CaCO_3$). If ocean water were even slightly acidic, calcium carbonate would dissolve and would be unavailable for these organisms.

Carbon dioxide plays a key role in controlling the pH of seawater. A substance that stabilizes a chemical system in this way is known as a **buffer**. Atmospheric CO_2 dissolves in ocean water producing carbonic acid (H_2CO_3) that dissociates into hydrogen ions (H^{+1}), carbonate ions (CO_3^{-2}), and bicarbonate (HCO_3^{-1}) ions. A chemical equilibrium develops in which carbon dioxide, carbonic acid, hydrogen ions, carbonate ions, and bicarbonate ions co-exist. Adding acid to the ocean shifts the equilibrium so that there are fewer carbonate ions, which in turn reduces the hydrogen ion concentration and raises the pH. Adding alkaline materials to ocean water raises the pH but the equilibrium shifts in such a way as to return the pH to the normal range.

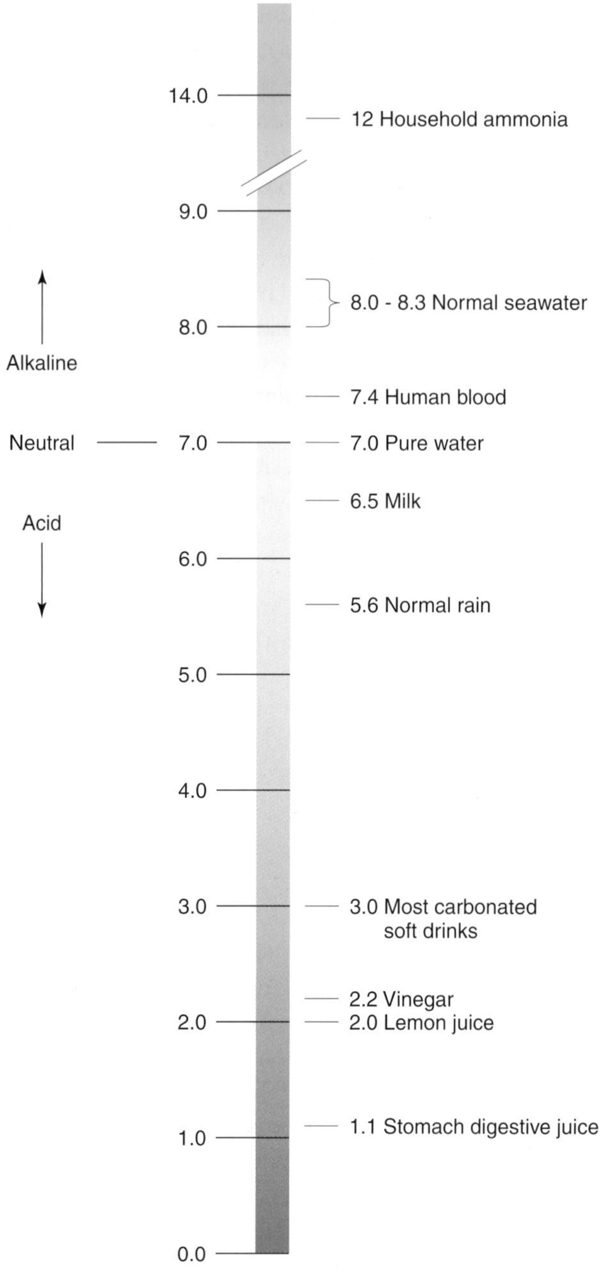

FIGURE 3.11
The acidity of water is expressed as pH, a measure of the hydrogen ion concentration. On this scale, pH increases from 0 to 14 as the hydrogen ion concentration decreases. Pure water has a pH of 7, which is considered neutral; a pH above 7 is increasingly alkaline and a pH below 7 is increasingly acidic.

Compared to seawater, freshwater lakes typically exhibit greater variability in pH in response to influxes of acid precipitation or runoff. Normally, rain and snow dissolve atmospheric CO_2 and are slightly acidic. Natural rainwater that is saturated with CO_2 has a pH of 5.6 and any rain or snow having a pH below this value is designated **acid rain** (or *acid snow*). Where air also contains oxides of sulfur (e.g., from coal burning) or oxides of nitrogen (e.g., byproducts of high-temperature industrial processes), precipitation can become much more acidic than normal. Carbonate rocks (i.e., limestone or dolostone) are alkaline so that lakes in carbonate basins usually neutralize acid rain and runoff whereas lakes occupying non-carbonate basins (e.g., granite) cannot buffer the acid rain or runoff they receive. Lowering the pH of lake waters is known to endanger aquatic life.

The mean pH of ocean water is declining with potentially serious implications for marine life. For information on ocean acidification, refer to this chapter's second Essay.

Physical Properties of Seawater

Some properties of water vary with salinity. Adding salts to water changes its temperature of initial freezing and the temperature at which it reaches maximum density. Because salt ions can not fit into the crystal lattice of ice, dissolved salts inhibit the formation of hexagonal ice crystals and this depresses the initial freezing point to temperatures below 0 °C. Seawater of average salinity (about 35) has an initial freezing temperature of –1.9 °C. Furthermore, seawater does not remain at the same temperature while freezing as does fresh water. Because salts are excluded from the ice structure as seawater freezes, the remaining unfrozen water becomes saltier and therefore freezes at still lower temperatures. This process is known as **brine rejection**. Unless cooled to very low temperatures, some concentrated liquid brine remains trapped in cells of sea ice, although the brine migrates downward over time toward the warmer water under the ice.

WATER DENSITY AND TEMPERATURE

Density is defined as mass per unit volume, which may be expressed as grams per cubic cm. When placed in water an object that is less dense than water will float to the surface whereas an object that is denser than water will sink. Freshwater density varies primarily with temperature whereas seawater density varies chiefly with temperature and salinity. Water is only slightly compressible (1-2%) so that pressure arising from the weight of overlying water does not significantly impact its density except in the deep ocean.

Most substances contract when cooled and expand when heated; their density increases with falling temperature and decreases with rising temperature. As the average kinetic molecular energy decreases (i.e., as the temperature falls), the same number of molecules occupies a progressively smaller volume. However, for fresh water, it is not quite that simple (Figure 3.12). As the temperature of fresh water falls steadily from say 25 °C (77 °F), the water contracts and its density increases. The density of fresh water reaches a maximum at about 4 °C (39.2 °F), but with additional cooling (below 4 °C), the water expands and its density decreases.

Ice-like molecular clusters continually form and break-up in liquid water and are responsible for the anomalous behavior of fresh water density at temperatures below 4 °C. As the temperature drops, the water molecules have less kinetic energy and move closer together so that the number of hydrogen bonds increases resulting in more ice-like clusters. These clusters occupy more volume than the unorganized water molecules in the liquid phase. Recall that ice crystals are open hexagonal (six-sided) structures with widely spaced water molecules (Figure 3.3). As the water temperature drops below 4 °C, the decrease in water density caused by increasing numbers of ice-like molecular clusters more than compensates for the increase in water density that would accompany a decline in kinetic molecular activity. Whereas most liquids contract when

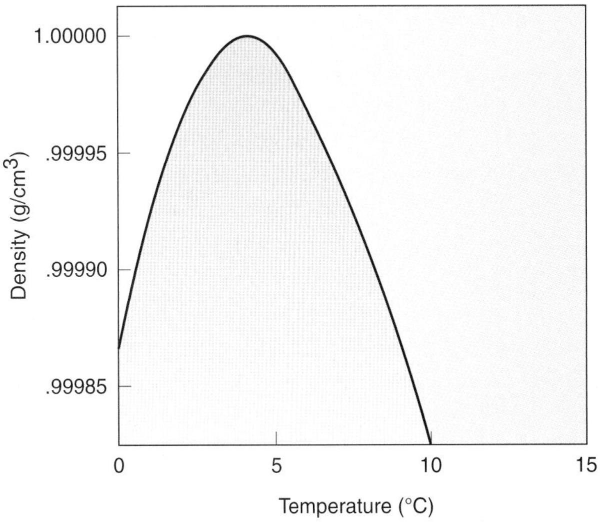

FIGURE 3.12
The density of fresh water reaches a maximum at about 4 °C (39.2 °F) and, with additional cooling (below 4 °C), liquid water expands until it freezes at 0 °C (32 °F).

they solidify, as water freezes, its molecules bond into an open hexagonal structure so that the density of ice is about 92% of liquid water.

The unique temperature-density behavior of fresh water explains why the less dense ice floats in more dense water and lakes freeze from the top down rather than the bottom up. If ice were denser than liquid water, ice that forms at the air-water interface would sink and in winter, lakes in cold climates would freeze solid from the bottom up, destroying all aquatic life. In autumn, lakes begin cooling at the air-water interface when the temperature of the overlying air falls below that of the lake surface. With the continued decline in air temperatures in fall, lake-surface waters cool, contract, become denser and sink to the bottom. This process of cooling and sinking of surface waters is repeated until the entire lake develops a uniform temperature of 4 °C (and the same density). Then, with additional cooling of surface waters below 4 °C, water density decreases and the coldest water remains at the surface. At 0 °C an ice cover begins to form. Once formed, ice contracts (becomes denser) as its temperature falls. But no matter how cold, ice remains less dense than liquid water.

The situation changes when salt is added to water. At constant temperature, the density of seawater increases with increasing salinity because the atomic mass of dissolved salts is greater than that of water molecules. Hence, less dense fresh water floats on more dense seawater. The salinity of seawater also affects the temperature of maximum density and the initial freezing temperature for the same reason: adding dissolved materials such as salt apparently interferes with the formation of ice-like clusters. As shown in Figure 3.13, the temperature of maximum density decreases linearly with increasing salinity. At any salinity less than 24.7, the maximum density of water occurs at a temperature above the initial freezing point (which also decreases linearly with increasing salinity). At a salinity of 24.7, the temperature of maximum density is the same as the freezing temperature (–1.33 °C). For salinities greater than 24.7, the temperature of maximum density is lower than the initial freezing point so that the water freezes before reaching maximum density. The density of seawater of average salinity (about 35) varies inversely with temperature; that is, seawater density always increases with falling temperature and decreases with rising temperature. Hence, seawater becomes denser with cooling and sinks. Except in a few important high latitude locations in the North Atlantic Ocean and the Southern Ocean around Antarctica (Chapter 6), relatively dense seawater usually does not sink to the deep ocean

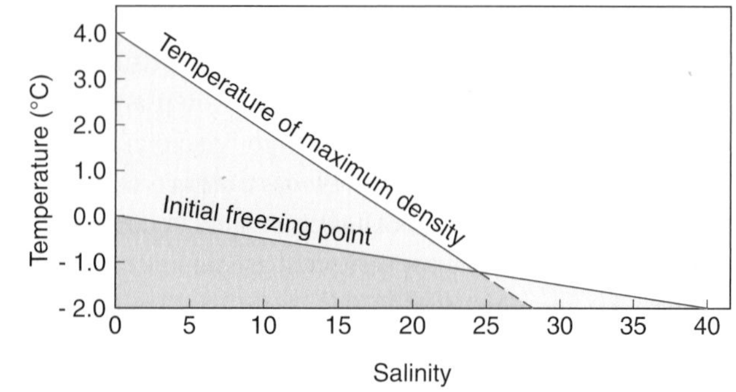

FIGURE 3.13
Water's temperature of maximum density and initial freezing point temperature decrease at different rates as the salinity increases. At a salinity of 24.7, the temperature of maximum density is the same as the freezing temperature (–1.33 °C).

bottom but descends to a level where the density of the surrounding water is the same.

Figure 3.14 shows the relative effects of temperature and salinity on seawater density. At a constant temperature of 15 °C, increasing the salinity from 33.7 to 34.9 raises the density from 1.025 g per cubic cm to about 1.026 g per cubic cm. Density changes an equivalent amount by cooling water with a constant salinity of 34.6 from 18 °C to 14 °C, a change of 4 Celsius degrees. Such variations in temperature are common at the ocean surface. As we will see in Chapter 6, differences in seawater density, caused by variations in temperature and salinity, are important controls of the vertical circulation of ocean water. (Note that ocean scientists usually write density in a shorthand notation. Because the density of seawater almost always starts with 1.0, they subtract 1 and then multiply by 1000. In

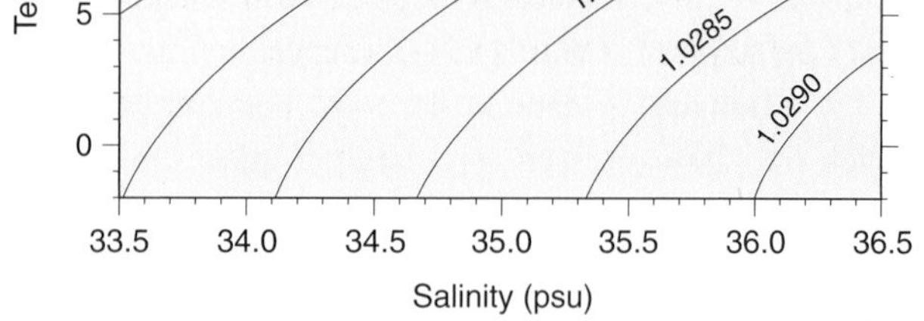

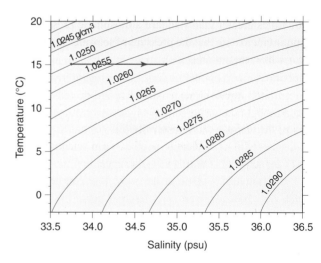

FIGURE 3.14
The density of seawater varies with temperature and salinity.

this way, a seawater density of 1.02178 g per cubic cm becomes 21.78 with no units.)

PRESSURE

Ocean water exerts **pressure** (i.e., force per unit area). It is useful to compare water pressure to air pressure. Atmospheric pressure can be thought of as the weight of a column of air acting over a unit area at the base of the column. By convention, *standard atmospheric pressure* is the average air pressure at sea level at 45 degrees latitude and an air temperature of 15 °C (59 °F). This pressure is equivalent to the weight exerted by a mass of 1.03 kg on a square cm (14.7 lbs per square in.) or 1013.25 millibars (mb). (A millibar is a unit of pressure where 1000 mb equals one bar.) Water is much denser than air so that a column of equivalent height produces much greater pressure.

A column of fresh water with a height of 10.33 m (33.9 ft) exerts a pressure at its base that approximately equals one standard atmosphere or 1.03 kg per square cm. When determining pressure, water can be approximated as an incompressible fluid; that is, water density does not vary significantly with increasing pressure. To a good first approximation, the pressure at any point in a water column is directly related to depth and the relationship is linear; that is, doubling the depth doubles the pressure. For example, at a depth of 103.3 m (339 ft), the water pressure is 10.3 kg per square cm, ten times the pressure at 10.33 m (33.9 ft) depth. The tremendous pressures encountered in deep ocean waters prevent direct exploration except by special submersible vehicles designed to withstand potentially crushing stresses. For example, the pressure at the deepest known place in the ocean, the Challenger Deep in the Mariana Trench in the Western Pacific where the ocean depth is about 11,000 m (36,100 ft), is more than 1000 times greater than standard atmospheric pressure. It is important to note that the pressure of seawater is slightly greater than that of fresh water due to the higher density of seawater.

Pressure has long been used as a measure of depth in the ocean because pressure is far easier to measure in situ than depth, and to a good approximation (error of less than 2%), pressure is equivalent to depth. Ocean scientists, like their counterparts in atmospheric science, commonly use the bar and its derivatives as a standard unit of pressure. The water pressure expressed in *decibars* (0.1 bar) is numerically equivalent to the water depth expressed in meters. The interchangeability of the two measures (i.e., water pressure in decibars and water depth in meters) greatly simplifies data analysis. For example, the water pressure at 100 m (330 ft) depth is about 100 decibars.

SEA ICE

Each winter, seawater freezes at high latitudes and even in some coastal areas of middle latitudes. When seawater is chilled below its initial freezing point, microscopic ice crystals form, later growing into hexagonal needles 1 to 2 cm (0.4 to 0.8 in.) long. At this stage, the sea surface takes on a dull appearance. As freezing continues, individual ice crystals freeze together and cover the surface like a blanket of wet snow. Eventually ice crystals begin to grow downward and form a thin, flexible, plastic-like ice layer, honeycombed with small cells that fill with seawater. Ice crystals themselves contain no salt, but the brines trapped in the small cells betweeen crystals are saltier than seawater. Typically 1 kg of newly formed sea ice consists of about 800 g of ice (salinity 0) and 200 g of seawater (salinity 35+). Thus, the average salinity of newly formed sea ice is about 7. As temperatures fall, more ice forms beneath the initial ice layer. The brines in the cells also partially freeze, making the remaining brines even saltier. If temperatures continue to fall, the salt in the brine eventually crystallizes as the last bit of water freezes.

The salt content of newly formed sea ice depends on temperature. At temperatures near freezing, sea ice forms slowly, which allows brines to flow out leaving little seawater in the cells. Such ice therefore contains little salt. At lower temperatures, ice forms more rapidly and traps seawater; this sea ice contains more salt but is less salty than the seawater from which it formed. Ice forms in the upper colder region of a brine cell while ice melts in the lower warmer region of the brine cell, causing the brine cell to migrate downward toward higher temperatures.

An ice layer up to 1 m (3 ft) or so thick can form in one winter; this is called *first-year ice*. First-year ice dominates the Southern Ocean around Antarctica. But in the Arctic pack ice of the central basin, sea ice melts little during summer and *multi-year ice* dominates. Figure 3.15 displays Arctic and Antartic ice cover for February 2011, as compared to median values. Over several seasons, the maximum thickness of multi-year ice is usually 2 to 3.5 m (about 7 to 11 ft). Winds, however, can pile up floes of ice, forming pressure ridges offshore and ice-shove ridges along the coast. Pressure ridges extend many meters above and below the ice pack and are hazardous to submarines navigating under the Arctic ice and impede icebreakers traveling through the ice pack. Many different terms are used to describe the various forms of sea ice. For more on sea ice terminology, see this chapter's third Essay. The impact of climate change on Arctic sea-ice cover is discussed in Chapter 12.

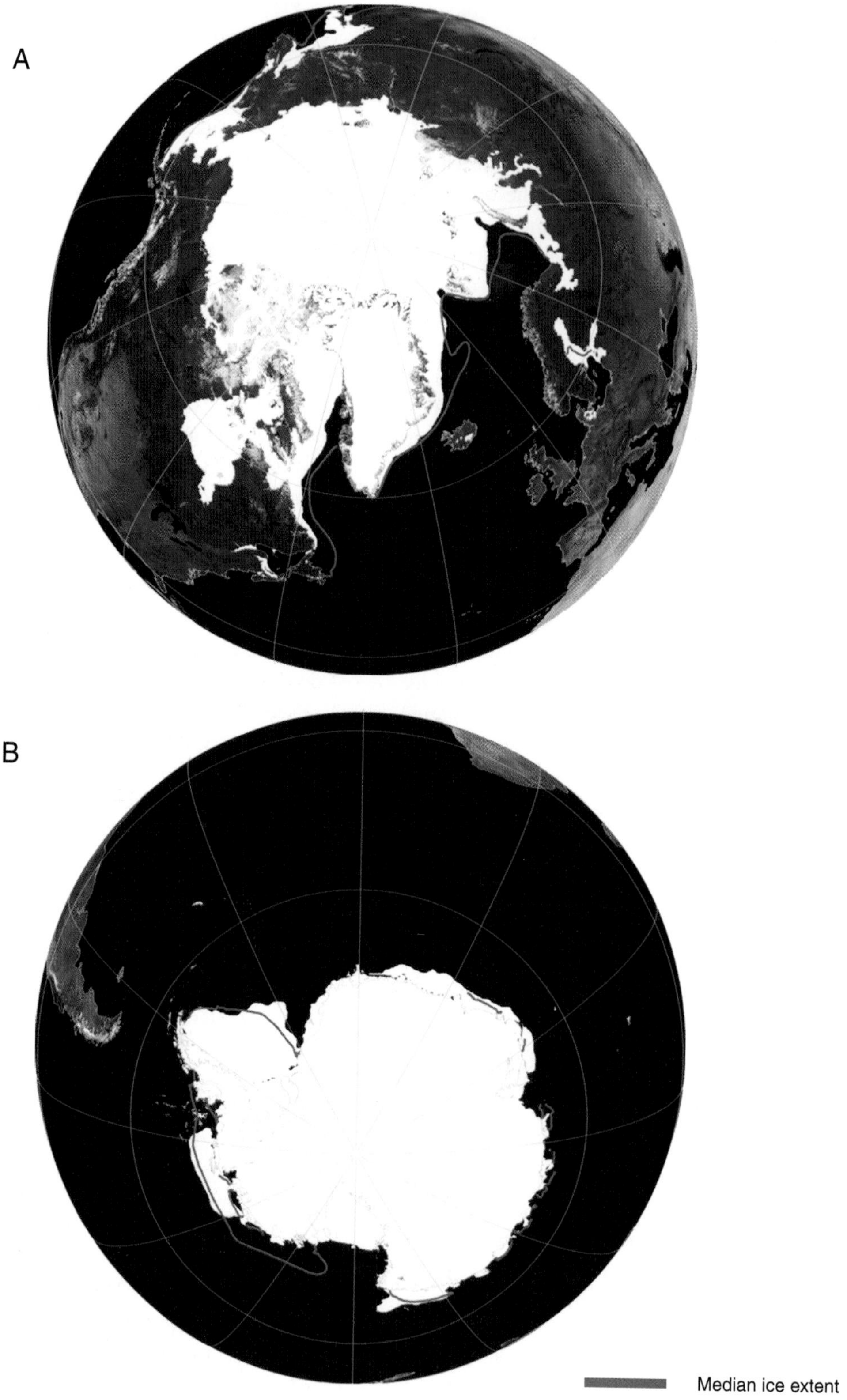

FIGURE 3.15
Sea ice extent in February 2011, based on satellite passive microwave data for (A) the Arctic where it was winter and (B) the Southern Ocean/Antarctica where it was summer. Total sea ice extent was 14.4 million square km in the Arctic and 2.5 million square km in the Antarctic. [Courtesy of National Snow and Ice Data Center, Boulder, CO.]

SOUND TRANSMISSION

Seawater is essentially transparent to sound just as the clear atmosphere is nearly transparent to sunlight. Whales use sound to communicate across ocean basins and to attract mates. Ocean scientists have developed techniques that take advantage of the sound propagating ability of seawater to determine ocean depth, locate underwater objects such as submarines and schools of fish, and to determine small changes in water temperature over great distances.

Sound propagates through some medium (e.g., air or water) as compression (push-pull) waves with the speed of propagation dependent on the properties of the medium. The speed of sound in ocean water (about 1500 m per sec or 5000 ft per sec) is more than four times the average speed of sound in air because water is less compressible than air. Knowing the speed of sound in seawater is the principle behind an *echo sounder* used by most seagoing vessels to determine the depth of water beneath the ship. This instrument sends a narrow beam of sound vertically to the seafloor where it is reflected back to the ship. The time interval between emission and return of the sound signal (the *echo*) is calibrated in terms of water depth. *Fish finders* are echo sounders that send and receive sound waves that are reflected by schools of fish.

SONAR (SOund NAvigation and Ranging) is similar to an echo sounder except that the operator of the instrument can alter the direction of the sound signal. Pulses of sound are sent out to locate targets such as submarines and the return echoes are displayed electronically on a monitor as distance or depth from the sounder. Complicating the use of SONAR for target location are changes in the speed of sound in water that can cause the sound wave to bend (refract). Sound waves bend toward regions where sound waves travel more slowly and away from regions where sound waves travel faster. The speed of sound in seawater is influenced by temperature, pressure, and salinity.

Variation of the speed of sound due to temperature/salinity differences in the ocean gives rise to the **SOFAR channel**, a zone centered at an ocean depth of about 1000 m (3300 ft) where the speed of sound is at a minimum value (Figure 3.16). (*SOFAR* is the acronym for *SOund Fixing And Ranging*). Sound waves are vertically trapped in the SOFAR channel and can travel thousands of kilometers horizontally with little loss of energy. Why does the speed of sound reach a minimum and produce the SOFAR channel? The speed of sound in seawater increases with increasing temperature, salinity, and depth (i.e., pressure). From the sea surface down to a depth of about 1000 m, temperature usually decreases and is the primary control of sound speed. Hence, sound speed decreases with depth to 1000 m. At greater depths, the water temperature tends to be uniformly low and variations in sound speed depend chiefly on pressure change with depth. With increasing depth and pressure, sound speed increases. As shown in Figure 3.16, sound waves are refracted upward (from

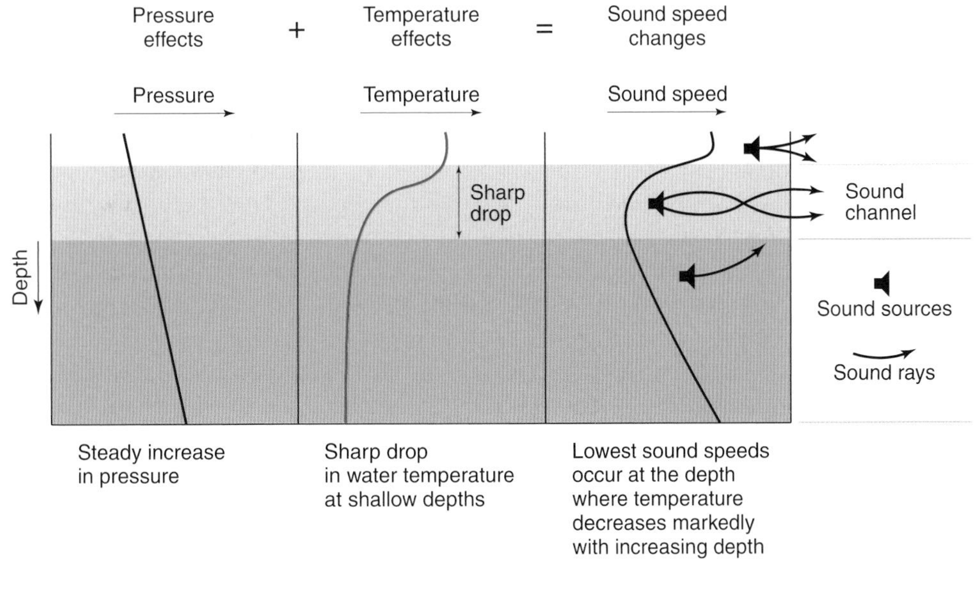

FIGURE 3.16
Refraction of sound waves as they travel through ocean waters gives rise to the SOFAR channel, a zone centered at a depth of about 1000 m (3300 ft) where the speed of sound is at a minimum value.

below the SOFAR) or downward (from above the SOFAR) toward the region of minimum sound velocity within the SOFAR channel.

During the 1990s, with increased interest in global climate change, scientists at the Massachusetts Institute of Technology and Scripps Institution of Oceanography developed and tested a method of monitoring temperature changes within the SOFAR channel. The approach, known as **acoustic thermometry**, is based on the dependency of the speed of sound on ocean water temperature and can measure very small temperature changes—to a few thousandths of a degree—over great distances in the ocean. An increase in sound speed between sound transmitters and receivers over a period of time would indicate warming of the ocean. Indeed, such warming has been observed in the Arctic basin using acoustic thermometry.

Conclusions

Water has unusual physical and chemical properties resulting from the water molecule's unique polar structure that gives rise to hydrogen bonding. Water is a powerful solvent, dissolving both salts and gases, and is an effective heat-storage and transporting medium. Consequently, the ocean plays a major role in Earth's weather and climate. Temperature, salinity, and pressure affect most physical properties of seawater such as density and sound velocity. The density of seawater, in turn, influences ocean currents. In the next chapter, we continue our examination of ocean properties by focusing on the types and sources of marine sediments.

Basic Understandings

- The unusual physical and chemical properties of water (H_2O) arise from the unique structure of the water molecule, consisting of two hydrogen (H) atoms bonded to an oxygen (O) atom. The bonding angle of 105° allows a slight positive charge to reside on each of the hydrogen atoms and a slight negative charge on the oxygen atom. This makes the water molecule a polar molecule. The positively charged (hydrogen) poles of a water molecule attract the negatively charged (oxygen) poles of other water molecules; this attractive force is known as hydrogen bonding and is the reason for the unique properties of water.
- Water is one of the few substances occurring naturally in all three phases within the usual range of temperature and pressure at or near Earth's surface. Ice has a three-dimensional internal framework consisting of a hexagonal ring of molecules that vibrate about fixed locations. Liquid water consists of scattered un-bonded molecules and transient clusters of hydrogen bonded molecules. When liquid water changes to vapor, all hydrogen bonds are broken.
- Temperature is directly proportional to the average kinetic energy of the atoms or molecules composing a substance. Heat is the name given to the energy transferred between objects at different temperatures and is always directed from the warmer to the colder object.
- When water changes phase, latent heat is either absorbed from the environment to break hydrogen bonds (melting, evaporation, sublimation) or released to the environment when hydrogen bonds form (freezing, condensation, deposition).
- The thermal response of a unit mass of a substance to a given input of heat energy, known as the specific heat of the substance, differs for various materials and is highest for water. Water's high specific heat has consequences for climate. Compared to the surface of an adjacent landmass, the surface of a body of water does not warm as much during the day (or in summer) and does not cool as much at night (or in winter). Places immediately downwind of the ocean have a maritime climate in which the average temperature contrast between summer and winter is less than it is at the same latitude but well inland in a continental climate.
- Water is an excellent solvent, readily dissolving solids, liquids, and gases. The polar nature of the water molecule favors the solution of both ionic and nonionic substances. Seawater is a solution of nearly constant salt composition with a concentration of about 3.5% in the open ocean.
- Salinity is based on measurements of the electrical conductivity of seawater and is determined as the ratio of the conductivity of the seawater sample to the conductivity of standard seawater. As a ratio, salinity has no units but numerically is essentially equivalent to grams of salt per kilogram of seawater (parts per thousand).
- Salts dissolved in seawater are derived from weathering and erosion of rock and sediment on

land, chemical reactions between cold seawater and newly formed oceanic crust, and volcanic eruptions. Rivers and streams deliver the products of terrestrial weathering and erosion to the ocean; this is the principal source of salt in the ocean.

- Gases are exchanged between the atmosphere and ocean at the ocean surface. If seawater is not saturated with a specific gas, the gas is transferred from air to ocean. If seawater is supersaturated with the gas, the transfer is in the opposite direction. When water is saturated with a gas, the rate at which the gas dissolves in water equals the rate at which the gas escapes to the atmosphere.
- The saturation concentration of a gas in water depends primarily on temperature in freshwater bodies such as lakes and rivers and a combination of temperature and salinity in seawater. In general, cold water can hold more dissolved gases than warm water, and fresh water can dissolve more gas than seawater. Also, the solubility of a gas in water depends on the gas species so that, for example, oxygen and carbon dioxide are more soluble than nitrogen.
- Below the ocean-atmosphere interface, biochemical processes control the proportions of certain dissolved gases. Within the photic zone, the sunlit upper portion of the ocean where photosynthesis takes place, dissolved oxygen is enhanced relative to carbon dioxide. Beneath the photic zone, decomposer activity and cellular respiration are the most important biochemical processes. For the ocean as a whole, CO_2 is the most abundant dissolved gas, accounting for 83% of the total.
- Normally, seawater is slightly alkaline. This is very important for marine organisms whose shells and skeletons are composed of calcium carbonate ($CaCO_3$).
- The density of fresh water is maximum at a temperature of 4 °C (39.2 °F) and freshwater ice is less dense than fresh water. This explains why ice floats and why lakes freeze from the top down. At constant temperature, seawater density increases with increasing salinity so that less dense fresh water floats on more dense seawater.
- As salinity increases, the initial freezing temperature and temperature of maximum density decrease linearly. At a salinity of 24.7 or greater, cooling seawater becomes more dense down to its initial freezing temperature. Once sea ice forms, it floats because its density is lower than that of the seawater.
- Temperature and salinity are the principal controls of seawater density. Lower temperature and increasing salinity cause density to increase. Changes in salinity are important near shorelines (due to freshwater runoff), in polar areas (due to cold water and sea ice formation and melting), and in regions of high precipitation or evaporation.
- Water pressure is the force exerted by a column of water per unit area at the base of the column. Water is only slightly compressible so that water pressure increases linearly with depth. Water pressure expressed in units of decibars is numerically equivalent to depth expressed in units of meters.
- Sea ice forms in polar regions and at some middle latitude coastal areas. As seawater freezes, salts are excluded from the sea ice and the resulting brines increase the salinity of underlying waters. First-year ice dominates the Southern Ocean around Antarctica whereas multi-year ice dominates in the Arctic Ocean.
- Ocean scientists have developed techniques that take advantage of the sound propagating ability of seawater to determine ocean depth, locate underwater objects such as submarines and schools of fish, and to determine small changes in average water temperature.
- Variations in the speed-of-sound due to temperature, salinity, and pressure differences in the ocean waters give rise to the SOFAR channel, a zone centered at an ocean depth of about 1000 m (3300 ft) where the speed of sound drops to a minimum value. Sound waves are vertically trapped in the SOFAR channel and can travel thousands of kilometers horizontally with little loss of energy.

Enduring Ideas

- Hydrogen bonding, a force of attraction between water molecules, arises from the polar nature of the H_2O molecule. It is responsible for water's unusual thermal and other physical and chemical properties.
- The relatively high specific heat of water helps to explain the great thermal inertia of the ocean and the contrast between maritime and continental climates.
- The salinity of surface ocean waters varies with precipitation, evaporation, and runoff from rivers.
- Photosynthesis, cellular respiration, and temperature affect the amount of gas that dissolves in surface ocean water. For example, more CO_2 can dissolve in colder water.
- At constant temperature, the density of seawater increases with increasing salinity. Both the temperature of maximum density and the initial freezing point of seawater decrease linearly with increasing salinity.

Review

1. Describe how hydrogen bonds form between neighboring water molecules. During which phase change of water are all hydrogen bonds broken?
2. Distinguish between temperature and heat.
3. Identify and describe three phase changes of water during which latent heat is released to the environment.
4. Define specific heat. When the same quantity of heat is added per unit mass, does a substance having a lower specific heat warm more or less than a substance having a higher specific heat?
5. Distinguish between a maritime climate and a continental climate in terms of annual temperature range and name some U.S. or Canadian cities having these climates.
6. Compare the salinity of surface ocean waters in regions where the precipitation is low and evaporation rates are high to the global annual average sea-surface salinity.
7. For the ocean as a whole, what is the most abundant gas dissolved in seawater? What is the role of this gas in photosynthesis and cellular respiration?
8. What is the pH of pure water? Is the average pH of ocean water slightly acidic or alkaline?
9. At constant temperature, how does the density of seawater change with increasing salinity? With constant salinity, how does the density of seawater change with falling temperature?
10. What is the significance of the SOFAR channel?

Critical Thinking

1. If ice were always denser than liquid water, speculate on some of the consequences.
2. Describe the seasonal changes in temperature of the ocean surface first in subtropical and then in polar latitudes. Explain the differences.
3. Describe how changes in the phase of water bring about a transfer of heat energy within the Earth system.
4. The ocean is a major player in Earth's climate system. Explain why.
5. River water ultimately empties into the ocean and yet the average salinity of river water is much less than the average salinity of seawater. Explain the difference.
6. How do biological processes influence the type and amount of gases dissolved in seawater?
7. What is the significance of a seawater salinity of 24.7?
8. What factors govern surface water salinity in the open ocean?
9. Explain how surface water salinity can act as a rain gauge over large areas of the ocean.
10. Identify two conservative properties of seawater. Explain why these are considered conservative.

ESSAY: Desalination

The Red Sea port of Jeddah, Saudi Arabia, is a growing city of more than 3.4 million people (Figure 1). Rapid population growth is taking place in spite of an arid climate in which the average annual rainfall is only about 6.1 cm (2.4 in.) and no significant freshwater sources exist nearby. There are no rivers or lakes and very little groundwater. For their freshwater supply, residents of Jeddah depend mostly on saline water piped in from the Red Sea and desalinated. Jeddah's five desalination plants produce more than 380 million liters (100 million gal) of fresh water per day.

FIGURE 1
Map of Saudi Arabia showing location of Jeddah.

Desalination is a process whereby dissolved solids (principally salts) are removed from saline water to make the water potable, that is, suitable for domestic and agricultural uses. According to the World Health Organization (WHO), water is considered potable when it contains less than 500 milligrams per liter (mg/L) of dissolved solids. (Most people report a disagreeable taste when dissolved salts approach 1000 mg/L and an insipid taste when the salt content is too low.) On average, seawater contains about 35,000 mg/L of dissolved solids—almost all salts. Most desalination worldwide utilizes one of two basic technologies: distillation or reverse osmosis. Multistage flash distillation and reverse osmosis account for about 88% of worldwide desalination capacity.

With *distillation*, water is purified through phase changes (usually from liquid to vapor and back to liquid). The simplest (and least expensive) distillation device is a plastic transparent dome placed over a reservoir of seawater. Solar radiation penetrates the dome, is absorbed (converted to heat), and is the source of latent heat for evaporating the water. Water vapor then condenses as freshwater drops on the underside surface of the dome and the drops drip into a collection trough. This so-called *solar still* works best in sunny tropical climates where solar radiation is intense year-round.

For much more rapid distillation of large quantities of salty water, desalination utilizes *multistage flash distillation (MFD)*. With MFD, the intake water is superheated, that is, under pressure the water temperature rises above its normal boiling point but the water remains liquid. The water then pours into another vessel where the pressure is somewhat lower and the water flashes into steam. Water vapor rises and condenses into collection devices. This same process is repeated many times as the remaining intake water flows into vessels at progressively lower pressure. The output is fresh water and concentrated brine. In most cases, MFD facilities utilize the waste heat from nearby fossil-fuel electric power plants as the energy source for desalination.

With *reverse osmosis*, pressure applied to saltwater forces water molecules through a thin semi-permeable membrane. Up to 97% of the substances dissolved in seawater including salts, other dissolved solids, and nonvolatile organics cannot pass through the membrane and are left behind. Pretreatment of the intake water with substances such as chlorine, hydrogen peroxide, or sulfuric acid may be necessary to prevent biological fouling or calcium carbonate scaling of the membrane surface (Figure 2).

Reverse osmosis is used at desalination facilities on the Outer Banks of North Carolina where fresh water is in limited supply. At the Kill Devil Hills facility, operating since 1989, brackish (somewhat saline) water is pumped from 14 wells at a rate of up to 1900 liters (500 gal) per well per minute and then filtered and pretreated with sulfuric acid. Water is pumped under pressure to three reverse osmosis units, each of which produces about 3.8 million liters (1 million gal) of water per day. Chlorine, fluoride, and a corrosion inhibitor are then added to the desalinated water. More recently, the new reverse osmosis desalination facility operated by Florida's Tampa Bay Water came up to full production supplying 95 million liters (25 million gal) per day, about 10% of the Tampa Bay region's drinking water supply.

One of the drawbacks of desalination is the discharge of concentrated brine into the ocean. For example, about 25% of the intake water at the Kill Devil Hills facility becomes a highly saline waste stream that empties into the Atlantic Ocean. The salinity of this wastewater stream typically is about twice that of ordinary seawater. Conservationists are concerned that this dense brine will sink to the ocean bottom and adversely impact marine ecosystems. Also, some marine organisms may be killed by the seawater intake system. Desalinated water used for irrigation may require pre-treatment because of elevated levels of boron (B) that is toxic to many crop species and deficiencies in nutrients required for plant growth (e.g., calcium, magnesium). A major limitation to desalination of large quantities of water is the relatively high cost of energy-intensive technologies (distillation more so than reverse osmosis). For example, California's desalinated water costs more than twice that of water from watershed transfers or groundwater pumping. But the economics is becoming more favorable with the development of more efficient desalination technologies and the rising cost of fresh water from conventional sources.

Desalination (via both distillation and reverse osmosis) is most common in the Middle East where energy is abundant and relatively inexpensive. In that arid region of the world where freshwater resources are in short supply, 1700 desalination plants convert a total of 20.9 billion liters (5.5 billion gal) of seawater to fresh water each day—estimated by Global Water Intelligence to be about 72% of the world's capacity. The world's largest reverse osmosis desalination facility, located in Ashkelon, Israel, on the southern Mediterranean coast, has been operating since December 2005. The 100 million cubic m per year production is used for both human consumption and irrigation.

With increasing demand for potable water by the rapidly growing human population in the coastal zone (especially where groundwater supplies are insufficient) plus the pressing need for more irrigation water, desalination of seawater is becoming more popular. (In some coastal areas, freshwater supply is the limiting factor for community growth and 69% of the world freshwater supply is used for irrigation.) From 1994 to 2004, the worldwide desalination capacity more than doubled, from 17.3 to 35.6 million cubic m per day. As of 2007, desalination of seawater supplies approximately 1% of the world's drinking water.

FIGURE 2
Sandia National Laboratories scientist Susan Altman conducts research to minimize biofouling on reverse osmosis membranes used for desalination in her laboratory. Sandia scientists investigate optimal ways for treating brackish water for human consumption. Their findings can be applied to facilities across the United States. [Photo by Randy Montoya]

ESSAY: Ocean Acidification

One of today's most pressing issues is the consequences for global climate of a continuing rise in atmospheric carbon dioxide (CO_2) and other heat-trapping greenhouse gases (Chapters 5 and 12). Another impact of rising levels of CO_2 is *ocean acidification*, a change in the chemistry of ocean waters with potentially significant and long-lasting implications for marine organisms and ecosystems (Figure 1). The present rate of change in ocean chemistry appears to be unprecedented in geologic time, which makes predicting the impacts on marine life even more challenging.

Atmospheric CO_2 concentration has been rising, at least, since the beginning of the Industrial Revolution. Fossil fuel combustion accounts for roughly 75% of the increase in atmospheric CO_2 while deforestation (and other land clearing) is likely responsible for the balance. At present, the atmospheric CO_2 concentration is increasing at a rate that, if sustained, could result in at least a doubling of the concentration by the close of this century. An estimated one-third of all CO_2 released to the atmosphere by human activities during the past 200 years has dissolved in ocean waters. Today, about 40% of CO_2 from fossil fuel combustion stays in the atmosphere, 30% is taken up by terrestrial vegetation via photosynthesis, and 30% is absorbed by the ocean. Ocean scientists are concerned about potential adverse effects of ocean acidification on the viability of marine organisms and marine food chains.

Ocean acidification occurs when increasing amounts of CO_2 dissolve in surface seawater. Carbon dioxide affects the pH of seawater by combining with water (H_2O) to produce carbonic acid (H_2CO_3). That is,

$$\mathbf{CO_2 + H_2O \leftrightarrow H_2CO_3}$$

Most of this weak acid instantly dissociates into hydrogen ions (H^{+1}) and bicarbonate ions (HCO_3^{-1}). Some bicarbonate ions, in turn, dissociate into carbonate ions (CO_3^{-2}) and hydrogen ions. (Carbonate ions also combine with hydrogen ions to form bicarbonate ions.)

$$\mathbf{CO_2 + H_2O \leftrightarrow H_2CO_3 \leftrightarrow H^{+1} + HCO_3^{-1} \leftrightarrow 2H^{+1} + CO_3^{-2}}$$

Increased CO_2 means more bicarbonate and more free hydrogen ions, which then bond with carbonate ions, also increasing bicarbonate, and thus the effect of these chemical reactions is declining pH (*ocean acidification*).

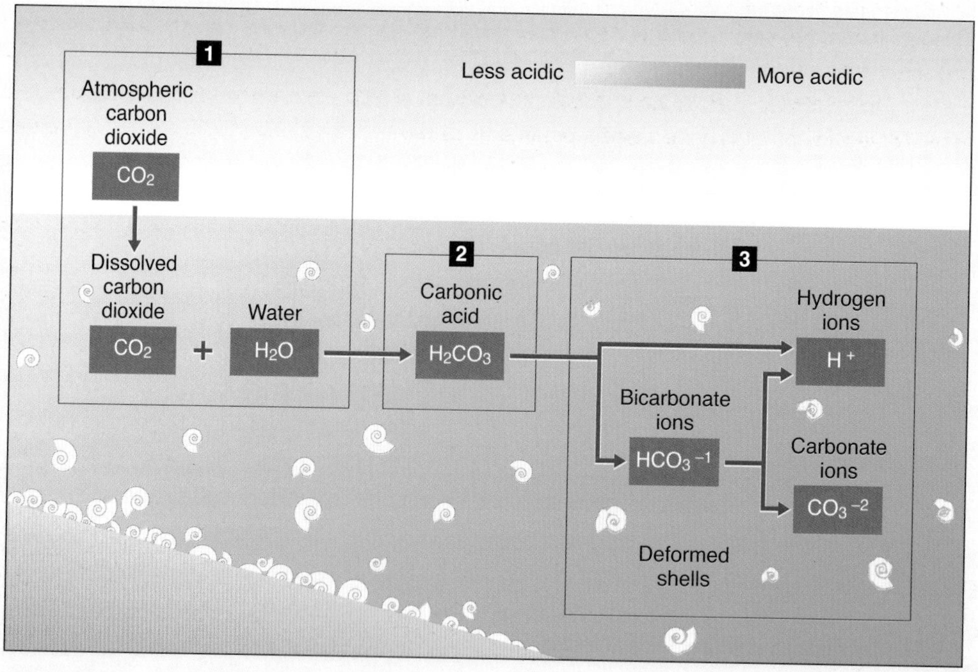

FIGURE 1
Visualization of the process of ocean acidification.

This directly affects *calcifying organisms*, those animals with a calcium shell such as mollusks (clams, oysters, snails and muscles) and some tubeworms. In more acidic seawater, calcium carbonate ($CaCO_3$) shells begin to dissolve. At the same time, the carbonate ions combine with hydrogen ions to form more bicarbonate ions, and so decrease the availability of carbonate ions for building the calcium carbonate ($CaCO_3$) component necessary for shells and skeletons. Also vulnerable to ocean acidification are corals that filter plankton from ocean water and secrete calcium carbonate. The reduction in carbonate ion concentration impairs the ability of reef-building coral to generate their calcium carbonate skeletons. Furthermore, ocean acidification may reduce the tolerance of some species to heat stress, to which polar ecosystems are particularly susceptible.

In 2010, Dan Bernie, of the Met Office Hadley Centre in the U.K., and colleagues used coupled climate, ocean, and terrestrial carbon models to predict the effect of more than 100 different CO_2 emissions scenarios on ocean pH. If no effort were made to stabilize CO_2 emissions, model simulations predict that ocean pH would decline from their current range (8.0- 8.3) to 7.67-7.81 by 2100. Models also predict that if CO_2 emissions were to peak in 2016 and then decline 5% each year, average ocean pH would stabilize at about 8.0.

A marine geochemist and Senior Scientist at Woods Hole (MA) Oceanographic Institution, Scott C. Doney has found that already "the acidity of the ocean has increased by 26% from pre-industrial times to today." At present, the average pH of surface ocean waters is estimated at 8.1, down from about 8.2 in 1800. A 0.1 unit decline in pH corresponds to a 30% increase in surface ocean acidity. If the average pH of ocean water drops an additional 0.3 to 0.4 units by 2100, as Bernie predicts will happen if there was no effort to stabilize CO_2, surface ocean acidity will increase 150% on average, a level not experienced in the ocean for many millions of years.

In fact, the global ocean pH is changing at an "unprecedented" rate, as Doney reports in the 18 June 2010 issue of *Science*, some 30 to 100 times faster than any such change in the recent geologic past "and the perturbations will last many centuries to millennia." He predicts that ocean acidification is likely to reduce the shell and skeleton growth of many marine calcifying species.

These organisms create calcium carbonate shells in three mineral forms, which differ in solubility. Aragonite (a form of calcium carbonate), which corals and pteropods (Figure 2) use for their shells, and magnesium calcite are more soluble than calcite (a stable form of calcium carbonate). This means they more readily dissolve in acidic waters. Also because CO_2 dissolves more quickly in cold water, cold-water calcifying organisms are the first to be affected by acidifying seawater. According to marine chemist Richard A. Feely of NOAA's Marine Environmental Laboratory in Seattle, WA, by the middle of this century, Arctic waters will be corroding the aragonite organisms and, by the end of the century, waters of the Southern Ocean and parts of the North Pacific Ocean will be corrosive to pteropods and other aragonitic organisms. On the other hand, shells of high latitude calcareous phytoplankton and zooplankton are composed of calcite, less soluble and less vulnerable to corrosion.

In related research, geochemist Justin Ries of the University of North Carolina, Chapel Hill, and colleagues grew 18 species of calcifying marine organisms in the laboratory for 6 months under various CO_2 levels that have been

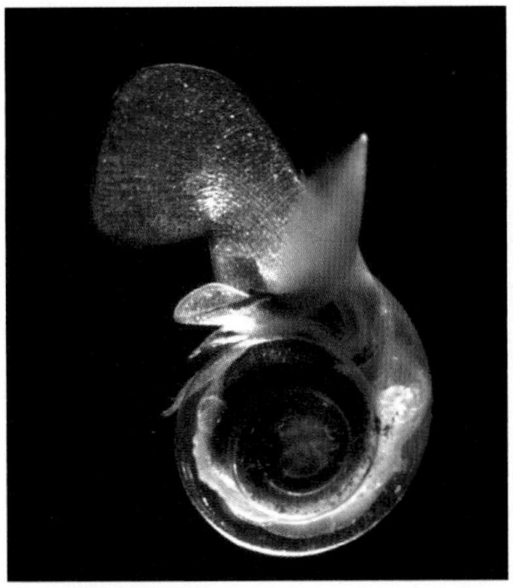

FIGURE 2
Ocean acidification affects organisms with aragonite calcium carbonate shells, such as the pteropod *Limacina Helicina.* [Photo courtesy of Russ Hopcroft, University of Alaska, Fairbanks/NOAA]

predicted for the next decade, the close of the century, and beyond. The researchers concluded that most species (i.e., periwinkles, oysters, sea urchins, and calcareous green algae) eventually produced less calcium carbonate with declining pH. Additionally, the greater hydrogen ion (H^{+1}) concentration of ocean acidification conveys more of those ions into the bodies of marine organisms, making their tissues more acidic. Calcifying marine organisms must use energy to restore and maintain their internal pH balance, taking away from their ability to grow and reproduce. Because of the rapid pace of ocean acidification, with a significant decline in pH expected to take place in 100 years or less, most species (especially those with slow reproductive rates) cannot adapt to the changing marine environment. *Adaptation* is an inheritable characteristic or trait that enhances an organism's chances for survival and reproduction in its habitat.

In the August 2010 issue of *Scientific American*, Marah J. Hardt, a research scientist and founder of Oceanlink, and Carl Safina, founding president of the Blue Ocean Institute, noted that the internal impacts of ocean acidification differ across the development stages of marine species. Laboratory studies reveal that with acidification, fertilization success may decline, early larval stages are often stressed, and adults have slower growth rates making them more susceptible to predators before they can reproduce. With reductions in strength, growth rate, immune function and/or reproduction rate, long-term population decline may be inevitable. We have more to say about the implications of ocean acidification in Chapter 4.

ESSAY: Sea Ice Terminology

Over the years, a complex terminology has evolved concerning sea ice, its development and various forms. The World Meteorological Organization (WMO) standardized definitions for many of these terms more than three decades ago. The purpose of this Essay is to define some terms that you may encounter while exploring sea ice.

Sea ice refers to any form of ice floating in ocean waters that originated from the freezing of seawater. As noted elsewhere in this chapter, the salt dissolved in seawater lowers slightly the freezing point of water. For example, seawater having a salinity of 34 (about 3.4% salt) begins to freeze when its temperature drops to about −1.9 °C (28.6 °F) whereas the freezing point of fresh water is 0 °C (32 °F). Note that icebergs are not sea ice; they are massive chunks of ice that break away from the snout of a glacier advancing from land to sea (or a large lake) or calve off the edge of an ice shelf. Icebergs consist of frozen fresh water and may be floating or grounded.

Sea ice is broadly classified based on age (and thickness) as *new ice* (less than 10 cm thick), *young ice* (10 to 30 cm thick), *first-year ice* (30 cm to as much as 120 cm thick), or *old ice* (second-year or multi-year ice). First-year ice is the product of no more than one winter's growth of sea ice. *Second-year ice* has survived one melt season (summer) whereas *multi-year ice* has persisted through at least two summers' melt. The latter is typically around 3 m (10 ft) thick and is more common in the Arctic Ocean than the Southern Ocean. Sea ice near Antarctica varies seasonally with summer ice cover usually ranging up to 150 km (93 mi) from shore whereas winter ice can extend from 450 km (280 mi) to more than 1700 km (1050 mi) from shore. Maximum sea-ice coverage occurs in the Weddell Sea and accounts for about 80% of the multi-year ice in the Antarctic.

During the initial stage of sea ice formation, frazil ice develops in the upper few centimeters of seawater. *Frazil ice* consists of millimeter-size platelets or discs that give the sea surface an oily appearance (Figure 1). Wave action can stir frazil ice to a depth of several meters. With continued freezing, frazil ice crystals coagulate to form a soupy mix that reflects little light and gives the surface a matte appearance, sometimes referred to as *grease ice*. Frazil ice and grease ice are forms of new ice but do not form distinctive ice floes. An *ice floe* is any contiguous piece of ice that can vary greatly in size from a few meters to more than 10 km across (Figure 1). Other types of new ice include frozen slush (an accumulation of water-saturated snow on an ice surface or floating in water) and *nilas* (a thin crust of elastic gray-colored ice on a calm sea that readily bends into interlocking fingers).

FIGURE 1
Frazil ice blowing into bands on surface next to ice floe. [Collection of Dr. Pablo Clemente-Colon, Chief Scientist, National Ice Center]

Grease ice aggregates into small chunks of ice, which then become pancake-shaped ice floes. *Pancake ice* consists of nearly circular pieces of ice from 30 cm to 3.0 m (1 to 10 ft) in diameter and having a thickness of up to 10 cm (4 in.). Collisions with other ice floes are responsible for the upturned rim that is characteristic of pancake ice (Figure 2). With falling temperature, the ice cover thickens. Winds, waves, and currents cause ice floes to override one another (known as *rafting*) further thickening the ice cover generally to 40-60 cm (15-24 in.). However, in some fierce winter storms in the Bering Sea, scientists have reported 3-m (10-ft) thick ice floes rafting onto other ice floes and building the ice to a thickness of 6 m (20 ft), of which about 90% is below sea level. Rafting also produces a rough surface on the ice floe.

Fast ice forms along the coast and is attached to the shore or shallow sea bottoms so that it cannot move laterally. This happens, for example, off the coast of northern Alaska. It may extend from a few meters to several hundred kilometers offshore and may be more than one year old. *Pack ice* is any area of sea ice that is not anchored to land and moves with the wind and ocean currents. Pack ice is described as very open (1/10 to 3/10 ice cover), open (4/10 to 6/10 ice cover), close (7/10 to 8/10 ice cover), very close (9/10 to less than 10/10), and compact (10/10 ice cover with no water visible). If ice floes are frozen together, the compact pack ice is described as consolidated. The *extent of sea ice* is defined as the area in which ice covers at least 15% of the ocean surface.

A significant portion of Antarctic sea ice thickens through a process whereby seawater floods the snow that has accumulated on top of the ice turning it to slush; at low temperatures the slush then freezes into snow ice. If the snow cover is sufficiently massive, the ice/snow interface is suppressed below sea level allowing seawater to intrude and turn the snow to slush. Alternately, water may migrate upward through brine channels within the sea ice and enter the snow.

Stresses produce fractures in sea ice that vary from a few meters to many kilometers in length. Where fast ice is attached to the shore, tide-induced vertical motions of the ice may produce *tide cracks*. These fractures are important for some forms of marine life; for example, they allow penguins and seals access to the ocean. A fracture that is wide enough to be navigable by surface vessels is known as a *lead* and may occur between the shore and the pack ice or between fast ice and pack ice or simply between large ice floes or in the ice pack at sea. At sea, leads provide breathing holes for whales. Leads also permit exchange of heat and moisture between the ocean and atmosphere. In addition, converging ice floes (perhaps wind- or current-driven) may be forced upward into a wall of fractured ice, known as a *pressure ridge*. Pressure ridges can extend several meters above the surrounding sea ice and ocean surface. The fractured ice that is forced downward meters below the ridge is known as an *ice keel*.

A non-linear shaped kilometer-scale opening in the sea ice cover that persists or reoccurs regularly (often annually), is called a *polynya*. These holes in the ice would not normally exist because of sub-freezing temperatures were it not for some mechanism that keeps them open. In some cases, persistent gravity-driven cold winds (*katabatic winds*) blow ice away from land (e.g., Antarctica), islands, and grounded ice leaving surface waters ice-free. In a second mechanism, warm water welling up from below melts the ice cover. Also, polynyas are common at the mouths of large rivers. Major polynyas in the Northern Hemisphere include the St. Lawrence polynya in the Bering Sea and polynyas off the northeast and northwest coasts of Greenland. These are called North East Water and North Water respectively.

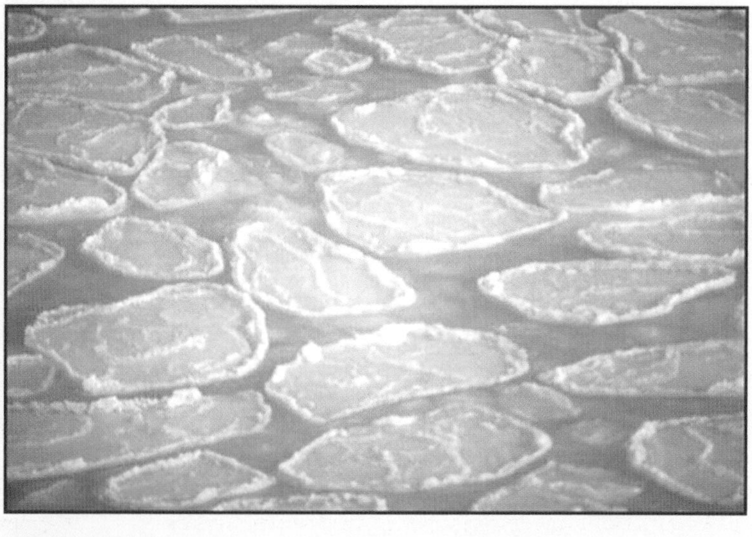

FIGURE 2
Pancake ice consists of pieces of ice 0.3 to 3.0 m across, up to 10 cm thick, with upturned rims caused by collisions with other pieces of ice. [Courtesy of Environment Canada]

CHAPTER 4

MARINE SEDIMENTS

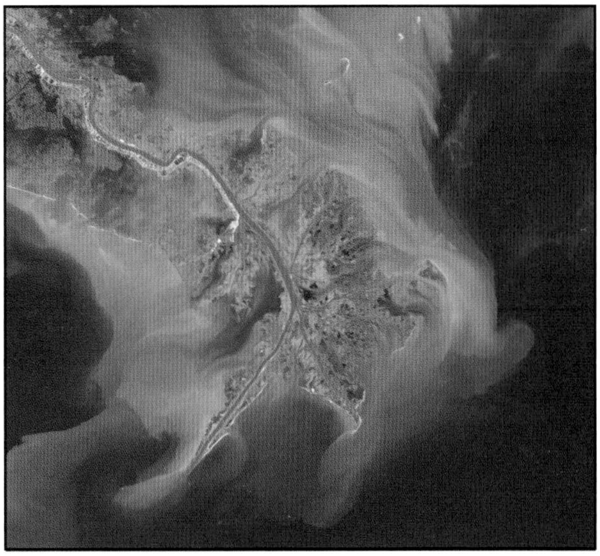

The mouth of the Mississippi River in Louisiana. The Mississippi River carries about 500 million tons of sediment into the Gulf of Mexico each year. [NASA Image by Robert Simmon, based on Landsat data provided by the University of Maryland Global Land Cover Facility]

Case in Point

Wastewater (i.e., water of unacceptable quality) enters rivers and streams via overland flow (e.g., agricultural runoff), groundwater, or direct input from drainage pipes (e.g., storm sewer systems). The net flow of water from land to sea in the global water cycle implies that all water-borne wastes eventually enter the ocean, either dissolved or suspended in water. Also as part of the global water cycle, rain and snow wash suspended particles from the atmosphere into the ocean. Furthermore, in some cases, industries and cities located along the coast discharge wastes directly into the ocean or barge them out to sea for dumping.

Some wastes discharged to the sea rapidly decompose physically, chemically, or biologically; others resist decomposition. These persistent chemicals enter food chains and move from one trophic level to the next higher trophic level increasing in concentration along the way. Problems caused by persistent chemicals are especially serious when they are toxic or hazardous to living organisms. In such cases, remediation requires either removal of the contaminants by dredging or capping them with layers of uncontaminated sediments to prevent their reentry into the ecosystem.

Perhaps the most infamous and now classic example of a toxic discharge to the sea that had serious consequences for human health took place in the small coastal village of Minamata, Japan. For many years, the Chisso Chemical Plant discharged industrial waste containing mercury into Minamata Bay. Because elemental mercury is insoluble in water, it was expected to sink to the bottom sediments and remain inert, causing no harm to the ecosystem or humans. Such was not the case.

Fish and shellfish taken from local waters were part of the staple diet of Minamata residents. But, in the early 1950s, they began noticing bizarre behavior in their cats, including twitching and stumbling, which we now recognize as signs of brain damage caused by mercury poisoning. In 1956, a five-year-old girl lapsed into a convulsive delirium and was diagnosed with neurological damage. A few weeks later, numerous people reported a variety of debilitating symptoms including numbness, headaches, loss of muscle control, and slurred speech. In some individuals, symptoms worsened and they developed violent trembling and paralysis. Children were born with physical deformities as well as severe mental retardation. Some victims died.

Upon investigation, fish and shellfish taken from Minamata Bay were found to contain high levels of methyl mercury in their tissues, a soluble and highly toxic form of mercury. Methyl mercury that enters the body attacks the central nervous system. In 1959, scientists demonstrated that bacteria in bottom sediments taken from Minamata Bay converted mercury to methyl mercury. Methyl mercury was subsequently taken up by aquatic organisms, readily moved up the food chain, and was consumed by people. In eating fish and shellfish from Minamata Bay, people were exposed to dangerous levels of a highly toxic material. More than 3500 people were seriously affected and about 50 died from methyl mercury poisoning—now referred to as Minamata disease. After many years of litigation, the Chisso Company finally accepted some responsibility for the Minamata tragedy and was required to make reparations to victims.

Discharge of mercury into Minamata Bay ceased in the late 1960s and the cleanup began. This involved dredging and removal of mercury-contaminated sediment deposits. These sediments were placed in reclamation areas where they were surrounded by dikes with impervious fabric liners and covered by clean sands to prevent run-off and further contamination of bay waters. By July 1997, the waters of Minamata Bay and its marine organisms were found to meet Japan's environmental standards. Minamata Bay was declared safe and reopened for human use including fishing and recreational activities but monitoring of mercury levels in the water, sediments, and marine life continues.

Driving Question:
What are the types and sources of sediment that enters the ocean?

Particles (sediments) blanket much of the ocean floor. Most of these particles are generated at the interfaces among Earth's various subsystems and are transported by rivers, wind, ice, and gravity to the ocean. Some particles originate in the sea (e.g., the excretions and secretions of marine organisms). Marine sediment deposits consist of rock fragments, soil particles, and the shells, bones and teeth of marine organisms, and even some material from outer space. Locked in the layers of ocean bottom sediments and marine sedimentary rock is a chronology of the history of Earth's subsystems over the past 200 million years, that is, since formation of the present ocean basins (Chapter 2). Analysis of sediments enables scientists to reconstruct past variations in parameters such as sea-surface temperature, salinity, carbon dioxide concentration, sea level, nutrient supply, and productivity. The main goal of geological oceanographers is to decipher the history of the ocean basins including past climate changes (Figure 4.1). We cannot begin to understand possible future climate change without understanding what has happened in the past and why.

The primary focus of this chapter is the types, sources, distribution, and environmental significance of marine sediments. We first describe how sediments are classified by size and the factors governing their rate of accumulation on the ocean floor. We then focus on the classification of marine sediments based on mode of origin, which largely determines their composition. There follows a description of the sedimentary deposits of the continental margin and the deep ocean floor and how geological processes modify these deposits. This provides

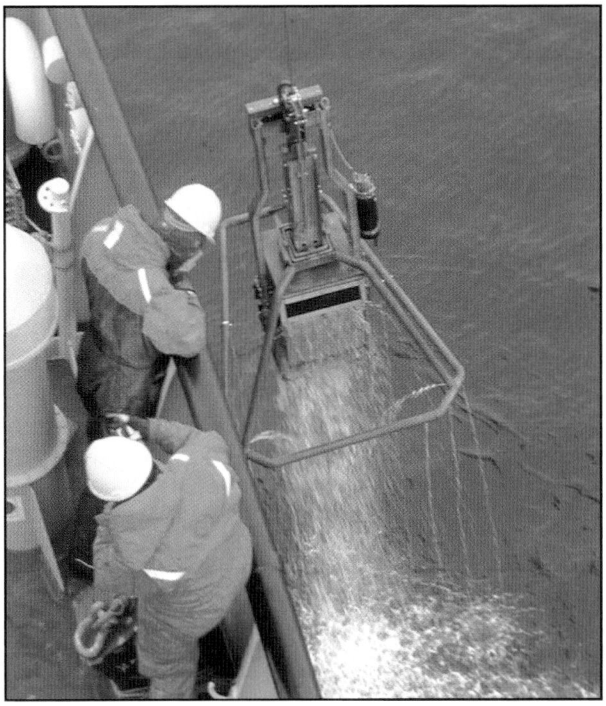

FIGURE 4.1
Scientists retrieve a box core used to gather samples of the ocean bottom sediments and fauna from the Canada Basin in the Arctic Ocean. This four-week international expedition, conducted during summer 2002, was designed to help reconstruct the history of climate. [NOAA Ocean Explorer; photo courtesy of Mike Vecchione, NOAA/Smithsonian Institution]

the background for a brief overview of sea floor resources including oil, natural gas, and minerals.

Classification of Marine Sediments by Size

Particles that accumulate on the sea floor are known as **sediment**. Marine sediments differ in source, composition, size, and the rate at which they accumulate on the sea floor. Of the many ways of classifying marine sediments, the two most popular are by particle size and source. In this section we describe the size range of sediments and the factors governing the settling rate of particles in ocean water.

SIZE AND SORTING

Marine sediments are classified by size into three broad categories: mud, sand, and gravel. As shown in Table 4.1, these categories are further subdivided by size using common descriptive terms. The smallest particle of mud (clay) has a diameter less than 0.0039 mm, too small to be seen without a microscope. At the other end of the scale, a boulder is more than 256 mm (10 in.) in diameter. Accumulations of sediment, called *sediment deposits*, on the sea floor also vary in the range of grain size, known as **sorting** (Figure 4.2). A well-sorted sediment deposit has a narrow range of grain sizes whereas a poorly sorted deposit has a broad range of grain sizes. In general, the greater the distance that sediment is transported by running water, ocean currents, or wind, the better sorted it becomes. For example, marine sediments that originated on land tend to be coarser and more poorly sorted on the continental margin than on the deep ocean floor. Grain size along with sorting can be used to infer the current direction and speed, the overall energy of the environment, and the transport mechanism (e.g., river, glacier, or wind).

TABLE 4.1
Wentworth Classification of Sediments by Size[a]

Sediment	*Type*	*Diameter (mm)*
Gravel	Boulder	>256
	Cobble	64-256
	Pebble	4-64
	Granule	2-4
Sand	Very coarse	1-2
	Coarse	0.50-1.0
	Medium	0.25-0.5
	Fine	0.125-0.25
	Very fine	0.0625-0.125
Mud	Silt	0.0039-0.0625
	Clay	<0.0039

[a]Adapted from C.K. Wentworth, *Journal of Geology* 30

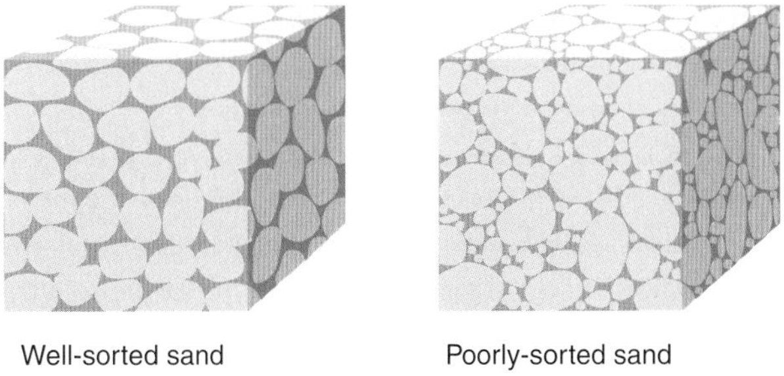

FIGURE 4.2
Well-sorted sediment deposit has a relatively narrow range of grain size whereas a poorly-sorted sediment deposit has a broad range of grain size.

Deposits of marine sediment are thickest on the continental margins (and near islands) where accumulation rates are relatively high. Most of these sediments are transported to the sea in suspension by rivers. As rivers flow downhill under the influence of gravity, some of the *kinetic energy* of moving water is used to erode the channel and transport sediment in suspension. Rivers also transport dissolved materials in solution and strong currents can push or roll larger sediment along the channel bottom. A river slows and diverges as it enters the ocean; dissipation of its kinetic energy reduces the river's ability to transport particles in suspension. More energy is required to transport larger particles than smaller ones. Hence, with diminishing kinetic energy, larger particles settle out of suspension almost immediately whereas finer particles are carried farther away from the coast before settling to the sea bottom.

Near the mouths of large sediment-transporting rivers, sediment accumulation rates can be enormous—perhaps as much as 8000 m (26,000 ft) per 1000 years. Waves and currents transport this material along the coast or offshore. Typical accumulation rates on the continental shelf and slope, on the other hand, range from 10 to 40 cm (4 to 16 in.) per 1000 years. The rain of sediment in the deep-ocean has been likened to the barely perceptible fall of dust in a classroom. Accumulation rates generally average from 0.5 to 1.0 cm (0.2 to 0.4 in.) per 1000 years. The thickness of deposits depends on the supply of particles and the length of time that sediments have been settling onto the ocean floor. As noted in Chapter 2, ocean floor sediments generally are thickest where the ocean crust is oldest so that deposits become thicker with increasing distance from the mid-ocean ridges (divergent or spreading plate boundaries). We would expect very little sediment on top of undersea mountains at a spreading center, but on the order of 1000 to 2000 m (3000 to 6000 ft) of sediment at distances farthest away from the spreading center at the edge of the ocean basins where the underlying oceanic crust approaches 200 million years old.

The length of time it takes for a particle to sink to the ocean bottom primarily depends on particle size. For example, a sand-sized particle may take a few days to sink to the bottom of the ocean whereas a clay-sized particle may take more than a century to cover the same distance. The longer it takes a particle to reach the ocean bottom, the greater the horizontal displacement of the particle by deep ocean currents and the more likely it is for soluble particles to dissolve in seawater. An important concept in understanding the ability of a fluid to transport a particle in suspension and the rate at which particles accumulate on the sea floor is terminal velocity.

TERMINAL VELOCITY

Terminal velocity is the constant speed attained by a particle falling through a motionless fluid such as water or air (Figure 4.3). The speed of a falling particle in calm water or air is regulated by (1) *gravity*, the force that accelerates the particle directly downward towards Earth's surface, and (2) the *fluid resistance* offered by the medium through which the particle is falling. A downward accelerating particle meets increasing, upwardly directed fluid resistance while gravity remains essentially constant. The magnitude of the upward resisting force eventually equals gravity; that is, the two opposing forces come into balance. When forces are balanced, the downward moving particle attains a constant speed. According to **Newton's first law of motion**, an object in constant straight-line motion or at rest remains that way unless acted upon by an unbalanced force. In this case, the fluid resistance (directed upward) balances gravity (acting downward) and the particle continues moving downward at constant speed. That speed is the particle's terminal velocity.

For a given medium, the terminal velocity increases with increasing particle size (assuming that the density and shape of particles vary little). This is the reason why sand-size particles settle to the ocean bottom faster than clay-size particles. Furthermore, the terminal velocity of a given particle varies with the medium. Due to its greater density, water is more viscous (offers more frictional resistance) than air. Hence, a particle of a given size has a greater terminal velocity in air than in water. For a particle to remain in suspension, turbulent motions in the medium must counter the terminal velocity. A fast-flowing river can be quite turbulent and can transport relatively large particles in suspension. The wind, on the other hand, ordinarily can transport only sand-size particles within a meter or so of the Earth's surface but can carry clay and silt-size sediment to altitudes of tens of thousands of meters and over horizontal distances of thousands of kilometers. Hence, as the wind or running water slows and becomes less turbulent, suspended particles settle out of suspension in an orderly sequence from larger to smaller particles.

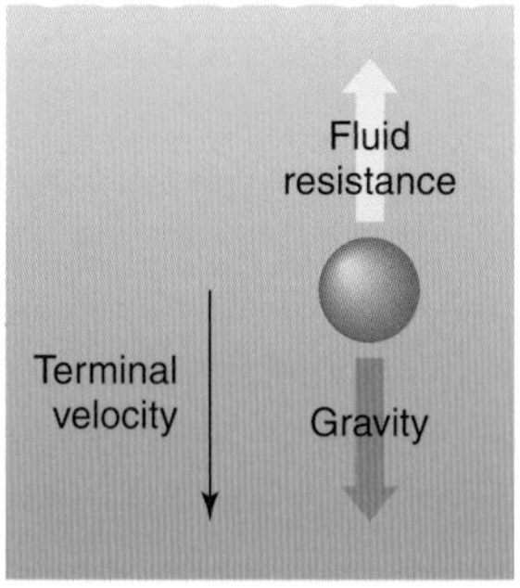

FIGURE 4.3
Terminal velocity is the constant downward-directed speed of a particle within a motionless fluid due to a balance between gravity (acting downward) and fluid resistance (directed upward).

Because of differences in terminal velocity, we would expect a deep ocean current to transport small slowly sinking particles a greater horizontal distance than a large rapidly sinking particle. For example, for a slow current of 1 km (0.6 mi) per day, the sand-size particle mentioned above, which takes a few days to sink to the bottom of the ocean, will be displaced horizontally only a few kilometers. In comparison, the clay-size particle, which may take more than a century to reach the bottom, can be displaced thousands of kilometers. For this reason, we might expect fine sediments on the sea floor to be quite different from the particles directly above them in the surface waters. Frequently, however, this is not the case because in seawater small particles tend to aggregate and form larger particles that rapidly settle to the sea bottom. The attractive force between particles having opposite electrical charges may be a reason why particles aggregate. Also, biological particles (e.g., fecal pellets) may stick together to form larger particles.

Classification of Marine Sediments by Source

Ocean scientists classify marine sediments based on their source as *lithogenous* (from rock), *biogenous* (from living organisms or their remains), *hydrogenous* (precipitated from seawater), and *cosmogenous* (from outer space). Cosmogenous sediments are much less common than the other three. In this section, we describe each of these sediment types and the processes that form them.

LITHOGENOUS SEDIMENT

Lithogenous sediment, originating from the weathering and erosion of pre-existing rock, accounts for about three-quarters of all marine sediments. As noted in Chapter 2, **weathering** refers to the physical disintegration and chemical decomposition of rocks that are exposed to the atmosphere while **erosion** is the transport of the products of weathering by running water, wind, glaciers, and gravity. Regions where the climate is warm and humid account for about two-thirds of river-borne marine sediments partly because of the greater rate of physical and chemical weathering in such climates.

Bruce Wilkinson of the University of Michigan estimates that about 86% of natural erosion takes place at elevations greater than 4000 m (13,125 ft) above sea level. This land represents only 2% of Earth's surface, but in steep terrain, rapidly flowing rivers and streams are powerful agents of erosion. So much so that, although the annual discharge of South American rivers is about twice that of Asian rivers, the South American rivers deliver less than half the volume of sediments to the ocean. Asia, already a larger continent, also has more land at higher elevations.

As the velocity of a current of water flowing over a sediment deposit increases, more grains of sediment move down-stream and the running water erodes the deposit. Based upon experimental work, the *Hjulström diagram* relates the current velocity needed to move a particle to the size (diameter) of the particle (Figure 4.4). The particle is assumed to be on a flat river bottom oriented horizontally at a depth of 1.0 m. Velocity, plotted on the vertical

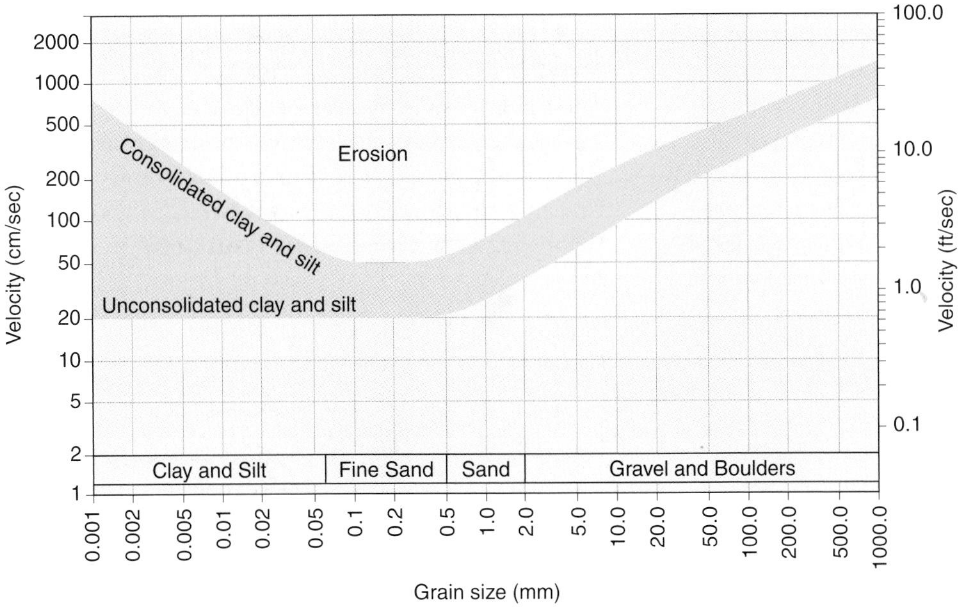

FIGURE 4.4
The Hjulström diagram relates the velocity of a current of water needed to move a particle in a down-current direction to the size (diameter) of the particle. [From Sundborg, A., 1956, *Geografiska Annaler*, Ser. A, v. 38, Fig. 16, p. 197.]

axis, increases upward, while grain size is plotted on the horizontal axis increasing from left to right. The shaded area in the diagram represents the *critical current velocity* above which particles are in motion (and undergoing erosion) and below which particles are stationary.

According to the Hjulström diagram, all particles with a diameter greater than 0.5 mm have a critical current velocity that steadily increases with increasing grain size, from sand to gravel to boulders. For finer grain sediment (especially less than 0.05 mm in diameter), the relationship between sediment size and critical current velocity is more complicated and depends on how well the particles consolidate (adhere to one another). For consolidated silt and clay, the critical current velocity increases with decreasing sediment size, so it would take water moving fast enough to push a boulder to displace much smaller particles of silt and clay. One of the principal reasons why fine particles become consolidated is the attraction between particles arising from opposite electrochemical charges, especially for clay and other soil particles. A similar effect takes place when marine organisms coat sediments with mucus webs making them "sticky" so that they adhere to one another.

Once the sediments reach the ocean, they are dispersed by waves and currents. Where highlands form the coastline, waves also undercut cliffs and, under the influence of gravity, rock debris slides and slumps into the sea. Explosive volcanic eruptions also contribute lithogenous fragments, known as **tephra**, which are rock fragments of various sizes, shapes and composition that fall through the air and accumulate in the ocean.

Human activity also plays an important role in delivering sediment to the ocean. In 2005, an international research team reported that through soil erosion, human activity (such as agriculture and deforestation) increased the amount of sediment transported by global rivers by 2.3 ± 0.6 billion metric tons per year. In fact, human activity is responsible for mobilizing about 15 times as much sediment as natural processes. At the same time, retention of sediments in reservoirs behind dams reduced the flux of sediment reaching the coast by 1.4 ± 0.3 billion metric tons per year. More than 100 billion metric tons of sediment is now sequestered in reservoirs most of which were constructed in the past 50 years or so.

The composition of lithogenous sediments produced through weathering depends on their source rock. The most abundant elements in the Earth's crust are oxygen (O) accounting for 46.6% by weight and silicon (Si) at 27.7% by weight (Table 4.2). The most common rocks composing Earth's crust are igneous, made up mostly of

TABLE 4.2
The Most Common Elements in the Earth's Crust

Element	*Weight Percent*
Oxygen (O)	46.6
Silicon (Si)	27.7
Aluminum (Al)	8.1
Iron (Fe)	5.0
Calcium (Ca)	3.6
Sodium (Na)	2.8
Potassium (K)	2.6
Magnesium (Mg)	2.1
All others	1.5

silicate minerals. The primary chemical building block of silicate minerals is the silicon-oxygen tetrahedron, consisting of one silicon atom bonded to four oxygen atoms (Figure 4.5). In the three-dimensional crystal lattice of a silicate mineral, silicon-oxygen tetrahedra link together through a sharing of oxygen atoms. Hence, the actual ratio of silicon atoms to oxygen atoms varies among silicate minerals. Quartz, one of the most common silicate minerals, has the chemical formula SiO_2. Quartz makes up about 12% of Earth's crust and is the principal component of most beach sand.

The silicon-oxygen tetrahedron combines chemically with other ions. Basalt, the chief constituent of oceanic crust, is an igneous rock that is rich in *ferromagnesian silicate minerals*, which contain iron (Fe) and magnesium (Mg) ions. They are dark in color and relatively dense. *Non-ferromagnesian silicate minerals* contain aluminum (Al), calcium (Ca), sodium (Na), or potassium (K) ions and are relatively light in both appearance and density. Granite, the chief constituent of continental crust, is an example of an igneous rock rich in non-ferromagnesian silicate minerals and is similarly light in color and comparatively less dense.

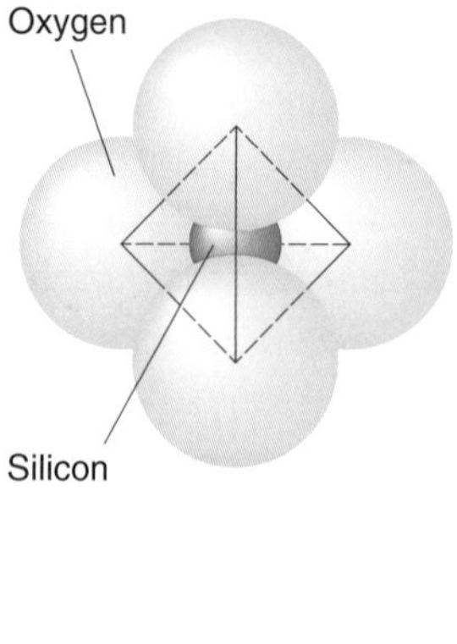

FIGURE 4.5
The silicon-oxygen tetrahedron is the building block of silicate minerals, the most common rock-forming minerals in the Earth's crust.

As a general rule, igneous rocks rich in ferro-magnesian silicate minerals weather more rapidly than those composed of non-ferro-magnesian silicate minerals. Differences in rates of weathering determine the dominant composition of lithogenous sediments delivered to the sea, which is mainly quartz grains and clay particles. Quartz is the most resistant of the common silicate minerals and clay particles are common weathering products of silicate minerals such as feldspars.

As noted earlier in this chapter, rivers transport the products of weathering in suspension, solution, or as part of the bed load. When a river reaches the coast, the denser and coarser particles settle out first and are deposited in deltas, estuaries, and shallow bays. Currents flowing nearly parallel to the shoreline supply sediment to beaches while the finer sediment can be carried to the continental shelf and beyond (Chapter 8).

FIGURE 4.6
A plume of Saharan dust spanning hundreds of kilometers is shown moving off the west coast of Africa into the North Atlantic on July 25, 2010. This image was captured by the MODIS instrument aboard NASA's Terra satellite. [Courtesy of NASA]

Almost every part of the ocean receives wind-borne dust. In fact, this is the primary mechanism whereby lithogenous particles reach the far-from-land deep ocean. Such fine particles compose much of the deep-sea red and brown clays located near the centers of major ocean basins. While high mountains and dry lake beds with barren surfaces are susceptible to erosion during episodes of strong winds, and volcanic ash is a source of wind-borne particles, the major source is subtropical deserts, such as the Sahara or Great Victoria in Australia. Because of the prevailing planetary-scale atmospheric circulation regime (Chapter 5), the latitudes around 30 degrees N and 30 degrees S are characterized by arid and semi-arid climates, resulting in large subtropical deserts. Furthermore, these regions of the ocean are nutrient-poor with low biological productivity so there are limited biogenous sediments to cover the clays on the sea floor.

Strong winds associated with weather systems that track across North Africa pick up dust particles from the dry topsoil and carry them to altitudes of 3000 m (10,000 ft). Trade winds, blowing from the northeast toward the southwest, then transport plumes of fine dust particles over the Atlantic and Caribbean and into Central America and the southeastern United States (Figure 4.6). This trans-oceanic journey takes between one and two weeks, mostly from June to October, peaking in July.

The analysis of two sediment cores from the North Atlantic sea floor, about 30 km (19 mi) from the mouth of the Senegal River, confirms that North Africa is a major source of wind-borne dust for the Atlantic basin. Additionally, in the 8 July 2010 issue of *Nature*, Stefan Mulitza of the University of Bremen in Germany and colleagues reported that the dust deposition dramatically increased following the arrival of commercial agriculture in the western Sahel (southern fringe of the Sahara) during the early 1800s. It also, unsurprisingly, increased during the long-term droughts that are endemic to the Sahel.

Wind-borne particles (*aerosols*) have both positive and negative impacts on marine organisms. Some aerosols are important sources of nutrients (e.g., nitrogen and phosphorous compounds) or trace metals (e.g., iron). The actual impact depends on a host of variables including aerosol composition, ocean chemistry, and types of organisms (e.g., different phytoplankton respond differently to chemically different particles).

Scientists have identified possible links between North African dust and red tides in the Gulf of Mexico as well as with threatened coral reefs in the Caribbean. In the summer of 2001, scientists reported that iron in North African wind-borne dust particles fertilized the waters of the Gulf of Mexico, increasing the frequency of **harmful algal blooms**, commonly known as **red tides**. Harmful algal blooms cause the die-off of fish, marine mammals, and birds as well as causing respiratory problems and skin irritations in humans. Enhanced levels of iron enable bacteria to convert nitrogen gas dissolved in seawater to an organic form that triggers an explosive growth in populations of potentially toxic algae.

North African dust may be harming coral reefs in the Caribbean through nutrient enhancement by similarly spurring the growth of populations of algae and phytoplankton that colonize the same environment as coral and interfere with its growth. Furthermore, North African dust may harbor a soil fungus that attacks coral reefs.

Glaciers also introduce lithogenous sediment by eroding bedrock and transporting rock fragments to the ocean. Sediments carried by glaciers are usually angular and can range in size from a fine powder to blocks of rock larger than automobiles. As glacial ice streams flow into the ocean, they float on the denser seawater, giving rise to stresses in the surrounding ice, which fractures into floating bergs. When the leading edge of a glacier breaks into icebergs upon entering the ocean (or large lake) it is known as **calving**. Transported by ocean currents and wind, icebergs eventually melt in warmer environments, releasing a poorly sorted mix of sand and boulders. Such ice-rafted glaciomarine sediments occur on about 20% of the sea floor, covering the Antarctic continental shelf and are common in the Arctic Ocean and on the nearby deep-ocean bottom. Layers of such glacial debris, alternating with non-glacial sediments, have also been identified in cores extracted from deep-ocean sediments of the North Atlantic. These deep-sea sediment layers record sudden releases of icebergs during the last Ice Age, known as Heinrich Events. For more on this, see the first Essay at the end of this chapter.

BIOGENOUS SEDIMENT

Biogenous sediment is sediment from a living source, such as excretions, secretions, and the remains of organisms. In the ocean, these include shells, fragments of coral, and parts of skeletons. The chemical composition of most biogenous sediments is either calcium carbonate ($CaCO_3$) or silica (SiO_2), which organisms secrete to form their shells. Biogenous sediments dominate 30% to 70% of the ocean's mid-depths, skeletal remains alone accounting for 25% to 50% of all particles suspended in seawater. Organisms consume much of the organic carbon in these particles as they sink to the sea floor or are exposed there following deposition.

On the ocean floor, calcareous sediments accumulate at a rate of between 1 and 4 cm (0.4 and 1.6 in.) per 1000 years, making them the most abundant of all biogenous sediment (Figure 4.7). They consist of calcium carbonate *tests* of *foraminifera* (single celled organisms), shells of *pteropods* (small, floating snails), and *coccoliths* (platelets secreted by one-celled algae known as *coccolithophorids)*. Figure 4.8A is a scanning electron microscope photograph of a common species of coccolithophorids with coccoliths that resemble tiny hubcaps. Robust shells characterize many foraminifera as well (Figure 4.8B), making it possible for them to reach the deep-ocean bottom intact, where they are preserved if covered by sediment. This is the origin of the calcareous mud that covers nearly half the deep-ocean bottom.

Many of the larger biogenous particles consist of **fecal pellets**, undigested organic matter that is concentrated and excreted by animals feeding in surface waters. Because of their relatively large size, fecal pellets have high terminal velocities, sinking hundreds of meters per day and reaching the ocean floor in less than a week. Smaller, uneaten particles are often trapped by the remains of shredded mucus webs, which are frequently produced by marine organisms. These mucus fragments cause particles to aggregate (clump together) into larger particles and they fall more rapidly to the sea floor, similar to the fecal pellets although not as fast. The remains of dead organisms from the upper layer of the ocean, fecal pellets, and various forms of non-living (inorganic) matter create a continuous flow of white particles through the ocean depths known as **marine snow**.

In sunlit surface waters, phytoplankton and other photosynthetic organisms convert inorganic carbon (CO_2) into usable organic carbon via photosynthesis and then are consumed by other organisms (Chapter 1). However, this carbon food source is limited in the deep ocean, where marine snow is the principal food source. Reporting in the November-December 2010 issue of *American Scientist*, Craig McClain of the National Evolutionary Synthesis Center in Durham, NC, noted that only about 3% of the total carbon fixed by photosynthesis in surface ocean waters sinks to the ocean interior.

Phytoplankton production and marine snow is greatest in coastal upwelling zones where nutrient-

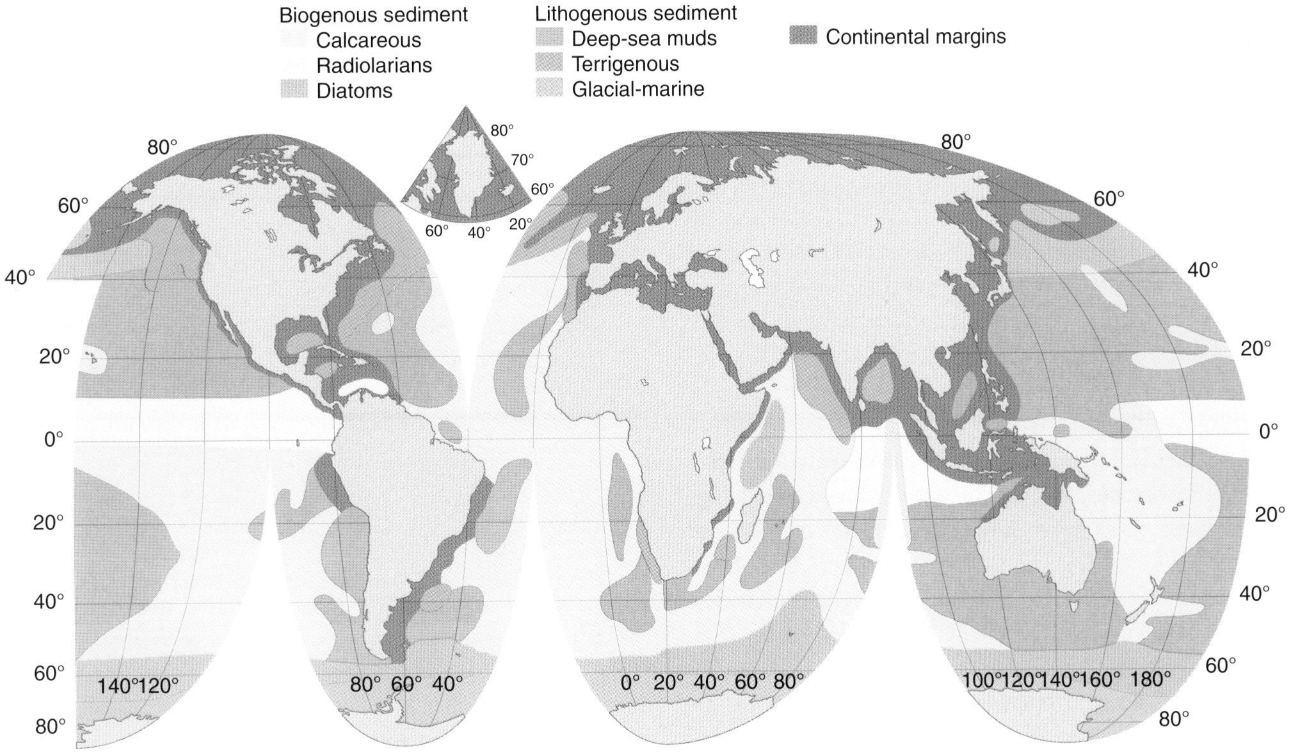

FIGURE 4.7
Approximate distribution of lithogenous and biogenous sediments on the floor of the world ocean.

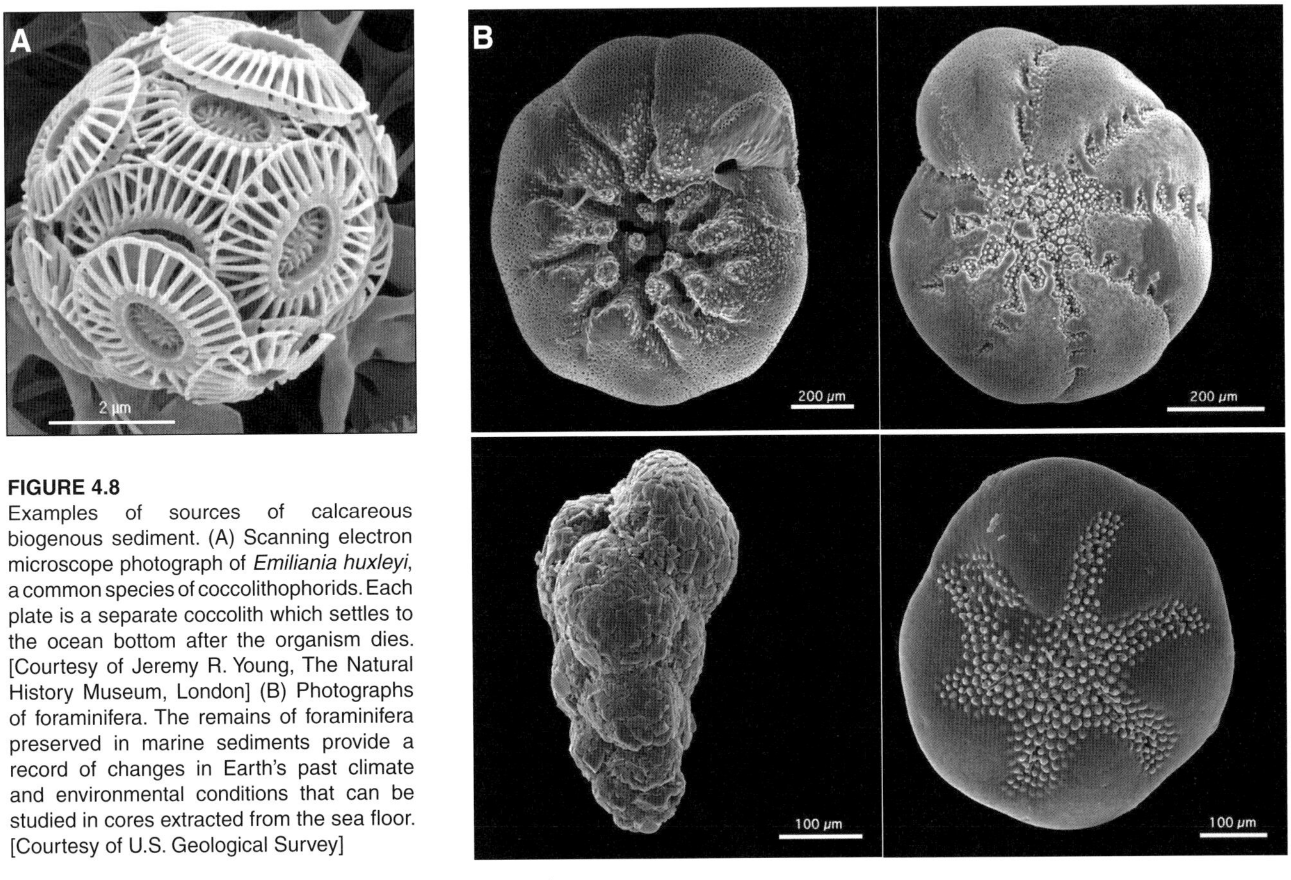

FIGURE 4.8
Examples of sources of calcareous biogenous sediment. (A) Scanning electron microscope photograph of *Emiliania huxleyi*, a common species of coccolithophorids. Each plate is a separate coccolith which settles to the ocean bottom after the organism dies. [Courtesy of Jeremy R. Young, The Natural History Museum, London] (B) Photographs of foraminifera. The remains of foraminifera preserved in marine sediments provide a record of changes in Earth's past climate and environmental conditions that can be studied in cores extracted from the sea floor. [Courtesy of U.S. Geological Survey]

rich deep-water is conveyed to the surface (Chapter 6). Furthermore, as the distance from shore and water depth increases, more particles of marine snow are consumed by various organisms. Hence, the supply of marine snow reaching the ocean floor decreases with increasing depth.

In the deep ocean, smaller and soluble particles are more likely to dissolve prior to reaching the bottom, bringing nutrients (nitrogen, phosphate and silicate compounds) and altering the chemical composition of deep-ocean waters. The most soluble particles dissolve rapidly and only accumulate on shallow ocean bottoms below regions with abundant shell-forming organisms. For example, organisms such as pteropods are plentiful in near-surface waters, but have particularly soluble calcium carbonate shells that readily dissolve in the more acidic deeper seawater. Consequently, pteropod shells are present only in sediment deposits on the tops of shallow seamounts, above 4500 m (15,000 ft) while, at greater depths, pteropod shells are not preserved.

Siliceous sediments are second in abundance to calcareous sediments on the ocean floor. These sediments come from the cell walls of *diatoms*, single-celled algae (Figure 4.9A), and *radiolaria*, single-celled zooplankton (Figure 4.9B). Siliceous particles dissolve at all ocean depths, though slightly more so in shallow warm water, so their presence on the sea floor indicates abundant organisms above. Siliceous mud is most common in the Pacific. Diatom-rich mud surrounds the Antarctic continent and occurs in the northernmost Pacific, while radiolarian-rich sediments dominate the ocean floor in equatorial latitudes, as radiolarian are most common in warm waters.

Phosphatic sediments (rich in phosphate) are rare in marine sediment deposits. Such deposits, primarily consisting of fish bones, teeth, and scales, occur mostly on shallow, isolated banks or near coastal areas where rates of biological productivity are especially high. Such near-shore phosphatic deposits may someday be exploited for their phosphate, which is used in manufacturing fertilizers.

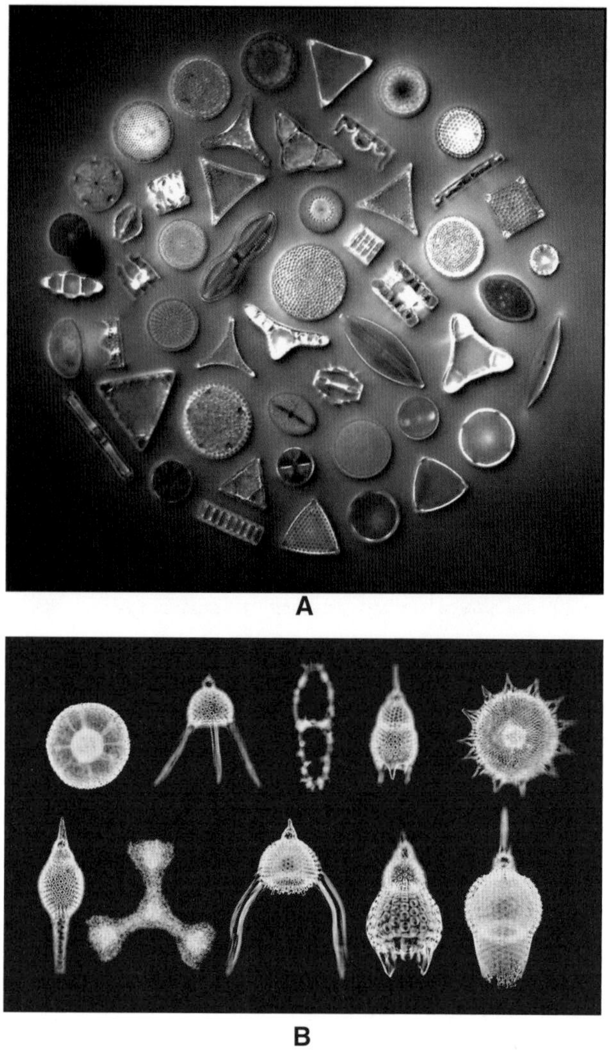

FIGURE 4.9
Examples of sources of siliceous biogenous sediment. (A) Diatoms as viewed through a microscope; they have silica exoskeletons. (B) These radiolaria have diameters of 0.5 to 1.5 mm. [Photos courtesy of Randolph Femmer /life.nbii.gov]

HYDROGENOUS SEDIMENT

Hydrogenous sediment encompasses particles that are chemically precipitated from seawater, in some cases forming coatings on other sea floor sediment. In addition, some hydrogenous sediment is the product of chemical reactions taking place in hot seawater discharged by hydrothermal vents on the sea floor (Chapter 2). Examples of hydrogenous sediment include some carbonates, halite (NaCl), gypsum ($CaSO_4 \cdot 2H_2O$), and manganese nodules. A rise in the temperature of shallow water may cause dissolved carbonate to precipitate as tiny pellets known as *ooliths*, perhaps 0.5 to 1.0 mm (0.02 to 0.04 in.) in diameter. Where evaporation rates are high and rainfall is low, salts precipitate from seawater in the sequence: carbonate salts, sulfate salts, and halite.

Most conspicuous of all hydrogenous sediments are **manganese nodules**, irregularly shaped, sooty black or brown nodules on the sea floor (Figure 4.10). Manganese nodules on average contain about 18% manganese by weight, 17% iron, and more importantly, small amounts (generally less than 1%) of copper, cobalt, and nickel. They range in size from tiny grains to large slabs weighing hundreds of kilograms; most, however, are the size of potatoes. Manganese nodules occur on the floor of all ocean basins

FIGURE 4.10
A box core from the floor of the tropical Pacific showing a relatively high density of manganese nodules. Box cores preserve the character of the undisturbed ocean bottom. [Courtesy of NOAA]

except the Arctic and in a variety of marine environments from abyssal plains to mid-ocean ridges. They are most abundant in an east-west belt about 5000 km (3000 mi) long on the floor of the tropical Pacific Ocean southeast of Hawaii and north of 10 degrees N.

Manganese nodules begin as coatings on hard objects (e.g., rock fragments, whale ear bones, shark teeth) that are exposed on the ocean bottom for lengthy periods of time. Burrowing organisms turn the nodules over, exposing all sides to seawater so that a coating forms on all sides of the object. Manganese nodules grow extremely slowly, ranging from about 1 to 10 mm (0.004 to 0.04 in.) per million years. With such slow growth rates, manganese nodules must remain unburied on the ocean bottom. Otherwise, sediments would cover the nodules, isolating them from contact with seawater and greatly inhibiting growth. In areas of the ocean receiving an abundant influx of sediment, only small (pea-sized or smaller) micronodules develop because they are buried before they can grow into larger nodules. Accordingly, rich manganese nodule deposits are found only in regions of the ocean far from shore, where input of lithogenous sediments is small and in unproductive areas where the rate of accumulation of biogenous particles is also low.

COSMOGENOUS SEDIMENT

Cosmogenous sediment comes from outer space, for example, as meteorite fragments. Perhaps 90% of the extraterrestrial solid particles entering Earth's atmosphere burns up prior to reaching the surface due to frictional heating. Most particles that survive the journey through the atmosphere and enter the ocean are so small that they dissolve before reaching the bottom. Nonetheless, cosmogenous particles are found mixed with other ocean bottom sediments. In addition, some cosmogenous particles that fall on the Greenland and Antarctic ice sheets enter the ocean via melting snow and ice.

Some cosmogenous sediments are remnants of the formation of planets in the solar system. These iron-rich particles likely have a chemical composition similar to Earth's core and mantle and their unique composition makes them readily recognizable in deep ocean sediment deposits. Other particles, formed from silicate rocks blasted off other planets or the Moon by meteorite impacts, are more difficult to identify as cosmogenous because they resemble lithogenous sediment.

A special type of sediment that is indirectly cosmogenous in origin consists of solidified droplets of rocks melted when huge meteorites struck the Earth. These small black fragments of silica-rich glass are known as **tektites**. Tektites are typically 2.5 to 5 cm (1 to 2 in.) across and have a teardrop or dumb-bell shape indicating that they were once fluid. They formed when droplets of molten rock produced by a meteorite impact were blasted into Earth's atmosphere and rapidly cooled and solidified. For example, the meteorite impact about 65 million years ago that is hypothesized to have contributed to the extinction of the dinosaurs generated great numbers of tektites that accumulated in and around the impact site, that is, Mexico's Yucatán Peninsula (Chapter 1). Tektites also have been identified in ocean bottom sediments off southern Australia and in the Indian Ocean.

Marine Sedimentary Deposits

The different types of marine sediments (i.e., lithogenous, biogenous, hydrogenous, and cosmogenous) occur in varying proportions and thicknesses on the ocean bottom as marine sedimentary deposits. In this section, we describe marine sediment deposits of the continental margin and the deep ocean basins.

CONTINENTAL-MARGIN DEPOSITS

Marine sediment deposits in the continental margin, called **neritic deposits**, are mostly lithogenous and occur in a wide range of sediment sizes. Most river-borne lithogenous sediments that reach the coast do not travel far seaward of the shoreline. About 95% of the coarsest sediments transported toward the ocean are trapped and deposited in bays, wetlands, estuaries, beaches or deltas. Only about 5% of river-borne sediment reaches the continental shelf or slope and very little sediment of terrestrial origin is transported beyond the continental margin into the deep-ocean basins. A notable exception is seaward of the mouths of major sediment-transporting rivers such as the Mississippi (United States), Ganges (Bangladesh), and Yangtse (China). Massive submarine avalanches and turbidity currents can transport sediments hundreds or thousands of kilometers out onto the continental rise and to the sea floor beyond.

As a sediment-laden river enters the ocean (or any open body of water), the water slows and sediments settle out of suspension. As long as currents do not carry them off as quickly as they are deposited, sediments accumulate on the sea bed at the mouth of the river. The body of accumulated sediments diverts the river along other paths, where it deposits more sediments eventually building a **delta**, so-called because it resembles the triangle of the Greek letter *delta* (Δ) when viewed from above (Figure 4.11). Water flows in a branching series of channels, known as *distributaries*, spreading fine, often nutrient-rich sediment over the surface of the delta. Unless artificially stabilized, the main current of the river can abruptly shift from one distributary to another.

The formation of many of today's deltas began when mean sea level stabilized near the end of the Pleistocene Ice Age, about 8500 years ago. Since then, river discharge, sediment load and compaction, subsidence of the sea (or lake) bottom, tides, and waves have mainly shaped deltas. Deltas are classified as *river-dominated* (e.g., Mississippi), *wave-dominated* (e.g., Niger), or *tide-dominated* (e.g., Ganges). In a river-dominated delta, the rate of input of sediments (from rivers or streams) exceeds the rate of removal of sediments (by waves and currents) and the delta develops the classic triangular shape. In a wave-dominated delta, the rate of sediment removal exceeds the input because strong wave action and currents limit sediment deposition, forming only a slight bulge on an otherwise straight coastline. Similarly, in a tide-dominated delta, the sediment is again moved by the ocean waters but the tidal currents rework it into long, narrow islands and submarine ridges.

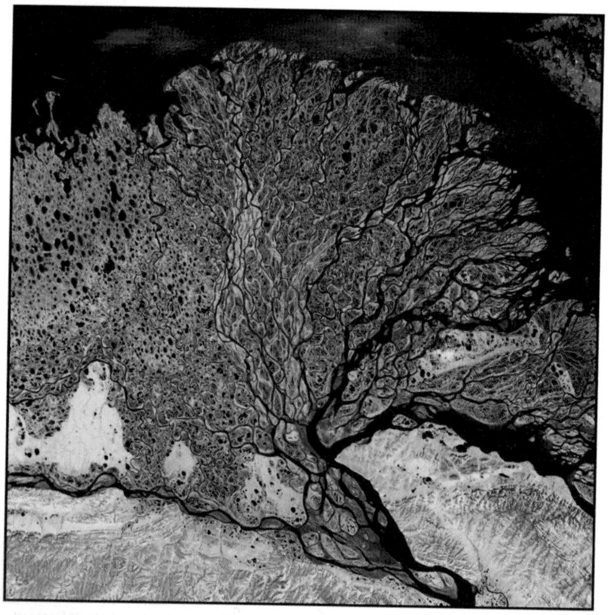

FIGURE 4.11
Landsat-7 image of the delta of the Lena River in Russia, one of the longest rivers in the world. [Courtesy of NASA}

Damming of rivers sequesters sediment in the reservoir behind the dam and disrupts delta formation. The retention of river-borne sediment in reservoirs reduces the need for dredging of harbors, and offshore coral reefs may benefit from less turbid waters and greater sunlight penetration. However, deprived of that sediment input, soil may compact, land surface subsides, and erosion by waves and currents increases. As detailed in this chapter's second Essay, this is happening to the Nile Delta of Egypt as a consequence of the Aswan High Dam.

Because sediment can accumulate so rapidly in coastal areas, typically at rates of several meters per thousand years, it is buried before it can react with the seawater or oxygen dissolved in it. Because some iron is oxygenated, these sediments exhibit a wide variety of vivid colors, especially in greens and blues, due to the various oxidation states of iron on the particles.

Common on deltas and coastal plains of the mid-Atlantic U.S. and Gulf of Mexico and in many other areas of passive continental margins are **wetlands**, low-lying areas covered by water, or with soils saturated for at least part of the year. Wetlands accumulate large amounts of organic matter and provide habitat for numerous populations of plants, animals and birds. Salt marshes help control coastal flooding, acting as a sponge during high water episodes.

As noted in Chapter 2, near the mouths of many major sediment-transporting rivers are **turbidity currents**, intermittent avalanches of dense, mud-rich waters that flow down submarine canyons (Figure 4.12A). While such flows are rarely directly observed in the ocean, they are well documented in lakes and reservoirs behind dams. Turbidity currents are powerful enough to break submarine cables, especially near the mouths of major rivers such as the Congo (Zaire) and Ganges.

Although occurring infrequently, turbidity currents convey large amounts of sediment that deposit thick beds, known as **turbidites**, over broad areas of

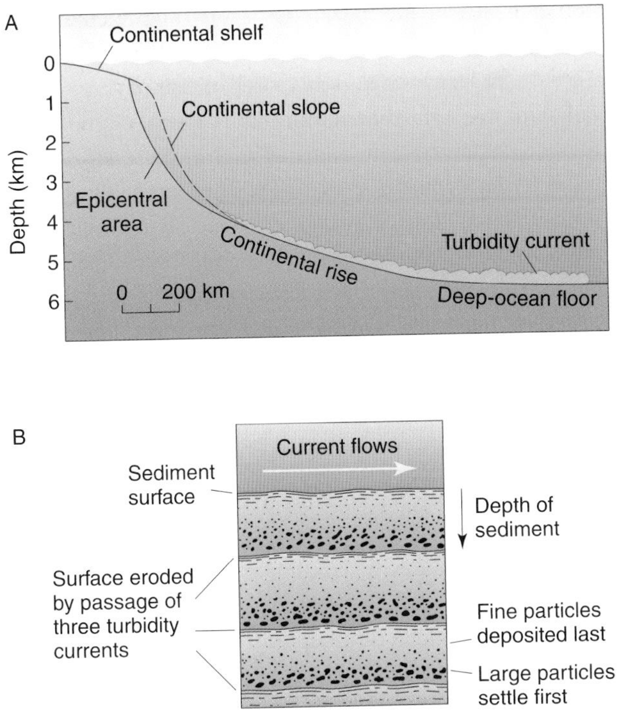

FIGURE 4.12
Turbidity currents can transport lithogenous sediment into the deep-ocean bottom. (A) Vertical profile of a turbidity current flowing on the continental slope and rise. (B) Vertical cross-section of turbidites deposited by three successive turbidity currents.

the deep ocean floor (Figure 4.12B). For example, in 1929 an earthquake triggered a submarine avalanche that produced a turbidity current off the Grand Banks, south of Newfoundland. Sediments were deposited on the nearby deep-ocean floor over an area that measured 100 km (60 mi) by 300 km (190 mi). Based on the time elapsed between the earthquake and submarine-cable breaks, scientists calculated the speed of the turbidity current at about 20 km per hr (12 mph), about the speed of a slow freight train. Submarine avalanches not only cause turbidity currents but also may trigger dangerous tsunamis (Chapter 7).

Turbidity currents are much denser than seawater because of the added suspended sediment load. Hence, they flow along the ocean bottom, often eroding channels just as rivers do on land. Such flow behavior explains why turbidites have unusual textures and contain abundant shells and other remains of shallow-water dwelling organisms. Submarine canyons near the mouths of major sediment-transporting rivers would fill with sediments without periodic scouring by turbidity currents.

Obstructions on the ocean bottom that deflect turbidity currents can prevent them from reaching the deep-ocean floor. In the Pacific Ocean basin, submarine ridges associated with volcanic island arcs and trenches block the flow toward the adjacent deep-ocean bottom. Without such obstructions, thick sediment deposits create the conspicuous continental rises in the Atlantic, northern Indian, and Arctic Oceans (Chapter 2). The tops of seamounts and submarine ridges typically are unaffected by turbidity currents and are free of turbidites.

DEEP-OCEAN DEPOSITS

Fine-grained sediments that gradually accumulate particle-by-particle on the deep-ocean floor form **pelagic deposits**. Most of these sediments are biogenous and their accumulation rates are considerably slower than neritic (near shore) sediments. On average, a 1-mm thick layer forms in about 1000 years. (A 1-inch thick layer forms in about 25,000 to 250,000 years.) In spite of these very low accumulation rates, sufficient time has passed since formation of the ocean basins that the average thickness of pelagic deposits is 500 to 600 m (1600 to 2000 ft).

Very small particles have relatively low terminal velocities and sink slowly through the ocean depths. Many of these particles are eaten by filter feeders or trapped in mucus nets secreted by marine organisms. Their relatively long suspension times in the water mean that they are widely dispersed by currents. There is also ample time for chemical reactions to occur. For example, iron minerals in suspended clay particles react chemically with dissolved oxygen in seawater, forming a rusty (iron oxide) coating. The abundance of such red- or brown-stained particles in deep-sea mud is responsible for the colors and common names of these deposits, that is, *red clay* and *brown mud.* Colors of deep-sea clays range from brick red (derived from the Saharan Desert) in the Atlantic Ocean to chocolate brown in the Pacific Ocean.

Pelagic deposits that are more than 30% biogenous by weight are called either calcareous ooze or siliceous ooze depending on composition. **Calcareous ooze**—made up of the tests (hard parts) of coccolithophorids, pteropods, and foraminifera—are the most abundant of pelagic oozes. Calcareous oozes generally are confined to ocean waters shallower than the **carbonate compensation depth (CCD)**, the depth of the ocean below which calcium carbonate ($CaCO_3$) shells and skeletons dissolve and do not accumulate (Figure 4.13). (The CCD is also called the *saturation horizon.*) The rate at which calcium carbonate dissolves in seawater increases with falling temperature; that is, cold water can hold more carbon dioxide which dissociates to form carbonic acid that dissolves $CaCO_3$. Ocean temperature usually drops

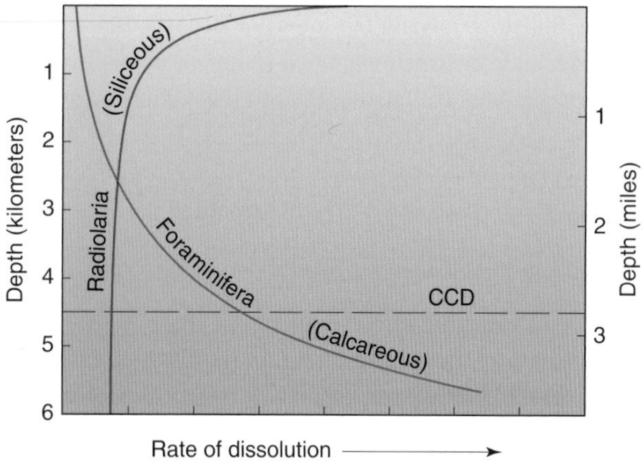

FIGURE 4.13
The rate at which radiolaria (siliceous) and foraminifera (calcareous) shells dissolve with increasing ocean depth in terms of percent weight loss. Siliceous shells more readily dissolve in surface waters whereas calcareous shells more readily dissolve at depth. The carbonate compensation depth (CCD) is defined as the depth at which the amount of calcareous particles declines to less than 20% of the mass of all particles present.

with increasing depth below surface waters so that more and more calcareous particles dissolve with increasing water depth. Above the CCD, waters are relatively warm and saturated with respect to the carbonate minerals calcite and aragonite. At the CCD, the rate of production of calcareous tests equals the rate of calcite dissolution. Below the CCD, waters are under saturated with respect to carbonate minerals, colder, and sufficiently acidic to dissolve carbonate shells and skeletons.

Worldwide, the carbonate compensation depth averages about 4500 m (14,800 ft). Calcareous oozes accumulate on ocean bottom features having depths shallower than 4500 m, such as the crest and slopes of the mid-ocean ridges. If the ocean basins were to suddenly dry up, the tops of volcanic peaks that had been above the CCD would appear light in color due to the accumulation of carbonate-rich sediments. Hence, the CCD is somewhat analogous to the snow line on terrestrial mountain ranges.

As discussed in Chapter 3, the increasing flux of atmospheric CO_2 into the ocean is lowering the pH of seawater. (Recall that ocean waters are slightly alkaline.) In response to this *ocean acidification*, the CCD has shifted upward an estimated 50 to 200 m (165 to 660 ft) since the beginnings of the Industrial Revolution and humankind's increasing reliance on fossil fuels. The expected continued increase in atmospheric CO_2 concentration is likely to be accompanied by a further thinning of the portion of the ocean in which the calcareous shells and skeletons of organisms do not dissolve.

Siliceous ooze is composed of tests of diatoms and radiolaria (described earlier in this chapter). Ocean water is under-saturated with silica so that these tests dissolve at all ocean depths (Figure 4.13). They occur on the sea floor below surface waters only where source organisms are particularly abundant. Siliceous and calcareous oozes consist of clay size particles and dominate the deep-ocean bottom sediment (Figure 4.7). Sand-size particles make up less than 10% of the deep-ocean sediments and the coarsest deep-ocean sediment deposits are the products of explosive volcanic eruptions, ice rafting, and turbidity currents.

As noted above, in the deep ocean, accumulations of sea floor sediments on average are hundreds of meters thick. However, in 2005, ocean scientists discovered an area in the South Pacific Ocean about 4000 km (2500 mi) east of New Zealand where pelagic sediments form an unusually thin veneer over basaltic oceanic crust—in some places sediments have a thickness of only 50 cm (20 in.). Approximately the size of the Mediterranean Sea, this so-called *South Pacific Bare Zone* owes its relative dearth of sediments to a number of factors. Surface waters are nutrient-poor limiting the populations of plankton and other organisms so that there are little organic remains to contribute to the rain of biogenous particles. The deepest water is relatively low in carbonate and silica so that the remains of organisms with carbonate or silica hard parts dissolve before reaching the sea floor. The region's great distance from a landmass means a limited supply of windblown dust. Little or no hydrothermal activity limits the amount of dissolved minerals that could precipitate as hydrogenous particles. Furthermore, surface currents that steer Antarctic ice bergs and their load of lithogenous sediment bypass the region.

Another recent discovery by ocean scientists calls for a revised description of the deep ocean abyss as a serene place where small particles gently rain down and settle on the ocean floor as pelagic deposits. Instruments lowered to the sea floor to measure currents and the resulting mobilization of bottom sediments detected a much more active environment than expected. Scientists found that occasionally bottom currents and **abyssal storms** scour the ocean floor, generating moving clouds of suspended sediment. Although called an abyssal storm, the water motion pales by comparison to the wind speeds in atmospheric storms.

Abyssal currents and storms apparently derive their energy from surface ocean currents. Wind-driven surface ocean currents flow about the margins of the ocean basins as huge gyres centered near 30 degrees latitude (Chapter 6). Viewed from above, these *subtropical gyres* rotate clockwise in the Northern Hemisphere and counterclockwise in the Southern Hemisphere. For reasons given in Chapter 6, surface currents flow faster, are narrower, and extend to greater depths on the western side of the gyres. They are known as *western boundary currents* and include, for example, the Gulf Stream of the North Atlantic Ocean basin. Abyssal currents are likewise most vigorous on the western side of the ocean basins, flowing along the base of the continental rise, which is on the order of several kilometers deep.

Abyssal storms may be caused by eddies (called *rings*) that occasionally break off from the main current of the Gulf Stream and other western boundary currents. During an abyssal storm, the eddy or ring may extend to the ocean bottom causing the typical velocity of a bottom current to increase 10-fold to about 1.5 km per hr (1 mph). While unimpressive as a wind speed, water is much denser than air so that its erosive ability and sediment-transport capacity is significant even at only 1.5 km per hr. At this higher speed, the suspended sediment load in the bottom current increases by a factor of ten. Abyssal storms scour the sea floor leaving behind long furrows in the pelagic deposits. After a few days to a few weeks, the current weakens or the eddy (ring) is reabsorbed into the main surface ocean circulation and the suspended load settles to the ocean floor. In this way, abyssal storms can transport tons of sediment great distances, thereby disrupting the orderly depositional sequence of layers of deep-sea sediments. Scientists must take this disruption into account when interpreting deep-sea sediment cores for their record of past environmental conditions (Chapter 12).

Marine Sedimentary Rock

As noted in Chapter 1, sediment is generated at the interfaces between the lithosphere, atmosphere, hydrosphere, cryosphere, and biosphere. Physical and chemical weathering processes break down exposed bedrock forming rock fragments (sediments) that are transported by rivers, glaciers, wind, and gravity to the ocean. Weathering also releases soluble constituents, such as calcium and sodium that dissolve in water and are transported in solution to the ocean.

As part of the *rock cycle* (Figure 4.14), over the millions of years that constitute geologic time, sediment that is deposited on the ocean floor is gradually converted

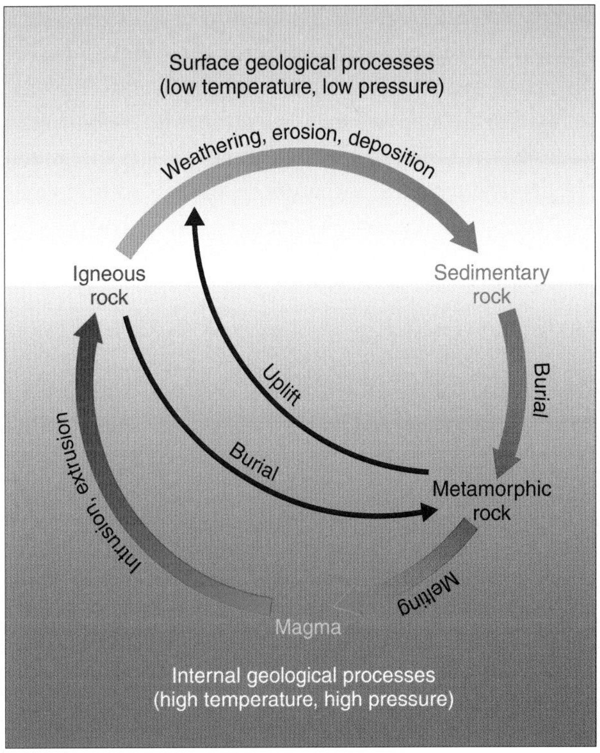

FIGURE 4.14
As shown in this schematic diagram, the conversion of marine sediments to sedimentary rock is part of the rock cycle.

to solid marine sedimentary rock through lithification. **Lithification** usually involves both compaction and cementing of sediments at relatively low temperatures (under 200 °C or 390 °F). Sediments are compacted by the increasing weight of sediments accumulating above that squeeze deeper sediments closer together. Siliceous and calcareous fluids migrating through the tiny openings between individual sediment grains precipitate minerals that fill the pore spaces, cementing grains to one another. The product of lithification is sedimentary rock such as shale, sandstone, or limestone depending on the composition and particle size of the constituent sediments. With deeper burial and further increases in temperature, pressure, and access to chemically active fluids, sedimentary rock may be converted to metamorphic rock such as slate, schist or gneiss.

Most sediment and sedimentary rock deposited on the ocean floor is transported with the underlying moving lithospheric plate, eventually entering a subduction zone, where they are driven down into the mantle and are metamorphosed or melted into magma. Magma produced in subduction zones is differentiated into lower density magma that is ejected by andesitic/rhyolitic volcanoes landward of the subduction zones. Some of the higher density magma might be extruded in backarc spreading areas but most enters the asthenosphere. While much of the marine sedimentary strata on the continents originated in extensive shallow seas, some marine sedimentary deposits were scraped off subducting oceanic plates and physically attached to a continental plate. This plus other tectonic activity such as mountain building explains why marine sedimentary rocks are found high in continental mountains. For an example of this, see this chapter's third Essay.

Resources of the Sea Floor

Resources extracted from the sea floor include oil, natural gas, sand, gravel, and minerals. While oil and natural gas account for more than 95% of the total monetary value of resources extracted from the sea floor, sand and gravel are the sea floor resources most commonly mined worldwide. Approximately 25% of the oil and 20% of the natural gas consumed in the U.S. each year comes from offshore sources including the Gulf of Mexico, Persian Gulf, and North Sea.

Global economics plays a central role in governing the amount of sea floor mining and is the principal reason why few resources are mined from the ocean, other than oil, natural gas, sand, and gravel. Production costs are usually much lower for onshore mining versus offshore mining. In the future, however, shifts in the international minerals markets and increased demand for strategically important minerals may make it feasible to begin mining more seabed resources.

OIL AND NATURAL GAS

Petroleum and natural gas are derived from the remains of marine plants and animals and occur in the pore spaces of marine sedimentary rock. Both oil and natural gas consist of hydrocarbons, compounds whose molecules contain only hydrogen and carbon atoms. Oil is a mixture of thousands of different hydrocarbons. By volume, natural gas is up to 99% methane (CH_4), plus small quantities of ethane (C_2H_6), propane (C_3H_8), and butane (C_4H_{10}).

Today's major deposits of oil and natural gas developed under very restricted and unusual biological and geological conditions many millions of years ago. Petroleum formation requires that substantial amounts of organic matter must accumulate at the bottom of a shallow quiet sea. Decomposers reduce dissolved oxygen (DO) to levels that can be tolerated by only anaerobic bacteria. (*Anaerobic* organisms can live and grow without free oxygen.) Products of anaerobic decomposition include methane and other light hydrocarbons. With continued accumulation of organic and other sediment on the sea floor, the deeper organic-rich sediments were subject to increasing temperature and pressure that spurred their lithification and the conversion of organic matter to oil and natural gas.

Conditions favorable to oil and natural gas production occurred primarily during the geologic periods known as the Ordovician (488-444 million years ago), Permian (299-251 million years ago), Jurassic (200-146 million years ago) and Cretaceous (146-66 million years ago). At those times, sea level was unusually high and ocean waters spread over low-lying portions of the continents producing large shallow seas. Enhanced biological productivity in those seas provided the raw material for the eventual formation of oil and natural gas deposits.

Oil and natural gas migrated upward from their source rock into more porous layers of sandstone or limestone, displacing the water that occupied the pore space in these strata. An overlying layer of less permeable rock trapped the petroleum in *reservoir rock*. Extraction wells must penetrate this reservoir rock to tap the oil or natural gas. Oil formed when sediments were buried to depths of at least 2 km (1.2 mi) and generally does not occur at depths greater than 3 km (1.9 mi). Natural gas is usually found in marine sedimentary rocks at depths of less than 7 km (4.3 mi).

MINERAL RESOURCES

Great quantities of sand, gravel, and shells are mined from the near-shore, shallow ocean bottom, especially near coastal cities. These resources are used primarily in road construction and the production of cement and concrete. Also, sand dredged from the shallow ocean bottom is used to replenish and restore nearby beaches eroded by storm waves (Chapter 8).

In some locales, valuable metallic and nonmetallic minerals such as iron, tin, platinum, gold, and diamonds occur mixed with coastal sands. Most of these resources are products of weathering and erosion of continental rock and sediment and are transported to the sea in suspension by rivers along with other lithogenous particles. Ocean waves and currents sort and concentrate metals and gemstones in coastal or submarine deposits, known as **placer deposits**. Placer minerals typically are relatively dense and resistant and were left behind as a *lag concentrate* as waves and currents removed the less dense sand grains. Although hundreds of such deposits are known from around the world, very few are actually being extracted. Today, dredging of placer deposits usually takes place in shallow waters just offshore and yields tin (Thailand and Indonesia), gold (Alaska, New Zealand, and the Philippines), and diamonds (Namibia and South Africa). Rivers also deliver minerals in solution to the sea including phosphorite, a deposit rich in phosphate that precipitates from seawater and accumulates on the sea floor under upwelling zones or other biologically productive areas, and manganese, which forms manganese nodules via biogeochemical cycling.

As noted earlier in this chapter, manganese nodules on the deep ocean floor have been studied extensively as potential sources of copper, nickel, and cobalt. Cobalt-rich manganese crusts occur at intermediate depths (2 to 3 km, or 1 to 2 mi) on the slopes of extinct volcanoes that form many Pacific islands. But after decades of exploration and development of deep-ocean mining techniques, there is still insufficient economic incentive for commercial production from the deep-ocean floor. Furthermore, environmental safeguards and financial obligations imposed by the *U.N. Convention on the Law of the Sea*, ratified in 1994, have greatly reduced interest in mining manganese nodules.

Development of the plate tectonics theory spurred interest in another source of marine mineral deposits: geological processes taking place at submarine plate boundaries (Chapter 2). Ocean basins include sites where mineral deposits form in place (a process called *mineralization*) rather than being delivered to the sea by rivers. **Seawater, magma, and new oceanic crust interact at plate boundaries (including spreading centers and subduction zones), exchanging heat and chemicals and producing *hydrothermal mineral deposits*.** Magma from Earth's interior heats the seawater that circulates through fractures and vents in the oceanic crust. Hot seawater dissolves metals from the magma and crust, and those metals react with sulfur, precipitating as sulfide minerals on the ocean floor. These hydrothermal deposits are potential sources of copper, zinc, silver, gold, and other metals.

As is the case with seabed placer deposits, development of hydrothermal mineral deposits depends on favorable economic conditions as well as consideration of potential environmental impacts. Nonetheless, sulfide metals associated with subduction zones (e.g., in the western Pacific) or hydrothermal vents are economically promising for several reasons, including: a greater percentage of precious metals, occurrence at shallower ocean depths (1000 to 2000 m or 3300 to 6600 ft), and location within the 370-km (200-nautical mi) jurisdiction of coastal or island nations. Worldwide, about 250 sites of massive sulfide deposits have been identified on the sea floor.

Over the past decade, interest in development of massive sulfide deposits on the sea floor has focused on an active hydrothermal vent system in the waters within the Exclusive Economic Zone (EEZ) of Papau New Guinea. This hydrothermal deposit features higher concentrations of gold, copper, zinc, and silver than extracted from terrestrial mines. Developers are encouraged by the availability of appropriate mining technology and rising metal prices. Because the proposed mine site is located within an EEZ, developers are not subject to the environmental restrictions of the *U.N. Convention on the Law of the Sea*. Within EEZs (discussed below), it is up to the individual nations to issue mining permits and impose environmental regulations. For economic and other reasons, Papau New Guinea is expected to be relatively lenient in this regard.

Conservationists are concerned that although mining of sulfide deposits will be restricted to small areas of the ocean floor, mining activities would disturb the benthic environment and produce sediment plumes that could smother or contaminate hydrothermal vent communities. Meanwhile, developers argue that the planned deep-sea mining will be less environmentally disruptive than similar mining conducted on land. For one, mining will focus on deposits associated with extinct vents thereby avoiding contact with lava and superheated water. Furthermore, developers point out that deep-sea mining of sulfide ores produces far less waste material than terrestrial mining. Typically, a targeted metal represents a tiny fraction of a huge rock mass. For example, 2 tons of ore extracted from the ocean floor should yield as much copper as 80 million tons of material mined on land.

In January 2011, *Nautilus Minerals*, headquartered in Toronto, Canada, announced that it had obtained from Papua New Guinea the world's first lease to mine minerals on the sea floor. The 20-year lease gives the company the green light to proceed with plans to begin mining for copper and gold in a 59 square km (22.7 square mi) area on the Pacific Ocean floor north of Rabaul, Papau New Guinea. *Nautilus Minerals* has committed to impact mitigation and restoration as well as responsible mine closure. Mining is scheduled to commence in 2013.

EXCLUSIVE ECONOMIC ZONE

Since 1958, the United Nations has worked to formulate international policies concerning exploitation of seabed resources, including fuels and minerals. But progress has been slow because of conflicts among the more than 150 nations involved in negotiations. A fundamental philosophical division exists between less-developed and developed nations. Many less-developed nations view ocean resources as the common heritage of all people, but they fear that the world's richest and most technologically advanced nations will reap the bulk of the harvest. While the U.S. and many other developed nations endorse the common heritage concept, they fear that too much power would be vested in the governments of less-developed nations if they were granted significant say in shaping ocean resource policy.

The 1982 *U.N. Convention on the Law of the Sea* granted jurisdiction over an **exclusive economic zone (EEZ)** to each of 151 coastal nations. In March 1983, the U.S. (later joined by fifty other nations) defined its jurisdiction over ocean resources (including minerals, fuels, and fisheries) to extend 370 km (200 nautical mi) offshore. This gave the U.S. jurisdiction over an ocean area that is 1.7 times that of the total land area of the U.S. and its territories. Within the EEZ, the federal government regulates all economic activity beyond the seaward edge of the individual state's jurisdictional area (its territorial sea) out to 200 nautical miles from its coast. For the U.S., the National Oceanic and Atmospheric Administration (NOAA) and the Interior Department share responsibility for managing seabed mineral resources.

The 370 km exclusive economic zone may or may not encompass the entire continental shelf associated with the landmass of a specific coastal nation (Figure 4.15). Hence, a provision of the 1994 *U.N. Convention on the Law of the Sea* allows a nation to expand its EEZ to the edge of the continental shelf, if it can establish that the new territory is a "natural prolongation" of its landmass. (Recall from Chapter 2 that the continental shelf is a submerged extension of a continent.) The challenge is to establish scientifically the location of the outer boundary of the continental shelf, that is, where the continental shelf ends and the slope begins. By one estimate, the U.S. stands to gain an additional 750,000 square km (290,000 square mi) of the Atlantic, Pacific and Arctic Oceans if the EEZ includes the shelf surrounding every one of the nation's islands and possessions, regardless of how small or remote.

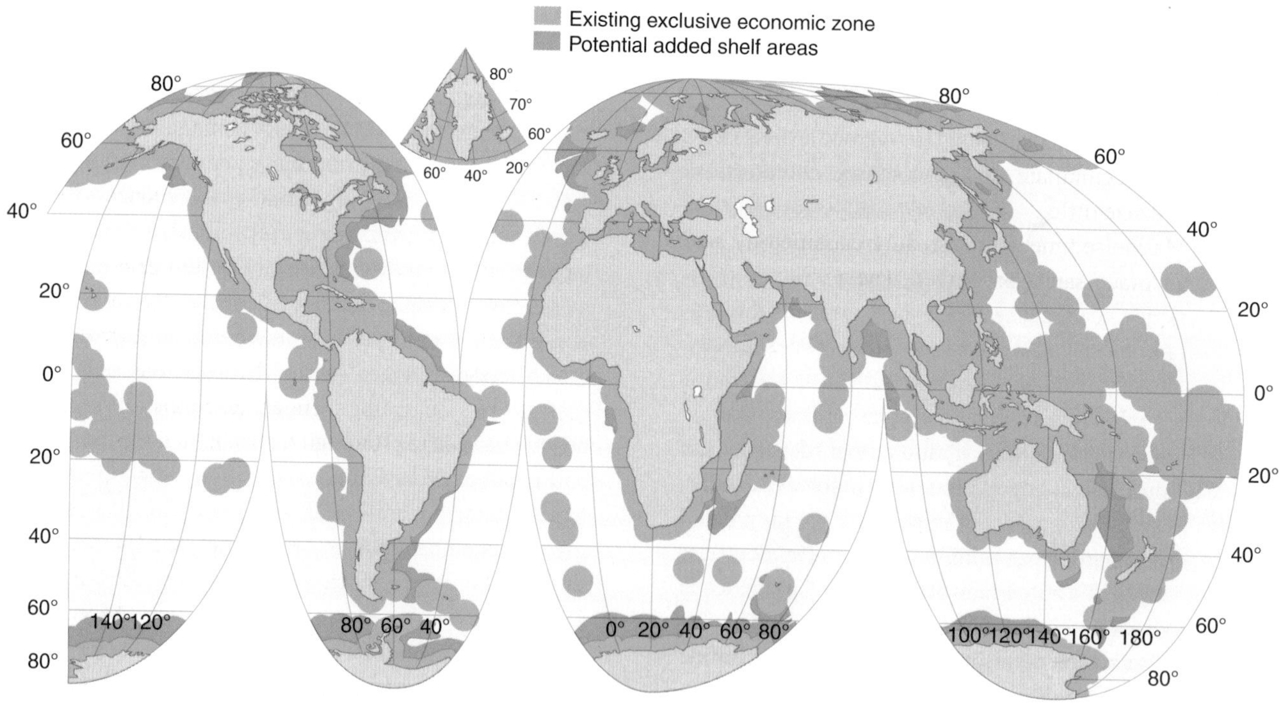

FIGURE 4.15
Current exclusive economic zones (brown) in the near future could be extended to the limits of the continental shelf (orange).

Conclusions

Sediments are produced by processes operating at the interfaces between the lithosphere, atmosphere, hydrosphere, cryosphere, and biosphere. Rivers, winds, glaciers, and gravity transport sediment from land to sea while some sediment originates in the ocean. Sediment deposits on the ocean floor record changes in the various subsystems of the Earth system over millions of years, that is, since formation of the present ocean basins. In later chapters, we revisit deep-sea sediments as we seek to understand how the various subsystems have changed through Earth history and what is revealed about past variations in Earth's climate.

Sedimentary processes (e.g., weathering, erosion, deposition, lithification) are key players in the functioning of the Earth system. For example, sequestering of carbon in sediments and sedimentary rocks is responsible for Earth's fossil fuel resources and accounts for the relatively small amount of carbon dioxide in Earth's modern atmosphere. Atmospheric carbon dioxide is one of many gases that contribute to the planet's greenhouse effect. We learn more about Earth's climate system in the next chapter where we consider the flux of heat energy and water between the ocean and atmosphere.

Basic Understandings

- Particles that accumulate on the sea floor are known as marine sediments. The rate at which they accumulate, and their source, composition, and size differ.
- Marine sediments are broadly classified by size as mud, sand, or gravel. The range of grain size within accumulations, known as sorting, also varies. A well-sorted sediment deposit has a narrow range of grain sizes whereas a poorly sorted one has a broad range of grain sizes. Sediment size, sorting, and composition provide information on past environmental conditions.
- Terminal velocity is the constant speed attained by a particle as it falls through a motionless fluid such as calm water or air. The speed of that falling particle is regulated by gravity and the resistance of the medium through which it falls. Once the magnitude of the resisting force equals the pull of gravity, the two forces come into balance and, as described by Newton's first law of motion, the falling particle attains a constant speed, its terminal velocity. Within a given fluid, the terminal velocity generally increases with increasing size of particles of the same substance.
- Marine sediments are classified based on their source as lithogenous (from rock), biogenous (from organisms or their remains), hydrogenous (precipitated from seawater), and cosmogenous (from outer space).
- Lithogenous particles, which mostly originate from weathering of terrestrial rock, are transported to the ocean by rivers, winds, glaciers, and gravity. Some lithogenous sediments are volcanic in origin.
- The largest lithogenous particles are deposited close to where they enter the ocean while smaller particles are transported further out to sea. Fine windblown sediment, clay and silt particles, can travel thousands of kilometers from its source.
- According to the Hjulström diagram, the critical current velocity required to mobilize particles with a diameter greater than 0.5 mm increases with increasing grain size (from sand to gravel to boulders). For finer grain sediment, less than 0.05 mm in diameter, the critical current velocity depends both on how the particles are consolidated and sediment size. For consolidated silt and clay, the critical velocity increases with decreasing sediment size.
- Biogenous sediment includes the excretions, secretions, and remains of organisms living in the ocean, usually in the sunlit waters. The chemical composition of most biogenous sediments is either calcium carbonate ($CaCO_3$) or silica (SiO_2).
- Hydrogenous sediments, such as carbonate salts, halite, and manganese nodules, are precipitated from seawater and occasionally coat other sea floor sediment. Other hydrogenous sediment is the product of chemical reactions from seawater circulating through hydrothermal vent systems.
- Neritic deposits, which occur on the continental margin, are mostly lithogenous. About 95% of the coarsest sediments transported to the ocean by rivers are trapped and deposited in bays, wetlands, estuaries, beaches or deltas. Only about 5% of river-borne sediment that pass the shoreline reaches the continental shelf or slope.
- Fine-grained sediments that slowly accumulate on the deep-ocean floor form pelagic deposits.

Most of these sediments are biogenous and their accumulation rate is considerably slower than neritic sediments.

- Occasionally bottom currents and abyssal storms scour the pelagic deposits on the ocean floor, generating moving clouds of suspended sediment and disrupting the orderly accumulation of sediments.
- Calcareous ooze, made up of the tests (hard parts) of coccolithophorids, pteropods, and foraminifera, are the most abundant of pelagic oozes. They occur where the deep ocean is shallower than the carbonate compensation depth (CCD).
- Siliceous ooze is composed of tests of diatoms and radiolaria. As seawater is under-saturated with silica, these tests dissolve at all ocean depths. Hence, they occur on the sea floor only in those portions of the ocean where diatoms and radiolaria are abundant in the surface waters and deposition occurs more quickly than dissolution.
- Over the thousands of millions of years of geologic time, the sediment deposited on the ocean floor gradually converted to solid marine sedimentary rock through lithification, which usually involves both compaction and cementing of sediments at temperatures under 200 ºC (390 ºF).
- Oil and natural gas are derived from the remains of marine organisms and their deposits are found in the pore spaces of marine sedimentary rock.
- Great quantities of sand, gravel and shells are mined from the near-shore, shallow ocean bottom, especially near coastal cities, primarily for road construction and cement and concrete production. In some locales, valuable metallic (e.g., tin) and non-metallic (e.g., diamonds) minerals occur mixed with coastal sands as lag concentrates.
- Since 1958, the United Nations has been working on international policies concerning the exploitation of seabed resources, including fuels and minerals. Progress has been slow because of long-standing conflicts among the more than 150 nations involved in negotiations.

Enduring Ideas

- Marine sediments, organic and inorganic particles that accumulate on the sea floor, are classified by size and source. Terminal velocity increases with increasing particle size so that larger-size sediment settles to the ocean bottom faster than smaller-size sediment.
- Lithogenous sediment is derived from weathering and erosion of rock and most commonly is composed of silicate minerals. It accounts for about 75% of all marine sediments.
- Biogenous sediment, encompassing the excretions, secretions, and skeletal and shell remains of organisms, is composed of calcium carbonate or silica. Hydrogenous sediment consists of particles that are chemically precipitated from seawater. Where evaporation rates are high and precipitation low, salts precipitate from seawater.
- Cosmogenous sediment includes particles of extraterrestrial origin such as meteorite fragments that survived the journey through the atmosphere without burning up from friction, and through the ocean without dissolving.
- Marine sedimentary deposits include deltas, wetlands, and turbidites. Calcareaous ooze and siliceous ooze are deep-ocean (pelagic) deposits. Geological processes operating as part of the rock cycle, convert these deposits into sedimentary rock layers.

Review

1. What is the relationship between the degree of sorting of a sediment deposit and the range in size exhibited by the individual sediments? Are well-sorted sediments more likely to be found in the continental margins or on the deep-ocean floor? Explain your answer.
2. Define terminal velocity and describe the relationship between terminal velocity and particle size of a given substance in a specified medium such as water or air.
3. What are the four major groups of marine sediments? Briefly describe the source of each sediment type.
4. What are the two most common types of biogenous sediment on the ocean floor?
5. A deep-ocean manganese nodule is an example of what type of marine sediment? Describe the ocean environment in which the largest nodules are found.
6. Compare the locations and accumulation rates of neritic and pelagic marine sediment deposits.
7. What are the two types of pelagic deposits that are more than 30% biogenous by weight? Which type is unlikely to occur on the ocean floor at depths greater than about 4500 m (14,800 ft) below sea level?
8. Define lithification and describe the processes involved in lithification.
9. How are economically important deposits of methane (natural gas) generated?
10. What is the significance of the U.S. exclusive economic zone (EEZ)?

Critical Thinking

1. What sequence of events could account for the presence of relatively coarse sediment on the deep-ocean floor?
2. Why are quartz grains and clay particles the most common lithogenous sediments transported to the ocean by rivers and streams?
3. How might a lower pH of seawater affect the abundance of calcareous sediment that accumulates on the sea floor?
4. Only a small fraction of the particles entering Earth's atmosphere from outer space reach the ocean floor as cosmogenous sediment. Explain why.
5. What is the significance of the carbonate compensation depth (CCD)? How might ocean acidification affect the CCD?
6. What would explain the near absence of sediment on some portions of the deep-ocean floor?
7. How might abyssal storms affect the interpretation of past climatic episodes obtained from analysis of deep-sea sediment cores?
8. Is the composition of biogenous sediment on the ocean floor always an indicator of the type of marine organisms living in the surface waters directly above? Explain your response.
9. List the advantages and disadvantages of mining in an EEZ versus the open-ocean beyond the EEZ.
10. Explain why the bulk of lithogenous sediment originates at relatively high land elevations.

ESSAY: Heinrich Events

Cores extracted from sediment deposits on the deep ocean floor of the North Atlantic Ocean contain layers of pebbles and other coarse rock fragments, called *Heinrich layers*, named for the German researcher Hartmut Heinrich who first described them in 1988. The size of these sediments is in marked contrast to the typically fine-grained pelagic sediments that blanket the deep-ocean floor. Heinrich sediments apparently were released by unusually large numbers of melting icebergs in the North Atlantic Ocean and settled to the sea floor (Figure 1). Heinrich layers were associated with massive discharges of icebergs from northeastern Canada (Hudson Bay and the St. Lawrence River) during the last Ice Age. From analysis of North Atlantic sediment cores, Heinrich identified six layers dating from the past 100,000 years. The layers were deposited during episodes of exceptionally low sea surface temperatures.

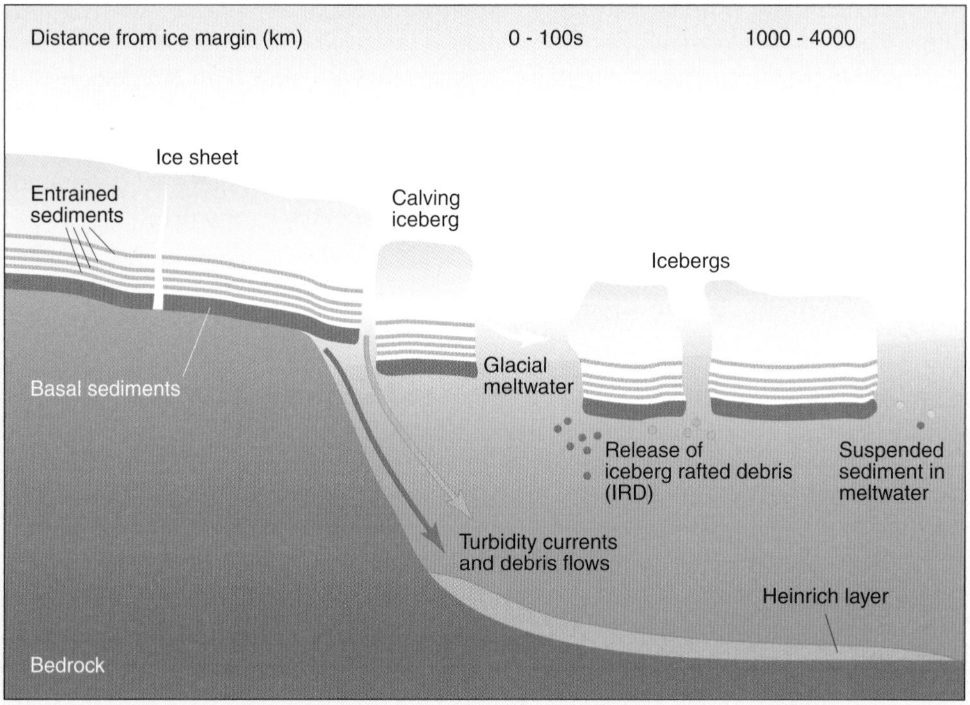

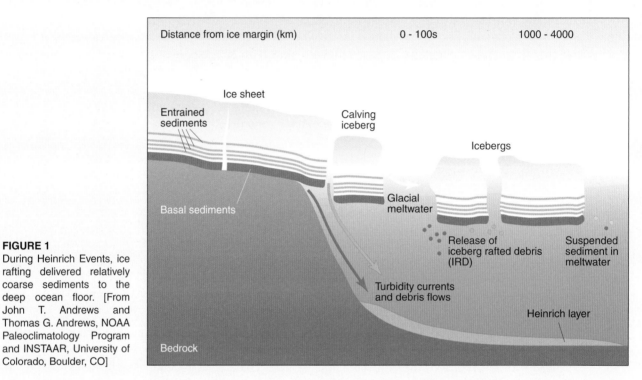

FIGURE 1
During Heinrich Events, ice rafting delivered relatively coarse sediments to the deep ocean floor. [From John T. Andrews and Thomas G. Andrews, NOAA Paleoclimatology Program and INSTAAR, University of Colorado, Boulder, CO]

During the last Ice Age, Earth's mean surface temperature may have fluctuated as much as 2 to 5 Celsius degrees (5 to 10 Fahrenheit degrees) with a regular periodicity, with deposition of Heinrich layers coinciding with the close of a cold episode. In addition to producing Heinrich layers, the massive influx of icebergs chilled the surface waters and the addition of large quantities of fresh water from the melting ice may have altered the ocean's deep current system that exerts a major influence on the climate of northern Europe (Chapter 6). Another effect of injections of melting icebergs was a rapid rise in sea level that drowned Caribbean coral reefs.

The cause of Heinrich events is still disputed but Doug MacAyeal of the University of Chicago has proposed an intriguing explanation. Scientists have known for several decades that major fluctuations in the planet's glacial ice cover arise from regular variations in Earth's orbital parameters that control the seasonal and latitudinal distribution of incoming solar radiation (Chapter 12). During episodes when Earth's orbital parameters favored warmer winters and cooler summers in central and northern Canada, some of the winter snows persisted year-round. In time, these climatic conditions gave rise to the Laurentide ice sheet that thickened and eventually spread over much of what is now Canada and the northern tier of the U.S. According to MacAyeal, over Hudson Bay the ice sheet initially was frozen to the bedrock but things changed as the ice sheet thickened. The growing ice sheet acted as an insulating blanket over Earth's surface and trapped enough geothermal heat conducted from Earth's interior that the bottom layer of the ice sheet thawed. Loaded with rock debris, this lubricated ice flowed rapidly into the North Atlantic releasing a massive surge of icebergs responsible for the Heinrich layers, changes in ocean circulation, and sea level rise. The now thinner ice sheet then re-froze to the bedrock and the cycle began anew as the ice sheet again thickened.

Smaller, shorter, and more frequent iceberg discharges into the North Atlantic Ocean are called *Dansgaard-Oeschger events* (named for the paleoclimatologists Willi Dansgaard and Hans Oeschger) or "flickers" because of their relatively short period. The Greenland ice core record contains indications of some 23 Dansgaard-Oeschger events during the period from 110,000 to 15,000 years ago. There is more on the Dansgaard-Oeschger events and their climatic implications in the first Essay of Chapter 12.

ESSAY: Future of the Nile Delta

Rains that fall in the eastern and central African highlands feed the White Nile (flowing from Lake Victoria, Uganda) and the Blue Nile (flowing from Lake Tana, Ethiopia). The two rivers merge at Khartoum, Sudan, as the Nile River flows north toward the Nile Delta and Mediterranean Sea. The scene in Figure 1, photographed from a NASA Space Shuttle looking northwest, shows the Nile River Delta, stretching about 160 km (100 mi) north and south, with 240 km (150 mi) of coastline bordering the Mediterranean. Just north of Cairo, Egypt, at the southern edge of the delta, the Nile splits into two branches (distributaries), the Rosetta flowing to the northwest and the Damietta flowing to the northeast, each about 240 km (150 mi) long. Nile water disperses across the delta through a fan-shaped network of smaller rivers, canals, and irrigation ditches. Today, the combined length of these delta waterways is more than 10,000 km (6200 mi).

FIGURE 1
The Nile River Delta. Photograph taken from NASA Space Shuttle in May 1996. [Courtesy of NASA]

For thousands of years, arid Egypt has depended on the Nile River for water to irrigate crops growing in the floodplain and delta. Mean annual rainfall is only 100-200 mm (3.9-7.9 in.) and falls mostly in winter. Suspended silt transported and deposited by the Nile and its branches during annual floods (August to October) provided the delta with the most fertile soils on the African continent. It is for this reason that the Greek historian Herodotus (*ca.* 484-425 BCE) described Egypt as "the gift of the Nile."

But the pressures of a rapidly growing population, especially during the 20th and 21st centuries, increasingly diverted Nile waters for agriculture. To meet the demand for irrigation water, and to control flooding and create hydroelectric power, Egypt constructed the 111 m (364 ft) high Aswan High Dam, located north of the Egypt-Sudan border and about 683 km (423 mi) directly south-southeast of Cairo (Figure 2). Construction took 10 years and the dam was completed in 1970. The huge reservoir behind the dam, Lake Nasser, can hold a two-year supply of the Nile's average annual flow and makes possible year-round irrigation, supporting two crops each year. The Aswan High Dam releases an average of 55 cubic km of water per year of which about 46 cubic km are diverted for irrigation. Before the Aswan High Dam, about 38% of the river's average annual flow reached the Mediterranean Sea but today so much is diverted for irrigation that very little reaches the Sea. Nile water now irrigates more than 3.3 million hectares (8.2 million acres) of cropland and this number is projected to climb to 4.6 million hectares (11.4 million acres) by 2020.

One consequence of diverting the Nile for irrigation is disrupting the natural balance between constructive and destructive forces. Prior to damming, the delta soils compacted naturally and the land subsided, but this was compensated by suspended silt that spread over the delta during floods. Prior to the Aswan Dam, perhaps 90% of the silt transported by the river spread over the delta, but today, almost all the silt is trapped in Lake Nasser. With much less sediment input to the delta, most of the delta is subsiding (nearly 1 cm per year in some areas near the coast), erosive forces are dominating, and the shoreline is much more vulnerable to wave erosion and is retreating—in some places more than 150 m (500 ft) per year. The Nile once delivered perhaps 100 million tons of silt to the Mediterranean, readily compensating the loss due to wave and current erosion along the coast. About 30% of the delta is now less than 1 m above sea level, and salt water is beginning to intrude the northern portion of the delta, contaminating well water up to 30 km (19 mi) inland.

During floods, the pre-dam Nile also delivered nutrients (phosphorus and nitrogen compounds) in solution as well as attached to sediments that fertilized the agricultural lands of the valley and delta. Nutrients also washed into the Mediterranean Sea, providing the primary production for phytoplankton blooms that fed a valuable fishery. Soon after the Aswan High Dam began operating in 1964, the fishery collapsed. In less than a decade, catches of sardines, fish, and shrimp

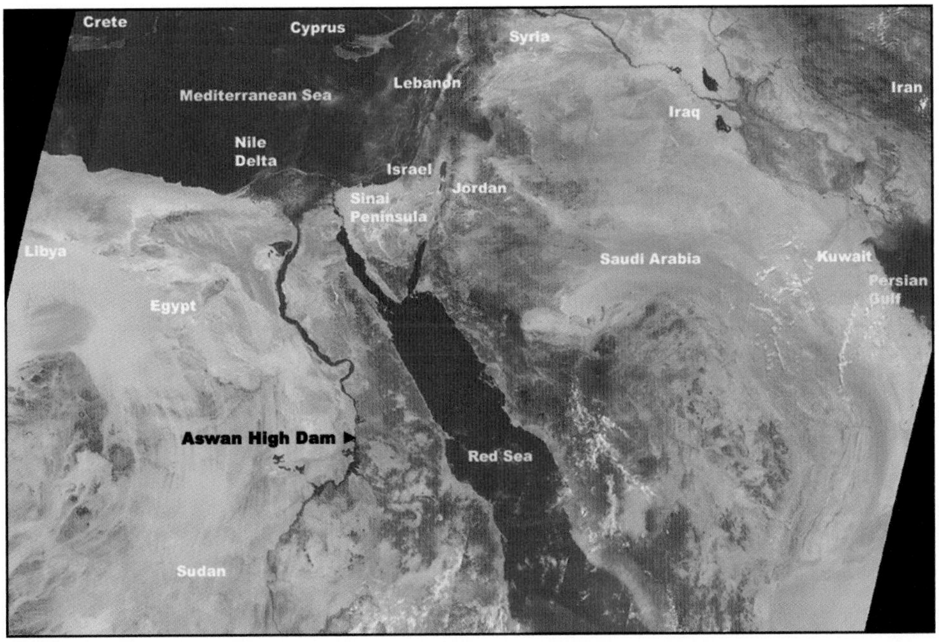

FIGURE 2
Map of Egypt and surrounding area. The Aswan High Dam is labeled, located just north of Lake Nasser. [NASA/GSFC/JPL, MISR Science Team]

plummeted nearly 80%. However, by the early 1980s, the eastern Mediterranean fishery began to recover, especially finfish such as grouper and pollock. Although the fish catch eventually returned to pre-dam levels, the slower recovering shrimp catch remains below pre-dam levels. Why the dramatic turn-around?

Reporting in the March-April 2004 issue of *American Scientist*, Scott W. Nixon, an oceanographer at the University of Rhode Island, argues that human activity was not only responsible for the decline of the eastern Mediterranean fishery (by damming the Nile) but also for the fishery's recovery. As natural fertilizer was greatly reduced, increasing amounts of synthetic (chemical) fertilizer were applied to the Nile Delta to restore and maintain soil fertility. In addition, increasing amounts of human waste from the rapidly growing population, in 2010 estimated to be 80.5 million with a third living in the delta region, added to the nutrient load that washed into the eastern Mediterranean. According to Nixon, fertilizer application and sewage outfalls supplied "more than enough nutrients to replace those captured by the Aswan High Dam."

Compounding the problem of land subsidence in the Nile Delta is climate change (Chapter 12). Warming causes seawater to expand, raising the level of the Mediterranean Sea. According to one estimate, if sea level rises 1 m (possible by 2050), one-third of the delta could be lost. The Egyptian government responded to this threat with several megaprojects intended to expand significantly the nation's habitable area. The Toshka Project is perhaps the most ambitious. Well on the way toward its scheduled completion in 2017, the project will divert 10% of Egypt's allocation of Nile water to 0.2 million hectares (0.5 million acres) of desert. (Only 2.92% of Egypt is arable land.) A huge pump with a discharge capacity of 1.2 million cubic m per hr draws water from Lake Nasser into a massive irrigation system, doubling the region's arable land and potentially accommodating a human population of more than 16 million in this formerly desolate area.

The complex interplay of events in the Nile Delta illustrates how the Earth system involves the interactions of many subsystems in response to both human and natural forces. To meet Egypt's growing demand for water and food in an arid climate, the Nile River was dammed, altering the natural balance of forces and thereby increasing the delta's vulnerability to land subsidence, coastal erosion, saltwater intrusion, and disruption of the eastern Mediterranean fishery. Compounding matters is large-scale climate change that is raising sea level which has required large-scale action.

ESSAY: Burgess Shale: A Glimpse into Ancient Marine Life

The earliest known fossil record of life on Earth dates back to the Precambrian Era, around 3.5 billion years ago, about a billion years after Earth formed and half a billion years after the ocean formed. Over the next 2 billion years, life consisted of simple soft-bodied unicellular marine organisms. Then, 650 million years ago near the close of the Precambrian, the earliest hard-shelled animals appeared. About 550 million years ago, in an evolutionary big-bang known as the *Cambrian Explosion*, a broad spectrum of complex multi-cellular life forms evolved in just 10 to 30 million years, a mere blink of an eye in the perspective of geologic time. While these organisms were still primarily invertebrates, this event represented a huge expansion of marine biodiversity and included the ancestors of shellfish, corals, crustaceans, and other inhabitants of the sea as well as creatures that would disappear entirely. What caused the Cambrian Explosion is still debated, although it is known that the transition between the Precambrian and Paleozoic Eras was a time of extreme fluctuation in climate.

A lack of fossils has hindered a detailed study of the Cambrian Explosion. Two conditions favor preservation for fossilization: rapid burial and hard parts (e.g. shells). Without hard parts, fossilization is only possible through rapid burial, which slows and prevents decay and, simultaneously, hinders scavenging. Almost none of the early unicellular life forms from the Precambrian appear in the fossil record. In fact, despite nearly continuous sedimentation and burial in the ocean, even fossils of the diverse community of animals from the Cambrian Explosion are rare, occurring in very few unique rock outcrops. One of these outcrops is that of the *Burgess Shale*, consisting of layers of shale and mudstone located high in the Canadian Rockies near the town of Field in eastern British Columbia. The Burgess Shale contains an abundance of fossils of soft-bodied animals, providing a rare window into marine life about 40 million years after the Cambrian Explosion. The Burgess Shale has yielded tens of thousands of unique specimens representing 170 species, the majority are benthic (sea-floor dwelling) organisms and many no longer exist. The most remarkable feature of the Burgess Shale fossils is that 60% to 80% of the fossils found were soft bodied creatures. How were they preserved?

During the Cambrian Period, when life was still restricted to the ocean and the land was barren, the paleo-North American continent was located in the tropics astride the equator. The continental margin of the paleo-continental coast was a warm, shallow tropical sea with a great algal reef next to a steep underwater cliff (or escarpment). The cliff, now known as the Cathedral Escarpment, fell away hundreds of meters into the ocean and, between it and the reef, sediment accumulated in submarine mud banks. In and on the mud banks lived a rich and diverse community of marine organisms consisting mostly of large, soft-bodied benthic invertebrates including coelenterates, echinoderms, mollusks, worms, and sponges. Among them were polychaete annelids, bristly scavenging worms (Figure 1) and more exotic spiny velvet worm creatures (Figure 2).

Periodically the accumulated mud became unstable and flowed down the escarpment as turbidity currents (described earlier in this chapter). Turbidity currents transported the animals in a slurry down the slope to the base of the reef escarpment, where most were buried fast and deep enough that they were sealed off from scavengers and decomposing bacteria, and preserved as fossils. This process was repeated many times, building a 10,000-m (33,000-ft) sequence of fine-grained, fossil-rich layers of sediment. In time, these sediments were lithified to shale and mudstone. Beginning about 175 million years ago (during the Jurassic Period), stresses associated with mountain building elevated and transported (via thrust faulting) the fossil beds from their ocean burial ground many kilometers eastward and upward. Fortunately, thrust faulting carried the shale beds above and out in front of the region where tectonic forces were building the Rocky Mountains. Otherwise, the shale would have been metamorphosed and the fossils destroyed. The Burgess Shale is located at the eastern edge of the Canadian Rockies just west of the Alberta plains in Canada's Yoho National Park. Following their close escape from tectonic metamorphism, these ancient shale and mudstones, with a fabulous treasure of fossils, were gradually exposed by erosive agents including glaciers, wind, running water, landslides, avalanches, and human activity.

In 1886, R.G. McConnell of the Geological Survey of Canada was the first geologist to visit the Burgess Shale site and described fossils he found as an "odd shrimp." Almost 100 years later, scientists identified the "odd shrimp" as the molted claw of a giant predator. Charles D. Walcott (1850-1927) heard reports about "stone bugs" (trilobites) and visited the site in 1909 while the director of the Smithsonian Institution (1907-1927), after he had served as director of the U.S. Geological Survey (1894-1907). Walcott is credited with discovering the Burgess Shale (named after nearby Mt. Burgess in the Canadian Rockies). During his visits, he collected many fossils, unlike any he had seen before, for the Smithsonian

FIGURE 1
Burgess shale fossil of a Canadia spinosa (a polychaete annelid). [Reproduced with permission from the Smithsonian Institution, Museum of Natural History, Department of Paleobiology]

FIGURE 2
Burgess shale fossil of a Hallucigenia sparsa (an onychophoran). [Reproduced with permission from the Smithsonian Institution, Museum of Natural History, Department of Paleobiology]

and attempted to classify them. Eventually, he identified 100 of the 170 recognized species. At present more than 65,000 specimens are housed at the Smithsonian Institution's National Museum of Natural History in Washington, DC.

In the late 1960s, paleontologist Harry Whittington of Cambridge, England, along with his graduate students Derek Briggs and Simon Conway Morris, began a thorough study of the Burgess Shale fossils (including those in storage). They were unable to classify all the fossil animals using the modern classification system and described those they didn't know as "unknown phyla," implying the diversity of marine fauna of half a billion years ago was greater than today. The specimens are odd and enigmatic creatures with elongated and gripping feeding devices and flattened and many-limbed body configurations, which would have been unknown if not for the Burgess Shale. None of their shapes seem any more viable than another, leaving it to the imagination to decide what caused so many species to become extinct while our ancestors were prolific. Continuing work has classified more fossil species, however, just as many new unclassifiable forms have been found. The biodiversity represented by the Cambrian Explosion remains greater than exists in the modern ocean.

Based on the Burgess Shale record, by the Middle Cambrian Period many species that appeared during the Cambrian Explosion suddenly disappeared, leaving few descendants based on the subsequent geologic record. Many scientists attributed the disappearance of the unique Burgess Shale-type species to a major extinction. This view changed in May 2010 when an international team of scientists led by Belgian paleontologist Peter Van Roy reported on fossils found near the Atlas Mountains and the city of Zagora, Morocco. These fossils (numbering about 1500 specimens) date from 480 million years ago and include Burgess Shale-type species. It now appears that the disappearance of the unique fossil fauna at the Burgess Shale site was the consequence of an absence of preservation rather than part of a mass extinction.

Similar fossils also have been found in Cambrian shale deposits near the town of Chengjiang in China's Yunnan Province. This mud/shale is about 15 million years older than the Burgess Shale and over 100,000 specimens have been collected there. Similar species found in both places suggests that the Cambrian Explosion was ocean wide.

The Burgess Shale is an exceptional example of fossil soft-body preservation and records a diversity of rare animals. Its beautifully preserved fossils constitute a valuable snapshot of Cambrian life, far more complete than deposits containing fossils of only hard parts. In 1981, the Burgess Shale was declared a UNESCO World Heritage Site.

CHAPTER 5

THE ATMOSPHERE AND OCEAN

The last of the upper limb of the Sun just before diving below the horizon. [NOAA photo by Commander John Bortniak, NOAA Corps (ret.)]

Case in Point

At middle latitudes, prevailing winds blow from west to east so that the moderating influence of the North Atlantic Ocean on climate is much more evident over Western Europe than Eastern North America. It is interesting to speculate on what would happen to the climate of Western Europe if this moderating influence were to weaken as it has at times in the past.

In Western Europe, the air temperature contrast between summer and winter is less than it is over most of North America. For reasons discussed in Chapter 3, sea surface temperatures (SST) change relatively little through the course of a year and this stable SST regime dampens the summer-to-winter temperature contrast of air flowing over the ocean to downwind Western Europe. Whereas summer average air temperatures are somewhat lower, winter average air temperatures are milder in Western Europe compared to upwind North America. The northward-moving warm Gulf Stream parallels the U.S. coastline from Florida to the Mid-Atlantic States and then the current turns east and northeastward across the North Atlantic. In winter, the relatively warm ocean surface moderates cold air masses as they surge from polar areas southeastward toward the British Isles and Western Europe.

Compare, for example, January and July temperatures at Cork, Ireland (51 degrees, 54 minutes N) and Saskatoon, Saskatchewan, Canada (52 degrees, 8 minutes N). At both places on average, January and July are the coldest and warmest months of the year respectively. Although located at about the same latitude, the two cities have markedly different climates. The average temperature contrast between July and January is about 36.5 Celsius degrees (65.7 Fahrenheit degrees) at continental Saskatoon but only about 11 Celsius degrees (20 Fahrenheit degrees) at maritime Cork. The reduced seasonal contrast at Cork is mostly due to much higher winter temperatures. January average temperature is 4.5 °C (40.1 °F) at Cork but −18.5 °C (−1.3 °F) at Saskatoon. July average temperatures are not much different at the two locations with 18 °C (64.4 °F) at Saskatoon and 15.5 °C (59.9 °F) at Cork.

Historical records indicate that during periods in the past, winters were much colder in the British Isles and other parts of Western Europe. The *Little Ice Age*, lasting from about CE 1400 to 1900, was one such period when relatively cold winters were more frequent than today. Sea ice cover expanded over the North Atlantic, mountain glaciers advanced, and growing seasons shortened over Western Europe bringing erratic harvests and much hardship for many people. Evidence of other cold episodes in the North Atlantic comes from climate signals unlocked from deep-sea sediment cores and annual ice layers in the Greenland ice sheet (Chapter 12). The past 10,000-year epoch (since the close of the last Ice Age) was punctuated by multi-century cold episodes that began abruptly (within decades) and occurred about every 1500 years. Across Northern and Western Europe, average winter temperatures during cold episodes were as much as 7 Celsius degrees (13 Fahrenheit degrees) lower than today.

A possible cause of the Little Ice Age and prior cold episodes in Western Europe is periodic weakening of the North Atlantic circulation. The Gulf Stream is part of a planetary-scale thermohaline circulation that transports enormous amounts of heat throughout the world ocean (Chapter 6). The flow in the Gulf Stream is at least 500 times that of the Amazon River. Researchers at Lamont-Doherty Earth Observatory of Columbia University propose that runoff of unusually great amounts of fresh water into the North Atlantic alters the salinity (and density) of surface ocean waters, produces less dense deep water, and weakens the thermohaline circulation. A weaker circulation would cause winter average temperatures in Western Europe to plunge abruptly by perhaps 5 Celsius degrees (9 Fahrenheit degrees). For Cork, this would mean winters more like those experienced at Spitsbergen, some 1000 km (600 mi) north of the Arctic Circle.

Driving Question:

What role does the ocean play in the long-term average state of the atmosphere?

To this point in our investigation of the ocean in the Earth system we have emphasized primarily the physical properties of ocean water and the ocean basin. In this chapter, our principal focus shifts to the flow of energy into and out of the Earth system especially as it involves its fluid subsystems and those processes operating at the interface between the ocean and atmosphere. In doing so, it quickly becomes apparent that the ocean plays a key role in the global radiation budget, the transport of heat between Earth's surface and atmosphere, the flow of heat from the tropics to higher latitudes, and the development of storm systems. For these reasons, the ocean is a major player in the state of the atmosphere (weather) and the climate system.

In this chapter, we examine radiational heating and cooling of the **Earth-atmosphere system** (Earth's surface plus overlying atmosphere), the interaction of incoming solar radiation with the atmosphere, ocean, and continents, the flow of infrared radiation to space, and the greenhouse effect. In response to heating imbalances within the Earth-atmosphere system, temperature gradients develop and heat is transferred via phase changes of water, conduction and convection, exchange of air masses, and ocean currents. Atmospheric and oceanic circulation operating at various spatial and temporal scales transports heat from where it is warmer to where it is colder. We begin our discussion by distinguishing between weather and climate.

Weather and Climate

Weather and climate are closely related concepts. We can think of **weather** as the state of the atmosphere at a particular place and time described in terms of such variables as temperature, precipitation, cloud cover, and wind speed. A place and time must be specified when describing weather because the atmosphere is dynamic; that is, its state is always changing from one place to another and with time. At the same hour the weather may be cold and snowy in Philadelphia, sunny and warm in Dallas, and cool and rainy in Seattle. *If you don't like the weather, wait a minute* is an old saying that is not far from the truth in many places. From personal experience, we know that tomorrow's weather may differ considerably from today's weather.

Climate is popularly defined as weather at a particular place averaged over a specific interval of time. By international convention, average values of weather elements such as temperature or precipitation are computed over a 30-year period beginning with the first year of a decade. At the close of a decade the averaging period is shifted forward ten years. As of this writing, the official averaging period is 1981-2010. Thirty-year average monthly and annual temperatures and precipitation totals are commonly used to describe climate. Other useful climatic parameters include average seasonal snowfall, length of growing season, and frequency of thunderstorms. Ultimately, climate governs the supply of fresh water, the geographical distribution of plants and native animals, and the type of crops that can be cultivated.

Climate encompasses extremes in weather in addition to average values of weather elements. Tabulation of extreme values usually covers the entire period of record (or at least for the period when the weather station was at the same location). Specifying the frequency of weather extremes provides information on the variability of climate at a particular place and gives a more complete and useful description of climate. Climate scientists study not only trends in average temperature and precipitation, but also changes in the frequency of extreme events such as excess heat, cold, drought, or rainfall.

Heating and Cooling Earth's Surface

As Earth orbits the Sun, its sunlit atmosphere and surface are absorbing solar radiation. Absorption of solar radiation heats the Earth-atmosphere system. At the same time the entire planet is emitting infrared radiation to space, which cools the Earth-atmosphere system. Over the long term, radiational cooling of the planet essentially balances radiational heating of the planet so that Earth remains in radiative equilibrium with surrounding space.

The Sun emits a band of electromagnetic radiation having wavelengths mostly between 0.25 and 2.5 micrometers. (One micrometer is a millionth of a meter or about one-tenth the thickness of a human hair.) Solar radiation is most intense at a wavelength of about 0.5 micrometer, in the green of the visible portion of the electromagnetic spectrum (Figure 1.12). The Earth-atmosphere system, on the other hand, emits to space a broad band of electromagnetic radiation having wavelengths mostly between 4 and 24 micrometers, in the infrared portion of the electromagnetic spectrum. The peak intensity of infrared radiation emitted by Earth's surface is at a wavelength of about 10 micrometers. In this section, we take a closer look at radiational heating and cooling of the Earth system.

SOLAR RADIATION

Earth's motions in space govern daily and seasonal variations in the amount of solar radiation striking Earth's surface. Once every day, Earth completes one rotation on its axis. At any instant, half the planet is illuminated by solar radiation (day) while the other half is in darkness (night). The tilt of Earth's spin axis is responsible for the seasons. Earth's spin axis is tilted 23 degrees 27 minutes from the perpendicular to the plane defined by the planet's annual orbit about the Sun (Figure 5.1). During Earth's annual revolution about the Sun, its spin axis remains in the same alignment with respect to the extremely distant background stars (the North Pole always pointing toward *Polaris*, the North Star) while its orientation to the Sun changes continually. Simply put, the Northern Hemisphere tilts away from the Sun in fall and winter and toward the Sun in spring and summer. In the Northern Hemisphere, astronomical winter begins on the winter solstice, on or about 21 December, and ends on the first day of spring (the vernal or spring equinox), on or about 21 March. Summer begins on the summer solstice, about 21 June, and continues until the first day of autumn (the autumnal

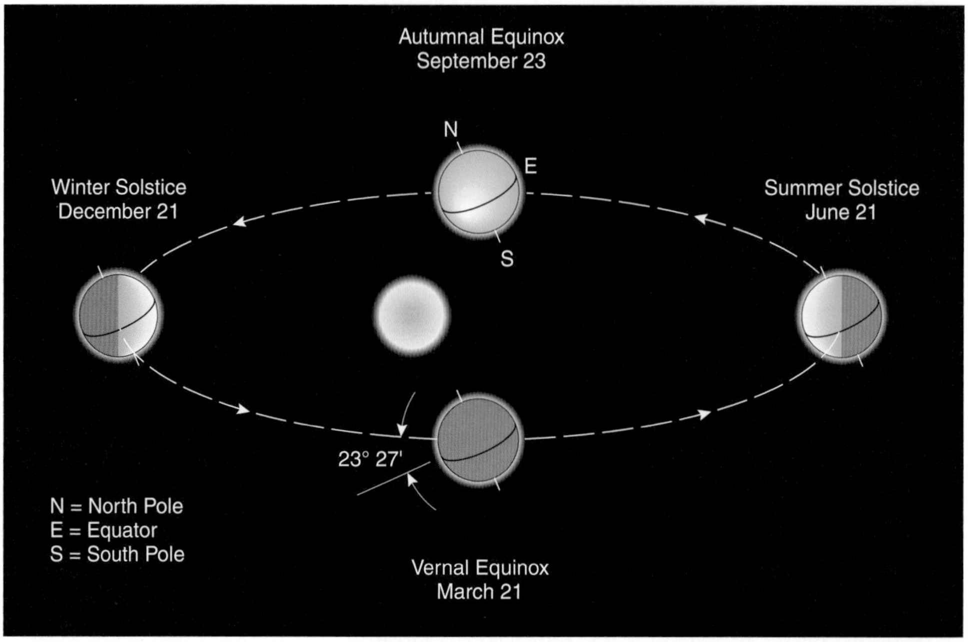

FIGURE 5.1
The seasons change because Earth's equatorial plane is inclined (at 23 degrees, 27 minutes) to the orbital plane. Seasons are given for the Northern Hemisphere. Note that the eccentricity of Earth's orbit is greatly exaggerated.

equinox), on or about 23 September. Precise dates of solstices and equinoxes vary because Earth completes one orbit of the Sun in 365.24 days, necessitating the leap year adjustment.

Regular changes in local maximum solar altitude and length of daylight accompany the annual periodic changes in the planet's orientation to the Sun. These changes, in turn, affect the amount of solar radiation that strikes Earth's surface at a particular location. **Solar altitude** is the angle of the Sun above the horizon and varies from 0 degrees (at sunrise or sunset) to as much as 90 degrees (where and when the Sun is directly overhead), and *length of daylight* is the number of hours and minutes between sunrise and sunset. At middle and high latitudes, the altitude of the noon Sun is higher, daylight is longer, and solar radiation is greater in summer than in winter.

With clear skies, the intensity of solar radiation striking Earth's surface at a point varies directly with solar altitude. With increasing solar altitude, more solar energy strikes a unit area of Earth's surface in a unit of time (Figure 5.2). Earth is so far from the Sun (a mean distance of 150 million km or 93 million mi) that solar radiation reaches the planet as nearly parallel beams of essentially uniform intensity. But the almost spherical Earth presents a curved surface to incoming solar radiation so that the noon solar altitude is higher in the tropics than at higher latitudes. Greater solar altitudes in the tropics translate into more intense radiation and higher temperatures at Earth's surface.

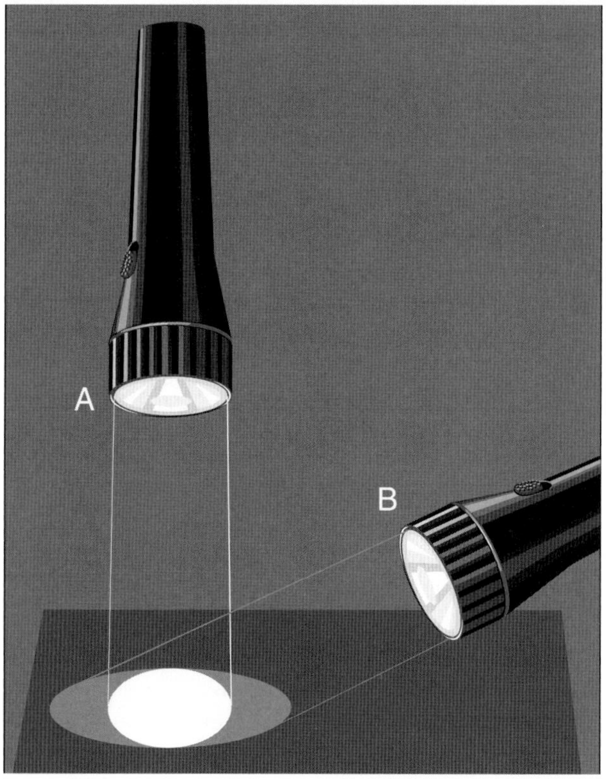

FIGURE 5.2
The intensity of solar radiation striking Earth's surface per unit area varies with the solar altitude. Consider this analogous situation: (A) A flashlight beam shines on a horizontal surface most intensely when the flashlight shines from directly overhead (analogous to a solar altitude of 90 degrees). (B) At an angle decreasing from 90 degrees, the flashlight beam spreads over an increasing area of the horizontal surface so that the light is less concentrated (less radiational energy received per unit area).

The daily path of the Sun through the sky on the solstices and equinoxes is shown schematically in Figure 5.3 for the equator, a middle-latitude Northern Hemisphere location, and the North Pole. At the middle-latitude location and North Pole, the altitude of the noon Sun is greatest on the summer solstice but at the equator the noon solar altitude is maximum on the equinoxes (when the Sun is directly overhead). The only places on Earth where the solar altitude ever reaches 90 degrees during the course of a year are within the latitude belt bounded by the Tropic of Cancer (23.5 degrees N) and the Tropic of Capricorn (23.5 degrees S).

At most latitudes that experience day and night each day, daylight is shortest on the winter solstice and longest on the summer solstice. The equator experiences essentially equal periods of daylight and darkness every day of the year. On the equinoxes, the length of daylight and night are about the same (12 hrs) everywhere on the planet except at the poles. Daylight is longer than night during spring and summer, and daylight is shorter than

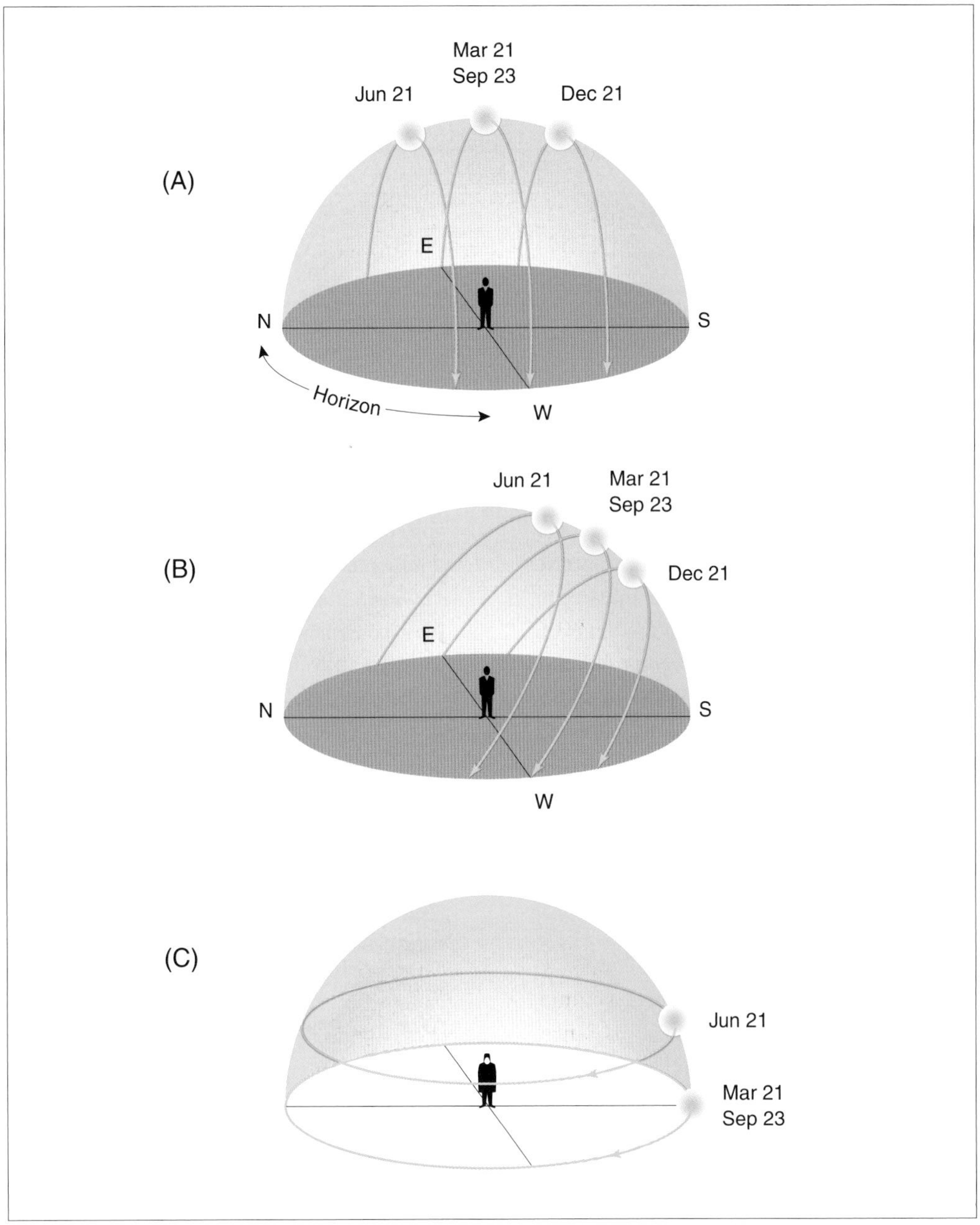

FIGURE 5.3
Path of the Sun through the sky on the solstices and equinoxes at (A) the equator, (B) a middle latitude location in the Northern Hemisphere, and (C) the North Pole.

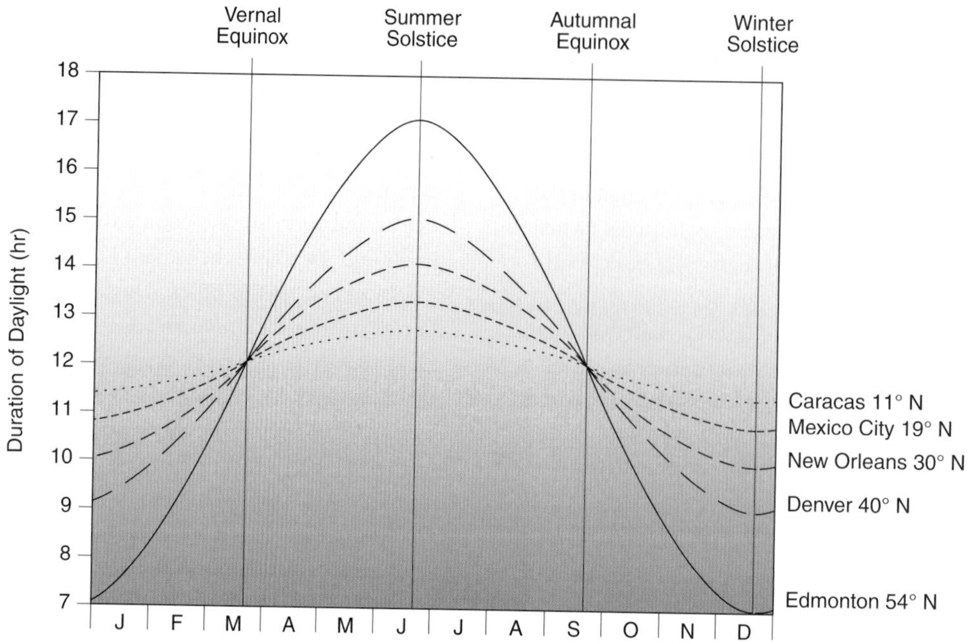

FIGURE 5.4
Variation in the number of hours and minutes of daylight through the year increases with increasing latitude.

night during autumn and winter. In the absence of an atmosphere, the difference in length of daylight between the summer and winter solstices increases from zero at the equator to a maximum (24 hrs) at the Arctic and Antarctic Circles (Figure 5.4). Regular variations in maximum solar altitude and length of daylight through the year are ultimately responsible for changes in the receipt of solar radiation and monthly average temperatures (Figure 5.5). Little annual variation in maximum solar altitude and length of daylight in the tropics translates into relatively uniform monthly mean temperatures through the year. In the tropics, the temperature difference between day and night often is greater than the summer-to-winter temperature contrast. At middle and high latitudes, however, maximum solar altitudes and length of daylight vary considerably through the year and are responsible for marked contrasts between average summer and winter temperatures (i.e., seasons).

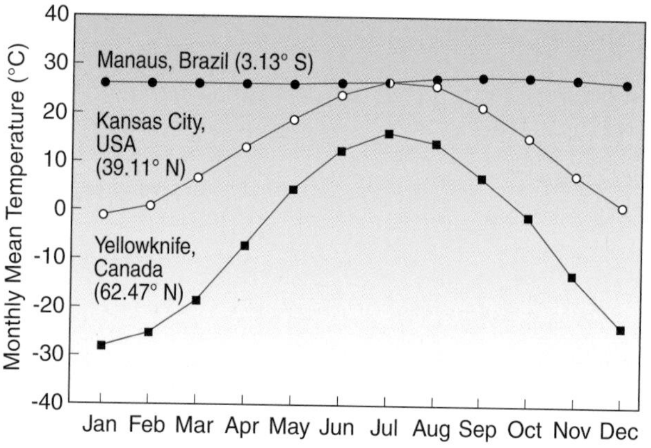

FIGURE 5.5
The seasonal contrast in monthly mean temperature generally increases with increasing latitude.

For reasons presented in Chapter 3, a large body of water such as the ocean influences the climate of downwind localities. Places downwind of the ocean experience a smaller contrast between average winter and summer temperatures and have a *maritime climate*. Places at the same latitude but well inland experience a greater temperature contrast between winter and summer and have a *continental climate*. Proximity to large bodies of water also affects the timing of the average warmest and coldest time of the year. Outside of the tropics, the annual temperature cycle lags (follows in time) the annual solar radiation cycle; that is, the warmest period of the year on average occurs after the summer solstice while the coldest period of the year occurs after the winter solstice. The Earth-atmosphere system takes time to adjust to seasonal changes in solar energy input. In the interior United States, the air temperature cycle lags the solar radiation cycle by an average of 27 days. But in coastal localities having a strong maritime influence (e.g., coastal California, Florida), the average lag time is up to 36 days.

Before modern navigational aids, mariners used the regular changes in the location of the Sun and other celestial bodies in the sky to locate themselves at sea. For an historical perspective on the challenges of navigating at sea, refer to this chapter's first Essay.

SOLAR RADIATION BUDGET

Solar radiation intercepted by Earth travels through the atmosphere and interacts with its component gases and aerosols. These interactions consist of scattering, reflection, and absorption. Solar radiation that is not absorbed or scattered or reflected back to space reaches Earth's surface where additional interactions occur.

With **scattering**, a particle disperses radiation in all directions: forward, backward, and sideways. Within the atmosphere, both gas molecules and aerosols (including the tiny water droplets and ice crystals that compose clouds) scatter solar radiation. Scattering explains the blue color of the clear daytime sky. Visible sunlight is made up of all colors (from violet at the short wavelength end of the solar spectrum to red at the long wavelength end). Air molecules preferentially scatter short wavelength visible light (blue-violet), causing it to enter the observer's eyes from many directions.

Reflection is a special case of scattering in which a large surface area redirects radiation in a backward direction. The fraction of incident radiation that is reflected by a surface is known as the **albedo** of that surface, that is,

$$\text{albedo} = \frac{\text{(reflected radiation)}}{\text{(incident radiation)}} \times 100\%.$$

Surfaces having a high albedo reflect a relatively large fraction of incident solar radiation and appear light in color. Surfaces having a low albedo reflect relatively little incident solar radiation and appear dark in color.

Within the atmosphere, the tops of clouds are the most important reflectors of incoming visible sunlight. Cloud top albedo depends primarily on cloud thickness and varies from under 40% for thin clouds to 80% or more for thick clouds. The average albedo for all cloud types and thicknesses is about 55%, and at any point in time, clouds cover about 60% of the planet. All other factors being constant, solar radiation striking the Earth's surface is more intense and surface air temperatures are higher when the daytime sky is clear rather than cloudy.

Scattering and reflection within the atmosphere alter the direction of solar radiation without its conversion to heat. **Absorption**, however, is a process whereby some of the radiation that strikes an object is converted to heat energy. Oxygen, ozone, water vapor, and various aerosols (including cloud particles) absorb solar radiation. Absorption by atmospheric gases varies with wavelength; that is, each gas absorbs strongly in some wavelengths and weakly or not at all in other wavelengths. Essential for life on Earth is the strong absorption of the Sun's ultraviolet (UV) radiation by oxygen and ozone (O_3) in the stratosphere, which shields organisms from exposure to potentially lethal intensities of UV. These absorption processes create the *stratospheric ozone shield*. For more on the stratospheric ozone shield and marine life, see this chapter's second Essay.

Solar radiation not scattered or reflected to space or absorbed by atmospheric gases or aerosols reaches Earth's surface where it is either reflected or absorbed. The portion that is not reflected is absorbed (i.e., converted to heat). High-albedo surfaces, such as snow-covered ground or pack ice, reflect a considerable amount of incident solar radiation whereas low-albedo surfaces, such as a water surface or an asphalt road, reflect much less incident solar radiation. The albedo of some common surfaces is listed in Table 5.1.

TABLE 5.1
Average Albedo (Reflectivity) of some Common Surface Types for Visible Solar Radiation

Surface	*Albedo (% reflected)*
Deciduous forest	15-18
Coniferous forest	9-15
Tropical rainforest	7-15
Tundra	15-35
Grasslands	18-25
Desert	25-30
Sand	30-35
Soil	5-30
Green crops	15-25
Sea ice	30-40
Fresh snow	75-95
Old snow	40-60
Glacial ice	20-40
Water body (high solar altitude)	3-10
Water body (low solar altitude)	10-100
Asphalt road	5-10
Urban area	14-18
Cumulonimbus cloud	90
Stratocumulus cloud	60
Cirrus cloud	40-50

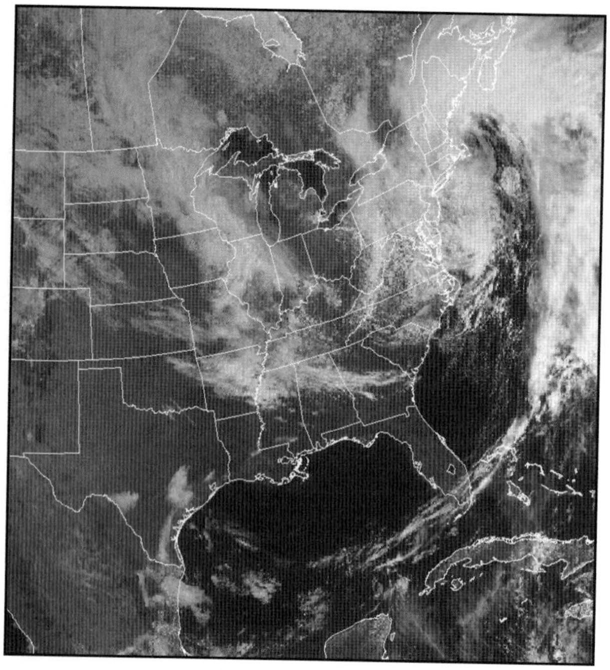

FIGURE 5.6
In this visible satellite image from 1 April 2011, the ocean surface appears dark because of its low albedo for visible solar radiation. White areas are clouds. [Courtesy of NOAA Aviation Weather Center]

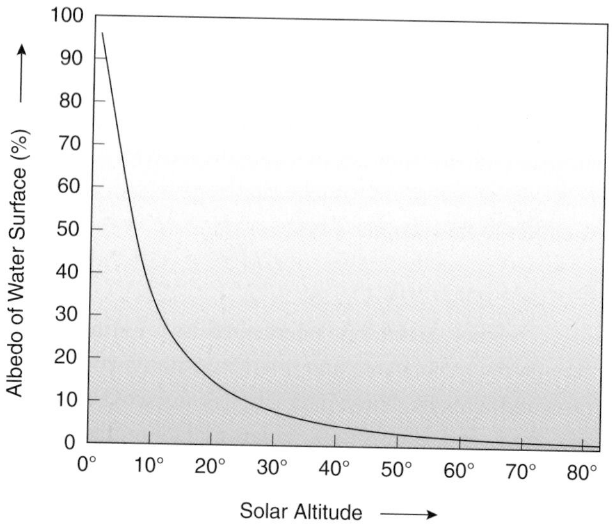

FIGURE 5.7
Under clear skies, the albedo of a flat and undisturbed water surface changes with solar altitude. A wave-covered water surface has a slightly higher albedo at high solar altitudes and a slightly lower albedo at low solar altitudes.

In the visible satellite image in Figure 5.6, the ocean surface appears dark because of its strong absorption of solar radiation, that is, its low albedo. The albedo of the ocean surface varies with the angle of the Sun above the horizon (solar altitude). Under clear skies, the albedo of a flat, tranquil water surface decreases with increasing solar altitude (Figure 5.7). The albedo is almost a mirror-like 100% near sunrise and sunset (when the solar altitude is near 0 degrees) but declines sharply as the solar altitude approaches 20 degrees. When the Sun is low in the sky (small solar altitude), light rays reflect off the water surface with little penetration. However, when the Sun is high in the sky, light rays penetrate the water to some depth and most of the sunlight is scattered below the surface with little scattered back to the atmosphere or space. With cloud-covered skies, only diffuse solar radiation strikes the water surface and the albedo varies little with solar altitude and is uniformly less than 10%. The roughness of a wave-covered water surface decreases its albedo. On a global basis, the albedo of the ocean surface averages only about 8%; that is, the ocean absorbs 92% of incident solar radiation. Considering that the ocean covers about 71% of the surface of the planet, the ocean is the principal sink (absorber) for solar radiation striking Earth's surface.

Measurements by sensors onboard Earth-orbiting satellites indicate that the Earth-atmosphere system reflects or scatters back to space on average about 30% of the solar radiation intercepted by the planet. This is Earth's **planetary albedo**. The atmosphere (i.e., gases, aerosols, clouds) absorbs only about 23% of the total solar radiation intercepted by the Earth-atmosphere system. In other words, the atmosphere is relatively transparent to solar radiation. The remaining 47% of solar radiation is absorbed by Earth's surface—mostly the ocean.

Earth's surface is the principal recipient of solar heating, and heat is transferred from Earth's surface to the atmosphere, which eventually radiates this energy to space. Hence, Earth's surface is the main source of heat for the atmosphere; that is, the atmosphere is heated from below. This is evident from the average vertical temperature profile of the troposphere (Figure 1.4). Normally, air is warmest close to the Earth's surface, and the temperature drops with increasing altitude, that is, away from the main source of heat.

SOLAR RADIATION AND THE OCEAN

Whereas the atmosphere is relatively transparent to (and absorbs little) solar radiation, the ocean absorbs most solar radiation within relatively shallow depths. As shown in Figure 5.8, the ocean's absorption of the visible portion of solar radiation is selective by wavelength. Water absorbs the longer wavelengths (i.e., reds and yellows) of visible light more efficiently than the shorter

FIGURE 5.8
Visible solar radiation is selectively absorbed by wavelength as it penetrates the surface waters of the open ocean.

wavelengths (i.e., greens and blues) so that green and blue penetrate to greater depths. Within clear, clean water, red light is completely absorbed within about 15 m (50 ft) of the surface, whereas green and blue-violet light may penetrate to depths approaching 250 m (800 ft). More green and blue light is scattered to our eyes, explaining the blue/green color of the open ocean. Suspended particles significantly boost absorption so that sunlight often is completely absorbed at shallower depths. In fact, some near-shore waters are so turbid (cloudy) that little if any sunlight reaches much below 10 m (35 ft). Suspended particles preferentially scatter yellow and green light, giving these waters their characteristic color.

The **photic zone** is the sunlit surface layer of the ocean, down to the depth where light is just sufficient for photosynthesis. The base of the photic zone is generally where the light is just 1% of the radiation incident on the surface. In clear ocean waters, this depth is usually from 100 to 200 m (330 to 650 ft) but is much shallower in highly productive or turbid waters. Although a small amount of light penetrates below the photic zone (into the so-called *twilight zone*), light is insufficient for plants to survive.

As the concentration of particles and dissolved organic matter in seawater increases, the color of light that penetrates deepest into the water shifts to yellow-green in coastal areas and to red in the most turbid estuarine waters. Hence, as light becomes dimmer with increasing depth, its color also changes. This color change affects plant production because each plant pigment is most efficient for a specific color of light. The combination of pigments in any type of phytoplankton determines its optimal depth distribution.

With some notable exceptions, marine life depends directly or indirectly on sunlight and organic productivity in the ocean's photic zone (Chapters 9 and 10). Even the diverse community of animals living at great depths on the ocean floor depends on organic particles produced within the photic zone that settle to the sea floor (Chapter 4); exceptions are organisms living near hydrothermal vents who depend upon *chemosynthesis* (Chapter 9).

INFRARED RADIATION AND THE GREENHOUSE EFFECT

If solar radiation were continually absorbed by the Earth-atmosphere system without any compensating flow of heat out of the system, Earth's surface temperature would rise steadily. Eventually, life would be extinguished and the ocean would boil away. Actually, global air temperature changes very little from one year to the next. **Global radiative equilibrium** keeps the planet's temperature in check to some extent; that is, emission of heat to space in the form of infrared radiation balances solar radiational heating of the Earth-atmosphere system. Although solar radiation is supplied only to the illuminated half of the planet, infrared radiation is emitted to space ceaselessly, day and night, by the entire Earth-atmosphere system. This explains why nights are usually colder than days and why air temperatures typically drop throughout the night.

While the clear atmosphere is relatively transparent to solar radiation, certain gases in the atmosphere impede the escape of infrared radiation to space thereby elevating the temperature of the lower atmosphere. This important climate control, the so-called **greenhouse effect**, refers to the heating of Earth's surface and lower atmosphere caused by strong absorption and emission of infrared radiation (IR) by certain gaseous components of the atmosphere, known as **greenhouse gases**. Solar radiation and terrestrial infrared radiation peak in different portions of the electromagnetic spectrum, their properties differ, and they interact differently with the atmosphere. As noted earlier, the atmosphere absorbs only about 23% of the solar radiation intercepted by the planet. The atmosphere absorbs a greater percentage of the infrared radiation emitted by Earth's surface, and the atmosphere, in turn, radiates some IR to space and some back to Earth's surface. Hence, Earth's surface is heated by absorption of both solar radiation and atmosphere-emitted infrared radiation.

The similarity in radiational properties between infrared-absorbing atmospheric gases and the glass or plastic glazing of a greenhouse is the origin of the term greenhouse effect. Greenhouse glazing, like the atmosphere, is relatively transparent to visible solar radiation but strongly absorbs infrared radiation. A greenhouse (where plants are grown) takes advantage of the radiational properties of glazing. Sunlight readily penetrates greenhouse glazing and much of it is absorbed (converted to heat) within the greenhouse. Objects in the greenhouse emit infrared radiation that is strongly absorbed by the glazing. The glazing, in turn, emits IR to both the atmosphere and to the greenhouse interior, thereby raising the temperature within the greenhouse. The analogy between the atmosphere and a greenhouse is not strictly correct, however. A greenhouse also functions as a shelter from the wind and this is the principal reason for the elevated temperature observed within most greenhouses. Nonetheless, *greenhouse effect* is such a commonly used term that we use it in this book.

The greenhouse effect is responsible for considerable warming of Earth's surface and lower atmosphere. Viewed from space, the planet (Earth-atmosphere system) radiates at about –18 °C (0 °F) whereas the average temperature at Earth's surface is about 15 °C (59 °F). The temperature difference is due to the greenhouse effect and amounts to:

[15 °C – (–18 °C)] = 33 Celsius degrees

or

[59 °F – (0 °F)] = 59 Fahrenheit degrees.

Without the greenhouse effect, Earth would be too cold to support most forms of plant and animal life. Water vapor is the principal greenhouse gas. Other greenhouse gases include carbon dioxide, ozone, methane (CH_4), nitrous oxide (N_2O), and halocarbons (once widely used as refrigerants and aerosol spray propellants). As shown in Figure 5.9, the percentage of infrared radiation absorbed by these gases varies with wavelength. An **atmospheric window** is a range of wavelengths over which little or no radiation is absorbed. A *visible window* extends from about 0.3 to 0.9 micrometers and the major *infrared window* is from about 8 to 13 micrometers. Significantly, this latter window includes the wavelength of the planet's peak infrared emission (about 10 micrometers). Through this window, most heat from the Earth-atmosphere system escapes to space as infrared radiation. IR sensors on Earth-orbiting satellites monitor this upwelling radiation which is calibrated in terms of the surface temperature of the radiating object: the higher the temperature, the more intense is the emission of IR radiation (Chapter 1).

Warming caused by atmospheric water vapor is evident even at the local or regional scale. Consider an example. Locations in the Desert Southwest and along the Gulf Coast are at about the same latitude and on a clear day receive essentially the same input of solar radiation. In both places, summer afternoon high temperatures commonly top 32 °C (90 °F). At night, however, air temperatures often differ markedly. Air is relatively dry (low humidity) in the Southwest so that infrared radiation readily escapes to space and air temperatures near Earth's surface may drop well under 15 °C (59 °F) by dawn. People who camp in the desert are well aware of the dramatic fluctuations in temperature between day and night. Infrared radiation does not escape to space as readily through the Gulf Coast atmosphere where the air is more humid. Water vapor strongly absorbs outgoing IR and emits IR back towards Earth's surface so that early morning low temperatures may dip no lower than the 20s Celsius (70s Fahrenheit). The smaller day-to-night temperature contrast along the Gulf Coast is due to more water vapor and a stronger greenhouse effect.

Clouds are composed of IR-absorbing water droplets and/or ice crystals, and also contribute to the greenhouse effect. All other factors being equal, nights usually are warmer when the sky is cloud-covered than when the sky is clear. Even high thin cirrus clouds through which the Moon is visible can reduce the nighttime temperature drop at Earth's surface by several Celsius degrees.

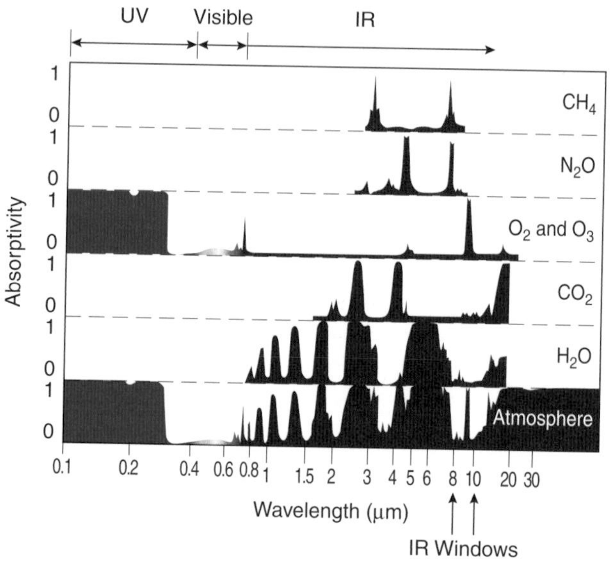

FIGURE 5.9
Absorption of radiation by selected gaseous components of the atmosphere as a function of wavelength. *Absorptivity* is the fraction of radiation absorbed and ranges from 0 to 1 (0% to 100% absorption). Absorptivity is very low or near zero in the *atmospheric windows.* Note the infrared (IR) windows near 8 and 10 micrometers.

Although water vapor is the principal greenhouse gas, changes in its atmospheric concentration (humidity) does not instigate warming or cooling trends in Earth's climate. Humidity varies in response to changes in temperature brought on, for example, by the buildup of other greenhouse gases such as CO_2. Water vapor's role in climate change is to amplify rather than trigger temperature trends.

Natural biogeochemical cycles continually transport greenhouse gases into and out of the atmosphere and ocean. In Chapter 3, for example, we saw how carbon dioxide cycles out of the atmosphere and into the ocean. Human activities alter the rate of biogeochemical cycling so that, for example, the atmospheric concentration of certain greenhouse gases is increasing. The consequent enhancement of the natural greenhouse effect is causing global-scale warming that has implications for all sectors of society. Since the beginning of the Industrial Revolution, combustion of fossil fuels (i.e., coal at first, oil and natural gas later) and clearing of forests and other vegetation have altered the global carbon cycle so that the atmospheric concentration of CO_2 is now about 39% higher than pre-industrial levels and continues to increase because of human activities.

The **Callendar effect** is the theory that global climate change can be brought about by enhancement of the natural greenhouse effect by increased levels of atmospheric CO_2 from anthropogenic sources, principally the burning of fossil fuels. The theory is named for the British engineer Guy Stewart Callendar (1898-1964) who investigated the link between global warming and fossil fuel combustion beginning in the late 1930s. We have more to say on the Callendar effect and its impacts on the Earth system in Chapter 12.

Heating Imbalances: Earth's Surface versus Atmosphere

Sensors onboard Earth-orbiting satellites detect imbalances in rates of radiational heating and radiational cooling. One important aspect of this heating imbalance involves Earth's surface versus the atmosphere.

Figure 5.10 shows how solar radiation intercepted by planet Earth interacts with the atmosphere and Earth's surface. Numbers represent global and annual averages. For every 100 units of solar radiation that enters the upper atmosphere, the Earth-atmosphere system

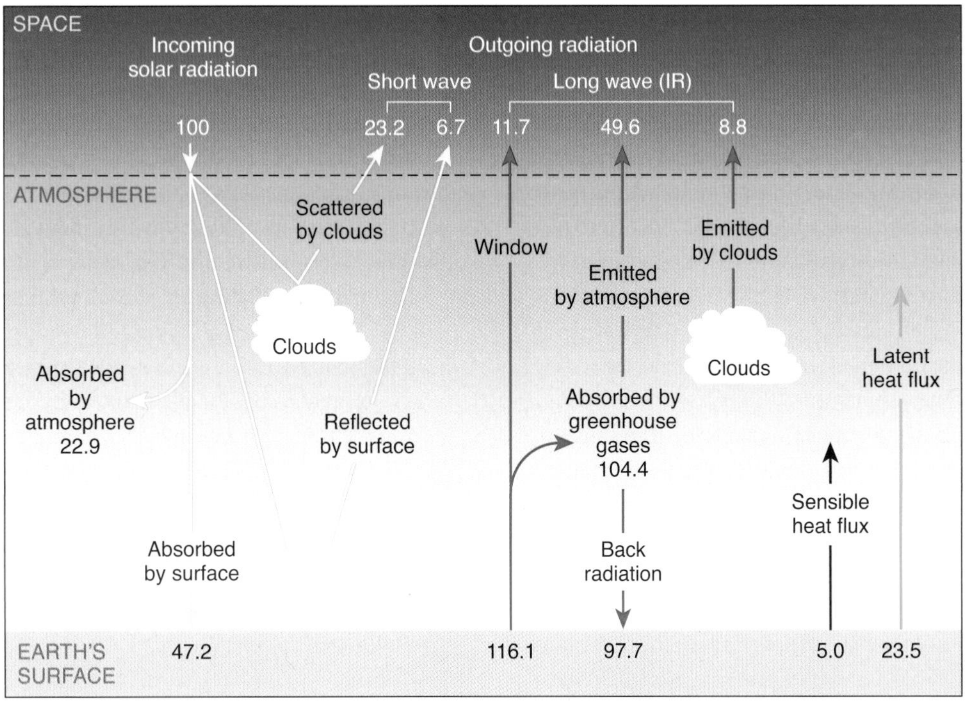

FIGURE 5.10
Globally and annually averaged distribution of 100 units of solar radiation entering the top of the atmosphere. Depicted from left to right are fluxes of solar radiation (yellow), infrared radiation (red), sensible heat (black), and latent heat (green). [Modified after K. E. Trenberth et al., 2009. "Earth's Global Energy Budget", *Bulletin American Meteorological Society*, vol. 90, issue 3, pp. 311-323]

TABLE 5.2
Global Radiation Balance

Solar radiation intercepted by Earth	100 units
Solar radiation budget	
Scattered and reflected to space (23.2 + 6.7)	29.9
Absorbed by the atmosphere (22.9)	22.9
Absorbed at the Earth's surface	47.2
Total	100 units
Radiation budget at the Earth's surface	
Infrared cooling (97.7 − 116.1)	−18.4
Solar heating	+47.2
Net heating	+28.8 units
Radiation budget of the atmosphere	
Infrared cooling (− 49.6 − 8.8 + 104.4 − 97.7)	−51.7
Solar heating	+22.9
Net cooling	−28.8 units
Non-radiative heat transfer: Earth's surface to atmosphere	
Sensible heating (conduction plus convection)	5.0
Latent heating (phase changes of water)	23.5
Net transfer	28.5 units

reflects or scatters 29.9 units to space, the atmosphere absorbs 22.9 units and Earth's surface (principally the ocean) absorbs 47.2 units. In response to radiational heating, Earth's surface emits 116.1 units of infrared radiation. Atmospheric gases and clouds absorb 104.4 units of infrared radiation and emit 97.7 units to Earth's surface (*greenhouse effect*). A total of 70.1 units of IR radiation are emitted out the top of the atmosphere to space, equal to the amount of solar radiation absorbed by the Earth-atmosphere system.

The global average annual distribution of incoming solar radiation and outgoing infrared radiation implies net warming of Earth's surface and net cooling of the atmosphere (Table 5.2). At Earth's surface, absorption of solar radiation is greater than emission of infrared radiation. In the atmosphere, on the other hand, emission of infrared radiation to space is greater than absorption of solar radiation. That is, on a global average annual basis, Earth's surface undergoes net radiational heating and the atmosphere undergoes net radiational cooling.

The atmosphere is not actually cooling relative to Earth's surface because radiation is not the only heat transfer mechanism at work. In response to the radiationally induced temperature gradient between Earth's surface and atmosphere, heat is transferred from Earth's surface to the atmosphere. A combination of latent heating (phase changes of water) and sensible heating (conduction and convection) is responsible for this transfer of heat. As shown in Figure 5.10, on a global annual average basis, 28.5 units of heat energy are transferred from Earth's surface to the atmosphere: 23.5 units (about 82% of the total) by latent heating and 5 units (about 18%) by sensible heating.

LATENT HEATING

Latent heating refers to the transfer of heat energy from one place to another as a consequence of phase changes of water. As discussed in Chapter 3, when water changes phase, heat energy is either absorbed from the environment (i.e., melting, evaporation, sublimation) or released to the environment (i.e., freezing, condensation, deposition). As part of the global water cycle, latent heat that is used to vaporize water at the Earth's surface is transferred to the atmosphere when

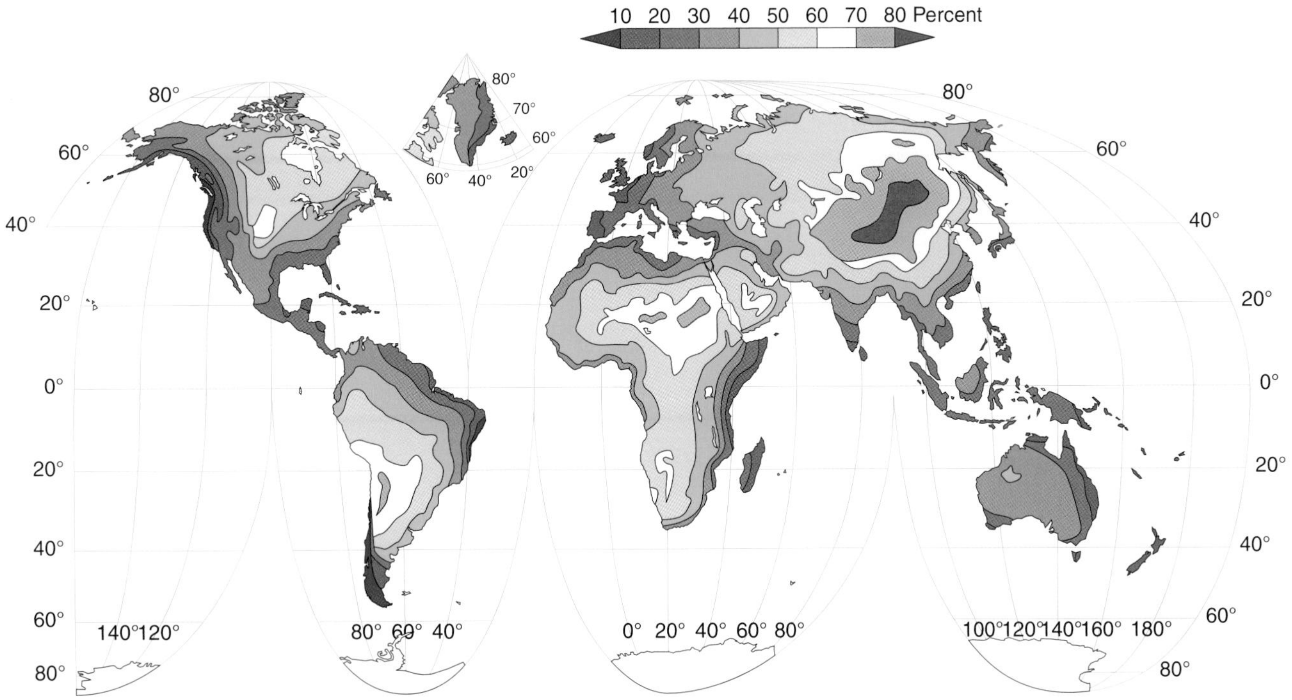

FIGURE 5.11
Percentage of precipitation over land that originated as evaporation on the continents, annually averaged over 15 years. In many land areas, the principal source of water for precipitation is evaporation from the ocean. [World Climate Research Programme, Global Energy and Water Cycle Experiment]

clouds form. Significantly for Earth's climate, ocean water covers a large portion of Earth's surface and is the principal source of water vapor that eventually returns to Earth's surface as precipitation. In general, only well inland does most precipitation originate as evaporation from the continents (Figure 5.11). Also, the ocean is a major source of salt crystals that spur condensation and cloud development in the atmosphere. These *cloud condensation nuclei* have a special chemical affinity for water molecules and readily promote cloud formation. When sea waves break, drops of salt water enter the atmosphere and evaporate leaving behind sea-salt crystals that function as nuclei.

As Earth's surface absorbs radiation (both solar and infrared), some of the heat energy is used to vaporize water from the ocean, glaciers, lakes, rivers, soil, and vegetation (*transpiration*). The latent heat required for vaporization (evaporation or sublimation) is supplied at the Earth's surface, and heat is subsequently released to the atmosphere during cloud development. Within the troposphere, clouds form as some of the water vapor condenses into liquid water droplets or deposits as ice crystals. During cloud formation, water changes phase and latent heat is released to the atmosphere. Through latent heating, then, heat is transferred from Earth's surface to the troposphere. In fact, latent heat transfer is more important than either radiational cooling or sensible heat transfer in the cooling of Earth's surface (Figure 5.12).

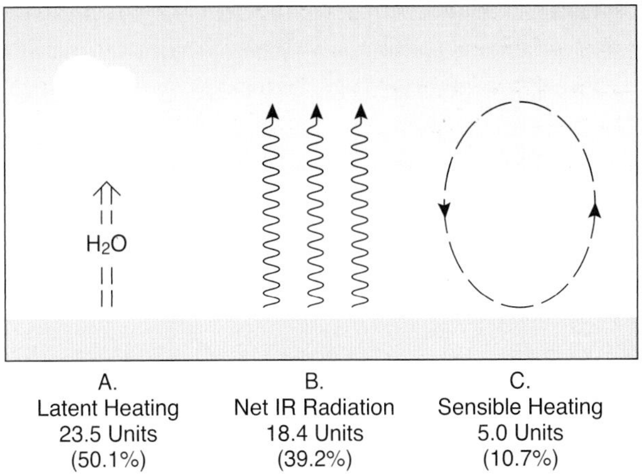

FIGURE 5.12
Earth's surface is cooled through (A) vaporization of water, (B) net emission of infrared radiation to the sky, and (C) conduction plus convection. Numbers are global annual averages based on 100 units of solar radiation entering the top of the atmosphere.

SENSIBLE HEATING

Heat transfer via conduction and convection can be monitored (sensed) by temperature changes; hence, **sensible heating** encompasses both of these processes. Heat is conducted from the relatively warm surface of the Earth to the cooler overlying air. Heating reduces the density of that air, which is forced to rise by cooler denser air replacing it at the surface (Figure 5.13). In this way, convection transports heat from Earth's surface into the troposphere. Because air is a relatively poor conductor of heat, heat convection is much more important than conduction as a transfer mechanism within the troposphere.

Often sensible heating combines with latent heating to channel heat from Earth's surface into the troposphere. This happens during thunderstorm development. Updrafts (ascending branches) of vapor-laden air in convection currents often produce *cumulus clouds*, which resemble puffs of cotton floating in the sky (Figure 5.14A). These clouds are sometimes referred to as *fair-weather cumulus* because they seldom produce rain or snow. On the other hand, if atmospheric conditions are favorable, convective currents can surge to great altitudes, and cumulus clouds merge and billow upward to form towering *cumulonimbus clouds*, also known as thunderstorm clouds (Figure 5.14B).

In retrospect, two important heat transfer processes (a combination of latent heating and sensible heating) took place last summer when that thunderstorm sent you scurrying for shelter.

At some times and places, heat transfer is directed from the troposphere to Earth's surface, the reverse of the global average annual situation. This reversal in direction of heat transport occurs, for example, when mild winds blow over cold, snow-covered ground or when warm air moves over a relatively cool ocean surface. Heat transport from the atmosphere to Earth's surface is the usual situation at night (especially when skies are clear) when radiational cooling causes Earth's land surface to become colder than the overlying air.

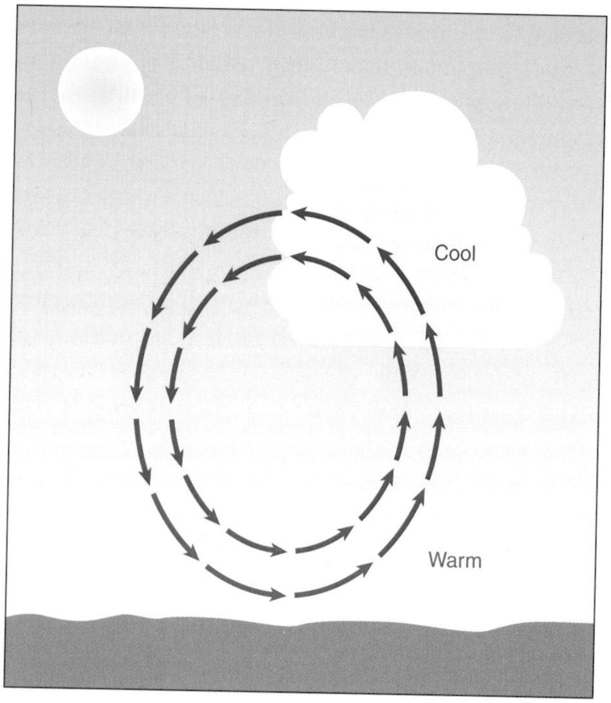

FIGURE 5.13
Convection currents transport heat from Earth's surface into the troposphere.

A

B

FIGURE 5.14
Latent heating and sensible heating are combined in the formation of (A) cumulus clouds and (B) cumulonimbus (thunderstorm) clouds.

The **Bowen ratio** describes how the heat energy received at Earth's surface (by absorption of solar and infrared radiation) is partitioned between sensible heating and latent heating. That is,

Bowen ratio = [(sensible heating)/(latent heating)]

At the global scale,

Bowen ratio = [(5.0 units)/(23.5 units)] = 0.2

The average Bowen ratio varies from one place to another depending on the amount of surface moisture. The wetter the surface, the less important is sensible heating and the more important is latent heating. The Bowen ratio ranges from about 0.1 (one-tenth as much sensible as latent heating) for the ocean to about 5.0 (five times as much sensible as latent heating) in deserts. Ocean waters cover much of Earth's surface so it is not surprising that the global Bowen ratio is a relatively low 0.2.

Heating Imbalances: Tropics versus High Latitudes

On a global scale, imbalances in radiational heating and cooling occur between not only Earth's surface and atmosphere but also between the tropics and higher latitudes. Because the planet is nearly a sphere, parallel beams of incoming solar radiation strike the tropics more directly than higher latitudes. (That is, solar altitudes are higher in the tropics and lower at higher latitudes.) At higher latitudes, solar radiation spreads over a greater area and is less intense per unit horizontal surface area than in the tropics.

Emission of infrared radiation by the Earth-atmosphere system also varies with latitude but less than solar radiation. Because air temperatures are generally lower at higher latitudes, IR emission also declines with increasing latitude. (Recall that radiation emission is temperature dependent.) Consequently, over the period of a year at higher latitudes, the rate of infrared cooling to space exceeds the rate of warming caused by absorption of solar radiation. At lower latitudes the reverse is true; that is, over the course of a year, the rate of solar radiational heating is greater than the rate of infrared radiational cooling (Figure 5.15). Averaged over the globe, incoming energy (absorbed solar radiation) must equal outgoing energy (IR emitted to space). That is, the areas under the two curves in Figure 5.15 are equal. The balance between energy entering and leaving the Earth-atmosphere system (*global radiative equilibrium*) is the prevailing (long-term average) condition on Earth.

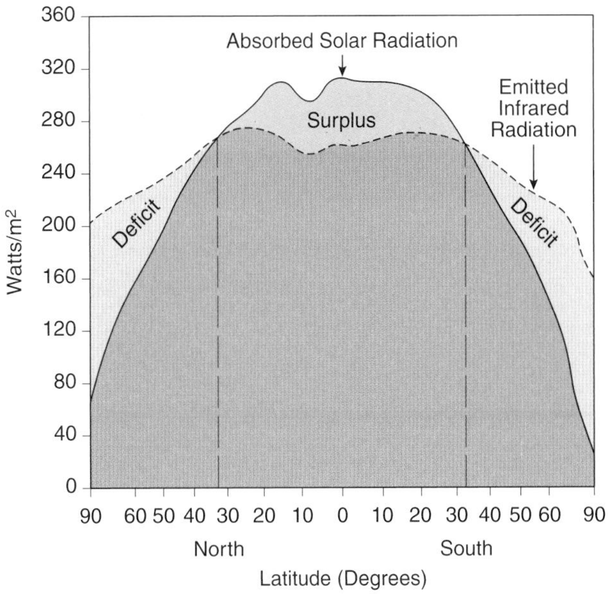

FIGURE 5.15
Variation by latitude of absorbed solar radiation and outgoing infrared radiation derived from satellite sensor measurements. [From NOAA/NESDIS]

Measurements by sensors onboard Earth-orbiting satellites indicate that the division between regions of net radiational cooling and regions of net radiational warming is close to the 35-degree latitude circle in both hemispheres. By implication, latitudes poleward of about 35 degrees N and 35 degrees S should experience net cooling over the course of a year, while tropical latitudes are sites of net warming. In fact, lower latitudes do not become progressively warmer nor do higher latitudes become colder because heat is transported poleward from the tropics into middle and high latitudes. **Poleward heat transport** is brought about by (1) air mass exchange, (2) storm systems, and (3) ocean circulation.

HEAT TRANSPORT BY AIR MASS EXCHANGE

North-south exchange of air masses transports sensible heat from the tropics into middle and high latitudes. An **air mass** is a huge volume of air covering thousands of square kilometers that is relatively uniform horizontally in temperature and humidity. The properties of an air mass largely depend on the characteristics of the surface over which the air mass forms (its *source region*) or travels (Figure 5.16). Air masses that form

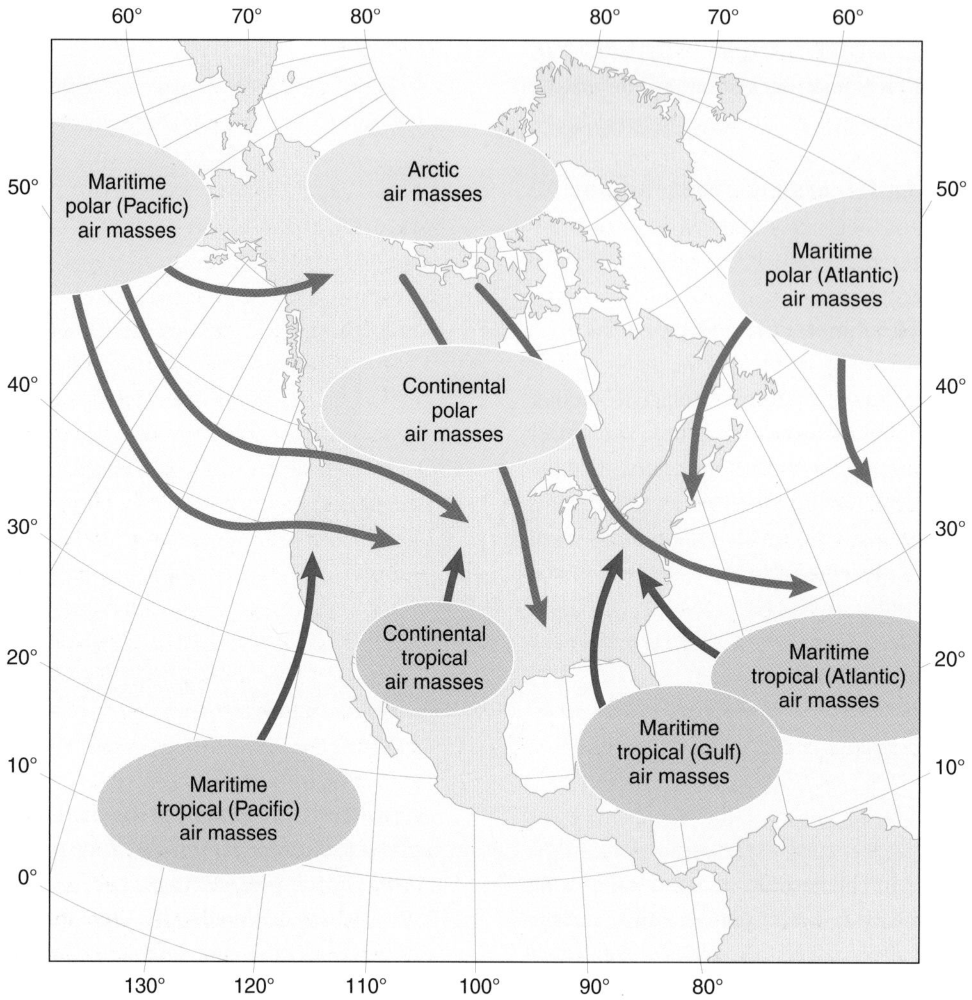

FIGURE 5.16
Source regions of air masses that regularly move over North America.

at high latitudes over cold, often snow- or ice-covered surfaces are relatively cold. Those air masses that form at low latitudes are relatively warm. Air masses that develop over the ocean are humid and those that form over land are relatively dry. Hence, there are four basic types of air masses: cold and humid, cold and dry, warm and humid, and warm and dry.

As shown in Figure 5.16, warm air masses that form in lower latitudes flow toward the pole while cold air masses flow toward the equator from source regions at high latitudes. Air masses modify (become cooler, warmer, drier, more humid) to some extent as they move away from their source region, gaining or losing heat energy and/or moisture in the process. In this north-south exchange of air masses, a net transport of heat energy takes place directed from lower to higher latitudes.

HEAT TRANSPORT BY STORMS

Acquisition and subsequent release of latent heat in storm systems (*cyclones* or *lows*) plays an important role in the poleward transport of heat. At low latitudes, water that evaporates from the warm ocean surface may be drawn into the circulation of a developing storm system. As the storm travels into higher latitudes, some of that water vapor condenses into clouds, thereby releasing latent heat to the troposphere. Latent heat of vaporization acquired at low latitudes is thereby delivered to middle and high latitudes. Because they entrain much more water vapor and latent heat, tropical storms and hurricanes are greater contributors to poleward heat transport than ordinary middle latitude (*extratropical*) storms. We discuss middle latitude and tropical cyclones in greater depth in Chapter 8.

HEAT TRANSPORT BY OCEAN CIRCULATION

The ocean contributes to poleward heat transport via wind-driven surface currents and the deeper thermohaline-driven circulation. Surface water that is warmer than the overlying air is a *heat source* for the atmosphere; that is, heat is transferred from sea to air via conduction, convection, and latent heating. Surface water that is cooler than the overlying air is a *heat sink* for the atmosphere; that is, heat is conducted from air to sea. Warm surface currents, such as the Gulf Stream, flow from the tropics into middle latitudes, supplying heat to the cooler middle latitude troposphere. At the same time, cold surface currents, such as the California Current, flow from high to low latitudes, absorbing heat from the relatively warm troposphere and greater solar radiation in the tropics. Chapter 6 has more on wind-driven ocean currents.

The ocean's **thermohaline circulation** is the density-driven movement of water masses. As noted in Chapter 3, the density of seawater increases with decreasing temperature and increasing salinity. More dense water tends to sink while less dense water rises. The thermohaline circulation transports heat energy, salt, and dissolved gases (e.g., carbon dioxide, oxygen) over great distances and to great depths in the world ocean and plays an important role in Earth's climate system. In the North Atlantic, for example, a warm surface ocean current flows north and eastward from the Florida Strait. At high latitudes, the surface waters cool, sink, and flow southward as cold bottom waters. This heat transporting mechanism is a key component of the ocean's meridional overturning circulation (MOC).

Meridional Overturning Circulation

The ocean undergoes large-scale overturning in a two-dimensional north-south (meridional) plane. This **meridional overturning circulation (MOC)** transports heat, salinity, and carbon dioxide throughout the world ocean. Thereby, the MOC contributes to poleward heat transport and is an agent of climate change.

The physical oceanographer Henry M. Stommel (1920-1992) was a major contributor to our understanding of global ocean circulation patterns. In 1958, he described the basic structure of ocean overturning. Water masses originate at high latitudes (Norwegian and Labrador Seas and near Antarctica) via cooling and evaporation at the air/sea interface. These relatively dense water masses sink and flow at depth beneath the western-intensified boundary currents of ocean basins toward the equator. Deep waters spread through the abyss, upwell to the surface, and flow poleward.

Some 25 years later, Wallace S. Broecker, a geoscientist at Columbia University's Lamont-Doherty Earth Observatory, and colleagues, likened the MOC to a "great ocean conveyor belt" linking the circulation of the world ocean and transporting cold water equatorward at depth (the lower limb of the conveyor belt) and warm water poleward at the surface (the upper limb of the conveyor belt). The deep flow was assumed to be smooth and continuous along the western boundary currents. A weaker interior circulation moves waters poleward and upward to the high latitude locales of deepwater formation.

Broecker and colleagues argued that ocean overturning was responsible for abrupt climate oscillations that punctuated the last major glacial episode. These climate oscillations were brought about by alternate weakening and strengthening of the conveyor belt circulation due to changes in production of deep water at high latitudes. During relatively mild (interglacial) episodes, continental ice sheets thinned and retreated discharging surges of fresh water into the ocean, decreasing or eliminating deepwater production, and weakening the conveyor belt. For Europe, this meant colder conditions and, in some locales, renewed glaciation. During relatively cold (glacial) episodes, continental ice sheets thickened and expanded so that less fresh water discharged into the ocean, enhancing deepwater production and strengthening the conveyor belt. For Europe, this meant milder conditions.

After several decades as a relatively popular paradigm, the validity of the conveyor belt model of the MOC was questioned. M. Susan Lozier of Duke University reports in the 18 June 2010 issue of *Science* that interior pathways may be more important than deep western boundary currents in transporting heat and other properties of ocean water, the deep western boundary currents are not continuous and appear to break up into migrating eddies at tropical latitudes, and wind-forcing may be more important than buoyancy (due to density differences) in transport by meridional ocean overturning. She notes that in the MOC, wind-forcing dominates on inter-annual to decadal time scales whereas buoyancy dominates on longer, centennial scales.

Recently, based on the output of coupled global climate models, some scientists predicted that global warming could cause the MOC to gradually weaken with implications for the climate of northern latitudes, particularly Europe. To test this hypothesis, a team of U.S. and British scientists from the *Rapid Climate Change Program* (begun in 2001) strung an array of moored instruments along the 26.5 degree N latitude circle across the North Atlantic Ocean. They reported their findings in *Science* in August 2007. Based on measurements made over a one-year period, scientists discovered considerable intra-annual variability in the MOC. The MOC varied by a factor of 8, from a low of 4.0 Sv to a high of 34.9 Sv, with an average of 18.7 ± 5.6 Sv. (1.0 Sv = 1.0 Sverdrup = 1.0 million cubic m per sec.) The challenge is to separate out any long-term trend from the substantial natural variability of the MOC but this would require a much more lengthy observational record. For now, there is no indication that the MOC is weakening.

Circulation of the Atmosphere: The Forces

As discussed earlier, the atmosphere circulates in response to temperature gradients that develop within the Earth-atmosphere system. These temperature gradients are due to differences in rates of radiational heating and radiational cooling between (1) Earth surface and atmosphere, and (2) the tropics and high latitudes. Circulation of the atmosphere and ocean transports heat from warmer locations to colder locations. In this section, we describe the principal forces operating in large-scale atmospheric circulation systems: the pressure gradient force and the Coriolis Effect. Other forces that influence atmospheric circulation are friction (important within about 1000 m or 3300 ft of Earth's surface), and gravity.

PRESSURE GRADIENT FORCE

Air exerts a force on the surfaces of all objects that it contacts. (A *force* is a push or pull on an object and is computed as mass times acceleration. A force is a vector quantity; that is, it has both magnitude and direction.) As noted in Chapter 3, we can think of **air pressure** at a given location on the Earth's surface as the weight per unit area of the column of air above that location. Unlike ocean water, air is highly compressible. The pull of gravity compresses the atmosphere so that the maximum air density and pressure are at the Earth's surface, and air density and pressure decrease rapidly with increasing altitude. The average air pressure at sea level is 1013.25 millibars (mb). At an altitude of only 5500 m (18,000 ft), air pressure is about half of its average value at sea level. The rapid drop in air pressure with altitude means that significant changes in air pressure accompany relatively minor changes in elevation. For example, the average air pressure at Denver, the mile-high city, is about 83% of the average air pressure at sea level.

Air pressure varies with both space and time and this drives the circulation of the atmosphere. Variations in air pressure are not always due to variations in elevation. In fact, atmospheric scientists are most interested in air pressure variations that arise from factors other than elevation. Hence, weather observers determine an equivalent sea-level air pressure value; that is, for weather stations located above sea level, they adjust air pressure readings to approximately what the pressure would be if the station were actually located at sea level. When this adjustment to sea level is carried out everywhere, air pressure is observed to vary from one place to another and fluctuate from day to day and even from one hour to the next. Spatial and temporal changes in air pressure at Earth's surface arise from variations in air temperature (principally), humidity (concentration of water vapor in air), and atmospheric circulation.

In the free atmosphere, air density varies inversely with both temperature and humidity. That is, air density increases with falling temperature and decreasing humidity. Cold, dry air masses are denser and usually produce higher surface pressures than warm, humid air masses. Warm, dry air masses, in turn, often exert higher surface pressures than equally warm, but more humid air masses. As one air mass replaces another at a specific location, the air pressure at that location may change. Falling air pressure often signals a turn to stormy weather whereas rising air pressure indicates clearing skies or continued fair weather.

On a weather map, a *HIGH* or *H* symbol designates the centers of places where sea-level air pressure is relatively high compared to the air pressure in surrounding areas. A high is also known as an *anticyclone* and is usually a fair weather system. A *LOW* or *L* symbol signifies the low-pressure center of regions where sea-level air pressure is relatively low compared to the air pressure in surrounding areas. A low is also known as a *cyclone* and often brings stormy weather.

The rate of decrease of air pressure from one place to another is known as an **air pressure gradient**. Air pressure gradients occur both vertically and horizontally within the atmosphere. A vertical air pressure gradient is a permanent feature of the atmosphere because air pressure always decreases with increasing altitude (at a rate dependent on the density of the air column). A horizontal air pressure gradient refers to pressure changes per unit distance along a surface of constant altitude (e.g., mean sea level). Horizontal air pressure gradients can be determined on weather maps from patterns of *isobars*, lines joining points having the same air pressure (adjusted to sea level). Usually isobars are drawn on weather maps at 4-millibar intervals.

In response to a horizontal air pressure gradient, the wind blows from where the pressure is relatively high toward where the pressure is relatively low. The force that causes air to move as the consequence of an air pressure gradient is known as the **pressure gradient force** and is always directed across isobars and toward low pressure. The magnitude of the pressure gradient force is inversely related to the spacing of isobars. The wind is relatively strong where the pressure gradient is steep (closely spaced isobars), and light or calm where the pressure gradient is weak (widely spaced isobars).

CORIOLIS EFFECT

If Earth did not rotate, surface winds would blow directly from the cold poles (where surface air pressure is relatively high) to the hot equator (where surface air pressure is relatively low) in response to horizontal pressure gradient forces directed toward the equator. And these winds would push ocean surface currents directly toward the equator. But because Earth rotates, anything moving freely over the planet's surface, including air and water, is deflected to the right in the Northern Hemisphere and to the left in the Southern Hemisphere (Figure 5.17). (No deflection occurs at the equator.) This deflection is known as the **Coriolis Effect**, named for Gaspard-Gustave de Coriolis (1792-1843), the French mathematician who first described the phenomenon quantitatively in 1835.

According to **Newton's first law of motion**, an object in constant, straight-line motion remains that way unless acted upon by an unbalanced force. Winds in the atmosphere (and currents moving in the ocean) exhibit this behavior. But these motions occur on a rotating Earth so that as air (or water) moves in a straight line, Earth rotates beneath the moving air (or water). Except at the equator, the wind is displaced from a straight-line path when its motion is measured with respect to the rotating Earth. While no force is causing this turning motion, this inconsistency can be incorporated in Newton's first law of motion by explaining the deflection to be the result of an imaginary force. This deflection is referred to as the Coriolis Effect and the apparent force invented to describe its magnitude and direction is called the *Coriolis Force*.

Reversal in the direction of the Coriolis Effect between the Northern and Southern Hemispheres is related to the difference in our perspective of Earth's rotation direction in the two hemispheres. To an observer looking down from high above the North Pole, the planet rotates counterclockwise, whereas to an observer looking down from high above the South Pole, the planet rotates clockwise.

For an Earth-bound observer, this reversal in the sense of Earth's rotation between the two hemispheres translates into a reversal in Coriolis Effect. That is, moving air (and water) is deflected to the right in the Northern Hemisphere, and to the left in the Southern Hemisphere. Because of these hemispheric differences, we refine the definition of the Coriolis Effect as always acting at 90 degrees and to the right of the direction of motion in the Northern Hemisphere, and at 90 degrees and to the left of motion in the Southern Hemisphere.

Although the Coriolis Effect influences the wind blowing in any direction, the amount of deflection varies significantly with latitude; that is, the magnitude of the Coriolis Effect varies from zero at the equator to a maximum value at the poles. This variation with latitude can be understood by visualizing the daily rotation of towers about their vertical axes when located at different latitudes. In a 24-hr day, Earth completes one rotation, as would towers located at the North and South Pole. In the same period, a tower at the equator would not rotate at all about its vertical axis because of its orientation perpendicular to Earth's axis of rotation. At any latitude

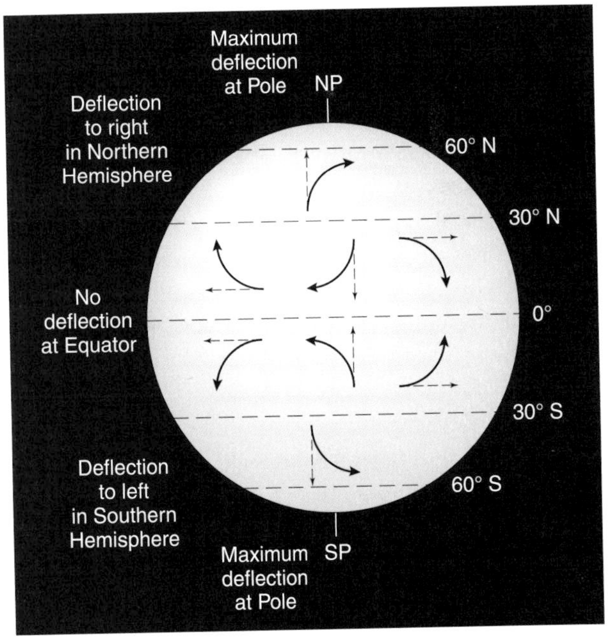

FIGURE 5.17
The Coriolis Effect arises from the rotation of Earth on its axis and causes deflection of winds (and ocean currents) to the right in the Northern Hemisphere and to the left in the Southern Hemisphere.

in between, some rotation of a tower occurs but not as much as at the poles.

The magnitude of the Coriolis Effect also varies with wind speed and spatial scale of atmospheric circulation. The Coriolis Effect increases as the wind strengthens because, in the same time interval, faster moving air parcels cover greater distances than slower moving air parcels. The longer the trajectory, the greater is the rotation of the underlying Earth. The Coriolis Effect significantly influences the wind only in large-scale weather systems, that is, systems larger than ordinary thunderstorms. Large-scale weather systems also have longer life expectancies than small-scale systems so that air parcels cover greater distances over longer periods of time, allowing the impact of Earth's rotation to manifest itself.

Circulation of the Atmosphere: Patterns of Motion

Air pressure gradients, the Coriolis Effect, and the physical properties of Earth's surface shape the circulation of the atmosphere. For convenience of study, atmospheric scientists subdivide atmospheric circulation into discrete weather systems operating at various spatial and temporal scales (Table 5.3). The large-scale wind belts encircling the planet (e.g., westerlies of middle latitudes, trade winds in the tropics) are **planetary-scale systems**. **Synoptic-scale systems** are continental or oceanic in scale; migrating cyclones are examples. **Mesoscale systems** include, for example, thunderstorms and sea breezes—circulation systems that are so small and short-lived that they may influence the weather in only a portion of a large city (Figure 5.18). A weather system covering only a very small area (e.g., a weak tornado) represents the smallest spatial subdivision of atmospheric circulation, **microscale systems**. This section focuses primarily on atmospheric circulation patterns operating at the planetary and synoptic scales.

PLANETARY-SCALE CIRCULATION

As shown in Figure 5.19, three broad wind belts encircle both the Northern and Southern Hemispheres. In the Northern Hemisphere, prevailing surface winds blow from the northeast (*trade winds*) between the equator and about 30 degrees N, from the southwest (*westerlies*) between about 30 and 60 degrees N, and from the northeast (*polar easterlies*) between about 60 degrees N and the North Pole. In the Southern Hemisphere, prevailing surface winds are southeasterly (*trade winds*) between the equator and about 30 degrees S, northwesterly (*westerlies*) between about 30 and 60 degrees S, and southeasterly (*polar easterlies*) between about 60 degrees S and the South Pole.

The planetary-scale atmospheric circulation plays an important role in the horizontal transport of water vapor between ocean basins (as part of the global water cycle). For example, the trade winds transport water evaporated from the tropical Atlantic Ocean basin across Central America to the tropical Pacific Ocean where it condenses into clouds that produce rain. This transport of water increases the salinity of Atlantic surface waters and freshens the Pacific surface waters.

In both hemispheres, trade winds blow out of the equatorward flank of the subtropical anticyclones and the westerlies blow out of the poleward flank of the subtropical anticyclones. A **subtropical anticyclone** (or high) is a massive semi-permanent high pressure system. Centered over the ocean basins near 30 degrees N and S, they are imposing features of the planetary-scale circulation that undergo seasonal changes in both location and strength. Viewed from above in the Northern Hemisphere, surface winds blow clockwise and

TABLE 5.3
Scales of Atmospheric Circulation

Circulation	*Space scale*	*Time scale*	*Example*
Planetary scale	10,000 to 40,000 km	weeks to months	Westerlies
Synoptic scale	100 to 10,000 km	days to a week	High, Low
Mesoscale	1 to 100 km	hours to a day	Thunderstorm
Microscale	1 m to 1 km	seconds to hours	Tornado

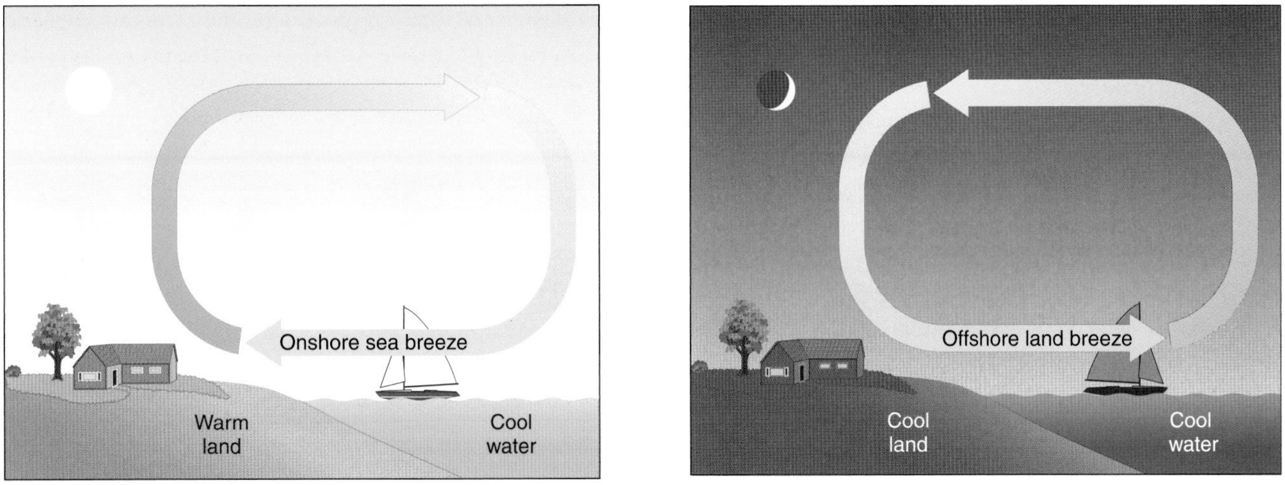

FIGURE 5.18
A sea breeze is a relatively cool mesoscale surface wind that develops during daylight hours and blows inland from the ocean in response to differential heating of land and sea. During the day, the land surface warms more than the sea surface inducing a horizontal air pressure gradient with high pressure over the ocean surface and low pressure over the land surface. At night, land becomes relatively colder than the sea surface. The circulation reverses and a land breeze blows offshore.

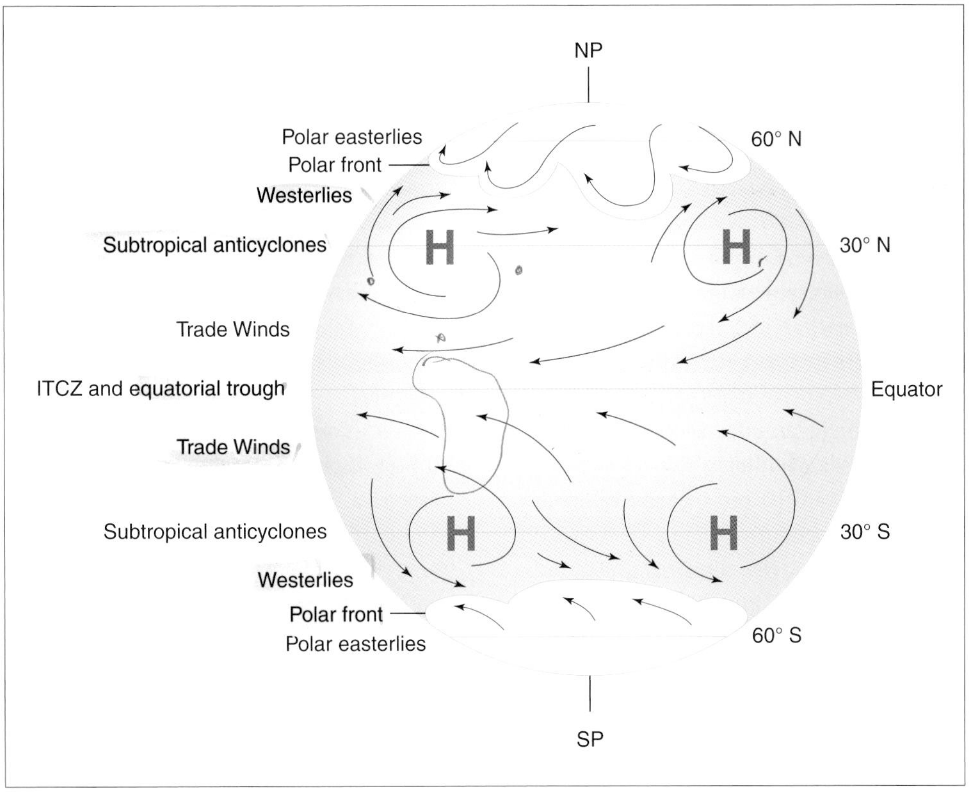

FIGURE 5.19
Schematic representation of the planetary-scale surface circulation of the atmosphere.

outward about the center of a subtropical high. Viewed from above in the Southern Hemisphere, surface winds blow counterclockwise and outward about the center of a subtropical high. In both hemispheres, climates are relatively dry on the eastern flank of the subtropical highs and moist on the western flank. Broad areas of ocean basins under the subtropical highs experience persistent episodes of warm dry weather. Low precipitation coupled with high temperatures cause high rates of evaporation and therefore relatively high salinity of surface waters (refer back to Figure 3.9).

Over a broad region encompassing the center of a subtropical high, the horizontal pressure gradient is weak so that surface winds are very light or the air is calm over extensive areas of the subtropical ocean. This situation played havoc with early sailing ships, which were becalmed for days or even weeks at a time. Ships setting sail from Spain to the New World were often caught in this predicament and crews were forced to jettison their cargo of horses when supplies of water and food ran low. For this reason, early mariners referred to this region as the **horse latitudes**, a name now applied to all latitudes between about 30 and 35 degrees N and S under subtropical highs.

On the poleward side of the subtropical highs, surface westerlies flow into regions of low pressure. In the Northern Hemisphere, there are two separate subpolar lows: the *Aleutian low* over the North Pacific Ocean and the *Icelandic low* over the North Atlantic Ocean. These pressure systems mark the convergence of the middle latitude westerlies with the polar easterlies. By contrast, in the Southern Hemisphere, the middle latitude westerlies and the polar easterlies converge along a nearly continuous belt of low pressure surrounding the Antarctic continent.

At middle and high latitudes of the Northern Hemisphere, prevailing winds at altitudes from about 5500 to 12,000 m (18,000 to 40,000 ft) blow generally from west to east in a wave-like pattern of ridges (clockwise turns) and troughs (counterclockwise turns) as illustrated in Figure 5.20. The temperature gradient between the relatively warm tropics and relatively cold polar areas initiates a northward flow of air aloft that is deflected to the right (to the east) by Earth's rotation (the Coriolis Effect). A similar situation occurs in the Southern Hemisphere where the Coriolis Effect deflects a southward-directed flow of air aloft to the left, resulting in a west wind in that hemisphere. The belts of westerlies encircle the planet and steer air masses, storms, and fair-weather systems generally from west to east.

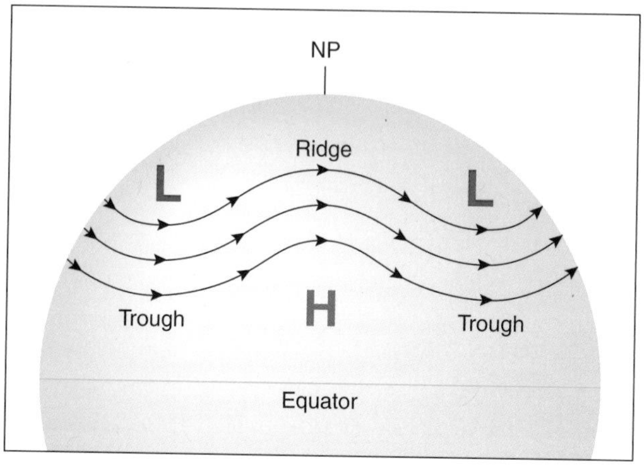

FIGURE 5.20
In the middle and upper troposphere, the Northern Hemisphere westerlies blow from west to east in a wave-like pattern of ridges (clockwise turns) and troughs (counterclockwise turns).

Also part of the weaving westerlies are narrow corridors of exceptionally strong winds in the upper troposphere known as *jet streams*. In the Northern Hemisphere, the middle-latitude jet stream is situated over a boundary (called the *polar front*) between colder air to the north and warmer air to the south and contributes to the development of storm systems.

SYNOPTIC-SCALE WEATHER SYSTEMS

The most important synoptic-scale weather systems are *highs* (or anticyclones) and *lows* (or cyclones). As noted earlier, highs usually are accompanied by fair weather whereas lows often bring clouds and precipitation. Whereas subtropical highs and subpolar lows are nearly stationary, synoptic-scale highs and lows move with the prevailing winds blowing at altitudes of several kilometers above Earth's surface, generally eastward across North America. Highs follow lows and lows follow highs. As a general rule, highs track toward the east and southeast whereas lows track toward the east and northeast. Important exceptions are tropical cyclones (e.g., hurricanes) that are imbedded in the trade wind flow and generally track from east to west over the tropical Atlantic and Pacific. Eventually these systems come under the influence of the middle-latitude westerlies and turn toward the north and northeast.

Highs originating over northwestern Canada bring cold, dry weather in winter and cool, dry weather in summer. Highs that develop farther south bring hot weather in summer and mild, dry weather in winter. Viewed from above in the Northern Hemisphere, surface winds in a high-pressure system blow in a clockwise

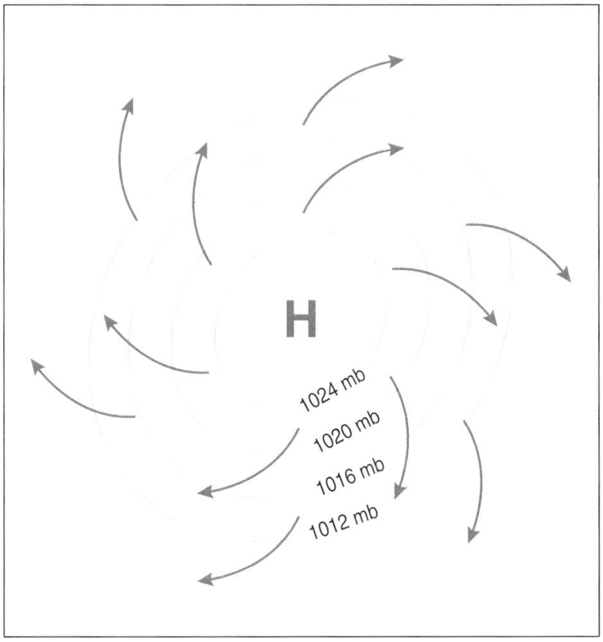

FIGURE 5.21
Viewed from above in the Northern Hemisphere, surface winds blow clockwise and outward in a high (anticyclone). The light blue lines are isobars drawn at 4-millibar intervals.

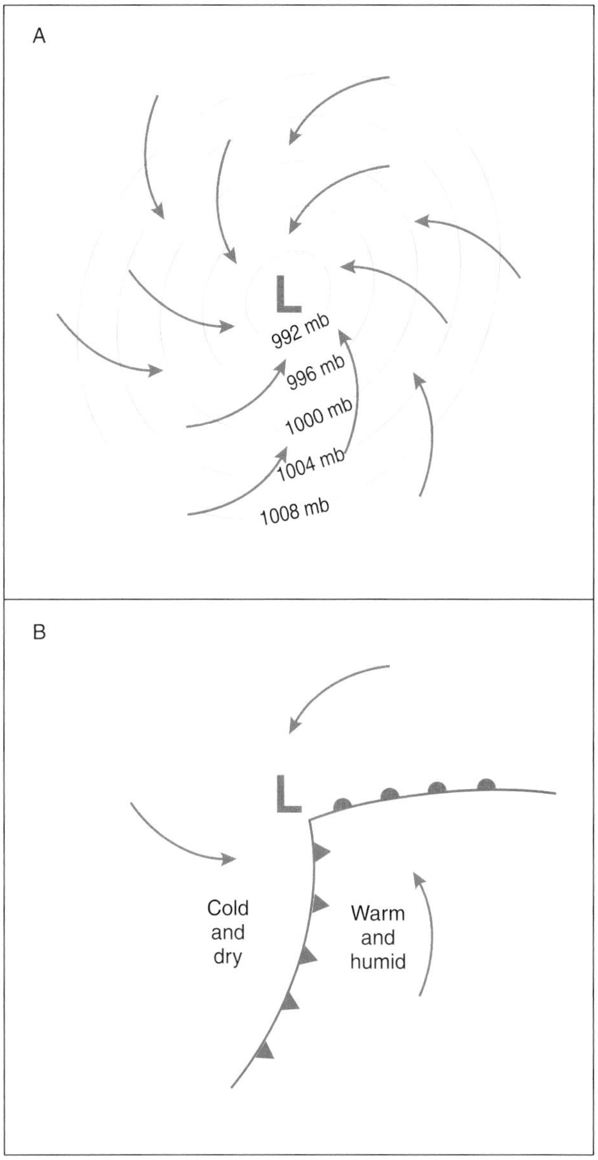

FIGURE 5.22
Viewed from above in the Northern Hemisphere, (A) surface winds blow counterclockwise and inward in a low (cyclone), (B) bringing together contrasting air masses to form fronts. The light blue lines are isobars drawn at 4-millibar intervals. The curved blue line with spikes is the leading edge of relatively cold and dry air (a cold front) while the curved red line with half circles is the leading edge of relatively warm and humid air (a warm front).

and outward spiral as shown in Figure 5.21. Reversal of the Coriolis Effect in the Southern Hemisphere means that surface winds in a high blow in a counterclockwise and outward spiral south of the equator. Surface winds spiraling outward induce descending dry air near the center of a high—hence, fair weather.

Viewed from above in the Northern Hemisphere, surface winds in a low blow in a counterclockwise and inward spiral as shown in Figure 5.22A. Because the Coriolis Effect reverses direction in the Southern Hemisphere, surface winds in a low blow in a clockwise and inward spiral south of the equator. For middle and high latitude lows, surface winds spiraling inward bring together contrasting air masses and induce ascending air along fronts (Figure 5.22B). A *front* is a narrow zone of transition between air masses that contrast in temperature and/or humidity. Air expands and cools as it ascends along a front causing water vapor to condense into clouds—hence, stormy weather.

Lows that track across the northern United States or southern Canada are more distant from sources of moisture (i.e., the Gulf of Mexico and the Atlantic Ocean) and usually produce less rain- or snowfall than lows that track farther south (such as lows that move along the Gulf Coast or up the Eastern Seaboard). Weather to the left side (west and north) of a storm's track (path) tends to be relatively cold, whereas weather to the right (east and south) of a storm's track tends to be relatively warm. For this reason, winter snows are most likely to the west and north of the path of a low-pressure system. We have more to say about storms and their impact on the coastal zone in Chapter 8.

Conclusions

In this chapter we have seen how the ocean plays a key role in the global radiation balance. The ocean, covering about 71% of Earth's surface, has a very low average albedo for incident solar radiation and is the principal sink for incoming solar radiation. The ocean also is the chief source of water vapor and latent heat for the atmosphere. Latent heat is absorbed when water evaporates from the ocean surface and is released to the atmosphere when water vapor condenses during cloud formation. This latent heat transfer is the principal means whereby heat is channeled from the Earth's surface to the atmosphere. Furthermore, ocean currents contribute to heat transport from the tropics to higher latitudes and breaking ocean waves are an important source of cloud condensation nuclei.

The atmosphere and ocean surface are frictionally coupled and winds supply the kinetic energy that drives surface ocean currents and generates ocean waves. Large-scale surface ocean currents emulate the prevailing planetary-scale atmospheric circulation. In the next chapter, we begin our examination of motion in the ocean with a look at wind-driven surface currents and the deeper thermohaline circulation.

Basic Understandings

- Weather is the state of the atmosphere at some particular time and place. Climate is defined as weather at a specific location averaged over a specified interval of time (30 years, by international convention). Climate also encompasses extremes in weather.
- As Earth orbits the Sun, its atmosphere and surface intercepts and absorbs a tiny fraction of the total radiation continually emitted by the Sun. Absorption of solar radiation heats the Earth-atmosphere system. At the same time, the planet continually emits infrared radiation to space, which cools the Earth-atmosphere system, the balance resulting in a nearly constant planetary average temperature.
- As Earth orbits the Sun, regular changes occur in the solar altitude and length of daylight, that affect the intensity of solar radiation striking Earth's surface at a point. Solar altitude is the angle of the Sun above the horizon. The amount of solar energy striking a unit area of Earth surface over an interval of time increases with increasing solar altitude. Daily periods of sunlight range from about 12 hrs every day of the year at the equator to 24 hrs or more of continuous daylight or darkness in polar regions over the period of a year. The regular variation of incoming solar radiation through the course of a year is reflected in the sequence of average monthly temperatures.
- The moderating influence of ocean water reduces the average temperature contrast between summer and winter. Locations immediately downwind of the ocean experience a maritime climate whereas locations well inland from the ocean have continental climates. Proximity to large bodies of water also affects the timing of the average warmest and coldest periods of the year. In the interior United States, the air temperature cycle lags the solar radiation cycle by an average 27 days. But in coastal localities having a strong maritime influence, the average lag time is up to 36 days.
- Solar radiation intercepted by Earth interacts with the atmosphere's component gases and aerosols. These interactions involve scattering, reflection, and absorption. Scattering and reflection alter only the direction of the radiation, but through absorption radiation converts to heat energy.
- Surfaces having a high albedo reflect a relatively large fraction of incident solar radiation and appear light in color. Surfaces having a low albedo reflect a relatively small fraction of incident solar radiation and appear dark in color. Within the atmosphere, cloud tops are the most important reflectors of solar radiation.
- Solar radiation that is not scattered or reflected to space or absorbed in the atmosphere strikes Earth's surface. There, some radiation is absorbed and some is reflected depending on the albedo of the surface. On a global basis, the average albedo of surface ocean waters is only about 8% so that the ocean, covering about 71% of Earth's surface, is the principal sink for solar radiation.
- On a global average annual basis, Earth's planetary albedo is about 30%. The atmosphere absorbs only about 23% of the total solar radiation intercepted by the planet, and Earth's surface (mostly ocean) absorbs 47%.
- Ocean waters absorb most incident solar radiation within relatively shallow depths.

Water absorbs the longer wavelengths (i.e., reds and yellows) of visible light more efficiently than the shorter wavelengths (i.e., greens and blues) so that green and blue penetrate to greater depths. The upper portion of the ocean in which solar radiation is detectable is known as the photic zone.

- The greenhouse effect refers to the heating of Earth's surface and lower atmosphere caused by strong absorption and emission of infrared radiation by certain gaseous components of the atmosphere. The principal greenhouse gas is water vapor. Without the greenhouse effect, Earth would be much too cold to support most forms of plant and animal life.
- The Callendar effect is the theory that global climate change can be brought about by enhancement of the natural greenhouse effect by elevated levels of atmospheric CO_2 from anthropogenic sources, principally the burning of fossil fuels.
- The global distribution of incoming solar radiation and outgoing infrared radiation implies net radiational warming of Earth's surface and net radiational cooling of Earth's atmosphere. In response to this temperature gradient, heat is transferred from Earth's surface to the atmosphere via latent heating and sensible heating. Latent heating refers to the transport of heat from one place to another as a consequence of phase changes of water. Sensible heating involves heat transport via conduction and convection.
- The Bowen ratio describes how heat available at Earth's surface is partitioned between sensible heating and latent heating. The Bowen ratio is lowest for wet surfaces and highest for dry surfaces, varying between 0.1 for the ocean to 5.0 for deserts. The global average Bowen ratio is 0.2.
- On a global average annual basis, radiational heating exceeds radiational cooling between about 35 degrees N and S latitude. Radiational cooling exceeds radiational heating poleward of about 35 degrees N and 35 degrees S. Imbalances in radiational heating and cooling imply net heating in the tropics and subtropics and net cooling at middle and high latitudes. But lower latitudes are not warming relative to higher latitudes because heat is transferred from the tropics toward the poles (poleward heat transport).
- Poleward heat transport is brought about by north-south exchange of air masses and latent heat in storm systems. Also, the ocean contributes to poleward heat transport via wind-driven surface currents and deeper thermohaline circulation that traverse the ocean basins.
- When air pressure readings everywhere are adjusted to sea level, to remove the influence of land elevation, air pressure varies from one place to another and fluctuates from day to day and even from one hour to the next. Spatial and temporal changes in air pressure at Earth's surface arise from variations in air temperature (principally), humidity (water vapor concentration in air), and atmospheric circulation.
- On a weather map, a HIGH or H symbol designates centers of places where sea-level air pressure is relatively high compared to the air pressure in surrounding areas. A high is also known as an anticyclone and is usually a fair weather system. A LOW or L symbol signifies the center of regions where sea-level air pressure is relatively low compared to the air pressure in surrounding areas. A low is also known as a cyclone and often brings stormy weather.
- In response to an air pressure gradient, the wind initially blows from where the pressure is relatively high toward where the pressure is relatively low. The pressure gradient force is always directed across isobars toward low pressure, and its magnitude is inversely related to the spacing of isobars.
- Because Earth rotates, anything moving freely over Earth's surface, including air and water, is deflected to the right in the Northern Hemisphere and to the left in the Southern Hemisphere. This deflection is known as the Coriolis Effect. The magnitude of the Coriolis Effect increases from zero at the equator to a maximum at the poles.
- Viewed from above in the Northern Hemisphere, surface winds blow clockwise and outward in a high, and counterclockwise and inward in a low. Viewed from above in the Southern Hemisphere, surface winds blow counterclockwise and outward in a high, and clockwise and inward in a low.
- At the planetary scale, three broad wind belts encircle each of the Northern and Southern Hemispheres. Trade winds characterize the

tropics, the westerlies encircle middle latitudes, and the polar easterlies prevail at high latitudes. The trade winds and westerlies are linked to the subtropical highs, massive semi-permanent systems that are centered over the ocean basins near 30 degrees N and S.

- Whereas subtropical highs and subpolar lows are nearly stationary, synoptic-scale highs and lows are steered by the prevailing winds blowing several kilometers above Earth's surface. Highs follow lows and lows follow highs, shaping day-to-day weather.

Enduring Ideas

- As Earth orbits the Sun, regular changes take place in local solar altitude and length of daylight resulting in seasonal variations in incoming solar radiation and the march of mean monthly temperatures on Earth's surface.
- Strong absorption and emission of infrared radiation by certain gaseous components of the atmosphere significantly elevates Earth's mean surface temperature. This is the greenhouse effect. The principal greenhouse gas is water vapor; others include carbon dioxide and methane. Human activities (burning of fossil fuels, land clearing) are enhancing the greenhouse effect, causing global warming, by raising the level of atmospheric carbon dioxide.
- The global distribution of incoming solar radiation and outgoing infrared radiation implies net radiational warming of Earth's surface and net radiational cooling of the atmosphere. In response, latent heating and sensible heating transfer heat from Earth's surface to the atmosphere.
- Also implied is net radiational warming of the tropics and subtropics and net radiational cooling at middle and high latitudes. In response, poleward heat transport is brought about by air mass exchange, latent heating in storm systems, and ocean circulation.
- The ocean is a major player in Earth's climate system. Because of its low albedo and large surface area, the ocean is the principal absorber of solar radiation intercepted by Earth. As the largest reservoir in the global water cycle, the ocean is the chief source of atmospheric water vapor and latent heat. As a major reservoir in the global carbon cycle, the ocean helps to regulate the level of the greenhouse gas carbon dioxide in the atmosphere.

Review

1. Distinguish between weather and climate.
2. Compare the length of daylight at a middle latitude location on the winter solstice with that on the summer solstice. Also compare the solar altitude at noon local time on these two days.
3. Outside of the tropics, what is the relationship between the annual solar radiation cycle and the annual temperature cycle? How is this relationship influenced by proximity to large bodies of water?
4. For incoming solar radiation, compare the average albedo of the ocean surface to the average albedo of cloud tops. What is the albedo of the Earth-atmosphere system (i.e., the planetary albedo)?
5. Define the photic zone and describe how its depth is affected by the concentration of suspended particles and dissolved organic matter in seawater.
6. What is the cause and significance of Earth's greenhouse effect?
7. Define latent heating and sensible heating and compare the importance of the two processes in transferring heat energy from Earth's surface to atmosphere on a global average annual basis.
8. Describe the processes involved in poleward heat transport within the Earth-atmosphere system.
9. How does the Coriolis Effect deflect large-scale winds and ocean currents in the Northern and Southern Hemispheres? Why is there no Coriolis Effect at the equator?
10. Describe the surface horizontal winds in high and low pressure systems as viewed from above in the Northern Hemisphere.

Critical Thinking

1. How does the ocean moderate the climate of downwind coastal areas?
2. The average albedo of the ocean surface is only 8%. What is the significance of this relatively low albedo for Earth's climate system?
3. Speculate on how higher sea surface temperatures (SST) might impact Earth's greenhouse effect.
4. Describe the ocean's role in poleward heat transport.
5. How do the subtropical anticyclones influence the salinity of surface ocean waters?
6. How does an enhanced greenhouse effect influence Earth's global radiative equilibrium?
7. How does less ice cover in the Arctic Ocean affect the flux of water vapor from the ocean to the atmosphere?
8. How does less ice cover in the Arctic Ocean affect heat transport from the ocean to the atmosphere?
9. In what way are hurricanes and other tropical cyclones important contributors to poleward heat transport?
10. Describe the track of a storm system (tropical or extra-tropical cyclone) that is most likely to cause considerable erosion along the U.S East Coast.

ESSAY: Location at Sea, A Historical Perspective

When humans first left their caves they faced the challenge of determining their location, how to get to where they wanted to go, and how to get back to where they started from. On land and inland waters, while not easy, the task was manageable. Using prominent landmarks such as shorelines or rock pinnacles as signposts, travelers developed a sense of direction and distance. Eventually common routes of travel for discovery, trade, and communication produced well-worn paths. Upon leaving land for the open ocean, the problem of location became much more difficult. Once out of sight of the shore, with water in all directions as far as the eye could see, ancient mariners depended mostly on luck to return them to port. It is not surprising that early seafaring was mostly a coastal affair.

Eventually humans developed a rudimentary understanding of winds and ocean currents and became familiar with the local geography of coastlines and the seas they regularly plied. Knowledge of the true shape of Earth allowed sailors to understand the relationship between their north-south position (latitude) and the location of the stars and Sun in the sky. By day, mariners determined their latitude by measuring the altitude of the Sun at its highest point in the sky (zenith) and knowing the calendar date. But they never knew with any certainty where their vessel was located in terms of the east-west direction. And when clouds obscured the sky or storm-tossed seas prevented an estimate of Sun or star angle, even the north–south position was in doubt. The inability to determine longitude with any certainty led to many shipwrecks, thousands of lives lost, and huge financial losses.

About CE 150, the Egyptian astronomer and geographer Ptolemy (ca. CE 85-165) proposed the concept of latitude and longitude as imaginary reference lines on the globe by which location could be specified. His zero degree parallel (latitude) was set at the equator because he knew that the Sun, Moon and planets pass almost directly over that location. His zero-degree longitude line (the *prime meridian*) was set in the Canary Islands. Perhaps for political reasons, others drew the prime meridian through Paris, Philadelphia, Moscow, and at least a half dozen other places. Since 1884, by international convention the prime meridian runs through Greenwich, England, a bit to the east of Ptolemy's mark. Longitude is measured in degrees to the east and west of that line to the common 180-degree meridian, essentially the *International Date Line* (the imaginary line on Earth that separates two consecutive calendar days).

The search for a connection between longitude and the regular motions of celestial bodies would challenge navigators and astronomers for more than 1000 years. By the early 18th century, governments of various maritime nations offered substantial prize money for the person who could solve the "longitude problem," that is, develop a reliable method for determining longitude at sea. With the Longitude Act of 1714, the British Parliament offered the biggest prize—several million dollars in today's money. A Board of Longitude was appointed consisting of a blue-ribbon panel charged with conducting trials of proposals and awarding prize money.

A promising approach to solving the longitude problem was to develop a way to determine precisely the time of the day. Earth is roughly a sphere that rotates through 360 degrees once every 24 hrs, that is, 15 degrees per hour. If one knew the current time at the prime meridian (from a clock) and the current time at one's location at sea (based on the Sun's position in the sky), it would be easy to calculate the number of degrees represented by the difference between the two times to yield the longitude. For example, at local noon on a ship, if the clock read 3 p.m., the longitude of the ship was 3 hours or 45 degrees west of the prime meridian.

The search for a solution to the longitude problem also involved considerable work on possible astronomical techniques (analogous to the way latitude was determined). The Board of Longitude favored astronomical solutions because they were considered more scientific than reliance on clocks. Nonetheless, the most successful efforts to solve the longitude problem focused on the design and construction of a very accurate clock that could keep time over a lengthy sea voyage and withstand the harsh conditions at sea without suffering mechanical failure.

The first in his series of five accurate and durable nautical chronometers was invented by the English clockmaker John Harrison (1693-1776) in 1733, and the final one completed about 39 years later. (Figure 1). In 1773, he was awarded the 20,000 Pounds Sterling prize offered by Great Britain for a clock that could be used to determine longitude to within one-half a degree (30 nautical miles at the equator) on a voyage from England to any port in the West Indies. It would be another 50 or so years for this technology to find its way onto virtually every ship. Meanwhile, most sea captains continued to rely on dead reckoning to determine their location. The clock-method for determining longitude was used for more than two centuries. Beginning in 1904, radio signals were broadcast worldwide to indicate the official time at the the prime meridian. Seafarers received these signals and also used a sexton to determine their local time. These two time values enabled seafarers to calculate the longitude of their ship. Since the start of World War II, ships have also relied on the radio-based navigation system known as LORAN (LOng RAnge Navigation) to determine location. In recent years, these techniques have been supplanted by satellite-based techniques including the *Global Positioning System (GPS)*, described in the first Essay in Chapter 6.

For more details on the longitude problem and how it was solved, refer to the delightful book *Longitude* by Dava Sobel (Penguin Books, 1995).

FIGURE 1
English clockmaker John Harrison (1693-1776), at top, was credited with solving the longitude problem by developing an accurate sea-worthy clock. Harrison's clock determined the longitude to within 0.5 degree on a voyage from England to a port in the West Indies. The last of John Harrison's chronometers is pictured below. [Bottom photo courtesy of the Collection of the Worshipful Company of Clockmakers; Photo by Racklever/License: Creative Commons Attribution ShareAlike 3.0]

ESSAY: The Stratospheric Ozone Shield and Marine Life

Ozone (O_3) is a relatively unstable molecule made up of three atoms of oxygen, but has certain physical and chemical properties that distinguish it from free oxygen (O_2). As shown in Figure 1, the greatest concentration of ozone occurs naturally in the stratosphere at altitudes near 40 km (25 mi). The formation and dissociation of stratospheric ozone shields marine and terrestrial organisms from exposure to potentially lethal intensities of solar ultraviolet (UV) radiation. Without the *stratospheric ozone shield*, life as we know it could not exist on Earth. It was not until about 1 billion years ago that the atmosphere contained sufficient oxygen (from which ozone is generated), that an effective ozone shield developed. From that point on, for example, marine organisms could thrive in surface waters. Another 560 million years would pass until there was sufficient ozone to allow life to leave the protection of the seas. However, in the latter portion of the 20th century, scientists became aware of a serious threat to the stratospheric ozone shield that poses a hazard for all forms of life including marine organisms.

Within the stratosphere, two sets of competing chemical reactions, both powered by solar ultraviolet radiation, continually generate and destroy ozone (Figure 2). During ozone production, UV strikes an oxygen molecule (O_2) causing it to split into two free oxygen atoms (O). Free oxygen atoms then collide with molecules of oxygen to form ozone molecules (O_3). At the same time ozone is destroyed. Ozone absorbs ultraviolet radiation, splitting the molecule into one free oxygen atom (O) and one molecule of oxygen (O_2). The free oxygen atom then collides with an ozone molecule to form two molecules of oxygen. The net effect of these opposing sets of chemical reactions is a minute reservoir of ozone that peaks at only about 10 parts per million (ppm) in the middle stratosphere. Ultraviolet radiation (at different wavelengths) powers both sets of chemical reactions so that much but not all UV radiation is prevented from reaching Earth's surface.

As a general rule, a 1% decline in stratospheric ozone concentration means a 2% increase in the intensity of UV that passes through the ozone layer and reaches Earth's surface, although it also depends on the clouds and dust in the atmosphere. The most dangerous portion of this radiation that strikes Earth's surface is *UVB*, which spans wavelengths from 0.29 to 0.32 micrometer. The effects on human health from exposure to too much UVB are well known and include increased incidence of skin cancer, premature aging of the skin, cataracts of the eye, and suppression of the immune system. We share those adverse effects with marine mammals, such as seals and sea lions which are also hairless and spend a portion of their life on the land. Less well known are the impacts on marine ecosystems.

An unprecedented increase in UV radiation penetrating ocean surface waters can disrupt the orderly functioning of marine food webs, adversely affecting predator-prey relationships, competition, species diversity, and the dynamics of trophic (feeding) levels. Of particular concern is the potential impact of elevated levels of solar ultraviolet radiation on autotrophs that form the base of marine food webs. Too much UV affects the growth and reproduction of autotrophs thereby

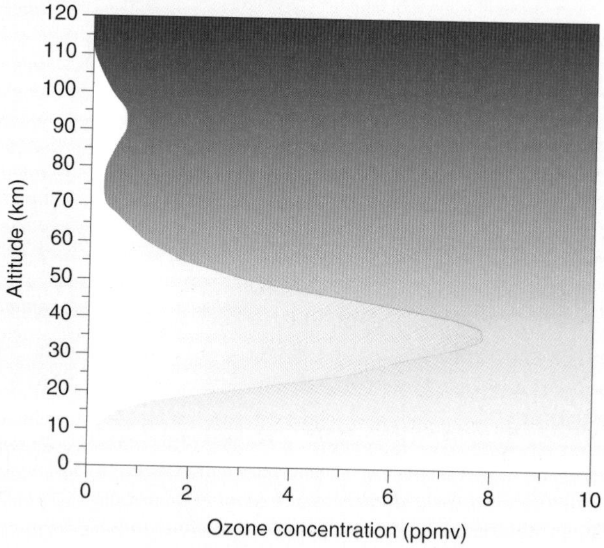

FIGURE 1
The concentration of ozone (O_3) peaks in the stratosphere. [Source: U.S. Standard Atmosphere, 1976]

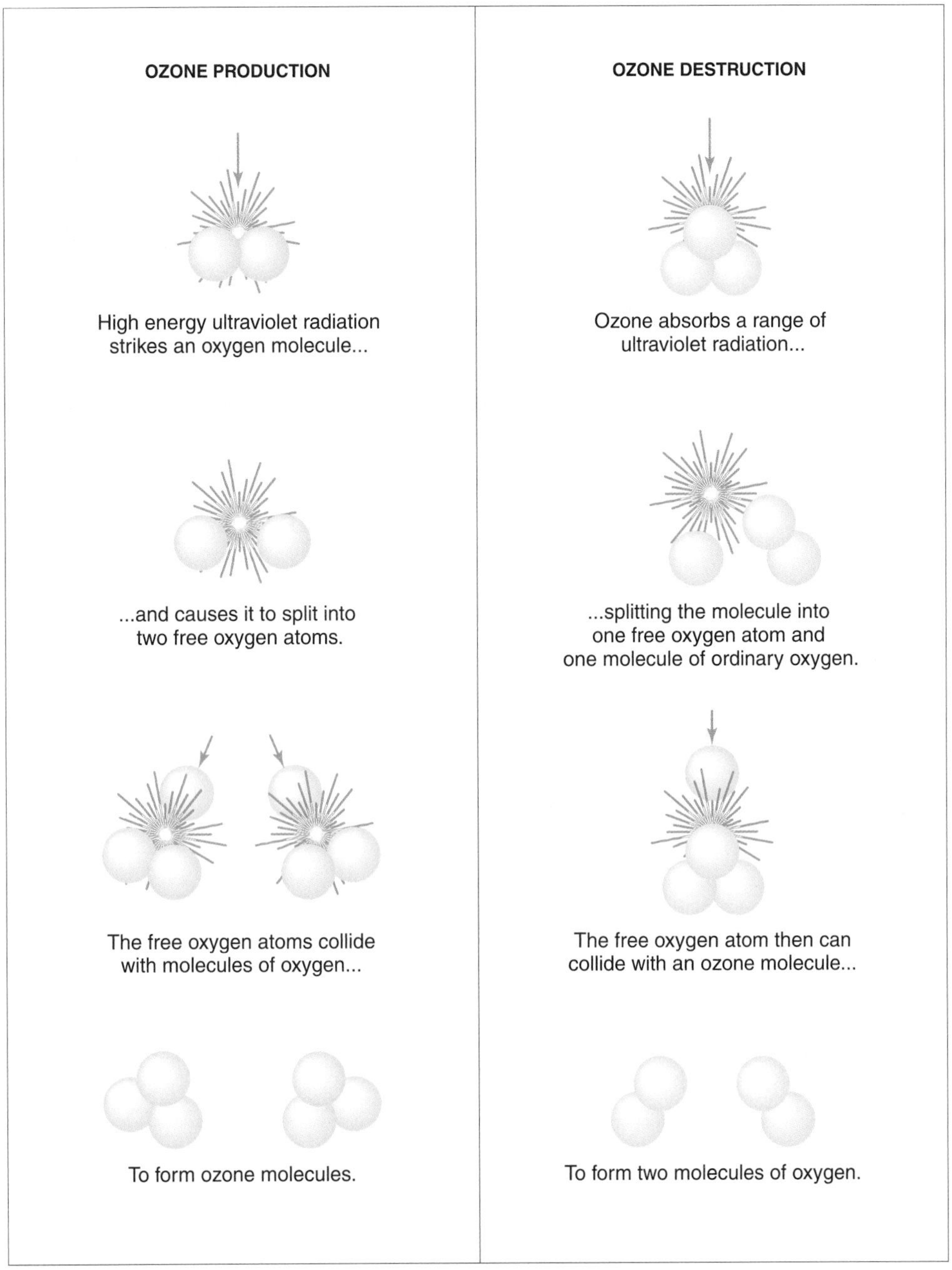
OZONE PRODUCTION
High energy ultraviolet radiation strikes an oxygen molecule...
...and causes it to split into two free oxygen atoms.
The free oxygen atoms collide with molecules of oxygen...
To form ozone molecules.
OZONE DESTRUCTION
Ozone absorbs a range of ultraviolet radiation...
...splitting the molecule into one free oxygen atom and one molecule of ordinary oxygen.
The free oxygen atom then can collide with an ozone molecule...
To form two molecules of oxygen.

FIGURE 2
Production and destruction of stratospheric ozone.

reducing the food (energy) supply for organisms occupying higher trophic levels (Chapter 9). Zooplankton (free-floating heterotrophs, such as bacteria and krill) feed on phytoplankton and are then food for small fish and larger arthropods, which are food for larger fish and marine mammals. All would be affected. Some phytoplankton can move out of harms way, but exposure to UVB impairs their orientation mechanisms and motility making even those species more susceptible to radiation damage. In addition, too much UV impairs larval development, harms young fish, shrimp, crabs, and amphibians, and threatens macro-algae and sea grasses.

In November 2010, Karina Acevedo-Whitehouse, a wildlife molecular epidemiologist at the Institute of Zoology in London, U.K., graduate student Laura M. Martinez-Levasseur, and colleagues presented evidence that exposure to UV may be adversely impacting whales, causing sunburned and blistered skin. Whales are exposed to solar ultraviolet radiation when they surface for air, feed their young, or just rest at the ocean surface. Researchers propose that a possible contributing factor to elevated levels of UV radiation is the depleted ozone shield.

Over a three-year period (2007-2009), the research team studied three cetacean species: blue whales (*Balaenopter musculus*), fin whales (*B. physalus*), and sperm whales (*Physeter macrocephalus*) in the Gulf of California. They took 156 high-resolution images of individual whales and skin biopsies from 142 whales (obtained from darts shot at the whales). Laboratory analysis indicated that 95% of all skin samples had sun-damaged cells and in 56% of the skin samples, damage extended through all layers of the skin down to the lowest basal layer. The amount of damage varied with species; blue whales were most susceptible of the three species probably because the blue whale has the lowest level of skin pigmentation (lightest in color) (Figure 3). During the three-year study period, the proportion of blue whales having skin blisters increased by 56%. Fin whales are the darkest of the three species and exhibited the least damage. Sperm whales had almost the same level of damage as the blue whales perhaps because they spend more time at the surface breathing between dives. So far there is no indication of cancers or melanomas.

FIGURE 3
Blue whale as seen from spotter aircraft. Blue whales are susceptible to skin damage due to their relatively light color and their need to come to the surface to breathe between dives. [Courtesy of NOAA]

The potential role of human activity in the depletion of the stratospheric ozone shield first came to public attention more than three decades ago when scientists discovered that a group of chemicals, known as CFCs (for *chlorofluorocarbons*), posed a serious threat to the stratospheric ozone shield. First synthesized in 1928, CFCs were widely used as chilling (heat-transfer) agents in refrigerators and air conditioners, for cleaning electronic circuit boards, and in the manufacture of foams used for insulation. F.S. Rowland and M.J. Molina of the University of California at Irvine first warned of the threat of CFCs to the stratospheric ozone shield in 1974. Five years later, use of CFCs as propellants in common household aerosol sprays such as deodorants and hairsprays was banned in the United States, Canada, Norway, and Sweden. For their pioneering work on the depletion of stratospheric ozone, Rowland, Molina (now at the University of California, San Diego), and P.J. Crutzen (of the Max Planck Institute for Chemistry, Germany) were awarded the 1995 Nobel Prize in chemistry.

Worldwide acceptance of the threat posed by CFCs and other ozone-depleting substances to the stratospheric ozone shield prompted the United Nations Environmental Programme (UNEP) in 1987 to draw up the *Montreal Protocol on Substances that Deplete the Ozone Layer*. This international treaty entered into full force on 1 January 1989 and by 16 September 2009, all 192 UN member states had ratified the original Montreal Protocol. The initial goal of the treaty was to cut CFC production in half by 1992 (compared to 1986 levels). However, the seriousness of the problem led to seven subsequent amendments (revisions) that included expanding the list of regulated substances (e.g., to include bromine-containing halons) and requiring the worldwide phase out of the manufacture and use of CFCs beginning January 1996.

Certain CFCs are inert (chemically non-reactive) in the troposphere, where they accumulated for decades. Atmospheric circulation transports CFCs into the stratosphere where, at altitudes above 25 km (15.5 mi), intense UV radiation breaks down CFCs, releasing chlorine (Cl), a gas that readily reacts with and destroys ozone. Products of this reaction are chlorine monoxide (ClO) and molecular oxygen (O_2). Chlorine (Cl) is a catalyst in chemical reactions that convert ozone to oxygen. (A catalyst is a chemical substance that facilitates reactions without being consumed in them.) In this way, each chlorine atom can destroy perhaps tens of thousands of ozone molecules.

First signs of a thinner stratospheric ozone shield came from Antarctica. For about six weeks during the Southern Hemisphere spring (mainly in September and October), the ozone layer in the Antarctic stratosphere (mostly at altitudes from 11 to 23 km, or 7 to 14 mi) thins drastically and then recovers during November (Figure 4). The *Antarctic ozone hole* is defined as the thinning of the ozone layer over the continent to levels significantly below what it was in 1979. The National Ozone Expeditions to the U.S. McMurdo Station in 1986-87 plus NASA aircraft flights into the Antarctic stratosphere

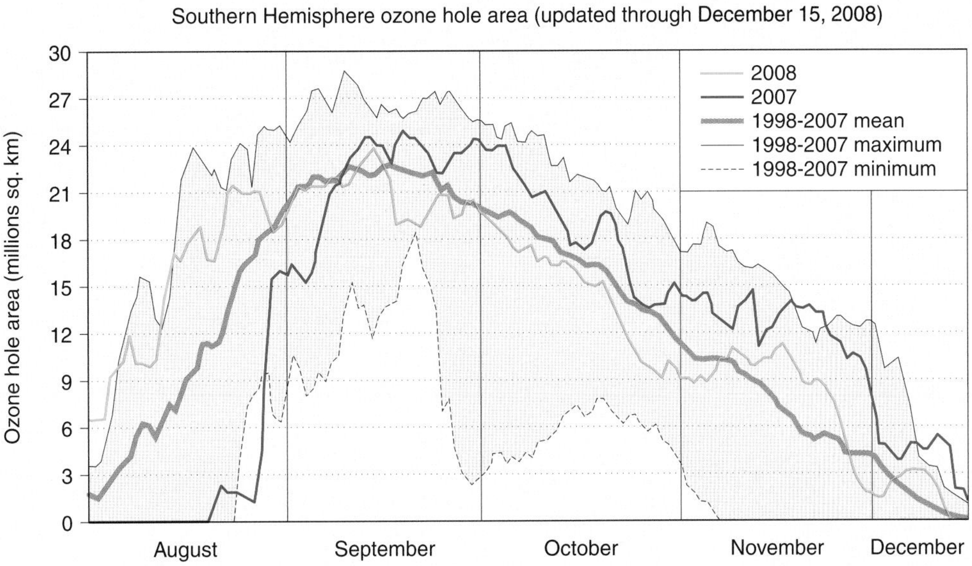

FIGURE 4
The Antarctic ozone hole is a widespread area of stratospheric ozone depletion that develops over Antarctica during the Southern Hemisphere spring. The mean area encompassed by the Antarctic ozone hole during 1998-2007 increased from near zero at the beginning of August, peaked in September, and declined to zero in December. [NOAA and NASA Goddard Space Flight Center]

in 1987 found relatively high concentrations of chlorine monoxide (ClO) in the Antarctic stratosphere. This discovery established a convincing link between ozone depletion and CFCs.

The Antarctic is particularly susceptible to ozone depletion because of the exceptionally low stratospheric temperatures arising from extreme radiational cooling during the long, dark winter. Stratospheric temperatures over Antarctica plunge below –88 °C (−126 °F). At such frigid temperatures, *polar stratospheric clouds (PSCs)* form. These clouds (composed of tiny particles of water ice, nitric acid, and sulfuric acid) promote ozone depletion by providing surfaces on which chlorine and bromine compounds that are inert toward ozone are converted to active forms that destroy ozone. Once the Sun reappears in spring, solar radiation supplies the energy that causes active forms of chlorine and bromine to begin destroying ozone.

Ozone depletion takes place while the Antarctic atmosphere is essentially cut off from the rest of the planetary-scale atmospheric circulation by the *circumpolar vortex*, a belt of strong winds that encircles the outer margin of the Antarctic continent. A month or so into the austral spring with returning solar radiation, however, the circumpolar vortex begins to weaken, allowing warmer ozone-rich air from lower latitudes to invade the Antarctic stratosphere. Polar stratospheric clouds vaporize and the stratospheric ozone concentration returns to normal levels; that is, the Antarctic ozone hole fills.

The 2000 ozone hole reached 28.3 million square km, exposing parts of southern Argentina and Chile and the Southern Ocean to elevated UV radiation. The 2006 ozone hole was the most severe, the thinnest and widest on record, reaching an average of 27.5 million square km from 21-26 September 2006 (Figure 5). (For some perspective on these numbers, the area of North America is 24.25 million square km.) Since then, the Antarctic ozone hole has been slowly decreasing and likely will be gone by 2080.

Scientists investigating stratospheric chemistry in the Arctic in early 1989 discovered ozone-destroying chlorine compounds and a slight thinning of ozone. An Arctic ozone hole comparable in magnitude to the Antarctic ozone hole is unlikely for two reasons. For one, in winter the Arctic stratosphere averages about 10 Celsius degrees (18 Fahrenheit degrees) warmer than the Antarctic stratosphere, making formation of polar stratospheric clouds unlikely. Secondly, the circumpolar vortex that surrounds the Arctic weakens earlier than its Antarctic counterpart. However, an exceptionally cold Arctic winter coupled with an unusually persistent circumpolar vortex could translate into significant ozone depletion in the Arctic.

Phase out of CFCs and other ozone-destroying chemicals slowed the release of chlorine into the atmosphere. However, CFCs have long atmospheric lifetimes so that these substances will continue to threaten the stratospheric ozone shield for some time to come in spite of the phase out. The 2006 UNEP *Scientific Assessment of Ozone Depletion Concentrations* reported that ozone-depleting chlorine levels in the stratosphere are on the decline, but recovery of stratospheric ozone to pre-1980 levels is not likely until the middle of this century in middle latitudes and a decade or two later at high latitudes. Meanwhile, scientists continue to monitor stratospheric ozone levels worldwide.

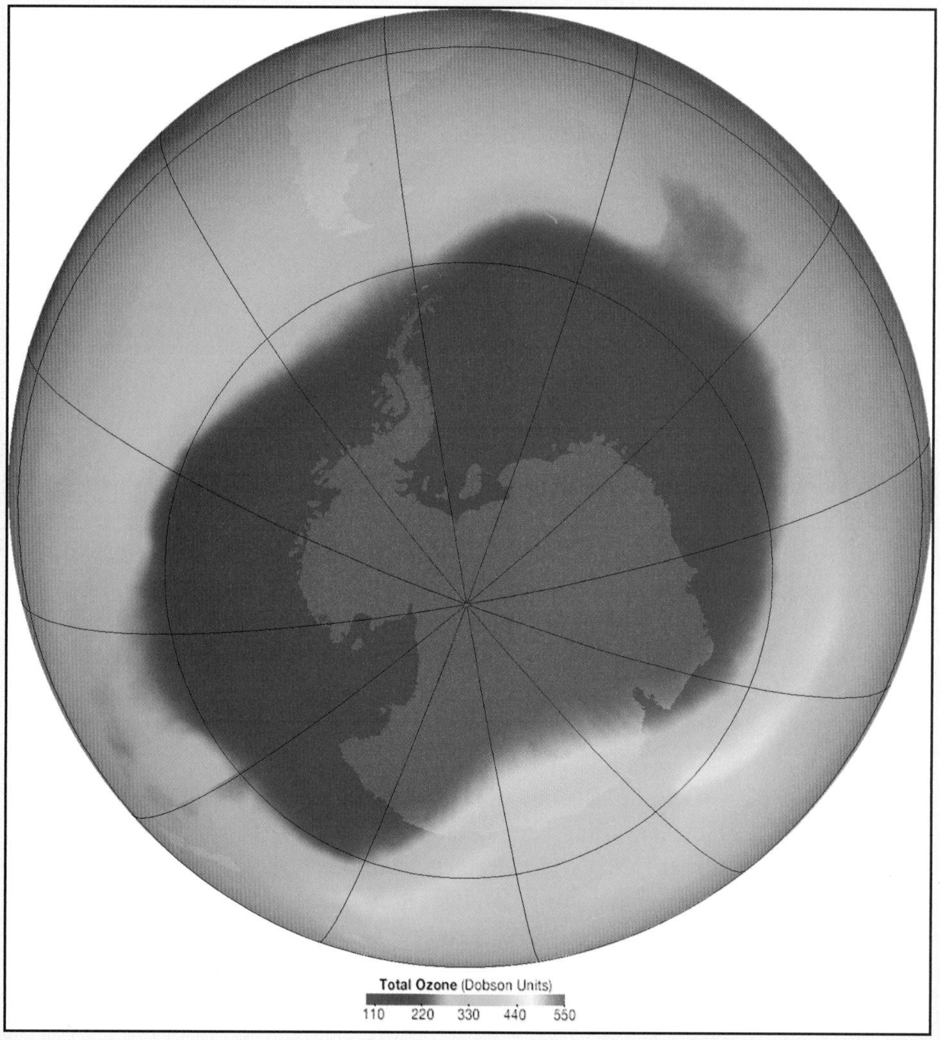

FIGURE 5
From September 21-30, 2006, the average area of the ozone hole was the largest ever observed at 27.5 million square km (10.6 million square mi). This image, created from measurements generated by the Ozone Monitoring Instrument on NASA's Aura satellite, shows the Antarctic ozone hole on September 24, 2006. The ozone hole on this day tied the single-day largest area of 29.5 million square km (11.4 million square mi), which was previously reached on September 9, 2000. The blue and purple colors indicate where there is the least ozone, and the greens, yellows, and reds display higher ozone levels. [Courtesy of NASA]

CHAPTER 6

OCEAN CURRENTS

Closeup of a Naval Oceanographic Office glider being deployed as part of a Gulf of Mexico Loop Current research cruise. [Courtesy of NOAA]

Case in Point

For much of human history, knowledge of the ocean and its currents was not recorded for many reasons. For one, most sailors could not write so their basic understandings of the sea were passed on orally. Even after the European nations (Portugal, Spain, Netherlands, England, France, and Russia) began systematic ocean exploration, their discoveries were closely guarded secrets, far too valuable to be made public. Hence, we know little of what sailors understood about ocean currents during the period of great ocean exploration from the 15th to 18th centuries.

Existence of the Gulf Stream, a major surface ocean current, was known as early as 1519. Bishop Resen of Copenhagen drew the first map of the Gulf Stream in 1605 based on records from trans-Atlantic voyages of the English explorer Martin Frobisher (1535-1594) and a crude sketch attributed to Icelanders. Subsequent charts showing the Gulf Stream were published in 1678 (by the Jesuit Athanasius Kircher) and 1685 (by a German named Happelius).

More widely known is Benjamin Franklin's study of the Gulf Stream. Franklin (1706-1790) served as colonial deputy postmaster general from 1753 to 1774. He found that British merchant ships arriving in the colonies from England took many days to weeks longer to make the voyage across the Atlantic than American vessels. British

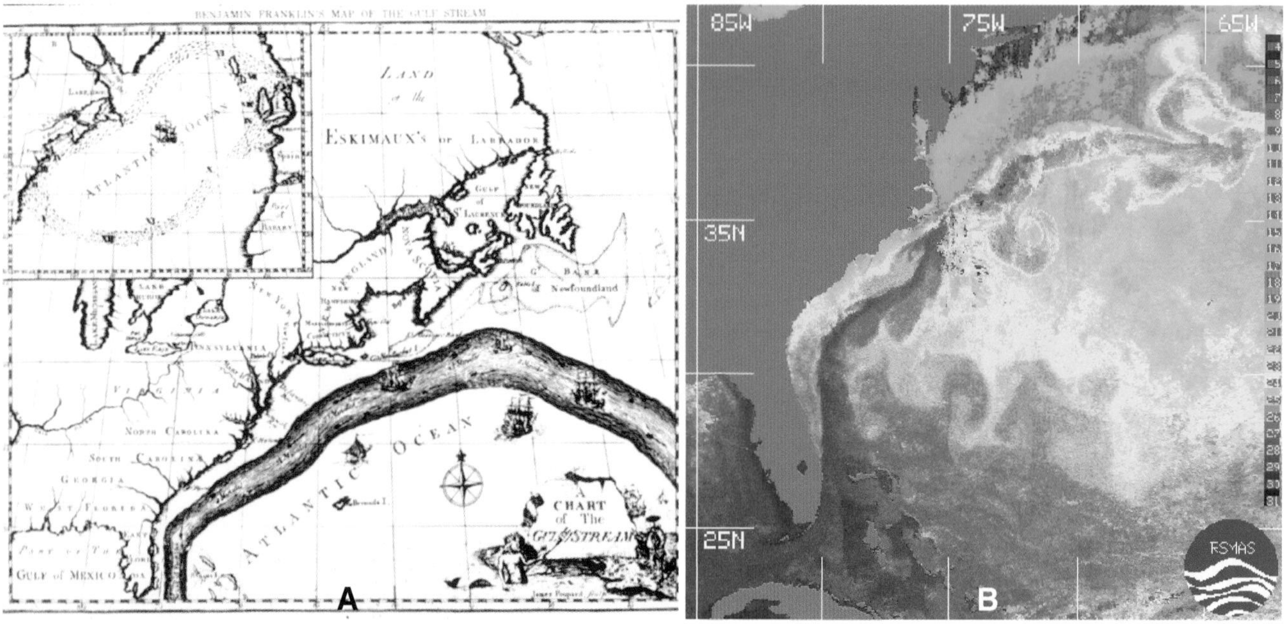

FIGURE 6.1
The location of the Gulf Stream (A) plotted on the 1769 Franklin-Folger map, and (B) on a much more recent infrared satellite depiction of sea-surface temperatures. Shades of red indicate the highest sea-surface temperatures. [Source: NOAA; University of Miami, Rosenstiel School of Marine & Atmospheric Science]

postal authorities asked Franklin for an explanation. Franklin consulted with his cousin, Timothy Folger, a Nantucket, MA, whaling ship captain, who told him about the Gulf Stream, the strongest surface current in the North Atlantic (often 10 to 12 km per hr or 6 to 7 mph—roughly the speed of a modern sailing ship). On eastbound voyages, American captains sailed with the current but on the return voyage they would avoid it, cutting up to weeks off the journey. At the time, British captains were unaware of the Gulf Stream.

Franklin published his cousin's chart showing the location of the Gulf Stream and presented it to the British. Interestingly, the location of the Gulf Stream plotted on the 1769 Franklin-Folger map shown in Figure 6.1A is remarkably similar to the location of the Gulf Stream as revealed by the satellite-derived sea-surface temperature (SST) pattern shown in Figure 6.1B. British authorities refused to take advantage of or even acknowledge the Franklin-Folger map. Franklin took measurements of sea surface temperatures during his many crossings of the Atlantic Ocean, thereby developing a navigational technique based on the location of the relatively warm Gulf Stream waters.

In the mid 19th century, the American naval officer Matthew Fontaine Maury (1806-1873) conducted the first systematic study of the ocean's surface currents and winds (Figure 6.2). Maury compiled information on currents and winds from the logbooks of sailors' observations

FIGURE 6.2
The American naval officer Matthew Fontaine Maury (1806-1873) compiled wind and ocean current data from thousands of ships' logs and produced the first reliable wind and current charts of the ocean. [Courtesy of U.S. Navy]

stored at the U.S. Navy's Depot of Charts and Instruments and published the first charts of the North Atlantic in 1847. He estimated current direction and speed by analyzing deflections in a ship's course caused by surface ocean currents. Failure to correct a ship's course for current-induced deflections meant that the ship's final position at the end of a run would differ from its intended destination. Combining thousands of such observations, Maury constructed a map of average surface currents over much of the ocean. A skilled navigator having knowledge of currents can correct a ship's course to compensate for current-induced deflections.

Studies of ocean currents continue today for many reasons, ranging from military operations to predicting oil spill trajectories at sea. Through the years, instruments used to monitor currents have become more sophisticated. Satellite observations and data from in situ instruments have largely replaced analysis of ships' logs. Global Positioning Systems (GPS) now measure a ship's location to within 5 to 10 m (16 to 33 ft) so that the deflection caused by ocean currents can be determined accurately and instantaneously. For more on GPS, see this chapter's first Essay.

Driving Question:
What causes ocean waters to circulate and what are the prevailing patterns of ocean currents?

Energy and matter (e.g., heat, water) are continually exchanged between the ocean and atmosphere, and these processes drive the ocean circulation. Evaporation, precipitation, runoff from the continents plus heating and cooling bring about changes in the temperature and salinity of surface waters. Density changes that accompany variations in temperature and salinity can cause water to sink or rise in the ocean. *Kinetic energy* (energy of motion) is transferred from near-surface winds to the ocean's surface layer, driving the currents that dominate the motion of the ocean's upper hundred meters or so. Winds are responsible for not only horizontal currents but also vertical water motions within the surface layer.

Most seawater (90%) is in the deep ocean, isolated from the atmosphere and its winds. Deep-ocean waters are cold and come to the surface primarily at high latitudes where they interact with the atmosphere. Differences in water density drive the sluggish circulation of deep water. Typically these waters flow at speeds of less than 1 cm per sec or 1 km per day—about 240 times slower than the Gulf Stream. Because of its volume and isolation from the atmosphere, the deep ocean is both a storehouse for heat (acting as a buffer for global-scale temperature change) and a reservoir for dissolved gases such as carbon dioxide that are sequestered for centuries to millennia.

The ocean features wind-driven surface currents and the deep ocean's slower density-driven thermohaline circulation. In this chapter, we examine the characteristics of these two circulation regimes, and the governing forces. We begin by describing the vertical structure of the ocean.

Ocean's Vertical Structure

Most of the ocean is divided into three horizontal depth zones based on density: the mixed layer, pycnocline, and deep layer (Figure 6.3). At high latitudes, a weak pycnocline may occur near or at the surface.

Wind-driven surface currents are restricted to the ocean's uppermost 100 m (300 ft.) or less. The strongest currents occur in the ocean's surface layer, although some surface currents such as western boundary currents like the Gulf Stream (discussed later) can be relatively strong to depths of several hundred meters. Surface currents are changeable, continually responding to variations in the wind, precipitation, and heating or cooling. Stirring of surface waters by the wind produces a well-mixed layer of uniform or nearly uniform density. For this reason, the surface ocean is known as the **mixed layer**. We are most familiar with the properties of the mixed layer because sensors onboard ships, aircraft, and Earth-orbiting satellites can readily monitor it.

The **pycnocline**, situated between the mixed layer and the deep layer, is where water density increases rapidly with depth because of changes in temperature and/or salinity. Recall that cold water is denser than warm water and seawater is denser than fresh water. Where declining temperature is responsible for the increase in density with depth, the pycnocline coincides with a **thermocline**; this is the case for most of the world ocean. On the other hand, if an increase in salinity is responsible for the increase in density with depth, the pycnocline coincides with a

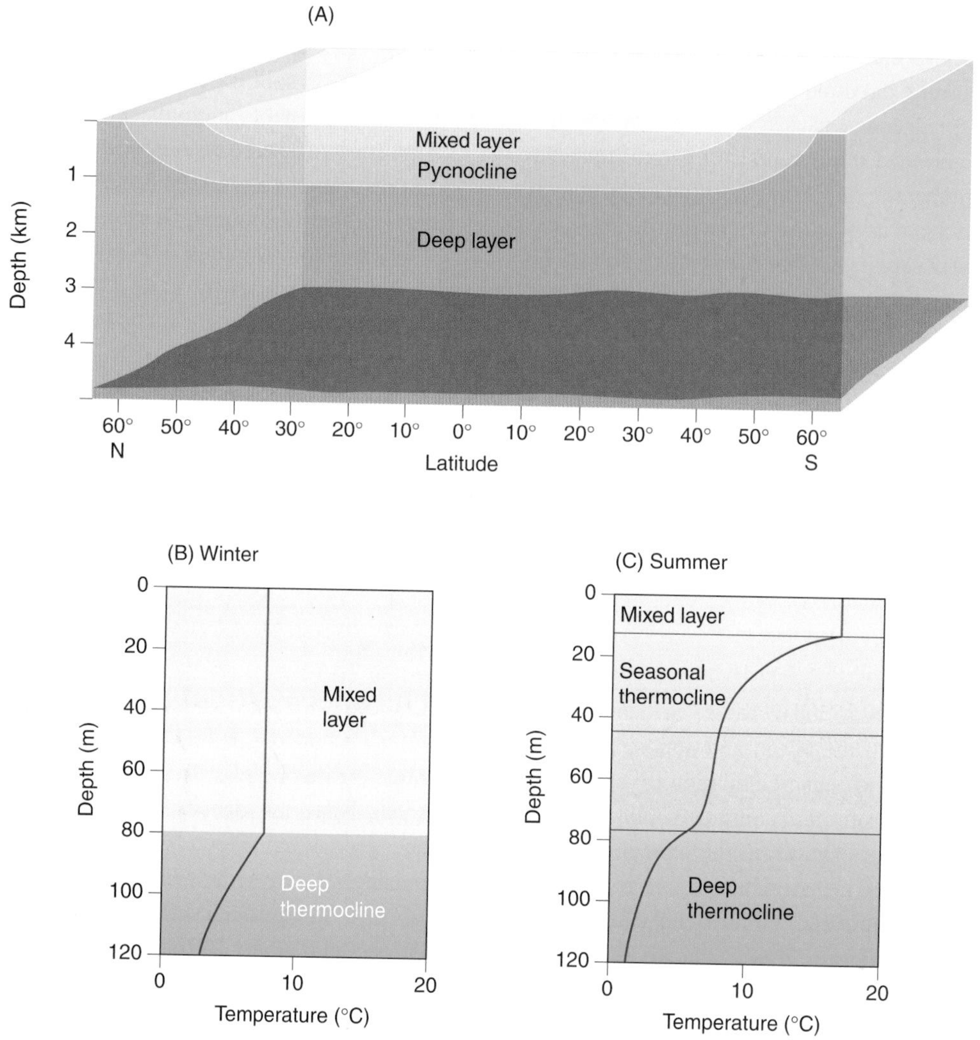

FIGURE 6.3
A model of density stratification in the ocean. (A) Except at high latitudes, a shallow mixed layer of relatively warm low-density water overlies the pycnocline characterized by a rapid increase in density with depth. The pycnocline, in turn, overlies the dark and cold deep layer accounting for 90% of the ocean's mass. In temperate latitudes, seasonal variations in the temperature profile are evident in (B) winter versus (C) summer. In these two seasonal cases, the pycnocline is equivalent to a thermocline and in summer a shallow thermocline develops above the main thermocline (pycnocline) in response to intense solar radiation.

halocline. Typically, the pycnocline extends to a depth of 500 to 1000 m (1600 to 3300 ft), but in middle latitudes seasonal pycnoclines may develop within the mixed layer (Figure 6.3C). The dark, cold deep layer below the pycnocline accounts for most of the ocean's mass. Within the **deep layer**, density increases gradually with depth and water moves slowly; in only a few locations (usually near the ocean bottom) are water movements fast enough to be considered currents.

The ocean's three-layer structure is an example of how gravity stratifies a fluid into layers such that the density of each layer is less than the density of the layer below it. More dense fluids sink and less dense fluids are buoyed up. The ocean's pycnocline is very stable thus suppressing mixing between the mixed layer and deep layer; that is, the pycnocline acts as a barrier to vertical motion within the ocean. The concept of stability is useful in understanding this property of the pycnocline.

Stability affects the vertical motion of ocean water. A system is described as stable if it tends to persist in its original state without changing. Following a disturbance (i.e., vertical motion), a **stable system** returns to its initial state or condition. As noted above, the usual stable state of the ocean features a layer of water (the mixed layer) that is warmest near its interface with the atmosphere and the mixed layer overlies water that becomes denser with increasing depth (the pycnocline). Strong storm winds may temporarily disturb this stable stratification bringing colder than usual water to the surface. Once the wind slackens, however, the original layered structure is soon restored.

To demonstrate the effects of differences in water density on the circulation of ocean water, you can conduct a simple experiment. You will need two glasses or glass beakers, food coloring, salt, and a medicine dropper. Fill one glass with fresh water and the other with salt water, both at room temperature. Add several drops of food coloring to the salt water and mix thoroughly. Let the glasses of water stand for a few minutes. Use the medicine dropper to put a drop of colored salt water into the glass of fresh water. Observe that the denser salt water sinks to the bottom of the fresh water. Now empty and wash the fresh-water glass. Again fill it with fresh water at room temperature. Add several drops of food coloring to the freshwater sample and mix thoroughly. Again, let the glasses of water stand for a few minutes. Use the medicine dropper to put a drop of the colored fresh water into the glass of salt water. You will observe that the fresh water rises to the surface because it is less dense than the salt water. Carefully add more fresh water with the dropper to form a two-layered system.

Suppose we add a drop of slightly salty water of intermediate density (using a different food coloring) to our two-layered system. The drop will sink through the fresh water and come to rest at the boundary between the two layers due to the differences in density between the water layers and the drop. The density distribution is stable when the densest water (in this case the saltiest) is at the bottom and the least dense fresh water is on top. Immediately after adding a drop of slightly salty water, we created an unstable density distribution. Because the drop is denser than the surrounding fresh water, it sinks. However, if a drop of slightly salty water is carefully injected into the very salty water, it rises.

This is an example of an unstable density distribution because the water drop is less dense than the surrounding waters and is buoyed upward. Hence, an **unstable system** shifts spontaneously towards a more stable density stratification.

Mixing the system so that water density is uniform throughout produces a neutrally stable density distribution. Following a disturbance, a **neutrally stable system** neither departs from its new state nor returns to its previous state.

Ocean in Motion: The Forces

Once the wind sets surface waters in motion as a current, the Coriolis Effect, Ekman transport, and the configuration of the ocean basin modify the speed and direction of the current. In this section, we consider the forces involved in the coupling of wind and ocean surface waters.

WIND-DRIVEN CURRENTS AND EKMAN TRANSPORT

Wind blowing over the ocean exerts a frictional drag that moves surface waters. Ripples or waves provide the surface roughness necessary for the wind to couple with surface waters (Chapter 7). A wind blowing steadily over deep water for 12 hrs at an average speed of about 100 cm per sec (2.2 mph) would produce a 2 cm per sec water current (about 2% of the wind speed).

If Earth did not rotate, frictional coupling between moving air and the ocean surface would push a thin layer of water in the same direction as the wind. This surface layer in turn would drag the layer beneath it, putting it into motion. This interaction would propagate downward through successive ocean layers, like pushing on cards in a deck, each moving forward at a slower speed than the layer above. However, because Earth rotates, the shallow layer of surface water set in motion by the wind is deflected to the right of the wind direction in the Northern Hemisphere and to the left of the wind direction in the Southern Hemisphere. We discussed the reason for the *Coriolis Effect* in Chapter 5. Except at the equator, where the Coriolis Effect is zero, each layer of water put into motion by the layer above shifts direction because of Earth's rotation.

Using arrows to represent the direction and speed of layers of water at successive depths, a simplified model of the three-dimensional current pattern caused by a steady horizontal surface wind emerges (Figure 6.4A). This model is known as the **Ekman spiral**, named for the Swedish physicist V. Walfrid Ekman (1874-1954) who first described it mathematically in 1905. Ekman based his model on observations made by the Norwegian explorer Fridtjof Nansen (1861-1930). Nansen was interested in learning about the currents of polar seas. In 1893, he allowed his 39-m (128-ft) wooden ship, the *Fram*, to freeze into the Arctic pack ice about 1100 km (685 mi) south of the North Pole. His goal was to drift with the ice and cross the North Pole thereby determining how ocean currents affect the movement of pack ice. The *Fram* remained locked in pack ice for 35 months but came no closer than about 394 km (245 mi) to the North Pole. As the *Fram* slowly

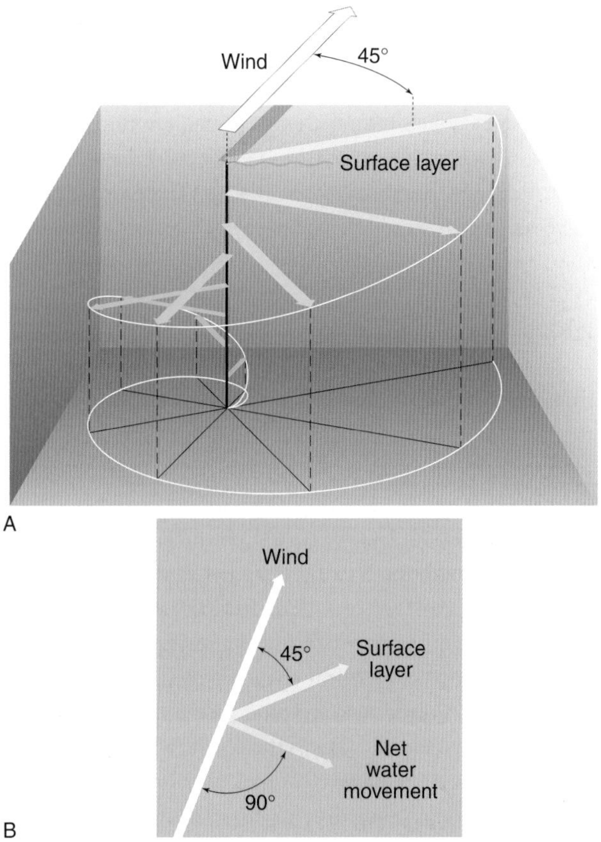

FIGURE 6.4
The Ekman spiral describes how the horizontal wind sets surface waters into motion. (A) As represented by horizontal vectors, the speed and direction of water motion change with increasing depth. (B) Viewed from above in the Northern Hemisphere, the surface layer of water moves at 45 degrees to the right of the wind. The net transport of water through the entire wind-driven column (Ekman transport) is 90 degrees to the right of the wind.

drifted with the ice, Nansen noticed that the direction of ice and ship movement was consistently 20 to 40 degrees to the right of the prevailing wind direction.

The Ekman spiral indicates that the direction of water movement and speed change with increasing depth. In an ideal case, a steady wind blowing across an ocean of unlimited depth and extent causes surface waters to move at an angle of 45 degrees to the right of the wind in the Northern Hemisphere (45 degrees to the left of the wind in the Southern Hemisphere). Each successively lower layer moves more toward the right and at a slower speed. At a depth of about 100 to 150 m (330 to 500 ft), the Ekman spiral has gone through less than half a turn. Yet water moves so slowly (about 4% of the surface current) that this depth is considered to be the lower limit of the wind's influence on the movement of ocean waters.

In the Northern Hemisphere, the Ekman spiral predicts net water movement through a depth of about 100 to 150 m (330 to 500 ft) at 90 degrees to the right of the wind direction (Figure 6.4B). That is, if one adds up all the arrows in Figure 6.4A, the resulting flow is at 90 degrees to the right of the surface wind direction. In the Southern Hemisphere, the net water movement is 90 degrees to the left of the surface wind direction. This net transport of water due to the coupling between the surface wind and water is known as **Ekman transport** and is an important type of flow in the ocean's mixed layer.

The real ocean departs from the idealized conditions represented by the Ekman spiral; that is, wind-induced water movements often differ appreciably from theoretical predictions. To arrive at 45 degrees as the angle between the directions of the surface waters and surface winds, Ekman made many simplifying assumptions. In the real ocean however the angle is closer to 15 to 20 degrees at the most, regardless of the depth of the water. The stable pycnocline inhibits the transfer of kinetic energy to deeper waters, restricting wind-driven currents to the mixed layer; that is, the pycnocline acts as a floor for Ekman transport. More fundamentally, because momentum is efficiently mixed within the mixed layer, what actually happens instead of an Ekman spiral is more like an Ekman slab.

Ekman transport piles up surface waters in some areas of the ocean and removes surface waters from other areas, producing variations in the height of the sea surface, causing it to slope gradually. One consequence of a sloping ocean surface is the generation of horizontal differences (gradients) in water pressure. These pressure gradients, in turn, give rise to geostrophic flow.

GEOSTROPHIC FLOW

The horizontal movement of surface water arising from a balance between the pressure gradient force and the Coriolis Effect is known as **geostrophic flow**. Geostrophic flow characterizes **gyres**, large-scale roughly circular surface current systems in the ocean basins. Depending on location, gyres are either subtropical or sub-polar and are the dominant type of flow within the ocean's mixed layer.

To a large extent, horizontal movement of ocean surface waters, including gyres, mirrors the long-term average planetary-scale circulation of the atmosphere. As shown in Figure 5.19, three prevailing surface wind belts encircle each hemisphere: trade winds (equator to 30 degrees latitude), westerlies (30 to 60 degrees), and polar easterlies (60 to 90 degrees). The trade winds, on the equatorward flank of a subtropical high pressure system, and the westerlies, on the poleward flank of a subtropical high pressure system, are portions of the atmospheric circulation in subtropical high pressure systems (anticyclones) centered near 30 degrees latitude. These massive highs feature generally light winds or calm air in their central regions and are semi-permanent, in that they persist throughout the year, but undergo seasonal shifts in relative strength and location.

These atmospheric subtropical anticyclones drive the oceanic **subtropical gyres**, which are centered near 30 degrees latitude in the North and South Atlantic, the North and South Pacific, and 30 degrees S in the Indian Ocean. Subtropical gyres in the Northern and Southern Hemispheres are similar except that they rotate in opposite directions because the Coriolis Effect acts in opposite directions in the two hemispheres. Viewed from above, subtropical gyres rotate in a clockwise direction in the Northern Hemisphere but in a counterclockwise direction in the Southern Hemisphere.

Ekman transport, induced by the prevailing winds about a subtropical anticyclone, causes waters to converge from all sides toward the central region of a subtropical gyre. This transport produces a broad mound of water as high as 1 m (3 ft) above mean sea level near the center of the gyre (Figure 6.5). As more water is transported toward the center of the gyre, the surface slope of the mound becomes steeper. At the same time, the steeper the slope, the greater the horizontal water pressure gradient produced. In response to the horizontal gradient in water pressure, water parcels move from where the pressure is higher toward where the pressure is lower, that is, downhill from the center of the gyre. Simultaneously, the Coriolis Effect shifts the direction of the moving parcels to the right in the Northern Hemisphere (to the left in the Southern Hemisphere). The water parcels accelerate until the Coriolis Effect balances the outward-directed *pressure gradient force*, and the water parcels flow around the gyre and parallel to contours of elevation of sea level as geostrophic flow.

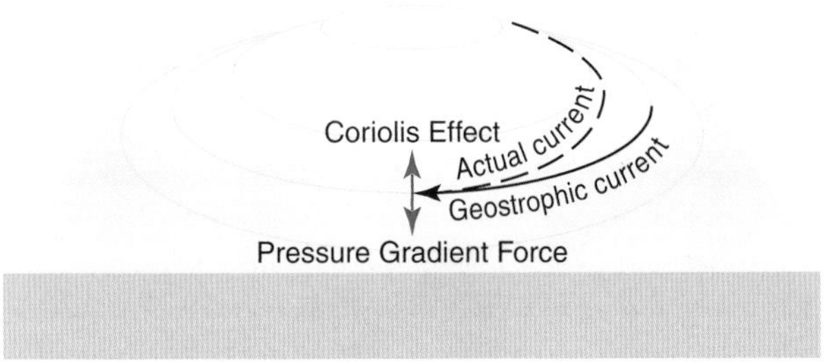

FIGURE 6.5
Ekman transport causes surface waters to converge toward the central region of a subtropical gyre from all sides, producing a broad mound of water. A balance develops between the Coriolis Effect and the force arising from the horizontal water pressure gradient such that surface currents flow parallel to the contours of elevation of sea level. This current is known as geostrophic flow.

It is interesting to note that persistent fair weather and high temperatures characterizing the subtropical anticyclones enhance the rate of evaporation while inhibiting precipitation within subtropical gyres, resulting in surface waters with salinities significantly higher than the global average (Figure 3.9). An example is the vast Sargasso Sea which lies under the Bermuda-Azores subtropical high of the North Atlantic and features an average surface water salinity of 36.5 to 37.0 (Chapter 10).

Sub-polar gyres, smaller than their subtropical counterparts, occur at high latitudes of the Northern Hemisphere; they are the Alaska gyre in the far North Pacific and the gyre south of Greenland in the far North Atlantic. The counterclockwise surface winds in the Aleutian and Icelandic sub-polar low pressure systems drive the sub-polar gyres. (The Aleutian low and Icelandic low are persistent features of the atmosphere's planetary-scale circulation.) Hence, viewed from above, the rotation in the sub-polar gyres is opposite that of the Northern Hemisphere subtropical gyres. Ekman transport causes surface waters to diverge away from the central region of the sub-polar gyres. The thinner surface layer permits more nutrient-rich waters from deeper in the ocean to move upward into the photic zone, thereby increasing biological productivity in these regions (Chapter 9).

Wind-Driven Surface Currents

The long-term average pattern of ocean surface currents is plotted in Figure 6.6. Some currents are relatively warm whereas others are cold. Some (i.e., western boundary currents) are faster moving than others. Winds associated with passing storm systems disturb the ocean surface and can cause the actual flow of ocean currents locally to deviate temporarily from the long-term average pattern.

Most wind-driven surface currents transport water within a specific ocean basin. One of the few areas of the world ocean where inter-basin transport takes place is the *Indonesian Through-flow*. The islands of Indonesia mark the boundary between the Indian and Pacific Oceans but only partially block the flow between the two ocean basins. Although details are still lacking, it is clear that warm, low-salinity waters from the Pacific are transported

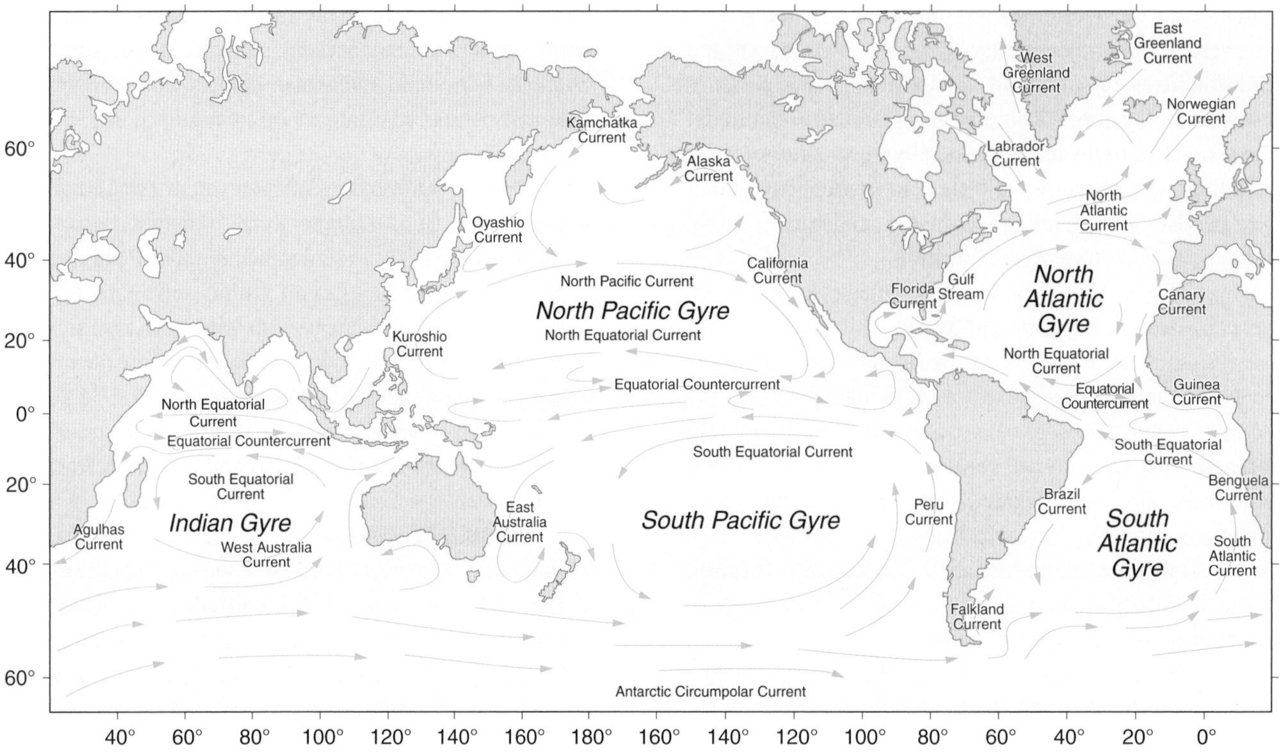

FIGURE 6.6
The long-term average pattern of wind-driven ocean-surface currents.

into the Indian Ocean via the many passages between the thousands of Indonesian islands. These waters replenish the large amounts of water removed by evaporation from the northern Indian Ocean. The summer Asian monsoon winds transport this water vapor from sea to land over India and Southeast Asia where it falls as torrential rains. After flowing westward across the Indian Ocean, waters enter the South Atlantic via the Agulhas Current flowing around southern Africa. In addition, partial blocking by the Indonesian Islands also causes warm surface waters to accumulate in the western equatorial Pacific Ocean, an important aspect of El Niño/La Niña (Chapter 11).

In this section, we examine the basic characteristics of wind-driven surface-ocean currents, components of the gyres that dominate the ocean basins. Western boundary currents are the strongest segments of the subtropical gyres and often spawn warm-and cold-core eddies known as rings. In coastal and equatorial regions, coupling of wind and surface waters can cause upwelling and downwelling.

SUBTROPICAL AND SUB-POLAR GYRES

Surface currents within subtropical gyres vary considerably in strength, width, and depth. The northeastward flowing Gulf Stream of the northwestern Atlantic and the Kuroshio Current of the northwestern Pacific are the swiftest surface currents with velocities averaging 3 to 4 km per hr (1.8 to 2.5 mph). Those currents are also relatively deep and narrow, usually measuring no more than 50 to 75 km (30 to 47 mi) across. On the eastern arms of these gyres, the southward flowing Canary and California Currents, respectively, are hundreds of kilometers wide and rarely flow at more than 1 km per hr (0.6 mph).

The westward flowing South Equatorial Current links the two subtropical gyres of the Atlantic Ocean. The eastward projection of Brazil splits the South Equatorial Current into two segments. The segment flowing southward forms the western arm of the South Atlantic gyre, the Brazil Current. The segment flowing northward merges with the North Equatorial Current, which then splits into two currents that rejoin as they exit the Gulf of Mexico between Florida and Cuba to become the Florida Current. This current then becomes the Gulf Stream that flows northeastward and passes Cape Hatteras, NC. In that region, the current speed may be as great as 9 km per hr (5.5 mph). Near Chesapeake Bay, the amount of water transported in the Gulf Stream exceeds 90 million cubic m per sec (90 Sv); the volume of water transported falls to about 40 Sv by the time the current reaches southern Newfoundland. (For comparison purposes, 90 Sv is about 4500 times the discharge of the Mississippi River—enough to fill the Lake Superior basin in about 1.5 days.) In the North Atlantic at about 40 degrees N and 45 degrees W, the Gulf Stream becomes the North Atlantic Current and flows toward the east. The Canary Current flows southward along the west coast of Spain, Portugal, and North Africa and then merges with the North Atlantic Equatorial Current, thus completing the North Atlantic subtropical gyre.

Like their Northern Hemisphere counterparts, currents in the South Pacific Ocean and South Atlantic Ocean are narrowest and flow most rapidly along their western margins but are broad and slow-moving along their eastern margins. The Indian Ocean subtropical gyre varies more than the others because of seasonal reversals in the monsoon winds. During the high-Sun season, surface winds blow from sea to land and during the low-Sun season, surface winds blow from land to sea.

In addition to the South Pacific and South Atlantic Ocean subtropical gyres, prevailing winds generate the Antarctic Circumpolar Current. At about 60 degrees S, this current also makes up, at least in part, the southern segments of the subtropical gyres of the Atlantic, Pacific, and Indian Oceans. This easterly-flowing current encircles the Antarctic continent rather than rotating as a basin-centered gyre and features the ocean's greatest water flow. Such a globe-circling current is possible only in the Southern Ocean; elsewhere, continents interrupt east-west currents. The Drake Passage, the relatively narrow strait between Cape Horn (the southern tip of South America) and the Antarctic Peninsula, deflects a portion of the Antarctic Circumpolar Current.

As part of the sub-polar gyre of the far North Atlantic, the cold Labrador Current flows southeastward between Canada and Greenland while the East Greenland Current flows southwestward between Greenland and Iceland. Farther east, the North Atlantic Current splits into the Norwegian Current which flows northeasterly between Iceland and Europe along the coast of Norway. In the sub-polar gyre of the far North Pacific, the Kamchatka current flows southwestward on the west side of the Bering Sea while the Alaska Current flows northwest off the south coast of Alaska.

EQUATORIAL CURRENTS

The tropical ocean encompasses broad areas of the Atlantic, Pacific, and Indian Ocean basins and is closely linked to the tropical atmosphere. The prevailing surface winds over the tropical ocean are the trade winds that blow from the northeast (toward the southwest) in the Northern Hemisphere and from the southeast (toward the northwest) in the Southern Hemisphere. The name for these winds was coined by sea captains who sailed for trading companies and took advantage of the persistent speed and direction of the trade winds when crossing the ocean. Trade winds drive both North and South Equatorial Currents westward, thus transporting warm ocean-surface waters in that direction. Equatorial Counter Currents and Equatorial Under Currents return some warm waters eastward. Counter Currents flow along the surface whereas Under Currents flow at greater depths below the surface.

The trade winds of the two hemispheres converge along a narrow east-west belt located near the equator known as the *intertropical convergence zone (ITCZ)*. The ITCZ is an important component of the planetary-scale atmospheric circulation that is particularly well defined over the tropical ocean. Warm and humid air ascends in the ITCZ giving rise to clusters of showers and thunderstorms that produce locally heavy rainfall (Figure 6.7). For mariners this region of the tropical ocean is known as the **doldrums,** feared by the captains of sailing ships because of light and variable winds. Seasonally, the ITCZ moves with the Sun, shifting northward during the Northern Hemisphere spring and southward during the Northern Hemisphere autumn but generally remaining north of the equator, especially over the Atlantic Ocean. The eastward-flowing Equatorial Counter Current, separating the surface current systems of the two hemispheres, also lies mostly just north of the equator.

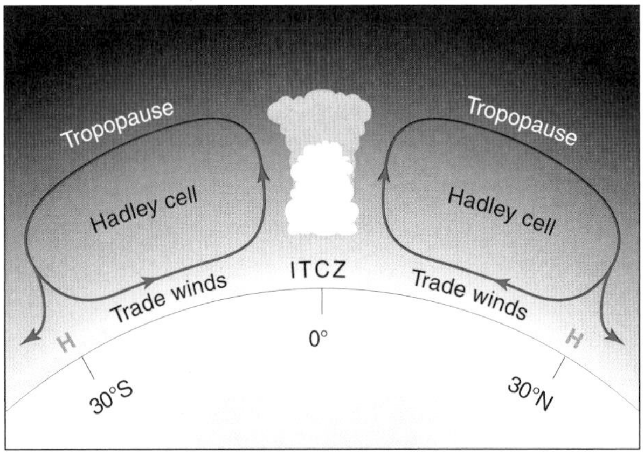

FIGURE 6.7
Vertical cross-section of the prevailing atmospheric circulation in the tropics showing the location of the Hadley cells and intertropical convergence zone (ITCZ). The ITCZ is marked by clusters of showers and thunderstorms and shifts north and south with the seasons.

The South Equatorial Current crosses the equator in the Atlantic and to a lesser extent in the Pacific. In this way, it transports surface waters and heat into the Northern Hemisphere. The return flow is through subsurface currents. The cape at the easternmost point of South America diverts part of the flow of the South Equatorial Current into the southward flowing Brazil Current. The remainder continues northwestward along South America's northeast coast into the Caribbean Sea.

BOUNDARY CURRENTS

In each of the subtropical ocean gyres, the circulation is asymmetrical in the east-west direction; that is, the gyre circulation appears to be shoved toward the western side of the ocean basin (Figure 6.8). On the western sides of the basin close to the east coasts of continents, currents are relatively fast, narrow, and deep. These are the **western boundary currents** and include, for example, the Gulf Stream, Kuroshio Current, and Brazil Current. Flows in the major western boundary currents transport 50 to 100 times the total water discharged by all the world's rivers. Conversely, on the eastern sides of the ocean basins, the currents are slower, much wider, shallower, and not as close to the coast. These **eastern boundary currents** include, for example, the Canary, California, and Peru Currents.

To explain why the western boundary currents are stronger than the eastern boundary currents and why the subtropical gyres are not centered in the ocean basin but shifted westward, we must invoke the principle of the **conservation of angular momentum**. This same principle applies to a figure skater performing a spin. The skater slows down when he/she extends the arms and spins faster when he/she pulls the arms closer to the body. The conservation of angular momentum in an ocean gyre involves three factors that must balance for the current to flow in a curved path. These factors are surface winds, the Coriolis Effect, and frictional drag of the coast, that is, the continental slope.

The large-scale surface winds in the North Atlantic subtropical high rotate the subtropical gyre in a clockwise direction (viewed from above) over much of the ocean basin. Near the equator, little horizontal spinning or rotation of the ocean water occurs because the Coriolis Effect is almost zero (Chapter 5). Along the western side of the ocean basin (U.S. Eastern Seaboard), the Gulf Stream transports this non-rotating ocean water northward. In traveling northward, the Coriolis Effect increases, shifting motion to the right in the Northern Hemisphere. Hence, the water that exhibited no spin at the equator picks up clockwise spin or rotation as it travels northward.

Once this ocean water reaches the northwest quadrant of the North Atlantic, it then flows eastward across the North Atlantic as part of the North Atlantic Current in balance with its acquired Coriolis Effect (clockwise spin). Upon entering the northeast quadrant of the North Atlantic, the ocean water turns again, but this time toward the south on the east side of the ocean basin. Now the Canary Current is transporting water with clockwise spin from high latitudes toward the equator where the water began its journey and the Coriolis Effect is weak. That is, in the transit from north to south, the water must lose

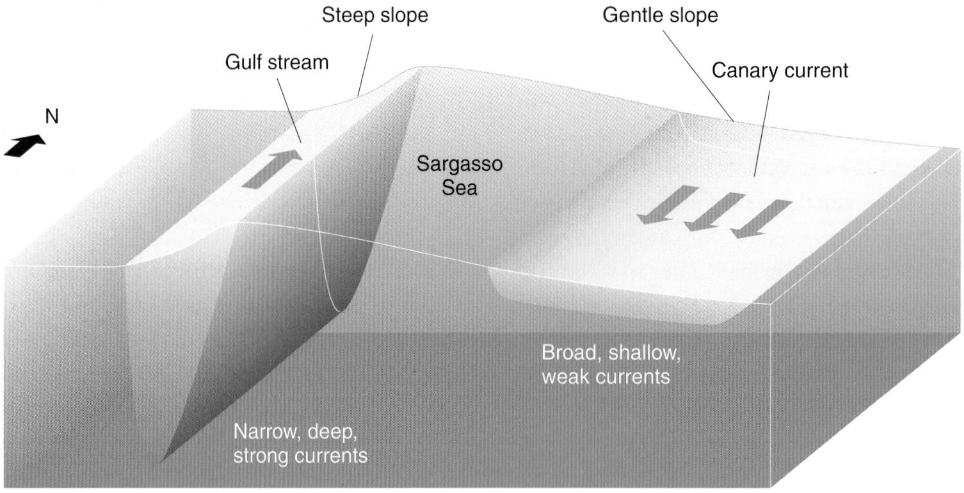

FIGURE 6.8
Surface ocean currents are faster, narrower, and deeper on the western side of an ocean basin. These are known as western boundary currents. Note that the vertical scale is greatly exaggerated.

the clockwise spin it gained at northern latitudes and stay in balance with the nearly zero Coriolis Effect near the equator (equivalently, the southward moving water must gain counterclockwise spin).

On the eastern side of ocean basins, surface winds add clockwise spin while movement southward subtracts clockwise spin to create an approximate balance. However, on the west side of the subtropical gyres, both wind and northward movement (increasing Coriolis Effect) add clockwise spin so that there is too much clockwise spin and hence an imbalance. The question then is how does the west side of the subtropical gyre gain counterclockwise spin to balance the excessive clockwise spin and not break into rotational eddies all over the North Atlantic?

The answer is that the western boundary current gains counterclockwise spin by accelerating, deepening, and pushing up against the coast thereby increasing frictional drag. *Shoving its shoulder* against the coast contributes a counterclockwise turning. The Gulf Stream (and the other western boundary currents) accelerates and produces enough friction against the coast to generate sufficient counterclockwise turning to balance the clockwise spin. To generate sufficient spin, however, the current becomes faster, deeper, and flows closer to the Eastern Seaboard. Conversely, on the eastern side of the ocean basin, the subtropical gyre does not need to generate much clockwise spin to balance angular momentum and so does not push against the coast, but rather remains relatively wide, slow, shallow, and away from the coast.

In our discussion of the ocean boundary currents, we have used a Northern Hemisphere example, but the same interactions occur in the Southern Hemisphere except that the spin is in the opposite direction because of the reversal of the Coriolis Effect. In the Southern Hemisphere, the circulation of subtropical gyres and winds are in the opposite direction but then so is the Coriolis Effect. Hence, the strongest boundary currents also occur on the west side of the subtropical gyre and flow poleward.

OCEAN RINGS

Sensors on Earth-orbiting satellites reveal that the ocean surface is much more dynamic than indicated by maps and charts that portray ocean conditions averaged over decades. Satellite images show that currents change in response to variations in oceanic and atmospheric conditions. For example, relatively swift western boundary currents occasionally spawn large turbulent rotating warm-core and cold-core eddies, also known as **rings**. A ring forms when a meander in a boundary current (or the Antarctic Circumpolar Current) becomes a loop that pinches off (separates) from the main current and moves independently as an eddy. Currents bordering a ring can rotate at more than 1.0 knot (1.0 nautical mi per hr or 1.86 km per hr or 1.15 mph), essentially isolating waters and organisms in rings from the surrounding waters. Rings extend to some depth in the ocean and should be thought of as cylindrical pools of water rather than simply surface features.

Rings form on either side of the Gulf Stream. On the north side, rings are typically 100 to 200 km (60 to 120 mi) across. These rings enclose warm waters from the Sargasso Sea located to the south and east of the Gulf Stream; hence, they are called **warm-core rings**. Viewed from above, these warm-core rings rotate in a clockwise direction. Because of the strong contrast in sea-surface temperatures, they are readily detected on infrared satellite images. Warm-core rings are also readily distinguished from the surrounding surface waters by their relatively low levels of biological production. The pool of water in a warm-core ring can extend to a depth of 1500 m (4900 ft) so that they cannot move onto continental shelves, which are shallow—typically 200 m (650 ft) deep or less. However, rings can come close enough to the shelf edge to modify coastal currents and transport unusual organisms onto the shelf. Occasionally boaters and fishers in normally cool coastal waters encounter organisms (e.g., sea turtles, tropical fish) that live in much warmer water having been transported in a warm core ring that spun off the Gulf Stream.

Rings that spin off the south side of the Gulf Stream entrain relatively cold and productive coastal waters and are called **cold-core rings** (Figure 6.9). These rings have diameters of about 300 km (185 mi) and, viewed from above, rotate in a counterclockwise direction. While not the case in Figure 6.9, cold-core rings can be more difficult for satellites to track because their originally cool surface waters are warmed by absorption of solar radiation, which can make them almost indistinguishable thermally from surrounding surface waters. However, cold water persists below the surface, sometimes extending down to the ocean floor at depths of more than 4000 m (13,000 ft), and can be detected in vertical profiles of temperature and salinity obtained by instrumented probes. Cold-core rings usually contain more nutrients and marine organisms than the biologically barren Sargasso Sea waters that surround them. Thus, they can also be identified in the subsurface by their unusually abundant marine life.

Rings move slowly (typically 5 to 6 km or 3 to 4 mi per day), drifting southwestward in the weaker

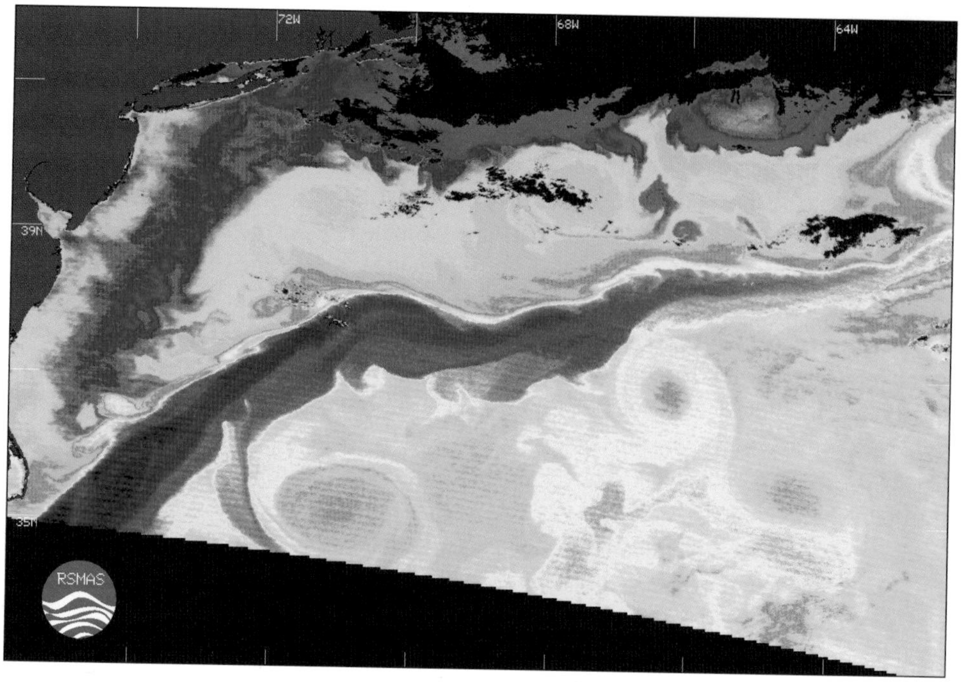

FIGURE 6.9
Gulf Stream sea-surface temperatures measured by NASA's Moderate-Resolution Imaging Spectroradiometer (MODIS). Two cold-core rings (indicated as yellow-rimmed circles with green shadings in the middle) are visible to the southeast of the Gulf Stream (red shadings). [Image courtesy of University of Miami; created by Bob Evans, Peter Minnett, and co-workers]

currents on either side of the northeast flowing Gulf Stream. The proximity of the Gulf Stream to the coast limits the southward movement of warm-core rings. Typically, after a few months to a year, a warm-core ring becomes caught between Cape Hatteras, NC, and the Gulf Stream; the ring is then reabsorbed back into the Gulf Stream. Cold-core rings are not as restricted in their movements as warm-core rings and may persist for several years; on average, individual cold-core rings last for about 18 months.

Some of the ocean's largest warm-core rings, with diameters up to 400 km (250 mi), form in the Gulf of Mexico. These rings spin off the Loop Current at highly irregular intervals ranging from several months to 1.5 years. As shown in Figure 6.10, the Loop Current enters the Gulf from the Caribbean by flowing through the Yucatán Strait between Cuba and Mexico, heads northwestward in the general direction of Louisiana, then makes a clockwise turn, and exits the Gulf through the Florida Strait (between Florida and Cuba). Rings drift westward across the Gulf at 2 to 5 km (1.2 to 3.1 mi) per day. Bordering currents of up to 4 knots (4.6 mph) can play havoc with offshore oil platform operations, damaging equipment and increasing the risk of accidents.

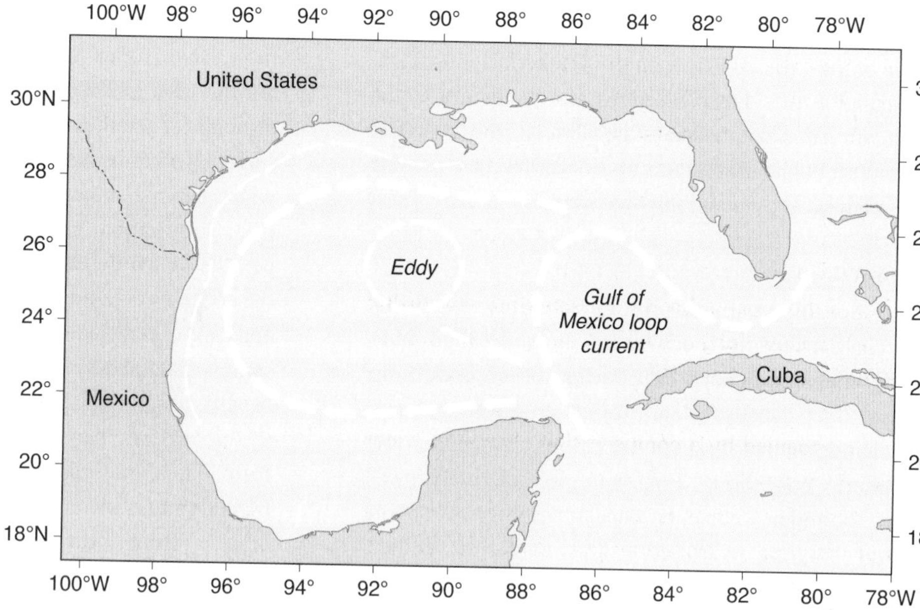

FIGURE 6.10
The Gulf of Mexico Loop Current. enters the Gulf through the Yucatán Strait between Cuba and Mexico and generally flows clockwise before exiting the Gulf between Florida and Cuba. Occasionally, warm-core rings (shown by the circle labeled "Eddy") spin off the Loop Current.

The Agulhas Current, the Indian Ocean's western boundary current, also spawns rings. This relatively fast southward-flowing current averages about 7 km per hr (4 mph). As it reaches Africa's southern tip, part of the current is caught up in the eastward flow around Antarctica and abruptly shifts direction back into the Indian Ocean. Some of the flow continues around South Africa as narrow (50 km or 30 mi wide) filaments that cool rapidly and mix with the surrounding waters in the large upwelling zone off Africa's Namibia coast. The Agulhas Current periodically sheds rings about 320 km (200 mi) across from its westernmost end. Rings in the Southern Hemisphere rotate in the opposite direction of those in the Northern Hemisphere. The Agulhas Current's warm-core rings rotate counterclockwise and contain Indian Ocean waters that are about 5 Celsius degrees (9 Fahrenheit degrees) warmer than nearby South Atlantic surface waters. They retain their identity as they move into the South Atlantic and transport heat, salt and organisms from the Indian Ocean into the South Atlantic. Over a two-year period in the mid-1990s, 14 rings were reported.

UPWELLING AND DOWNWELLING

In some coastal areas of the ocean (and large lakes such as the Great Lakes of North America), the combination of persistent winds, Earth's rotation (Coriolis Effect), and restrictions on lateral movements of water caused by shorelines and shallow bottoms induces upward (upwelling) and downward (downwelling) movements of water.

As explained above, the Coriolis Effect plus the frictional coupling of wind and water (Ekman transport) cause net movement of surface water at about 90 degrees to the right of the wind direction in the Northern Hemisphere and to the left of the wind direction in the Southern Hemisphere. **Coastal upwelling** occurs where Ekman transport moves surface waters away from the coast; surface waters are replaced by water that wells up from below (Figure 6.11). Where Ekman transport moves surface waters toward the coast, the water piles up and sinks in the process known as **coastal downwelling** (Figure 6.12). Upwelling and downwelling illustrate *mass continuity* in the ocean; that is, water is a continuous fluid so that a change in distribution of water in one area is accompanied by a compensating change in water distribution in another area.

Upwelling is most common along the west coasts of continents (eastern sides of ocean basins). In the Northern Hemisphere, upwelling occurs along west coasts (e.g., coasts of California, Northwest Africa) when winds blow from the north (causing Ekman transport of surface water away from the shore). Figure 6.13 shows the influence of upwelling on sea-surface temperatures along the central and northern California coast. Winds blowing from the south cause upwelling along continents' eastern coasts in the Northern Hemisphere, although it is not as noticeable because of the western boundary currents. Upwelling also occurs along the west coasts in the Southern Hemisphere (e.g., coasts of Chile, Peru, and southwest Africa) when the wind direction is from the south because the net transport of surface water is westward away from the shoreline. In the Southern Hemisphere, winds blowing from the north cause upwelling along the continents' eastern coasts.

Upwelling and downwelling also occur in the open ocean where winds cause surface waters to diverge (move away) from a region (causing upwelling) or to converge toward some region (causing downwelling).

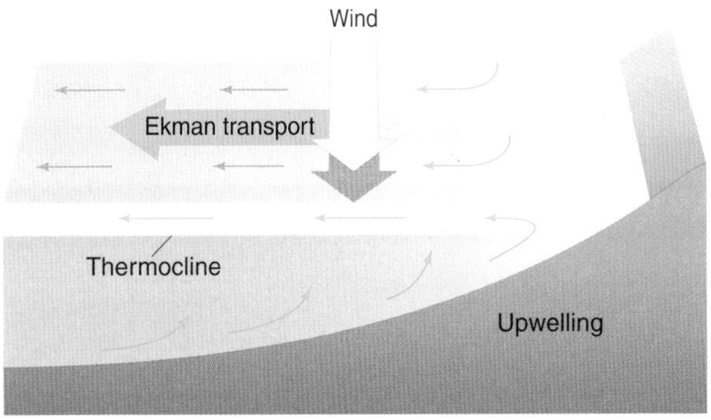

FIGURE 6.11
Where Ekman transport moves surface waters away from the coast, surface waters are replaced by water that wells up from below in the process known as upwelling.

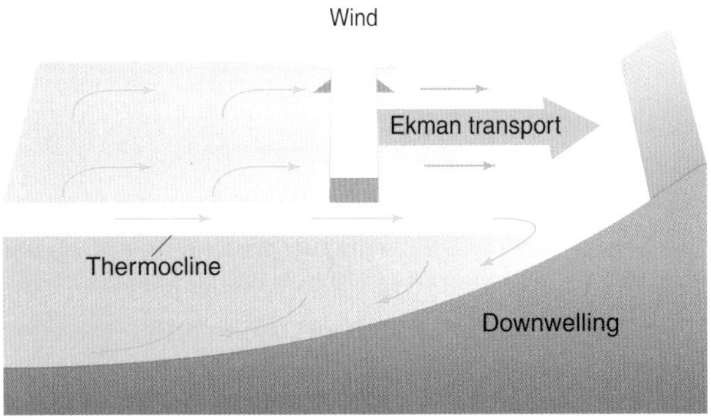

FIGURE 6.12
Where Ekman transport moves surface waters toward the coast, the water piles up and sinks in the process known as downwelling.

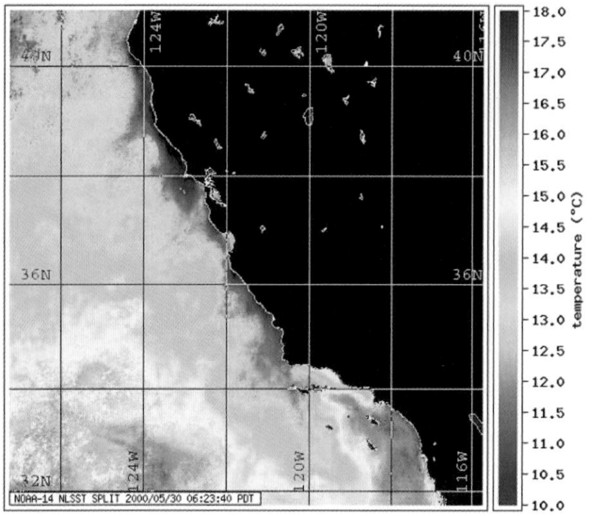

FIGURE 6.13
An infrared satellite image shows the relatively cold upwelling waters along the coast of central and northern California in response to the Ekman transport caused by winds blowing from the north. According to the color scale, the lowest sea surface temperatures are plotted as blue and purple. Note how in places, strong upwelling extends well offshore from the coast. [NOAA Ocean Explorer]

For example, upwelling takes place along much of the equator (Figure 6.14). Recall that the deflection due to the Coriolis Effect reverses direction on either side of the equator. Hence, westward-flowing, wind-driven surface currents near the equator turn northward on the north side of the equator and southward on the south side. Surface waters are moved away from the equator and replaced by upwelling waters; this is known as **equatorial upwelling**.

Upwelling and downwelling influence sea-surface temperature and biological productivity (Chapter 9). Upwelling waters originate at considerable depth and are usually colder than the surface waters they replace. You may experience this phenomenon at the beach on a windy day when the warm surface water is blown offshore and replaced by chilly water from below. Coastal upwelling also transports waters rich in dissolved nutrients (nitrogen and phosphorus compounds) from the ocean depths into the photic zone where sunlight penetrating the water supports the growth of phytoplankton populations. The world's most productive fisheries are in areas of coastal upwelling (especially in the eastern boundary regions of the subtropical gyres); about half the world's total fish catch comes from upwelling zones (Chapter 10). On the other hand, in zones of coastal downwelling, the surface layer of warm, nutrient-deficient water thickens as water sinks. Downwelling reduces biological productivity and transports heat, dissolved materials, and surface waters rich in dissolved oxygen to greater depths. This occurs, for example, along the west coast of Alaska in the eastern boundary region of the Gulf of Alaska sub-polar gyre.

Alternate weakening and strengthening of upwelling off the coast of Ecuador and Peru are associated with El Niño and La Niña episodes in the tropical Pacific (Chapter 11). During an El Niño event, upwelling wanes, and cold nutrient-rich water remains so deep that weak upwelling brings only warm, nutrient-poor water into the photic zone. In extreme cases, nutrient-deficient waters coupled with over-fishing cause fisheries to collapse bringing about severe economic impacts.

Coastal upwelling and downwelling also influence weather and climate. Along the central and northern California coast, upwelling lowers sea-surface temperatures (Figure 6.13). Relatively cold surface waters chill the overlying humid marine air to saturation so that thick fog develops, especially in summer. Also, seasonal upwelling and downwelling reduce the annual temperature range along the west coasts of the Americas. During El Niño and La Niña, changes in sea-surface temperature patterns

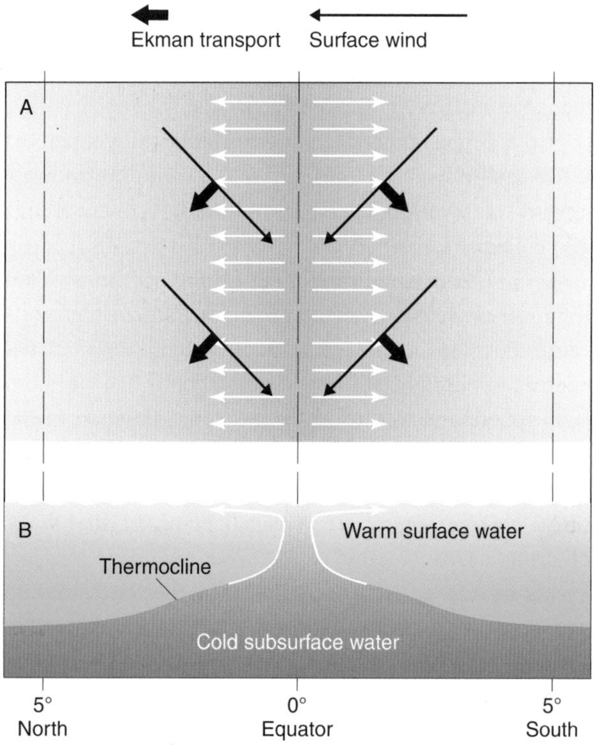

FIGURE 6.14
Equatorial upwelling. (A) In this plan view of the ocean from 5 degrees S to 5 degrees N, the trade winds of the two hemispheres are shown to converge near the equator. The consequent Ekman transport away from the equator gives rise to upwelling as shown in (B) a vertical cross section from 5 degrees S to 5 degrees N.

associated with weakening and strengthening of upwelling off the northwest coast of South America and along the equator in the tropical Pacific affect the distribution of precipitation in the tropics and elsewhere. Upwelling cold water also inhibits formation of tropical cyclones (e.g., hurricanes and tropical storms) there. As we will see in Chapter 8, tropical cyclones derive their energy primarily in the form of latent heat from warm surface waters.

Monitoring the Ocean Depths

While ocean scientists have measured surface and near-surface temperature, salinity, and currents for more than a century, probing the ocean depths to determine the vertical structure and circulation of the ocean has been much more challenging. Accurate measurements of deep currents and water temperature and salinity at various depths not only require instruments that can withstand the stresses of the ocean environment, but also appropriate platforms that provide a means of delivering the instruments to the desired depth and retrieving the data once measurements are made. In the past few decades, however, systematic surveys of the deep ocean have greatly expanded our knowledge of the properties of seawater and water movements. Deep-sea casts from oceanographic research ships have produced much of these data.

Ocean scientists use the term *cast* (as in net cast or CTD cast) for measurements made at various depths in the ocean. This term likely originated with boat pilots (such as Mark Twain) who used a weighted measured line thrown overboard to determine water depth. Comparatively, atmospheric scientists use the term *sounding* to denote a sequence of measurements obtained at various altitudes in the atmosphere.

New promising ocean sensing technologies include submersible, instrumented profiler *floats* that obtain vertical profiles of temperature, pressure (a measure of depth), and conductivity (a measure of salinity) (Figure 6.15). A widely spaced array of 3000 plus instrumented floating profilers are deployed across the ocean at 3-degree latitude/longitude intervals. The primary purpose of the float array, known as *Argo (Array for real-time geostrophic oceanography)* is to monitor the climate (long-term average conditions) of the wind-driven layer, pycnocline, and below in the top 2 km (1.2 mi) of the ocean on denser spatial and temporal scales than previously available. The long-term observations provided by Argo are the basis for mapping the large-scale average oceanic flow.

Research ships, commercial vessels, and low-flying aircraft drop profiler floats into the sea. Typically, a float is programmed for a 10-day cycle during which it sinks to a prescribed depth of 1000 m (3300 ft), drifts with the current for 9 days before sinking to a maximum depth of 2000 m (6600 ft), and then returns to the surface monitoring ocean water properties along the way. At the surface, the float relays its collected data via satellite to computer databases. Tracking the position of the float over time also records water movements.

While the Argo global-scale array is an international effort, various institutions in the U.S. have deployed about half the floats. Among the other 22 participating nations are Australia, Canada, Japan, France, and the United Kingdom. The U.S. portion of the Argo program includes floats that are under the auspices of the University of Washington, Scripps Institution of Oceanography, and the Woods Hole Oceanographic Institution. Argo data are archived at NOAA's Atlantic Oceanographic and Meteorological Laboratory in Miami, FL.

In Chapter 3, we described another technology that uses changes in the speed of sound to measure temperature over great distances within the ocean (*acoustic thermometry*). In addition, the distribution of various tracer materials, such as dissolved oxygen, CFCs (chlorofluorocarbons), or radioactive substances from both atmospheric and oceanic testing of nuclear devices have been used to track movements of subsurface waters. For more on oceanographic casting methods, refer to this chapter's second Essay. For a description of a unique way to track ocean currents, go to this chapter's third Essay.

Large-Scale Ocean Circulation

As noted earlier, the upper portion of the ocean is put into motion by wind forcing at the air-sea interface. Below the pycnocline, where the deep ocean is shielded from the direct action of the wind, ocean currents are driven by density differences in water masses. These density contrasts are caused by variations in water temperature and salinity, the densest water being cold and salty (Chapter 3). This deep-ocean circulation driven by variations in density is the **thermohaline circulation**, *thermo* meaning heat and *haline* referring to salinity.

The thermohaline circulation is a major player in the Earth system. The temperature difference between the warm surface waters and the cold deep waters governs ocean stratification and, although the deep

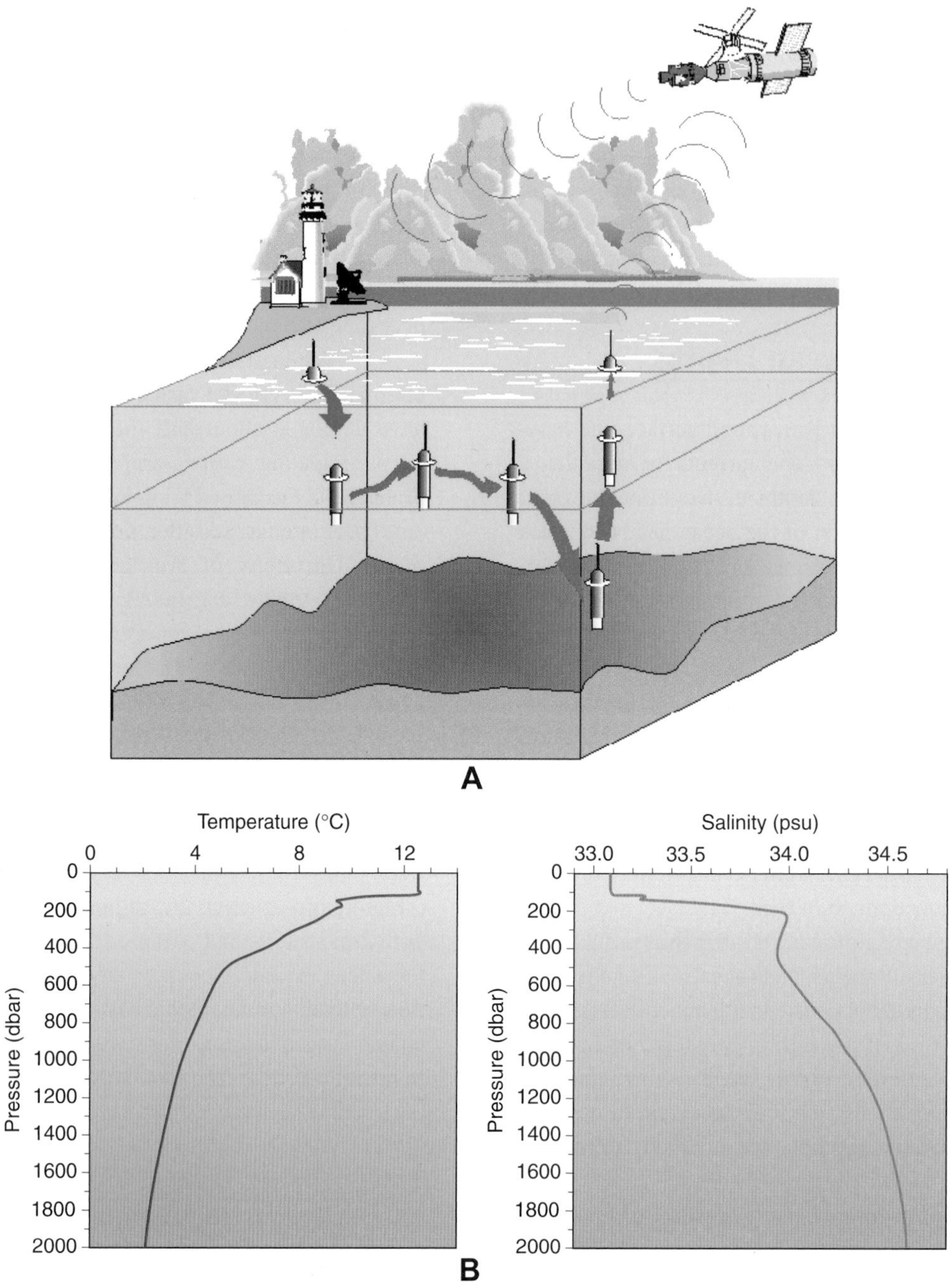

FIGURE 6.15
An Argo float (A) obtains continuous profiles of ocean temperatures and salinity to a maximum depth of about 2000 m. The instrument surfaces and sends data to a satelite for downloading. (B) This is a sample plot of float-derived temperature and salinity profiles obtained at a location in the eastern North Pacific, west of Northern California, on 1 February 2008. [Courtesy of the U.S. Global Ocean Assimilation Experiment]

ocean currents are relatively weak, the volume of water involved is much greater than that of surface waters so the magnitude of water transport is similar. Especially because of the fast flow of warm surface water towards the poles, the thermohaline circulation affects inter-annual climate variability to long-term climate change (Chapters 5 and 12).

As discussed in Chapter 5, thermohaline flow and wind-forcing are involved in the ocean's prominent **meridional overturning circulation (MOC)**, the large-scale overturning of the ocean from the north to south and back. In the North Atlantic Ocean, the MOC consists of two western-intensified boundary currents, the surface warm Gulf Stream flowing north, and the cold Deep

Western Boundary Current flowing toward the equator. The Gulf Stream is a heat source for the atmosphere until, in the subpolar North Atlantic, surface waters become cold and dense, sinking to the ocean bottom. This cold deep water then flows southward along the continental slope of the eastern U.S. as the Deep Western Boundary Current far below, but not far to the east of the Gulf Stream on the surface.

FORMATION OF DEEP WATER

Density gradients in the deep ocean are established by dense waters sinking in just a few locations. These regions are generally characterized by very low temperatures and, in some cases, high salinity. In the Northern Hemisphere, deep water is formed in the Greenland-Norwegian Sea and the Labrador Sea during late winter when the surface water has reached its lowest temperature and greatest density. The primary mechanism is **open ocean convection**, where cold winds cool the surface water until its density is greater than that of the water below, creating an unstable water column through which it sinks.

Deep waters that formed in the Greenland-Norwegian Sea are separated from the main basin of the Atlantic Ocean by a submarine ridge running east from Greenland, through Iceland, to Europe. Dense waters overflow through gaps in this ridge, cascading down-slope into the Atlantic Ocean and entraining overlying waters along the way. In contrast, convection in the Labrador Sea funnels water directly into the mid-depth Atlantic. Unlike other sources of deep water in the Atlantic, deep water formation in the Labrador Sea undergoes considerable inter-annual variability. Some winters lack any deep convection whereas other winters are characterized by vigorous overturning.

A third source of deep waters in the North Atlantic Ocean consists of very salty outflows from the Mediterranean Sea. These waters originate in the northwestern Mediterranean in winter as cold, dry *Mistral winds* cool the surface waters and enhance evaporation. This dense salty water fills the deep Mediterranean basin and then spills over the sill at the Strait of Gibraltar.

In the Southern Hemisphere, deep waters form at several locations around the Antarctic continent, primarily in the Weddell Sea south of South America. Unlike the Northern Hemisphere, much of the deep water formation occurs underneath floating sea ice and details are not well known. Likely, cold winds acting on openings in the sea-ice cover further cool the water and sea ice formation creates denser water. As noted in Chapter 3, during **brine rejection**, the salt excluded when seawater freezes increases the salinity of the underlying water. The water made denser by cooling and brine rejection sinks along the continental slope of Antarctica into the deep ocean. As this cold, dense water sinks, it entrains additional water, increasing the total flow. In addition to the Weddell Sea, cold dense bottom waters are created in the Ross Sea, south of New Zealand, as well as other scattered locations along Antarctica's continental shelf.

WATER MASSES

Difficult to observe directly, much of what we know about deep ocean circulation is inferred from water properties. Their currents are very weak, on average slower than 1 mm per sec. Only at the surface of the ocean, when interacting with the atmosphere, are water properties altered. So, after sinking, the temperature and salinity are conserved. These large, homogeneous volumes of water, with a characteristic range of temperature and salinity, are **water masses**. They are typically identified by their source region (Figure 6.16), therefore, the deep water convection in the North Atlantic forms a water mass called *North Atlantic Deep Water (NADW)* while the deep water mass formed around Antarctica is called *Antarctic Bottom Water (AABW)* (Table 6.1).

In addition to deep water, the other principal type of water mass in the world ocean is *intermediate water*. Above the deep waters and below the wind-driven circulation, at a depth of 1 to 2 km, intermediate waters are formed by a variety of processes. The most extensive intermediate waters, *Antarctic Intermediate Water (AAIW)*, is formed by open ocean water cooling in the Southern Ocean. Other intermediate waters are formed in marginal seas, such as *North Pacific Intermediate Water (NPIW)* formed in the Sea of Okhotsk in northeast Russia, *Red Sea Intermediate Water (RIW)*, and *Mediterranean Intermediate Water (MIW)*.

As noted above, AABW, the densest waters in the Southern Ocean, forms in the Weddell Sea. Trapped close to the Antarctic continent by submarine ridges, AABW does not flow into the other ocean basins. Above AABW, encircling the entire Southern Ocean, is *Circumpolar Deep Water (CDW)*. CDW results from the mixing of AABW with overlying deep waters of the Atlantic, Pacific, and Indian Ocean basins. The Antarctic Circumpolar Current efficiently mixes and spreads this water mass around Antarctica. CDW flows from this southern starting point northward as bottom water in all three ocean basins (Atlantic, Pacific, and Indian).

The Atlantic Ocean is the primary site of deep water formation in the Northern Hemisphere because of

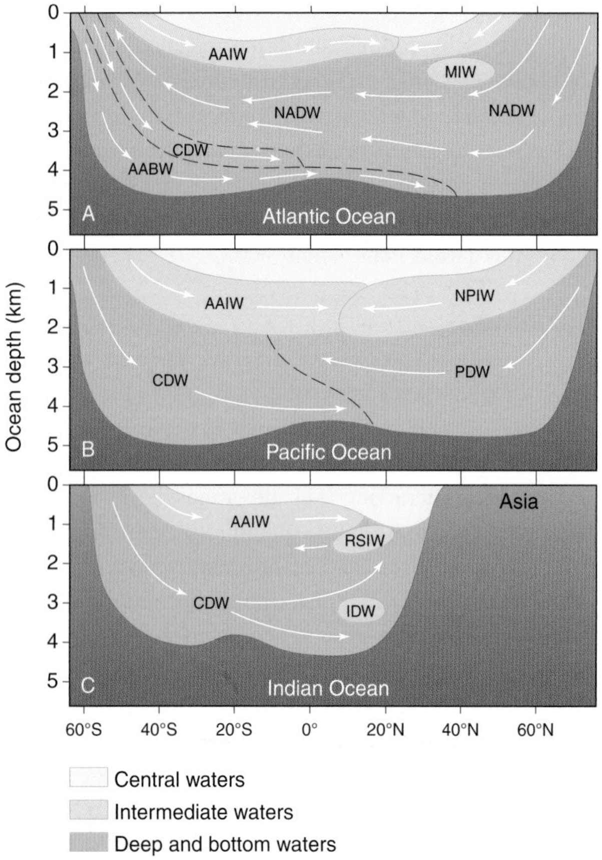

FIGURE 6.16
Vertical distribution of water masses in the (A) Atlantic, (B) Pacific, and (C) Indian Oceans.

TABLE 6.1
Water Masses of the Atlantic, Pacific, and Indian Oceans

Atlantic Ocean
Antarctic Intermediate Water (AAIW)
Mediterranean Intermediate Water (MIW)
Circumpolar Deep Water (CDW)
North Atlantic Deep Water (NADW)
Antarctic Bottom Water (AABW)
Pacific Ocean
Antarctic Intermediate Water (AAIW)
North Pacific Intermediate Water (NPIW)
Pacific Deep Water (PDW)
Circumpolar Deep Water (CDW)
Indian Ocean
Antarctic Intermediate Water (AAIW)
Indian Deep Water (IDW)
Red Sea Intermediate Water (RSIW)
Circumpolar Deep Water (CDW)

surface waters exposed to cold winter winds poleward of 60 degrees and the high salinity of the Atlantic. NADW is characterized by high salinity, a property which can be used to trace its spread throughout the world ocean. Deep water formed in the North Atlantic spreads south across the equator and then is entrained into the circumpolar flow around Antarctica. From the Southern Ocean, NADW spreads into both the Indian and Pacific Oceans. Flowing northward underneath the NADW in the Atlantic is a wedge of deep water formed in the Southern Ocean. CDW and AABW reach into the Northern Hemisphere, gradually mixing and entraining into the NADW above.

In contrast to the Atlantic Ocean, deep water does not form in the Pacific Ocean. Although surface waters in the northern Pacific are cooled to very low temperatures, their low salinity (due to precipitation and runoff) prevents them from sinking to the bottom. In addition, the shallow Bering Strait, at a depth of only about 50 m (165 ft), isolates the deep Pacific Ocean from the deep Arctic Ocean. The only water mass locally produced in the Pacific is *North Pacific Intermediate Water (NPIW)*, formed in moderate quantities in the Sea of Okhotsk. With no deep water formation, the deepest portions of the Pacific basin are filled with *Circumpolar Deep Water (CDW)*, which originates in the Southern Hemisphere. Over time, CDW mixes with the overlying waters, forming a third water mass: *Pacific Deep Water (PDW)*, not directly formed by cooling at the surface, PDW flows slowly south with a portion upwelling into the pycnocline while another portion returns to the Southern Ocean.

The circulation in the Indian Ocean resembles that of the Pacific Ocean. The smallest of the three major ocean basins, most of the Indian Ocean is located in the Southern Hemisphere and no deep water masses form along its tropical and subtropical northern boundaries. Small amounts of intermediate water form from the saline outflow of the Red Sea, but this flow is most notable for its effect on salinity. Like the Pacific, deep water comes from CDW flowing north along the bottom, gradually upwelling and mixing to form *Indian Deep Water (IDW)*.

IMPLICATIONS FOR CLIMATE

The ocean is a key component of Earth's climate system (Chapters 3 and 5). Its relatively great thermal inertia implies a capacity to store large quantities of heat, moderating fluctuations in air temperature (e.g., maritime versus continental climates). The ocean also significantly influences the global radiation budget and poleward heat transport. The meridional overturning circulation (MOC) moves heat energy, salt, and carbon dioxide, which influence climate over a broad range of spatial and temporal scales. Changes in the operation of the MOC have important implications for future climate as it has in the past.

An essential element in the MOC is formation of bottom waters in the polar and subpolar reaches of the North Atlantic Ocean. The combination of cooling of surface waters, evaporation, and sea-ice formation produces cold, salty North Atlantic Deep Water (NADW) that sinks and then flows southward toward Antarctica. Continued production of NADW is essential to maintain the ocean's meridional overturning circulation and northern Europe's moderate climate.

The MOC could weaken if the North Atlantic surface-water salinity were to drop too low for formation of deep ocean water masses. An influx of large amounts of fresh water into the North Atlantic could trigger such an event, as may have happened in the past. For example, some climate scientists argue that weakening of the MOC was responsible for the Little Ice Age, CE 1400 to 1900, when the climate of northern Europe was markedly colder. Mountain glaciers advanced, growing seasons became shorter, and sea-ice cover increased. Paintings from the era show Dutch skaters on frozen canals, an impossibility in today's climate. Cores extracted from deep-sea sediment deposits contain evidence of earlier cold episodes that may also be linked to weakening of the MOC, including the Heinrich events described in the first Essay of Chapter 4. There is more on the ocean and climate change in Chapter 12.

Conclusions

The ocean's surface mixed layer, its currents driven by winds, is the part of the ocean that is most directly involved in Earth-system processes. It transports heat globally, supplies water vapor to the atmosphere, dissolves and transports salts, nutrients, and gases, supports fisheries, and plays major roles in day-to-day weather and short-term climate variability. The deep-ocean is isolated from the atmosphere by the mixed layer and pycnocline except at high latitudes; it is mostly cold and dark. The relatively slow thermohaline circulation is driven by density contrasts between water masses and involves about 90% of the ocean's waters. The deep circulation is important in long-term climate change, plays an important role in sequestering or buffering greenhouse gases such as carbon dioxide, and is important in transporting dissolved nutrients. We continue our investigation of the dynamic nature of the ocean in the next chapter where we examine waves and tides.

Basic Understandings

- Based on density, the ocean is divided into three horizontal zones: the surface mixed layer of uniform or near uniform density, the intermediate pycnocline where density increases markedly with depth because of changes in temperature and/or salinity, and the deep layer where density increases gradually with depth. The pycnocline is very stable and inhibits blending of ocean waters between the mixed layer and the deep layer.
- The wind-driven currents in the surface mixed layer and the relatively slow thermohaline circulation in the deep ocean dominate ocean circulation.
- Wind-driven currents are maintained by kinetic energy transferred from the horizontal winds to ocean surface waters. Once the wind sets surface waters in motion as a current, the Coriolis Effect and configuration of the ocean basin modify the speed and direction of the current.
- The frictional coupling of the surface winds, which puts surface waters into motion, the Coriolis Effect, and the frictional coupling of successively deeper layers of water, causes the horizontal movement of water to change direction and slow with increasing depth, producing the Ekman spiral. In an ideal case, a steady wind would cause surface waters to move at an angle of 45 degrees to the right of the wind in the Northern Hemisphere and to the left in the Southern Hemisphere.
- The Ekman spiral brings about net water movement through a depth of about 100 m (330 ft) at 90 degrees to the right of the wind direction in the Northern Hemisphere and 90 degrees to the left of the wind direction in the Southern Hemisphere.
- Because of the wind circulation in semi-permanent subtropical highs, Ekman transport causes surface waters to converge toward the central region of subtropical gyres, which produces a mound of surface water near the center of the gyre.
- Surface water parcels flow outward and down slope away from the center of the gyre while the Coriolis Effect causes water parcels to turn to the right in the Northern Hemisphere and the left in the Southern Hemisphere. Eventually, the

outward-directed pressure gradient force balances the Coriolis Effect and water parcels flow around the subtropical gyre, parallel to the contours of elevation of sea level. This horizontal movement of surface water is known as geostrophic flow.

- Ocean gyres resemble the long term planetary-scale surface wind patterns. Surface water currents form subtropical gyres roughly centered in each ocean basin near 30 degrees latitude. Viewed from above, currents in these gyres flow in a clockwise direction in the Northern Hemisphere and a counterclockwise direction in the Southern Hemisphere, similar to the subtropical anticyclones.
- In the sub-polar gyres of the far north, current directions are opposite that of the Northern Hemisphere subtropical gyres.
- Trade winds of the Northern and Southern Hemispheres drive equatorial currents and warm surface water westward across the tropical ocean.
- Winds associated with passing storm systems disturb the ocean surface and can cause the flow of ocean surface currents locally to deviate temporarily from long-term average patterns.
- On the west sides of the subtropical gyres, surface currents are fast, narrow, and deep. These are the western boundary currents and include the Gulf Stream, Kuroshio Current, and Brazil Current. On the eastern sides of the subtropical gyres, surface currents are slower, much wider, shallower, and farther from the coast. These eastern boundary currents include the Canary, California, and Peru Currents.
- The contrast in characteristics of the western versus eastern boundary currents is due to the conservation of angular momentum, the variation of the Coriolis Effect with latitude, and the current's frictional interaction with the coast.
- Western boundary currents spawn warm- and cold-core rings (eddies). Rings develop when a meandering current forms a large loop that pinches off and separates from the main current. In the Northern Hemisphere, warm-core rings form on the current's landward side and rotate clockwise whereas cold-core rings form on the current's ocean side and rotate counterclockwise.
- In coastal areas, the combination of persistent winds blowing parallel to the coast, the Coriolis Effect, and restrictions on lateral movements of water caused by shorelines and shallow bottoms induces vertical water movement. Where winds generate Ekman transport of surface waters away from the coast, colder water wells up from below, a process called upwelling. Upwelling supplies nutrient-rich waters to the sunlit surface zone of the ocean, spurring biological productivity.
- Upwelling also occurs along the equator in response to Ekman transport associated with the trade winds of the two hemispheres.
- Where winds generate Ekman transport of surface waters toward a coast, water piles up and sinks, known as downwelling.
- Over the past few decades, systematic surveys of the deep ocean have greatly expanded our knowledge of the properties of seawater and water movements. Casts denote a sequence of measurements (e.g., temperature, conductivity) obtained vertically through the ocean depths such as obtained by Argo profiler floats.
- Deep-ocean currents are driven primarily by slight differences in seawater density. Variations in water temperature and salinity, with cold salty water being the densest combination, cause the density differences that drive the thermohaline circulation.
- A water mass is a large, homogeneous volume of water with a characteristic range of temperature and salinity. Water masses are typically identified based on their source region and their ocean depth.
- The meridional overturning circulation (MOC) links the surface and thermohaline circulation regimes and transports heat and salt on a planetary scale. It is an important component of Earth's climate system.

Enduring Ideas

- Ocean circulation is driven by processes that continually exchange energy (e.g., heat) and matter (e.g., water) between the ocean and atmosphere. Precipitation, evaporation, and runoff from continents, along with heating and cooling, bring about changes in the temperature and salinity of surface waters. Variations in seawater density that accompany changes in temperature and salinity can cause water to sink or rise in the ocean.
- Kinetic energy is transferred from surface winds to the ocean's mixed layer, driving the surface currents that characterize the motion of the upper 100 m of the ocean.
- Frictional coupling of the surface wind with the ocean surface waters and the Coriolis Effect transport water horizontally at a 90 degree angle to the wind direction to the right in the Northern Hemisphere and 90 degrees to the left in the Southern Hemisphere. This motion is known as Ekman transport and can result in upwelling or downwelling in coastal or equatorial regions.
- Upwelling delivers nutrient-rich cold water into the sunlit photic zone, spurring biological productivity.
- Surface currents flow around the subtropical gyres, parallel to contours of sea level elevation. Gyres resemble the long-term planetary-scale atmospheric circulation pattern.
- Ocean surface currents are narrower, deeper, and flow faster on the western than eastern side of the subtropical gyres. These are the boundary currents.

Review

1. Describe the vertical structure of the ocean in middle latitudes based on the variation in temperature with increasing depth.
2. How does water density vary with depth within the pycnocline? Under what conditions is a pycnocline coincident with a thermocline?
3. In the Northern Hemisphere, how does the direction of Ekman transport of ocean water compare to the surface wind direction?
4. Distinguish between the directions of the trade winds in the Northern Hemisphere versus the Southern Hemisphere. What is the significance of the intertropical convergence zone?
5. Contrast the surface ocean currents on the eastern and western sides of the subtropical gyres.
6. Compare the rotation direction and productivity in warm- and cold-core rings of the Northern Hemisphere. Do cold-core rings associated with the Gulf Stream entrain waters from coastal areas or the Sargasso Sea?
7. Describe the role played by Ekman transport in coastal upwelling. Distinguish between coastal upwelling and equatorial upwelling.
8. How does the speed of water motion in the thermohaline circulation compare to that of wind-driven surface ocean currents?
9. Define water mass. How are water masses classified?
10. Describe the significance of the meridional overturning circulation (MOC) in Earth's climate system.

Critical Thinking

1. Suppose that the surface wind is blowing from the north along the Oregon coast. Predict the direction of Ekman transport of surface waters. Would you expect coastal upwelling or downwelling?
2. Compare the direction of Ekman transport in the subtropical gyres of the Northern Hemisphere versus the Southern Hemisphere. Explain the difference.
3. How might rings (warm-core and cold-core eddies) influence the intensity of storm systems over the ocean?
4. How and why do coastal upwelling and downwelling affect sea-surface temperature?
5. Define thermohaline circulation and describe how it is maintained in the ocean.
6. What role is played by the Coriolis Effect in western boundary currents?
7. Describe the relationship between trade winds and biological productivity along the equator.
8. How does a semi-permanent, subtropical atmospheric high pressure system influence the salinity of the underlying surface ocean waters?
9. Describe the weather conditions that would favor upwelling along the Wisconsin coast of Lake Michigan.
10. How does the wind direction over the Arctic Ocean compare to the direction of motion of the underlying ice floes?

ESSAY: Global Positioning System

Satellite navigation was the brainchild of the U.S. Department of Defense. Similar to the 18th century governments who offered prize money for anyone who could solve the longitude problem (Chapter 5), the U.S. government spent over $12 billion to develop a satellite-based system that would determine location precisely, at all times and in all types of weather. The *Global Positioning System (GPS)* was intended originally for military use and dates to the 1978 launch of the first Navstar satellite. GPS can provide near pinpoint location accuracy globally for a wide variety of military and civilian applications. Not to be outdone by its then Cold War adversary, the former Soviet Union developed a similar system called GLONASS. This system has suffered maintenance problems but now appears to be on the verge of rejuvenation via agreements with India for launch of additional satellites and talks with the U.S. about changing certain radio frequency standards which would allow a certain amount of interoperability with the U.S. GPS as well as the relatively new commercial Galileo system of the European Union.

Civilian use of GPS began in the 1980s, mostly on large ships. At least 24 Navstar satellites are needed to provide continuous service. Once this was achieved in the early 1990s, inexpensive, small portable receivers were developed and marketed, and civilian use of GPS soared. Today the civilian sector accounts for about 92% of GPS equipment sales. (Worldwide, sales of GPS equipment were projected to rise from $17 billion in 2003 to $50 billion by 2010.) GPS is widely used for navigation by ships, planes, delivery trucks, and automobiles, and is increasingly being used to provide location in cellular telephones.

The principle behind GPS is *trilateration* (related to triangulation) from satellites that are in view of the receiver (Figure 1). At least four satellites are needed for maximum precision. The time it takes for a ranging signal to travel from a satellite to a receiver is determined using accurate clocks. Atomic clock times are embedded in the code of a signal that is continuously transmitted by the satellites. When the timed signals arrive at the receiver they are compared with a code generated by the clock of the receiver. The time difference between the codes is the important factor. Radio signals travel at a finite speed (300,000 km per sec, or 186,000 mi per sec, the speed of light) so that elapsed time can be converted to distance. All GPS satellites are identifiable by their coded signals and their orbits are regular allowing a refinement of the triangulated distances. Using an inexpensive widely available receiver, a person can readily locate his/her position within a few meters. More sophisticated military receivers can achieve location accuracy in centimeters.

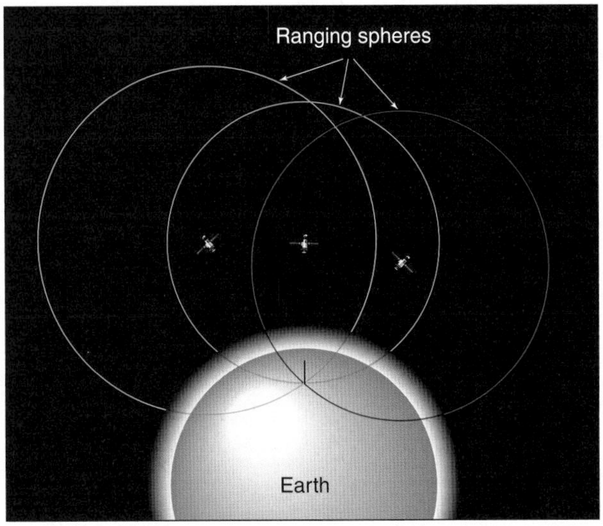

FIGURE 1
To determine its location on Earth's surface, a GPS receiver measures the travel times of radio signals sent by three satellites of known orbital location. Travel times are converted to distances. This trilateration technique is more precise when a fourth satellite is used to synchronize the clocks on the GPS satellites and receiver.

Using one satellite, a person's location must be some point on a sphere centered on the satellite with the radius of the sphere equal to the distance between the person's location and the satellite. Knowing our distance from a second satellite, our location must be some point on a sphere having a radius equal to the distance to that satellite. The two spheres have different radii and geometry indicates that the intersection of two spheres is a circle. Our location must be somewhere on that circle. A third satellite describes yet another sphere that intersects with the other two spheres, narrowing our location to one of two possible points, that is, where all three spheres intersect. Usually one of the two points is either in space or in Earth's interior. Hence, the other point is our location on the surface of Earth. A computer in the GPS receiver has algorithms that can distinguish between locations likely to be correct and locations that are spurious.

Use of a fourth (or additional) satellite is intended to compensate for the imprecision caused by slight differences in timing between the precise atomic clocks flown onboard the GPS satellite and the less accurate quartz clocks in the GPS receiver. Measurements made by the fourth satellite are used to synchronize the two clocks. Actually, most GPS receivers will choose the best signals from many satellites and use four of them to fix the position so that assumptions about which one of two points of intersection is correct never have to be made. A mariner at sea level does not need to know altitude, and this further simplifies the procedure.

Plans are currently underway to increase the location accuracy of GPS by broadcasting new signals from the navigation satellites. One of the anticipated benefits of these new signals is a reduction in interference caused by Earth's ionosphere (a region of the upper atmosphere containing a relatively high concentration of electrically charged particles). These improvements to GPS soon should be in place and are essential for completion of the transition of civilian aviation to GPS-based navigation and air traffic control.

ESSAY: Profiling the Ocean Depths

In situ sampling at various depths in the ocean provides profiles of water transparency, temperature, salinity (from electrical conductivity), and density. Sunlight is essential for photosynthesis and an easy way to estimate the maximum depth in the water column reached by visible light utilizes a simple instrument known as a Secchi disk. The *Secchi disk* is a plate-like device about 20 cm (7.9 in.) in diameter—although they come in different sizes depending upon the anticipated range of water clarity (Figure 1). The disk is hung from a weighted line marked off in increments of length and is divided into alternate black and white quadrants to make it easier to see at depth. As close to mid-day as possible, the Secchi disk is gradually lowered into the water (oriented parallel to the sea surface) on the sunny side of the boat until the observer can no longer see the disk (i.e., cannot distinguish between the black and white quadrants) and the depth is recorded. The disk is then raised slowly until it just reappears, and this depth is recorded. The Secchi depth is the average of these two depth readings and is a reasonable estimate of the compensation depth of 0.1% to 1% light penetration. It is a semi-quantitative method that is still used today. Of course, with modern electronics, very precise measurements of the transparency of ocean water to sunlight are made from profiling instruments.

Some of the first temperature and salinity casts made in the early 20th century employed collection bottles and reversing thermometers that were attached to a wire. The specially designed bottles obtained seawater samples at designated water depths and their contents were sent to a ship or onshore laboratory for chemical analysis. Thermometers recorded seawater temperature which when combined with salinity readings yielded water density and depth. Two liquid-in-glass thermometers were used at each sampling depth, one accounting for pressure at depth and the other protected from the effects of pressure at depth. A collection bottle plus thermometers were attached to the wire at fixed depth intervals. To obtain data, a messenger (metallic weight) slides down the wire, tripping the shallowest collection bottle so that it flips over, captures a seawater sample, and closes. At the same time, the thermometers reverse thereby fixing their readings. This action releases another messenger that slides down the wire to the next deeper bottle where the process of data collection is repeated. Although this casting method is time consuming, reversing thermometers and collection bottles are still used today for calibrating other instruments.

FIGURE 1
A Secchi disk is a plate-like device hung from a weighted line. It is a tool for measuring the maximum depth that visible light can reach in a water column. [Photo by Leszek Bledzki]

In the 1970s, electronic sensors were introduced for more rapid and reliable measurements of CDT: conductivity (a measure of salinity), depth, and temperature. A CDT instrument package is linked to an oceanographic research ship via a hydrographic cable. As the CDT is lowered and raised in the water, its sensors provide a continuous stream of data at specific depths. An additional component is a series of a dozen or more Niskin (a refinement of the long-used Nansen device) collection bottles plus thermometers arranged in a rosette pattern (Figure 2). Ship-based researchers are able to open and close collection bottles at designated depths of the cast.

In the mid 20th century, a mechanical instrument called a *bathythermograph (BT)* was developed that could obtain a nearly continuous temperature profile with depth from a slowly moving ship. As this torpedo-shaped device sinks into the water, a temperature-sensitive element (typically a deformation or bimetallic thermometer) inside the BT moves a stylus (a needlelike marking device) across a metal- or smoke-coated slide recording temperature as a function of depth (obtained from pressure bellows). The resulting temperature trace was interpreted after the instrument was retrieved from the water.

Since the 1970s, newer electronic technologies have permitted more rapid castings that produce an almost continuous profile of temperature. An *expendable bathythermograph (XBT)* is a non-recoverable instrument package deployed from fast moving ships or airplanes that obtains a nearly continuous ocean temperature profile to a depth of approximately 1800 m (5900 ft). This torpedo-shaped device consists of a thermistor (electronic thermometer) housed in an expendable casing. A thin conducting wire connects the XBT to the ship or plane and transmits the temperature signal to an onboard recorder. The electrical conductivity of seawater is used to complete the electrical circuit. As the device sinks into the ocean at a known rate, data are transmitted electronically at regular intervals, permitting the recording of ocean temperature as it changes with depth. A slightly different profiling instrument is an *expendable conductivity-temperature-depth profiler (XCTD)* that provides essentially continuous profiles of ocean conductivity (salinity) in addition to temperature and pressure.

NOAA's Atlantic Oceanographic and Meteorological Laboratory in Miami, FL, oversees the XBT program. Volunteer commercial ships deploy as many as four XBTs daily along selected shipping lanes. The data are compiled by a computer onboard ship and then transmitted via satellite relay to the Laboratory for global distribution. More than 70 ships voluntarily produce 26 monthly transects across the three major ocean basins.

In 1985, the ten-year international *Tropical Ocean Global Atmosphere (TOGA)* program commenced. One of TOGA's projects was deployment of *TAO (Tropical Atmosphere/Ocean)*, an array of moored buoys (small, non-piloted, instrumented platforms) in the tropical Pacific Ocean. Data from this array have been extremely valuable in detecting and predicting such atmospheric/oceanic episodes as El Niño and La Niña (Chapter 11). This instrument array, renamed *TAO/TRITON* in 2000, presently consists of approximately 70 deep-sea moorings that measure several atmospheric variables (air temperature, wind, relative humidity) as well as oceanic parameters (sea-surface and subsurface temperatures at 10 depths

FIGURE 2
A rosette of Niskin water collection bottles being lowered over the side of an oceanographic research ship. The purpose of this experiment is to determine the concentration of chlorophyll and biological productivity. [NOAA Ocean Explorer]

in the upper 500 m or 1640 ft). Several newer moorings also have salinity sensors, along with additional meteorological sensors. Five moorings along the equator also measure ocean velocity using a Subsurface Acoustic Doppler Current Profiler. The data are collected and relayed in near real-time to shore via satellite. Real time data displays from the TAO/TRITON array are available on the Internet from NOAA's *Pacific Marine Environmental Laboratory (PMEL)* in Seattle, WA.

Over the last decade or so, a variety of autonomous instrumented profilers have been developed and deployed to measure large-scale subsurface currents and make repeated near real-time vertical measurements of ocean variables. Early versions of these free-drifting profilers were identified by the term *PALACE (Profiling Autonomous Lagrangian Circulation Explorer)* whereas a subsequent version was called *APEX (Autonomous Profiling Explorer)*. Elsewhere in this chapter, we describe the Argo array.

An Argo float (profiler) is about 1 m (3.3 ft) long and less than 20 cm (8 in.) in diameter. A profiler moves vertically between the ocean surface and a maximum depth of about 2000 m (6600 ft) by pumping hydraulic oil between an internal reservoir and an external bladder to change its bulk density. The profiler ascends when the relatively lower density oil (as compared to the density of sea water) flows into (and expands) the bladder (decreasing the density) and descends when oil flows in the opposite direction. During the float's ascent, onboard sensors record the temperature, pressure (depth), and conductivity (salinity) nearly continuously. Upon returning to the ocean surface, the float telemeters these data to a satellite for subsequent relay to data collection stations (Figure 6.15). The float's position is determined by satellite, and subsurface ocean current information is inferred from the horizontal displacement of the float between subsequent surfacings. The geostrophic assumption is used to compute ocean currents at depth. Following a programmed interval at the surface, hydraulic oil is pumped back to the internal reservoir, and the float returns to depth for the next cycle. The anticipated lifetime of one Argo float is 100 cycles.

A glider is a more recent type of float designed to cover a greater distance and have a longer sampling life than an Argo float. A glider, pictured on the front page of this chapter, is shaped like a torpedo and changes its buoyancy in order to rise and sink in the ocean. Unlike an Argo float, it can move independent of the surrounding ocean currents, making it more versatile and highly maneuverable. We have much more to say about gliders in Chapter 13.

ESSAY: Bottles and Rubber Duckies, Tracking Currents with Flotsam and Jetsam

The classic message-in-a-bottle is one of the simplest ways to determine ocean surface currents. Bottles, examples of *jetsam* the traditional term for floating objects intentionally tossed from a ship, are dropped into the ocean at a known time and place and left to float with the surface current. Inside the bottle is a message, typically requesting the finder to notify the sender of where and when it was found.

Prince Albert I of Monaco (1848-1922), an early amateur oceanographer, employed this elementary technique to map the currents of the North Atlantic subtropical gyre in the late 1800s. From his yacht, the *Hirondelle*, he dropped hundreds of bottles throughout the North Atlantic. Additionally, he collected accounts of bottles dropped from commercial ships making trans-Atlantic voyages. One such bottle, recovered in the Caribbean off the coast of the Yucatán Peninsula in 1890 contained a common message: *"This bottle was thrown from the S.S. Cephalonia of the Cunard Line on the 24th of November, 1887, by W.C. Lippard about 400 miles from Boston. Anybody finding this will please send it to the Boston newspapers."*

A drawback of the message-in-the-bottle method of tracking ocean currents is that typically only one or two percent of the bottles are recovered. Hence, many bottles must be tossed into the ocean for only a small amount of data. Instead of jetsam, flotsam, traditional term for goods lost over the side of a boat, can be serendipitously used if spilled in large quantities at a known time and place.

In May 1990, a heavy storm in the North Pacific washed 21 cargo containers off the container vessel *Hansa Carrier*. Five containers held shipments of Nike® shoes, an estimated 80,000 shoes were lost as the containers fell into the sea. Unfortunately, waste at sea is routine and this particular loss would not have been noteworthy except that 10 months later, hundreds of shoes began washing up on the shores of Oregon, Washington, and Vancouver Island. Stories of the large flotilla of shoes caught the attention of oceanographer Curtis C. Ebbesmeyer, who had made a career of tracking currents using buoys and jetsam. Ebbesmeyer used code numbers from the shoes to confirm that they originated from the *Hansa Carrier* spill. Knowing the exact date and location the shoes were lost, Ebbesmeyer and fellow oceanographer W. James Ingraham Jr. used the locations of the found shoes to test predictions made by a computer model of North Pacific Ocean currents.

Another large flotsam dump occurred in January 1992 when a freighter crossing the Pacific from Hong Kong to Tacoma, WA, lost several cargo containers near the International Dateline during a heavy storm. One container held 28,800 plastic floating bathtub toys manufactured by Floatees. The container broke open as it fell into the sea and seven months later beachcombers near Sitka, Alaska, began finding plastic ducks, turtles, frogs, and beavers, numbering in the hundreds. As late as 2007, nearly 1000 of these floating toys continued to be recovered along beaches. Data from this spill indicate that the typical "trip" around the sub-polar gyre of the North Pacific takes 2 to 4 years. The toys would not be confined to the North Pacific, however. Eventually some began drifting into the Arctic Ocean and by the year 2000 the first plastic ducks showed up in the North Atlantic as well.

Although worldwide an estimated 10,000 cargo containers are lost overboard each year, only a fraction of flotsam ends up washing up on beaches. Much of it collects at the center of the large subtropical gyres. Subtropical gyres are characterized by convergent Ekman transport. The wind forcing pushes surface waters toward the middle of the gyre. Convergence causes a large-scale sinking of the water, i.e. *Ekman pumping*. Although the water sinks, objects floating on the surface are left behind causing a gradual accumulation of flotsam. This is true for all floating objects, flotsam, jetsam, and those naturally occurring. The Sargasso Sea, at the center of the North Atlantic subtropical gyre, derives its name from the high concentration of floating sargassum weed which accumulates because of convergent Ekman currents.

While the use of flotsam to study ocean currents is a novel way of obtaining useful results from unfortunate accidents, it also highlights the long-term hazard of plastics in the marine environment. After decades of exposure at sea, the plastic ducks and frogs continue to float and pose hazards to marine life. The North Pacific lacks sargassum weed, thus the most distinctive objects accumulating in the subtropical gyre are floating plastic objects. This has given rise to a rather unfortunate name for this region, the "North Pacific Trash Vortex."

CHAPTER 7

OCEAN WAVES AND TIDES

Sandbars and sculpted sands left behind by a receding tide. [Courtesy of Randolph Femmer /life.nbii.gov]

Case in Point

The USS *Ramapo* was en route to San Diego, CA, from the Philippines on the early morning of 7 February 1933 when it encountered an intense storm with near hurricane-force winds. The 146 m (478 ft) Navy tanker was struck on the stern by a freak wave higher than 34 m (112 ft). In September 1965, while crossing the Atlantic, one of the world's largest ocean liners, the *Queen Elizabeth 2 (QE2)*, was hit by a giant wave. The *QE2* had only just changed course to avoid a hurricane and, despite rough seas, thought the danger was passed until the captain saw a wave about 29 m (95 ft) above the sea surface, a height later verified by an instrumented buoy. On the morning of 16 April 2005, a wave the height of a seven-story building crashed into the bow of the cruise ship *Norwegian Dawn* off the coast of Georgia.

Unusually high and potentially destructive **rogue waves**, also called freak waves, develop in the open ocean and coastal areas. Although they can arise when conditions are tranquil, they are most common during stormy weather and, even in intense storms with high waves, rogue waves are much larger and strike without warning. The *Ramapo, QE2,* and *Norwegian Dawn* were large and lucky enough to survive their clashes with rogue waves but many ships, including modern supertankers are not so fortunate. They can be extensively damaged or sink with fatalities. Rogue waves can also be destructive in coastal areas where they can destroy off-shore oil-drilling platforms.

When does an ocean wave qualify as a rogue wave? This depends not only on the height of the wave, but also the **significant wave height** of the waves in the surrounding ocean. The significant wave height is found by averaging the height of the tallest third of the waves and, if the wave is a rogue wave, its height is 2.2 times greater. Another definition sets the minimum height of a rogue wave at 25 m (82 ft). Rogue waves also exhibit a distinctive form unlike other ocean waves, a deep trough and steep forward face. Mariners describe their encounter with a rogue wave as "sailing into a hole in the sea" and then "hitting a wall of water."

Rogue waves have been reported from all ocean basins. Until recently, ocean scientists assumed that they were a relatively rare phenomenon, estimated to be a once-in-10,000-year event at any location in the ocean. However, wave data collected by instruments on a gas-drilling platform off the southern coast of South Africa indicate that during any hour, the probability of a rogue wave is about 3.1%. Similarly, there is a 3.7% probability in the South Atlantic Ocean east of Rio de Janeiro, Brazil.

What causes rogue waves? Mariners have long attributed the sudden occurrence of exceptionally high waves to a random superposition of waves with different wavelengths, resulting in *constructive interference*. With constructive interference, the crests of different waves are additive as they coincide to produce a giant wave. The shape of the sea floor and coastline configuration may also help direct the movement of swells, amplifying the interaction.

Rogue waves are relatively frequent off the coast of South Africa where storm waves driven by southwest winds from the Southern Ocean run into the swift Agulhas Current flowing from the northeast. With the waves and current moving in opposite directions, the wavelength shortens and wave crests build. Another explanation for rogue waves formed by the Agulhas Current is that eddies (rings) may focus wave energy, just as an optical lens focuses light. Such focusing might also occur near the Gulf Stream in the "Bermuda Triangle" where ships have been known to vanish mysteriously.

With improved understanding of the genesis of rogue waves, scientists hope to develop models to predict the likely location of these dangerous waves. With adequate warning, mariners could alter their course, saving lives and cargo.

Driving Question:
What are sea waves and tides and what causes them?

In thinking about the ocean, most people visualize a vast water surface disturbed by continually changing and sometimes seemingly chaotic patterns of waves. Ceaselessly, waves break against the shore; waves and sea spray buffet a small fishing boat bobbing in the water just offshore; and waves the height of a three-story building batter a supertanker plying through storm-tossed waters a thousand kilometers at sea. Potentially destructive waves are not limited to the ocean. A huge wave may have contributed to the sinking of the ore carrier *Edmund Fitzgerald* on Lake Superior on 10 November 1975 with the loss of its 29 member crew.

What are sea waves and what causes them? Most sea waves are the product of an interaction between the ocean and atmosphere in which *kinetic energy* (the energy of motion) of the wind is transferred to surface waters. Hence, these waves are often called **wind-waves**. Wind-waves are part of the current-generating processes described in the previous chapter. In this chapter, we discuss the formation and life cycle of these waves, their generation as wind-waves or "sea," plus their existence at sea as free waves and swell before expending their energy as breakers on a distant reef or shoreline. We also describe internal waves and tsunamis.

Coastal residents are very familiar with not only sea waves but also the periodic rise and fall of sea level known as tides. Tides are also a type of wave but with very long wavelengths approaching the dimensions of an ocean basin. Because of their importance to marine interests, tides were among the earliest ocean characteristics to be monitored on a regular basis. In fact, elaborate mechanical computers were developed early on to predict tides. In this chapter, we examine tide generation, types of tides and tide prediction.

Water Waves

A **wave** is the physical oscillation in a solid, liquid, or gaseous medium as energy is transmitted through that medium. A **sea wave** is an oscillation or undulation on the ocean surface that propagates (moves) along the interface between the atmosphere and ocean (Figure 7.1). A schematic drawing of an idealized sea wave is shown in Figure 7.2. As a wave propagates, the sea surface alternates up and down about an equilibrium, which is the still-water level. The **wave crest** is the highest part of the wave above the equilibrium level and the **wave trough** the lowest. **Wave height** is the vertical distance between crest and trough (half the wave height is the *amplitude*) while the horizontal distance between successive wave crests (or equivalently between any other two corresponding points of two consecutive waves) is the **wavelength**. The time needed for one wavelength, two successive wave crests, to pass a fixed point is the **wave period**. The number of waves that pass a fixed point over time is **wave frequency**, which is the inverse of the wave period.

WIND-WAVE GENERATION

While energy can be introduced to the ocean by coastal and underwater landslides, tectonic movements of the sea floor (responsible for tsunamis), and astronomical forces (responsible for tides, a type of shallow-water wave), the primary wave genesis is wind.

If the air over the ocean were always calm, the sea surface would be smooth and motionless, a condition sailors refer to as *flat calm*. Wind disturbs the equilibrium as some of its *kinetic energy* is transferred to the ocean surface as a wave-generating force. Waves begin as small ripples, called **capillary waves**, with wavelengths less than 1.7 cm (0.7 in.). At these short wavelengths, water's surface tension is the restoring force that smoothes and flattens the small waves back to equilibrium.

Surface tension is the force acting at the interface between liquid and air, due to the difference in attraction between liquid molecules and gaseous molecules. In the case of a liquid water/air interface, hydrogen bonding within the water creates this surface tension (Chapter 3). Compared to the ocean surface, the density of water molecules is relatively low in the air above. Hence, the H_2O molecules within the surface water layer are more strongly attracted to each other through hydrogen bonding (and to the layer of water molecules below) than to those above the surface. Surface tension explains why we can pour water into a glass to just above the rim without the water spilling. Among liquids, the strength of water's surface tension is second only to mercury.

Stronger winds disturb the equilibrium of the water surface even more, producing larger waves with longer wavelengths and greater wave heights. For these larger waves, *gravity* is the restoring force, leveling the wave crests to fill in the wave troughs. But even as gravity pulls the crested water down, momentum carries the water

FIGURE 7.1
Sea waves on the Pacfic Ocean. Image taken during the deployment of a NOAA Deep-ocean Assessment and Reporting of Tsunamis (DART) system buoy. [Courtesy of NOAA]

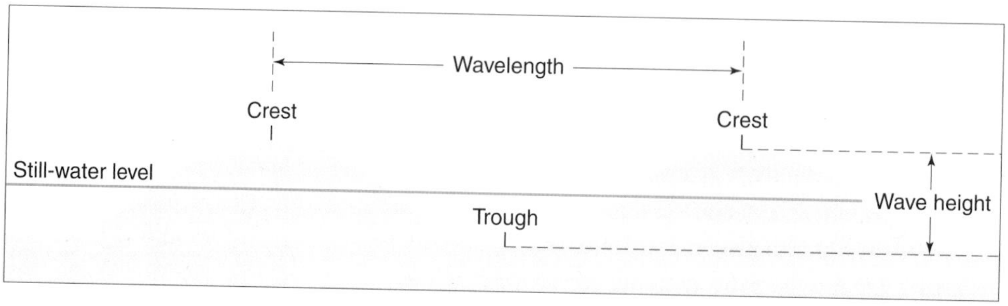

FIGURE 7.2
Vertical cross-section of an idealized sea wave.

further downward and forms a new trough. The lowered water is then buoyed upward raising to the next crest. Together, wind and gravity are responsible for most waves observed on the ocean surface.

As a wave propagates along the water surface, the disturbed water particles travel in circular orbits, moving forward, down, back and up (Figure 7.3). In this way, waves (the changing shape of the water surface) move away from the disturbance that caused them. As long as the wave is traveling through relatively deep water, the wave height is the same as the diameter of the surface water particle's orbit. As a complete wave (from crest to trough to crest) passes a fixed point, the water particle completes one orbit, returning to approximately its original location. Within the orbit, the top, forward speed is slightly greater than the bottom, backward speed so that, when the water particle leaves the wave, it has traveled slightly forward with the wave. Below the water surface, the oscillations of water particles become smaller (the orbits decrease in diameter). The depth at which orbits no longer form is the **wave-base**, at approximately half the wavelength.

The propagation and shape of a wave change as it moves through the water. If a rock is thrown into a quiet pond, the disturbed water generates a ring of waves that spreads outward radially. As the waves spread, they travel in packs with approximately the same speed and period known as *wave trains*. Within wave trains, a single packet of energy, the largest wave crest, belongs to the last wave, on the inside of the ring. As they propagate, this wave crest grows in height, moving at twice the speed of the train as it passes through it, then disappears at the front on the outside of the ring. The lead wave disappears because it transfers half its energy ahead, to initiate motion in the undisturbed water and half to maintain the motion in the wave following. A new wave continuously forms at the back so the number of waves in the wave train remains constant.

Celerity is the speed of a single wave and is found using the wave's wavelength (λ), the distance between crests (or troughs), and the wave period (T), the time it takes for two consecutive crests (or troughs) to pass the same point. (The wave period is constant.)

$$C = \lambda/T$$

An individual wave travels twice as fast as the leading edge of the wave train. The *group speed* (*V*) is the speed of the wave train, that is.

$$V = C/2$$

The wave energy travels at this group speed. Transitional and deep ocean waves exhibit this same behavior as they also propagate at twice the speed of the wave energy.

Formation and evolution of ocean waves depend on the interaction of wind speed, duration of the wind, and **fetch**, the distance the wind blows in the same direction over a water surface. Together, these factors determine

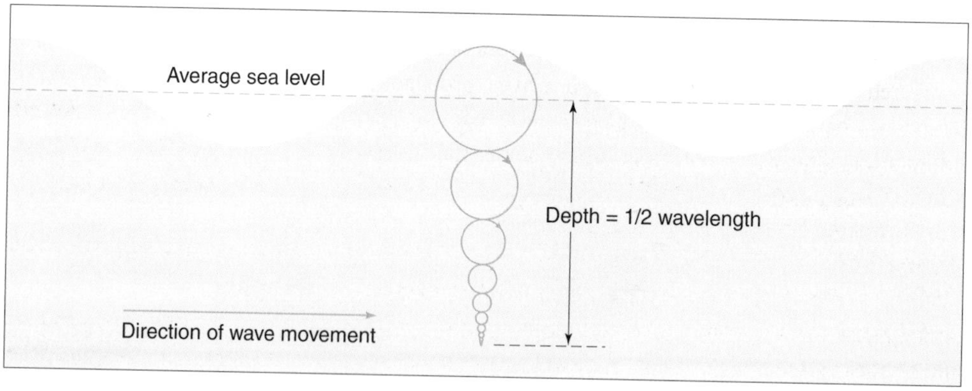

FIGURE 7.3
For waves passing through relatively deep water, the nearly circular orbits of water particles weaken with increasing depth and essentially disappear at a water depth of one-half the wavelength.

the amount of available kinetic energy transferred from the atmosphere to the surface waters. Increasing wind speed translates into more transferable kinetic energy and, therefore, higher waves. As the wind strengthens, it becomes more turbulent and creates *eddies*, energetic irregular whirls of air motion. We experience eddies as wind gusts and lulls on a windy day. Turbulent eddies exert forces acting parallel and perpendicular to a water surface because they have both horizontal and vertical components. Parallel forces drag surface water particles laterally whereas perpendicular forces are responsible for the vertical movement. Changes in the elevation, the perpendicular forces, mark changes in potential energy while the orbital activity of the water particles involves kinetic energy. The greater the wave height, the greater the combination of potential and kinetic energy, as potential energy is continuously converted to kinetic energy and back.

At low wind speeds (less than a few km per hr), weak eddies produce capillary waves on the water surface and, as noted earlier, the surface tension restores these tiny waves to smooth equilibrium. However, these ripples roughen the water surface, thereby increasing the wind's drag on the ocean surface, increasing the wave height, which increases the roughness, and causes even greater wind resistance and wave growth. Even with the same wind speed, higher waves can be produced with a longer fetch. *Duration* refers to the length of time the wind blows from the same direction. For the same wind speed and fetch, waves will heighten and elongate the longer the wind blows from the same direction.

In general, waves build in height and length as long as the wind supplies more energy than is dissipated by breaking waves. Conversely, when the wind supplies less than is dissipated, waves decay (become weaker and smaller). When the amount of dissipated energy equals the amount of wind supplied energy, no further wave build up occurs.

An ocean surface disturbed by storm winds becomes a confused mass of sharp-crested waves of various heights and lengths, all moving in different directions. This condition is known as **sea**. These are *forced waves*, generated by storm winds. As these waves propagate beyond the area of strong storm winds, they become more rounded than the waves forming directly under the storm winds and are known as **swells**. A swell consists of *free waves* where the only force acting on them is gravity (Figure 7.4).

In deep water, waves with longer wavelengths and periods travel faster than waves with shorter wavelengths and periods so the longer waves of a swell outdistance the shorter waves. With little dissipation from friction, a swell can travel thousands of kilometers from its source. This is why strong waves, from a distant storm, strike the coastline even when the local weather is tranquil.

FIGURE 7.4
Easterly swell at Lyttelton Harbour, New Zealand. [Photo courtesy of Phillip Capper/License: Creative Commons Attribution 2.0]

Wave interference also influences the growth and decay of ocean waves. Waves generated by winds associated with two or more storm systems may interfere constructively or destructively (Figure 7.5). In **constructive wave interference**, two or more wave crests coincide to form composite waves with heights greater than any of the original waves. As discussed in this chapter's Case-in-Point, constructive wave interference may contribute to the formation of *rogue waves*. On the other hand, the crests and troughs of waves can interact, partially canceling all of the waves. The product is a composite wave with a height smaller than that of the original waves. This interaction is known as **destructive wave interference**. Both types of wave interferences are always happening and contribute to the continually changing pattern of waves on the open ocean. Furthermore, waves from many sources may intersect at any angle, making possible many interference patterns.

DEEP-WATER AND SHALLOW-WATER WAVES

Waves in water deeper than their wave-base (about half the wavelength) are known as **deep-water waves**. Recall that **celerity** is the speed of an individual wave and that the celerity of deep-water waves depends on wavelength and gravity.

$$C = \sqrt{(1.56 \times \lambda)}$$

C is the celerity in m per sec and λ is the wavelength in meters. For example, a 10 m (33 ft) long wave travels at

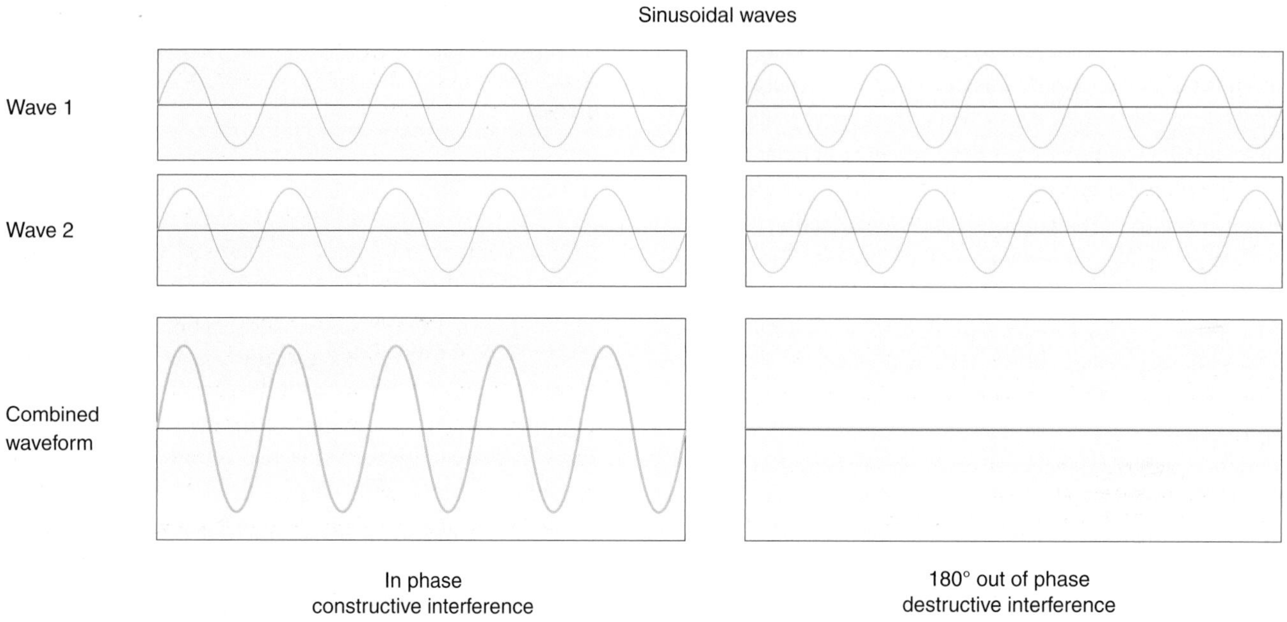

FIGURE 7.5
In constructive wave interference (left), two waves in phase with one another combine to produce a wave with double the height of each original wave. During destructive wave interference, two waves out of phase in effect cancel each other out. In the ocean, waves from many sources may interact at various angles, leading to interference patterns much more complex than this example.

3.9 m per sec (9 mph) whereas a 100 m (330 ft) long wave travels at 12.5 m per sec (28 mph).

There is no interaction between the orbital motions of the water particles in deep-water waves and the ocean bottom. However, as waves approach the coastline, they encounter increasingly shallow (shoaling) waters. Eventually waves enter waters shallower than their wave-base where, affected by the ocean bottom, they are no longer the free waves of the open ocean. Because the wave period is constant, a wave interacting with the ocean bottom slows, its wavelength shortens, and its height increases. (With a constant wave period, the water and energy have no place to go but up!) With a **shallow-water wave**, which occurs in water depths less than one-twentieth the wavelength, the orbits of water particles flatten with increasing depth, changing from circular to elliptical and ultimately to a back-and-forth motion near the ocean bottom (Figure 7.6). **Transitional waves** refer to waves entering water with a depth between one-twentieth and half of the wavelength.

When waves enter waters less than half their wavelength, the wave celerity depends on water depth and gravity but not wavelength. Shallow-water wave celerity can be computed using the formula:

$$C = \sqrt{(g \times D)}$$

where C is the celerity in m per sec, g is gravity (9.8 m/sec^2), and D is the depth in meters. For example, waves in 10 m (33 ft) of water travel at 9.9 m per sec (22 mph) whereas waves in 2 m (7 ft) of water travel at 4.4 m per sec (9.8 mph). Hence, shallow-water waves slow as they enter shoaling waters.

As the water shoals, particles of water in the building wave crest orbit forward faster than the wave

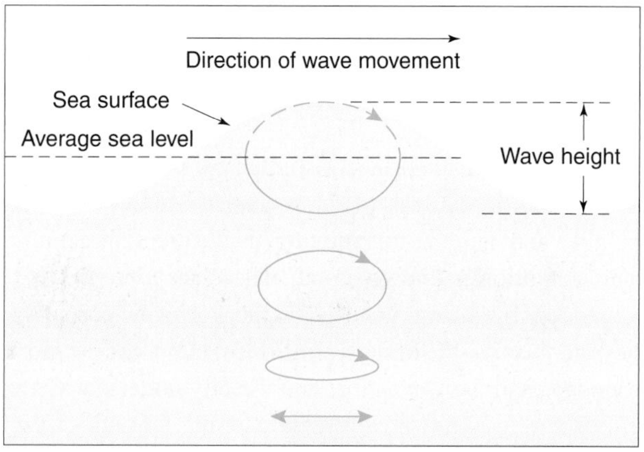

FIGURE 7.6
In shallow-water waves, the orbits of water particles gradually flatten with increasing depth, changing from circular to elliptical and ultimately to a back-and-forth motion near the ocean bottom. Figure not drawn to scale.

FIGURE 7.7
A wave propagating through shallow water becomes steeper and unstable. When the ratio of wave height to wavelength approaches 1 to 7, the crest plunges forward as a breaker. [Photo by J.M. Moran]

propagates toward shore. The wave becomes steeper and eventually unstable, which causes the crest to plunge forward, forming a **breaker** and dissipating the wave energy (Figure 7.7). Waves break when the wavelength is approximately seven times the wave height and the wave-crest angle (a measure of the steepness of the wave) is close to 120 degrees (Figure 7.8). Breakers 2 m (7 ft) high exert a pressure equivalent to 15,000 kg per square m (3000 lbs per square ft), significantly impacting both natural and artificial features along the coast. A nearly continuous train of waves breaking along a shore is called **surf**.

FIGURE 7.8
Closeup of a breaking wave in Santa Cruz, CA. [Photo by Mila Zinkova/License: Creative Commons Sharealike 3.0]

Through the years, mariners have developed methods of describing the state of the sea and relating that state to near-surface winds. For more on this, see this chapter's first Essay.

SEICHE

If you've ever sloshed water back and forth in a bathtub, you have experienced a seiche, a wave phenomenon first studied in Lake Geneva, Switzerland, in the 1700s. A **seiche** (pronounced *say-sh*) is a rhythmic oscillation of water in an enclosed basin (e.g., bathtub, lake, or reservoir) or a partially enclosed coastal inlet (e.g., bay, harbor, or estuary). During a seiche, the water level in a basin rises at one end while simultaneously falling at the other end. A seiche episode may last only a few minutes or a few days. A seiche is a **standing wave**. This is in contrast to the wind-driven waves (described above) that are **progressive waves** in that they move through a body of water. With wind-driven waves, crests and troughs travel along the water surface whereas with standing waves, crests alternate vertically with troughs but at fixed locations. Gravity is the restoring force for both progressive and standing waves.

For a typical seiche in a simple enclosed basin, the water level near the center does not change (Figure 7.9). This location, called a *node*, is where water moves fastest horizontally. At either end of an enclosed basin, where vertical motion of the water surface is greatest and horizontal movement of water is minimal, are the *antinodes*. The motion of the water surface during a seiche is somewhat like that of a seesaw. The pivot point of the seesaw does not move vertically (analogous to a node) while the people seated at either end of the seesaw move up and down (analogous to an antinode). Partially enclosed basins usually have a node at the mouth (rather than near the center) and an antinode at the landward end. Furthermore, some basins are complex and have several nodes and antinodes, therefore seiches can be uninodal or multinodal.

Wind, air pressure gradients, earthquakes, and astronomical tides can induce a seiche. For example, wind blowing persistently in the same direction across the broad expanse of a bay causes water to pile up at the downwind shore. When the wind slackens, the piled water is released and the water surface oscillates back-and-forth from one end of the bay to the other as a seiche until the water calms. Because it is directly proportional to the basin length, the period of a seiche is considerably longer in a large coastal inlet than in a small pond, and can range from minutes to hours. Also, for the same basin, the period

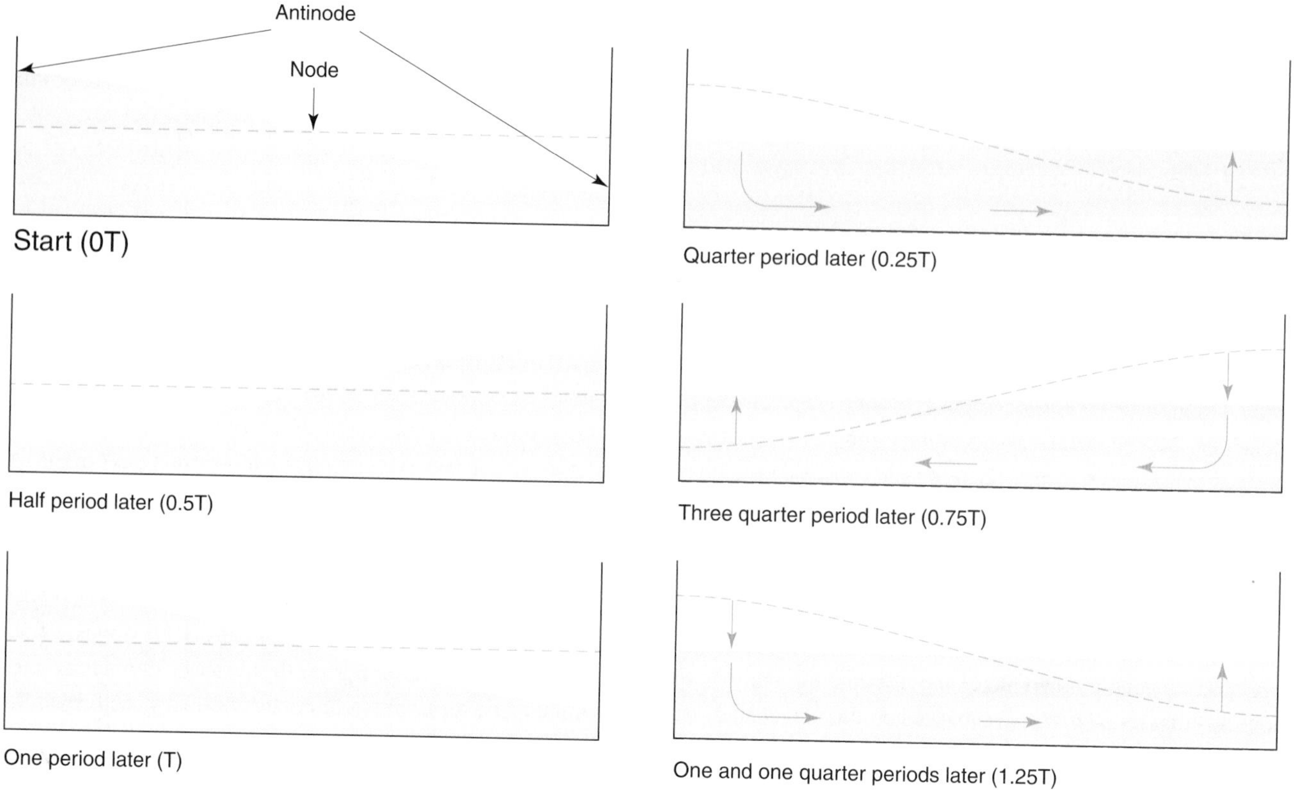

FIGURE 7.9
A seiche in an enclosed basin with a single node. The maximum vertical motion of the water takes place at the antinodes and the maximum horizontal motion occurs at the node.

is inversely proportional to water depth so deeper water has a shorter period.

Usually a seiche in a harbor or lake causes little concern because the vertical movements of water level are typically small, only a few centimeters. Under certain conditions, however, a seiche may grow, flooding coasts and damaging moored vessels, as has happened in San Diego Bay and Los Angeles/Long Beach, CA. A seiche grows as a consequence of **resonance**, meaning that the period of the disturbance (such as an earthquake or wind) matches the period of oscillation of the basin. By timing your rhythmic disturbance of the water in a tub to match the period of the tub (about 1 second), you can cause the seiche to build until water splashes over the lip of the tub and onto the floor. Through resonance, vibrations from the Northridge, CA, earthquake on 17 January 1994 caused several swimming pools to overflow throughout Southern California. In bays open to the ocean, if the period of tidal forcing equals the period of the bay, resonance can greatly increase the *tidal range* (difference in water level between high tide and low tide).

WAVE MEDIATED ATMOSPHERE-OCEAN TRANSFER

Sea waves help bring about transfers of energy and matter between the atmosphere and ocean, a major interaction among subsystems of the Earth system. By transferring momentum from the wind to ocean surface waters, the largest wind-generated waves are important in driving ocean currents. Waves with shorter wavelengths play a major role in heat transfer from the ocean surface to the atmosphere through *latent heating* and *sensible heating* (Chapter 5). Latent heat derived from ocean's surface water can power storm systems, such as thunderstorms and tropical cyclones (Chapter 8).

Waves with shorter wavelengths (especially breaking waves) also deliver tiny salt particles to the atmosphere where they function as cloud condensation nuclei spurring development of clouds (Chapter 5). Droplets of ocean spray transfer microscopic marine algae and viruses into the atmosphere where winds can transport them long distances. Breaking waves capture myriads of air bubbles that are carried tens of meters below the ocean surface. Gases in these bubbles dissolve in seawater; this process, known as **bubble injection**, is an important source of oxygen and carbon dioxide dissolved in surface waters (Chapter 3).

Internal Waves

Internal waves form within the ocean along interfaces where the change in density with depth is relatively abrupt. Fridtjof Nansen (1861-1930) is credited with discovering internal waves in the Arctic Basin in the late 1800s.

Fundamentally, waves are pulses of energy that travel along surfaces separating fluids having different densities. Hence, wind-driven waves and seiches occur at the interface between air and water. Similar interfaces also occur beneath the ocean surface. As described in Chapter 6, the ocean is not homogeneous; water density changes with depth principally in response to variations in temperature and salinity. Favorable sites for development of internal waves include the base of the mixed layer and interfaces (pycnoclines) between water masses having different densities. Internal waves also form in estuaries along pycnoclines between fresh river water and salty ocean water (Chapter 8).

The smaller the density contrast between two fluids in contact, the slower the internal wave propagates and the greater the wave height. Density contrasts between different water masses are about 1000 times less than that between air and water so that internal waves propagate more slowly than surface waves but have much greater wave heights. Simply put, with internal waves there is a much smaller density difference for gravity to act upon. Typically, internal waves have lengths in the hundreds of meters and heights of several meters but sometimes are considerably higher (100 m (330 ft) or more). Besides astronomical tides, slumping on the ocean floor, turbidity currents, and water masses slipping over one another also can generate internal waves. Even ships moving across the sea surface can generate internal waves along shallow pycnoclines.

Ocean Tides

Astronomical tides are the regular rise and fall of the sea surface caused by the gravitational attraction between the rotating Earth and the Moon and Sun. Tides can be thought of as progressive planetary-scale waves that propagate across ocean basins; wave crests are high tides and wave troughs are low tides. The length of a tide wave is considerably greater than the depth of the ocean so it behaves as a shallow-water wave.

In the hypothetical case of an Earth covered with an ocean of infinite depth and with no continents, tides can be visualized as waves having lengths of one half of the circumference of the planet. Astronomical tides are *forced waves* in that they always follow the driving force of the Moon and Sun. Tide crests would be located directly below the celestial body (Moon or Sun) responsible for the tide-generating force. On our hypothetical Earth, the speed of propagation of the tide crest depends, at least in part, on the rotation of the planet relative to the Sun or Moon. On the equator where the planet's circumference is about 38,600 km (24,000 mi), the tide crest would travel at about 1600 km per hr (1000 mph).

On the real Earth, continents break up the ocean into separate basins and the ocean has a variable depth. Tides are shallow-water waves so that wave celerity depends on water depth. For an average ocean depth of 4000 m (13,000 ft), tidal celerity is about 200 m per sec (400 mph). Tides speed up where the ocean is relatively deep and slow over ridges where the ocean is shallower. At sea, the amplitude of the tide wave is well under a meter but increases as it enters shallower coastal waters. Tides are measured mostly at coastal locations as local changes in water level through time. The average vertical difference in height between water levels at high and low tides is called the **tidal range** and generally varies between less than a meter to several meters (Figure 7.10). The time between successive high tides is the **tidal period**.

FIGURE 7.10 Low tide at a mooring in a harbor along the coast of Maine. Note the water-mark on the dock indicating the level of the water at high tide. [Photo by J.M. Moran]

Tides are important because of their effects on ecosystems, local navigation, moorings, coastal structures, legal boundaries, fisheries, and recreation. Today, considerable scientific research focuses on the global nature of tides, including their influence on other physical processes in the ocean such as circulation, mixing, and wave generation. Furthermore, ocean and atmospheric scientists are interested in how storm-driven waves and surges combine with tides to affect the potential for coastal flooding and the rate of coastal erosion (Chapter 8). In the following section, we focus on some fundamentals of ocean tides and their prediction.

TIDE-GENERATING FORCES

The tide-generating force is produced by the combination of (1) the gravitational attraction between Earth and the Moon and Sun, and (2) the rotations of the Earth-Moon and Earth-Sun systems. Forces combine to deform Earth's ocean surface into a roughly egg-shape with two bulges. One ocean bulge faces towards the Moon and the other is on the opposing side of the planet, facing away from the Moon (Figure 7.11). A similar interaction between Earth and Sun produces two other ocean bulges that line up towards and away from the Sun.

According to the **law of universal gravitation** (Isaac Newton, 1642-1727), the *gravitational attraction* between two bodies is directly proportional to the product of the masses of the two bodies and inversely proportional to the square of the distance between them. Simply put, the greater the mass, the greater is the force of attraction whereas the greater the distance, the smaller the force of attraction. Although the Sun is 10^7 times more massive than the Moon, the Moon is much closer to Earth and for that reason exerts a greater gravitational pull on Earth. In fact, the tide-generating force of the Moon on Earth is more than twice that of the Sun on Earth. For now, we will ignore the influence of the Sun and focus on the Moon.

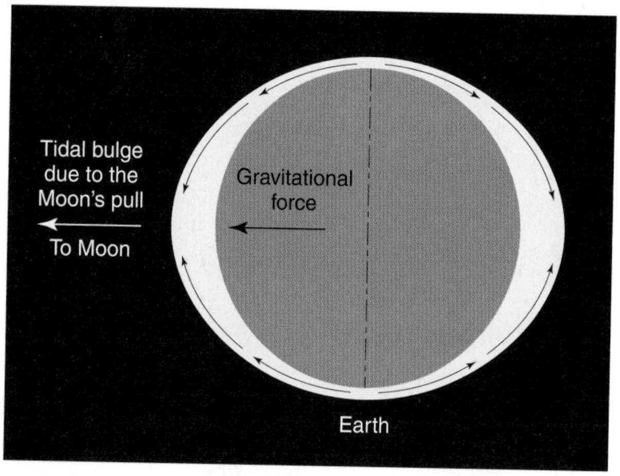

FIGURE 7.11
Two ocean tidal bulges produced by the gravitational attraction of the Moon combined with the rotation of the Earth-Moon system on an idealized water-covered Earth.

The gravitational pull of the Moon on Earth is primarily responsible for the bulge in the ocean surface that is oriented toward the Moon. (Actually, Earth's rotation drags the bulge ahead of the position directly under the Moon.) On the opposing side of the planet, the gravitational pull is weaker and the rotation of the Earth-Moon system is primarily responsible for the tidal bulge. Earth and Moon revolve around a common center of mass. Because Earth is much more massive than the Moon, the center of mass of the system is within 4700 km (2900 mi) of Earth's center, that is, 1700 km (1060 mi) below Earth's surface. This has been likened to a seesaw with an adult seated at one end and a child at the other end. The pivot point (center of mass of the adult-child system) must be moved toward the adult for the two individuals to balance the seesaw.

Newton's first law of motion predicts that a net force must operate in any rotating system and this net force in the rotating Earth-Moon system gives rise to the tidal bulge on the side of the planet opposite the Moon. (For the Earth-Moon system, that net force is the changing direction of gravitational attraction.) Recall from Chapter 4 that according to Newton's first law of motion, an object in constant straight-line motion remains that way unless acted upon by an unbalanced (net) force. In a rotating system, the net force confines an object to a curved (rather than straight) path. Consider an analogy. Suppose that you are a passenger in an auto that rounds a curve at high rate of speed. You feel a force that pushes you outward from the turning auto. Actually, you are experiencing the tendency for your body to continue moving in a straight path while the auto follows a curved path. In the same way, the rotation of the Earth-Moon system causes the ocean to bulge outward on the side of the planet opposite the Moon.

So far in our discussion of tide-generating forces, we have used the *equilibrium model of tides*, which assumes a frictionless Earth entirely covered by water. With this model, ocean bulges would always align with the celestial body that caused them. Furthermore, any location on the planet that is moved by Earth's rotation through these bulges would experience rising and falling sea level (i.e., tides). If only one celestial body (Moon or Sun) were present, each day a low-latitude locality would experience two high tides (when bulges pass) and two low tides (when halfway between the bulges). If the positions of Earth and the other celestial body remained the same in space, the period of these waves would be the time it takes for one half a rotation of Earth, about 12 hrs.

While Earth is rotating, however, the Moon is revolving around Earth. The Moon revolves around Earth once each lunar month (averaging 29.5 days, between new moons) and in the same direction as Earth's daily rotation. Hence, Earth must make more than a full rotation in order for a specific location on the planet to line up again with the advancing Moon. Catching up with the advancing Moon requires 24 hrs plus 1/29.5 of a day, which is approximately 24 hrs and 50 min. This Moon-based day (24 hrs, 50 minutes) is also called the **tidal day**. Because the tidal day is longer than the solar day, the times of high and low tide change by about 50 minutes from one solar day to the next.

The ocean's tidal bulges produced by the Moon remain in the same alignment relative to the Moon, but change their latitudinal (north-south) positions on Earth from day to day as they follow the Moon during its monthly revolution about Earth. The plane of the Moon's orbit is inclined by 5 degrees to Earth's equatorial plane so that during one lunar month, the Moon's latitudinal position moves from directly over the equator northward as much as 28.5 degrees N (5 degrees beyond the Tropic of Cancer), back to the equator, on southward to 28.5 degrees S (5 degrees beyond the Tropic of Capricorn), or as little as 18.5 degrees N and S (5 degrees less than the Tropic positions). This total range in the Moon's declination cycles every 18.6 years.

When the Moon is at its maximum latitudinal position, the center of one tidal bulge is just north of the Tropic of Cancer and the center of the other tidal bulge is on the other side of the planet just south of the Tropic of Capricorn. Hence, a location along or near the Tropic of Cancer or Tropic of Capricorn experiences only one significant tidal bulge in 24 hrs.

Sun-related ocean tidal bulges are produced in the same way as those caused by Earth-Moon interactions. That is, the gravitational attraction between Earth and Sun plus Earth's annual revolution around the mutual center of mass of Sun and Earth generate a second set of similar but smaller tidal bulges that are aligned with the Sun. As noted earlier, because of the Sun's much greater distance from Earth, the Sun's tidal pull on the ocean is less than half (about 46%) of the Moon's tidal pull. These tidal bulges follow the Sun (just as Moon-related bulges track the Moon), their locations changing as Earth rotates and follows its yearly orbit about the Sun (Figure 5.1). The period of the solar tide is 24 hrs.

The tide-generating force diminishes rapidly with increasing distance so that celestial bodies at greater distances from Earth than the Sun are too far away to exert a significant tidal pull on Earth's ocean. The *tide-generating force* (arising from a combination of gravitational and rotational forces) is inversely proportional to the cube of the distance between Earth and any other celestial body. Hence, doubling the distance between two bodies reduces the tide-generating force by a factor of 2^3 or 8 times.

TYPES OF TIDES

Based on the number of high and low tides and their relative heights each tidal day, tides are described as diurnal, semi-diurnal, or mixed (Figure 7.12). When the Moon is directly over Earth's equator, its associated tidal bulges are centered on the equator. In theory, all locations on the planet except at the highest latitudes would rotate through the two tidal bulges and experience two equal high tides and two equal low tides per tidal day; this is known as a **semi-diurnal tide**. A semi-diurnal tide has a period of 12 hrs and 25 min, and theoretically has a wavelength of more than half the circumference of Earth.

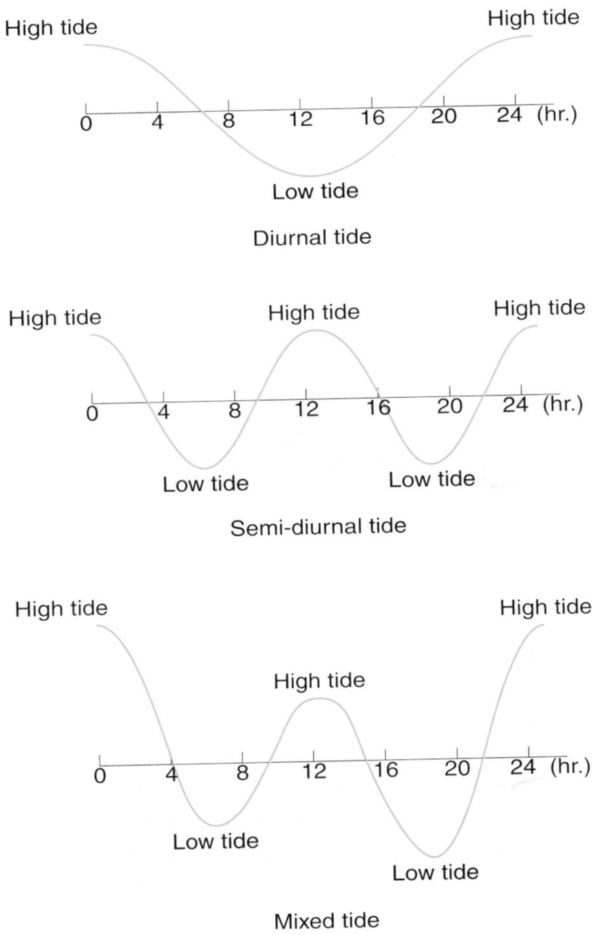

FIGURE 7.12
Types of astronomical tides observed in coastal locations.

On an ocean-covered Earth, different types of tides occur when the Moon is either north or south of the equator. Whereas semi-diurnal tides would theoretically occur at the equator at all times, most locations north or south of the equator would experience two unequal high tides and two unequal low tides per tidal day; this is called a **mixed tide** and the difference in height between successive high (or low) tides is called the **diurnal inequality**. When the point on Earth directly under the Moon is near the Tropic of Cancer or Tropic of Capricorn, the diurnal inequality is at its maximum and the tides are called *tropic tides*. When the Moon is above or nearly above the equator, the diurnal inequality is at its minimum and the tides are known as *equatorial tides*. When the Moon and its associated tidal bulges are either north or south of the equator, most points at high latitudes in theory would be impacted by one tidal bulge and would experience one high tide and one low tide per tidal day. This so-called **diurnal tide** has a period of 24 hrs and 50 min.

The separate sets of ocean bulges related to the Moon and Sun act at times together and at other times in opposition. About every two weeks, the positions of the Sun, Moon, and Earth fall along a straight line (Figure 7.13A). At these times of new and full Moon phases as viewed from Earth, the lunar- and solar-related ocean bulges also line up (and add up) to produce tides having the greatest monthly tidal range (that is, the highest high tide and lowest low tide); these are called **spring tides**. Between spring tides, at the first and third quarter phases of the Moon, the Sun's pull on Earth is at right angles to the pull of the Moon (Figure 7.13B). At this time, tides have their minimum monthly tidal ranges (that is, the lowest high tide and the highest low tide); these are called **neap tides** or *fortnightly tides*. Furthermore, the Moon's orbit about Earth is an ellipse (rather than a circle) so that the Moon is closest to Earth (stronger tide-generating force) at *perigee* and farthest from Earth (weaker tide-generating force) at *apogee*. The Moon completes one perigee-apogee-perigee cycle every 25.5 days. Similarly, tides due to the Sun vary annually because Earth's yearly orbit around the Sun is an ellipse so that Earth is closest to the Sun around January 3 each year (*perihelion*) and furthest from the Sun on July 4 each year (*aphelion*).

TIDES IN OCEAN BASINS

To this point in our discussion of ocean tides, we have assumed a water-covered Earth that rotates on its spin axis through the tidal bulges. In a more realistic situation, many non-astronomical factors modify tides, including ocean bottom topography, the presence of continents, coastline configuration, the Coriolis Effect, winds, and water depth. Tidal bulges move relatively unimpeded around the globe only in the Southern Ocean

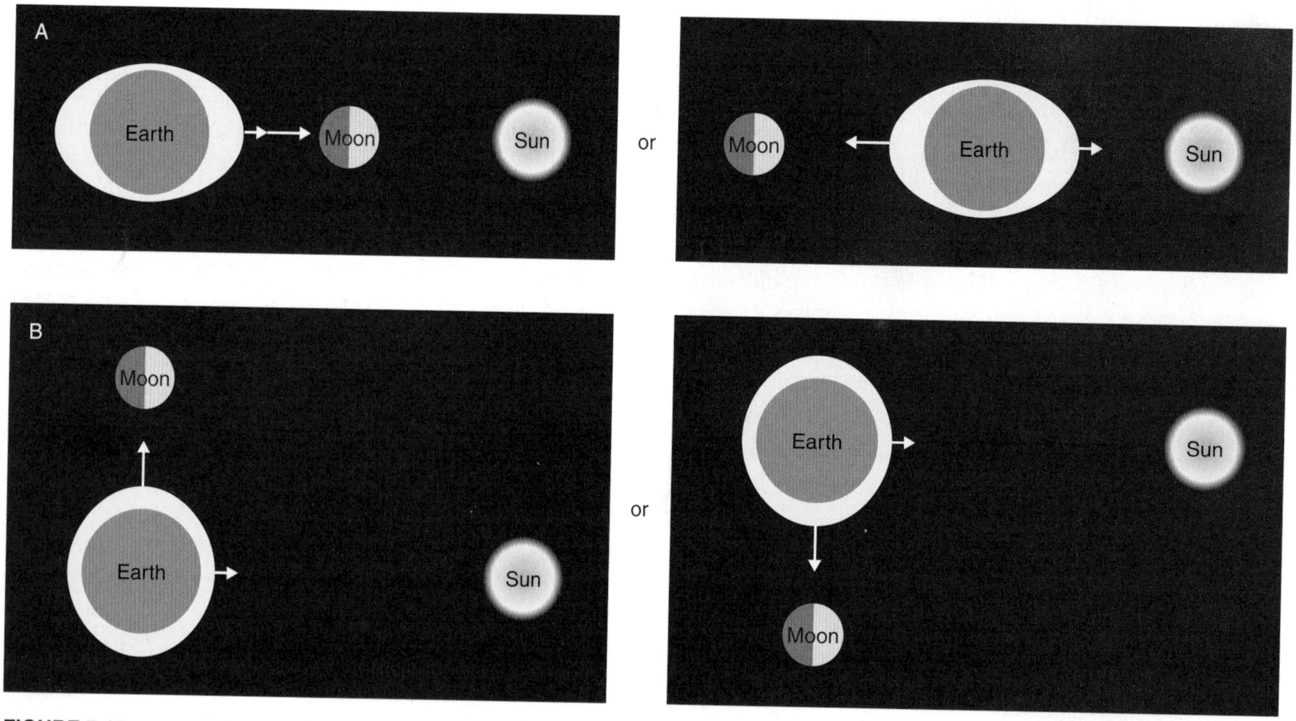

FIGURE 7.13
Configurations of Earth, Moon, and Sun responsible for (A) spring tides, and (B) neap tides.

near Antarctica. Consider, for example, the idealized case of tides in a Northern Hemisphere ocean basin of uniform depth that is completely surrounded by land (Figure 7.14). Assume also that the Moon is providing the only tide-generating force and initially is situated directly above the ocean basin.

As Earth rotates from west to east, the tidal bulge shifts toward the western boundary of the ocean basin and the water surface slopes gently downward toward the east. The western boundary of the basin experiences high tide while the eastern boundary experiences low tide. Tide waves are shallow-water waves. Hence, as noted earlier in this chapter, the orbits of water particles flatten with increasing depth, changing from circular to elliptical and ultimately to a back-and-forth motion near the ocean bottom.

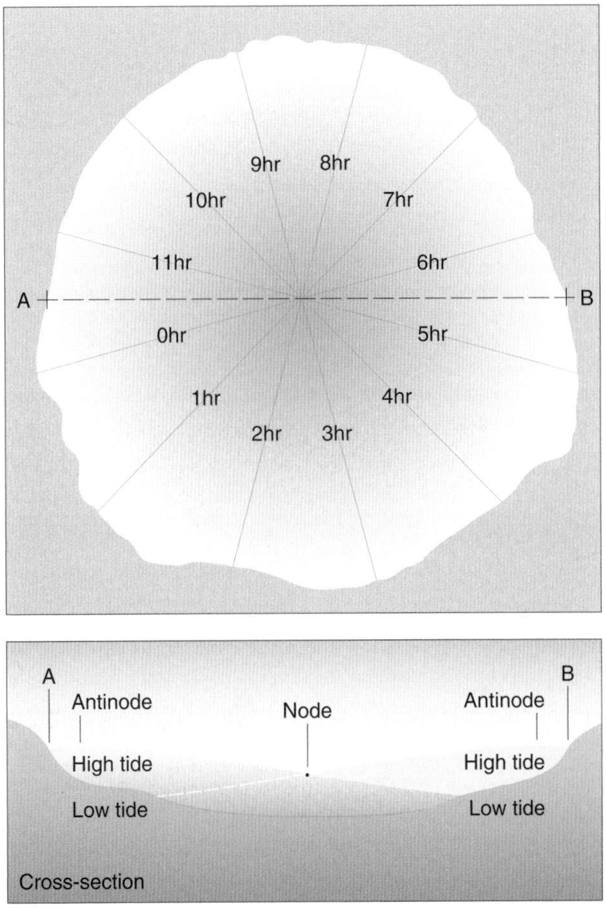

FIGURE 7.14
In this idealized Northern Hemisphere ocean basin bordered on all sides by land (top), a tide wave rotates in a counterclockwise direction (viewed from above). Lines radiating outward from the central node are cotidal lines that join points where high tide occurs at the same time of day. This is a semi-diurnal tide. Also shown (bottom) is a vertical cross-section from point A to point B. Note that the tidal range varies from zero at the node to a maximum at the antinodes (along the coast).

The horizontal motion of water particles in a tide persists for long periods (because the tide wave is constantly being forced) so that the tide wave is subject to the Coriolis Effect. As the tidal bulge at the western boundary begins to move down slope toward the east, the Coriolis deflects water particles to the right (in the Northern Hemisphere) so that the tidal crest (high tide) rotates into the southern portion of the basin. Now, the water surface slopes downward toward the north. The tidal crest continues to rotate around the basin in a counterclockwise direction (viewed from above). When the tide is high on one side of the basin, it is low on the opposite side. In Southern Hemisphere basins, reversal of the Coriolis Effect causes the tide wave to rotate in a clockwise direction (viewed from above).

This so-called *dynamic model of tides* applies reasonably well to seas and large embayments as well as the open ocean. Ocean scientists graphically represent the rotary motion of the standing tide wave in a basin by a series of cotidal lines radiating outward from a central node like the spokes in a bicycle wheel (Figure 7.15). A **cotidal line** joins points where high tide occurs at the same time of day; they are usually drawn at one-hour intervals. The tidal range varies from zero at the node to a maximum at the antinode along the coast.

Diurnal tides make one complete circuit per tidal day whereas semi-diurnal or mixed tides complete two circuits per tidal day. The period of the rotary tide waves is 12 hrs 25 min for semi-diurnal tides and 24 hrs 50 min for diurnal tides. For an ocean of average depth (about 4000 m or 13,000 ft), a tide wave progresses as a shallow-water wave at about 645 km per hr (400 mph). (As with wind-generated shallow-water waves, this is the speed of the wave crest, not the water.) Actual wave celerity along the coast varies greatly due to changes in bathymetry. Along the west coast of the U.S., tide waves travel at about 565 km per hr (350 mph), whereas along the west coast of Africa in the Northern Hemisphere, speeds are about 360 km per hr (225 mph). For coastal residents, however, these relatively high speeds are not apparent because for a diurnal tide, a 2-m (6.5-ft) rise in water level may take slightly more than 6 hrs.

The natural period of oscillation of a shallow basin may match the period of the tide-generating force causing resonance. The extraordinary tidal range (up to 16 m or 53 ft during spring tide) in the Bay of Fundy (Figure 7.16) is a two-step amplification involving the Gulf of Maine (12-hr period) and the Bay of Fundy (6.2-hr period). The Gulf of Maine resonates with the semi-diurnal lunar tide and the Bay of Fundy resonates with the Gulf of Maine.

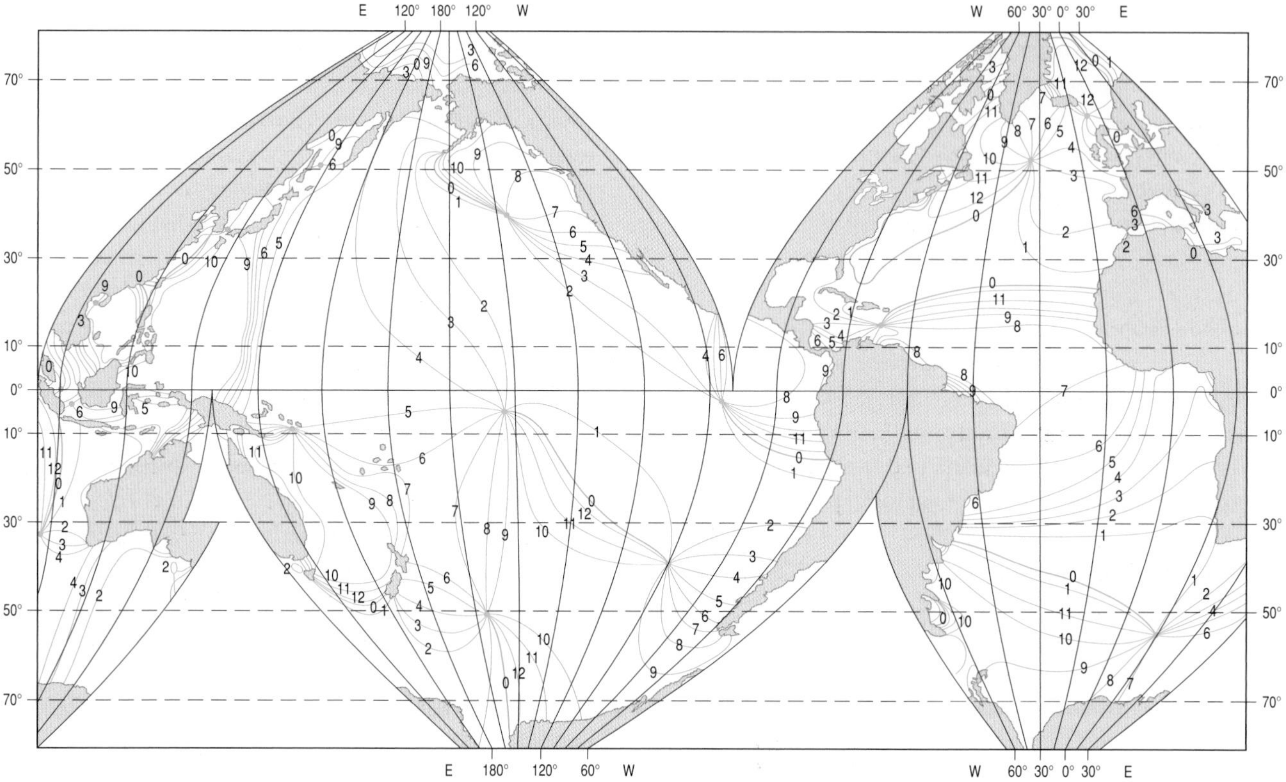

FIGURE 7.15
Phase relations of tides in the Pacific and Atlantic oceans. The map shows the cotidal lines of the semidiurnal tide referred to the culmination of the Moon in Greenwich. The tidal amplitude approaches zero where the cotidal lines run parallel (such as between Japan and New Guinea). Much of the tidal motion has the character of rotary waves. In the south and equatorial Atlantic Ocean the tide mainly takes the form of north-south oscillation on east-west lines. This complex reality should be compared to the simple concept which is the basis for existing calculations of the lunar orbital evolution and which pictures the tide as a sinusoidal wave progressing around the Earth in the easterly direction. [Image modified after Defant, A., 1961: Physical Oceanography, Vol. II., Pergamon, 598 pp., description provided by NASA]

FIGURE 7.16
ASTER satellite image of the Bay of Fundy, Nova Scotia, Canada, at high tide (left) on 20 April 2001, and low tide (right) on 30 September 2002. The world's highest astronomical tides (tidal range of 16 m) occur in Minas Basin at the eastern extremity of the Bay of Fundy due to resonance of the Bay of Fundy-Gulf of Maine system; that is, the natural period of oscillation of about 13 hrs is close to the 12 hrs 25 min period of the lunar tide of the Atlantic Ocean. [Courtesy of NASA/GSFC/METI/ERSDA/JAROS, and U.S./Japan ASTER Science Team.]

Although tide characteristics may vary during the month at a specific locale, all coastal areas have primarily diurnal, semi-diurnal, or mixed tides. Non-astronomical factors help explain why locations on the U.S. Atlantic coast have predominantly semi-diurnal tides whereas many places on the Gulf Coast have mostly diurnal tides, and localities on the Pacific Coast have mostly mixed tides. Portions of Canada's Atlantic coast also have mixed tides.

TIDAL CURRENTS

Alternating horizontal movements of water accompanying the rise and fall of astronomical tides in coastal areas are known as **tidal currents**. Along the boundaries of an ocean basin (the location of antinodes), tidal ranges and hence tidal currents reach their maximum speed. Irregularities along the coast can modify the rotary motion of tide waves so that tidal currents move more directly into and out of rivers and harbors (Figure 7.17). Tidal currents flow in one direction during part of the tidal cycle and in the opposite direction during the remainder of the tidal cycle. When tidal currents are directed toward the land, water levels rise in harbors and rivers; these are called **flood tides**. Tidal currents flowing seaward with falling sea levels are called **ebb tides**. Between flood and ebb tides are slack water periods with little or no horizontal movement.

In some coastal areas where the tidal range is relatively large and the flood tide enters a narrow bay or channel, a tidal bore forms and moves upstream in a river or shallow estuary. A **tidal bore** is a wall of turbulent water, usually less than a meter in height. Tidal bores are well known at the mouth of the Amazon River in Brazil, on the Severn River in England, in Turnagain Arm off Cook Inlet, Alaska, in the Qiantang River in China, and in the upper embayments and rivers of the Bay of Fundy, Nova Scotia.

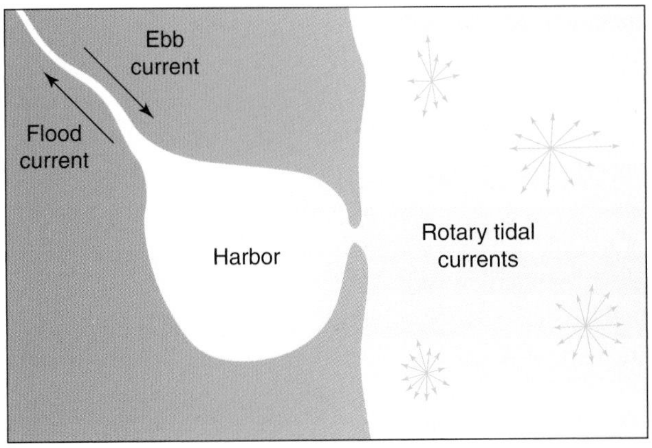

FIGURE 7.17
Ebb tide and flood tide in a harbor viewed from above.

OBSERVING AND PREDICTING TIDES

Observing the changing water levels caused by astronomical tides is relatively simple and has long been important for major ports. Knowing tide levels helps pilots and ship captains avoid running aground in shallow stretches of harbor channels. In the 1850s, the port of New York began operating a real-time tide gauge that indicated to ship operators the tide level and whether it was rising or falling. Today government agencies (e.g., NOAA, Canadian Hydrographic Services, and the British Admiralty) maintain tide observing and prediction systems that advise mariners based primarily on the output of numerical models and computers. On the other hand, fewer sources of information exist on tidal currents in harbors because they are much more difficult and expensive to monitor than tide levels. Whereas tide levels are nearly uniform over broad areas, tidal currents change quickly with variations in wind and river discharge. Furthermore they are affected by complex shorelines and bottom topography. Nonetheless, NOAA's *Physical Oceanographic Real Time Services (PORTS)* provides real-time information on tides and tidal currents to ship operators for 18 major U.S. ports. In the U.S., tide and tidal current predictions are made by NOAA's National Ocean Service CO-OPS (*Center for Operational Oceanographic Products and Services*) for more than 3000 locations and are available to the public online. Consider how tide predictions are made.

Periods of motions of the Earth, Sun, and Moon in space (i.e., orbit and rotation) are fixed and known precisely. The predictability of the movements of tide-generating celestial bodies means that astronomical tides can also be predicted with great accuracy. Tides are waves so that local tides can be resolved mathematically into their various components, or tidal constituents, called **partial tides**. Each partial tide describes a unique individual or combination of periods of motion (e.g., spring-neap cycle, perigee-apogee cycle). Partial tides are predicted individually and added together to forecast the height and timing of future local tides. Although as few as four partial tides can account for 70% of the total tidal range in the open ocean, some 60 components are commonly used (to account for both astronomical and non-astronomical factors including the effects of shallow water). More than 100 components must be considered to predict accurately the tides along a complex coastline, such as that of Cook Inlet, Alaska, or in the upper portion of a tidal river such as at Philadelphia, PA. Predictions of local astronomical tides are most reliable when based on data collected for at least 18.6 years (i.e., a *nodal cycle*), a period that encompasses the astronomical configurations

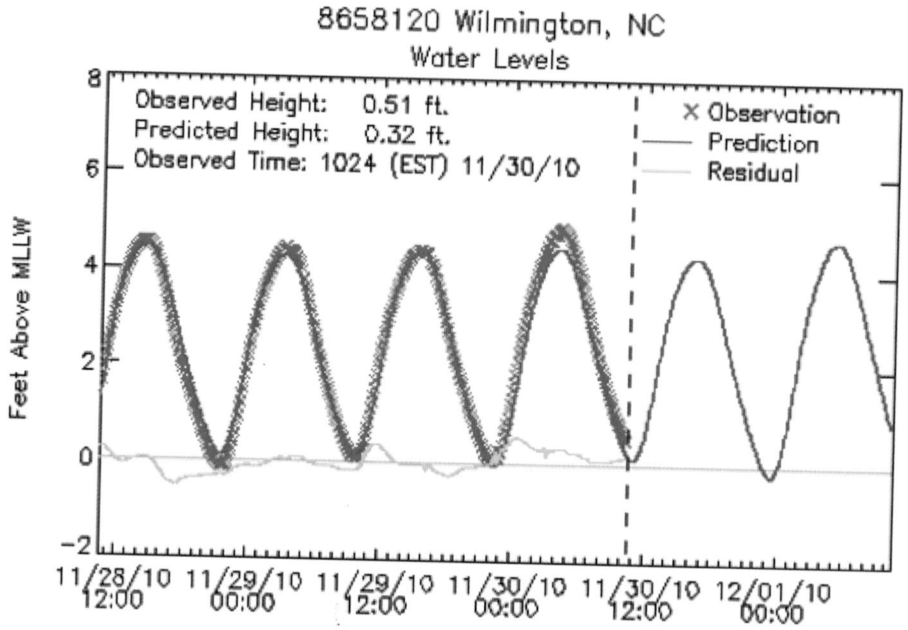

FIGURE 7.18
Observed (red) and predicted (blue) water level (in feet above Mean Lower Low Water) for the tide station at Wilmington, NC from 28 November to 1 December 2010. Vertical dashed line marks the "present" time. [From NOAA, National Ocean Service, Center for Operational Oceanographic Products and Services.]

of the Earth-Moon-Sun system that generate the tides. However, only a single year of tide station observations usually suffices for very reasonable tide predictions once they are adjusted for the nodal cycle. In most cases, local winds and atmospheric pressure variations are the primary causes of the difference between the actual and predicted tide (Figure 7.18). In upper tidal estuaries, seasonally high river discharge may also cause significant differences between actual and predicted tides and tidal currents.

OPEN-OCEAN TIDES

An Earth-orbiting satellite routinely measures tides over the deep-ocean. Such a satellite is equipped with a radar-altimeter that bounces microwave signals off the sea surface and precisely measures sea level. With such data, it is possible to determine what happens to tide waves and their energy as they travel across deep-ocean basins. The *TOPEX/Poseidon satellite*, a joint venture of the U.S. and France, accurately measured global sea level every 10 days. However, this mission finally ended in January 2006 after 13 years and 5 months of service when the spacecraft lost its ability to maneuver. The *Jason satellite* which was launched in December 2001, now measures global sea level. For more on measuring sea level from space, refer to this chapter's second Essay.

As shallow-water waves, tides lose energy through frictional drag with the ocean floor, especially in shallow seas and along continental margins. Satellite measurements show that about three quarters of the global tidal energy dissipates in shallow seas bordering northern Europe, in the Yellow Sea off Asia, in the shallow seas around Australia, near Argentina, and in Canada's Hudson Bay.

Open-ocean tides are important in mixing deep-ocean water. Ocean scientists long assumed that wind was the principal mixing agent of the open ocean, but satellite altimeter data now show that tidal mixing in the deep ocean is about as important as the wind. Perhaps as much as half of the tidal energy in the ocean is dissipated in mixing processes when tidal currents in the deep ocean flow over seamounts, ridges, and other rugged features on the ocean floor or weave through passages between islands.

Tidal currents flowing over topographic irregularities on the ocean floor generate internal waves that propagate away from their source. These internal waves arise from the fact that water density increases gradually with increasing depth. As tidal currents encounter a seamount or submarine ridge, relatively dense water is forced upward into slightly less dense water. Then to the lee of the obstacle gravity pulls the denser water downward. However, the descending water gains momentum and overshoots its equilibrium level and descends into denser water. The water then ascends thereby forming an oscillating wave that propagates horizontally. Because these waves are generated by tides, they occur at tidal frequencies and are called **internal tides**. Internal tide waves can travel thousands of kilometers beyond the obstruction that formed them and can have very large wave heights. They also break like surf on a beach but under water, locally mixing waters above and below the internal wave. Internal tides are important in mixing cold bottom waters with warmer surface waters as part of the global oceanic circulation (Chapter 6).

Ocean scientists have evidence that internal tides influence the gradient of the continental slope (Chapter 2). The inclination of the continental slope varies from very gentle (as small as one degree) to precipitous (up to 25 degrees where submarine canyons cut into the slope). About 80% of the continental slope is inclined at less than 8 degrees and the average inclination is about 4 degrees. According to geological studies, however, the sediments supplied to the continental slope (mostly by rivers) would support a stable average slope of perhaps 15 degrees or greater. Data acquired from model studies, dives in piloted submersible vessels, and moored instruments show that internal tides produce strong currents that prevent accumulation of sediment that would add to the steepness of the continental slope. In fact, the internal waves ascending the continental slope apparently behave very much like ordinary sea waves entering the shoaling waters of a coastal area (with changes in amplitude, wavelength, and water velocity). Whereas the influence of internal tides is widespread along the continental slope, turbidity currents and tectonic forces can be important locally and regionally in shaping the slope.

A vast amount of energy is involved in ocean tides and waves so it is not surprising that there is some interest in developing technologies to tap this energy to generate electricity. Although the potential is enormous, very little of this energy resource has been developed to date. For more on tidal power, see this chapter's third Essay.

The Tsunami Hazard

On 1 April 1946 in a span of 27 minutes, two earthquakes shook the Scotch Cap Lighthouse on Unimak Island in Alaska's Aleutian chain (Figure 7.19). Shortly afterward, a huge sea wave appeared and obliterated the lighthouse, killing the five-man crew and hurling debris 35 m (115 ft) above sea level. All told, this destructive sea wave, known as a **tsunami** (from *tsu-nami*, the Japanese word for large harbor wave and pronounced *sue-nah-mee*), was responsible for more than 165 fatalities and $26 million (1946 dollars) in property damage.

A tsunami can be a deadly natural hazard. Over the past 2000 years, tsunamis probably claimed the lives of hundreds of thousands of people living in the Pacific Rim. During the 20th century, the 141 most damaging tsunamis killed more than 70,000 people. During the 1990s, 82 tsunamis were reported worldwide, 10 of them claiming more than 4000 lives total. On average, 57 tsunamis occur each decade but improved global communications systems, which allow for more complete and reliable detection and reporting, gives the impression that tsunamis are increasing in frequency worldwide. During the 1990s, the highest tsunami reported reached 31 m (102 ft), one of a series of huge waves that struck the small island of Okushiri, Japan, on 12 July 1993, taking 239 lives. On 26 December 2004, a tsunami devastated portions of the coastal zone of the Indian Ocean, claiming an estimated 230,000 lives, the greatest death toll ever recorded by a tsunami. Injuries

FIGURE 7.19
The Scotch Cap Lighthouse on Unimak Island, Alaska before (left) and after (right) the earthquake and tsunami of April 1946. A magnitude 8.0 (Mw) earthquake with the source to the south of Unimak Island generated a tsunami that destroyed the five-story lighthouse, located 9.8 m above sea level. [Courtesy of US Coast Guard]

numbered in the hundreds of thousands and at least 2 million people were made homeless. On 11 March 2011, a tsunami struck the northeast coast of Honshu, Japan's main land mass and largest island (Figure 7.20). This tsunami was generated by a submarine earthquake, the most powerful earthquake to strike Japan in at least 130 years. As of this writing, there were 15,093 deaths and 9,093 missing in Japan as a result of the tsunami. (The 2004 and 2011 tsunamis are described in the Case-in-Point of Chapter 1).

Although sometimes referred to as a "tidal wave," a tsunami has nothing to do with astronomical tides. About 85% of all tsunamis originate in the Pacific Ocean (about 25% near Japan), generated by strong undersea earthquakes occurring along convergent tectonic plate boundaries and associated subduction zones (Chapter 2). As the lower plate subducts into the mantle, the upper plate abruptly shifts horizontally and upward. These tectonic movements disturb the overlying water column, generating a long-wavelength sea wave. The height of the tsunami generated depends on the amount of movement of the sea floor. In addition to subduction zone earthquakes, volcanic eruptions, meteorite impacts, submarine landslides, and even calving glaciers can produce tsunamis. The violent eruption of the Indonesian volcano Krakatoa on 26-27 August 1883 spawned huge tsunami waves (one reported to be the height of a 12-story building) that killed at least 36,000 people.

The tsunami hazard is much less in the Atlantic Ocean than in the Pacific Ocean because of the absence of convergent plate boundaries and subduction zones. In 2006, scientists surveying the ocean floor off the northern coast of Puerto Rico found evidence of past submarine landslides. Some of these landslides were large enough to trigger a tsunami similar to the one that struck Puerto Rico in 1918 with a height of 2.5 m (8 ft). The potential for tsunamis triggered by submarine landslides also exists on the unstable volcanic slopes of the Canary Islands off the northwest coast of Africa.

Once formed, a tsunami radiates outward in all directions. Tsunamis typically originate near the coastline so that their initial impact is on nearby beaches. In the opposite direction, the open ocean, a tsunami propagates as a series of waves with wavelengths that can exceed 750 km (465 mi). Because tsunamis are *shallow-water waves*, with lengths much greater than the depth of the ocean, they disturb little with their passing. Average ocean depth of 4000 m (13,000 ft) and wave celerity is about 645 km per hr (400 mph)—approaching the speed of a commercial jetliner—but with a wave height less than 1 m (3 ft) and taking 10 to 30 minutes to pass, a tsunami would be unnoticed moving under a ship at sea. However, a tsunami can have a long reach, transporting destructive energy many thousands of kilometers away.

In the deep water of the open ocean, tsunami wave energy is distributed through a huge volume of water. As a tsunami enters shoaling coastal waters, as any

FIGURE 7.20
An aerial view shows damage to Wakuya, Japan, 15 March 2011, four days after the devastating 9.0 earthquake and resulting tsunami. [Courtesy of U.S. Department of Defense]

other ocean wave, it slows and its energy focuses into a decreasing volume of water. The wavelength shortens to 15 to 20 km (9 to 12 mi) and wave height increases, sometimes building to extraordinary heights as the tsunami approaches the coastline. The 2004 Indian Ocean tsunami had a maximum height of at least 30 m (100 ft) when it first came ashore in Sumatra's Aceh Province and then surged inland up to 4.5 km (2.8 mi).

Often preceding the arrival of a tsunami in a coastal area is a trough that draws water off the beach and exposes the sea floor. This unusual sight can entice curious people to investigate, only to be caught by the impending tsunami crest. The first wave is usually followed by a succession of powerful tsunami waves. Tsunamis normally inundate land as huge breaking waves, a flood similar to a flood tide, or wall of water, taking lives, crushing buildings, bringing down trees and otherwise obliterating the landscape. Flexure of the tectonic plate following an earthquake, where the ocean bottom rises in the offshore subduction zone as the coastal land depresses, may enhance the inland propagation of a tsunami.

In the United States, coastal residents of Hawaii and Alaska are at greatest risk from a tsunami, but they have struck elsewhere along the West Coast. Hawaii has reported more than a dozen damaging tsunamis since 1895, the most destructive of which occurred on 1 April 1946. A tsunami severely damaged most of Hilo's waterfront and caused 159 deaths. (This was the same tsunami that destroyed the Scotch Cap Lighthouse mentioned earlier.) On 23 May 1960, another tsunami, having a maximum wave height of 10.7 m (35 ft), struck the Hilo Bay area, taking 61 lives and wrecking 540 homes and businesses. This tsunami was triggered by a magnitude 8.25-8.5 earthquake off the west coast of South America and took 15 hours to reach the Hawaiian Islands. The tsunami that struck the Seward, Alaska, area on 28 March 1964 claimed more than 100 lives. Although tsunamis are rare along the West Coast of the U.S. outside of Alaska, geological evidence suggests that the Cascadia subduction zone located off the Pacific Northwest Coast is the site of tsunami-generating earthquakes about every 300 to 700 years. Geological evidence uncovered in the late 1980s pointed to the occurrence of a major tsunami along the coast of British Columbia, Washington, and Oregon around 1700. More recently, in April 1992, a magnitude 7.1 earthquake at the southern end of the subduction zone triggered a small tsunami that caused property damage near Cape Mendocino, CA.

The Hawaiian tsunami of 1946 provided the impetus for establishment of the *Pacific Tsunami Warning Center* in Ewa Beach on Oahu, Hawaii, in 1948. The tsunami that struck Prince William Sound during the 1964 Alaskan earthquake led to formation of the *West Coast and Alaska Tsunami Warning Center* in Palmer, Alaska, in 1967. Scientists at these centers warn the public of an approaching tsunami based on earthquake occurrences detected by seismic networks around the Pacific Rim and tide gauge readings in coastal areas. However, these methods have not been very successful—the false alarm rate has been around 75% since the 1950s. Undersea earthquakes do not always generate a tsunami and warnings provided by coastal tide gauges generally arrive too late for adequate warning.

Saving lives and protecting property through improved early warning of an approaching tsunami is the goal of NOAA's *Deep-ocean Assessment and Reporting of Tsunamis (DART®)* system. DART is aimed at gathering data for early detection of tsunamis in the open ocean and real-time forecasting for coastal regions in the path of a tsunami. Anchored to the ocean floor at depths up to 500 m (1640 ft), each DART station has a *tsunameter*, consisting of a pressure sensor that measures slight variations in pressure exerted by the overlying water column in response to a passing tsunami wave. The DART is sensitive enough to detect water pressure changes associated with a tsunami wave as small as one centimeter. When a pressure change is sensed, an acoustic signal transmits data from the tsunameter to a companion moored buoy floating on the ocean surface, and then to a satellite for communication to the two Tsunami Warning Centers operated by NOAA's National Weather Service (Figure 7.21).

The first four DART stations were in place in August 2000, but when the 2004 Indian Ocean tsunami struck, NOAA was operating only 6 DART stations, all in the Pacific Ocean. As of October 2010, 50 DART stations had been deployed at sites in regions having a history of generating destructive tsunamis. DART stations are located mostly in the Pacific Rim, but they are also in between New Zealand and Southeast Asia, Hawaii, and between the equator and Chile, as well as in the Caribbean, the Gulf of Mexico, and the western North Atlantic (Figure 7.22). As of this writing, the U.S. operates 40 DART stations, Australia 7, while Chile, Indonesia, and Thailand each have one.

DART sensors are a critical element of NOAA's *Tsunami Program*, a partnership with representatives of various U.S. federal agencies (i.e., NOAA, FEMA, and the U.S. Geological Survey) and Pacific Coast states. In recent years, this program has also installed new shore-

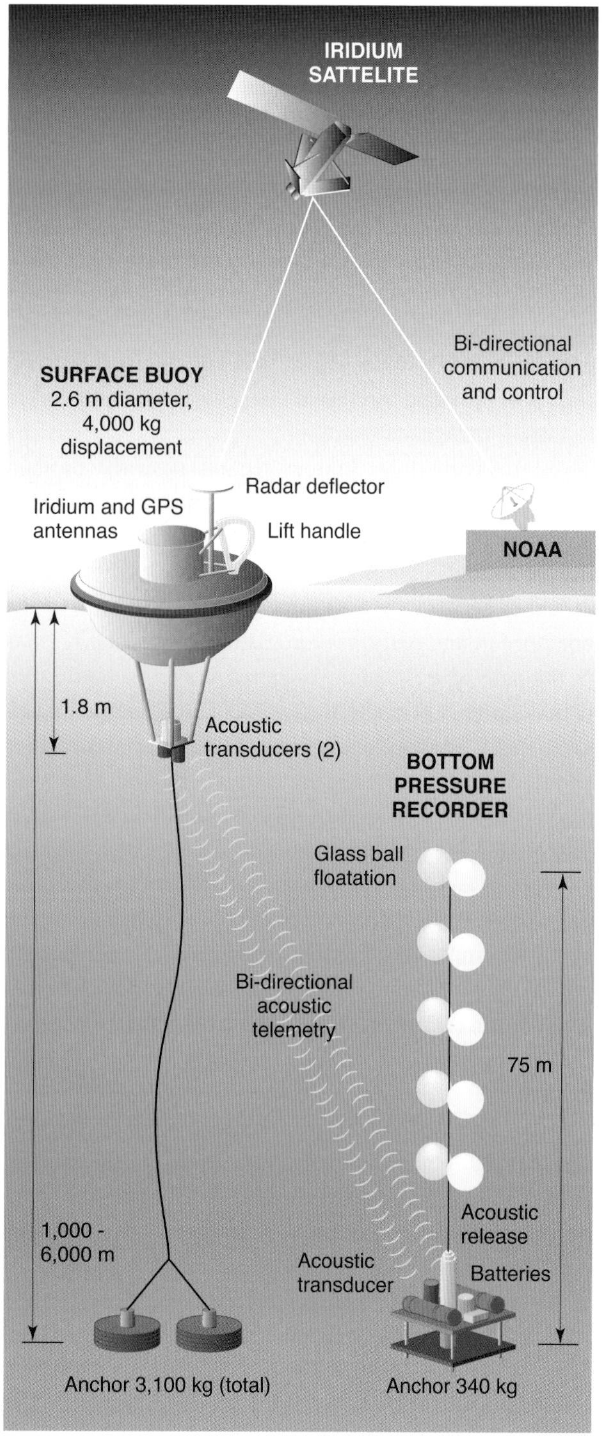

FIGURE 7.21
NOAA's DART system gathers data for early detection of tsunamis in the open ocean. A tsunameter, anchored to the sea floor at each DART station, contains a pressure sensor that can measure slight variations in pressure caused by a passing tsunami wave. When such a pressure change is sensed, data is transmitted via an acoustic signal from the tsunameter to a moored buoy on the ocean surface, and then to the Iridium Satellite for communication to the two NOAA Tsunami Warming Centers.

based tide gauges and upgraded seismometers operated by the USGS and NSF. The goal is to provide the public with adequate advance warning of both local- and remote-source tsunamis, and substantially reduce false alarm rates. They also plan to create greater public awareness of tsunami hazards and construct maps that identify coastal areas most susceptible to inundation (Figure 7.23).

In September 2010, the National Research Council (NRC) reported that since the 2004 Indian Ocean tsunami, the ability of the U.S. to detect and forecast tsunamis worldwide has improved. The NRC report attributed this improvement to upgrades in the Global Seismic Network and coastal sea-level stations, expansion of the DART network, and production of more and better quality hazard and evacuation maps. Nonetheless, according to the NRC, significant problems persist. For one, a tsunami generated by a large near-shore earthquake can strike the coast in a matter of minutes, making it difficult, or impossible, for the public to receive the warning and effectively evacuate. (During the 2004 Indian Ocean tsunami, the first of three tsunami waves inundated Banda Aceh, Indonesia, within only 40 minutes of the earthquake.) The NRC recommends development of education programs that familiarize the public with natural cues to an approaching tsunami (i.e., earthquake tremors, recession of water from the shore) so that they will take action even if they receive no official warning. The NRC also recommends better communication and coordination among the two Tsunami Warning Centers, emergency responders, and the media so that the public receives consistent information regarding the tsunami threat and the appropriate response.

Prior to the devastating 2004 Indian Ocean tsunami, no tsunami warning system operated in the Indian Ocean. Without warnings, tsunamis waves traveled across the Bay of Bengal and crashed against the shore, taking 16,269 lives in India and about 31,000 lives in Sri Lanka. Since then an international effort, coordinated by the *Intergovernmental Oceanographic Commission* of the *United Nations Educational, Scientific, and Cultural Organization (UNESCO)*, has been attempting to rectify the situation by installing an improved seismic network for detecting large earthquakes, a number of tsunameters, and a real-time coastal tide gauge network. The Indian Ocean network is intended to be part of an integrated global tsunami warning system, which includes the U.S.

One outcome of the extensive scientific documentation of the 2004 Indian Ocean tsunami was an opportunity to use real-time data to test existing tsunami numerical models and make appropriate refinements. With information about the location and magnitude of

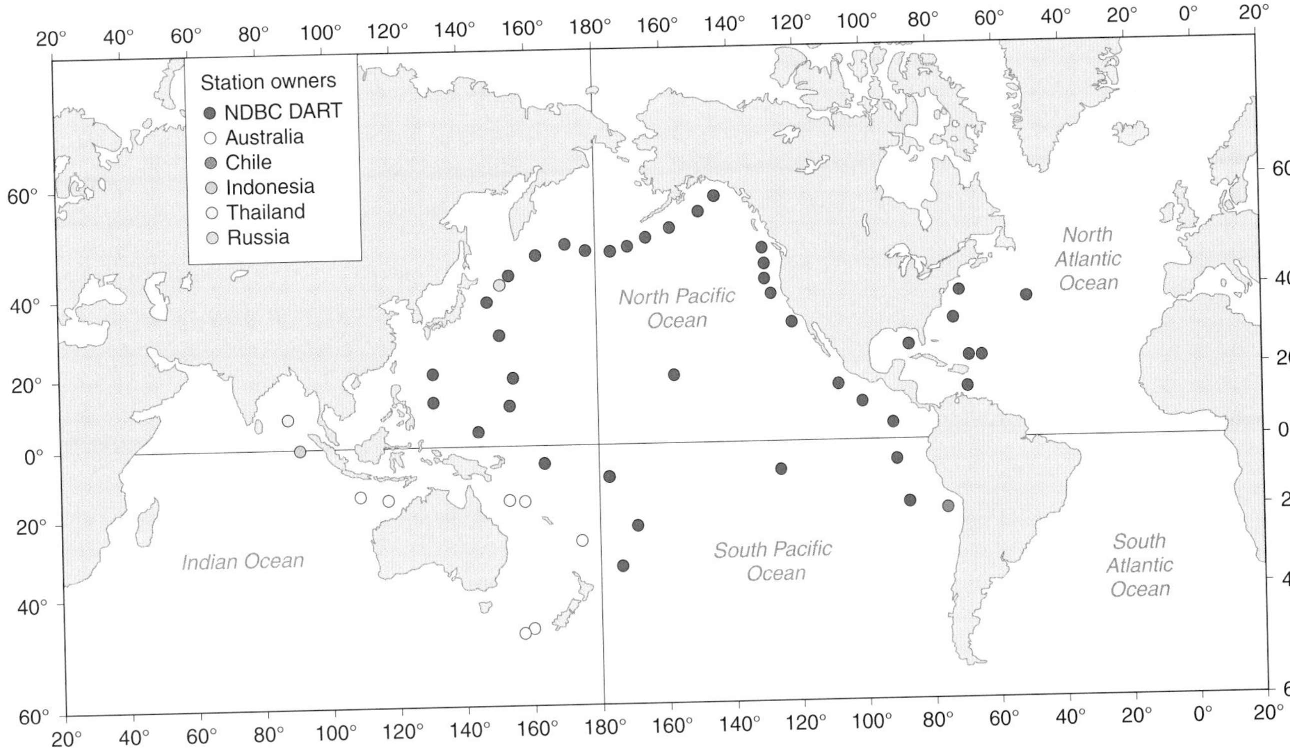

FIGURE 7.22
Location of completed and planned DART and Tsunameter Stations for real-time detection of tsunamis. [NOAA, National Data Buoy Center]

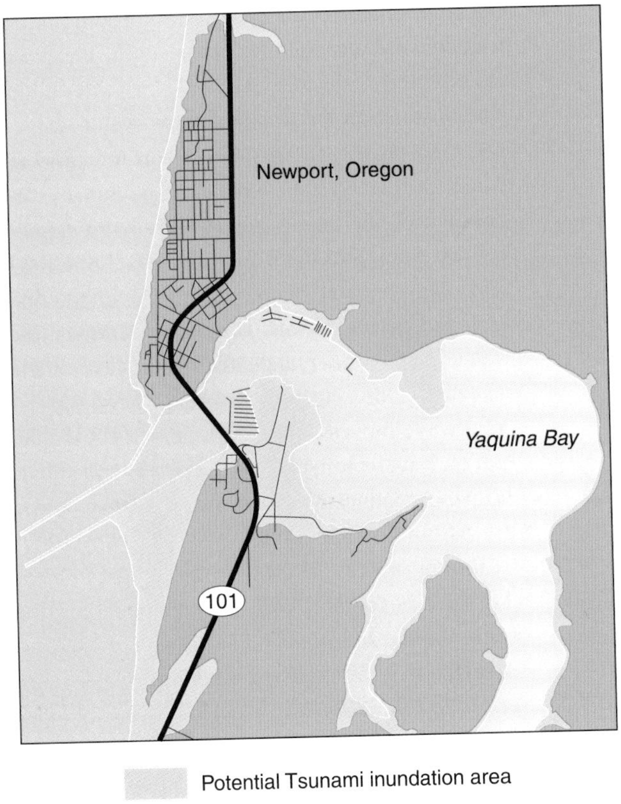

FIGURE 7.23
Tsunami-risk map for Newport, OR, identifying areas of greatest potential for tsunami flooding. Such maps inform local planning for tsunami hazard mitigation. [Produced by the Oregon Department of Geology in collaboration with the Oregon Graduate Institute of Science and Technology]

tectonic activity on the sea floor, and the volume of displaced seawater, Japanese and American researchers have designed models to simulate the propagation of a tsunami through the open ocean and the pattern of flooding as the wave moves onshore. They can also forecast when and where a tsunami is likely to strike along the coast and how far inland its waters might flood. Because a satellite altimeter, Jason-1, happened to be in position over the Indian Ocean (refer to the second Essay of this Chapter), and tracked the propagation and height of the 2004 Indian Ocean tsunami, the models could be tested and refined. These models were put to the test only hours after the deadly 2011 Tohoku earthquake and tsunami struck Japan, predicting the path of the tsunami across the Pacific and identifying areas likely to be impacted hours in advance.

Conclusions

Winds supply the kinetic energy that forms waves on the ocean surface, which then propagate horizontally until they expend their energy by breaking on a distant shore. As noted in Chapter 6, this is an important aspect of ocean-atmosphere interaction that provides energy to drive surface currents and mix surface waters. Wave behavior depends on wind speed, fetch, and duration as well as water depth. In terms of interaction with the ocean bottom, a distinction is made between deep-water waves and shallow-water waves. A tsunami is a shallow-water wave that passes unoticed by a ship at sea, but can build to tremendous heights in coastal areas.

The gravitational attraction of the Moon and Sun combined with the rotations of the Earth-Moon and Earth-Sun coupled systems generate planetary-scale shallow-water tide waves in the ocean basins. The consequence is a periodic rise and fall of sea level that is so familiar to coastal dwellers as ocean tides. Tides also cause currents that are among the strongest of all currents in coastal regions and are responsible for some water movements in the deep ocean. In Chapter 8, we take a closer look at waves and tides in the coastal zone, where the majority of the human population lives.

Basic Understandings

- When the kinetic energy of moving air (wind) is transferred to the ocean surface, it generates surface waves. These wind-driven ocean waves are oscillations of the sea surface propagating horizontally along the ocean/atmosphere interface, away from where they were generated. Energy is carried with the wave, not the water (except where the wave breaks).
- Water's surface tension is the restoring force for capillary waves, which have wavelengths less than 1.7 cm (0.7 in.). For longer sea waves, the restoring force is gravity. Open ocean deep-water waves propagate away from their energy source with little energy loss through friction. Energy is not significantly expanded until encountering the ocean bottom and shore, and finally dissipates in breaking in the surf zone.
- Wave height, the vertical distance between wave trough and wave crest, depends on the wind speed (and turbulence), the duration the wind blows from the same direction, and the distance over which it blows (the fetch).
- Wave interference also affects the growth and decay of sea waves. Wind-driven waves associated with two or more storm systems may interfere constructively or destructively. In active wave-forming areas, waves (called a sea) have a mixture of wavelengths and periods, as well as sharp crests, which gives the ocean surface a chaotic appearance. As waves move away from the area where they formed, they become more rounded and arranged by wavelength; these are called swell.
- The distinction between shallow-water waves and deep-water waves depends on the ratio of wavelength to ocean depth. Unaffected by the ocean bottom, deep-water waves have depths greater than half their wavelengths. Shallow-water waves are affected by frictional interaction with the ocean bottom when the depth of the water body is equivalent to half or less of the wavelength of a surface wave.
- As a wave moves into water depths of less than half its wavelength, the wave shortens and slows. With no change in period, wave energy forces the water to build and the wave becomes steeper as the top moves faster than the bottom. Soon, the wave is unstable and breaks, expending the energy in the surf zone.
- A seiche is a rhythmic oscillation of water in an enclosed basin or partially enclosed coastal inlet. During a seiche with one node, the water level in a basin rises at one end and simultaneously falls

at the other. A seiche grows through resonance, amplification that occurs when the period of the disturbance (e.g., earthquake, wind) matches the natural period of oscillation of the basin. When it does, seiches can grow to great heights, causing flooding or damaging moored vessels.

- Astronomical tides are planetary-scale, shallow-water waves responsible for the regular rise and fall of the ocean surface. Measured at coastal locations, the difference in height between water level at high and low tide is called the tidal range and the time between successive high tides is the tidal period.
- Tides are caused by the gravitational attraction between the Earth, Moon, and Sun, combined with the rotation of the Earth-Moon and Earth-Sun coupled systems. Although its mass is considerably smaller than that of the Sun, the Moon exerts a greater influence because it is so close to Earth.
- Through gravity, the Moon (or Sun) produces two bulges of water on Earth's ocean surface, one directly below the Moon or Sun and one on the opposite side of the planet. Earth then rotates through the tidal bulges, which causes the regular rise and fall of water level.
- Depending on factors such as location and phase of the solar and lunar cycles, there may be two equal high tides and two equal low tides per tidal day (semi-diurnal tide), two unequal high tides and two unequal low tides per tidal day (mixed tide), or one high tide and one low tide per tidal day (diurnal tide). Many non-astronomical factors modify tides, including the presence of continents, the Coriolis Effect, winds, and variations in coastline configuration, water depth, and bathymetry.
- In a Northern Hemisphere ocean basin, the Coriolis Effect causes the shallow-water tide wave to rotate around the basin in a counterclockwise direction as viewed from above. When the tide is high on one side of the basin, it is low on the opposite side. The Southern Hemisphere reversal of the Coriolis Effect causes the tide wave to rotate in a clockwise direction as viewed from above.
- The rotary motion of a tide wave in an ocean basin is graphically represented by a series of cotidal lines radiating outward from a central node. A cotidal line joins points where high (or low) tide occurs at the same time of day.
- Alternating horizontal movements of water accompanying the rise and fall of tides in coastal areas are called tidal currents. The maximum tidal range and tidal currents are along the boundaries of an ocean basin (coasts). In coastal embayments and harbors, tidal currents are constrained to flow back and forth as flood tides (flowing toward land) or ebb tides (flowing away from land).
- Treated as a wave, the tide at a particular location can be resolved into many component tidal constituents, or partial tides. Partial tides are predicted individually according to their causes and added together to forecast the local astronomical tide.
- Below the ocean surface, internal waves form and propagate on interfaces of different densities. Because density contrasts between water masses are about 1000 times less than that between air and water, internal waves propagate more slowly than surface waves and with much greater wave heights.
- A tsunami is a shallow-water ocean wave that develops most often when a submarine earthquake, or a landslide or volcanic eruption, disturbs the ocean. While in the open ocean, tsunami wave height is less than a few meters, with a period of 10 to 30 minutes, a wavelength ranging from 100 to 200 km (60 to 120 mi), and a speed of hundreds of kilometers per hour. With such small heights, long periods, and long wavelengths, tsunamis are imperceptible to ships at sea. However, when they reach shore and the wavelength drastically shortens, the wave can build to tremendous heights, inundating coasts, taking many lives and causing considerable damage.
- Improved early warning of an approaching tsunami is the goal of the NOAA Deep-ocean Assessment and Reporting of Tsunamis (DART) system. Each DART station has a *tsunameter*, consisting of a pressure sensor anchored to the ocean floor at depths to 500 m (1640 ft), measuring the slight variations in pressure exerted by the overlying water column in response to a passing tsunami wave.

Enduring Ideas

- An ocean wave is an oscillation or undulation of the sea surface that propagates along the interface between the atmosphere and ocean. As a wave progresses, the sea surface oscillates up and down (from wave trough to wave crest) about an equilibrium, the still-water level.
- The principal source of kinetic energy for ocean wave generation and propagation is the surface wind. Tectonic movements of the sea floor, coastal landslides, volcanic eruptions, and astronomical forces also generate sea waves. Gravity is the restoring force for all but the smallest ocean waves (capillary waves).
- Wave height depends on wind speed, duration of the wind, and fetch. In the open ocean with uniformly deep water, waves with longer wavelengths and periods travel faster than waves with shorter wavelengths and periods so that the longer waves of a swell outdistance the shorter waves.
- The celerity of deep-water waves (for water deeper than wave-base) depends on wavelength and gravity. The celerity of shallow-water waves (for water depths of less than half the wavelength) depends on water depth and gravity but not wavelength.
- A seiche is a standing wave, a rhythmic oscillation of water in an enclosed basin or partially enclosed coastal inlet. It can be initiated by wind, air pressure gradients, or astronomical tides, and can grow as a consequence of resonance.
- The gravitational attraction among the Earth, Moon, and Sun combined with the rotation of the Earth-Moon and Earth-Sun coupled systems gives rise to the astronomical tides. Based on the number of high and low tides and their relative heights each tidal day, tides are diurnal, semi-diurnal, or mixed.
- A tsunami is a rapidly propagating shallow-water ocean wave that is generated when a submarine earthquake, landslide, or volcanic eruption disturbs deep ocean water. A tsunami entering shoaling coastal waters may build to extraordinary heights, push inland, take many lives, and cause considerable damage.

Review

1. What is the immediate energy source for most waves on the ocean surface? What type of energy is transferred from this source to the surface waters?
2. Distinguish between wave height and wavelength.
3. For deep-water waves, the diameter of the orbit of a water particle at the ocean surface is equivalent to the __________.
4. For deep-water waves, the depth of no wave motion (wave-base) is approximately equal to __________ the wavelength.
5. Define fetch. For the same wind speed, higher sea waves are generated with increasing or decreasing fetch?
6. What is a seiche? How does a seiche differ from wind-driven ocean waves?
7. Astronomical tides are caused by the gravitational attraction between the rotating Earth and what other two bodies in space? Which one of these celestial bodies produces greater gravitational pull with Earth and why?
8. Do astronomical tides behave as deep- or shallow-water waves? Explain your answer.
9. Distinguish between semi-diurnal and mixed tides.
10. Tsunamis most frequently originate in which ocean basin? What type of tectonic activity is responsible for most tsunamis?

Critical Thinking

1. Compare the restoring force for capillary waves versus that for longer wavelength ocean waves.
2. How does frictional interaction with the ocean bottom influence sea waves?
3. Explain how wave interference influences the height of sea waves.
4. How does bathymetry affect a shallow-water wave such as a tsunami?
5. Distinguish among a forced wave, standing wave, and progressive wave. Also provide an example of each.
6. How and why does the tidal range vary between a spring tide and a neap tide?
7. Explain how tectonic activity can affect the inland progression of a tsunami.
8. Compare the types of tides usually experienced along the U.S. Pacific coast and Atlantic coast. What might explain the difference?
9. What is the significance of resonance for the tidal range in coastal inlets?
10. Why are tsunamis more common in the Pacific Ocean basin than the Atlantic Ocean basin?

ESSAY: The State of the Sea

Experienced sailors are usually adept at describing the state of the sea. The term *sea state* refers to the overall appearance of the ocean surface, including the dimensions of wind-generated waves and the presence of surface phenomena such as white caps, foam, or spray. Because the wind is largely responsible for the appearance of the sea surface, approximate wind speed can be inferred from the sea state. The sea state also provides an indication of wind direction because wave crests tend to move in approximately the same direction as the near-surface wind.

In the 19th century, prior to the invention of accurate instruments for measuring wind speed (anemometers), mariners developed a method to estimate the wind speed from the observed sea state. This method evolved from a wind force scale proposed in 1805 by Francis Beaufort (1774-1857), later an Admiral in the British Royal Navy. Initially, Beaufort devised a method for estimating the force of the wind using the amount of sail needed by a fully rigged sailing vessel, a British frigate of the day. He divided his scale into 13 increments, ranging from 0 (calm) to 12 (hurricane). Hence, for example, a wind of force 1 (light air) was described as "just sufficient to give steerage way" for a frigate, whereas a force 12 (hurricane) was "that which no canvas could withstand." By 1838, the *Beaufort scale* was modified to specify the sea state and was adopted by the British Navy. On this scale, force 1 winds produce "ripples with appearance of scales; no foam crests" (wave heights of about 7.5 cm or 3 in.) whereas a force 12 wind produces a chaotic sea state described as "air filled with foam; sea completely white with driving spray; visibility greatly reduced."

In the 20th century, the first reliable anemometers allowed mariners to quantify the Beaufort scale in terms of wind speed. For example, a force 1 wind is 1.6 to 4.8 km per hr (1 to 3 mph) whereas a force 12 wind is 119 km per hr (74 mph) or higher (see Figure 1 for examples). In 1926, the Beaufort scale was extended to include the effects of wind on land (Table 1). In addition to wind speed, the state of the sea depends on swell and the fetch and duration of the wind. Other factors such as water depth, heavy rain, or ice can also affect wave height. Furthermore, the Beaufort scale applies to waves on the open sea; waves in enclosed waters tend to be smaller and steeper.

Experience tells us that waves do not have a single wave height or wavelength, but occur in a spectrum of sizes. This variation is partly due to the inherent variability of wind speed and direction over a continuum of space and time scales. One of the most useful statistics for describing wave characteristics is *significant wave height*, the average height of the highest one-third of waves observed in a patch of ocean. Because smaller waves are usually not visible against the background of larger waves, we can assume that significant wave height approximates the visually observed mean wave height. Statistical analysis of waves indicates that the largest individual wave that one might encounter in a storm would be roughly twice as high as the significant wave height. The *dominant wave period* represents the period of waves exhibiting the maximum wave energy.

Spurred by the need for better weather and climate forecasts, the meteorological and oceanographic communities have expanded their monitoring of wind and wave conditions over the open ocean. These observations utilize more sophisticated techniques than visual observations from ships of opportunity traversing limited regions of the ocean. Today, sensors on moored automated buoys and orbiting satellites as well as ships at sea gather near-surface wind and wave data.

Moored buoys, deployed by various nations in their coastal waters, serve as instrumented platforms for making automated weather and oceanographic observations. The *National Data Buoy Center (NDBC)*, part of NOAA's National Weather Service, operates approximately 106 moored buoys in the coastal and offshore waters of the western Atlantic Ocean, Gulf of Mexico, and the Pacific Ocean from the Bering Sea to southern California, around the Hawaiian Islands, and in the South Pacific, as well as the Great Lakes. Buoys are equipped with accelerometers or inclinometers that measure the heave acceleration or the vertical displacement provided to the buoy by waves passing during a specified time interval. An onboard computer uses statistical wave models to process these measurements and generate wind-sea and swell data that are then transmitted to shore stations. These data include significant wave height, average wave period, and dominant wave period during each 20-minute sampling interval. Selected buoys also measure directional wave data, such as mean wave direction.

Measuring the winds at sea has always been difficult and the historical database has very sparse coverage. Today, Earth-orbiting satellites equipped with radar scatterometers use the sea state to estimate near-surface wind speed and direction and provide the first global map of the surface wind field over the ocean. A *scatterometer* is a sensor that measures the return reflection or scattering of a microwave (radar) signal sent to Earth's surface. A rough sea surface reflects back (backscatters) to the antenna on the satellite a stronger signal than does a smooth sea surface for off vertical angles. Backscatter is proportional to roughness because more water surface faces directly at the radar when the water is rougher (with waves) than when the water is smooth. Computer algorithms estimate wave height and then the wind speed from differences measured in the strength of the return signals. If two emitted beams are spaced with a precise angular distance, slight variations in the return signals from the roughened surface permit determination of wind direction. As a feasibility study, the first space-borne scatterometer was onboard the U.S. Skylab Mission in 1973-74. From June to October 1978, the U.S. SEASAT-A Satellite Scatterometer (SASS) demonstrated that accurate winds could be obtained remotely from space. The NSCAT (NASA Scatterometer) was flown onboard the Japanese satellite ADEOS, launched into orbit in 1996 and operational until mid-1997. SeaWinds sensors accompanied the U.S. QuikSCAT (Quick Scatterometer) launched in 1999 and the Japanese Midori 2 satellite in December 2002.

FIGURE 1
The state of the sea corresponding to Beaufort Force 1 wind speed: 1 to 3 mph (top) and 10 wind speed: 55 to 63 mph (left). [Courtesy of NOAA's National Weather Service, Milwaukee/Sullivan, WI]

TABLE 1
Beaufort Scale of Wind Force

Beaufort Number	*General description*	*Sea and land observations for estimating wind speed*	*Wind speed 10 m above ground in km/hr (mph)*
0	Calm	Sea like mirror; smoke rises vertically.	< 1 (< 1)
1	Light air	Slight ripples at sea; smoke but not wind vane shows direction of wind.	1-5 (1-3)
2	Light breeze	Small, short wavelets; wind felt on face, leaves rustle, wind vane moves.	6-11 (4-7)
3	Gentle breeze	Large wavelets, scattered whitecaps; leaves and small twigs move constantly, small flags extended.	12-19 (8-12)
4	Moderate breeze	Small waves, frequent whitecaps; dust and loose paper raised, small branches moved.	20-28 (13-18)
5	Fresh breeze	Moderate waves; small leafy trees swayed.	29-38 (19-24)
6	Strong breeze	Large waves, some spray; whistling heard in utility wires, large branches in motion.	39-49 (25-31)
7	Near gale	White foam from breaking waves; whole trees in motion.	50-61 (32-38)
8	Gale	Moderately high waves of great length; twigs break off trees.	62-74 (39-46)
9	Strong gale	Crests of waves begin to roll over, spray may impede visibility; slight structural damage.	75-88 (47-54)
10	Storm	Sea white with foam, heavy tumbling of sea; trees uprooted, considerable structural damage.	89-102 (55-63)
11	Violent storm	Very rare, unusually high waves; widespread damage.	103-118 (64-73)
12	Hurricane	Very rare, much foam and spray greatly reduce visibility; disastrous.	119 (74) and higher

ESSAY: Monitoring Sea Level from Space

Flown onboard a polar-orbiting satellite, a *radar altimeter* measures the time it takes a pulse of microwave energy to traverse the distance down to the sea surface and back. Directed downward by the satellite, the microwave pulse strikes the ocean surface and part of the signal reflects back to the satellite's radar instrument. Using the known speed of the microwave signal and the time between initial transmission and receipt, the distance is calculated between the satellite and the ocean surface. With the latest satellite altimeters, ocean scientists can determine the height of the ocean surface with an accuracy of a few centimeters. Taken into account during this remote sensing are a variety of factors external to the ocean surface such as the satellites' orbit, atmospheric influence on the microwave pulse, variations in gravity, and astronomical tides.

Many variables affect sea level, with changes in local gravitational attraction being one of the most important. The local gravitational attraction between the solid Earth and overlying liquid ocean depends on both the distance and mass of Earth's lithosphere. The greater the mass, the greater the gravitational attraction, whereas the greater the distance, the smaller the attraction. Ocean water piles up, increasing sea surface height, over areas of stronger gravitational attraction, while it drifts away from areas with weaker gravitational attraction, decreasing sea surface height. Density differences in Earth's lithosphere can also cause significant variations in sea surface height. Sea level drops about 4 m (13 ft) for every 1000 m (3300 ft) of ocean depth, owing to local gravitational effects of Earth's crust on the overlying ocean water. Ocean bottom features such as ridges and trenches can cause sea height to vary by tens of meters. These mounds and depressions in the sea surface are not readily apparent when sailing or flying because the ocean's horizontal dimensions are so great that the slopes are small.

Variations in sea-surface elevation are measured relative to Earth's geoid. The *geoid* is the equipotential surface of the planet's gravitational field, which corresponds to global mean sea level in the absence of any disturbance such as winds, currents, or tides. (Gravity everywhere acts downward and perpendicular to a geopotential surface.) The geoid varies less than 200 m (650 ft) across the globe. For example, sea level rises about 1 m in the subtropical gyres and drops about 2 m across the Antarctic Circumpolar Current. The somewhat irregular surface of the geoid contrasts with the *reference ellipsoid*, an idealized mathematical representation of Earth's surface as a smooth oblate spheroid, somewhat flattened at the poles and bulging at the equator.

Although efforts at remote sensing of sea-surface topography by satellite were conducted in the 1970s and 1980s, the technique was not fully developed until the early 1990s. The *TOPEX/Poseidon satellite*, a joint mission of NASA and the *Centre National d'Etudes Spatiales (CNES)* in France, provided the first continuous global coverage of ocean surface topography (sea level) at 10-day intervals. Launched in 1992, it operated until January 2006 and completed almost 62,000 orbits of the planet. In polar orbit at an altitude of 1336 km (830 mi), the radar (microwave) altimeters onboard bounced microwaves off the ocean surface to obtain precise measurements of the distance between the satellite and the sea-surface. These data, combined with data from the Global Positioning System (GPS), generated images of sea-surface height. Focusing on temperature instead of gravitational attraction, elevated topography (hills) indicates warmer than usual water whereas areas of low topography (valleys) indicate colder than usual water. Such images can be used to calculate ocean surface currents and identify and track the development of El Niño and La Niña (Chapter 8).

In December 2001, NASA and CNES, launched *Jason 1*, successor to *TOPEX/Poseidon*, continuing to gather data on ocean circulation. Among *Jason 1*'s primary objectives were improving climate forecasting, measuring changes of sea level in response to global climate change, and studying large-scale ocean circulation and heat transport. Sensors on *Jason 1* map wave heights, wind speed, and ocean surface topography to an accuracy of 3.3 cm (1.3 in.) over about 95% of Earth's ice-free ocean surface every 10 days. *Jason 1* serendipitously tracked the propagation and height of the December 2004 Indian Ocean tsunami (Figure 1). On 15 June 2008, *Jason 2*, the successor to *Jason 1*, was launched into the same orbit as *Jason 1*. This latest *Ocean Surface Topography Mission (OSTM)* is an international collaborative effort involving NASA, NOAA, CNES, and the *European Organisation for the Exploitation of Meteorological Satellites (EUMETSAT)*. *Jason 2* will extend the continuous climate record of precise sea surface height measurements using the next generation of instruments.

The *Jason* series of satellite radar altimeters has proven to be a versatile and powerful tool for remote sensing of the ocean. Analysis of satellite altimeter derived sea-surface elevations provides the most complete data set on the bathymetry of the ocean floor. Physical oceanographers also use satellite altimetry to measure changes in the height of the ocean surface due to wind and density differences, enabling them to calculate patterns of surface currents (*geostrophic flow*) from observed sea-surface slopes (Chapter 6). Altimeter data are also used to determine wave heights, estimate surface wind speed, and track surface water masses, such as the tropical Pacific basin during El Niño and La Niña (Chapter 11).

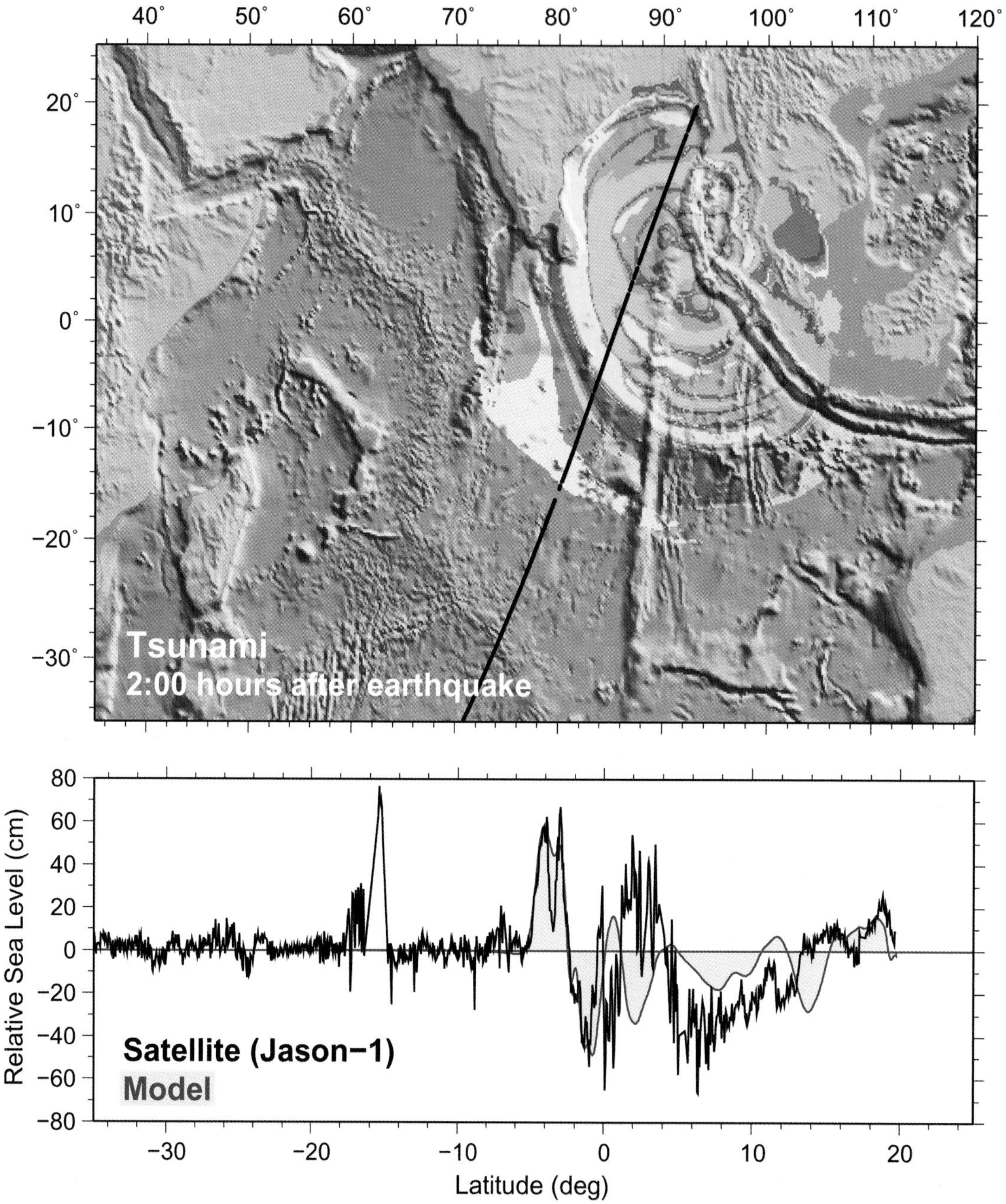

FIGURE 1
Sea-surface height measurements obtained by the altimeter onboard the Jason-1 satellite enabled NOAA scientists to track and measure the height of the December 2004 Indian Ocean tsunami. The image represents the wave pattern two hours following the earthquake. The graph shows variations in sea level height measured along the dark black line on the top map. [Courtesy of NOAA]

ESSAY: Power Generation from Ocean Tides and Waves

Ocean tides and waves have long interested scientists, engineers, and conservationists as energy sources for generating electrical power and "clean" energy alternatives to lessen our dependence on fossil fuels. Tidal power has received the most attention—both the regular rise and fall of the sea surface as well as strong tidal currents. Indeed, tidal power was used for milling grain as early as the 11th century in Britain and France and the 17th century in Boston.

In the last half of the 20th century, large tidal-power installations were built in France, Canada, Russia, and China. These plants operate much like hydroelectric facilities except that water flows in both directions. Suitable coastal sites must have tidal ranges greater than 5 m (16 ft) and relatively fast tidal currents that flow through a constricted inlet into a large basin. A dam (called a barrage) is constructed across the inlet with turbines submerged at the base of the dam. The ocean is on one side of the barrage and the basin is on the other side. The incoming flood tide is channeled through the turbines and the basin fills with water. During the slack water period the gates on the barrage are closed. Then, with the outgoing ebb tide, the gates are opened and the exiting water again drives the turbines that generate electricity.

The downside of this technology is the fact that electrical generation is not continuous at the same level of output 24-hrs a day, diminishing during slack periods when tidal currents are weak. Very few locations possess the ideal combination of coastal configuration and tidal range. Furthermore, where such a combination does exist (e.g., Bay of Fundy, Nova Scotia), the barrage is likely to disrupt the local ecology, fishery, and tourism.

As shown in Figure 1, the most successful of the tidal-power installations is La Rance Plant on the Brittany coast of France just south of Saint-Malo where the tidal range varies from 9 m (30 ft) to more than 13 m (43 ft). This 240 megawatt (MW) plant has 24 turbines that generate about 90% of Brittany's electricity. The 18 MW facility at Annapolis Royal, Nova Scotia, has been less successful and never operated commercially. A small, tidal power plant was built on the sea floor near the Arctic tip of Norway to generate power from the area's strong tidal currents and provide electricity for Hammerfest, one of the world's most northerly towns. The power plant uses large windmill-like turbines mounted on the sea floor and transmits electricity to the local power grid via submarine cables. Initially the plant generated enough electricity for about 300 homes but is planned for expansion to power perhaps 1000 homes.

Many engineering schemes have been proposed for generating electrical power from ocean surface waves. A few have been built and operated briefly as demonstration projects. One demonstration plant was built in the mid-1990s on the remote island of Islay in the Western Islands of Scotland, an area known for its high waves. The plant operated for a short time before being destroyed by high waves in a storm. Small-scale wave-powered generators are often used to power navigation aids on offshore buoys.

FIGURE 1
Aerial view of the tidal barrage of La Rance Plant and Saint-Malo.

The world's first commercial wave energy conversion project is based on a new technology known as the *Pelamis Wave Energy Converter*. From a site off the north coast of Portugal, the system initially produced 2.25 MW of electricity; anticipated expansion is expected to bring electrical output to a level that will meet the needs of 1500 homes. The Pelamis system consists of four cylindrical tube segments, each 120 m (395 ft) long and 3.5 m (11.5 ft) in diameter (Figure 2). The segments float semi-submerged and moored in waters 50-60 m (165-200 ft) deep and about 5 km (3.1 mi) from the coastline. Hinged joints link together the segments and it is the wave-induced motion of these joints that is the immediate source of energy. This motion, in turn, drives hydraulic rams that pump high pressure oil that powers hydraulic motors that, in turn, run electrical generators. An ocean-floor cable delivers electricity to the shore. The Pelamis device is moored in such a way that its tube segments orient toward oncoming waves thereby optimizing the harvesting of energy. Production of 30 MW would require many interconnected tubes covering an ocean area of about 1 square km.

Because of its long coast facing the waves of the open Atlantic, wave energy conversion may be Portugal's most promising alternative energy source. Costs are significantly less than both wind power and solar photovoltaic power. Interest in this technology is also growing in the United Kingdom.

FIGURE 2
A Pelamis Wave Energy Converter during the final tests at the port of Peniche, Portugal. [Photo by Dipl. Ing. Guido Grassow/License: Creative Commons Attribution ShareAlike 3.0]

CHAPTER 8

THE DYNAMIC COAST

Haystack Rock at high tide at Cannon Beach in Oregon. [Photo by Michael Theberge/NOAA]

Case in Point

A *wetland* is a low-lying, generally flat area that periodically is covered by water or has soil that is saturated with water (Figure 1). A coastal wetland is part of the dynamic transition zone between land and ocean that provides many valuable services and resources. Wetlands act as a buffer, absorbing energy from coastal storms (including tropical storms and hurricanes), thereby preserving shorelines and protecting human populations and infrastructure. The roots of wetland vegetation anchor the soil and sediment in place thereby reducing erosion by high waves and storm surges. They function as nurseries for juvenile fishes, shellfish, crabs, and shrimp. They are critical habitats for numerous migratory bird populations. Some birds, such as geese and ducks, winter over in coastal wetlands while smaller birds might spend the summer whereas others remain year round. However, these services and resources are increasingly threatened as sea-level rise, projected to accompany climate change during this century, causes inundation and drowning of coastal wetlands.

Not all coastal wetlands are equally vulnerable to inundation and drowning. In the December 2010 issue of the journal *Geophysical Research Letters*, scientists from the U.S. Geological Survey (USGS) emphasize that some coastal wetlands can adapt to rising sea level through natural modifications of their physical environment. They report that the fate of coastal wetlands depends on sediment supply, tidal range, and the rate of sea-level rise.

Coastal wetlands are continually flushed by currents associated with rising and falling tides and are maintained by sediments transported by water flowing through the wetland from upstream sources. With each tidal cycle (diurnal or semi-diurnal), sediment spreads across the surface of the wetland causing it to gain elevation. Where the tidal range (the difference in height between consecutive high tide and low tide) is relatively great, the sediment delivery system and the concentration of transported sediment favors the building of more elevation that may counter sea-level rise.

Sea-level rise is projected to accompany global warming through the 21st century and beyond. Mean sea level is rising in response to melting land-based glaciers and thermal expansion of seawater. A key question regarding the future of coastal wetlands is how the rate of sea-level rise might compare to the rate of sedimentation in coastal wetlands. USGS scientists attempted to answer this question using models set for rapid and slow sea-level rise scenarios.

FIGURE 8.1
Wetlands protect coastal communities from erosion. [Courtesy of NOAA]

They concluded that with rapid rise, most coastal wetlands worldwide would disappear by the end of this century. Even with slow rise, wetlands having low sediment supply and low tidal range will be vulnerable to inundation and drowning. On the other hand, wetlands with high sediment supply and high tidal range are more likely to survive. Along the U.S. Atlantic coast, the most vulnerable coastal wetlands are Plum Island Estuary (New England's largest estuary) and the Albemarle-Pamlico Sound of North Carolina. Limited sediment supply is the principal reason for their vulnerability.

Driving Question:

Why and in what ways is the coastal zone a particularly dynamic and vulnerable portion of the Earth system?

The **coast**, the primary focus of this chapter, encompasses the relatively narrow region transitional between land and ocean. Although on a given day the coast can appear stable and benign, the many components of the Earth system (ocean, atmosphere, geosphere, cryosphere, and biosphere) continually interact in the coastal zone making it an exceptionally active environment. Forces operating at the land/sea interface include tides, breaking waves, glacial processes, sea-ice conditions, near-shore currents, sediment discharged by rivers and streams, changing sea level, tectonic and other geological processes, and the impact of living organisms including humans. These interactions are responsible for the wide variety of coastline types worldwide including, for example, rocky beaches, cliffed headlands, barrier islands, mangrove swamps, estuaries, sand dunes, coral reefs, and fjords.

Rapid human population growth is stressing the coastal zone. In 2009, the National Oceanic and Atmospheric Administration (NOAA) reported that the coast is home to about 52% of all Americans. Some 159 million U.S. residents currently live in coastal counties—about 39 million more than in 1980—and the population is expected to swell by 7.1 million by 2015. Los Angeles County, CA, leads the U.S. in coastal population with 6% of the total. According to NOAA, a county is described as coastal if it is on a coast (i.e., Atlantic, Gulf, Pacific, Great Lakes) or at least 15% of the county's land area is in a coastal watershed. The 673 coastal counties represent 17%

of the nation's land area and have three times the nation's average population density. Population growth is most rapid from Texas through Virginia with Florida leading the nation in population growth (and hurricane potential). Of the top 10 counties with the greatest percentage population growth, 6 of them are in Florida.

Increasing human population density in the coastal zone translates into greater demands for electrical power, fresh water, wastewater treatment, and roads. Environmental stressors that accompany population growth include waste disposal, recreational activities, tourism, infrastructure needs, industrial expansion, homes, and power plants, and the exploitation of fisheries, minerals, oil, and natural gas deposits. Adverse impacts on coastal ecosystems include shrinking plant and animal habitats, surface- and ground-water pollution, and more frequent algal blooms due to enhanced runoff of fertilizers and other nutrients. Furthermore, population growth along the Atlantic and Gulf coasts exposes more people to the hazards of hurricanes and powerful extratropical storms.

In this chapter, we examine the various types of coasts and the processes responsible for their evolution and distinguishing features, and how human activity is altering the coastline. Special attention focuses on beaches, barrier islands, salt marshes, and estuaries. We describe hazards posed to coastal residents by tropical and extratropical storm systems and the key role played by air-sea interactions in powering these weather systems. We begin with coastline forming processes. Coastlines also border the Great Lakes. For more on the Great Lakes and how their properties and processes compare with those of the ocean, refer to this chapter's first Essay.

Coastline Formation

As shown schematically in Figure 8.2, the farthest inland extent of storm waves is one way of delineating the **coastline**, in some cases marked by sand dunes or wave-cut cliffs. Another definition is the mean high water level. Land exposed at low tide up to the coastline is the **shore** and the average low tide line is the **shoreline**. The **intertidal zone**, the shore area between high and low tides, is alternately under water and exposed to the atmosphere as the tide rises and falls. Moisture, temperature, and salinity fluctuate dramatically so that only organisms with special adaptations survive drastic changes in the intertidal zone (Chapter 10).

A fundamental distinction among coastlines is whether they occur in tectonically active or inactive

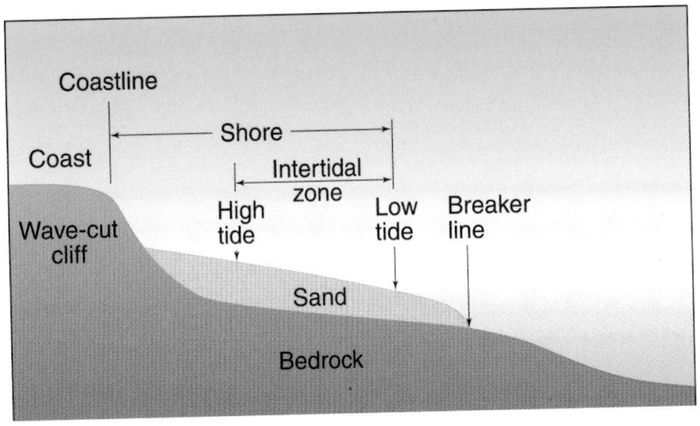

FIGURE 8.2
Idealized cross-section of a coastal zone, in this case bordered on the land side by wave-cut cliffs.

(passive) continental margins (Chapter 2). Mountain building and volcanic activity are dominant geological processes in tectonically active coastlines such as those that rim much of the Pacific Ocean basin. These processes can elevate coasts so rapidly that waves and currents cannot erode or deposit sediment fast enough to significantly modify long stretches of the shoreline. The few beaches that exist are small, isolated and usually restricted to near wave-cut cliffs or the mouths of rivers, which supply sand and gravel. For example, subduction is causing uplift along the Pacific Northwest coast. The Juan du Fuca plate is created at the submerged Juan de Fuca Ridge and is moving eastward, subducting beneath Washington, Oregon, and Northern California. Along the southern Oregon coast, the rate of tectonic uplift is greater than the global rise in sea level. Wave-cut benches are visible in the cliffs well above present high-water level. Meanwhile, along the northern Oregon coast, the rate of sea-level rise is keeping pace with tectonic uplift.

Erosion, transport, and deposition of sediment are dominant geological processes along passive shorelines such those that border the Atlantic Ocean and Gulf of Mexico. In passive continental margins, sediments (mostly sand and silt) are deposited by rivers and streams entering the ocean and supplied by breaking waves undercutting shoreline cliffs that subsequently slump or slide into the ocean. Waves and currents transport these sediments along the shore, forming beaches, spits, and offshore barrier islands.

Rocky shorelines that are common in New England and Atlantic Canada formed during the Pleistocene Ice Age (1.8 million to 10,500 years ago) when huge lobes of glacial ice scoured the land numerous times, eroding soils and bedrock. As the ice retreated, sea level rose (Figure 8.3). The time since the ice retreated has been too short for accumulation of significant quantities of

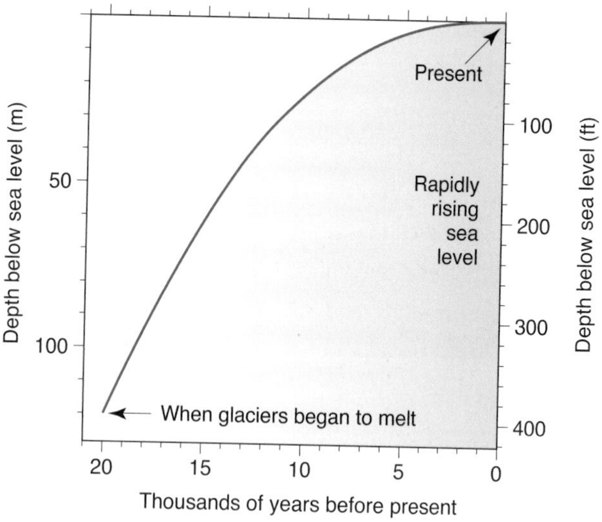

FIGURE 8.3
Smoothed post-glacial curve showing rise in global sea level from about 20,000 years ago to the present.

sands and gravels on most rocky shorelines. Furthermore, the retreating glacial lobes left behind many lakes and bogs that trap the little sediment produced, preventing the remainder from reaching the shore. Hence, not enough sediment is available for waves and tides to smooth the complex coastlines. Such recently de-glaciated coastlines feature many small harbors and inlets, with rocky islands offshore.

As waves enter coastal waters of varying depth, they usually undergo **refraction**, bending of a wave in response to changing wave speed. Along the coast, wave refraction influences patterns of erosion, sediment transport, and deposition. Consider some examples. Suppose that the shoreline is straight and coastal waters shoal uniformly in a direction perpendicular to the coastline. If wave crests approach the shoreline at an angle other than 90 degrees, the wave bends. The segment of the wave moving through deeper water progresses toward shore at its original speed while the segment of the wave entering shoaling water begins to "feel" the bottom and slows (Chapter 7). Consequently, the wave crest is refracted toward the shallower water. Refraction causes approaching wave crests to come into near alignment with the depth contours (bathymetry) of the ocean bottom and to closely conform to the shape of the shoreline.

On the other hand, suppose that the coastline is irregular and coastal waters do not uniformly shoal toward the shore. This situation is shown in Figure 8.4 viewed from above. Well offshore in deep water, approaching wave crests (continuous white lines) are essentially straight and parallel. Waves eventually enter water that is not uniformly deep and are refracted toward shallower water. Hence, waves approaching an irregular coastline are refracted, converging toward headlands (rocky cliffs that project seaward) and away from coves or bays. Convergence concentrates wave energy causing undercutting and erosion of headlands whereas spreading out (divergence) of wave energy in bays weakens wave action and favors deposition of sediment and expansion of the beach. Eventually, the forces of erosion cut back the headlands and sediment fills in coves and bays so that the coastline straightens. Figure 8.5 shows the remnants of a headland that was extensively eroded by wave action.

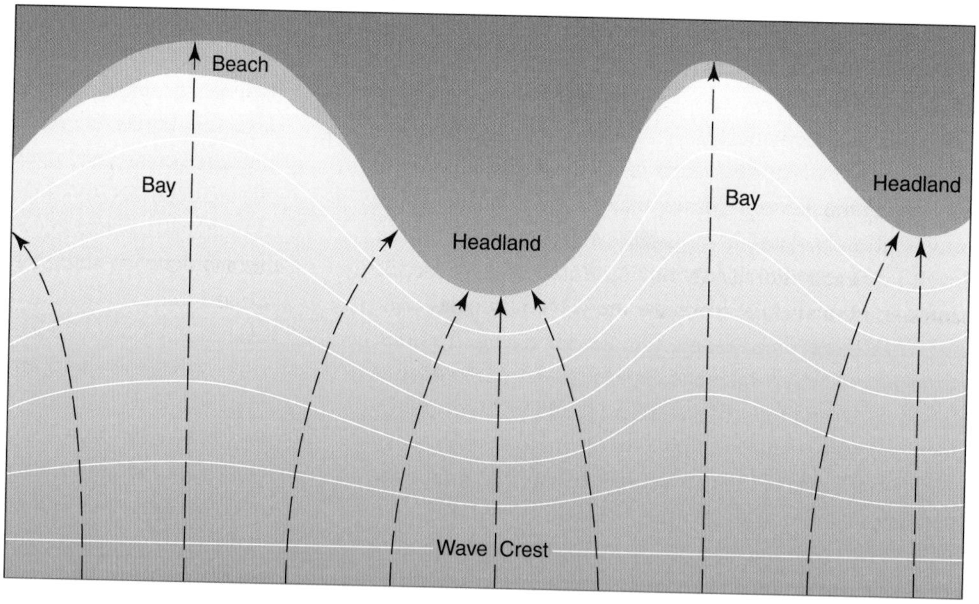

FIGURE 8.4
Viewed from above, waves slow in shoaling water and undergo refraction as they approach an irregular coastline. Solid white lines are wave crests; dashed lines are orthogonals to wave crests.

FIGURE 8.5
These rocky knobs along the Washington coast are what remains of a headland eroded by wave action. [Photo by J.M. Moran]

Coastal Features

Erosive and depositional processes impact to some degree both tectonically active and passive coastlines. The immediate source of energy for these processes is breaking waves. As we saw earlier in this book, the Sun drives atmospheric circulation (Chapter 5) and winds transfer kinetic energy to the ocean surface forming waves (Chapter 7). Hence, solar energy ultimately powers erosive and depositional processes operating along the coast. In this section, we focus on selected features of the coast, that is, beaches, barrier islands, salt marshes, and wetlands. We also consider how and why humans alter the coastal zone.

BEACHES

A **beach** is an accumulation of wave-washed sediment (usually sand and gravel) at the landward margin of the ocean (or a lake). At the mean high-water mark, most beaches feature a prominent **berm**, a platform of sand that is nearly flat-topped and slopes steeply seaward. A beach is a changing system in which erosive and depositional forces continually do battle. Waves, tides, and currents erode cliffs, wash away dunes, and transport sand to and from the beach. Forces acting to add or remove beach sediment are always at work and are seldom in balance.

A beach functions as a buffer that minimizes erosion by storm-driven ocean waves. As storm waves push ashore, water movement spreads the sand flattening the beach surface. In this way, wave energy is distributed over a broader area of the beach thereby dissipating energy and reducing erosion. Following a storm, sand transported to the beach by currents fills in eroded areas and returns the beach to a steeper profile. Especially at middle and high latitudes, significant differences are noted in the appearance of a beach in winter versus summer.

In winter, erosive forces tend to prevail over depositional forces as powerful storm waves cut back the beach, moving sands onto shallow submerged bars near the shoreline. Hence, winter beaches have a steeper slope. Strong storm winds blow sands from dry areas of the beach to dunes building behind the beach or into a lagoon. In summer, the weather often is more tranquil and depositional forces tend to prevail over erosive forces. Waves transport sand from offshore bars back onto the beach and the beach widens. Summer beaches therefore feature a gentler slope.

Sediment is delivered to the shore by rivers and streams and eroded from seaside cliffs by ocean waves. As rivers and streams enter the sea, their current slows and diverges, depositing much of their suspended sediment load near their mouths (Chapter 4). Breaking waves attack the base of seaside cliffs, carving out a notch. The cliff slope becomes too steep and rock debris slumps or slides into the ocean (Figure 8.6). A number of factors govern the propensity of a cliff to slumping or sliding including composition (e.g., consolidated versus unconsolidated sediment), geological structure (e.g., presence or absence of fractures), and rainfall. Furthermore, the relentless pounding of waves breaking against the coast fragments exposed rock. Coastal currents then transport these sediments to and from the beach. Even the rock debris that accumulates at the base of a seaside cliff is eventually removed, exposing the cliff to more erosion.

Earlier in this chapter, we noted that waves approaching the shore bend (are refracted) toward shallower water causing their crests to closely conform to the shoreline. Despite the effects of refraction, wave crests rarely line up perfectly parallel to the beach, often striking the beach at oblique angles up to 10 degrees or so (the angle between wave crests and the shoreline). Breaking waves striking the shore at an oblique angle give rise to sediment-transporting currents along the shore. If a breaking wave approaches the shore directly, a thin sheet of water rushes straight up the sloping beach face as *swash*, stops briefly, and then flows straight down slope as a *backwash* (Figure 8.7). If, on the other hand, a breaking wave approaches the shore at an oblique angle, the swash follows a diagonal up the beach face, then straight down slope (Figure 8.8). Together, the upwash and backwash constitute the zigzag or saw tooth motion that rolls particles along the beach face, known as *beach drift*. The oblique approach of waves to the shore produces a component of water motion that flows parallel to the shore

FIGURE 8.6
Wave-eroded cliffs along the Oregon coast. [Photo by J.M. Moran]

known as a **longshore current** (Figure 8.9). A longshore current transports sediments as **littoral drift** (a river of sand) along the coast, either nourishing or cutting back beaches. In some cases, littoral drift transports sediment offshore into the head of a submarine canyon where it is permanently lost to the beach.

The **beach sediment budget** is summarized in the following word equation:

Sediment Budget = Sediment Input − Sediment Output

In summary, sediment input processes include delivery by rivers, streams and wind, erosion of seaside cliffs, longshore current supply, and onshore transport. Sediment output processes encompass removal by longshore currents, wind erosion, and offshore transport. If the sediment balance is positive, (e.g., in summer) the beach grows; with a negative balance, (e.g., in winter) the beach is cut back. If input and output are equal, the beach is in a steady state.

Along the Atlantic coast, different geological processes account for the significant differences in beach characteristics. From south Florida to the south shore of Long Island, NY, beaches are mostly along the shorelines of barrier islands. As discussed later, a **barrier island** is an elongated, narrow accumulation of sand oriented parallel to the coast and separated from the mainland by a lagoon, bay,

FIGURE 8.7
Swash on a sloping beach face at Gaviota State Park about 33 miles west of Santa Barbara, CA. [Photo by Edward Hopkins]

FIGURE 8.8
Wave crests approaching the beach at an oblique angle at the Assateague Island National Park, VA. [Courtesy of NOAA Corps]

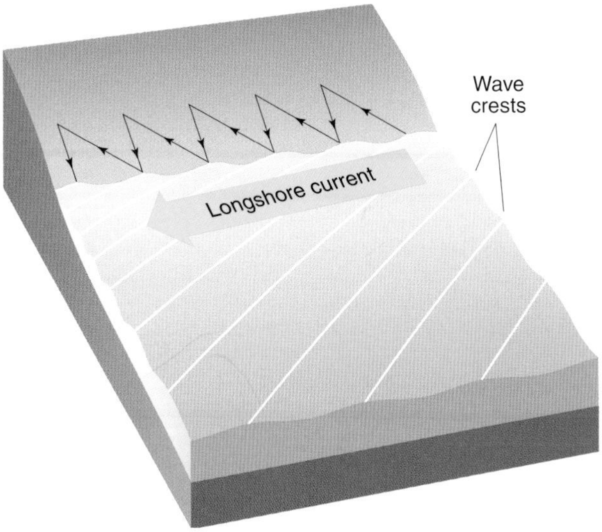

FIGURE 8.9
Viewed from above, waves approaching the coastline at an oblique angle produce a component of water motion roughly parallel to the shore (longshore current) that transports sediment (littoral drift). Zig-zag arrows represent the flow of water up and down the beach face.

or estuary. Chains of barrier islands lie offshore of a broad coastal plain that extends to a relatively smooth continental shelf. This Atlantic coastline of *barrier island beaches* is segmented by estuaries such as Chesapeake Bay.

North of Long Island, the coastal plain is absent and the Appalachian Mountain bedrock stretches beyond the shore. In this region, slow weathering of resistant bedrock and numerous episodes of glaciation and deglaciation formed the beaches. Much of coastal New England is rocky and irregular, overlain in places by unconsolidated glacial sediments, and featuring small pocket beaches nestled between rocky headlands. Small islands, ledges, shoals, and submerged basins are scattered offshore. Most New England beaches are *mainland beaches*, occurring on the mainland, not offshore islands. The beaches can also be on **sand spits**, finger-like ridges of sand or gravel that protrude from the shore into a body of water, or a **tombola**, a spit linking an offshore island to the mainland or another island.

Seasonal variations in beach width are part of longer trends. In 2011, the USGS published a report on the Mid-Atlantic (Virginia/North Carolina border to Long Island, NY) and New England coasts, measuring shoreline changes over a 150-year period at more than 21,000 sites over a distance of more than 1050 km (650 mi). They concluded that 68% of beaches are eroding with a coastal change (eroding and prograding beaches) average of −48.8 cm/yr (−1.6 ft/yr). In the most extreme case, a beach was cutting back −18.3 m/yr (−60 ft/yr). Over the past 25-30 years, the percentage of beaches undergoing erosion declined to 60%, perhaps because of beach restoration projects which are discussed later in this chapter. A combination of many factors governs the rate of erosion, including sand supply, storms, sea level change, and human activity. Substrate is also important, as demonstrated by the greater erosion with Mid-Atlantic sandy beaches than the primarily rocky coastline of New England.

BARRIER ISLANDS

Nearly 300 barrier islands fringe much of the Atlantic and Gulf Coasts primarily from Texas to Long Island, NY, the best developed of any barrier island chain in the world. (The number of barrier islands worldwide exceeds 2500.) They occur where the supply of sand is abundant, the sea floor is gently sloping, waves are energetic and the continental margin is passive. Barrier islands vary in length from a few hundred meters to more than 100 km (62 mi) whereas the width is often a kilometer or less. Padre Island is the longest of the nation's barrier islands extending more than 180 km (112 mi) along the Texas Gulf Coast (Figure 8.10). Tidal inlets segment barrier islands. As described in this chapter's second Essay, three major tidal inlets link the Gulf of Venice to the lagoon where the island city of Venice, Italy, is located.

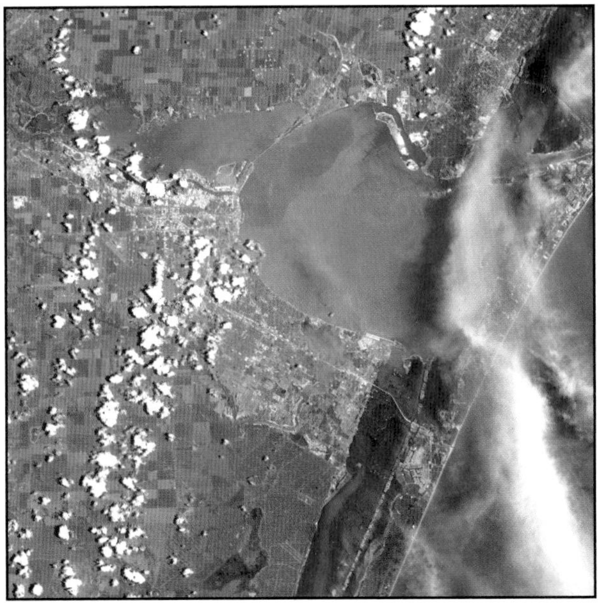

FIGURE 8.10
Padre Island is shown offshore of Corpus Christi Bay and the city of Corpus Christi, TX on the Bay's southern end. [NASA image by Jesse Allen, using data provided courtesy of the NASA/GSFC/MITI/ERSDAC/JAROS, and U.S./Japan ASTER Science Team]

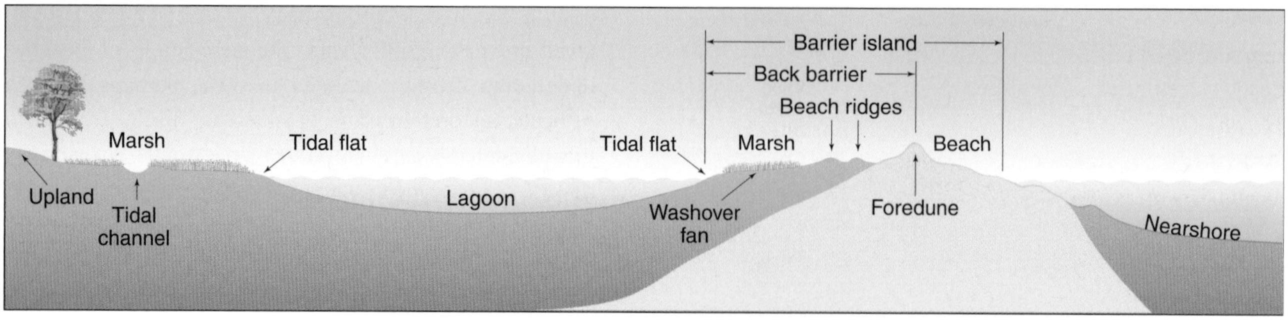

FIGURE 8.11
A cross-sectional profile of a typical undeveloped barrier island with the mainland to the left and the open ocean to the right.

Beginning at the ocean side of a typical undeveloped barrier island and working our way toward the mainland, we would encounter successively a beach, dunes, marsh, flats, lagoon, flats, marsh, and upland areas (Figure 8.11). The near-shore and beach zones are high-energy environments with the potential for considerable impact from waves, longshore currents, and onshore winds. Just inland from the beach is a dune field consisting of one or more parallel ridges of wind-blown sand from the beach (Figure 8.12). The next zone is the back-island flat, an extensive plain consisting of grassland (marsh) or woodland, depending on the age and size of the island and its exposure to storms and high winds. This area is prone to **washover** by high waves generated by the winds of powerful coastal storms. The flat grades into a salt marsh that borders a lagoon, estuary, or bay; it is a low-energy environment where fine sediments accumulate. A **lagoon** is a partially enclosed body of water separating a barrier island from the mainland. Normally, lagoons receive essentially no input of river water and tidal fluxes maintain salinities near that of seawater. (In this way, lagoons differ from estuaries—discussed later in this chapter.) Where evaporation rates are relatively high, the salinity of water in some isolated lagoons may exceed the usual seawater salinity. In an extreme example, Laguna Madre between Padre Island and the Gulf coast has salinities as high as 60 psu.

FIGURE 8.12
Sand dunes located just inland from a beach on a barrier island on North Carolina's Outer Banks. [Photo by J.M. Moran]

Scientists recognize several different ways whereby a barrier island may form. Many (especially along the southeast Atlantic and Gulf Coasts) date from the mid Holocene Epoch (about 5000 years ago) when sea level was somewhat higher than it is now. (The Holocene Epoch covers the past 10,500 years.) Higher sea level submerged beaches isolating the tops of dune ridges as offshore islands while flooding lower areas landward of the dune ridges, which became lagoons. Some barrier islands (e.g., small Gulf Coast islands) are the product of the vertical growth of a submarine sand bar that emerged from the sea as an island. Along an irregular coastline, sand spits can lengthen into barrier islands through longshore sand transport. During intense storms, the spit many be breached, forming barrier islands. This is the origin of barrier islands along some segments of the New England coast.

A barrier island is a continually changing system. Waves breaking on the shores of a barrier island dissipate their energy by shifting sands and modifying the shape of the island. As sea level gradually rises, episodes of storm-induced washover become more frequent. Sand is removed from the ocean-side of the island and transported to the lagoon-side and in this way a barrier island migrates toward the mainland—eventually becoming a mainland beach.

Barrier islands absorb the brunt of powerful storm-driven sea waves thereby providing some protection for coastal beaches, estuaries, wetlands, and shoreline structures. But many barrier islands have been developed, especially for cottages and resorts, and their sands are now temporarily stabilized under layers of asphalt or concrete. Some coastal cities, including Atlantic City, NJ, Miami Beach, FL, and Virginia Beach, VA, are built entirely on barrier islands. Such exposed locations are particularly vulnerable to the ravages of the high winds and floodwaters associated with tropical cyclones (discussed later in this chapter). On developed barrier islands, much of the energy of storm waves may be expended in demolishing buildings and roads instead of shifting sands. Conflict is inevitable between rigid structures such as roads and buildings and the inherently changeable platforms (i.e., barrier islands) upon which they are built.

RIP CURRENTS

In the *surf zone* (where shoreward moving waves break in shoaling coastal waters), water moves toward and usually along the shore. Whereas the shoreward transport of water occurs over broad areas of the surf zone, the return seaward flow of water tends to concentrate in narrow widely spaced belts often corresponding to depressions in the seafloor or breaks in sand bars. The narrow seaward flow of water must balance the broad shoreward flow of water so that the offshore flow occurs as a relatively swift surface or near-surface current with speeds up to 2 knots (1.7 mph) or so. This is known as a **rip current**. (Sometimes, a rip current is referred to as a *rip tide*, but this is a misnomer; rip currents have nothing to do with astronomical tides.) Usually the strength of a rip current increases as wave height and wave period increase.

A rip current flows at nearly right angles to the shoreline and spreads seaward a few meters to hundreds of meters beyond the line of breaking waves (Figure 8.13). Viewed from above, a rip current is marked by choppy water made turbid by suspended sand and floating debris being transported seaward. Conventional thinking was that rip currents were transient phenomena lasting from a few minutes to a few hours. However, field studies conducted by University of Florida scientists indicate that a rip current may persist for weeks or even months at essentially the same location along the shore, varying in strength during that period.

A rip current is hazardous for people swimming in the surf zone and can develop offshore of any surf beach including those bordering the Great Lakes. Most people are unable to swim against the strong current and are swept into deeper offshore waters. According to the U.S. Lifesaving Association, rip currents claim more than 100 lives on average each year along the nation's beaches.

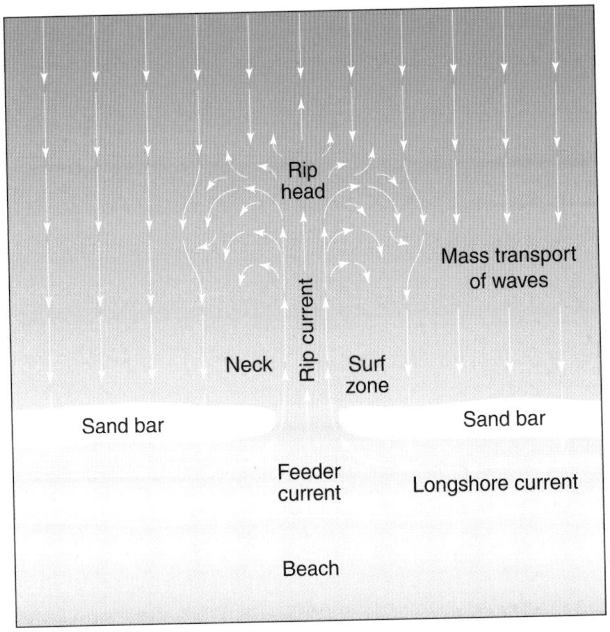

FIGURE 8.13
A rip current viewed from above.

More than 80% of all rescues performed by surf beach lifeguards involve rip currents. In Florida, since 1989, drowning deaths due to rip currents have averaged 19 annually—more than the combined yearly state average death toll from hurricanes, tornadoes, and lightning. As noted above, however, a rip current is narrow. If caught in the current, swimmers are advised not to fight the current but to swim parallel to the shore until they escape into calmer waters and can swim back to shore.

In coastal areas, NOAA's National Weather Service issues a daily rip current outlook that rates the risk of rip currents as low, moderate, or high. These outlooks plus increased public awareness of the hazards of rip currents contributed to a recent decline in rip current fatalities at the surf beaches of southeast Florida. From 1979, when the Miami National Weather Service office began issuing outlooks, to 1991, the average annual death toll declined from 12 to 5.

SALT MARSHES AND WETLANDS

A **salt marsh** is composed of salt-tolerant plants that colonize shores that are protected from wave action. Along the New England coast, the dominant plant species in undisturbed salt marshes are cordgrass (along the shore), marsh hay, black rush, and shrubs (inland often bordering trees). Tidal creeks meander through most salt marshes allowing seawater to flood the marshes on rising tides and

FIGURE 8.14
A tidal creek meanders through a Massachusetts salt marsh. [Photo by J.M. Moran]

FIGURE 8.15
Red mangorve "islands" in an an intracoastal waterway near Ft. Pierce, FL. [Courtesy of Randolph Femmer /life.nbii.gov]

drain on ebbing tides (Figure 8.14). Vegetation slows the movement of water through a salt marsh thereby inducing deposition of suspended fine sediment and the plants anchor the sediment in place. A salt marsh offers many benefits including protecting the coastline from storm wave erosion, filtering runoff thereby reducing the input of pollutants and excess nutrients into estuaries, and providing habitat for organisms such as crabs and oysters.

In the tropics and subtropics, such as South Florida and Puerto Rico, coastal wetlands support thick growths of salt-tolerant woody trees called **mangroves** (Figure 8.15). Mangroves have extensive root systems that trap sediments. They also provide habitat for both marine and land animals. Some organisms are especially adapted for survival in mangrove thickets, such as mangrove oysters, which attach themselves to roots and branches. Mangrove forests protect low lying coastal areas from storm surges and tsunamis. Unfortunately, in many areas, development activities are destroying the forests. For example, during a recent 100-year period, Tampa Bay, FL, lost an estimated 44% of its coastal wetland area, including mangrove forests and salt marshes.

Wetlands are among the most productive natural ecosystems on Earth, providing food-rich nursery grounds for estuarine and other coastal organisms, including many commercially important fish. Roots of wetland plants (both grasses and trees) trap sediment and organic matter. Where sea level is rising, wetlands can build up fast enough that their surface remains at sea level, assuming an adequate supply of sediment. If sea level is rising too rapidly and the sediment supply is limited, marshes are eroded as is presently the case on the Mississippi River delta and around the Chesapeake and Delaware Bays. Where sea level is stable, wetland plants can trap enough sediment (if available) to build the marsh surface above normal sea level. Where this happens, the shore advances and grasslands or forests eventually replace the coastal wetland.

HUMAN ALTERATIONS

Human alteration of the coastal zone has many purposes including preservation of recreational beaches, flood control, maintenance of harbors and navigation channels, and protection of homes, roads, and other structures built near the beach, in the dunes behind them, or on the top of seaside cliffs. When erosion threatens to damage or destroy shoreline structures, three protective strategies are available: armor, artificial beach nourishment, and strategic retreat. *Armor* includes construction of dams, breakwaters, jetties, groins, and sea walls. *Artificial beach nourishment* refers to the addition of sand to a beach, and *strategic retreat* involves inland displacement of a structure threatened by rising sea level and storm waves.

Coastal armor has both costs and benefits, but research suggests that the overall costs outweigh the benefits. Armor can disrupt littoral drift and alter the natural flow of sediments to and from a beach. Sediments settle out of suspension in the reservoir behind a dam, reducing the river's sediment load below the dam and the sand supply for the beach. Longshore currents transport existing sands away from the beach, sometimes to the extent that all that remains is rocky rubble (cobbles and boulders).

A **breakwater** is a long, narrow offshore structure, usually constructed of large blocks of rock or concrete and oriented parallel to the shoreline (Figure 8.16). Its purpose is to provide calm waters for docking boats or to protect beaches from erosion by absorbing the energy of breaking waves or reflecting waves back to sea. Some breakwaters are built on open coastlines where no natural harbors exist. However, sediments settle to the bottom of the relatively calm waters on the landward side of the breakwater and costly dredging may be required to keep the harbor open for navigation.

Breakwaters also disrupt littoral drift. Most sediment accumulates in the harbor and is unavailable for deposition on down-current beaches. Just such a situation occurred at Santa Barbara, CA, after a breakwater was built in 1928. To keep the harbor open and maintain longshore sand transport, sand is continually dredged from the harbor and piped as slurry (sediment and water mixture) to the down-current side of the harbor. A pipeline now transports sand formerly conveyed by natural longshore currents.

A **jetty** is similar to a breakwater except that it is oriented perpendicular to the shoreline and extends seaward up to a kilometer or more. In some cases, a jetty serves the same function as a breakwater by protecting harbors from storm waves. Often jetties are also used to keep a barrier island inlet open for navigation into and out of a harbor on the mainland side of the island. Jetties constructed on either side of the ocean-side entrance of a tidal inlet constrict the flow of tidal currents. Constriction accelerates the flow of water, reducing sediment deposition and perhaps eroding the channel bottom. On the other hand, jetties disrupt littoral drift across the inlet. Up current from a jetty, beaches widen due to enhanced sediment deposition (*beach nourishment*) whereas beaches disappear down current from a jetty due to lack of sand input and wave erosion. Consider an example.

Ocean City, MD, is a beach resort that has undergone considerable development since the 1950s. The resort is located at the south end of Fenwick Island, a barrier island. Although steps were taken in the late 1970s to curb further development by limiting dredging and filling of wetlands (intended to create more land for dwellings), the resort remains a popular destination. Most residents and visitors are probably unaware that in August 1933 a hurricane made a direct hit on the Middle Atlantic region, the only one to do so during the 20th century. (For a direct hit, the storm must approach the coast from the southeast.) Although at landfall, the hurricane was of moderate intensity, the air pressure pattern caused strong onshore winds that made it the most damaging hurricane in the history of the area. Hurricane-force winds devastated the coast from Virginia to New Jersey, fatalities numbered 47, and property damage was $40 million (in Depression era dollars). The hurricane lashed Fenwick Island and opened an inlet at its southern end linking the backwater bay with the Atlantic Ocean. Subsequently, the U.S. Army Corps of Engineers erected two rock jetties to maintain the inlet for navigation. However, the jetties also altered the southward flowing littoral drift causing the beach to widen north of the jetty at Ocean City while the beach at Assateague Island, south of the inlet, was deprived of sand and eroded back.

Other structures called **groins** resemble jetties but are smaller and usually more closely spaced along a beach (Figure 8.17). Their purpose is to widen a beach by trapping sand moving downstream as littoral drift. But once again, sand trapped by groins is not available to beaches down current so that local beach widening is at the expense of beach erosion elsewhere.

A **seawall** is a concrete or rock embankment intended to protect beaches, roads, buildings, and

FIGURE 8.16
Breakwater beacon at the mouth of Delaware Bay, DE. [Photo by R.S. Weinbeck]

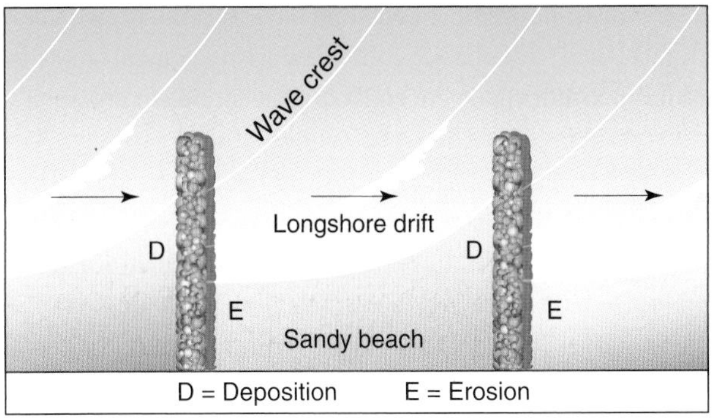

FIGURE 8.17
Viewed from above, closely-spaced groins disrupt the longshore drift so that sand is deposited on the up-current side (D) and eroded from the down-current side (E). In this way, groins trap sand along a beach and the beach widens locally.

FIGURE 8.18
Seawall protecting property from erosion by storm waves. [Courtesy of NOAA]

shoreline cliffs from erosion by storm waves (Figure 8.18). However, seawalls also cut off a source of sediment for the beach by preventing erosion of bluffs. Furthermore, evidence suggests that wave energy, rather than being mostly absorbed, is actually reflected by a seawall, stirring up the water and sand at the base of the structure, accelerating erosion, and causing beaches to narrow. Deeper water in the near shore area means that rather than breaking well offshore, storm waves break against the seawall, further exacerbating erosion and ultimately destroying the seawall.

Beaches are the economic life-blood of many coastal communities, attracting visitors for a variety of recreational activities. Perhaps 70% of all vacationing Americans spend time at a beach. Breakwaters, jetties, groins, and seawalls are often constructed to preserve beaches, protect seaside property, and support the local economy. But for reasons discussed above, these strategies to *armor* or *harden* the coastline have serious drawbacks. In fact, these drawbacks convinced many coastal states to prohibit armoring of the coast to prevent beach erosion.

An alternate and widely used strategy to maintain existing beaches or restore badly eroded beaches is **artificial beach nourishment** (Figure 8.19). Sand is usually dredged from nearby lagoons or inlets or deep waters just offshore and then spread on the beach. In some cases, sand is trucked to the beach from an inland sand pit. On the plus side, artificial beach nourishment attempts to re-establish the natural balance between sediment inputs and outputs. On the negative side, beach nourishment is expensive. The new sand is not the same as the original and tends to erode at a faster rate so that beach nourishment is often short-lived (typically lasting 5 years or less) and must be repeated regularly to maintain the beach.

Today's large-scale beach nourishment projects are rooted in the Ash Wednesday storm of 1962 which caused considerable beach erosion along the New Jersey shore. Congress authorized the U.S. Army Corps of Engineers to repair the damage by dredging sand and pumping it on the beach. As of this writing beach nourishment projects have taken place on more than 125 East Coast beaches at a total cost in excess of $2.5 billion. In many cases, the process had to be repeated.

The transient nature of artificial beach nourishment is illustrated by what happened at Folly Beach, SC, a barrier island southeast of Charleston, SC. In September 1989, Hurricane Hugo's storm surge severely eroded the beach. In 1992, the U.S. Army Corps of Engineers spent $15 million to restore the beachfront through artificial beach nourishment. But only two months after the project was completed, a coastal storm washed away 80% of the new sand. Similar experiences are common along U.S. coastlines.

When erosion threatens shoreline buildings, strategic retreat is an alternative to artificial beach nourishment or armoring the coast with seawalls, groins or other structures. *Strategic retreat* refers to the physical relocation of a building inland. In 1999, this strategy was used to preserve the historic Cape Hatteras Lighthouse. (For more on this, refer to this chapter's third Essay.) In some cases, entire settlements are moved inland. For example, the town of Port Valdez in Alaska's Port Valdez fjord (terminus of an oil pipeline) was relocated after being heavily damaged by a tsunami triggered by the 1964 Alaska earthquake (Chapter 7). Originally, the town was situated at the head of the fjord and was moved to one side for greater protection.

FIGURE 8.19
Artificial beach nourishment project in Hawaii. [Courtesy of State of Hawaii, Department of Land and Natural Resources]

Human activity also has modified salt marshes and other coastal wetlands either intentionally or unintentionally. For centuries most people considered wetlands to be wastelands and reclaimed them by ditching, draining, or filling. In this way, former wetlands were developed for airports, agriculture, housing, and shopping malls. Over the past 250 years, more than half of the original wetland acreage in the United States was converted to other purposes. But we now know that wetlands provide habitats that are necessary to support coastal marine ecosystems and efforts are underway to reclaim damaged wetlands. For example, restoration efforts are underway to return the area of saltpans in southern San Francisco Bay to its original wetland condition by removing dikes, grading, and replanting grasses. Comparable projects are underway in the Netherlands where *polders* (reclaimed continental shelf areas) are also being restored to their original condition. Wetlands are complex ecosystems, however, and it is difficult to re-establish a wetland that will function as well as an undisturbed one.

Estuaries

An **estuary** is a partially isolated body of water where fresh water (from rivers and streams) mixes with seawater—from the lower extent of the photic zone to the most inland reach of tidal action. As pointed out in Chapter 1, estuaries are among the most productive ecosystems on Earth because of their special combination of physical and biological characteristics. They receive a continual supply of nutrients and organic matter delivered by river water and ocean tides. In Chapter 10, we focus on biological production in estuaries; in this chapter, we consider the origins of estuaries and their circulation patterns.

Most estuaries developed during the post-glacial (Holocene) sea-level rise that inundated the land drowning the mouths of rivers. Other estuaries developed through glacial erosion of mountain valleys, tectonic processes, or from the action of longshore sediment transport. In high latitude mountainous regions during the Pleistocene Ice Age, tongues of glacial ice occupied former river valleys and eroded those valleys below present sea level. With subsequent deglaciation and concurrent rise in sea level, ocean water flooded the valleys. Such a steep-walled enclosure of water is known as a **fjord** (Figure 8.20). Along some tectonically active coasts, the sea surged over down-faulted blocks of crust. And in other coastal locations, sand spits grew via littoral drift and partially isolated a body of water from the ocean.

FIGURE 8.20
A fjord along the mountainous coast of Norway. [Photo by J.M. Moran]

Although some mixing of fresh river water and seawater occurs in all estuaries, the characteristic circulation of an estuary depends on the amount of river inflow versus the strength of tidal currents. Based on circulation type, ocean scientists distinguish among salt-wedge estuaries, partially mixed estuaries, and well-mixed estuaries. At one extreme (salt-wedge estuaries), inflow of river water is the dominant factor and mixing is minimal and at the other extreme (well-mixed estuaries), tides are the dominant factor and mixing is extensive.

In estuaries where river inflow is relatively strong and tidal currents are weak, the low-salinity river water is quite distinct from the high salinity ocean water. That is, the two layers of water remain stratified. Where they meet, the wedge of less dense fresh water overlies the wedge of denser seawater; the thin vertical transition zone between the two layers is a *halocline*. This is known as a **salt-wedge estuary** (Figure 8.21). Seaward flow of the fresh water is much stronger than the landward movement of the seawater. The only mixing that takes place is due to internal waves that develop and propagate along the halocline (Chapter 7). Breaking of these internal waves mixes relatively small amounts of saltwater (from below) with the fresh water (above) so that strong stratification persists. During peak flow, a salt-wedge estuary forms near the mouth of the Columbia River but when river discharge is low, the salt-water wedge migrates upriver. Other examples of salt-wedge estuaries occur at the mouth of the Mississippi and Hudson Rivers. The Amazon River in Brazil is an extreme case of mixing in that the river's discharge is so great that seawater does not penetrate into the river's mouth and mixing of seawater and river water is confined to the continental shelf.

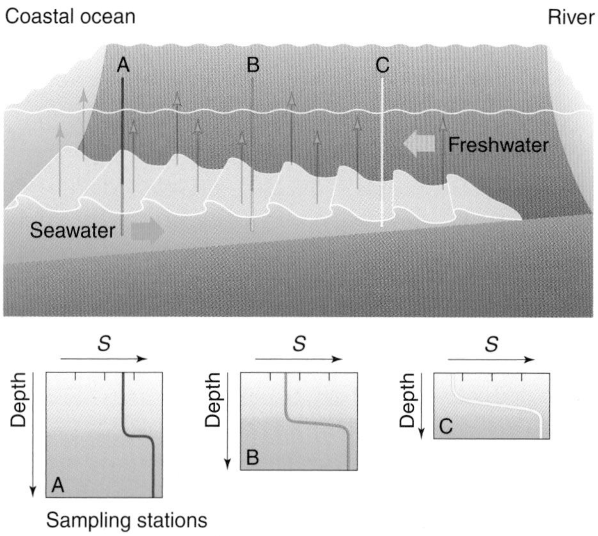

FIGURE 8.21
Circulation in a salt-wedge estuary. The charts at the bottom show the variation in salinity (S) with depth at three water-sampling stations.

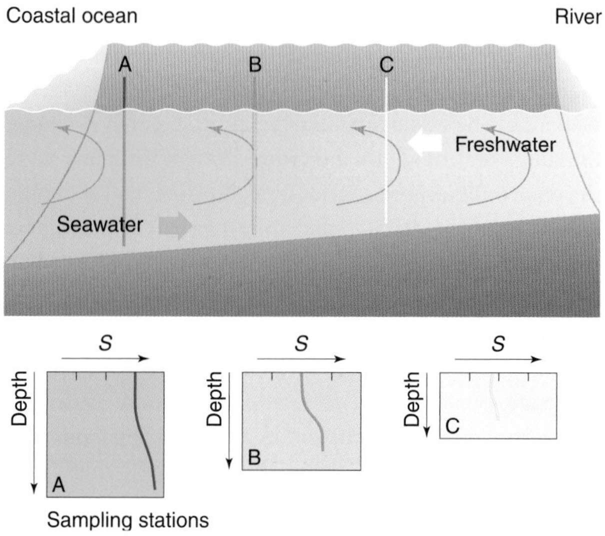

FIGURE 8.22
Circulation in a partially mixed estuary. The charts at the bottom show the variation in salinity (S) with depth at three water-sampling stations.

With declining river discharge and strengthening tidal currents, vertical mixing within an estuary becomes more vigorous so that stratification and the halocline weaken. In these **partially mixed estuaries**, salinity typically varies by less than 10 psu from the bottom water to the surface water (Figure 8.22). That is, the flow of fresh water seaward (above the halocline) and the flow of seawater landward (below the halocline) are both stronger than in a salt-wedge estuary. Chesapeake Bay is a classic example of a partially mixed estuary and is the largest estuary in the United States. It is almost 300 km (185 mi) long, 65 km (40 mi) at its broadest, and averages about 20 m (66 ft) deep. The estuary was formed during the post-glacial rise in sea level that flooded the ancient Susquehanna River Valley. The Bay receives about half its water from the Atlantic Ocean and the other half from the more than 150 rivers and streams draining a 166,000 square km (64,000 square mi) land area encompassing portions of New York, Pennsylvania, West Virginia, Delaware, Maryland, Virginia, and the District of Columbia. Other examples of partially mixed estuaries include San Francisco Bay and Puget Sound.

In a **well-mixed estuary**, strong tidal currents dominate the inflow from rivers and thoroughly mix the fresh water and saltwater. With no vertical stratification and no halocline, the only salinity gradient is a horizontal gradient from fresh water up estuary to seawater in the ocean. Delaware Bay (Figure 8.16) is an example of a well-mixed estuary.

In broad estuaries, the Coriolis Effect influences the circulation and mixing. In the Northern Hemisphere, both seaward and landward flowing currents are deflected to the right of their initial direction. That is, the two currents are deflected toward opposite banks of a broad estuary, such as Chesapeake Bay, resulting in fresh river water flowing to the sea on one side and saltwater flowing up river on the other side. This results in a tilted halocline and lateral mixing in contrast to the vertical mixing in estuaries that are too narrow to be significantly influenced by the Coriolis Effect.

Sediment distribution within an estuary depends on the relative strength of river discharge compared to tidal currents. River-borne sediments dominate salt-wedge estuaries whereas the sediment in well-mixed estuaries comes mostly from offshore sources. Partially mixed estuaries have a mixture of sediments from both sources. The two-layered circulation pattern in salt-wedge and partially mixed estuaries trap fine river-borne sediment. River-transported sediments eventually encounter the weaker current associated with the halocline. Coarser sediments settle to the bottom of the estuary while finer sediments (clays) remain suspended and concentrate near the halocline, forming a *turbidity maximum*.

In arid coastal regions that receive less fresh water input than is lost through evaporation, vertical circulation patterns are opposite those in most estuaries (Figure 8.23). In these systems, called *anti-estuarine circulations* or *reverse estuaries*, landward-flowing surface

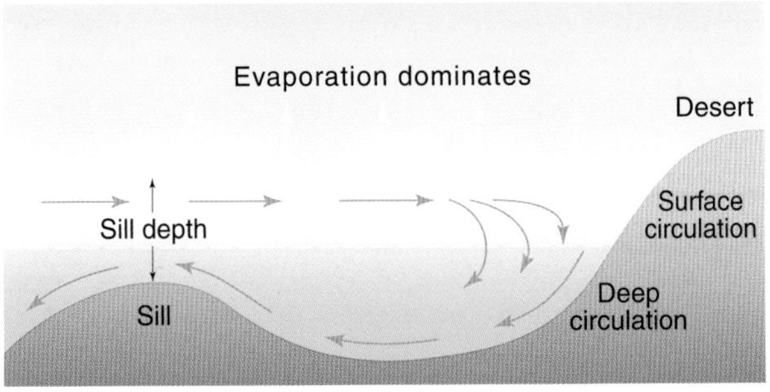

FIGURE 8.23
Anti-estuarine circulation characterizes the Mediterranean and Red Seas.

waters replace waters lost through evaporation. Intense evaporation produces highly saline waters that sink and flow seaward. Such circulation systems characterize the Mediterranean and Red Seas.

Coastal Storms and Storm Surge

People who live in the coastal zone are vulnerable to a variety of natural hazards and those hazards can be exacerbated by human activity. Among these coastal hazards are sea-level rise, tsunamis, land subsidence, beach or shoreline erosion, salt-water intrusion into aquifers, and storm surge. In this section we describe the storm surge that can be produced by coastal storms. In Chapter 12, we consider the threat of coastal flooding caused by climate change and rising sea level.

Atmospheric scientists distinguish between two types of storm systems that pose a threat to the coast: tropical cyclones and extratropical cyclones. Both of these low-pressure weather systems are characterized by surface winds that blow in a counterclockwise and inward direction as viewed from above in the Northern Hemisphere (Chapter 5). A **tropical cyclone** originates in a uniform mass of warm humid air over the tropical ocean but may track into higher latitudes; it includes hurricanes, tropical storms, and their precursors. An **extratropical cyclone** forms in middle latitudes and spends its entire life cycle in middle and high latitudes; its circulation brings together contrasting warm and cold air masses to form fronts and frontal weather (clouds and precipitation).

Potentially the most devastating impact of tropical and extratropical storms on the coastal zone is ocean water driven ashore by strong onshore winds (blowing from sea to land) associated with an intense storm system centered over or near the ocean. Strong winds (coupled with low air pressure) pile up a dome of seawater 80 to 160 km (50 to 100 mi) at its maximum width that sweeps across the coastline bringing floodwaters inland that can take lives and cause considerable property damage (Figure 8.24). A **storm surge** can erode beaches, overwash barrier islands, cut new tidal inlets, wash out roads and railway beds, and demolish marinas, piers, cottages, and other coastal structures.

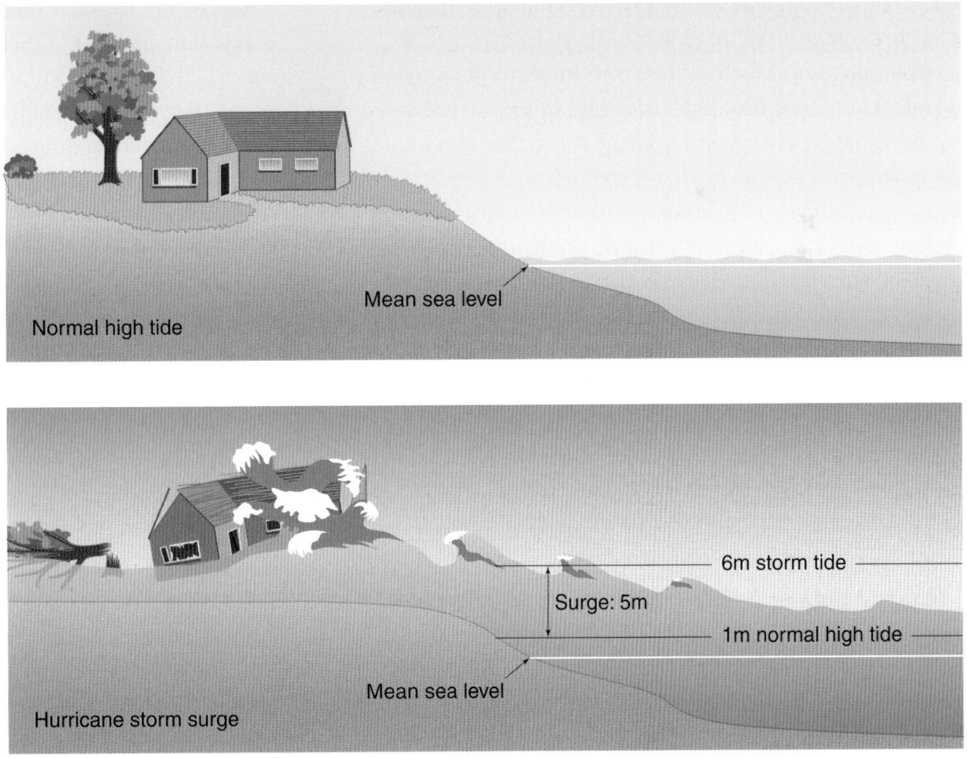

FIGURE 8.24
A storm surge is a dome of seawater 80 to 160 km (50 to 100 mi) across that is driven onshore by strong winds associated with tropical or extratropical cyclones. The storm tide is the sum of the storm surge and the prevailing astronomical tide. [Adapted from NOAA]

FIGURE 8.25
Damage produced at Gulfport, MS, by the storm surge of Hurricane Katrina in August 2005. [Courtesy of NOAA]

Many factors influence the potential for a storm surge, including storm intensity, timing of astronomical tides, bathymetry, and shoreline configuration. In general, a storm surge of 1 to 2 m (3 to 6.5 ft) can be expected with a weak hurricane, whereas the storm surge accompanying a violent hurricane may top 5 m (16 ft). Lower air pressure at the center of the storm contributes to the mounding of water, with sea level rising about 0.5 m (1.6 ft) for every 50-millibar drop in air pressure. A storm surge is superimposed on astronomical tides so that the impact is greatest when the storm surge coincides with high tide and especially spring tide (Chapter 7). In addition, wind-driven waves having heights of 1.5 to 10 m (5 to 33 ft) top the storm surge, further exacerbating the impact on coastal areas (Figure 8.25). All other factors being equal, the more gradual the slope of the ocean bottom in coastal areas, the greater is the storm surge. Low-lying coastal plains are most vulnerable to a storm surge especially if no barrier islands are present to dissipate the energy of the surge.

A numerical model developed in 1979 by NOAA predicts the location and height of a storm surge. Weather forecasters report considerable success with the **SLOSH (Sea, Lake, and Overland Surges from Hurricanes)** model. Analysis of SLOSH output and local topographic maps enables forecasters to identify areas most likely to be inundated by floodwaters when a hurricane threatens. Hypothetical hurricanes are simulated using various combinations of storm intensity, forward speed, and expected landfall location to predict the maximum high water level for various SLOSH basins along the coast. Public safety officials use this information in developing evacuation plans. SLOSH model coverage includes the entire U.S. East and Gulf Coasts, as well as parts of Hawaii, Guam, Puerto Rico, and the Virgin Islands.

Understanding storm surges and why some coastal areas are more prone to storm surges than others requires a closer look at the basic characteristics of tropical and extratropical cyclones.

Tropical Cyclones

The most catastrophic storm surges are associated with hurricanes that make landfall, that is, hurricanes that track from sea to land. In the most deadly natural disaster in U.S. history, more than 8000 people perished, mostly by drowning, when a hurricane storm surge devastated Galveston, TX, on 8 September 1900. Ocean waters flooded the entire city located on a barrier island. Some 3600 homes were destroyed; a dam of wreckage that built up some six blocks inland from the beach probably spared the business district total destruction. To protect the city from future storm surges, the U.S. Army Corps of Engineers elevated the city an average of 3 m (10 ft) and erected a massive concrete seawall along the city's Gulf shore.

On the evening of 28 August 2005, the effects of the approaching Hurricane Katrina were beginning to be felt along the north central Gulf Coast as the storm's outer rain bands swept ashore. Within 24 hrs, Katrina's ferocious winds, storm surge, and heavy rains would devastate the Gulf Coast of Louisiana and Mississippi, obliterating coastal communities and devastating New Orleans. According to the National Weather Service, Katrina was the third most intense hurricane (minimum central air pressure of 920 mb) to make landfall on the United States since reliable records began in 1851. Katrina claimed at least 1500 lives and, in terms of economic loss, Katrina ranks as the most destructive hurricane to strike the United States.

Katrina began in the southeastern Bahamas on 23 August 2005 as the twelfth tropical depression of the season; the next day the system strengthened to a tropical storm and was assigned its name. Katrina intensified as it slowly moved northwestward and then westward through the Bahamas. On 25 August, Katrina made landfall in south Florida as a category 1 hurricane on the Saffir-Simpson Hurricane Wind Scale (described later in this chapter), producing maximum sustained winds of 130 km per hr (80 mph) and heavy rain (more than 12.5 cm or 5 in. in southeast Florida). Tracking almost due west and fueled by the warm waters of the Gulf of Mexico, the storm quickly intensified and gradually turned toward the northwest and then north. By the afternoon of 26 August, Katrina was a major hurricane. As a precaution, public officials issued mandatory evacuation orders for the people of New Orleans, LA, Gulfport, MS, and sections of Mobile, AL. Earlier, personnel on platforms and oil rigs in the Gulf had been evacuated.

By the morning of 28 August, Katrina (Figure 8.26) was a category 5 hurricane with maximum sustained surface winds of 282 km per hr (175 mph). Katrina weakened to a strong category 3 prior to making landfall on the morning of 29 August in southern Plaquemines Parish just south of the town of Buras, LA, at the mouth of the Mississippi River. Maximum sustained winds were 202 km per hr (125 mph) to the east of the storm center with hurricane force winds extending 195 km (120 mi) from its center, driving a 6 to 9 m (20 to 30 ft)

FIGURE 8.26
NOAA geostationary satellite image of Hurricane Katrina taken on Aug. 28, 2005, at 11:45 a.m. EDT, as the powerful storm churned in the Gulf of Mexico as a category 5 storm with sustained winds near 175 mph, a day before the storm made landfall on the U.S. Gulf Coast. [Courtesy of NOAA.].

storm surge that reached well inland (to the 30-ft contour line) along the Louisiana and Mississippi Gulf Coast and as far east as Mobile, AL, flooding parts of the city. Several hours later, after passing over Breton Sound and Lake Borgne, the storm made another landfall at the mouth of the Pearl River on the Louisiana-Mississippi border. By then, Katrina had weakened somewhat but was still a category 3. At 11 p.m., EDT, the north/northeast moving system was centered near Columbus, MS, downgraded to a tropical storm. Along much of Katrina's path, total rainfall exceeded 20 to 25 cm (8 to 10 in.).

Topography made New Orleans particularly vulnerable to Hurricane Katrina. The city occupies a bowl between the Mississippi River and Lake Pontchartrain. Much of the bowl, home to more than 1 million residents and businesses, is up to 1.8 m (6 ft) below sea level. The people of New Orleans depend on earthen levees and concrete floodwalls to keep the water out and pumps to remove flood waters. However, the combination of strong winds, storm surge, and heavy rainfall caused numerous breaches in the levee system, the pumps failed, and up to 80% of New Orleans was flooded to depths that in many places reached 6 m (20 ft).

Although warned in advance of the approaching potentially catastrophic storm, thousands of people either were unable or chose not to evacuate New Orleans. As weather conditions deteriorated and floodwaters began to rise, many people fled to the Louisiana Superdome, the city's shelter of last resort. On 29 August, winds gusting to more than 160 km per hr (100 mph) ripped off part of the roof of the Superdome and knocked out electricity. An estimated 30,000 people spent many days in overcrowded, deplorable conditions without adequate food, water, sanitary facilities, or air conditioning. Other residents of New Orleans fled the rapidly rising waters, climbing to the upper floors, attics or roofs of their homes where they awaited rescue by boat or helicopter. The damage to New Orleans was so extensive that recovery is still ongoing.

HURRICANE CHARACTERISTICS

Hurricane is likely derived from *Haracan*, the name of the storm god of the Taino people who inhabited Caribbean islands at the time of Spanish exploration of the New World. A **hurricane** is an intense cyclone that originates over tropical ocean waters, usually in late summer or early fall (when sea-surface temperatures are highest), and has a maximum sustained wind speed of at least 119 km per hr (74 mph). By convention in the United States, *sustained wind speed* is a one-minute average measured at the standard anemometer height of 10 m (33 ft).

A hurricane develops in a uniformly warm and humid air mass. Typically, the central pressure at sea level is considerably lower and the horizontal air pressure gradient is much greater in a hurricane than an extratropical cyclone. A hurricane is usually a much smaller system, averaging a third the diameter of a typical extratropical cyclone. Rarely do hurricane-force winds extend much more than 120 km (75 mi) beyond the system's center. The circulation in a hurricane weakens rapidly with altitude and usually becomes anticyclonic (clockwise when viewed from above in the Northern Hemisphere) in the upper troposphere at altitudes above about 12,000 m (40,000 ft).

At the center of a hurricane is an area of almost cloudless skies, subsiding air, and light winds (less than 25 km per hr or 16 mph), called the **eye** of the storm. The eye generally ranges from 10 to 65 km (6 to 40 mi) across, shrinking in diameter as the hurricane intensifies and winds strengthen. At a hurricane's typical rate of forward motion, the eye may take up to an hour to pass over a given locality. People are sometimes lulled into thinking the storm has ended when skies clear and winds abruptly slacken following a hurricane's initial blow. They may be experiencing passage of the hurricane's eye; heavy rains and ferocious winds soon will resume but blow from the opposite direction.

Bordering the eye of a hurricane is the **eye wall**, a ring of thunderstorm (cumulonimbus) clouds that produce heavy rains and very strong winds. The most dangerous and potentially most destructive part of a hurricane is the portion of the eye wall on the side of the advancing system where the wind blows in the same direction as the storm's forward motion. On that side, hurricane winds combine with the storm's forward motion producing the system's strongest surface winds. In the Northern Hemisphere, this dangerous semicircle of high winds and high ocean waves is on the right side of the hurricane when facing in the direction of the system's forward movement. Cloud bands, producing heavy convective showers and hurricane-force winds, spiral inward toward the eyewall.

WHERE AND WHEN

Three conditions are necessary for a tropical cyclone to form: high sea-surface temperatures, adequate Coriolis Effect, and weak winds aloft. In addition, relatively humid air in the mid-troposphere (near 4900 m or 16,000 ft) favors tropical cyclone development. To a large extent, these requirements dictate where and when tropical cyclones form.

Tropical cyclone formation requires a sea-surface temperature (SST) of at least 26.5 °C (80 °F) through an ocean depth of about 45 m (150 ft) or more. Such exceptionally warm ocean waters sustain the system's circulation by the latent heat released when water vapor, evaporated from the ocean surface, is conveyed upward and condenses within the storm system. Temperature largely governs the rate of evaporation of water, so the higher the SST, the greater is the supply of latent heat for the storm system. Furthermore, the spray from breaking ocean waves readily evaporates, adding to the supply of water vapor that condenses and releases latent heat in the developing tropical cyclone.

As a tropical cyclone makes landfall or moves over colder water, however, it loses its warm-water energy source and weakens. The strong winds of a tropical cyclone stir up surface ocean waters. Cyclonic winds induce Ekman transport and divergence of surface waters under the storm system, bringing cold water to the surface. Until the normal thermal structure of the ocean is restored, lower than usual SST can inhibit development of subsequent tropical cyclones over the same region of the ocean. Water welling up to the surface is also nutrient-rich spurring biological production (stimulating algal blooms) often in portions of the ocean (e.g., Sargasso Sea in the Atlantic east of Bermuda) that normally are low in productivity. In 2004, Steven M. Babin and his colleagues at Johns Hopkins University reported that their analysis of satellite imagery revealed an enhancement of chlorophyll concentration at the ocean surface in the wake of each of the 13 hurricanes that passed over the Sargasso Sea from 1998 to 2001.

In recent years, atmospheric scientists discovered that changes in SST associated with warm- and cold-core ocean rings influence tropical cyclone development. Recall from Chapter 6 that rings are eddies that break off western boundary currents such as the Gulf Stream. Tropical cyclones may intensify over warm-core rings and weaken over cold-core rings. For example, the Loop Current (Figure 6.10) in the Gulf of Mexico is a source of warm-core rings that may intensify tropical cyclones passing overhead.

The second condition required for tropical cyclone development is a significant Coriolis Effect; that is, the influence of Earth's rotation must be sufficiently strong to initiate a cyclonic circulation. As noted in Chapter 5, the Coriolis Effect weakens toward lower latitudes and is zero at the equator. With very rare exceptions, tropical cyclones do not form within 5 degrees of the equator (a distance of about 480 km or 300 mi).

The first two conditions that favor tropical cyclone formation (i.e., high SST and sufficient Coriolis Effect) occur only over certain portions of the ocean. The main ocean breeding grounds for tropical cyclones along with average storm trajectories are plotted in Figure 8.27. Most hurricanes form in the 8- to 20-degree latitude belts. Major breeding grounds are: (1) the tropical North Atlantic west of Africa (including the Caribbean Sea and Gulf of Mexico), (2) the North Pacific Ocean west of Mexico, (3) the western tropical North Pacific and China Sea, where a hurricane is called a *typhoon* (from the Cantonese *tai-fung*, meaning great wind), (4) the South Indian Ocean

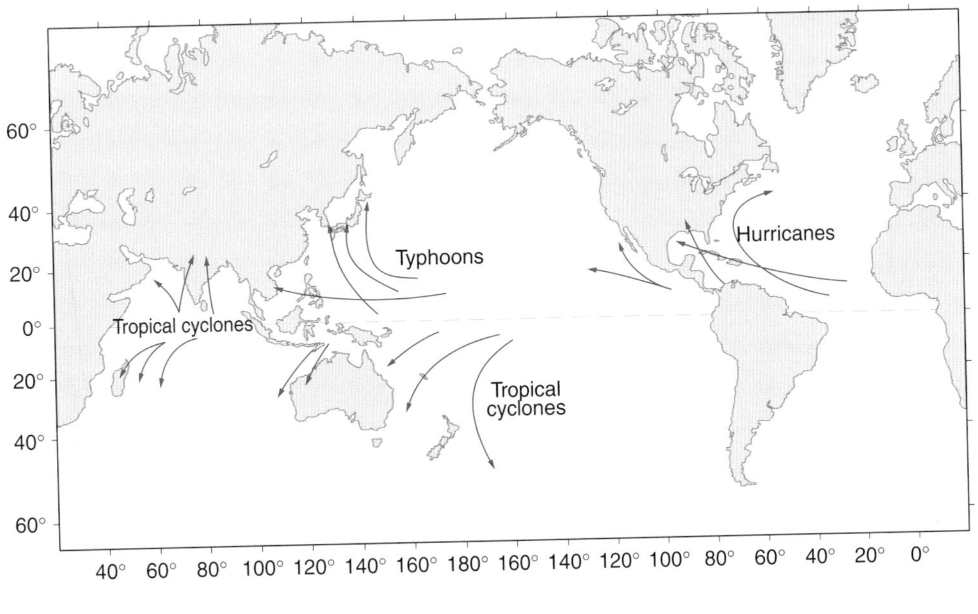

FIGURE 8.27
Tropical cyclone breeding grounds are located only over certain regions of the world ocean. Arrows indicate average hurricane trajectories.

east of Madagascar, (5) the North Indian Ocean (including the Arabian Sea and Bay of Bengal), and (6) the South Pacific Ocean from the east coast of Australia eastward to about 140 degrees W. In the Indian Ocean and near Australia, hurricanes are simply called *cyclones*.

The requirement of high sea-surface temperatures explains the seasonal occurrence of tropical cyclones (Figure 8.28). For reasons presented in Chapter 3, the temperature of surface ocean waters lags the regular seasonal variations in incoming solar radiation. Sea-surface temperatures reach a seasonal maximum roughly 6 to 8 weeks after the date of most intense solar radiation. Most Atlantic hurricanes develop when surface waters are warmest, that is, in late summer and early autumn; the official hurricane season runs from 1 June to 30 November, with the peak hurricane threat for the U.S. coastline between mid-August and late October.

The third condition for tropical cyclone development is relatively weak winds aloft (in the middle and upper troposphere) over oceanic breeding grounds. Weak winds aloft allow a cluster of cumulonimbus clouds to organize over tropical seas, the first step in the evolution of a hurricane. By contrast, strong west-to-east winds aloft shear off the tops of westward tracking thunderstorms, preventing the systems from building vertically and organizing. Strong vertical *wind shear* is the principal reason hurricanes rarely form off the east or west coasts of South America (although Caribbean hurricanes occasionally impact the north coast of Venezuela).

According to the *International Best Track Archive for Climate Stewardship (IBTrACS)* team at the World Data Center for Meteorology, operated under the auspices of NOAA's National Climatic Data Center, for the 30-year period from 1980-2009, the global annual average of named tropical cyclones is 86.5; slightly more than half of those storms, 45.4, develop into a hurricane, typhoon, or cyclone (Category 1-3), and of those, 21.9 reach major storm status (Category 4-5). The western Pacific Ocean, with its vast expanse of warm surface waters, is the most active area for tropical cyclones, with an average of 26.3 systems each season (1980-2009). More than half of those storms, 15.9, develop into a typhoon (Category 1-3), and of those, 8.0 reach major storm status (Category 4-5)

Hurricanes spawned over the tropical Atlantic, Caribbean Sea, and Gulf of Mexico pose the most serious threat to coastal North America. According to the National Hurricane Center, based on the period 1966-2010, a seasonal average 11.4 named tropical storms form over these waters. Of these systems, on average, 6.3 intensify into hurricanes and of these 2.4 become major hurricanes. On average, two hurricanes strike the U.S. coast each year. The 2005 Atlantic hurricane season set a record with 27 named tropical cyclones (and one unnamed subtropical cyclone); the previous record was 21 set in 1933. In 2005, 15 Atlantic tropical cyclones became hurricanes (a record) and 4 major hurricanes struck the U.S. (also a record). The 1995 Atlantic hurricane season was the third most active in recorded history with 19 tropical storms, 11 of which

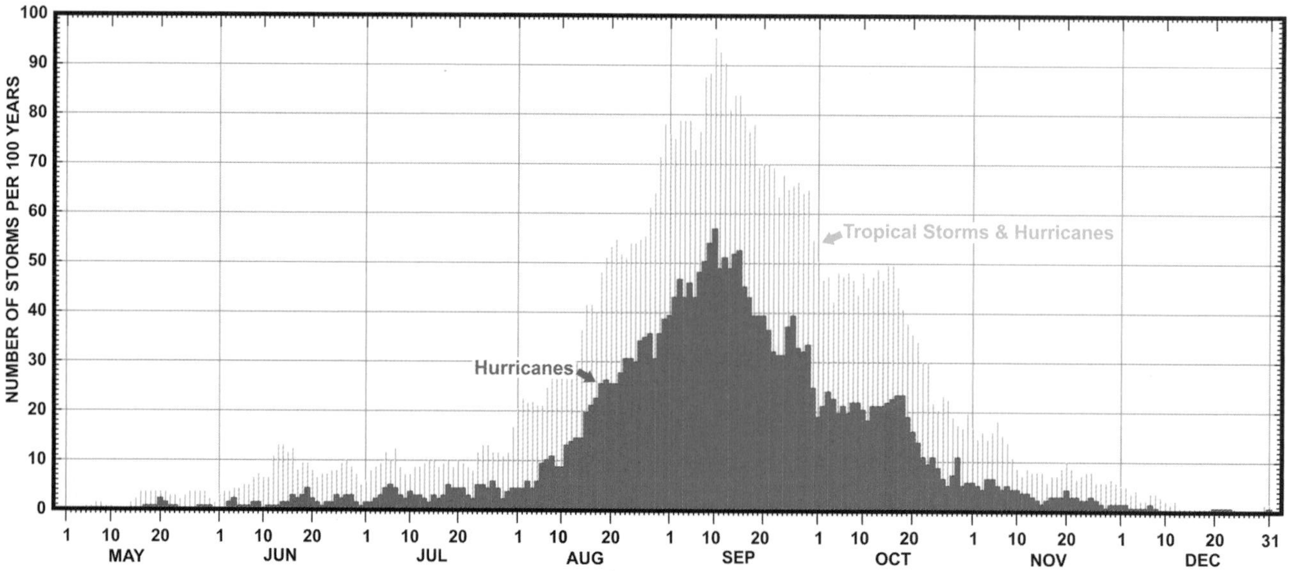

FIGURE 8.28
The frequency of Atlantic basin tropical storms and hurricanes by date (number of storms per 100 years). The date of peak frequency occurs about 2.5 months following the summer solstice. [Courtesy of NOAA]

attained hurricane strength. However, the annual number of hurricanes may have little bearing on the number of landfalling hurricanes and their impact. Whereas only 4 hurricanes occurred during the 1992 Atlantic season, one of them, Hurricane Andrew, was the second most costly in terms of property damage in U.S. history, amounting to $34.3 billion (in year 2000 dollars).

Hurricanes have hit every Atlantic and Gulf Coast State from Texas to Maine. Florida is the most hurricane-prone of all the states, with 66 hurricanes crossing its coastline between 1900 and 2010. In the same period, Texas was second with 41 hurricanes, while Louisiana had 31, and North Carolina had 29.

The primary breeding ground for hurricanes in the Atlantic shifts east and west with the seasons. Early in the hurricane season (May and June), hurricanes form mostly over the Gulf of Mexico and the western Caribbean. By July, the main area of hurricane development begins to shift eastward across the tropical North Atlantic. By mid-September, most hurricanes form in a belt stretching from the Lesser Antilles (in the eastern Caribbean Sea) eastward to south of the Cape Verde Islands (off Africa's West Coast). After mid-September, hurricanes again originate mostly over the Gulf of Mexico and the western Caribbean.

According to the National Hurricane Center, the Pacific Ocean off Mexico and Central America ranks second to the western North Pacific in average annual number of tropical cyclones (15.1 per year between 1966 and 2010); the majority of these (8.4) develop into hurricanes. The Pacific Coast of the United States is rarely a target of hurricanes, although one or two tropical cyclones typically make landfall on the Mexican Pacific Coast each year. Prevailing winds (northeast trades) are directed offshore and usually steer tropical cyclones that form west of Central America away from the coast. Also, the southward flowing cold California Current plus upwelling just off the southern California (and Baja California) coast produce sea-surface temperatures that normally are too low to sustain hurricanes that travel toward the northeast. However, during unusual atmospheric/oceanic circulation regimes, hurricanes have struck coastal Southern California and even traveled over the Desert Southwest.

The Hawaiian Islands are sometimes threatened by tropical cyclones that develop over the central tropical Pacific or track into that region from the Pacific hurricane breeding grounds west of Mexico. Fortunately, in an average year, only 3 to 4 tropical cyclones affect the central Pacific and since 1957 only 4 hurricanes have impacted the islands. However, the most recent one, Iniki in September 1992, devastated the island of Kauai. Seven people lost their lives. Total property damage was estimated at $2.3 billion (in year 2000 dollars), making this the most costly natural disaster in the history of the State of Hawaii.

HURRICANE LIFE CYCLE

The first sign that a hurricane may be forming is the appearance of an organized cluster of thunderstorm clouds over tropical seas. This region of convective activity is labeled a **tropical disturbance** if a center of low pressure is detected at the surface. If atmospheric/oceanic conditions favor hurricane development and if those conditions persist, the surface air pressure falls and a cyclonic circulation develops. Water vapor condenses within the storm, releasing latent heat of vaporization, and the heated buoyant air rises. Expansional cooling of the ascending air triggers more condensation, release of even more latent heat, and an additional increase in buoyancy. Rising temperatures in the core of the storm, coupled with an anticyclonic outflow of air aloft, cause a sharp drop in surface air pressure, which in turn, induces more rapid convergence of humid air at the surface. The consequent uplift surrounding the developing *eye* leads to additional condensation and release of latent heat.

Through this process, a tropical disturbance intensifies and its winds strengthen. When maximum sustained wind speeds reach 37 km per hr (23 mph) or higher, the developing system is called a **tropical depression**. When maximum sustained wind speeds reach at least 63 km per hr (39 mph), the system is classified as a **tropical storm** and assigned a name, such as Ann or Bill. Once maximum sustained winds reach 119 km per hr (74 mph) or higher, the storm is officially designated a hurricane. As a hurricane weakens and decays, the system is downgraded by reversing this classification scheme.

Hurricanes that form over the Atlantic Ocean near the Cape Verde Islands usually drift slowly westward with the trade winds (along the southern flank of the Bermuda-Azores subtropical high) across the tropical North Atlantic and into the Caribbean. At this stage in the storm's trajectory, it is not unusual for the system to travel at a mere 10 to 20 km per hr (6 to 12 mph) and take a week to cross the Atlantic. Once over the western Atlantic, however, the forward speed of the storms usually increases and the storm begins curving northward along the western flank of the subtropical high, and then northeastward as the system enters the middle latitude westerlies. Precisely where this curvature takes place determines whether the hurricane enters the Gulf of Mexico (perhaps then

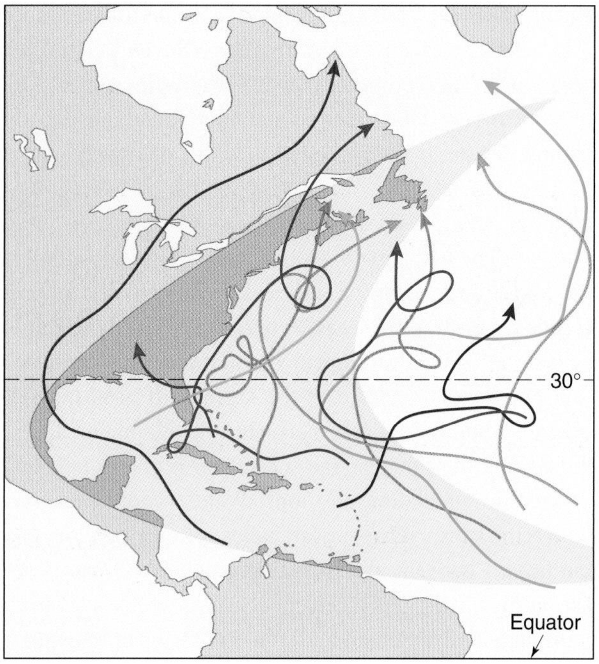

FIGURE 8.29
Tropical cyclone trajectories are sometimes erratic as shown by these samples. As indicated by the darker blue shaded area, however, most Atlantic tropical cyclones initially drift westward and then curve toward the north and northeast when they reach the western Atlantic. [From NOAA, *Hurricane*, Washington, DC: Superintendent of Documents, 1977.]

tracking up the lower Mississippi River Valley or over the Southeastern States), moves up the Eastern Seaboard, or curves northeastward into the Atlantic.

Upon reaching about 30 degrees N, an Atlantic hurricane may begin to acquire extratropical characteristics as colder air circulates into the system and fronts develop. From then on the storm resembles an extratropical cyclone and completes its life cycle, usually over the North Atlantic. Many hurricanes, however, depart significantly from the track just described. Some of the hurricane tracks plotted in Figure 8.29, for example, are very erratic. A hurricane can describe a complete circle or reverse direction. In addition, some hurricanes, fueled by warm Gulf Stream waters, maintain tropical characteristics far north along the Atlantic Coast. Coastal New England and Atlantic Canada, for example, have been the targets of hurricanes.

HURRICANE HAZARDS

In addition to storm surge (discussed earlier in this chapter), hazards of hurricanes are heavy rain and inland flooding, strong winds, and tornadoes. Hurricanes and tropical storms can produce very heavy rainfall with amounts typically in the range of 13 to 25 cm (5 to 10 in.). Even when the system tracks well inland, heavy rains often persist and may trigger costly and life threatening flooding. According to research conducted by scientists at NOAA's National Hurricane Center in Miami, FL, freshwater flooding was responsible for almost 60% of the 600 U.S. deaths attributed to tropical cyclones or their remnants during the period from 1970 to 1999. In those three decades, far more people (351) died from inland flooding than from coastal storm surge flooding (only 6 deaths). On the other hand, the majority of the more than 1500 fatalities associated with Hurricane Katrina in August 2005, were caused by the storm surge. The storm surge remains as the most serious potential impact of a landfalling hurricane and is the primary reason people are evacuated from low-lying coastal areas and barrier islands.

From 1970 to 1999, about 12% of tropical cyclone fatalities were wind-related. Winds pushing on the outside walls of buildings exert a pressure that increases dramatically as winds strengthen. Furthermore, debris carried by the wind and hurled against structures exacerbates the damage potential of strong winds. Hurricane winds diminish rapidly once the system makes landfall, so that most wind damage is confined to within about 200 km (125 mi) of the coastline. Two factors account for the abrupt drop in wind speed once a hurricane makes landfall. A hurricane over land is no longer in contact with its energy source, warm ocean water. In addition, the frictional resistance offered by the rougher land surface slows the wind and shifts the wind direction toward the low-pressure center of the system. This wind shift causes the storm to begin to fill; that is, its central pressure rises, the horizontal air pressure gradient weakens, and winds slacken.

Although wind speed decreases once a hurricane makes landfall, the system may produce tornadoes. (A *tornado* is a small column of air that whirls rapidly and violently about a nearly vertical axis and is made visible by clouds, dust and debris). Usually only a few tornadoes occur with a hurricane but in 1967, Hurricane Beulah reportedly spawned as many as 115 tornadoes across southern Texas. Tornadoes are most probable after the hurricane enters the westerly steering current and curves towards the north and northeast; they form mostly to the northeast of the storm center, often outside the region of hurricane-force winds.

In the early 1970s, H.S. Saffir (1917-2007), a consulting engineer, and R.H. Simpson, former director of the National Hurricane Center, designed the Saffir-Simpson Hurricane Intensity Scale, a rating system

TABLE 8.1
Saffir-Simpson Hurricane Wind Scale

Category	*Wind speed km/hr (mph)*	*Damage potential*
1	119-153 (74-95)	Very dangerous winds will produce some damage
2	154-177 (96-110)	Extremely dangerous winds will cause extensive damage
3	178-209 (111-130)	Devastating damage will occur
4	210-249 (131-155)	Catastrophic damage will occur
5	> 249 (> 155)	Catastrophic damage will occur

for hurricanes. First used in hurricane advisories in 1975, this scale identified hurricanes in five categories (1 to 5) by increasing intensity. The categories were based on maximum sustained wind speed and the potential for property damage, and provided the range of central air pressure and storm surge potential. In fact, during the 1970s and 1980s, central pressure was used as a proxy for wind speed. Since 1990, aircraft reconnaissance has directly measured both air pressure and wind speed, negating the need to include pressure in the scale. Storm surge ranges in the original scale could be misleading because factors such as bathymetry, topography and hurricane size, forward speed and angle to the shoreline had different local impacts that were not immediately evident from the reported category. Updated in early 2010, the **Saffir-Simpson Hurricane Wind Scale** (Table 8.1) removed air pressure and storm surge potential, providing a more scientifically accurate description of hurricane impacts that reduces public confusion.

Of the 185 hurricanes that struck the U.S. Atlantic or Gulf Coasts between 1901 and 2010, 69 (37.3%) were classified as major; that is, they rated 3 or higher on the Saffir-Simpson scale. Property damage potential rises rapidly with ranking on the Saffir-Simpson scale. In fact, destruction from a category 4 or 5 hurricane can be 100 to 300 times greater than that caused by a category 1 hurricane. The 25 major hurricanes that made landfall along the Gulf or Atlantic Coasts between 1949 and 1990 accounted for three-quarters of all property damage from all landfalling tropical storms and hurricanes during the same period.

Scientists at the National Hurricane Center watch for the development of tropical cyclones over the Atlantic, Caribbean, Gulf of Mexico, and the Eastern Pacific. The Center operates the SLOSH model for prediction of storm surges, prepares and distributes hurricane watches and warnings for the public, conducts research on hurricane forecasting techniques, and sponsors public awareness programs.

EVACUATION

From 1981 through 2000, only ten major hurricanes (category 3 or higher) struck the U.S. mainland. This was similar to the 1971 through 1990 period when only nine major hurricanes made landfall on the U.S. coast. From 10 August 1980 to 17 August 1983, no hurricanes struck the United States. The infrequency of major hurricanes during the 1970s and 1980s lulled many coastal residents of the Southeast U.S. into a false sense of security and encouraged development and population growth in areas that could be devastated by a major hurricane. More and more resort hotels, high-rise condominiums, and expensive homes were constructed perilously close to the shoreline and even among coastal sand dunes (Figure 8.30).

About 45 million permanent residents inhabit hurricane-prone portions of the nation's coastline, and the population continues to climb. Until the recent upturn in hurricane frequency, most Atlantic and Gulf Coast residents never experienced the full impact of a major hurricane. Some of them may have weathered with relative ease a weak hurricane or the fringes of a strong system. But such an experience may lull them into complacency so that they are less likely to prepare adequately should a major hurricane threaten. Compounding the problem of the growth of the resident population in the Southeast is the arrival of holiday, weekend, and seasonal visitors to seaside resorts. The human population in some of these locales swells ten- to one-hundred-fold during vacation periods. Many of these resorts are in low-lying coastal areas or on barrier island beaches that are subject to rapid inundation by a storm surge.

A

B

FIGURE 8.30
The relative infrequency of hurricanes in the 1970s and 1980s lulled some residents of the U.S. Gulf and East Coast into a false sense of security and inspired construction of resort hotels (A) and vacation homes (B) within meters of high-tide level. [Photo by J.M. Moran]

Evacuation of people from barrier islands, as well as other low-lying coastal areas, is the traditional strategy in the event of a major storm threat (Figure 8.31). Since the 1985 hurricane season, evacuation has proven its value in lives saved. But the potential downside of evacuation was illustrated in September 1999 when Floyd, an unusually massive hurricane, threatened much of the Eastern Seaboard. More than 2 million residents of the coastal area from south Florida to South Carolina took to the roads and fled inland. In many areas the result was gridlock—too many vehicles on too few highways. As it turned out, Floyd spared most of the evacuated area and made landfall near Cape Fear, NC, on 16 September 1999 as a category 2 system. Torrential rainfall, totaling as much as 38 to 50 cm (15 to 20 in.) over portions of eastern North Carolina and Virginia caused extensive inland flooding, 56 deaths, and property damage of at least $6 billion.

Successful evacuation of coastal communities hinges on sufficient advance warning of a hurricane's approach but hurricanes are notorious for sudden changes in direction, forward speed, and intensity. (In retrospect, perhaps 75% of all coastal storm evacuations prove to be unnecessary.) A hurricane's erratic behavior is especially troublesome for people living in isolated localities (e.g., a barrier island linked to the mainland by a single bridge) and congested cities where highway systems have not kept pace with population growth. In such places the time required for evacuation may be lengthy. The Federal Emergency Management Agency (FEMA) estimates evacuation times during the peak tourist season as up to 50-60 hrs for New Orleans, LA, Ocean City, MD, and Fort Myers, FL, up to 30-39 hrs for the Florida Keys, the Outer Banks of North Carolina, Cape May County, NJ, and Atlantic City, NJ, and up to 20-29 hrs for Long Island, NY, and Galveston, TX.

As coastal communities continue to grow, the time required for evacuation of their population lengthens. Evacuation must begin earlier when a storm is farther away and greater uncertainty surrounds its likely track. Such uncertainty necessitates a broader zone of evacuation that translates into greater economic losses associated with evacuation (e.g., closed businesses). Cognizant of the problem of lengthy evacuation times, some coastal communities are considering the option of *vertical*

FIGURE 8.31
In hurricane-prone areas, special road signs mark evacuation routes. [Photo by J.M. Moran]

evacuation, moving people to shelters on the upper floors of well-constructed public buildings. Vertical evacuation may be a viable alternative to sending people fleeing on congested roads and highways for large urban areas where evacuation options are limited.

Extratropical Cyclones

Of all the extratropical cyclones to affect the U.S. coastal zone, nor'easters generally receive the most attention. With strong winds and heavy precipitation, they have the greatest potential for impacting large numbers of people. A **nor'easter,** named for the direction from which the most powerful winds blow, is an intense extratropical cyclone that tracks northward along the East Coast of North America. Unlike hurricanes, nor'easters begin as relatively weak weather systems over land, forming along boundaries (*fronts*) between air masses with contrasting temperatures and humidity. Like hurricanes, they often intensify over warm ocean waters. Nor'easters are most frequent from October through April, when the seasonal contrasts in air mass characteristics are greatest. The storm tracks toward the northeast and, if centered just offshore, strong onshore winds on the northern and northwestern flanks of the system (blowing from the east and northeast) can cause a storm surge and high waves resulting in considerable coastal erosion, flooding, and property damage.

The essential ingredients for a nor'easter to form and intensify along or near the coast are: low-level temperature contrast between land and sea, areas of rising air, abundant moisture (supplied by the Gulf of Mexico or Atlantic Ocean), and a pool of Arctic air (a cold high-pressure system) to the north or northwest. Two jet streams, corridors of exceptionally strong winds in the upper troposphere, the subtropical jet stream over the southeast U.S. and the polar jet stream to the north, work in tandem to induce and sustain regions of rising air throughout the troposphere. Copious amounts of snow come with a pronounced southerly loop in the polar jet stream, allowing for south-north oriented flow along the Eastern seaboard. This circulation pattern not only channels cold air into the coastal storm, it also slows the forward speed of the system and intensifies it. The heaviest snow usually falls in one or more bands located 160 to 320 km (100 to 200 mi) northwest of the storm center. These conditions most often come together during winters characterized by the warm phase of the El Niño-Southern Oscillation and the negative phase of the North Atlantic Oscillation/Arctic Oscillation (Chapter 11).

Although rarely attaining the strength of even a category 1 hurricane, a nor'easter can pack a powerful punch. In an intense nor'easter, winds can produce wave heights of 1.5 to 10 m (5 to 33 ft) on top of a 5 m (16 ft) storm surge. If the path of the nor'easter's center follows high tide, flooding further increases. Because, as noted earlier, the diameter of an average extratropical cyclone is three times that of an average hurricane, a nor'easter impacts a much greater swath of coastline than a typical East Coast hurricane. Whereas a hurricane typically tracks directly from sea to land, impacting only about 100 to 150 km (60 to 90 mi) of coastline, a nor'easter often moves parallel to the coast so its onshore winds can sweep over more than 1500 km (900 mi) of coastline. In fact, a less intense but slower-moving nor'easter causes more damage than a more intense but faster-moving nor'easter. Winds in the slow-moving system blow in the same direction for a longer period than the fast-moving system, and persistent winds make a destructive storm surge and high waves more likely.

Strengthened by latent heat acquired from evaporation of relatively warm ocean water, some nor'easters develop very rapidly. By convention, a rapidly intensifying extratropical cyclone is labeled a *bomb* if its central pressure drops at least 24 millibars in 24 hrs. Few cyclones actually meet this criterion, and most of those that do, develop over warm ocean currents such as the Gulf Stream or the Kuroshio Current. During early January 1989, an extreme example of an East Coast bomb developed. The incipient cyclone was first identified off of Cape Hatteras, NC, at about 7:00 p.m. (EST) on 4 January with a central air pressure of 996 mb. Only twenty-four hours later, the storm was centered 700 km (435 mi) south of Newfoundland with a central pressure of 936 mb. The storm had intensified (pressure decreased) by 60 mb, 2.5 times the criterion for a bomb.

One of the most noteworthy nor'easters of the 20th century occurred on 12-15 March 1993 when a major storm system tracked from the Gulf of Mexico up the Eastern Seaboard (Figure 8.32). Although the system's size and minimum central pressure were not particularly unusual, its human impact was enormous because of its track along the densely populated East Coast. With this track, the storm drew into its circulation a tremendous amount of moist air from the warm ocean and, with the large pool of cold air inland, produced heavy snow to the north and west of its track from Alabama to Maine (Figure 8.33). In addition to property damage caused by heavy snow and high winds, a *squall line* (an elongated group of strong to severe thunderstorms) associated with the storm produced

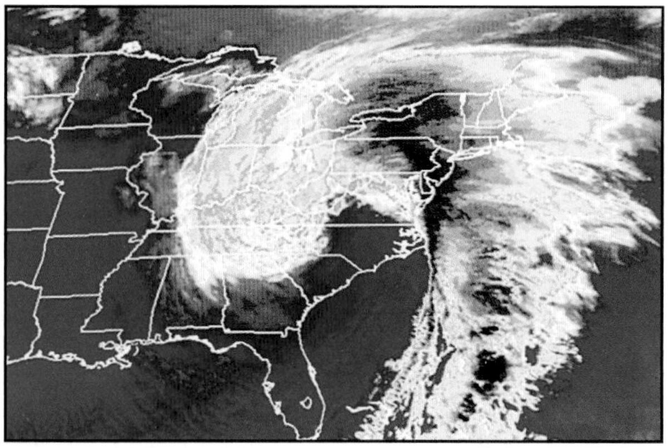

FIGURE 8.32
One of the greatest winter storms in United States' history, a nor'easter, struck the eastern third of the nation between March 12 and 15, 1993. Colorized infrared mage taken March 13, 1993. [Courtesy of NOAA]

FIGURE 8.33
A downed tree and snow-covered car attest to the magnitude of snowfall in the Asheville, North Carolina area during the nor'easter of March 1993. Some snowfall totals were: Mt. Mitchell, NC, 50 in.; Grantsville, MD, 47 in.; Syracuse, NY, 43 in. [Courtesy of NOAA]

27 tornadoes in Florida and a 3 m (10 ft) storm surge in the Gulf of Mexico, flooding the Apalachicola area. All told, this nor'easter claimed 270 lives, more than three times the combined death toll of Hurricanes Hugo and Andrew, and caused damage estimated at $7 billion (in 2002 dollars), the costliest extratropical cyclone in U.S. history.

While the impact of the March 1993 nor'easter remains unmatched, the winter of 2009-2010 was historic for many areas of the mid-Atlantic. Four nor'easters each brought widespread 25-50 cm (10-20 in.) swaths of snow, with even greater amounts locally. Two of the storms arrived back-to-back (4-7 February and 9-11 February), shutting down school systems for over a week and closing Federal government offices for 4 successive days. Seasonal snowfall records were shattered in Washington, DC, Baltimore, MD, and Philadelphia, PA. Baltimore received a total of 196 cm (77 in.), a far cry from the average seasonal total of 46 cm (18.2 in.).

The powerful winds and surge of nor'easters can significantly impact coastal areas. As described earlier in this chapter, the 1933 hurricane altered the landscape of Ocean City on Fenwick Island, MD. Many people visiting Ocean City are probably unaware of the coastal erosion and extensive flooding from the great nor'easter of 5-8 March 1962. This slow moving and intense storm system produced waves as high as 9 m (30 ft) superimposed on a 2 m (7 ft) storm surge. During five successive high-tide cycles over two days, storm waves washed over almost all of Fenwick Island and property damage at Ocean City was estimated at $7.5 million. In view of the considerable development that has occurred on the island since 1962, can you imagine the impact of a similar storm today?

In January 2006, NOAA introduced a scale that categorizes the impact of major snowstorms (including nor'easters) on the Northeast. The focus on Northeast snowstorms stems from their potential disruption of transportation systems and economic effects on the rest of the nation. The **Northeast Snowfall Impact Scale (NESIS)** (Figure 8.34) rates snowstorms on the amount of snowfall, the area affected by the snowstorm and the population in the path of the storm. With this scale, assessment of snowstorm impact is available in days instead of weeks. Snowstorms are assigned to one of five categories: 1 (notable), 2 (significant), 3 (major), 4 (crippling), or 5 (extreme). The highest rating on the NESIS is assigned to storms producing heavy snowfall

Category	NESIS Value	Description
1	1 - 2.499	Notable
2	2.5 - 3.99	Significant
3	4 - 5.99	Major
4	6 - 9.99	Crippling
5	10.0+	Extreme

FIGURE 8.34
The Northeast Snowfall Impact Scale (NESIS). [Image adapted from table at http://www.ncdc.noaa.gov/snow-and-ice/nesis.php]]

over large geographical areas encompassing major metropolitan centers. For example, the 12-15 March 1993 nor'easter rated 5 on the Northeast Snowfall Impact Scale, and the two storms of February 2010 were each assigned category 3. Development of the scale was the joint effort of Louis Uccellini, director of NOAA's National Centers for Environmental Prediction (NCEP) and Paul Kocin, formerly of *The Weather Channel*. The NESIS joins the Saffir-Simpson Hurricane Wind Scale and the Enhanced Fujita scale, which is for tornadoes, as tools for better informing the public on weather extremes.

Conclusions

Essentially all components of the Earth system come together and interact in the coastal zone: ocean, atmosphere, cryosphere, geosphere, and biosphere (including humans). The land/sea interface is a particularly dynamic portion of the Earth system where numerous forces shape the coast. Beaches, barrier islands, cliffed headlands, and estuaries are among the many features of the coastline. People are increasing their presence in the coastal zone and their activities sometimes conflict with the forces of nature including rising sea level, storm surges, and littoral drift.

Having examined the physical properties of the ocean, we next focus on life in the ocean. The next chapter covers marine ecosystems and the processes that support life in the ocean. Chapter 10 considers marine organisms and how they have adapted to the varying physical conditions of their environment.

Basic Understandings

- The coast encompasses the relatively narrow region transitional between land and ocean. Although the coast may appear stable and unchanging, the many components of the Earth system (ocean, atmosphere, biosphere, geosphere, and cryosphere) interact in the coastal zone, making it an exceptionally active environment.
- The farthest inland extent of storm waves defines the coastline, which can be marked by sand dunes or wave-cut cliffs, or the high water level. The average low tide is the shore line while the land exposed from low tide line up to the coastline is the shore, or the intertidal zone. It is alternately under water and exposed to the atmosphere as the tide rises and falls.
- Mountain building, volcanic activity, and earthquakes are dominant geological processes in tectonically active coastlines, such as along the rim of the Pacific Ocean basin. Sediment erosion, transport, and deposition are dominant geological processes along passive shorelines such as those surrounding the Atlantic Ocean and the Gulf of Mexico.
- As sea waves move into coastal waters, they undergo refraction, as they bend toward shallower water in response to changing wave speed caused by changing water depth. In this way, waves approaching an irregular coastline converge toward headlands (rocky cliffs that project seaward) and diverge away from coves or bays. Convergence concentrates wave energy, causing erosion of headlands, whereas spreading (divergence) of wave energy in bays reduces wave action, favoring deposition of sediment and expansion of a beach.
- Erosive and depositional forces continually battle on the landward margin of ocean and lake beaches where wave-washed sediment, usually sand and gravel, accumulates.
- When breaking waves approach the shore at an oblique angle, the zigzag motion of the swash moves particles along the beach face, which is known as beach drift. They also produce a longshore current, a component of water motion that is parallel to the shore and transports sediments as littoral drift along the coast, either nourishing or cutting back beaches.
- An abundant supply of sand, energetic waves and a gently sloping sea floor form a barrier island, an elongated, narrow accumulation of sand parallel to the coast and separated from the mainland by a lagoon, estuary, or bay. Barrier island beaches border the Atlantic Coast from south Florida to Long Island, NY. From the ocean side of a typical undeveloped barrier island and working toward the mainland, there follows beach, dunes, flats, marsh, and lagoon.
- Mainland beaches occur from Long Island, NY, northward through coastal New England and Atlantic Canada.
- Along the coast, the broad shoreward flow of water is balanced by the narrow seaward flow of

water, which occurs as a swift surface or near-surface current. This is known as a rip current.

- A salt marsh is a nearly flat, low-lying, protected coastal area where fine-grained sediment accumulates and salt-tolerant plants grow. Tidal creeks meander through most salt marshes, allowing seawater to flood the marshes on a rising tide and drain on the ebb tide.
- Artificial structures (including dams, breakwaters, jetties, groins, and sea walls) disrupt littoral drift and alter the natural flow of sediments to and along a beach. An alternate strategy to maintain existing beaches or restore badly eroded beaches is artificial beach nourishment. With this approach, sand is dredged from deep waters and spread on the beach. Artificial beach nourishment is an expensive and temporary solution but provides substantial protection for coastal property and businesses.
- An estuary is a partially isolated body of water where fresh water (from rivers and streams) mixes with seawater. Most estuaries developed during the post-glacial (Holocene) rise in sea level. Other estuaries developed through glacial erosion of mountain valleys, tectonic processes, or longshore sediment transport. Based on the circulation regime, a distinction is made among salt-wedge estuaries, partially mixed estuaries, and well-mixed estuaries.
- A tropical cyclone, which includes tropical storms and hurricanes, originates in a uniform mass of warm humid air over the tropical ocean but may track into higher latitudes. An extratropical cyclone, which brings together contrasting warm and cold air masses to form fronts and frontal weather, forms in middle latitudes and spends its entire life cycle in middle and high latitudes.
- A dome of ocean water driven ashore by strong onshore winds associated with an intense storm system centered over or near the ocean, known as a storm surge, is potentially the most devastating impact of tropical and extratropical cyclones in the coastal zone. The storm intensity, timing of tides, bathymetry, and shoreline configuration all influence a storm surge.
- A hurricane is an intense tropical cyclone that originates over tropical ocean-waters, usually in late summer or early fall, with a maximum sustained wind speed of at least 119 km per hr (74 mph). In addition to storm surge, hurricanes bring flooding rains, strong winds, and tornadoes.
- To form a tropical cyclone there must be (1) relatively high sea-surface temperature, (2) adequate Coriolis Effect, and (3) weak winds aloft. The energy source for tropical cyclones is latent heat released when water vapor that had evaporated from the ocean condenses in the system. As a tropical cyclone intensifies, the successive stages in its life cycle are designated tropical disturbance, tropical depression, tropical storm and, finally, hurricane.
- The Saffir-Simpson Hurricane Wind Scale rates hurricanes from category 1 to 5 by increasing intensity and potential property damage from a hurricane landfall.
- A nor'easter is an intense extratropical cyclone that tracks along the East Coast of North America and is named for the direction from which its most destructive winds blow. Unlike hurricanes, nor'easters may originate over land, forming along the boundary between air masses that contrast in temperature and humidity. They are most frequent from October through April.

Enduring Ideas

- The coastal zone is a dynamic environment where the ocean, atmosphere, geosphere, cryosphere, and biosphere continually interact. The intertidal zone is alternately under water and exposed to the atmosphere causing drastic changes in temperature and salinity. Organisms have special adaptations to survive.
- Tectonically active coastlines, characterized by mountain building, volcanic activity, and earthquakes, are near plate boundaries. Passive coastlines, which have erosion, transport, and deposition of sediment as dominant geological processes are distant from plate boundaries.
- Longshore currents transport sediments as littoral drift along a beach. Hard coastal modifications such as dams, breakwaters, jetties, groins, and sea walls disrupt littoral drift and may either nourish or cut back a beach.
- A tropical cyclone (tropical storm, hurricane) originates in a uniform mass of warm, humid air over tropical ocean waters. The system's primary source of energy is latent heat released when water that evaporated from the ocean surface condenses into clouds. SST and latitude governs the geographical and seasonal distribution of tropical cyclones.
- Potentially the most devastating aspect of both tropical and extratropical cyclones is a storm surge in the coastal zone. The height and inland reach of a storm surge depends on storm intensity (air pressure and wind speed), timing of tides, bathymetry, and shoreline configuration.

Review

1. Explain why the coastal zone is a particularly dynamic part of the Earth system.
2. What are the dominant geological processes operating in tectonically active coastlines and passive coastlines? Of North America's Atlantic and Pacific coasts, which one is more tectonically active?
3. Why does a wave crest bend (refract) as it approaches the shoreline at an angle other than 90 degrees?
4. How and why does the width of an East Coast beach change from winter to summer?
5. Briefly describe the origins of a longshore current.
6. Describe rip currents and why they are potentially hazardous for swimmers.
7. Are barrier islands found on tectonically active or passive continental margins? Justify your choice.
8. Distinguish among the types of estuaries in terms of fresh water and seawater flow and vertical stratification.
9. Identify the hazards to the coastal zone posed by landfalling hurricanes.
10. What is the "fuel" that powers both tropical cyclones and extratropical cyclones?

Critical Thinking

1. Explain how wave refraction affects erosion and sediment deposition along an irregular coastline.
2. Why does an irregular coastline tend to straighten with time?
3. What is the origin of the sediment that forms a beach and how do human activities disrupt the supply of that sediment?
4. Identify some of the costs and benefits of coastal armor.
5. List some of the advantages and disadvantages of artificial beach nourishment.
6. Explain how an undeveloped barrier island is a continually changing system.
7. What role is played by the ocean in the development of tropical cyclones?
8. Why do hurricanes rarely strike Southern California?
9. Under what conditions might an extratropical cyclone cause more coastal erosion than a hurricane?
10. Speculate on some of the ways whereby a rising sea level might impact the coastal zone.

ESSAY: The Great Lakes and the Ocean

The Great Lakes of North America are among the great lakes of the world, which include Great Bear Lake and Great Slave Lake of Canada, Lake Baikal of Russia, Lakes Victoria and Malawi of Africa, Lake Titicaca in South America and others. These lakes, and particularly the Great Lakes, are often referred to as "models of the ocean." How accurate is this statement?

Combined, the Great Lakes constitute one of the largest reservoirs of surface fresh water on the planet, accounting for about 18% of the total. (Only the polar ice sheets and Lake Baikal contain more fresh water.) The Great Lakes are a chain of lakes (Figure 1). Lake Superior, the largest and deepest of the five lakes (Figure 2), drains into Lake Huron via the St. Marys River. Lake Michigan waters also empty into Lake Huron through the Straits of Mackinac. From Lake Huron, water flows via the St. Clair and Detroit Rivers into Lake Erie, the smallest and shallowest of the five lakes. Water then flows from Lake Erie through the Niagara River and Welland Canal into Lake Ontario and empties into the St. Lawrence River, the Gulf of St. Lawrence, and ultimately the North Atlantic Ocean. Water draining into Lake Superior along its western shore travels some 3600 km (2230 mi) to the Gulf of St. Lawrence. *Residence time* of water in the Great Lakes ranges from 2.6 years in Lake Erie to 191 years in Lake Superior.

The Great Lakes are a legacy of the Pleistocene Ice Age that began about 1.8 million years ago and ended about 10,500 years ago. Prior to the Pleistocene, rivers likely flowed through valleys where lake basins are now located. Huge

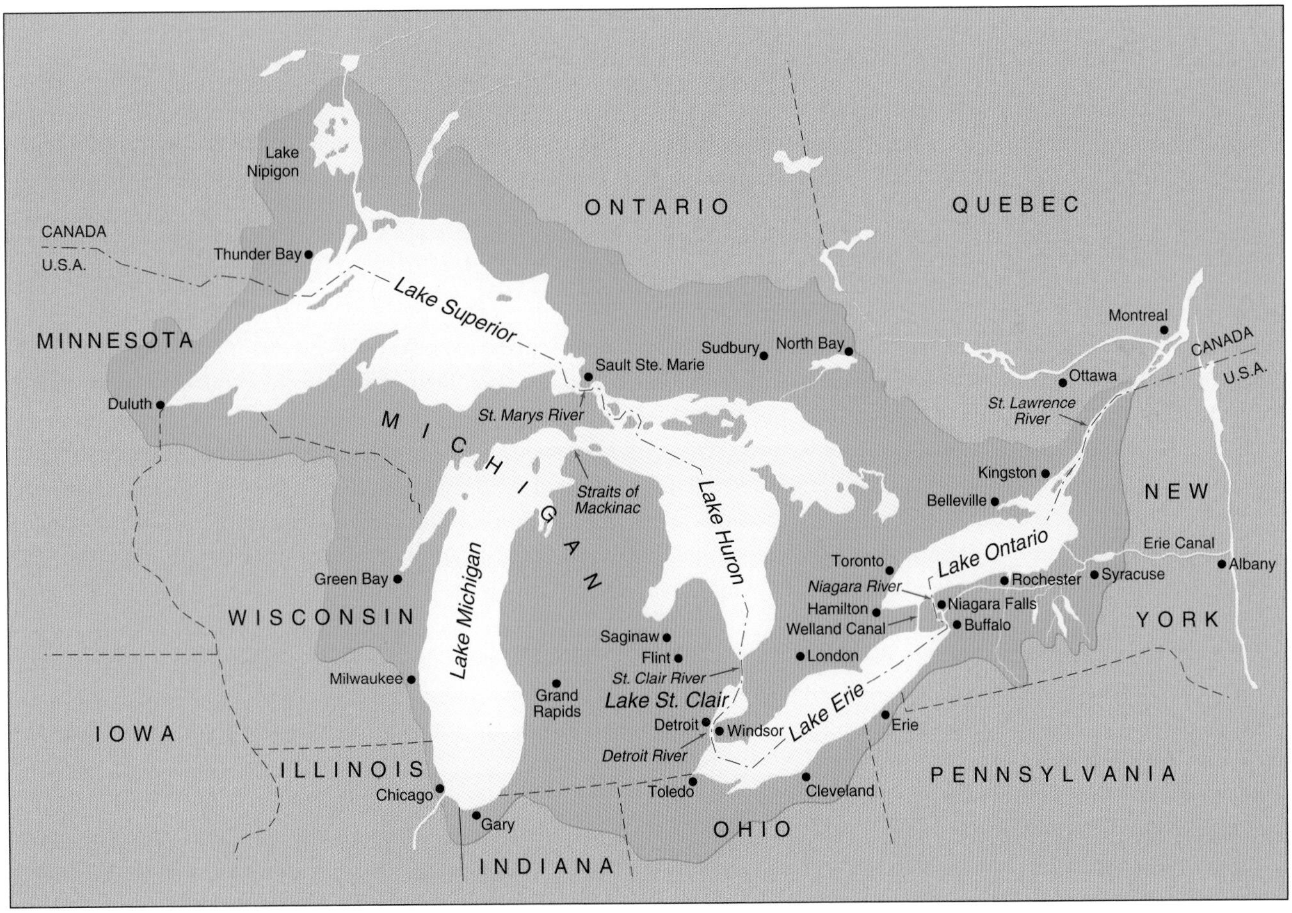

FIGURE 1
The Great Lakes of North America with the drainage basin shaded in green. Weather and climate over the lakes and the drainage basin control lake levels.

FIGURE 2
Shoreline of Lake Superior, the largest of the Great Lakes, near Munising, MI. [Photo by J.M. Moran]

lobes of glacial ice advancing southward from what is now Canada widened and deepened the river valleys, and the weight of the ice further depressed the evolving lake basins. Ocean basins, on the other hand, are much older (hundreds of millions of years) and are the products of tectonic processes (Chapter 2).

The Great Lakes are small compared to ocean basins; their dimensions are similar to those of the continental shelf, that is, 10s to 100s of kilometers (10s to 100s of miles) in the horizontal, and 10s to 100s of meters (10s to 100s of yards) in the vertical. Like the ocean, the Great Lakes are sufficiently large that the Coriolis Effect plays a role in the circulation of their waters (Chapter 5). The Coriolis Effect along with winds and stratification generate circulation phenomena that are the same as those observed in the ocean (e.g., wind-driven Ekman transport, coastal upwelling and downwelling, and geostrophic flow).

Unlike small lakes but like the ocean, the Great Lakes are sufficiently large and deep that much of their water mass is unaffected by frictional interaction with the coast and bottom. If wind sets the surface water in motion at the nearly frictionless center of the lake basin, inertia keeps the water in motion. (Recall from Chapter 4 that according to *Newton's first law of motion*, an object in constant straight-line motion or at rest remains that way unless acted upon by an unbalanced force. *Inertia* is the name given to this tendency for an object to continue moving or to remain at rest.) With no other forces operating, the flow is governed by the Coriolis Effect. In the Great Lakes, Coriolis deflection is to the right, so that a surface current describes a loop, called an *inertial circle*, having a period of about about 17 hrs at 45 degrees North—near the middle of Lake Michigan. (Actually, the current follows a straight path while the basin rotates with the spinning Earth). For a current speed of 10 cm per sec (0.2 mph) at 45 degrees N, the radius of an inertial circle is 1 km (0.6 mi), whereas for a current speed of 100 cm per sec (about 2 mph), the radius is 10 km (6 mi).

Ekman transport operates in the Great Lakes. If the wind blows parallel to the coast, depending upon the direction of the wind, Ekman transport causes either coastal upwelling or coastal downwelling (Chapter 6). In response to upwelling or downwelling, the thermocline tilts near the coast creating a horizontal gradient in density (or pressure). The balance that develops between this horizontal density gradient and the Coriolis Effect produces *geostrophic flow* (Chapter 6). In the Great Lakes, geostrophic flow is observed as a coastal current or *jet*.

When winds slacken, the release of the tilted thermocline (caused by upwelling or downwelling) generates two types of free internal waves involving the Coriolis Effect: inertial waves or oscillations and Kelvin waves. Inertial oscillations occur in the thermocline across the entire lake but with maximum currents at the center of the lake. A *Kelvin wave* is an *edge wave*, meaning that maximum wave height is at the coast but diminishes to lake level within 5 to 10 km (3 to 6 mi) offshore. Kelvin waves depend upon the Coriolis Effect and progress cyclonically around the lake with the "edge" to the right of the direction of motion. Crests and troughs of a Kelvin wave are not readily observed except by temperature readings at municipal water intakes. A Kelvin wave crest brings deeper cooler water to the intake whereas a Kelvin wave trough delivers shallower warmer water. The period of a Kelvin wave is typically much longer than the period of an inertial oscillation. Large scale eddies are also observed in the Great Lakes coastal waters that resemble those on the shelf and in the open ocean.

Both inertial oscillations and Kelvin waves also occur in the ocean. Kelvin waves develop in the equatorial Pacific when the trade winds weaken during El Niño and the water piled up in the western tropical Pacific sloshes back toward South America (Chapter 11). The sloshing consists of Kelvin waves with maximum height on the equator where the Coriolis Effect changes sign. When these Kelvin waves strike the South American coast, they split into two waves, one progressing northward and the other southward along the coast. The northward moving Kelvin wave eventually reaches the U.S. West Coast.

An obvious difference between the Great Lakes and the ocean is salinity: Great Lakes waters are fresh whereas seawater is saline. In the Great Lakes, water density depends on temperature and maximum density is at 4 °C (39.2 °F). On the other hand, seawater density depends on both temperature and salinity and its temperature of maximum density is below its initial freezing point (Chapter 3). An important implication of this contrast in physical properties is the occurrence of seasonal turnover in lakes but not the ocean. Spring and fall turnover of lakes in temperate latitudes is important for organisms living in lakes because turnover replenishes the dissolved oxygen supply of those water bodies. Lake turnover also recycles nutrients, especially nitrogen and phosphorus compounds, from bottom sediments to the overlying water, where they become available to aquatic plants, especially algae.

Bright summer sunshine penetrates a lake to shallow depths, warming the surface layer of water. Meanwhile, the dark deep water does not benefit from solar heating and remains cold, with an average temperature as low as 4 °C. This stable stratification (layering) of warm, less dense surface water and colder, denser water at depth persists through most of the summer, with little mixing between the layers. The upper layer of the stratified lake is known as the *epilimnion* and the lower layer is known as the *hypolimnion*, with the transition zone between the two layers being the *thermocline*.

In a stratified lake, oxygen that is supplied by the atmosphere and photosynthetic organisms replenishes the dissolved oxygen supply of only the epilimnion. In the hypolimnion, cellular respiration by decomposers and other organisms removes dissolved oxygen. With very little transfer of oxygen from the epilimnion, dissolved oxygen levels in the hypolimnion steadily decline. If lake stratification were to persist, the dissolved oxygen concentration would fall to levels that would severely stress cold-water fish species (e.g., trout, whitefish) living in the hypolimnion. Fortunately, these fish and other inhabitants of the hypolimnion usually do not perish because lake stratification eventually breaks down, lake waters turn over, and the dissolved oxygen content of the hypolimnion is replenished.

As summer gives way to autumn, the Sun becomes lower in the sky, daylight shortens, and air temperatures drop. Heat is lost from the warm surface waters to the overlying cool air and ultimately to space. Eventually the temperature of the epilimnion cools to that of the hypolimnion, the thermocline disappears, and the lake has a uniform temperature and density from top to bottom. Winds blowing across the lake transport surface water from the upwind shore toward the downwind shore. Oxygen-depleted water wells up from below on the upwind shore and oxygen-rich surface water sinks at the downwind shore. This *fall turnover* of lake-waters brings oxygen-depleted water from the lake bottom to the surface where it is exposed to the atmosphere and replenished in dissolved oxygen.

Cooling during late autumn and early winter eventually drops the temperature of surface waters to 4 °C and lower. The lake begins to stratify with the coldest and least dense water at the surface. With continued cooling, the lake surface temperature eventually drops to 0 °C and a skim of ice forms. While large areas of the Great Lakes (especially sheltered inlets such as Green Bay) may develop an ice cover, stirring of waters by strong winds brings slightly warmer water to the surface and keeps much of the lake surface ice-free through the winter. For the Great Lakes, the entire water column must overturn to reach maximum density (at 4 °C) before the lake can cool below 4 °C, stratify, and freeze. For Lake Superior, the deepest of the Great Lakes, this is a maximum depth of 400 m (1300 ft). Lake-water temperatures typically vary from 0 °C just under the ice to 4 °C at the lake bottom. Ice forms a barrier that prevents exchange of oxygen between the lake-water and atmosphere, and cellular respiration within the lake causes a gradual decrease in dissolved oxygen levels. When ice finally melts in spring, the lake waters may be seriously depleted in dissolved oxygen.

With the arrival of spring and lengthening daylight, air temperatures rise, and lake surface waters warm. Eventually, the temperature of the surface water reaches 4 °C and once again the temperature and density of the lake become uniform from top to bottom. Just as in autumn, strong winds trigger turnover of lake waters and the mixing that accompanies *spring turnover* brings oxygen-depleted water to the surface where oxygen is replenished directly from the atmosphere and via photosynthesis. As spring gives way to summer, lake stratification is reestablished. What is intriguing and different from ocean water is that in spring the Great Lakes typically begin to stratify in shallower water near shore and stratification progresses toward the center of the Lakes at about 1 km (0.6 mi) per day.

Wind-driven waves are similar on the ocean and Great Lakes although the scales are different. Swells and larger-size waves are more common on the ocean because of more persistent winds and longer fetch (Chapter 7). Through the years, many powerful storms have swept over the Great Lakes, producing huge waves that sank thousands of vessels with a considerable loss of life. Even very large ships may be unable to withstand the fury of storm waves on the lakes. For example, on 10 November 1975, a fierce storm tracking over Lake Superior produced winds gusting to 145 km per hr (90 mph). Waves of 3.5 to 5 m (12 to 16 ft) sank the *Edmund Fitzgerald*, at the time the largest ore carrier on the Great Lakes at 222 m (729 ft) long, 23 m (75 ft) wide, and carrying 26,116 tons of taconite pellets. The *Edmund Fitzgerald* sank in 160 m (525 ft) of water at the eastern end of Lake Superior about 27 km (17 mi) from Whitefish Bay, MI, with the loss of its crew of 29. Speculation is that huge waves lifted the bow and stern enough to cause the ship to break amidships and sink before the crew could react.

The ocean and the Great Lakes also share some similarities in long-term changes in water level but there are also fundamental differences. Astronomical tides are evident in the ocean, but are not very noticeable on the Great Lakes. On the other hand, seiches are common on the Great Lakes but are not prominent in the open ocean or on the continental shelves, although they are observed in harbors and embayments. For both tides and seiches the time scale is hours and the height scale is centimeters to meters.

Over the past 3000 to 3500 years, Great Lakes levels have fluctuated in a quasi-periodic manner about an established datum (based on mean lake levels during the modern record of 1918-1998). Departures from this so-called *International Great Lakes Datum (IGLD)* were never much more than a meter during the 3000 to 3500 year period. As of this writing, Great Lake levels are near historic lows (Figure 3). Great Lakes levels are governed by precipitation, runoff from spring snowmelt, and air temperature. Temperature affects the rate of evaporation of water and controls ice cover that also influences evaporation. Should the present global warming trend continue, lake levels are expected to fall whereas sea level is predicted to rise. We have more to say about these prospects and their implications in Chapter 12.

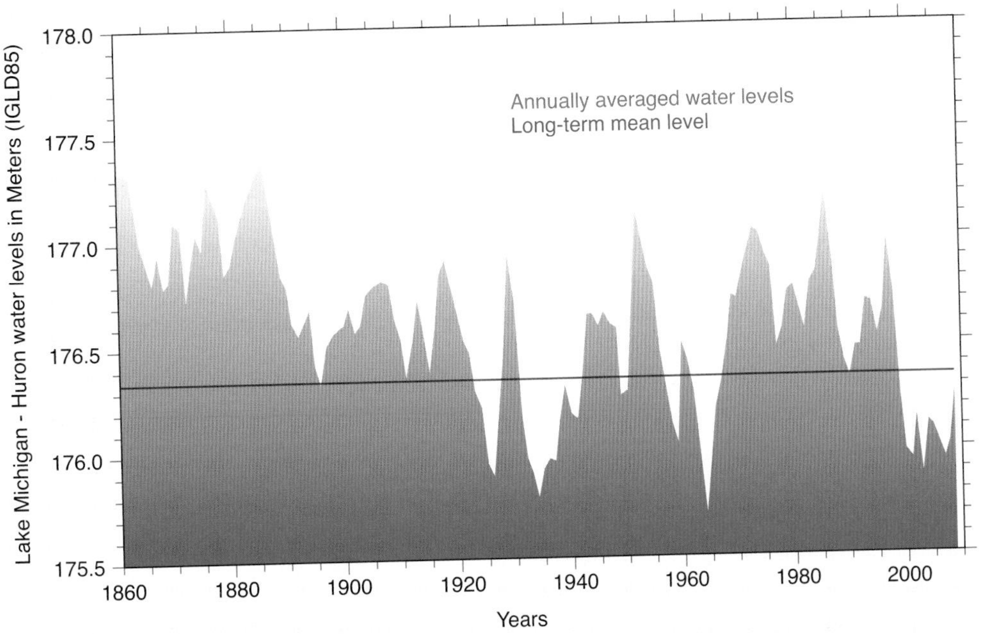

FIGURE 3
Record of water levels (in meters) of Lakes Michigan and Huron beginning in 1860. Horizontal red line is the long-term mean. These data are collected and archived by NOAA's National Ocean Service. [Courtesy of NOAA's Great Lakes Environmental Research Laboratory (GLERL)]

ESSAY: Venice, Italy, and the Encroaching Sea

Many of the world's most populous cities are located along the coast and efforts to protect them from storm-generated flood surges are becoming more challenging, especially as *mean sea level (msl)* rises in response to global climate change (Chapter 12). This is not a new problem, as demonstrated by Venice in northern Italy.

At the head of the Adriatic Sea, Venice is situated on a cluster of 120 salt-marsh islands in a large coastal lagoon, one of the most important wetlands in the Mediterranean Basin (Figure 1). Some 200,000 birds use the lagoon as a winter migration destination and breeding area. Bridges connect some of the islands and a causeway and ferries link Venice to the mainland. A long narrow barrier island separates the lagoon from the sea (Gulf of Venice) except for the three major tidal inlets. At one time, the sea and the lagoon protected the city from foreign invaders but today, the encroaching sea threatens Venice's art and architectural treasures, and perhaps the city's continued existence.

Slow subsidence, combined with rising sea level, has increased the frequency of flooding. Venice has been slowly sinking under its own weight for centuries, exacerbated by the withdrawals of large quantities of groundwater for industrial use from the 1930s to 1970s. Groundwater occupies the tiny pore spaces in sediment and, as the water is pumped out, the sediment compacts and the ground subsides. Land subsidence is also occurring because of the oxidation of organic-rich soils in fertile farmland that had been reclaimed from marshland at the southern end of the Venice lagoon. A drainage

FIGURE1
ASTER satellite image of Venice, Italy, located in the saltwater lagoon between the mainland and barrier islands. The image was acquired on 9 December 2001. ASTER (Advanced Spaceborne Thermal Emission and Reflection Radiometer) is flown on NASA's Terra Satellite orbiting Earth at an altitude of 705 km (430 mi). [Courtesy of NASA/GSFC/METI/ERSDAC/JAROS, and U.S./Japan ASTER Science Team]

system that keeps the soil relatively dry promotes oxidation and, consequently, loss of soil mass. Reclamation of marshland began in the late 19th century and continued into the late 1930s. Since then land subsidence has reached 2 m (6.5 ft) and is currently taking place at a rate of 1.5 to 2 cm per year. According to estimates by the *Venice Organic Soil Subsidence (VOSS)* project, in 50 years oxidation could remove all the organic soil, causing additional land subsidence of 0.75 to 1.0 m (2.5 to 3.3 ft).

Ten major floods have inundated Venice over the past 70 years, while smaller floods have become much more frequent. During the often stormy late fall and early winter, November through January, some areas of the city are under water almost every day. Portions of Venice, including the historic Piazza San Marco, are flooded about 100 days per year, which a century ago, flooded only about 7 days per year. Flooding forces businesses to close and inconveniences residents and tourists alike but, more significantly, the salt water corrodes the brick underpinnings of historic buildings along the Grand Canal and elsewhere in the city.

To protect Venice from future flooding, the Italian government embarked on a $7 billion project to construct 78 floodgates at the three tidal inlets linking the Venice lagoon to the Adriatic Sea. Most of the time, the hollow gates, which can measure up to 5 m (16 ft) thick, 20 m (66 ft) wide, and 27 m (89 ft) in length, will be filled with seawater and lie horizontally on the sea floor within a foundation (caisson) so as not to impede tidal currents or navigation. When a storm surge is predicted to elevate the tide to a height of 110 cm (43 in.) or more, the floodgate system will be activated. (For perspective, the water level reached 194 cm, or 6.4 ft, during the disastrous flood of November 1966.) Compressed air will be pumped into the gates to expel the water, thus allowing them to swing upwards and block the flow of floodwaters through the inlets. This system is designed to protect Venice from a 60 cm (24 in.) rise in mean sea level and a 3 m (10 ft) flood. Locks (already completed) will allow passage of ocean going container-cargo and cruise ships, as well as the fishing fleet, when the gates are up. Similar flood control structures have been built on the Thames River in England to protect London and at the mouths of estuaries to protect the low-lying Dutch coastal plain.

Although the idea for the Venice floodgates dates back to the mid 1980s, it was not until May 2003 that a stone-laying ceremony signaled the beginning of construction on the *Moses Project* (known in Italian as *Mose* for Modulo Sperimentale Elettromeccanico). As of this writing, the project is slated for completion in 2012 but environmental protests have slowed construction. Opponents of the Venice floodgate scheme argue that closing the inlets will cause substantial environmental damage, especially to the fragile lagoon ecosystem. Most of Venice's sewage empties untreated into the many canals dissecting the city, or directly into the lagoon. Normally, tidal currents dilute and transport these wastes out to sea but, when the gates are closed for extended periods, waste will accumulate in the city's canals and lagoon. Some scientists speculate that the gates could be closed for as many as 50 days per year.

ESSAY: Moving the Cape Hatteras Lighthouse

When beach erosion threatens to damage or destroy coastal structures, three protective strategies are available: *armor* (e.g., groins, seawalls), *artificial beach nourishment*, and *strategic retreat* (relocation of structures inland). At various times, all three strategies were employed to keep the encroaching sea from claiming the historic Cape Hatteras Lighthouse (Figure 1).

In 1870, a 63-m (208-ft) brick lighthouse was built at Cape Hatteras on one of the barrier islands that form North Carolina's Outer Banks. At the time, the lighthouse with its distinctive white and black diagonal stripes was about 460 m (1500 ft) from the shoreline. But the beacon's location on a barrier island and the steady rise in sea level (presently about 2 mm or 0.08 in. per year) made it increasingly vulnerable to attack by powerful sea waves generated by the winds of tropical and extratropical cyclones that often track near Cape Hatteras. In less than 50 years, about 335 m (1100 ft) of the east-facing beach in front of the lighthouse had washed away. In the 1930s and early 1940s attempts to stabilize the beach with groins failed and after World War II, the lighthouse was taken out of service (replaced by a beacon mounted on a metal tower located well inland) and assigned to the National Park Service (NPS).

The Cape Hatteras Lighthouse soon became one of North Carolina's most popular tourist attractions and public support for preserving the lighthouse grew. Artificial beach nourishment was tried many times in the 1960s and 1970s but with short-lived success. The growing threat to the lighthouse was made clear in 1982 when an intense winter storm brought storm waves to the base of the lighthouse. During the height of the storm, quick acting NPS personnel tore up slabs of asphalt from a newly constructed parking lot and piled them around the base of the lighthouse to protect the foundation. Subsequently, sand bags were piled on top of the asphalt slabs.

Coastal engineers advised against a plan to build a seawall around the lighthouse. They pointed out that the structure's brick and stone foundation is less than 2 m (7 ft) deep and rests on yellow pine timbers that must remain submerged in fresh water to prevent decomposition. Surrounded by a seawall, the lighthouse eventually would become an island, saltwater would replace fresh water in the subsurface and rot the pine timbers, the foundation would collapse, and the lighthouse would fall into the sea.

FIGURE 1
This photo of the Cape Hatteras Lighthouse was taken in April 1995. Notice the sandbags at the base of the tower. [Photo by J.M. Moran]

By 1987, the Cape Hatteras Lighthouse was only 49 m (160 ft) from the sea. The only option to save the lighthouse was a strategic retreat inland, a plan endorsed by the National Park Service in 1989. In effect, retreat is a preservation/protection strategy that evades rather than confronts the forces of nature. After years of public debate, the lighthouse was physically moved on rollers and iron rails 884 m (2900 ft) to a new location (and a new foundation) some 488 m (1600 ft) from the edge of the Atlantic Ocean. The lighthouse began its journey on 17 June 1999 and arrived at its destination twenty-two days later. The Cape Hatteras Lighthouse again opened to the public on 26 May 2000.

Is this the end of the Cape Hatteras Lighthouse story? Not likely. Sea level will probably continue to rise, tropical and extratropical cyclones will generate storm surges, and barrier islands will migrate toward the mainland. In fact, on 18 September 2003, Hurricane Isabel made landfall on the Outer Banks as a category 2 system causing considerable coastal erosion and flooding. The storm surge cut a new inlet and came precariously close to the Cape Hatteras Lighthouse, damaging roads and parking lots.

CHAPTER 9

MARINE ECOSYSTEMS

School of fish near a coral reef. [Courtesy of NOAA]

Case in Point

In December 1987, many people in Montreal, Quebec, became ill with abdominal cramps, vomiting, diarrhea, severe headaches, hallucinations, seizures, and memory loss after eating farmed mussels. Scientists worked around the clock for days at a government laboratory in Halifax, Nova Scotia, to identify the toxin (poison) responsible for the illness and its source. The mussels, from Prince Edward Island in eastern Canada, caused a malady, later identified as *amnesic shellfish poisoning*, that took the lives of three people while 105 others experienced acute poisoning.

A mussel is a filter-feeding mollusk that siphons large volumes of water, filtering out organisms such as diatoms and other tiny particles for food. The farmed mussels had fed on a particular species of diatom belonging to the genus *Pseudo-nitzschia*, which produces a neurotoxin known as domoic acid. While harmless to the mussels, this naturally occurring amino acid causes debilitating symptoms in the human neural system and is particularly damaging to the memory centers of the brain. Tests for this toxin enabled fishery managers to immediately halt the harvesting and sale of mussels from contaminated areas (Figure 9.1). Cooking does not destroy algal toxins in shellfish so that commercial supplies of mussels, clams, oysters and scallops are now rigorously tested for the presence of toxins.

FIGURE 9.1
A harmful algal bloom (HAB) may pose a hazard to human health when a naturally occurring toxin enters the food chain necessitating the closure of commercial shellfish beds. [Courtesy of Woods Hole (MA) Oceanographic Institution]

Blooms of phytoplankton are common in surface ocean waters and normally consist of many species. Sometimes, however, a bloom of a single species occurs and can harm the environment, cause the demise of marine organisms, and result in economic loss and human health problems, including illness and death. Of the thousands of species of microscopic algae that inhabit the ocean, only a few dozen produce potent toxins. These **harmful algal blooms** (HABs) may take on a distinct color, such as brown or red and be named accordingly. For example HABs that produce a red hue in Gulf of Mexico waters are often referred to as **red tides** (Figure 9.2); HABs that produce a brown hue in Long Island Sound waters are called *brown tides*. Not all HABs are toxic or caused by algae; many toxic blooms are caused by dinoflagellates, microscopic one celled organisms. Different types of poisoning are caused by different HABs. Amnesic shellfish poisoning is only associated with toxic blooms of *Pseudo-nitzschia* whereas blooms of *Karenia brevis* (Florida red tide) produce brevetoxins that cause neurotoxic shellfish poisoning.

A potentially deadly illness that is linked to an algal toxin is *paralytic shellfish poisoning (PSP)*. In this case the dinoflagellate, a photosynthetic swimming algae, is *Alexandrium catanella*. In sufficient numbers, *A. catanella* causes the water to take on a luminescent red appearance. They produce saxitoxen, another neurotoxin, which causes PSP in humans (and other animals) with tingling and numbness of lips, fingers, face and extremities, vomiting, diarrhea, slurred speech, headache, and rapid heartbeat. Death is caused by respiratory paralysis.

Harmful algal blooms seem to be increasing in frequency world-wide, although it may be that we are just detecting them more often because of increased study over the past several decades. Some HAB producing organisms show up in the ballast water discharged by ships far from their port of origin and may contribute to more widespread occurrence of HABs. In the coastal waters of China, for example, HABs increased 10-fold between 1975 and 1995. Another possible reason for more frequent HABs is elevated nutrient levels (nitrogen and phosphorus) in coastal waters due to runoff from agricultural fields, suburban lawns, and sewage from urban areas.

Blooms of toxic algae have caused fish kills and increased mortality of marine mammals and birds around the world. In late February 2002, the deaths of 20 dolphins that washed ashore in Southern California were attributed to the domoic acid secreted by diatoms. The 2005 deaths of significant numbers of the endangered manatee population of southwest Florida were attributed to *Karenia brevis* blooms.

In addition to toxin-producing blooms, a second type of harmful algal bloom, known as a **high biomass bloom**, significantly reduces the supply of *dissolved oxygen (DO)* in seawater (Chapter 3). As excess

FIGURE 9.2
A large red tide formed by the dinoflagellate *Noctiluca*. [Courtesy of Peter Franks, Scripps Institution of Oceanography, University of California, San Diego]

amounts of nutrients are discharged into coastal waters, phytoplankton blooms become extremely dense. When the algae die, their remains sink to the ocean bottom where they are eaten by *aerobic decomposers*, organisms that breathe oxygen. Eventually the water becomes severely depleted of dissolved oxygen, making it impossible for other organisms (except *anaerobic decomposers*) to survive. Such conditions cause massive fish kills.

Harmful algal blooms are known in the waters of almost every coastal state of the U.S. According to a 2007 NOAA report, HABs are responsible for direct economic losses in the United States averaging $75 million annually. Costs are incurred in public health, closure of commercial fisheries, and management and monitoring. The most frequent toxic HABs (*Karenia brevis* or Florida red tide) occur off the coast of Florida every year. It has been estimated that impacts of these blooms cost $19 to $32 million annually. In 2009, the threat to human health associated with HABs forced the closing of Maine's shellfish fisheries from April until, in some places, late September. At its peak, 97% of the fisheries were closed, at great economic loss. The decomposing residue of algal blooms that washes up on beaches produces a foul odor and unpleasant appearance that drives away visitors costing the tourism and recreation industries millions of dollars each year. In 1991, a *Pseudo-nitzschia* bloom forced the closure of Washington State beaches to commercial shellfish harvesting and recreation at an estimated revenue loss of $15 to $20 million for local communities.

Advance warning of a harmful algal bloom increases options for managing and mitigating potential impacts. To this end, NOAA's National Ocean Service operates the *HAB operational forecast system (HAB-OFS)* in the Gulf coast region of Florida (since 2004) and Texas (since 2010). The HAB-OFS utilizes satellite imagery, instrumented buoys, field observations, public health reports, and models to assess the potential for development of HABs, and their movement, extent, and potential human impacts. Bulletins, issued once or twice weekly depending on confirmed or expected bloom conditions, inform state and local coastal managers, and other individuals with resource management responsibilities. A public conditions report is issued alerting the general public to current bloom conditions and potential impacts.

Driving Question:
What are the basic components and structure of marine ecosystems and what is their source of energy?

With this chapter, our primary focus shifts to life in the ocean. The physical and chemical properties and processes examined in earlier chapters influence all marine organisms and their interactions. In this chapter, we describe how interactions among the biosphere, atmosphere, and ocean govern the distribution and abundance of life in the ocean.

A fundamental subdivision of the Earth system is the *ecosystem*. This chapter examines the components of marine ecosystems including producers, consumers, and decomposers. We emphasize the source of energy (photosynthesis, chemosynthesis) and the transfer of energy among organisms occupying the various feeding (trophic) levels within food webs. We then take a closer look at the processes operating within marine ecosystems including energy supply for growth and reproduction, factors influencing biological production in the ocean, the role of nutrients and trace elements as limiting factors, and the importance of microbes in marine ecosystems. In view of the contemporary interest in the carbon cycle and global climate change, we also examine the physical and biological processes governing the flux of carbon into and out of the ocean over a range of time scales. We begin by identifying the essential requirements for marine life.

Requirements for Marine Life

The ocean is home to a huge number of different types or species of organisms. *The Census of Marine Life 2010* discovered over 6000 potentially new species and increased the estimate of known species to 250,000. The Census extrapolated that there are at least 1 million different marine species and tens to hundreds of millions of microbe types. The wide variety of marine habitats makes possible this great biodiversity. Some marine ecosystems, such as coral reefs (Chapter 10), are structurally complex and provide niches for many types of organisms that have evolved different adaptations to their environment. The discovery of hydrothermal vents and their associated thousands of species of marine organisms dates only to 1979 (Chapter 2).

Many of the fundamental requirements for life in the ocean are the same as those needed by terrestrial organisms. Essential for all life on Earth are a source of energy (e.g., sunlight), liquid water, the appropriate mix of chemical constituents (e.g., nutrients), and the right combination of environmental conditions (e.g., range of temperature). Major differences between life in the ocean and life on land include the much greater space, a larger life zone, variety of marine habitats, and interactions between plants and herbivores. A *life zone* refers to space where the necessary and sufficient ingredients for the maintenance of life forms are found. On land, the tops of the tallest plants are only a few tens of meters above the ground and only a few species of birds can fly higher than 100 m (330 ft). In the ocean, microscopic plants live at depths as great as about 200 m (650 ft) and animals inhabit all depths. Even the deepest trenches (about 11,000 m or 36,000 ft deep on the ocean floor) support life. Another important difference is that terrestrial plants typically spend their entire life cycle rooted in the soil at a fixed location whereas those in the open ocean are suspended in a continually moving fluid. Additionally, in terrestrial systems, plants are generally larger than herbivores. In the ocean, the opposite is true. Terrestrial plants also have a large percentage of their biomass tied up in inedible parts (i.e., tree trunk) limiting herbivory. Again, the opposite holds true in marine systems.

All living organisms require energy, which they obtain either as producers (via photosynthesis or chemosynthesis) or as consumers eating other organisms. Photosynthetic organisms use light energy (solar radiation) to convert simple inorganic compounds (carbon dioxide and water) into complex energy-rich organic substances, which provide food. Chemosynthetic organisms live in the absence of light and derive energy from substances such as hydrogen sulfide (H_2S).

In addition to water and energy, marine life forms require nutrients and trace elements in sufficient quantities—just as humans require certain minerals and vitamins in small quantities in their diet. Nutrients that are required by marine organisms in relatively large quantities include phosphorus and nitrogen compounds, calcium, and magnesium—all of which are essential for plant growth. Iron and several other elements, so-called *trace elements* or *micronutrients*, are also required, but in very small quantities. Iron is especially important because it plays a role in the formation of *chlorophyll*, the most common photosynthetic pigment in plant cells, and it is essential for certain plant enzymes. When any one of these essential factors is not available in sufficient quantities, plant growth is limited.

According to the **law of the minimum**, the growth and well being of an organism is limited by the essential resource that is in lowest supply relative to what is required. This most deficient resource is known as the **limiting factor**. For example, trace amounts of iron may be a limiting factor in high nutrient-low chlorophyll (HNLC) areas of the ocean. For more on this topic and the possible implications for mitigating the current upward trend in atmospheric CO_2, see this chapter's first Essay.

Structure of Marine Ecosystems

An **ecosystem** is a fundamental subdivision of the Earth system in which communities of organisms interact with one another and with the physical conditions and chemical substances of their habitats. All ecosystems have both living (*biotic*) and nonliving (*abiotic*) components plus energy sources. Biotic components include producers, consumers, and decomposers whereas abiotic components constitute their physical and chemical environment. As we shall see, ecosystems vary enormously in structure and complexity.

Where sunlight is available, the basis of ecosystems is a group of organisms collectively known as **producers**, photosynthesizing organisms. On land, they are primarily green plants. In the ocean, producers include familiar seaweeds and sea grasses in shallow waters, algae, and some bacteria. Producers are also called *autotrophs* (meaning *self-nourishing*) because they manufacture the food they need by harvesting energy (either light energy or chemical energy) from the abiotic environment. The manufactured food constitutes primary production, the base of most common ecosystems. More precisely, **primary production** is the amount of organic matter synthesized by organisms from inorganic substances and is usually measured in units of grams of carbon per square m per unit time. Autotrophs store energy-rich organic matter in their cells for use in all life processes, such as growth, maintenance of cells, and reproduction.

Organisms that feed on autotrophs are called **consumers** or *heterotrophs* (meaning *other-feeders*). The organic material produced in the growth of consumers is known as **secondary production**. Heterotrophs may be herbivores, carnivores, or omnivores. *Herbivores* feed exclusively on plants, whereas *carnivores* eat herbivores or other carnivores. *Omnivores* eat both plants and animals. In this section we take a closer look at marine producers, consumers, and decomposers (organisms that break down dead plants and animals).

PRODUCERS

In the sunlit portion of the ocean, producers are plants, algae, and some bacteria. Multi-celled plants are uncommon in the ocean. Seaweeds (which look like plants, but are actually large, multicellular algae) and sea grasses (true vascular plants) grow attached to some substrate on the ocean bottom. They can survive only in shallow, relatively clear water where light levels are sufficient for photosynthesis. Floating, microscopic unicellular algae, photosynthetic bacteria, and other groups of organisms capable of photosynthesis are responsible for more than 99% of biological production in the photic zone. Collectively these floating organisms are referred to as **phytoplankton**, from the Greek words *phyto* for plant and *plankton* for wanderer.

Compared to terrestrial ecosystems, marine ecosystems have higher rates of production for a given amount of biomass. Phytoplanktonic organisms are very small and make up only 1% of the total mass of plants on Earth, but they are responsible for considerable global primary production. As is typical of very small organisms, phytoplankton reproduce rapidly during favorable environmental conditions (as often as eight times a day) so that populations of phytoplankton species can fluctuate dramatically. Some types have very complex life cycles with numerous stages and different, poorly known forms. This is especially true of certain toxic and otherwise harmful algal species, discussed in this chapter's Case-in-Point.

Here we describe four major groups of phytoplanktonic organisms: diatoms, coccolithophorids, dinoflagellates, and bacteria. Although all are single-celled, some may form colonies or chains of individual cells. Some types of phytoplankton (i.e., dinoflagellates and bacteria) are not simply classified because they have characteristics of both plants and animals. They may carry out photosynthesis and they may also act as heterotrophs. Organisms that can be both autotrophic and heterotrophic are called **mixotrophs**. The various types of phytoplanktonic organisms range in size from less than 2 micrometers to about 2 mm (visible to the naked eye).

One of the most abundant groups in phytoplankton, **diatoms** are encased in shell-like cases (called *frustules*) made of silica (Figure 4.9A). These cases can be very ornately decorated. Diatoms contain photosynthetic pigments (primarily *chlorophyll*) and contribute significantly to primary production. When diatoms die, they sink to the bottom where their frustules form a major component of marine biogenous sediment deposits in some areas of the ocean bottom (Chapter 4).

FIGURE 9.3
Coccolithophorid bloom in the Bering Sea off Alaska, north of the Aleutian Island chain. Bright white areas are clouds or snow covered land. [NASA SeaWiFS image]

Coccolithophorids are also single-cell photosynthesizing organisms; they are covered with tiny calcium carbonate plates (Figure 4.8A). Coccolithophorids are unusual in that they appear to thrive in nutrient-poor waters, forming large blooms that turn surface waters greenish blue as viewed from space (Figure 9.3). The plates of coccolithophorids are major contributors to marine calcareous sediment deposits including, for example, the White Cliffs of Dover on the southeast coast of England (Figure 9.4). About 70 million years ago a white mud, composed of mostly coccolithophorid fragments, accumulated at the bottom of a shallow sea and gradually converted to chalk, a fine-grained calcareous rock. Although the chalk was deposited at a very slow rate (about 0.5 mm per year), up to 500 m (1600 ft) of chalk eventually accumulated in some places. Tectonic forces elevated the chalk layers above sea level and later, toward the end of the Pleistocene Ice Age, formation of the English Channel made Britain an island and exposed the White Cliffs, rising up to 90 m (300 ft) or more above sea level.

FIGURE 9.4
White Cliffs of Dover. [Photo by Sue Morley]

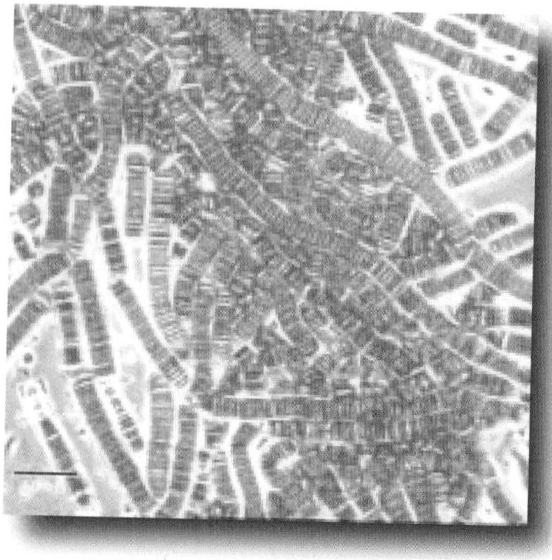

FIGURE 9.5
Cyanobacteria are the most abundant photosynthetic organisms in the ocean. [NASA]

Dinoflagellates have two flagella (thread-like structures) and no solid covering so that they are not preserved in marine sediment deposits. Many species of dinoflagellates are mixotrophs, many are bioluminescent (Chapter 10), and some species form red tides (Figure 9.2). Species of dinoflagellates that do not contain photosynthetic pigments are exclusively heterotrophs and are considered to be zooplanktonic (described below).

The smallest and probably the most numerous of phytoplanktonic organisms are certain groups of bacteria. They are distinct from the other single-celled organisms just mentioned in that they lack internal cellular structure such as nuclei, chloroplasts and mitochondria. The many types of bacteria play important roles in the functioning of marine ecosystems. Some are autotrophs; others are heterotrophs; and still others are decomposers, breaking down organic material from dead organisms (known as *detritus*) and recycling the nutrients back into the ecosystem. Cyanobacteria have a blue-green pigment and are mixotrophs (Figure 9.5). Some bacteria play an important role in the ocean's nitrogen cycle.

Whereas photosynthesis drives the most familiar marine ecosystems, scientists have discovered extraordinary ecosystems in the deep ocean in which chemical energy, rather than sunlight, drives the basic biological processes. That is, marine organisms in these ecosystems harvest energy from chemical reactions. This method of primary production is referred to as **chemosynthesis**. For example, bacteria oxidize sulfides, such as hydrogen sulfide (H_2S), to obtain energy to fix carbon from carbon dioxide into simple sugars. Regardless of whether an ecosystem is driven by photosynthesis or by chemosynthesis, the total amount of carbon converted from carbon dioxide into organic matter is known as *primary production*. The process is also referred to as *carbon fixing* because carbon is changed, or fixed, from a simple (inorganic) to a more complex (organic) form, usually energy rich sugars and carbohydrates.

CONSUMERS

Small herbivores are the chief consumers of phytoplankton; small and large carnivores, in turn, prey on herbivores. Single-cell and multi-cellular consumers that drift passively with ocean currents or are weak swimmers are referred to collectively as **zooplankton** (animal plankton). They include bacteria and dinoflagellates as small as 2 micrometers (called *microplankton*). Very large organisms, such as jellyfish and other gelatinous animals, are also zooplankton; these animals may measure as much as a meter (3 ft) across. Active swimmers can pursue and capture their prey, but most zooplanktonic organisms are suspension or **filter feeders**. They use tiny hairs (called *cilia*) or mucus-covered surfaces to capture particles suspended in the water.

The most ubiquitous members of the zooplankton are **copepods**, a diverse group of tiny crustaceans covered with an exoskeleton made of chitin, a colorless and amorphous substance (Figure 9.6). Copepods include both herbivorous and carnivorous species. In some areas of the ocean, herbivorous copepods inhibit the growth of phytoplankton populations. While nutrient availability typically limits the growth of phytoplankton populations,

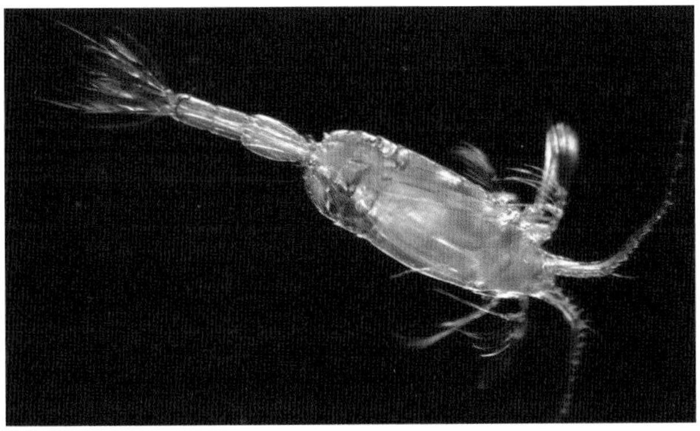

FIGURE 9.6
Copepods are tiny crustaceans that are an important food source for juvenile fish and shellfish. [NOAA Office of Ocean Exploration]

copepods are exceptional herbivores in that they can reproduce at such a high rate that they provide grazing pressure that limits phytoplankton growth. Copepods play an important intermediate role in marine food chains, feeding on phytoplankton and, in turn, being eaten by larval and juvenile fish.

Larger members of the zooplankton community include **euphausiids** and other shrimp-like crustaceans called **krill** (Figure 9.7). Krill, up to 6 cm (2.5 in.) in length, are so numerous in the ocean that their biomass is equal to the combined weight of all people on Earth! A major reason for their abundance in the Southern Ocean is the upward transport of nutrients in the North Atlantic Deep Water (NADW) around Antarctica. Krill are a direct or indirect food source for a variety of marine organisms including whales, fish, squid, seals, penguins, and seabirds. In fact, krill are regarded as the keystone species in the Southern Ocean. A *keystone species* is an essential member of an ecosystem that contributes to the ecosystem's integrity and stability; its loss would likely lead to the extinction of other forms of life in the ecosystem, thereby decreasing its biodiversity. Larval forms of fish, crabs, lobsters, starfish and other larger animals are also members of the ocean's zooplankton community.

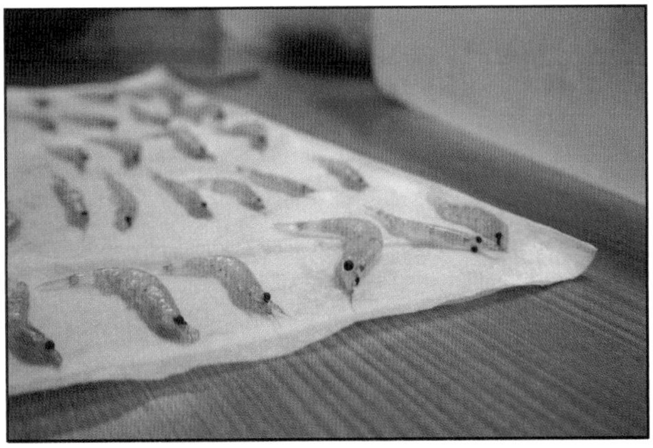

FIGURE 9.7
Euphausiids (krill) occur in large swarms in the Southern Ocean and feed on algae on the under surface of sea ice. This photo shows krill sorted by size. [Photo by Dr. David Demer, NOAA/NMFS/SWFSC/AMLR]

Zooplankton is the main food source for larger consumers in the sea. These include larval and juvenile fish and even the much larger basking sharks (e.g., whale sharks) and baleen whales (Chapter 10), which filter zooplankton out of the water by passing seawater through comb-like structures in their mouths. It is important to remember, however, that larger animals are rare in the sea compared to the enormous number of smaller zooplanktonic organisms, protists, and bacteria. In fact, the size of the average animal in the open ocean is no larger than a mosquito!

DECOMPOSERS

Decomposers, or detritivores (detritus eaters), are consumers that feed on dead organic matter, either on the ocean bottom or in the water column. Decomposers are essential components of all ecosystems because they break down organic matter and recycle nutrients back into the ecosystem. Whereas most decomposers are bacteria size, some decomposers such as worms may be complex organisms living on the sea floor and in ocean bottom sediments.

TROPHIC STRUCTURE OF ECOSYSTEMS

As noted in Chapter 1, a **food chain** is a linear path of feeding relationships among organisms (e.g., producers, consumers) along which energy and materials move within an ecosystem. For example, in a marine food chain, phytoplankton are eaten by herbivores, which in turn are eaten by carnivores, which are then eaten by other carnivores and so on. In reality, such simple food chains rarely exist in the ocean and marine heterotrophs are more likely to feed on a wide variety of prey organisms. Most ecosystems consist of several levels of consumers including fish, marine mammals, and sea birds. Humans tend to feed at the highest levels in marine ecosystems. This complex of linked food chains is more realistically described as a **food web** and each feeding position occupied by an organism within a food web is called a **trophic level** (Figure 9.8).

Within food webs, energy is transferred in only one direction: from lower trophic levels to higher trophic levels (e.g., from producers to consumers). But this energy transfer is inefficient. As organisms occupying one trophic

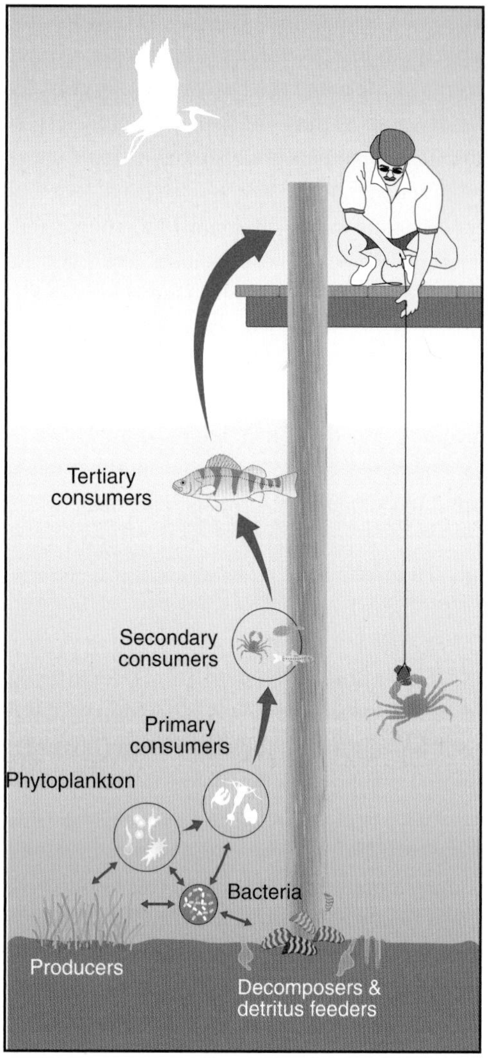

FIGURE 9.8
This simple food web is typical of an estuary. [Adapted from NOAA]

level feed on the tissues of its prey, only a small fraction of the energy is passed to the next higher level of the trophic system. We can visualize the energy distribution within a food web as a pyramid with producers forming a broad base containing most of the ecosystem's energy. Each successively higher trophic level contains considerably less energy than the one below it. That is, energy transfer within food webs is much less than 100% efficient. **Ecological efficiency** is defined as the fraction of the total energy available at one trophic level that is transformed into work or some other usable form of energy at the next higher trophic level.

Worldwide, ecological efficiency varies depending on types of organisms and ecosystems but is never very high. In some ecosystems, no more than 5% of the energy at one trophic level becomes incorporated in the tissues of organisms that occupy the next higher trophic level. An ecological efficiency of 10% is typical for the open ocean. In some highly productive marine ecosystems having few trophic levels and where strong upwelling provides an abundant supply of nutrients to surface waters, ecological efficiency can be as high as 20%.

Biomass is easier to measure than energy so that scientists usually follow and describe the flow of energy within food webs in terms of biomass. (*Biomass* is the total mass or weight of organisms.) Applying the *10% rule of ecological efficiency*, some 10,000 grams of primary producers will supply food for 1000 grams of herbivores (e.g., copepods). They, in turn, will feed 100 grams of primary consumers (e.g., fish larvae) and only 10 grams of secondary consumers—salmon, for example. If humans consume the salmon, then only 1 gram of human biomass will result from the original primary production of 10,000 grams of phytoplankton. As we have noted, however, few marine ecosystems are as simple as this example; most marine food webs consist of many different organisms feeding on multiple food sources. Many of the fish listed on restaurant menus (e.g., tuna, shark, salmon) are carnivores that feed at higher trophic levels and require considerable food energy from the ecosystem. Furthermore, a complex food web with diverse food sources tends to be more stable and less vulnerable to environmental change than is a simple ecosystem. The loss of one species as a food source may not destroy an entire complex food web as it might a simple food chain. On the other hand, a short, simple food chain is more efficient at transferring energy.

Why is ecological efficiency so low? There are many reasons. For one, not all the biomass at each trophic level is harvested. Many organisms have characteristics (adaptations) that enable them to avoid predation thereby reducing the harvest. Not all the harvested biomass is ingested. Parts of the prey organism (e.g., skeleton) may not be eaten even though they contain food energy. Not all the ingested biomass is digested (assimilated). If consumed, indigestible materials are excreted and subsequently broken down by decomposers. Furthermore, not all the assimilated biomass is converted to usable energy. Energy losses occur when assimilated food enters the organism's cells and is processed to liberate energy for maintenance, growth, and reproduction. This process of energy liberation, known as **cellular respiration**, occurs in all living cells. However, cellular respiration is inefficient. Less than half the energy in sugars (the direct source of energy for cellular respiration) is converted to a usable form. The rest is transformed to heat that cannot be used to perform work (biological activities) and eventually is released to the environment. Carbon dioxide and water, the other products of cellular respiration, also escape to the environment.

Low ecological efficiency has implications for *mariculture* such as rearing salmon in *fish farms*. Farmed salmon are generally fed fishmeal, made by grinding up marine fish caught in the wild. Given an ecological efficiency of 10%, it takes at least 10 grams of fishmeal (actually, more than 10 grams of whole fish since it is dried prior to grinding) to produce one gram of salmon. Hence, considerable biomass cannot be converted to usable energy.

BIOACCUMULATION

Food webs are pathways not only for energy but also for toxins, especially those that persist in the environment. They persist because they do not break down physically, chemically, or biologically. Examples include polychlorinated biphenyls (PCBs) and heavy metals such as mercury. Such substances tend not to dissolve in water but to concentrate in fatty tissue. These toxins may interfere with the metabolism and reproduction of sensitive organisms and in some cases pose health hazards for humans. We described a tragic example of bioaccumulation in the Case-in-Point of Chapter 4 when mercury contamination of sediments in Minamata Bay, Japan entered the marine food web and caused human illness and death.

Some persistent toxins move from one trophic level to the next, building in concentration along the way as one organism consumes another. This process of continually increasing concentration within a food web is called **bioaccumulation**. Bioaccumulation occurs because, while consumers can ingest (eat) these toxins, they cannot digest them. The digestible portion of the food consumed is converted into biomass and energy but the toxins remain, often stored in the animal's fat. Hence, while an organism on a lower trophic level may contain only a small amount of toxin, the consumer on a higher trophic level ingests many such organisms over a long period of time and the amount of toxin slowly increases (Figure 9.9). At higher trophic levels, the toxin is so great that it interferes with biological processes.

Bioaccumulation is especially pronounced in aquatic (marine and fresh water) food webs, which usually consist of 4 to 6 trophic levels rather than the 2 or 3 levels that are common in terrestrial ecosystems. In addition to contamination via bioaccumulation, some fish

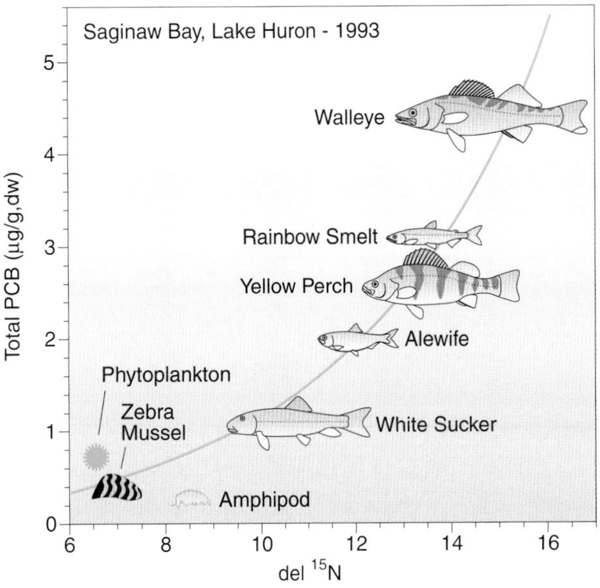

FIGURE 9.9
Field studies in the fresh water ecosystem of Saginaw Bay, Lake Huron, MI found bioaccumulation of PCBs in successively higher trophic levels. Scientists use "del ^{15}N" as a measure of the distance between trophic levels where ^{15}N is an isotope of nitrogen. [Adapted from NOAA Great Lakes Environmental Research Laboratory]

and shellfish can absorb certain toxins directly through their gills from the surrounding water. The combination of absorption plus bioaccumulation causes a toxin to become anywhere from 1000 to more than a million times more concentrated in upper-trophic level organisms (compared to the concentration of the toxin in the environment). Hence, people and fish-eating birds (e.g., eagles, osprey) that feed on fish at higher trophic levels run the greatest risk of consuming harmful levels of toxins with their meal. Especially susceptible individuals (e.g., children, pregnant women, nursing mothers) are advised to restrict their intake of fish that may have been exposed to toxins.

Persistent toxic substances can enter the ocean in the discharge of rivers or can settle out of the atmosphere. As we saw in this chapter's Case-in-Point, a toxin may even be a substance that is naturally produced by a marine organism.

Ecosystem Processes

Sunlight provides energy for photosynthetic organisms, which inhabit sunlit surface ocean waters and provide the food for other marine organisms. Chemosynthesizing organisms, which live in dark, deep-ocean waters and within the rock and sediment on the seafloor, obtain their energy through complex chemical reactions involving a mix of chemicals in the hydrothermal fluids circulating through rock fractures. In this section, we examine in more detail the sources of energy for growth and reproduction of marine organisms, marine production, how nutrients and trace elements can be limiting factors, and the important role of microbes in marine ecosystems.

ENERGY FOR GROWTH AND REPRODUCTION

Photosynthesis is represented by the following chemical equation:

$$6H_2O + 6CO_2 + \text{ENERGY} \rightarrow C_6H_{12}O_6 + 6O_2$$

Photosynthetic pigments in marine plants and bacteria absorb light energy and use it to combine water and carbon dioxide to form simple carbohydrates ($C_6H_{12}O_6$) and release oxygen gas. That is, photosynthesis converts solar energy into chemical energy. However, less than 0.5% of the available solar energy becomes chemical energy; much of the rest of the solar energy is converted to heat. Chlorophyll is the most common photosynthetic pigment and is the ingredient that makes the leaves on most terrestrial plants green. (Chlorophyll reflects radiation in the green portion of the visible spectrum and absorbs radiation at all other wavelengths.) Other pigments dominate in some plants and are responsible for the brown and red colors of some algae. The complex chemical structure of these pigments enables them to trap the energy of light photons and transfer it to electrons in the sugar or carbohydrate molecules formed during photosynthesis.

Because solar radiation is essential for photosynthesis, this process occurs only during daylight and primarily in relatively shallow waters (Chapter 5). Most sunlight is absorbed very near the ocean's surface; typically 80% is absorbed in the upper 10 m (33 ft) of water. The blue and green portions of the visible spectrum penetrate to the greatest depth in the water column (Figure 5.8) and photosynthesis is most efficient within this range (peaking at wavelengths of about 0.44 and 0.65 micrometers). The depth to which sunlight penetrates ocean water depends on the clarity of the water. Where waters are turbid because of abundant phytoplankton or suspended particles, or discolored by dissolved materials, the depth of the photic zone is reduced. In muddy rivers the photic zone may be only a few centimeters deep, whereas in very clear areas of the open ocean, it may be more than 200 m (660 ft) deep. Coastal waters, with their abundant dissolved and suspended sediments, tend to be more turbid and absorb light at much shallower depths.

Energy-releasing chemical reactions can occur in the ocean without sunlight. Many microbes synthesize organic matter, obtaining energy from chemical reactions fueled by energy-rich substances, such as hydrocarbons (petroleum or methane gas), hydrogen sulfide, and various metals. These energy-rich substances are discharged in the waters of hydrothermal vents (Chapter 2), from sea-floor petroleum seeps, decomposing whale carcasses, or decaying salt-marsh vegetation. For example, hydrogen sulfide (H_2S) takes the place of the Sun in supplying energy for chemosynthesis:

$$H_2S + 2O_2 \rightarrow H_2SO_4 + \text{Energy}$$

Hydrogen sulfide (H_2S) combines with dissolved oxygen (O_2) to produce sulfuric acid (H_2SO_4) and releases energy. Chemosynthetic organisms can use this energy to produce carbohydrates (just as in photosynthesis):

$$24H_2S + 6O_2 + 6CO_2 \rightarrow C_6H_{12}O_6 + 24S + 18H_2O$$

Hydrogen sulfide (H_2S) combines with dissolved oxygen (O_2) and carbon dioxide (CO_2) to produce carbohydrates ($C_6H_{12}O_6$), sulfur (S), and water. Hence, chemosynthesis supports ecosystems near hydrothermal vents on the ocean bottom just as photosynthesis in the surface waters supports ecosystems in much of the rest of the ocean.

On the ocean floor whale carcasses are known to nourish chemosynthetic-based ecosystems. Whales are negatively buoyant so that after death, their remains sink to the sea bottom (known as *whale falls*). In 1987, Craig Smith, an oceanographer at the University of Hawaii at Monoa, used the deep-sea research submersible *Alvin* (Chapter 13) to investigate the remains of a 20-m whale he discovered in 1240 m (4070 ft) of water at the bottom of the generally barren Santa Catalina Basin off Southern California. Smith observed that the whale skeleton, partially buried in sediment, was surrounded by diverse communities of life forms including worms, clams, snails, and limpets plus patches of white microbial mats.

As reported in the February 2010 issue of *Scientific American* by C.T.S. Little of the University of Leeds in England, subsequent studies by Smith and others have discovered many other locations on the ocean floor where whale falls support chemosynthetic communities. Scientists have identified more than 400 whale-fall related species, some 30 of which were previously unknown. Based on his study of whale-fall ecosystems during the

1990s, Smith found that at least some whale falls go through three overlapping stages of decomposition during which scavengers initially consume the whale's soft tissue, then snails, worms and other animals appear and feed on the leftovers. Finally, anaerobic bacteria break down lipids contained in the bones. In this way, a single large whale could support ecosystems for upwards of a century. According to Smith, in some cases, whale falls may be close enough on the ocean floor for larvae to move from one chemosynthetic site to another (e.g., whale carcass to hydrothermal vent), a stepping stone model for dispersal of chemosynthetic organisms in the deep ocean.

PRODUCTION IN THE PHOTIC ZONE

As noted earlier, cellular respiration makes energy available in a form that organisms use for growth, maintenance, and reproduction. In plants, photosynthesis is the dominant process during daylight hours, but both plants and animals respire continually. Whereas photosynthesis is confined to the photic zone, respiration takes place at all depths in the ocean.

Net production is the amount of organic matter produced during photosynthesis that exceeds the amount consumed via cellular respiration. Because respiration occurs at all depths and at all times in the ocean while photosynthesis occurs only in light, net photosynthetic primary production is confined to surface sunlit waters. The total amount of carbon fixed into organic matter through photosynthesis in a given unit of time is the **gross primary production** (expressed in units of grams of carbon per square m per day or year). But respiration by producers releases some of the energy stored in these carbon-containing compounds for their own metabolic processes such as growth and reproduction. The amount of carbon remaining is called **net primary production** and this is the total food and energy supply available for the rest of the ecosystem.

No net primary production occurs below the **compensation depth**—usually the ocean depth where the light level diminishes to about 1% of what it is at the surface. The compensation depth varies with location and time of day. At that depth, the amount of carbohydrates and oxygen produced by photosynthesis balances the amount consumed by respiration. Photosynthetic organisms cannot survive for long below the compensation depth because they cannot fix enough carbon or produce sufficient organic carbon based food to meet their own respiratory needs.

When organisms die, their tissues decay and nutrients are returned to the environment in their non-organic states. If decomposition occurs within the surface zone, living organisms quickly recycle nutrients. Primary production using recycled nutrients is called **regenerated production** (Table 9.1). Where decomposition occurs below the pycnocline, nutrients cannot readily mix back up into the photic zone. Therefore, recycling is much slower because little new growth occurs below the photic zone. Primary production based on nutrients brought into the ecosystem by processes such as upwelling or winter mixing is called **new production**. Pulses of phytoplankton growth may occur when storm winds and waves transport nutrients from below the pycnocline upward into the sunlit surface waters.

TABLE 9.1
Types of Marine Production

Gross Primary Production	=	New Production	+	Regenerated Production
Net Primary Production	=	Gross Primary Production	–	Respiration

The present estimate of total gross primary production in the world ocean is 50 gigatons (Gt) of carbon per year, where one Gt equals one billion tons. Given the area of the ocean, this figure translates into about 145 grams of carbon per square m per year—about the same production as terrestrial forests, grasslands, and cropland. To put these numbers into some perspective, a standard paper clip weighs about 0.5 gram so that each square meter of the ocean (summed to the depth where the light level drops to 1% of its surface value) fixes the equivalent of about 290 paper clips of carbon. Actual values of primary production (in grams of carbon per square m per year) range from less than 50 in the open ocean, up to 500 in coastal upwelling zones, and to as high as 1250 in shallow estuaries.

Primary production requires sunlight, nutrients, and phytoplankton (above the compensation depth) and varies with both location and season. Primary production is very low in much of the tropical ocean. The chief reason for this desert-like condition is lack of nutrients. In the tropics, incoming solar radiation is intense throughout the year so that the ocean is stratified year-round thereby inhibiting new nutrients from moving into the photic zone from below. Brief exceptions are caused by the strong winds of a tropical cyclone that mix water to great depths and transport nutrients into sunlit surface waters. In addition, tropical coral reefs are highly productive.

Overall, the tropical ocean has considerable biological diversity but relatively little biomass.

Low productivity in the central subtropical ocean is also due to lack of nutrients. Recall from Chapter 6 that Ekman transport in the subtropical gyres causes surface waters to converge producing net downwelling near the center of the gyres. Downwelling depresses nutrient-rich water to depths well below the photic zone.

At temperate latitudes, production varies with the season. In winter, sunlight is weak and surface waters become more dense and sink. The winds of winter storms mix water to the extent that the shallow seasonal pycnocline disappears and most autotrophs are carried to depths below the photic zone. In spring, more intense solar heating reestablishes the pycnocline and stratification, halting the deep mixing of autotrophs. With abundant nutrients, more sunlight, and a more stable water column, conditions are favorable for a dramatic increase in phytoplankton populations, an event known as the **spring bloom** (Figure 9.10). Once the spring bloom is well underway, there is an explosive growth of zooplankton populations that graze on the phytoplankton. (This is sometimes referred to as a *zooplankton bloom).* In some areas of the ocean, such as the North Atlantic, the spring bloom ends when the zooplankton have reduced the phytoplankton to levels that are not sustainable. In other areas of the ocean, the spring bloom is limited by nutrient supply. Often, a secondary bloom occurs in early fall while sunlight is sufficiently intense and the first storms of the season transport new nutrients upward into the sunlit surface layer.

Phytoplankton populations grow abundantly where upwelling transports nutrient-rich waters into the photic zone. Persistent upwelling occurs along the equator and on the eastern side of ocean basins, due to prevailing winds and offshore Ekman transport (Chapter 6). Such upwelling zones support about one quarter of all fish production in the ocean. For example, upwelling along the west coast of South America supports the rich sardine and anchovy fisheries off Chile, Ecuador, and Peru. These fisheries collapse during El Niño episodes when upwelling weakens and results in surface waters that are relatively low in nutrients and unable to promote phytoplankton growth (Chapter 11).

In polar regions lack of sunlight in fall and winter limits production although ocean stratification is generally absent and nutrients are abundant. (High latitude seas such as the Bering Sea are seasonally stratified.) In summer, after some of the ice cover melts, the long polar day triggers high primary production.

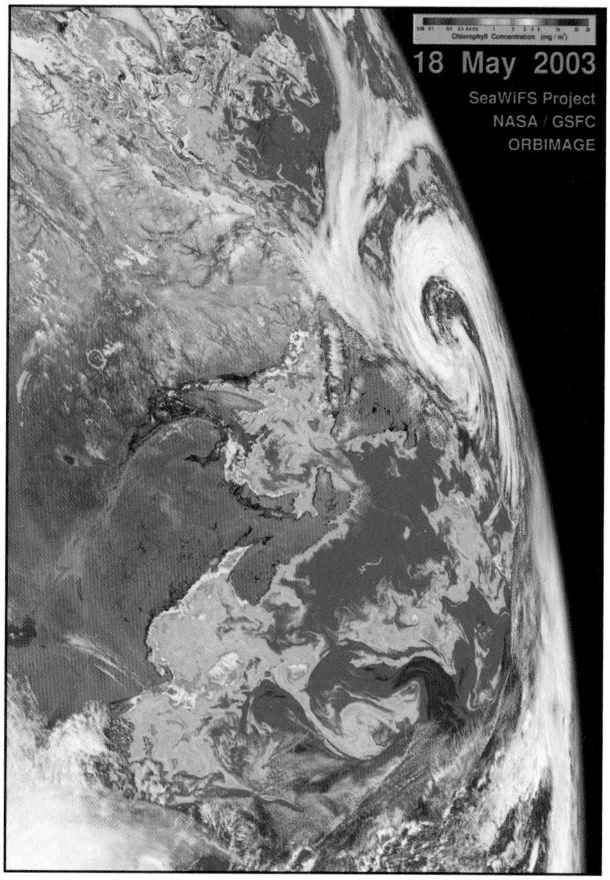

FIGURE 9.10
This satellite image shows the spring bloom around Arctic Canada in May 2003. Areas of high chlorophyll concentration are shown in red, orange, and yellow whereas areas of lower production are in shades of green and blue. [NASA]

NUTRIENTS AND TRACE ELEMENTS AS LIMITING FACTORS

For much of the world ocean, photosynthetic production continues until the primary nutrients (nitrogen and phosphorus compounds) are exhausted. However, some areas of the ocean have sufficient nitrogen, phosphorus, and sunlight but still have low levels of production. Often these areas are in the middle of ocean basins, far from land, and are known as **HNLC regions** (for high nutrients, low chlorophyll). Note the areas of very low production in the central ocean basins shown in Figure 9.11.

HNLC regions have long puzzled biological oceanographers. If the nutrient supply is sufficient, why is primary production so low? They suspected that this condition might arise from the lack of some essential element that usually occurs in trace quantities. Research conducted in the 1980s and 1990s demonstrated that iron is often a limiting trace metal in marine ecosystems. Iron is very insoluble in oxygen-rich ocean waters and usually

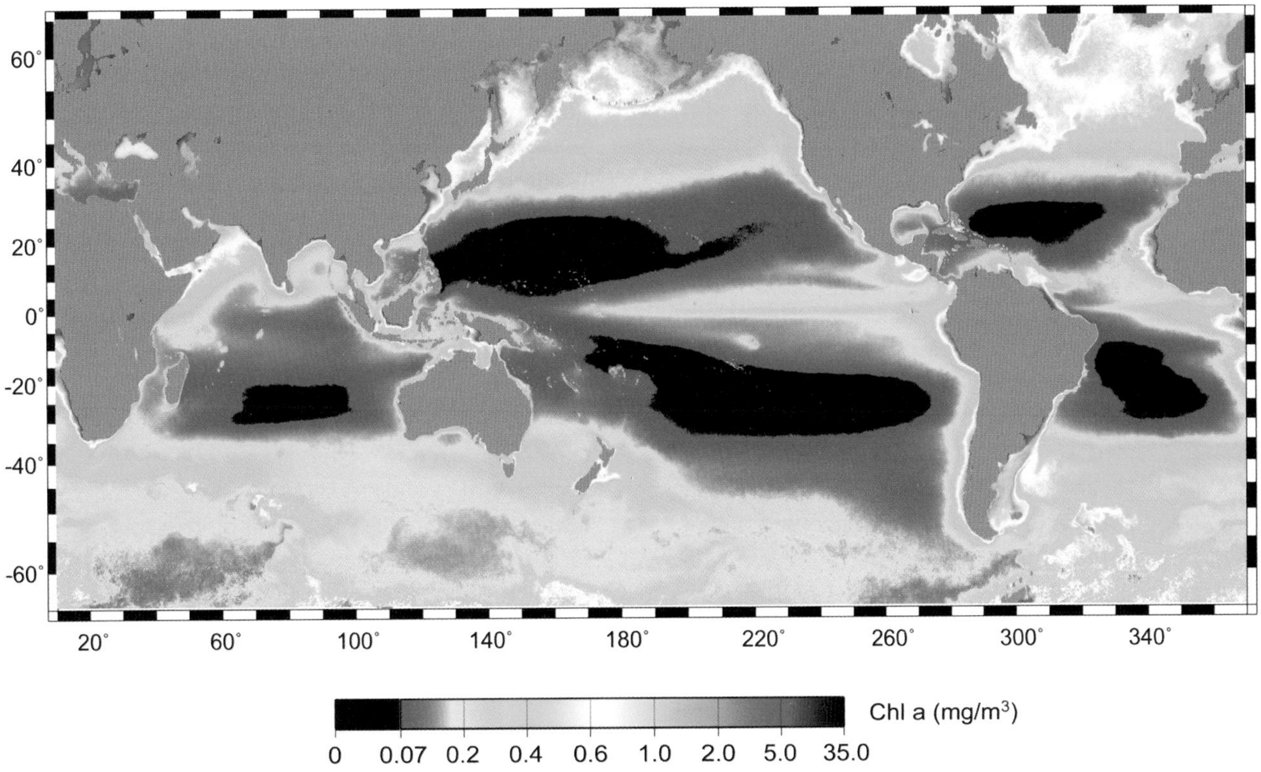

FIGURE 9.11
World map of chlorophyll concentration in March 2008. Areas of relatively high production are shown in red, orange, and yellow whereas areas of relatively low production are in dark blue and black. [Courtesy of NOAA]

attaches to particles and quickly sinks to the ocean bottom. Winds transport iron-rich dust particles from land sources (e.g., mountains and deserts) to the ocean where they settle into surface waters and often promote phytoplankton blooms (Chapter 4). However, HNLC regions of the ocean may be so distant from terrestrial sources of wind-borne iron particles that input of iron-rich dust is insufficient to make these areas productive. For more on this topic, refer to this chapter's first Essay.

MICROBIAL MARINE ECOSYSTEMS

The traditional view of marine ecosystems is that large organisms such as fish dominate food webs. However, ocean scientists now recognize that such large organisms are actually very rare in the ocean. In fact, microbes, single-celled organisms including bacteria, apparently dominate life in the ocean along with viruses and a group of organisms known as *archaea*. A single drop (1 milliliter) of seawater contains as many as one million bacteria and 1 billion viruses! The biomass of bacteria and other microbes in the ocean is so large that they are probably responsible for processing and recycling more carbon and nutrients than any other component of the ecosystem.

Now scientists are unraveling the role of these microbes in the ocean. In the 1970s the American oceanographer, John Steele, then at Woods Hole (MA) Oceanographic Institution, found that he could not account for the production of all the fish catches reported in government statistics from around the world. His estimate of total global primary production simply did not provide enough food to support the reported commercial fish catches, let alone all the other organisms in the sea. Where was the extra primary production coming from? Until that time, bacteria and protists had been considered to be important in the ocean only for their role in decomposing and recycling materials in ecosystems. Using improved microscopes and microbiological techniques, scientists discovered that microbes act as both producers and consumers. Furthermore, they concluded that microbes contribute enormously to global oceanic primary production and are especially important in open-ocean waters. Consequently, ocean scientists are revising their traditional view of the marine ecosystem in which plants are the major producers.

To observe these single-celled organisms, scientists must use light microscopes or electron scanning microscopes (for the smaller forms). The Dutch scientist

Antony van Leeuwenhoek (1632-1723), who ground his own magnifying lenses, was the first to observe microbes in fresh water. Some types of bacteria can be grown in the laboratory by placing a few drops of water on a special jelly (agar, made from marine algae). Hence, a few types of microbes, such as bacteria, are relatively well known.

The ocean is home to three major groups of microbes: bacteria, archaea, and viruses. Bacteria are single-celled prokaryotes, lacking a membrane to enclose their DNA, and, in fact have essentially no internal cellular structure. (Contrast against eukarayotes, which include single-celled organisms as well as all multicellular organisms, such as plants and animals.) They are ubiquitous throughout the planet, inhabiting everything from soil, radioactive waste, open water and living organisms. Neither prokaryotes nor eukaryotes, viruses require a host to manufacture and reproduce their DNA.

The genetic structure of archaea differs from that of bacteria. They have some different cell structural properties and unusual physiological adaptations that are likely related to the fact that archaea often inhabit extreme environments such as hydrothermal vents, Antarctic sea ice, or extremely salty pools. In the 1980s, Farooq Azam of the Scripps Institution of Oceanography in La Jolla, CA discovered archaea living in ocean water. Subsequent research has shown that archaea play important roles in decomposing simple organic compounds and dissolving organic carbon in the ocean.

Archaea living within the pore spaces of marine sediments appear to be involved in the generation of gas hydrate deposits. Gas hydrates are a potential source of methane (the chief constituent of natural gas), may contribute to global climate change, and may be a factor in the potential for submarine landslides. For more on methane hydrates, refer to this chapter's second Essay. Eleven different types of microbes (nine bacteria and two archaea) dominate life in the ocean in numbers and probably in mass. For example, one photosynthetic bacterium, *Prochlorococus*, accounts for up to half of primary production in the tropical ocean, making it the most abundant and most important primary producer on Earth. This particular organism was discovered in 1988.

Viruses are even stranger. Some question whether they are even 'living' in the traditional sense of the word because they cannot reproduce their own DNA or metabolize (create or use energy) without a host organism. Their structure is even simpler than that of prokaryotes. They are studied using molecular techniques and may be many times more abundant in ocean waters than bacteria.

Many different types of unicellular *eukaryotes* of very small sizes also inhabit the ocean. These have complex internal cellular structure, similar to the cells of more complex organisms, and are collectively known as *protista*. The major groups of protists in the ocean include ciliates and dinoflagellates. Many of these single-celled organisms are capable of weak swimming, propelled by hair-like structures.

Scientists now know that bacteria are most abundant in surface ocean waters to depths of about 150 m (490 ft). As we have seen, in addition to their importance as primary producers, bacteria and archaea are significant consumers in marine food webs. They can directly absorb and digest *dissolved organic carbon (DOC)* that cannot be used as food by larger organisms. DOC comes from many sources, including the contents of ruptured phytoplankton cells and microbial cells (as they die or are partially eaten by zooplankton), and from the liquid excretions of zooplankton.

Bacteria and archaea are food sources for the smallest heterotrophs, including dinoflagellates and ciliates. All of these microbes and protists may be too small to be a significant food source for most other zooplankton. Thus, the organic carbon they consume seldom reaches the rest of the marine food web, but is recycled within a semi-separate food web, the so-called **microbial loop** that operates parallel to the larger food web (Figure 9.12). The microbial loop is important in regulating the flow of carbon and energy in the marine ecosystem, particularly in recycling within the photic zone. The microbial loop is a self-sustaining system that scientists are just beginning to understand. It is clear,

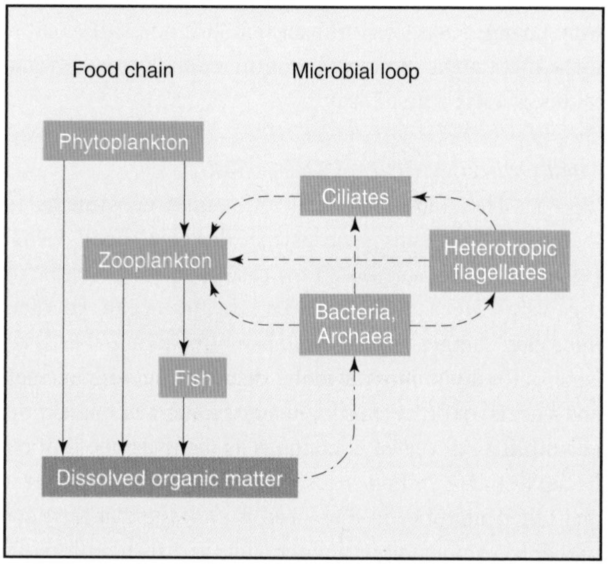

FIGURE 9.12
A simple microbial loop linked to a schematic marine food chain.

however, that the ocean is home to large numbers of bacteria, though of only a few types compared to terrestrial ecosystems. Viruses apparently infect bacteria and other microbes, thereby holding their populations in check.

Microbes occur not only in ocean water but also in greater abundance and diversity on and below the ocean floor. In fact, recent studies suggest that perhaps as much as 70% of all microbes in the Earth system inhabit the geosphere beneath the ocean water. These microorganisms depend for their survival on the chemical energy in minerals that compose oceanic crust and sediment.

In 2008, Katrina J. Edwards, a biochemist at the University of Southern California in Los Angeles, and colleagues reported on their analysis of crustal rock that crystallized from lava over the past 20,000 years along the East Pacific Rise off the northwest coast of South America. They found that while deep water in the area contained 8000 to 90,000 microorganisms per cubic cm, pore space in the underlying seafloor basalt contained 3 million to 1 billion microbes per gram (a comparable measurement to cubic cm). They also discovered that while deep water contained on average 12 types of microorganisms, the ocean bottom basalt was host to about 440 different types.

Another recent study found that microorganisms can thrive in sediments that were deposited millions of years ago but are now situated hundreds of meters below the sea floor. R. John Parkes, a microbiologist at Cardiff University in Wales, and colleagues analyzed sediment recovered off the shore of Newfoundland at a depth of 860 to 1626 m (2800 to 5330 ft) below the sea floor. They found that on average sediments held about 1.5 million microorganisms per cubic cm, about 60% of which were living. Based on the fact that microbes are known to survive at temperatures up to about 120 °C (248°F) observed at a depth of about 4000 m (13,100 ft) below the seafloor, Parkes estimates that ocean bottom sediments could host about 70% of all living microbes on Earth.

Ocean's Role in the Global Carbon Cycle

One of today's most widely discussed environmental concerns is global climate change and the role played by human-induced changes in the global carbon cycle, one of the chief biogeochemical cycles operating in the Earth system. The ocean is a major reservoir in the carbon cycle and an important player in Earth's climate system. Carbon dioxide, a greenhouse gas, is transferred between the atmosphere and ocean as part of the global carbon cycle.

In terms of changes taking place over a time frame of a few thousand years, the ocean—primarily the deep ocean—is the most significant reservoir of carbon in the Earth system. The ocean holds about 50 times the amount of carbon dioxide contained in the atmosphere and about 20 times the amount of carbon stored in the biosphere (primarily as organic matter in soils). Over time frames of hundreds of millions of years the largest carbon reservoir consists of sedimentary rock (e.g., limestone, shale) and fossil fuels (i.e., coal, oil, and natural gas). Together these large reservoirs contain almost 1400 times more carbon than in ocean water. In this section, we focus on the cycling of carbon among the ocean, atmosphere, and biosphere.

On time scales of years to millennia, two different sets of processes, biological and physical, govern the cycling of carbon into and out of the ocean. With both sets of processes, some carbon is cycled with the global thermohaline circulation and is isolated from the atmosphere for centuries to millennia whereas some carbon cycles at a faster pace as in primary production.

PHYSICAL PUMP

An important process controlling the ocean subcycle of the global carbon cycle is primarily physical in nature and is known as the **physical pump** or *solubility pump*. The latter name refers to the fact that CO_2 is more soluble in cold water than warm water. Hence, greater amounts of atmospheric CO_2 dissolve in and sink with the cold dense water masses that form at high latitudes of the North Atlantic Ocean as well as around Antarctica (Chapter 6). In this way, the physical pump acts in concert with ocean heat transport and heat loss to the atmosphere to convey carbon to the deep ocean where it may be sequestered for millennia.

Surface ocean waters take up atmospheric carbon dioxide primarily at middle and high latitudes. The major Northern Hemisphere western boundary currents (Gulf Stream and Kuroshio Current) that originate in the tropics are chilled during their flow northward. (Similar changes occur in Southern Hemisphere currents.) Because of lower temperatures and relatively high biological production in spring and summer, surface ocean waters take up large amounts of CO_2. These dense CO_2-rich waters sink as North Atlantic Deep Water (NADW) in the Greenland-Norwegian and Labrador Seas and as Antarctic Bottom Water (AABW) primarily in Antarctica's Weddell Sea.

As noted in our description of the thermohaline circulation in Chapter 6, situated above AABW is a large water mass that encircles the entire Southern Ocean: *Circumpolar Deep Water (CDW)*. CDW is produced by mixing of AABW with the overlying deep waters of the Atlantic and Pacific Ocean basins. The Antarctic Circumpolar Current spreads this water mass around Antarctica. From this southern starting point, CDW flows northward as bottom water in all three major ocean basins (Atlantic, Pacific, and Indian). Meanwhile, deep water formed in the North Atlantic spreads south across the equator and becomes entrained into the circumpolar flow around Antarctica. From the Southern Ocean, NADW spreads into both the Indian and Pacific Oceans. Flowing northward underneath the NADW in the Atlantic is a wedge of deep waters formed in the Southern Ocean. CDW flows into the Northern Hemisphere in the Atlantic, gradually mixing into the NADW lying above.

Circumpolar Deep Water circulates slowly throughout the ocean on a time scale of 1000 years. It gradually warms and diffuses upward into the surface layer where CO_2 is released (*outgassed*) to the atmosphere because carbon dioxide is less soluble in warmer water. The largest natural releases of CO_2 from the ocean's depths take place in the narrow zones of coastal and equatorial upwelling. As cold water from below moves up into the surface zone, it warms and releases some dissolved carbon dioxide. During El Niño, the presence of a thick layer of warm water in the central and eastern tropical Pacific inhibits upwelling of cold, carbon dioxide-rich waters, thus suppressing the outgassing of CO_2 from deep ocean water (Chapter 11).

BIOLOGICAL PUMP

The second set of processes whereby carbon cycles within the ocean involves biological activity and is known as the **biological pump**, the downward transport of particulate organic carbon (Figure 9.13). Eighty percent of primary production in the ocean takes place in the photic zone of the open ocean and roughly equals the amount of organic matter produced by all land plants annually. (The portion of the photic zone where photosynthesis takes place is called the *euphotic zone*). Unfortunately, the carbon sub-cycle in the ocean is not well understood, and poorly measured, especially in the intermediate- and deep-ocean zones. Consequently we examine how carbon is fixed as organic matter and then decomposed by bacteria and recycled through the ocean's ecosystems. Consider first what happens in the open ocean.

As we have seen, in the photic zone, phytoplankton and other photosynthetic organisms convert inorganic carbon into organic carbon and organisms grow and reproduce rapidly. Having very brief life spans, their cells divide several times a day and individuals die or are eaten by herbivores. In either case, the organic particles they produce, which include *marine snow*, fecal pellets of zooplankton, or aggregates of dead phytoplankton cells, sink out of the surface layer. The carbon contained in this material plus the carbon in the sinking bodies of zooplankton and other marine animals, is called **particulate organic carbon (POC)**. Up to 50% of the carbon fixed in the photic zone settles out of the surface layer in this form. POC may be eaten by zooplankton or decomposed by bacteria or other microbes. Through this process, some POC quickly converts back to dissolved inorganic carbon. POC that is eaten by zooplankton is repackaged into larger fecal pellets that sink more quickly through the water column (greater terminal velocity). Some organic matter reaches the ocean bottom and is buried by the subsequent rain of particles (Chapter 4).

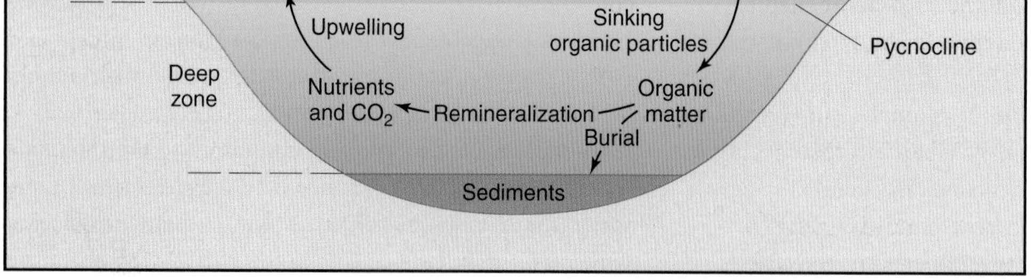

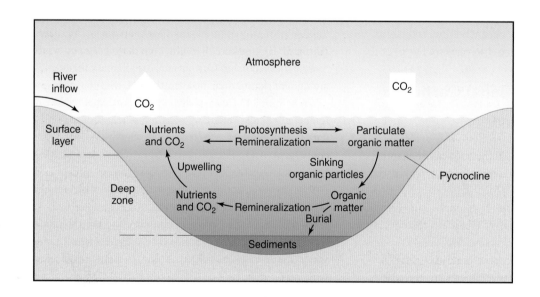

FIGURE 9.13
Biological processes that control the distribution of carbon in the ocean constitute the biological pump.

Only about 10% of the POC that sinks below the photic zone reaches depths greater than 1000 m (3300 ft). The intermediate zone of the ocean, from just below the photic zone down to a depth of 1000 m is known as the **twilight zone** (also known as the *disphotic zone*). Within the twilight zone, light levels are too weak for photosynthesis, but sufficient for visual predators to locate prey. This is also the depth where the concentration of dissolved oxygen reaches a minimum because of the higher demand for dissolved oxygen by decomposers of POC in the main pycnocline. The decline in dissolved oxygen begins just below the photic zone and the dissolved oxygen concentration reaches a minimum just beneath the pycnocline.

Particles of organic matter that settle to the sea floor and are not recycled by bottom-dwelling organisms are buried in sediment deposits, removing the carbon they contain from the water and the biosphere for perhaps millions of years. However, only about 3% of the carbon fixed in the photic zone is removed from the ocean in this way for longer than 1000 years; most organic carbon finds its way back to the surface waters and atmosphere in less time via the thermohaline circulation.

Another way in which organic carbon is recycled in the ocean is by *remineralization*, that is, by being converted back to soluble inorganic forms. Respiration by all organisms releases CO_2 into the water. If this happens at shallow depths, the CO_2 enters the surface layer and potentially escapes to the atmosphere relatively quickly (within a few years). Dissolved carbon dioxide in the deep ocean remains there for centuries or longer. It may eventually return to the surface via the physical pump described earlier. When a single-celled organism dies, its cell may rupture, spilling the contents into the seawater. Also, zooplankton and other organisms excrete liquid in addition to fecal pellets. These organic carbon-containing substances are ideal food for microbes, which ingest them, returning CO_2 in solution to the water. This process begins in the surface layer and continues through the water column as bacteria and other microbes feed on and decompose organic particles of all sizes. The ocean depth at which decomposition occurs determines the length of time that carbon is retained. Larger and heavier particles are more likely to pass through the mid-ocean depths without significant alteration by microbes until they reach the ocean floor.

Cycling of carbon into and out of the ocean is a key player in Earth's climate system (Chapter 12). Human activities, especially the burning of fossil fuels, the heating of limestone to make cement, and deforestation are adding carbon dioxide to the atmosphere. Based on analysis of ancient air bubbles trapped in polar ice cores, the pre-industrial atmospheric CO_2 concentration was about 280 parts per million by volume (ppmv). By 2010, it was about 390 ppmv and could top 550 ppmv by the close of this century if present trends continue. If there were no photosynthesis or biological pump operating in the ocean, the present atmospheric CO_2 concentrations would be 1000 ppmv and Earth's climate would be much warmer because of a greatly enhanced greenhouse effect (Chapter 5). If the efficiency of the biological pump were perfect, on the other hand, only 100 ppmv of CO_2 would be in the atmosphere.

Carbon dioxide fluxes at the interface between the ocean and atmosphere vary greatly from one region of the ocean to another. For example, across much of the central and eastern equatorial Pacific Ocean, upwelling transports cold deep-ocean waters rich in CO_2 to the sea surface. With intense solar radiation, sea-surface temperatures rise, and carbon dioxide is released to the atmosphere. In the North Atlantic, on the other hand, chilling high salinity surface waters forms dense waters that sink, carrying CO_2 from the surface down into the deep ocean. In addition, nutrient-rich waters ensure high primary production in the North Atlantic. Hence, in this region of the ocean, physical and biological processes combine to transport relatively large amounts of carbon dioxide from the atmosphere into the deep ocean. Similar conditions exist seasonally in the Southern Ocean. Decades-long studies of the ocean carbon sub-cycle have shown how the ocean "breathes," that is, where it absorbs CO_2 from the atmosphere and where it returns CO_2 to the atmosphere. This global pattern of breathing changes with the seasons and is affected by ocean circulation.

A mystery that puzzled ocean scientists for some time is the appearance of vast phytoplankton blooms at locales within an otherwise barren portion of the open ocean. These blooms along with high nutrient levels, huge numbers of diatoms, and other marine organisms occur within cold-core eddies (Chapter 6). In the summers of 2004 and 2005 (for a total of 6 months), a team of scientists led by Dennis McGillicuddy, an oceanographer with the Woods Hole Oceanographic Institution (WHOI), conducted ship-based investigations of eddies in the Atlantic's Sargasso Sea. They relied on altimeters onboard NASA satellites to locate eddies and sailed two heavily instrumented research vessels into eddies to obtain basic observational data. In a report published in *Science* magazine in 2007, the scientific team concluded that the rotating transient eddies interact with wind-driven currents to promote upwelling of nutrients that fuel phytoplankton blooms in the upper layers of water. It had been assumed that within eddies, carbon particles (e.g., from the remains of phytoplankton blooms) in surface waters sink to the deep ocean floor where they are sequestered in sediments. However, the team found that most of the biologically fixed carbon was recycled within the surface waters and not exported to deep water. This discovery points to a mechanism that supplies a major portion of the nutrients that sustains primary production in the subtropical ocean.

To this point, we have been discussing the operation of the carbon sub-cycle in the open ocean. Consider now the carbon sub-cycle in the coastal ocean. The coastal ocean is highly productive, receiving organic matter from many sources including tidal currents that flow through wetlands, the discharge of rivers and streams, wind-borne particles, and erosion of sediments deposited on continental margins. The total amount of organic carbon of terrestrial origin in the coastal ocean is estimated to be equivalent to about one-fifth of the primary production of open-ocean surface waters. Burial rates of carbon are relatively high in the coastal zone because of a combination of high production and shallow water. This supply of carbon is exacerbated by the discharge of nutrients by rivers and streams that stimulate production.

Ecosystem Observations and Models

Studies of the oceanic carbon sub-cycle on various temporal and spatial scales rely on data obtained remotely by satellite-borne instruments. One of these satellite sensors is NASA's *Sea-viewing Wide Field-of-View Sensor (SeaWiFS)*. SeaWiFS maps ocean color, that is, the distribution of photosynthetic pigments and particularly chlorophyll-a, the most significant photosynthetic pigment in marine algae. With frequent global observations, we can follow annual and seasonal changes in the distribution of chlorophyll-a concentration.

Global satellite images obtained from SeaWiFS show the narrow bands of highly productive coastal upwelling waters, especially along the western sides of the continents (Figure 9.11), and the moderately productive waters of the central and eastern equatorial Pacific and the margins of the Southern Ocean. Using such data, the ocean can be divided into similar biological provinces. In addition, the *Moderate-resolution Imaging Spectroradiometer (MODIS)* onboard NASA satellites and the *Advanced Very High-Resolution Radiometer (AVHRR)* sensor onboard NOAA's polar-orbiting satellites measure sea-surface temperatures on a global scale.

By combining remote sensing data from satellite sensors such as SeaWiFS and the AVHRR, one biological and the other physical, scientists can estimate the global (or regional) annual primary production (exclusive of that due to autotrophic microbes or chemosynthesis). In this way, they can deduce the amount of carbon fixed through photosynthesis. This information adds to our understanding of the role of the oceanic carbon sub-cycle in global climate change (Chapter 12). For more on SeaWiFS, refer to this chapter's third Essay.

For information on ocean properties at depths below the range of satellite sensors, ocean scientists rely on in situ observations taken at specific locations over a lengthy period, spanning years, decades, or longer. A major impetus for developing such time series of data was the *Joint Global Ocean Flux Study (JGOFS)*. This multi-disciplinary international program (with participants from 20 nations) was established in 1987 under the auspices of the Scientific Committee of Oceanic Research (SCOR). The principal goal of JGOFS was to "assess more accurately, and understand better the processes controlling, regional to global and seasonal to inter-annual fluxes of carbon between the atmosphere, surface ocean and ocean interior, and their sensitivity to climate changes." After JGOFS ended in 2005, the *Ocean Carbon & Biochemistry (OCB) Program* was created by NSF, NASA, and NOAA in 2006 to continue research on the ocean's role in the global carbon cycle. In late 1988, JGOFS established long-term ocean time series projects at sites near Bermuda and Hawaii, now called BATS and HOT, respectively. At these stations, observations continue to be routinely collected on numerous physical, chemical, optical, biological, and atmospheric variables from the surface through the ocean depths.

These time series data permit scientists to construct numerical models ranging in complexity from those applicable to a single location, to three-dimensional models of the global ocean. The principal goals of these models are to identify processes and explain observations, combine many types of data to improve estimates of biogeochemical fluxes, and improve predictability of future ocean states. Another major goal of ocean modelers is to produce coupled models that integrate biogeochemical and physical processes. However, physical data that drive the models (e.g., solar radiation, wind, cloud cover) currently are more readily available than biogeochemical data. Biogeochemical aspects of ocean modeling are essential and data are being collected, processes are being studied, and the gap between physical modeling and biogeochemical modeling is closing.

When the model results match observational data, researchers can be reasonably confident that they have included most of the major processes affecting ecosystems and biogeochemical cycles and their model more accurately reflects what is happening in the real ocean. In cases where models cannot reproduce the observed data, important processes are still not adequately understood or represented. In some cases, more research on the processes is required and in others, more environmental data must be collected to permit more realistic simulations of ocean processes in time and space.

Conclusions

Although many of the fundamental ingredients of life in the ocean are the same as those required by terrestrial life, the ocean has a larger vertical life zone, a greater variety of habitats than the continents, and a very different interaction between plants and herbivores compared to that in terrestrial systems. Marine ecosystems are composed of producers, consumers, and decomposers and their energy is supplied via photosynthesis or chemosynthesis. For several reasons, energy transfer between trophic levels in marine food webs is inefficient. Persistent toxic chemicals may enter marine food webs and bioaccumulate, threatening the well being of organisms (including humans) feeding at higher trophic levels.

We have seen that primary production in the ocean's photic zone varies spatially and temporally because of seasonal changes in sunlight, weather, fluctuations in nutrient supply, the availability of trace elements, stratification of ocean waters, and upwelling. We have also examined the physical and biological processes that govern the role of the ocean in the global carbon cycle. We return to this topic in Chapter 12 where we discuss prospects for an enhanced greenhouse effect and global climate change. Our examination of ocean life continues in the next chapter.

Basic Understandings

- Many of the fundamental requirements for life in the ocean are the same as those needed by terrestrial organisms. Essential for all life on Earth are energy sources, liquid water, the appropriate mix of chemical constituents (e.g., nutrients), and a favorable combination of environmental conditions (e.g., range of temperature).
- Major differences between the ocean and land include the much greater space, variety of marine habitats, and interactions between plants and herbivores. Also, terrestrial plants require a substrate to support them, usually soil. In the ocean, stratification plays this role for phytoplankton and is an important part of their ecosystem, to keep them in the euphotic zone so that they can produce more than they respire.
- When any one of the essential factors (e.g., nutrients) is not available in sufficient quantities, plant growth is limited. According to the law of the minimum, the growth and well being of an organism is limited by the essential resource that is in lowest supply relative to what is required. This most deficient resource is known as a limiting factor.
- An ecosystem is a fundamental subdivision of the Earth system in which communities of organisms interact with one another and with the physical conditions and chemical constituents of their surroundings. Ecosystems vary in complexity but all have biotic (living) and abiotic (non-living) components.
- The simplest organisms in marine ecosystems consist of producers (or autotrophs), most commonly photosynthesizing plants, algae and bacteria. Autotrophs manufacture their food from the physical and chemical environment. That food constitutes primary production. More precisely, primary production is the amount of organic matter synthesized by organisms from inorganic substances and is usually measured in units of grams of carbon per square m per unit time.
- Organisms that feed on autotrophs are called consumers or heterotrophs. The organic material produced in the growth of consumers is known as secondary production. Heterotrophs may be herbivores, carnivores, or omnivores. Organisms that function as both autotrophs and heterotrophs are called mixotrophs.
- Marine producers are algae and bacteria, which use sunlight to produce food through photosynthesis. Floating, microscopic unicellular algae, photosynthetic bacteria, and other groups of organisms capable of photosynthesis are responsible for much of biological primary production in the ocean. Collectively these organisms are referred to as phytoplankton. Four

principal groups of phytoplanktonic organisms are diatoms, coccolithophorids, dinoflagellates, and bacteria.

- While photosynthesis drives most marine ecosystems, for ecosystems in the deep ocean, chemical energy, rather than sunlight, drives basic biological processes. This method of primary production is referred to as chemosynthesis.
- Single-celled and multi-cellular consumers that drift passively with ocean currents or are weak swimmers collectively are known as zooplankton. They include bacteria and dinoflagellates as small as 2 mm, as well as very large organisms, such as jellyfish and other gelatinous animals.
- Decomposers or detritivores (detritus eaters) are consumers that feed on dead organic matter, either on the ocean bottom or in the water column. They are essential components of all ecosystems because they break down organic matter and recycle nutrients back into the ecosystem.
- Food chains and food webs are sequences of feeding relationships among organisms (e.g., producers, consumers) through which energy and materials move within an ecosystem. Within food chains and food webs, energy transfer occurs in one direction, from lower trophic (feeding) levels to higher trophic levels (from producers to consumers). But, for a variety of reasons, this energy transfer is inefficient. As an organism occupying one trophic level feeds on its prey, a small portion of the energy stored in the tissues of the prey organism is passed on to the next higher trophic level. In the open ocean, ecological efficiency is about 10% where ecological efficiency is defined as the fraction of the total energy at one trophic level that is transformed into work or some other usable form of energy at the next higher trophic level.
- Food webs are pathways not only for energy but also for toxins that persist in the environment because they do not break down physically, chemically, or biologically. Some persistent toxins move from one trophic level to the next higher trophic level, building in concentration as one organism consumes another. This process of continually increasing concentration within a food web is called bioaccumulation.
- Sunlight supplies the energy needed by photosynthetic organisms to convert organic matter from water and CO_2, substances that are readily available from the atmosphere and ocean. Because solar radiation is essential for photosynthesis, this process occurs only during daylight and primarily in relatively shallow waters.
- Other energy-releasing chemical reactions can occur in the deep ocean in the absence of sunlight. Many microbes synthesize organic matter, obtaining energy from chemical reactions fueled by energy-rich substances, such as hydrocarbons (petroleum or methane gas), hydrogen sulfide, and various metals.
- The total amount of carbon fixed into organic matter through photosynthesis in a given unit of time is the gross primary production (expressed in units of grams of carbon per square m per day or year). But respiration by producers releases some of the energy stored in these carbon-containing compounds for their own metabolic processes. The amount of carbon remaining is the net primary production and this is the total food and energy supply available for the rest of the ecosystem. No net primary production occurs below the compensation depth—usually the depth where the light level diminishes to about 1% of what it is at the ocean surface.
- Photosynthetic primary production requires sunlight, nutrients, and phytoplankton above the compensation depth and varies with both location and season. Primary production is very low in much of the tropical and subtropical ocean chiefly because of a lack of nutrients or trace elements. At temperate and polar latitudes, productivity varies seasonally and geographically.
- Phytoplankton grows abundantly where upwelling brings nutrient-rich waters into the photic zone. Areas of persistent wind-induced upwelling occur along the equator and on eastern sides of ocean basins, due to prevailing winds and Ekman transport.
- In the ocean, tiny microbes far outnumber large organisms. The ocean is home to three major groups of microbes: bacteria, archaea, and viruses.
- An important process controlling the ocean's role in the global carbon cycle is primarily physical in nature. This is called the physical pump or the solubility pump. The latter name refers to the fact that CO_2 is more soluble in cold water than warm

water. Hence, greater amounts of atmospheric carbon dioxide dissolve in and sink with the cold and dense water masses that form at high latitudes of the North Atlantic as well as around Antarctica.

- The second set of processes by which carbon cycles through the ocean constitutes the biological pump. Some of the carbon that is fixed as organic particles in the photic zone sinks to the deep ocean bottom where it is sequestered for an extended period. However, consumption and remineralization within the twilight zone decrease the efficiency of sequestration.
- Studies of the oceanic carbon sub-cycle on large temporal and spatial scales depend on data obtained by remote sensing from satellite-borne instruments. One of these satellite sensors is NASA's Sea-viewing Wide Field-of-View Sensor (SeaWiFS). SeaWiFS maps surface ocean color, that is, the distribution of photosynthetic pigments and particularly chlorophyll-a, the most significant photosynthetic pigment in marine algae.

Enduring Ideas

- All organisms that live in the ocean require a source of energy. That energy is made available via either photosynthesis (in sunlit waters) or chemosynthesis (in dark, deep waters).
- An ecosystem is a fundamental subdivision of the Earth system in which biotic components interact with one another and their abiotic environment. Biotic components include producers, consumers, and decomposers.
- Energy and materials are transferred within and among ecosystems along food chains or more complex food webs, linear paths of feeding relationships among organisms (e.g., producers, consumers). Energy transfer is always in one direction (producers to consumers) and, for a variety of reasons, relatively inefficient.
- Bioaccumulation is the process whereby persistent toxins become more concentrated as they move from one trophic (feeding) level to the next. This process tends to be more pronounced in aquatic (marine and fresh water) food webs.
- The ocean's primary role in the global carbon cycle entails both physical and biological processes. The physical pump hinges on the greater solubility of carbon dioxide in cold water. In the biological pump, carbon is fixed as organic matter and then decomposed by bacteria and recycled through marine ecosystems.

Review

1. In what general ways do marine ecosystems differ from terrestrial ecosystems?
2. Distinguish among producers, consumers, and decomposers in ecosystems.
3. Define "limiting factor" and briefly explain its significance.
4. Distinguish between photosynthesis and chemosynthesis.
5. Photosynthesis is confined to what portion of the ocean?
6. What is the 10% rule of ecological efficiency?
7. How is cellular respiration one of the reasons for the relatively low ecological efficiency?
8. Define "bioaccumulation" and explain its potential significance for human health.
9. Distinguish between gross primary production and net primary production.
10. Distinguish between the physical pump and the biological pump in the carbon sub-cycle.

Critical Thinking

1. Summarize the reasons why marine ecological efficiency is relatively low.
2. If no photosynthesis or biological pump operated in the ocean, what is likely to happen to the concentration of atmospheric carbon dioxide?
3. All other factors being equal, how does sea-surface temperature affect the rate at which carbon dioxide dissolves in seawater?
4. How might the law of the minimum apply in the event of a major change in climate?
5. How do humans attempt to increase ecological efficiency?
6. Explain why bioaccumulation of toxins often is a more serious problem in marine food webs than in terrestrial food webs.
7. What are the basic requirements for primary production at ocean depths above the compensation depth?
8. Identify factors that reduce the flux of carbon from surface ocean waters to the ocean depths.
9. What are the implications of iron fertilization for ocean acidification?
10. Humans feed at what level of a marine food web?

ESSAY: Ocean Iron Fertilization and Climate Change

Ocean iron fertilization (OIF) has been proposed as a means of boosting the ocean's uptake of carbon dioxide from the atmosphere. OIF refers to the intentional introduction of iron into surface ocean waters to stimulate an algal bloom that will remove CO_2 via photosynthesis. Marine organisms then feed on the bloom, producing carbon-containing waste particles that drift to the sea bottom where the carbon is sequestered for centuries. A decrease of CO_2 in surface waters increases the flux of CO_2 from the atmosphere into the ocean.

Despite receiving abundant sunshine, broad areas of open ocean are nearly devoid of algal blooms. While it may be due to the low concentrations of dissolved phosphorus and nitrogen compounds, other areas have high concentrations of those essential nutrients but low concentrations of algae. Such waters occur in the Antarctic, subarctic, and equatorial Pacific Ocean.

The mysterious lack of algae blooms was explained by John H. Martin (1935-1993), an oceanographer based at Moss Landing (CA) Marine Laboratories. Initially, Martin studied how classic plant nutrients (nitrogen, phosphorus) limit growth and then focused on the role of trace metals in seawater and algae. Martin's study was made difficult by potential contamination from sampling seawater onboard ship and during laboratory chemical analyses. To avoid contamination, he used special clean rooms with filtered air, ultra-clean plastics. and even prepared his own ultra-pure chemical reagents. He found that most trace-metal concentrations reported in the scientific literature were too high because of contamination.

Martin discovered that while algal growth is inhibited by the presence of copper and zinc in sea water, trace amounts of these metals are essential for their growth. He then proposed that the high-nutrient low-chlorophyll (HNLC) areas, which cover a third of the ocean, are deficient in iron and other trace elements resulting in low productivity. He proposed that large amounts of dissolved iron would stimulate algal growth, decreasing CO_2 in the surface water, and so increasing the CO_2 drawn from the atmosphere.

Lecturing at the Woods Hole (MA) Oceanographic Institution in July 1988, Martin said, "Give me a half tanker of iron, and I will give you an ice age." Martin's declaration came at a time when scientific interest was growing in the carbon dioxide-enhanced greenhouse effect and global temperature change. Ocean iron fertilization (OIF) attempts to boost the ocean's biological pump. But for this to work as a climate change mitigation strategy, two additional consequences are required. The enhanced biological pump must lead to a lower dissolved CO_2 concentration in surface ocean waters thereby increasing the rate of transfer of CO_2 from the atmosphere and the carbon transported to the deep ocean must remain sequestered in the ocean depths, not be quickly upwelled. Unfortunately, Martin's favored HNLC locales are regions of upwelling (e.g., the tropical Pacific Ocean and Southern Ocean) where models suggest that any carbon pumped down would be just as quickly upwelled and vented to the atmosphere.

Before he could carry out his ocean iron fertilization experiment, Martin died. But, In 1993, Kenneth Coale and his colleagues at Moss Landing Marine Laboratories, conducted Martin's experiment in a patch of HNLC water near the Galápagos Islands in the eastern equatorial Pacific. From their ship, the *R.V. Melville*, they released a half ton of iron over an ocean area of 64 square km (25 square mi). Within a day, the clear water turned soupy green as algae levels increased three-fold. However, the hoped for sequestering of CO_2 in the deep ocean did not happen because zooplankton consumed much of the algae, and released CO_2 back into surface waters.

In 1995, Coale and colleagues conducted another OIF experiment in the eastern Pacific Ocean that succeeded in both increasing algal biomass and reducing the CO_2 concentration of surface waters. The different result was attributed in part to changes in experimental procedure. Subsequently, three open-ocean iron fertilization experiments were conducted in the Southern Ocean and all three increased algal biomass and reduced the amount of carbon dioxide dissolved in surface waters. However, evidence that particulate organic carbon (POC) sank into the deep ocean was inconclusive.

The international oceanographic community has conducted more than a dozen OIF experiments since 1993. Although adding to our understanding of the role of iron in marine ecosystems, the outcomes of these experiments were inconclusive about OIF's effectiveness in sequestering carbon. The next generation of OIF experiments will aim at answering key questions including the fate of the algal blooms, how much carbon is consumed by zooplankton and how much sinks to the ocean floor. The goal is to sequester carbon below the 100-year horizon, equivalent to a water depth of 500 m (1600 ft). Water below this level does not come into contact with the atmosphere for at least a century. Numerical models predict that HNLC regions of the ocean are not likely to sequester more than several hundred million tons of carbon each year. Furthermore, the scale of these field experiments makes accurate assessment very challenging.

A volcanic eruption in August 2008 cast further doubt on the efficacy of reducing the concentration of atmospheric CO_2 by seeding the ocean with iron. Fortuitously, scientists on a research cruise observed a huge spike in the plankton population, with the chlorophyll concentration increasing 150% in only a few days over a broad area of the northeast Pacific between Canada and the Aleutian Islands. Reporting in the October 2010 issue of *Geophysical Research Letters*, chemical oceanographer Roberta Hamme and colleagues from the University of Victoria, British Columbia, estimated that a 5- to 10-fold increase in the dissolved iron concentration was responsible. They traced the iron to an ash cloud produced during the 7-8 August eruption of the Kasatochi volcano in the Aleutians, some 2000 km (1240 mi) away. Apparently, strong winds dispersed the iron-borne ash over a wide area of the ocean. Despite the large bloom that developed, the uptake of carbon was relatively modest.

Many scientists and public policymakers vigorously oppose geoengineering schemes, such as ocean iron fertilization, aimed at combating global climate change. They argue that such quick fixes deflect public attention away from the root cause of the global environmental issue, that is, humankind's dependence on fossil fuels. There are also ethical considerations, and the many uncertainties involved with altering the Earth system on such a grand scale. Open-ocean and deep-ocean ecosystems, as much as the Earth system and possible feedbacks, are too poorly understood to accurately predict unintended consequences of OIF.

ESSAY: Methane Hydrates

Methane hydrate (also called *methane ice*) is a clathrate, an ice-like solid formed from methane (CH_4) and water, the single molecule of methane enclosed in a frozen cage of water molecules (Figure 1). Cages are linked together forming a crystalline solid similar to water ice except that the crystal structure is stabilized by the caged gas molecule. Many gases besides methane can form hydrates, such as carbon dioxide, hydrogen sulfide, and other light hydrocarbons, but methane hydrate is the most common. Methane is the chief component of natural gas.

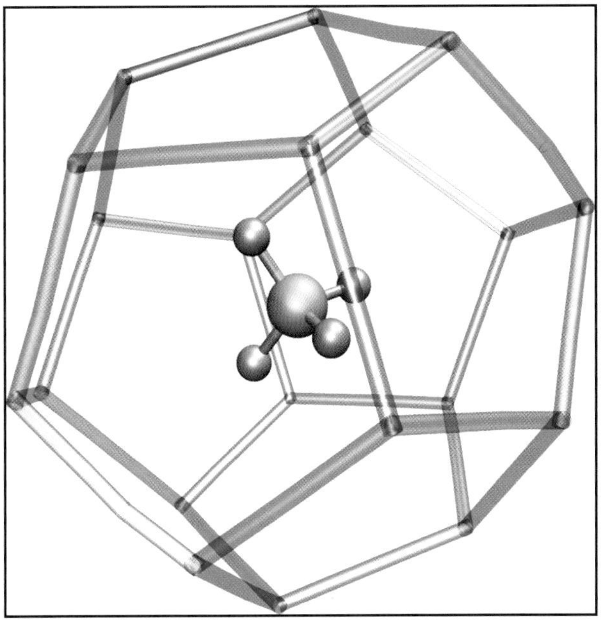

FIGURE 1
Methane hydrate is a solid substance in which water molecules form an open crystalline latice enclosing, without chemical bonding, molecules of methane. [Courtesy of U.S. Department of Energy National Energy Technology Laboratory]

As early as 1811, the British chemist Sir Humphry Davy (1778-1829) made the first gas hydrates in his laboratory, but not until 1965, under permafrost in Siberia, was a natural deposit found. Five years later, scientists accidentally drilled through methane hydrates on an expedition and although nothing happened, research-drilling efforts began avoiding suspected methane hydrate deposits, fearing that a pressurized pocket of methane could explode. But only recently have geoscientists started documenting the presence of methane hydrates in submarine and underground deposits, and exploring their potential contribution to climate change. As concerns over pressurized gas diminished, mounting scientific curiosity has emboldened researchers to try boring through gas hydrate fields. Starting in 1992, researchers with the international *Ocean Drilling Program (ODP)* intentionally breached methane hydrate deposits several times without incident. Through ongoing exploration and research, marine geologists have identified deposits at hundreds of sites along most continental margins of the world ocean (Figure 2).

Marine methane hydrates, not stable at surface temperatures and pressures, form in pore spaces within seafloor sediments at ocean depths greater than 400 m (1300 ft) where temperatures are sufficiently low, and/or pressures sufficiently high, to squeeze water and methane into hydrates. When methane hydrates, within sediment cores containing methane hydrate, or as solid chunks, are brought to the surface, the reduction in pressure and rise in temperature causes the gas hydrates to become unstable and decompose, much as ice melts when warmed. Without precautions, gas hydrates melt and fizz away before reaching the surface. Because of its volatility, chunks of methane hydrate will burn at the touch of a match and leave a puddle of water.

Though their precise origin is unknown, scientists suspect that *archaea* living within the sediments use rich organic materials and water to generate methane from carbon and hydrogen. Where there is dense sediment preventing rich detritus from oxidizing, archaea rapidly generate large quantities of methane. At the appropriate temperatures and pressures, the methane molecules are captured within the frozen hydrate structure. By contrast, conventional deposits of methane form through a different process. Seafloor sediments are buried far deeper, where temperatures are much higher, and the organic material in the sediments simmers until it transforms into petroleum and, eventually, methane.

Although it was hypothesized that bacteria might colonize methane ice mounds, multicellular animals were unexpected. Methane hydrates usually are buried deep but in the Gulf of Mexico they are exposed on the ocean bottom, and occasionally burst through in mounds often 1.8 to 2.4 m (6 to 8 ft) across. In 1997, a team of scientists using a mini research submarine on a NOAA-funded research cruise discovered a new species of centipede-like worms living among the mounds of methane ice 550 m (1800 ft) deep about 240 km (150 mi) south of New Orleans. Flat, pinkish, and 2.5 to

5 cm (1 to 2 in.) long, ice worms were living in dense colonies burrowed into the methane hydrate (Figure 3). The worms were observed using their two rows of oar-like appendages to move about the yellow and white honeycombed surface of the icy mound. Researchers speculate that the worms may be grazing on, or living symbiotically with, the chemosynthetic bacteria that grow on the methane. Scientists have also managed to keep a number of the exotic worms alive at shore side laboratories.

Methane hydrates in the Gulf of Mexico greatly affected recovery efforts following the Deepwater Horizon oil spill in 2010. Engineers attempted to capture the oil and natural gas (mostly methane) surging from the ocean floor by deploying an oil containment dome, but this and other initial efforts failed because the mixing of methane and seawater produced methane hydrate. We have more to say about the Deepwater Horizon oil spill in Chapter 15.

Terrestrial deposits of methane hydrates occur at relatively shallow depths at high latitudes where low temperatures combine with moderate pressues (rather than high pressure) to keep them stable. For example, petroleum companies have encountered methane hydrates while drilling through permafrost (perennially frozen ground) in Alaska, Canada, and Siberia. In this environment, the flux of geothermal heat from Earth's interior generates methane from deeply buried organic matter.

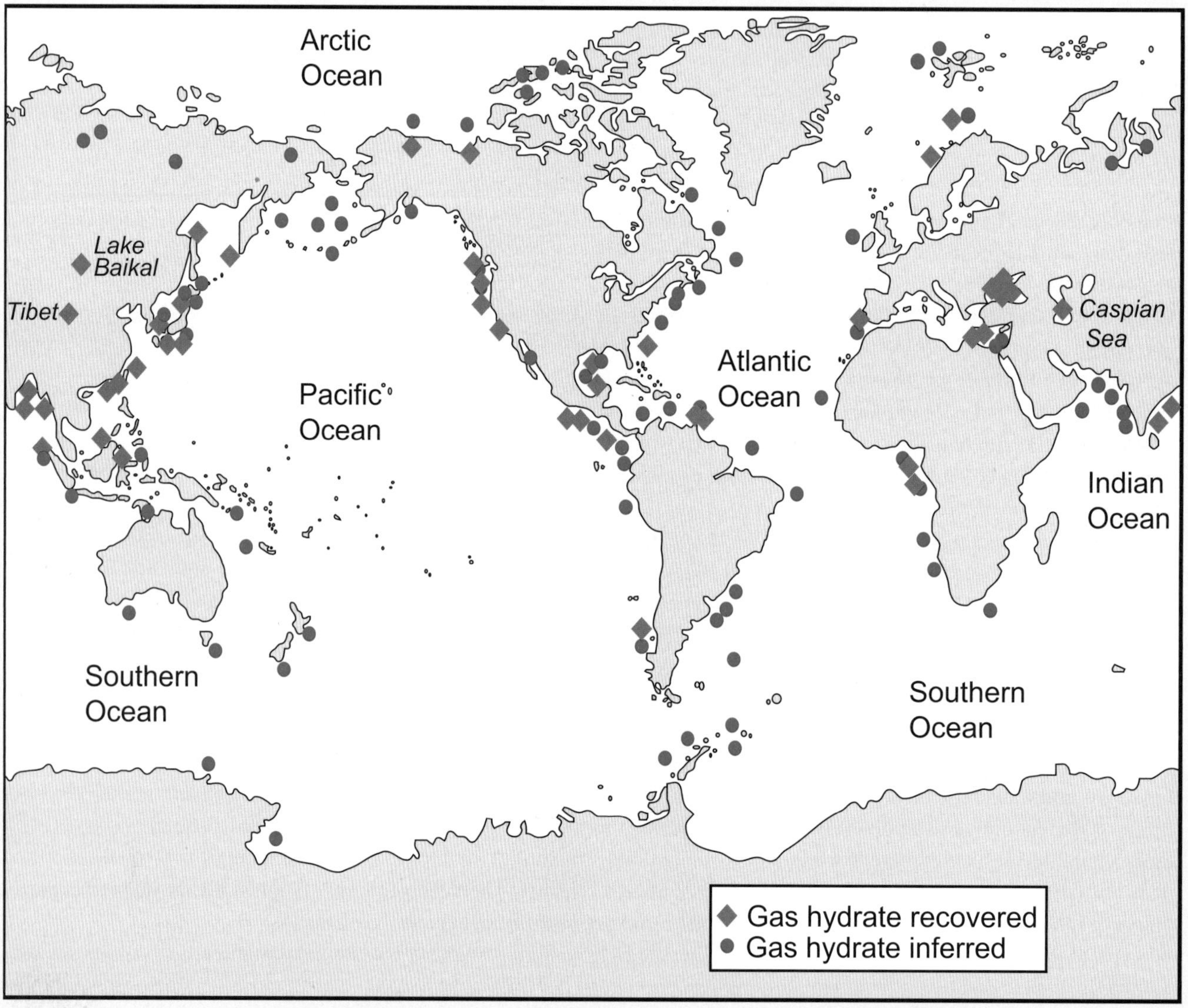

FIGURE 2
Worldwide location of recovered and inferred methane hydrate deposits. [From Collett, T. S., A. H. Johnson, C. C. Knapp, and R. Boswell, 2009, Natural Gas Hydrates: A Review, in T. Collett, A. Johnson, C. Knapp, and R. Boswell, eds., Natural gas hydrates; Energy resource potential and associated geologic hazards: AAPG Memoir 89, p. 146; 219.]

The gas migrates upward through openings in the sediments until it reaches the subsurface area where temperatures and the presence of water favor formation of methane hydrates. Methane hydrates formed below permafrost are much more concentrated than submarine deposits.

The greatest impact of methane hydrates on the Earth system could be their contribution to global climate change. Atmospheric methane is a greenhouse gas and hydrates may be a major source or sink (Chapter 5). A warming climate, increasing the temperature of the sea floor by a few degrees Celsius, would begin melting hydrates. The released methane could raise Earth's surface temperature significantly, although it would take centuries or millennia for the submarine hydrate-trapped methane to reach the atmosphere. Terrestrial methane hydrates would be released to the atmosphere as quickly as the permafrost over them, and they, melted. On the other hand, cooling could bind more methane in hydrates, removing it from the atmosphere at the same rate they could be released, and ultimately lower Earth's surface temperature. As both temperature and pressure play a role in the stability of methane hydrates, the possible feedbacks are complex.

FIGURE 3
Ice worms living in a methane hydrate deposit at a location on the floor of the Gulf of Mexico about 240 km (150 mi) south of New Orleans, LA. [Courtesy of NOAA/Pennsylvania State University]

One possible feedback, as global temperatures rise, would cause methane hydrate to begin melting, which releases methane to the atmosphere and enhances the greenhouse effect. The increase in warming would lead to melting glacial ice that would raise sea level. However, a higher sea level increases the pressure on ocean bottom sediments. This would enhance the formation of methane hydrates at the ocean bottom, thereby reducing the amount of methane entering the atmosphere. A reduction in the greenhouse effect would encourage cooling and the formation and expansion of glaciers, which would cause the sea level to drop again. While somewhat speculative, evidence from the geologic past shows that dramatic shifts in the ocean methane concentration coincided with episodes of sudden global warming. For example, release of methane from gas hydrates may have contributed to the *Paleocene/Eocene Thermal Maximum (PETM)*, the global warming at the close of the Paleocene Epoch, 55 million years ago.

Methane hydrates may also play a role in causing submarine landslides. Gas hydrates freeze sediments together, influencing the strength of sedimentary deposits and, possibly, triggering submarine landslides on continental slopes. Evidence of methane hydrate instability pockmarks the ocean floor along the Blake Ridge off the coast of South Carolina and Georgia. Numerous craters and depressions, 500 to 700 m (1600 to 2300 ft) wide and 20 to 30 m (65 to 100 ft) deep, may have formed when gas hydrates melted, releasing methane and caving the surrounding sediment. In other cases, melting at the base of the hydrate layer destabilized seafloor slopes and is thought to be a factor in submarine landslides off Alaska, the U.S. Atlantic coast, British Columbia, Norway, and Africa. Such inherent instability could spell problems for future drilling platforms resting on top of hydrate-rich deposits, such as in the Gulf of Mexico. Additionally, if a gas hydrate collapse is large enough, it could trigger a tsunami (Chapter 7).

Methane, as natural gas, is already used for energy and methane hydrate deposits could become a future energy resource. As a hydrate ice component, methane is highly concentrated, the breakdown of a single unit volume at sea level pressure produces about 160 volume units of methane gas. Methane in submarine hydrates is conservatively estimated to contain at least 10,000 gigatons of carbon, about twice the amount of all other fossil fuel reserves on Earth. The United States Geological Survey (USGS) estimates that gas hydrates on the continental margin of North Carolina alone contain 350 times the energy consumed by the U.S. in one year. At present, because most deposits are too thin to be recovered economically, it cannot be brought to market at competitive prices. Another major challenge is efficiently recovering the methane hydrate from the ocean bottom without losing the methane to the environment during hydrate decomposition.

The first successful extraction of methane from a gas hydrate deposit took place in March 2002 at a site in the Mackenzie River delta in northwestern Canada. Scientists drilled a 1200 m (3900 ft) well through permafrost, pumped hot water down the well, melted the crystals, and released natural gas to the surface, where it was flared off. Research continues on methods to mine methane from hydrates in continental margin deposits worldwide. Japan is already drilling exploratory wells in the Nankai Trough subduction zone. However, it may be decades before methane hydrates contribute to the world's energy supply.

ESSAY: Ocean Color and Marine Production

Satellite remote sensing of ocean-surface color is contributing to our understanding of the global carbon cycle, other biogeochemical cycles, and global climate change. Ocean-surface color is an index of the global distribution and concentration of photosynthetic organisms living in the upper layers of the ocean. The *Coastal Zone Color Scanner (CZCS)* flown onboard NASA's Nimbus-7 satellite pioneered the technique between 1978 and 1986. SeaWiFS (Sea-viewing Wide Field-of-view Sensor), the successor to CZCS, is on the SeaStar satellite, launched on 1 August 1997 into a Sun-synchronous orbit 705 km (438 mi) above Earth's surface. SeaWiFS is an important component of NASA's Earth Science Enterprise, designed to investigate the Earth system from space. These satellite sensors eliminate many of the sampling problems associated with ship-based measurements of marine production—greatly increasing spatial and temporal coverage. (However, remote sensing instruments must be periodically calibrated by surface data.) A satellite sensor can view every square kilometer of cloud-free ocean surface at least once every 24 hrs.

Ocean color sensors indirectly measure variations in the distribution and concentration of phytoplankton within the upper tens of meters of the open ocean (at somewhat shallower depths in more turbid coastal waters). As noted elsewhere in this chapter, phytoplankton are microscopic, single-celled organisms that form the base of most marine food webs. Chlorophyll-a (and other pigments) in phytoplankton absorbs selected wavelengths of visible solar radiation that penetrate seawater and that energy is used for photosynthesis. (The spectrum of *visible light* is composed of all colors, from violet at the short wavelength end to red at the long wavelength end.) Chlorophyll, the most common of the phytoplankton pigments, absorbs primarily in red and blue portions of the visible spectrum and reflects green light. Some of that reflected light is scattered to the ocean surface and then into the atmosphere. The more phytoplankton present, the greater is the concentration of plant pigments, and the greener is the water.

Of the scattered radiation intercepted by an ocean-color sensor, only about 10% to 20% actually originates in ocean waters; the rest is sunlight scattered back by atmospheric aerosols and air molecules (primarily nitrogen and oxygen). Ocean waters scatter very little solar radiation having wavelengths in the red (0.67 micrometer) and near infrared (0.75 micrometer). Using these two spectral channels, the atmosphere's contribution to scattered sunlight can be subtracted from the total radiation scattered by the ocean-atmosphere system. The product is a color-coded image of phytoplankton concentration.

SeaWiFS estimates near-surface phytoplankton concentration from measurements of scattered sunlight over a swath width of 2800 km (1740 mi) with a pixel resolution on the order of 1.1 km by 1.1 km. The sensor measures energy in eight spectral bands (six visible, one near-infrared, and one far-infrared) chosen because of their sensitivity to chlorophyll in phytoplankton. From data acquired over a succession of many orbits, SeaWiFS provides a composite global view of variations in marine production. SeaWiFS imagery clearly shows upwelling zones off the coasts of Peru, northwest Africa, and the U.S. West Coast, among the most productive regions of the world ocean (Figure 1). As described in Chapter 6, in these upwelling zones, coupling of the wind with surface ocean waters moves water away from the shore (Ekman transport) and nutrient-rich water wells up from below, spurring marine production. In fact, about 25% of the world fish harvest comes from coastal upwelling zones.

FIGURE 1
SeaWiFS image showing high concentrations of chlorophyll along the California, Oregon, and Washington coastlines. [Courtesy of NASA]

Marine phytoplankton plays an important role in regulating biogeochemical cycles. The atmospheric carbon dioxide concentration has been rising since the beginning of the Industrial Revolution (currently about 1% per year) principally because of the burning of fossil fuels. About half of all CO_2 released to the atmosphere from fossil fuel combustion during that time period has been taken up by ocean waters. Through photosynthesis, phytoplankton removes carbon dioxide dissolved in water and releases oxygen as a by-product. A better understanding of the distribution and concentration of photosynthetic organisms in the sea will improve our understanding of the flux of carbon dioxide into and out of the ocean and the possible implications for global climate change. Furthermore, ocean-color remote sensing is now combined with data from other passive and active satellite-borne sensors for a wide variety of practical applications from locating productive fishing grounds to helping plan the lowest fuel consumption routes for ocean shipping.

CHAPTER 10

LIFE IN THE OCEAN

Dolphins—a type of marine mammal. Marine mammals are warm blooded, air breathing animals that bear live young. [Courtesy of NOAA]

Case in Point

Codfish were once so plentiful in the western North Atlantic, especially on the Grand Banks southeast of Newfoundland, that early explorers reported that their ships were slowed by the cod and they could catch them easily using hand nets from small boats. For nearly 500 years the cod fishery was the basis of the economy of Newfoundland and other parts of Atlantic Canada. Some experts argue that the availability of easily transported salt cod (codfish preserved in salt) provided the food that sustained western European sailors as they explored the North Atlantic and colonized its shores.

Natural fluctuations in cod populations meant periods of low catches but fish stocks always recovered. All that changed after the middle of the 20th century when fishers began using huge factory trawlers to fish intensively for cod on the offshore banks, in waters beyond the limits of Canada's national jurisdiction. In 1977, Canada declared a 323-km (200-mi) exclusive fisheries zone. At that time, Canadian factory trawlers replaced foreign vessels. Although strict regulation ensued, major damage had been done to the cod fishery.

By 1992 cod stocks were so depleted that the Grand Banks fishery was closed. Almost immediately, some 40,000 people lost their jobs in Newfoundland where many small coastal villages depended entirely on the cod fishery for employment. Despite more than 15 years of relief from commercial fishing, the cod population has shown no sign of recovery; in fact, their numbers continue to decrease. Declines in the cod fishery elsewhere prompted the Canadian government in April 2003 to close much of the remaining cod fishery in other areas of the western North Atlantic off Newfoundland and Atlantic Canada. Apparently, the ecosystem that formerly supported cod stocks was altered in such a way as to favor other fish and even lobsters in place of cod. Given our present limited understanding of how this ecosystem functions, it is unclear if the fish stock and the ecosystem that sustained it for centuries can be restored.

In the early years of the 21st century, roughly the same scenario played out as the fisheries authorities of the European Union struggled to find a way to protect the dwindling cod stocks of the North Sea. Heavily subsidized fishing fleets, modern fish finding equipment, and better nets permitted fishers to catch so many fish that stocks fell below the levels needed to maintain the fishery. Efforts to regulate the fisheries were met with strong opposition from the fishers and the fishing industry throughout most of the European Union. Although fish catches were reduced, scientists argue that these cuts were too little and too late to restore stocks sufficiently to permit a sustainable fishery.

The impact of the decline in cod stocks on the fishing industry in northwestern Europe is expected to be serious with the loss of many thousands of jobs and devastation of fishing communities—comparable to what happened earlier in Newfoundland. The effects on the consumer will be significant including, for example, loss of the traditional fish and chips from the menus in British pubs and salt cod from dinner tables in Portugal. Furthermore, extensive disruption of the ecosystem may also lead to the demise of other species, including fish such as haddock that replaced cod on restaurant menus.

The problem of the declining cod fishery also extends into New England where federal courts and NOAA's National Marine Fisheries Service (NMFS) initially dealt with the problem of the declining cod fishery by reducing the length of time that fishers could spend at sea catching cod and other fish. In 2010, NMFS started a rights based management system for managing cod off New England, in which fisherman are no longer limited by days at sea.

Driving Question:

How have the large and diverse populations of marine organisms adapted to environmental conditions in the ocean?

In the previous chapter we examined the components, structure, functioning, and energy flow of marine ecosystems. In this chapter we continue our look at life in the ocean by describing the various flora and fauna, marine habitats, and some of the ways marine organisms have adapted to the diverse environmental conditions in the ocean.

Marine plants and animals are distributed non-uniformly throughout the ocean, often occurring in scattered areas. We open this chapter by describing marine habitats, oceanic life zones, and the characteristics of plankton and nekton in the open ocean. Central to this discussion is the variety of adaptations evolved by marine organisms including specialized means of floating and moving in the ocean, defense mechanisms to protect against predators, plus feeding and reproductive strategies. We then examine marine plant and animal life at the boundaries of the ocean including organisms living in the intertidal zone, wetlands and estuaries, the deep-sea floor, and coral reefs. Our discussion closes with brief descriptions of marine fishes, mammals, and reptiles plus sea birds.

Marine Habitats

The ocean contains about 99.5% of Earth's potentially inhabited living space; the remaining 0.5% is on land.

Living space in the ocean not only vastly exceeds that on land but also consists of many habitats unfamiliar to land dwellers. Ocean scientists know most about life in the ocean's relatively shallow surface layer (especially in the coastal zone and over the continental shelf), but much less about life in mid-depths, the deepest ocean waters, and the deep-sea floor.

One way we are learning more about life in the ocean is through findings of the *Census of Marine Life (CoML).* This 10-year $650 million international project, which culminated in the publication of a comprehensive report in October 2010, was the first integrated investigation of the biodiversity, distribution, and abundance of life in the ocean. The overall goal of the project was to determine what changes have occurred in ocean life and what the future might hold for ocean life. The Census was essentially a roll call of marine plants and animals including fish, sea birds, marine mammals, invertebrates, and plankton. The greatest advances were realized in biodiversity of which very little was known prior to the Census. Participating were about 2700 scientists from more than 600 institutions in some 80 nations around the world. Their efforts were coordinated under 14 Census field projects (e.g., Mid-Atlantic Ridge Ecosystem Project) while logging more than 9000 days at sea.

Census researchers logged nearly 30 million observations, all of which are available via the *Ocean Biogeographic Information System* database created by the project. Researchers described more than 1200 new species, but estimate that perhaps 80% of all ocean species remain undiscovered. The majority of undiscovered species likely live in the tropics, deep-seas, and Southern Hemisphere. An average of about 10,750 known marine species inhabits 25 key ocean regions. Crustaceans (e.g., crabs, shrimp, lobsters, and krill) account for the greatest percentage (about one-fifth) of all known marine species. Of the 25 ocean regions surveyed, the five most diverse are Australian waters (33,000 known species), Japanese waters (33,000), Chinese waters (22,365), Mediterranean Sea (16,848), and the Gulf of Mexico (15,374).

Living organisms inhabit all parts of the ocean, even extreme environments such as Arctic and Antarctic sea ice and hydrothermal vents on the sea floor. Marine habitats are diverse and include estuaries, coral reefs, polar oceans, and deep-sea trenches. Even sediments and bedrock on the sea floor harbor abundant life. We begin our survey of marine organisms and their habitats by identifying the ocean's principal life zones and then examining the plankton and nekton of the open ocean. (For information regarding life on, within, and beneath sea ice, refer to this chapter's first Essay.)

OCEANIC LIFE ZONES

There are several ways to define *oceanic life zones*. The most commonly used system defines marine habitats in terms of distance from shore and depth (Figure 10.1). In general, open ocean waters constitute the **pelagic zone**, from the ancient Greek *pelagos* meaning the deep or open ocean. Pelagic organisms include **plankton** (passive floaters or weak swimmers such as copepods, larval fish, and jellyfish) and **nekton** (strong swimmers including most fish, squid, turtles, and marine mammals). The environment of the sea floor at all depths is called the **benthic zone**. Benthic organisms (*benthos*) live either on the ocean bottom or within sediment deposits. They include attached (or sessile), burrowing, and mobile organisms, such as sea stars, crabs, worms, clams, sea cucumbers, sea anemones, urchins, snails, and barnacles.

The pelagic and benthic zones can also be subdivided according to water depth. The upper part

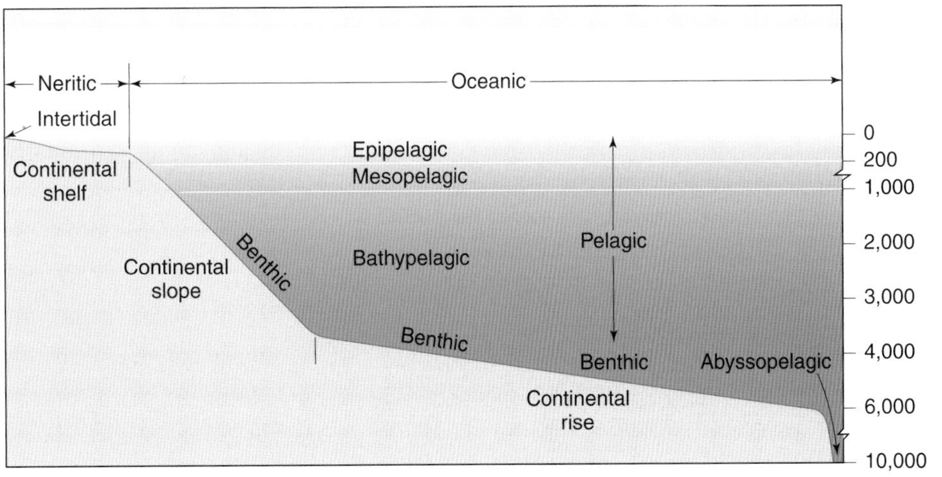

FIGURE 10.1
Life zones in the ocean.

of the pelagic zone, up to 200 m (650 ft) deep, is the *epipelagic zone* and roughly corresponds to the photic zone. Below that, from 200 to 1000 m (650 to 3300 ft), is the *mesopelagic zone*, and the deep waters of the open ocean from 1000 to 4000 m (3300 to 13,000 ft) constitute the *bathypelagic zone*. (*Bathy* is a Greek prefix meaning deep). Even greater depths are referred to as the *abyss* or *abyssopelagic zone*.

The area along the shore between high- and low-tide lines is the **intertidal zone**, also known as the *littoral zone*. Two familiar intertidal habitats are rocky shores and sandy beaches. Intertidal zones are home to ecosystems such as salt marshes and mangrove swamps. The area seaward from the shore, across the continental shelf, to the shelf break at a water depth of 120 to 200 m (390 to 650 ft) forms the **neritic zone**. The neritic zone includes the *intertidal zone* and is commonly referred to as the *coastal zone* (Chapter 8).

Oceanic life zones are also defined in terms of nutrient supply (e.g., phosphorus and nitrogen compounds) and productivity. Marine organisms are relatively sparse in the vast open-ocean areas far from land. Many of these areas are lacking in one or more nutrients essential for photosynthetic plants and bacteria. Such nutrient-poor waters having low primary production are described as *oligotrophic*. Waters are exceptionally clear and appear luminous blue in sunlight due to the lack of organisms and suspended particles. The Mediterranean Sea and large areas of open-ocean within the subtropical gyres (e.g., the Sargasso Sea) are examples of *oligotrophic waters*. On the other hand, ocean waters over continental shelves are especially rich in marine plants and animals. These relatively shallow waters are much more productive than most open-ocean waters because of coastal upwelling and river discharge of nutrients. Nutrient-rich waters having high primary productivity are described as *eutrophic*. In coastal areas where runoff from agricultural lands or sewage discharge enhances the nutrient supply, waters may become excessively eutrophic. Excessive amounts of nutrients promotes an abundant growth of phytoplankton (an algal bloom), which die, sink to the bottom, and deplete waters of dissolved oxygen as they decompose. Each summer, extreme depletion of dissolved oxygen in the Gulf waters off the Louisiana coast eliminates fish and shellfish. For details on the cause of this "dead zone," refer to this chapter's second Essay.

Algal blooms are not always associated with a heavy influx of nutrients. In some cases, blooms arise from disruption of predator/prey relationships within ecosystems. For example, in 1999, massive algal blooms began occurring in San Francisco Bay, a large estuary. Observations made over the previous 20 years reported no algal blooms and, prior to the 1999 bloom, nutrient loading in the Bay had been declining. Scientists attribute the explosive growth of algae populations to the collapse of the population of bivalves (mollusks), organisms that prey on algae. Without the bivalves to keep their populations in check, algal blooms were inevitable. The loss of bivalves, in turn, was linked to greater coastal upwelling of nutrient-rich waters that favored greater numbers of flatfish and crustaceans, consumers of bivalves.

PLANKTON IN THE PELAGIC ZONE

Most plankton are very small members of phytoplankton, zooplankton or microbial communities. In Chapter 9, we described the roles of these organisms in the functioning of marine ecosystems. We also saw earlier in this book how their remains are incorporated into marine sediments and sedimentary rocks, thereby linking the biosphere and geosphere through the rock cycle (Chapter 4). In this section we describe strategies and adaptations plankton developed for life in the pelagic zone. Here we define **adaptation** as a genetically controlled trait or characteristic that enhances an organism's chances for survival and reproduction in its environment.

Phytoplankton are slightly denser than seawater and would gradually sink below the sunlit photic zone were it not for characteristics that counteract the tendency to sink. Some species have flattened shapes; others are spiny or occur as long chain-like colonies. These shapes give the cells a relatively large surface area compared to their volume. This adaptation significantly slows sinking, especially in less dense, warm waters, and allows them to remain in the photic zone above the pycnocline. Not surprisingly, the most ornate plankton species inhabit warm tropical waters. Diatoms have a similar adaptation (Figure 10.2). The delicately ornamented outer shell (*frustule*) of diatoms consists of porous silica (SiO_2) that allows for direct uptake of nutrients and gases from seawater as well as rapid expulsion of wastes. One of the most common types of diatoms is shaped like a pillbox; others are elongated or form chains of individual cells. Diatoms are most abundant in cool, nutrient-rich waters, especially at high and middle latitudes and where silica concentrations are relatively high.

The complex shapes and spiny structures of zooplanktonic organisms also increase their surface area to volume ratio, thereby adding to their buoyancy. This is especially obvious in the larvae of many organisms, which have beautifully complex shapes. (The larvae's complex

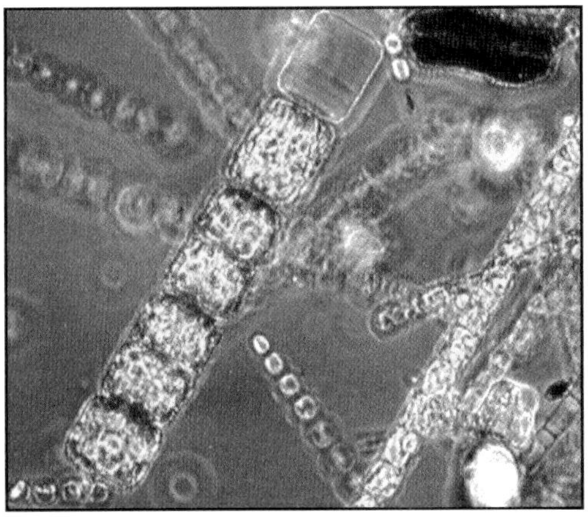

FIGURE 10.2
Although microscopic in size, diatoms can be beautiful and occur in a variety of shapes that provide them with buoyancy in the ocean. [Courtesy of NASA]

shape may also deter predators.) Another adaptation that increases the buoyancy of planktonic animals is storage of fats in large, oily globules within their body cavities. Fat is less dense than the surrounding seawater.

Some large planktonic animals have gelatinous bodies, consisting of 95% to 98% water, and are almost neutrally buoyant (having nearly the same density as the surrounding water). Jellyfish, for example, are gelatinous zooplankton having contractile tissues (primitive muscles) that enable them to pulsate their bell-shaped bodies to maintain their position and move slowly in the ocean (Figure 10.3). They swim weakly upward and then sink slowly downward, trapping organisms under their bodies with their tentacles. Jellyfish rely on jet propulsion for locomotion and have been doing so for about 550 million years. The bell-shaped body of a jellyfish is a gelatinous network of nerves and tissue. An extremely thin layer of muscle lines the internal wall of the bell. The jellyfish contracts these muscles ejecting water through the open base of the bell thereby propelling the jellyfish forward. In relatively large jellyfish (which tend to have broader and flatter bodies), jet propulsion is augmented by a paddle-type driving force.

The Portuguese man-of-war (*Physalia physalia*) is a colonial jellyfish-type organism (a siphonophore). It consists of a collection of specialized polyps of four types: a gas-filled float that holds the colony at the surface oriented with the wind, stinging tentacles that trap food, bag-like polyps that secrete enzymes to digest the trapped prey, and reproductive polyps. The specialized stinging cells (nematocysts) in the tentacles produce a powerful toxin, which can be dangerous to humans.

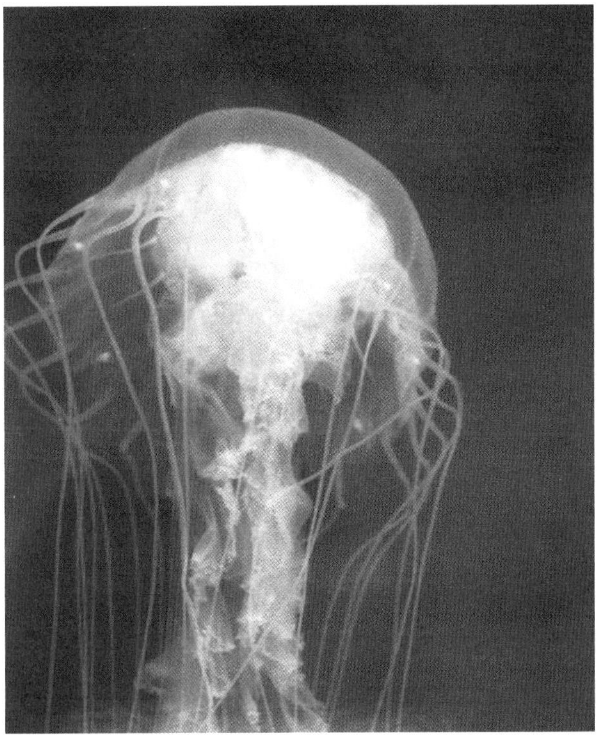

FIGURE 10.3
The bane of swimmers in the Chesapeake Bay, the stinging sea nettle, *Chrysaora quiquecirrha*. [NOAA Photo Library, photo by Mary Hollinger, NODC biologist, NOAA]

NEKTON IN THE PELAGIC ZONE

Larger, free-swimming pelagic animals, collectively called **nekton**, include fish of all sizes, squid, sea turtles, and marine mammals. Some migrate long distances. For example, California gray whales annually migrate from their breeding grounds in warm, shallow waters off the Pacific coast of Baja California, Mexico, to feeding grounds on the Arctic Ocean's continental shelf between Alaska and Siberia, a distance of about 13,000 km (8100 mi). On the other hand, the distribution of many pelagic animals is limited by their tolerance for water temperature, salinity, or food availability.

Except for the smallest zooplankton and jellyfish, marine animals are much denser than seawater. Hence, these animals must swim actively (requiring considerable energy) or utilize other mechanisms to maintain themselves at suitable depths in the ocean. Many adaptations for buoyancy are found in nektonic organisms. The simplest of these are gas bladders. The beautiful chambered nautilus (a primitive relative of squid and octopus) has an external, spiral shell with internal chambers (Figure 10.4). These animals cannot control the air pressure in the chambers—it is always the same as at sea level—so they

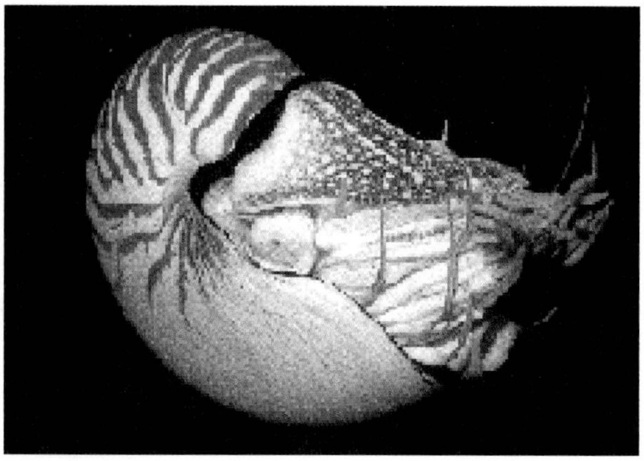

FIGURE 10.4
The nautilus is the only living cephalopod with a shell. They are ancient animals related to squid and octopi. [Courtesy of OceanLink and Bamfield Marine Sciences Center]

cannot swim too deep or the shell will collapse under the increasing water pressure. However, they can adjust the amount of fluid, and hence the air volume, in the shell chambers, thereby having some control over buoyancy. In fact, it appears that the nautilus may have been the first large swimming predator to appear in the ocean perhaps 500 million years ago. They are referred to as *living fossils*, that is, organisms that have remained essentially unchanged since their appearance in the geologic past.

Squids are invertebrates; they are mollusks and related to mussels, clams, and snails. They are extremely fast and highly maneuverable swimmers, living at mid-depths of the ocean and migrating to surface waters at night. These fast-growing animals are very successful predators. Most are from 10 to 100 cm (4 to 40 in.) in length and have acute vision. The mysterious giant squid, up to 23 m (75 ft) long, lives in deep waters where it is prey for sperm whales. A few specimens have been caught in fishing nets or have drifted ashore. Their movements are abrupt and powerful and they use their ten long tentacles, covered with suckers, to trap prey and bring it to the mouth, which has a sharp beak-like structure for tearing food apart.

The elusive giant squids are neutrally buoyant and use a special mechanism to prevent sinking. Muscles and cavities in their bodies contain abundant ammonia ions. Since ammonium chloride is less dense than the sodium chloride in seawater, they gain a buoyancy advantage. There is a disadvantage as well. Because they are neutrally buoyant, giant squid gradually float upward if they become trapped in warm water. Their hemocyanin (a form of invertebrate blood) carries oxygen less efficiently in warm water and the squid suffocates and dies. Much of what is known about these mysterious creatures has been learned from dying animals recovered from warm surface waters. Other types of squid must swim continuously in order to maintain their preferred depth in the water column. They are equipped with an internal jet propulsion system. By contracting muscles, the squid draws water into its body through a pipe-like structure called a siphon and then pumps the water out through the siphon as a jet.

Many fishes have gas-filled swim bladders that control buoyancy and regulate the amount of gas in the bladder through a connection to the gut. This makes them sensitive to changes in water pressure with depth, however. Deep-sea fishes with swim bladders are rarely caught alive. If they are brought to the surface too fast, their bladders expand and burst because they cannot adjust to the decreasing pressure fast enough. Fat is less dense than seawater so that another means of increasing buoyancy is to increase the proportion of fat in the body. To achieve neutral buoyancy, about 33% of a fish's tissue must be fat. Some sharks come close to this—their large liver is rich in a fatty substance that is much less dense than seawater. Still, many sharks must swim their entire lives to avoid sinking.

Most pelagic fish rely on active swimming to maintain their level in the ocean and obtain food. Some fish swim all the time, seeking prey over large areas of the ocean perhaps covering hundreds of kilometers each day. The streamlined shape of their body reduces frictional resistance and increases their swimming efficiency. Fish with this feeding strategy, such as tuna, require a great deal of energy, efficient muscles, and a rich oxygen supply. Most of these fish usually feed at a high trophic level and are referred to as *top predators*.

Life Strategies and Adaptations

Marine organisms have evolved many different adaptations to obtain food and avoid being eaten by predators. These include vertical migration, special coloration, eyes sensitive to low light levels, bioluminescence, use of sound, and specialized feeding strategies.

VERTICAL MIGRATION

The photic zone is where photosynthetic primary producers live and food is most plentiful. However, it is also a dangerous place for marine organisms because predators can easily see them. Many types of zooplankton avoid this threat by daily **vertical migration**. Each day at dusk, they come to the surface zone to feed on phytoplankton. As daylight comes, they return to the relative safety of darker, deep waters (the *twilight zone*). These migrations, typically over a vertical distance of about 200 m (650 ft), require expenditure of enormous amounts of energy. In a single daily migration, most small zooplanktonic animals will travel a distance equal to tens of thousands of times their body length. Some fish follow the zooplankton in their daily migration and prey on them. Vertical migration also plays a role in the carbon cycle in that carbon consumed by zooplankton feeding on phytoplankton near the surface at night is transported to deeper water as the animals respire during their return to depth at dawn.

World War II submarine sonars first detected vertical migration as a sound-scattering layer (known as the *deep scattering layer* or *DSL*) that moved up through the water after sunset and down again at sunrise. Sound waves were reflected by the millions of zooplanktonic organisms moving together up and down in the water column to and from the surface. Submarine commanders were able to use this knowledge to hide their vessels beneath the sound-scattering layer.

LIGHT AND VISION

Near the ocean surface, light is abundant and predators have no problem locating prey. Therefore, species of smaller fish that are prey for larger ones have adaptations that make them less visible and therefore less likely to be eaten. The most common adaptation is **adaptive coloration,** or camouflage, where the animal's color pattern closely matches its background substrate. Hence, they blend in with their surroundings thereby avoiding detection. Many fish exhibit **countershading**, that is, their dorsal side (or back) is a dark color, making it difficult for predators above to see them against the dark, deep water. Their ventral side (or underbelly) is lighter colored, making the fish more difficult for predators to see them from below against the more brightly lit surface waters. Squids and octopi have yet another defense against predators. They can eject a cloud of black or brown "ink" into the water and escape without being seen by a predator.

Marine animals that rely on sight to locate their prey in the dim light of the twilight zone have large, sensitive eyes. In this vast area of the ocean, from 200 to 1000 m (660 to 3300 ft) deep, there is just enough light for predators to locate their prey. In general, the deeper in the twilight zone, the dimmer is the light, and the larger are the eyes of predators living there. Many of these deeper-living fish tend to be reddish in color; red light is absorbed near the surface of the ocean so that red objects are virtually invisible at greater depths.

Below the twilight zone, in the greatest depths of the ocean, light is absent and vision is not useful for locating prey. There is a major exception to this rule, however. Many marine animals (including for example, squid, dinoflagellates, and some species of fish) emit light, a phenomenon known as **bioluminescence** (Figure 10.5). The light usually is a product of a chemical reaction that takes place in specialized cells (*photocytes*) or organs (*photophores*). When a substance known as luciferin reacts with oxygen in the presence of the enzyme luciferase, the chemical product gives off blue-green light. In some squid and fish, bacteria living within certain organs are responsible for bioluminescence. Note that bioluminescence differs from *phosphorescence* or *fluorescence* in which light involves energy received from other sources causing emission. Bioluminescence is mainly a marine phenomenon and is thought to have evolved a number of different times because it is a characteristic of a broad array of organisms.

Marine animals use light emission to attract mates or prey, frighten or confuse predators, or to disguise themselves. Various deep-sea animals exhibit characteristic patterns of light flashes or luminescent shapes to help them attract a mate. Some fish use bioluminescence to distinguish between male and female. Short bursts of bioluminescent light apparently are especially effective at disorienting prey. Some bathypelagic fishes have strange shapes and many use bioluminescence in their feeding strategies.

The female *anglerfish*, common name for members of the *Ceratiidae* fish family, has an unusual light-producing organ that ensures its survival in the

FIGURE 10.5
Bioluminescence is a property of a wide variety of marine organisms such as the jellyfish *Aequorea aequorea* pictured here. [Courtesy of NSF and Osamu Shimomura, Marine Biological Laboratory, Woods Hole, MA]

FIGURE 10.6
The female anglerfish. [SEFSC Pascagoula Laboratory; Collection of Brandi Noble, NOAA/NMFS/SEFSC]

food-limited deep-sea (Figure 10.6). A modified dorsal fin slides along a canal in the back of the fish. The structure resembles a fishing rod with a lighted blue lure at the tip of the rod above the forehead. The light is produced by symbiotic bacteria. When an unwary prey is attracted to the fish by the light, it is consumed with the sudden snap of the female anglerfish's huge mouth.

Light can also be a defensive adaptation for animals in the pelagic environment. The ink ejected by some squid species is luminescent, thereby further confusing a potential predator. Some shrimp emit a luminescent cloudy substance in the direction of an approaching predator while moving rapidly out of sight into the darkness. Also, some marine animals (e.g., squid) are able to camouflage themselves by using bioluminescence to match their body's coloration with that of their background. Even phytoplankton use bioluminescence defensively. Some organisms, such as certain species of dinoflagellates emit light when disturbed. For example, copepods feeding on dinoflagellates stimulate their light emission. But sometimes this attracts the attention of nearby fish to the copepods, allowing the dinoflagellates to escape predation.

SOUND

Marine organisms have other ways of locating prey including sound. Analogous to hearing, many marine animals have evolved ways of sensing the vibrations produced by other organisms moving through the water. A variety of structures and specialized organs have evolved for this purpose. As pointed out in Chapter 3, seawater is essentially transparent to sound. Many marine mammals, including whales, routinely communicate over great distances, even the width of an ocean basin. Scientists are beginning to understand some of their vocabulary. Whales and dolphins possess highly developed auditory senses. In addition to communication and locating prey, they may use sound for navigating over long distances. Fish too are capable of producing sound by using their swim bladders like drums. Predatory seals, dolphins, and whales take advantage of this sound in their search for prey.

Concern is growing over increasing noise levels in the ocean caused by human activities. Sounds such as those produced by large ships, military weapons testing, scientific research, and undersea oil and gas exploration significantly disturb many types of whales. Noise in the ocean may be the reason for some whale strandings in shallow waters or on beaches. Noise need not be very loud to cause problems; low frequency sound waves have higher energy and travel much farther than high frequency sound waves, especially in the ocean's SOFAR channel (Chapter 3).

FEEDING STRATEGIES

Tiny marine animals and the smaller one-cell plants and bacteria mostly move with the waters around them because they cannot swim against ocean currents. Some marine animals have evolved special mechanisms for capturing drifting organisms such as appendages covered with fine bristles that act like strainers to trap small prey. Such structures are very common among copepods and are often very ornate, especially in warm water where they also add buoyancy and act as parachutes to slow sinking out of the photic zone. The general name for marine animals equipped with features that strain food particles out of large volumes of water is **filter feeder**. Even blue whales, the largest animal known to have lived on Earth, are filter feeders. It and thirteen other species of baleen whales use plates of fibrous baleen (a substance similar to human finger nails) hanging in a row from their upper jaw to filter plankton from seawater pumped through their mouths by contractions of their massive tongue muscles (Figure 10.7). By feeding directly on primary production phytoplankton, baleen whales shorten the food chain to only two trophic levels.

Filter feeders eat most of the food items trapped in their feeding structures; that is, they are not selective. Other organisms are selective feeders and eat only specific prey organisms. Selective feeding may require special adaptations. For example, some starfish eat mussels, using strong suction to pry open their shells.

Some planktonic organisms, such as the planktonic snails called pteropods (also known as sea butterflies), produce a large sticky net of mucus, which functions much like a spider web to trap small water-borne

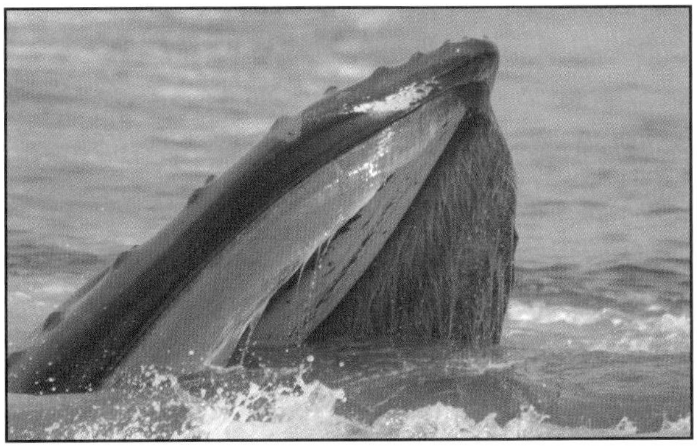

FIGURE 10.7
This humpback whale feeds by straining water and trapping food in the baleen plates in its mouth. [Courtesy of Dr. Brandon Southall, NMFS/OPR]

organisms and other food particles. The entire net is then eaten and digested along with the trapped organisms.

Life at the Ocean's Edge

After the pelagic zone, the next largest environment for marine life is the benthic zone, the sea floor at all depths, from the intertidal zone to the deep ocean. The benthic environment is essentially two-dimensional (limited vertically), unlike the vast three-dimensional pelagic zone. Marine organisms that live in the benthic zone collectively are called **benthos**.

Three basic life strategies characterize the benthic zone. Organisms may live attached to a firm surface, they may construct burrows or tunnels or simply dig into sediment deposits, or they may move freely on the sea floor. These habitats dictate their feeding strategies. Generally, attached animals are filter feeders, mobile animals are active predators, and burrowing organisms ingest sediment, digesting the organic material it contains and excreting the rest. Some also dig into the bottom sediments for protection from predators.

Attached organisms include seaweed (technically known as *macro-algae*) and sea grasses, the only truly marine form of the more complex plants (angiosperms). Macro-algae have root-like structures called *holdfasts* that attach to the ocean bottom while they absorb nutrients and water directly from the surrounding seawater. Some familiar attached animals are barnacles (Figure 10.8a), mussels (Figure 10.8b), clams (Figure 10.8c), oysters, corals, and anemones. In general, benthic animals that are not attached move relatively slowly. Starfishes, urchins, and snails are examples of mobile benthic predators. Others are crabs and

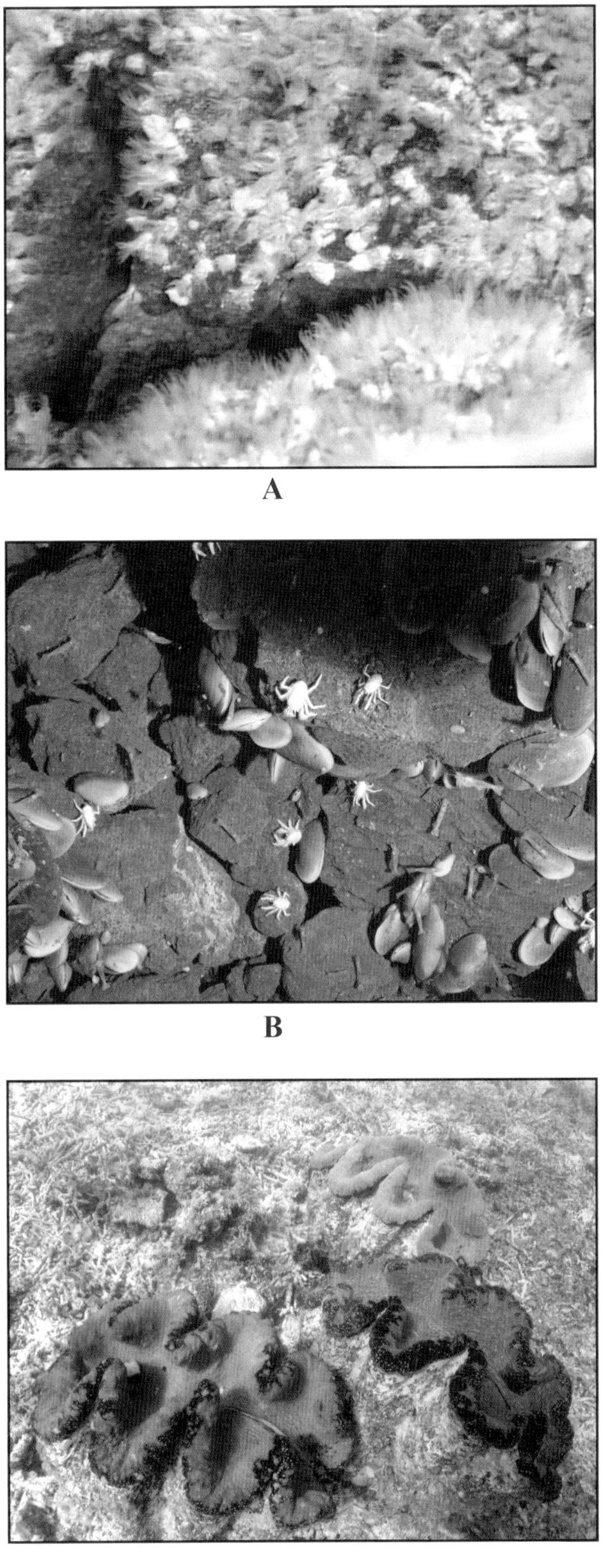

A

B

C

FIGURE 10.8
Examples of familiar benthic animals include (a) barnacles, (b) mussels, and (c) clams. [Courtesy of NOAA]

lobsters, which can move quickly in short spurts to catch their prey or to avoid predation.

Burrowing animals are less familiar to us because they largely disappear beneath the surface of sediment deposits, even in shallow tide pools. They live on sandy and muddy bottoms and include various forms of mollusks and worms. Often their presence can be detected by small holes in the sediment surface and piles of mud or sand ejected from their tunnels during burrowing.

INTERTIDAL ZONE

As mentioned earlier in this chapter, the intertidal (or littoral) zone is the area along the shore between low- and high-tide levels. Most intertidal zones are readily accessible so that properties and components are well known to scientists and amateur naturalists. Because the intertidal zone is where the atmosphere, hydrosphere, biosphere, lithosphere, and cryosphere interface and interact, this zone is among the most changeable portions of the Earth system (Chapter 8). Waves, winds, and tidal currents continually disturb the intertidal zone. With the rise and fall of the tides, the intertidal zone is alternately exposed to seawater and the atmosphere with accompanying drastic shifts in temperature and salinity. In winter, winds, tides, and currents may drive floating pans of sea ice ashore where they abrade and erode the intertidal zone. Special adaptations enable animals and plants to survive in this stressful environment.

Although all intertidal zones share some common characteristics, a distinction is made among intertidal habitats based on the type of substrate (e.g., rocky, muddy, sandy). The most energetic of these habitats are rocky intertidal shorelines where organisms are frequently subject to strong wave action, especially in winter. Rocky shorelines, especially in middle latitudes, are home to complex marine ecosystems with organisms adapted for life in particularly harsh conditions. Where the distance between low and high tide levels is relatively great, organisms may be exposed to dry air and sunshine for many hours at a stretch. Depending on the season, they may have to withstand extremely high or low air temperatures when they are not covered by seawater.

Some intertidal zones feature extensive mud flats. A **mud flat** is a nearly level area of fine silt along the shore. Streams and creeks deliver the sediment and wave action is typically weak. Mud flats are often ideal habitats for submerged aquatic vegetation, salt marshes, and many forms of benthic animals. In addition, sandy beaches are intertidal habitats (Chapter 8). Seaweeds are important components of the intertidal zone and provide habitat for animals. For example, many snails, sea urchins, and other marine herbivores graze on seaweed and may control its population in some areas. Seaweeds are distributed according to their light requirements and their ability to resist wave action as well as drying and/or freezing conditions during low tide.

All seaweeds are algae and are green, brown, or red, depending on the color of the dominant photosynthetic pigment. In general, brown and red algae are more robust and often have a rubbery feel, whereas green algae are more delicate. Bright green sea lettuce, with thin leaves up to one meter (3.3 ft) long, lives from just above the low tide mark (so it is not exposed to air for very long periods) to depths of about 10 m (33 ft). It requires abundant light. Red algae photosynthesize most efficiently at lower light levels; hence, they are more likely to be found in deeper water or in tide pools that are shaded by rock outcrops or cliffs. They also prefer the higher temperatures of tropical waters.

Brown algae are the most common seaweeds found along the rocky coasts of the northeastern and western United States. They flourish in cold water and usually have a heavier structure that resists strong waves. They often have many small, gas-filled bladders that help keep their blades (not true leaves) floating near the water surface. Another adaptation of brown algae for life in the intertidal zone is the presence of gelatinous substances that helps them retain seawater at low tide when they are exposed to drying air. These substances have many commercial uses as stabilizers in ice cream and as emulsifiers in many food products. Algin extracted from brown algae is also used to thicken textile dyes, in paints, and even by dentists in taking impressions of teeth. Various red, green, and brown seaweeds are used extensively as food in coastal cuisines around the world. For example, sheets of dried red alga are commonly used in Asia in soups and to wrap sushi.

Open-ocean waters are too deep for plants to attach to the bottom and survive. Some plants do survive, however, in an area in the western Atlantic Ocean where Gulf Stream waters flow around the western side of the North Atlantic subtropical gyre (Chapter 6). This area is known as the Sargasso Sea, after the abundant, brown seaweed (*Sargassum*) that floats on the surface. *Sargassum* weed originates in the Caribbean Sea, where it grows attached to hard surfaces in the intertidal zone. It is broken off by storm waves and becomes trapped in the current system of the North Atlantic subtropical gyre where it accumulates as large, floating rafts.

Sargassum weed is supported by bubble-like gas filled structures that keep the plants floating near the surface where light for photosynthesis is abundant.

Surface waters of the subtropical gyre are nutrient poor and cannot support large phytoplankton and fish communities. However, rafts of *Sargassum* weed provide a unique habitat for animal life. The small animals that can survive in the nutrient-poor waters maintain a close relationship with the seaweed, either attached to it or swimming within the floating mats. Some small fishes and crabs have color patterns that allow them to blend in with the background of *Sargassum* weed. Many invertebrates such as snails, crabs, and anemones live attached to or moving about the weed. All of these animals are feeding generalists (omnivores), eating whatever comes their way in the water. Sea turtles, especially young ones, also find a protective environment within the *Sargassum* weed of the Sargasso Sea.

As with seaweed, animals are distributed in the intertidal zone according to their tolerance for drying and changes in temperature and salinity. Highly mobile animals, such as crabs, move up and down with the changing water levels as the tide floods and ebbs. Animals such as barnacles (Figure 10.8a) and mussels (Figure 10.8b) attach themselves to hard surfaces and can shut their shells completely, thereby retaining enough seawater inside to survive until they are covered again by the flood tide. For this reason, barnacles are the dominant animals in the upper, more exposed portion of the intertidal zone. Although they resemble mollusks, barnacles are actually crustaceans. They feed only when covered by water, opening their shells and extending their feathery legs through their shell top to filter food from the water. Barnacles are so firmly cemented to rock surfaces that waves cannot loosen them. Mussels are mollusks that usually live slightly lower in intertidal zones and are firmly attached to rocks by sticky threads.

Other intertidal mollusks include limpets and chitons that attach to rocks with a strong suction created by a muscular foot-like appendage. Both move so slowly that they are unable to keep pace with tidal oscillations. They are fairly resistant to drying out and are only active when they are covered with water. Their food is plant material that they scrape off rock surfaces with a radula, a tongue-like structure consisting of rows of chitinous teeth. Still lower in the intertidal zone is a greater diversity of animals including starfishes, anemones, urchins, and crabs.

While walking along a rocky shore at low tide you may have encountered tide pools sheltering a diverse animal community. A **tide pool** is a volume of water left behind in a rock basin or other intertidal depression by an ebbing tide. If a tide pool is exposed to direct sunshine, organisms inhabiting the pool must contend with great fluctuations in water temperature. Rainfall can also alter the salinity of tide-pool waters. If a tide pool is shaded or in a coastal cave, temperature and salinity variations are much less so that a greater variety of more delicate animals can survive. Typically, tide pools harbor species that have evolved to deal with wide fluctuations in environmental conditions; these organisms include certain algae, sea anemones, starfish, snails, small crustaceans, barnacles, mussels, and fish.

Just below low-tide level on rocky shores live the most diverse communities of plants and animals. The complex substrate and the seaweed provide protection for many small animals. Here soft bodied and delicate animals like anemones and sea cucumbers can thrive. Predators, such as lobsters, crabs, small fishes, octopi, and sea otters are important in this ecosystem.

SEA GRASS BEDS AND SALT MARSHES

On mud flats and other soft-bottomed habitats (such as along sandy and estuarine shorelines), the most important plants are *sea grasses*, also known as *submerged aquatic vegetation (SAV)* (Figure 10.9). These are *angiosperms*, flowering and seed bearing vascular plants with true roots. They are limited to water depths where light is sufficient for photosynthesis and prefer water that is clear. Sea grass beds are highly productive with some rivaling the primary production of intensively developed agricultural land in the amount of carbon fixed (500 to 1000 gm per square m per year). Sea grass beds export large amounts of organic matter to nearby coastal waters. However, water pollution, sediment runoff from land, and dredging are among the many human activities that threaten sea grass beds. Where nutrient input (nitrogen and phosphorus compounds) is excessive, such as in runoff from agricultural lands,

FIGURE 10.9
A sunlit sea grass meadow in the Florida Keys National Marine Sanctuary. Sea grasses are also known as submerged aquatic vegetation. [NOAA photo by Heather Dine]

excessive numbers of algae may grow on the blades of sea grasses, shading and eventually killing them.

Perhaps the most extensively studied sea grass ecosystem is the Chesapeake Bay. When Europeans arrived in the Bay area in the early 17th century, sea grasses covered an estimated 240,000 hectares (600,000 acres) of the Bay. By 1978, the sea grass area had shrunk to only 16,000 hectares (41,000 acres). Destruction of sea grass beds is a concern for several reasons. Sea grass anchors sediment and dampens wave action thereby controlling erosion and turbidity. It is a food source for many organisms including waterfowl and small mammals and serves as a primary nursery for crabs and many species of fish.

Loss of sea grass habitat and increased harvest pressure caused the bottom-dwelling blue crab (*Callinectes sapidus*), a mainstay of the Chesapeake Bay fishery, to decline drastically from 400 million for the latter half of the 20th century to 130 million in 1997. Underwater sea grass beds serve juvenile blue crabs, those less than a year old, as nurseries, a source of food, and protection from predators. Without adequate sea grass habitat, the blue crab is more vulnerable to predation by striped bass, known locally as rockfish. Fortunately, restoration efforts have had some success and by 2009, the area of sea grasses along the shallows and shorelines of Chesapeake Bay had increased to an estimated 34,763 hectares (85,899 acres), up 12% from the previous year, though less than half of the Bay wide restoration goals. Adult blue crabs have rebounded as well, from 120 million in 2008 to 315 million in 2010, due to increased habitat and, even more helpful, strict rules to curtail harvesting pregnant females. There is more on efforts to restore the Chesapeake Bay ecosystem and blue crabs in Chapter 15.

Salt marshes commonly occur along sheltered shorelines and are ecologically similar to sea grass beds in estuaries. Salt marsh grasses and bushes differ from sea grasses by being salt-tolerant true land plants, pollinating in the air. Salt marshes are flooded at high tide, but the grasses are never completely covered by salt water. Salt marshes are also home to abundant marine life, and are refuges for waterfowl and other wildlife. They are true transition zones between marine and terrestrial ecosystems (Figure 10.10).

In estuaries, the types of plants and animals vary from marine species near the mouth of the estuary to brackish water species where the salinity is lower, to those that prefer a nearly freshwater environment at the head of the estuary (Chapter 8). The salinity of estuarine waters occasionally changes drastically as when the discharge of fresh water entering the estuary increases abruptly following a period of heavy rainfall. This change in salinity can impact the types of species and their distribution in the estuary depending on how fast they can move or their tolerance.

FIGURE 10.10
The edge of a salt marsh meeting oak and other higher ground flora on Daniel Island, SC. [NOAA photo by Captain Albert E. Theberge, NOAA Corps (ret.)]

In the tropics, between about 30 degrees N and 30 degrees S, mangrove swamps are common along muddy, low-lying coastlines and in estuaries. A **mangrove swamp** consists of tropical plant species including trees that grow in low marshy areas and can tolerate salt water flooding of their roots and lower stems (Figure 10.11). They generally compete successfully with local marsh grasses. Mangroves form dense growths and the aerial roots found in most mangrove species provide a complex and protective habitat for many organisms. For example, mangrove oysters and other organisms attach to the roots, which trap sediment and organic material. Mangrove swamps are also important

FIGURE 10.11
Red mangroves are common in Florida. Mangrove roots serve as critical habitat for many species and nutrient filters. [Courtesy of NOAA/OAR/National Undersea Research Program (NURP)]

nursery grounds for many marine species. However, many hectares of mangrove swamps are lost each year, cleared to create farmland or for coastal development projects such as housing, shopping malls, resorts, airports, and industrial parks. Furthermore, some coastal residents have removed mangrove trees because they obstructed their view of the ocean. Ironically, mangrove swamps may be the final line of defense against a storm surge (Chapter 8).

Application of new remote sensing techniques has provided an accurate assessment of the status of mangrove forests. In 2010, NASA and USGS scientists published the first high-resolution, satellite-based map of the global distribution of mangroves. Writing in the *Journal of Global Ecology and Biogeography*, USGS scientist Chandra Giri and colleagues report that mangrove species cover about 53,190 square mi, about 12% less than prior estimates. Mangroves are most common in Asia (42%), Africa (21%), and North and Central America (15%). According to the International Union for the Conservation of Nature, about 16% of mangrove species worldwide are in danger of extinction.

KELP FORESTS

Seaward of the intertidal zone, kelp forests grow where waters are cool and nutrient-rich (Figure 10.12). **Kelp** includes various species of brown algae that grow to enormous size. In clear waters, individual plants grow at depths of about 30 to 40 m (100 to 130 ft) and their stipes (stem-like structures) and leaf-like fronds reach all the way to the surface. Found in cool waters worldwide, they are especially abundant in coastal upwelling zones off California and the Pacific Northwest.

Kelp plants grow attached to rocky bottoms by root-like structures (*holdfasts*). Holdfasts are quite small so that in many species, the weight of the plant is supported at the surface by a bulbous gas filled float below the fronds. Hence, strong waves easily destroy kelp beds and huge amounts of detached kelp wash onto beaches following storms. Fortunately, kelp can grow rapidly—up to tens of centimeters per day. Because of this, it can be easily harvested and will regrow if only the tops of the blades are removed. Kelp's primary production is high, rivaling that of some of the richest farmlands.

A dense stand of kelp is like a submarine forest, and it supports a rich community of animals that lives below its canopy. Small fishes, urchins, crabs, and lobsters are common on the sea floor around the holdfasts of kelp plants. Smaller algae, sea anemones, mollusks, and other invertebrates may attach to these structures. Further up the stipe where light is a little brighter, red and brown algae may live on the surface of the kelp. Sea urchins climb up the kelp plant eating these algae, or even the kelp itself. The canopy protects large schools of young fishes such as herring and sardines. Also, many organisms such as snails graze upon kelp fronds.

FIGURE 10.12
A forest of giant kelp near Baja California. These large brown algae can grow as much as 0.6 m (2 ft) per day in water depths up to 45 m (150 ft). [Image courtesy of NOAA]

Along the Pacific coast of North America, kelp forests are home to large numbers of sea otters. Sea urchins are the favorite food of otters and urchins abound in the kelp forest habitat. In fact, urchins can destroy a kelp forest through grazing if they become too numerous. Sea otters feeding on urchins keep their populations in check. However, when sea otters were hunted (for their fur) to near extinction in the 18th and 19th centuries, sea urchin populations exploded and many kelp beds were lost. A similar relationship exists among kelp, urchins, and lobsters in North Atlantic kelp habitats, and over-fishing of lobsters has had similar results.

In 2007, an international team of ocean scientists reported that kelp forests may be much more widespread in the tropics than previously believed. (Widely scattered reports of kelp in tropical latitudes had circulated for many years.) A computerized kelp-habitat model was developed that incorporates the essential environmental conditions for kelp; that is, cold nutrient-rich water welling up from the deep, adequate sunlight, and a suitable ocean bottom for their holdfasts. The model predicted that kelp forests could be found in various patches of the tropical ocean covering a total area of 23,500 square km (9100 square mi) worldwide. The scientists confirmed the computer prediction when they discovered eight patches of dense kelp forests near the tropical Galápagos Islands.

CORAL REEFS

A **coral reef** is among the ocean's most spectacular features. After centuries to millennia of slow growth, they can become so large that they are visible from space. In many parts of the tropical ocean, reefs stand hundreds of meters above the sea floor and extend hundreds of kilometers along the shoreline. Carbonate-secreting colonial animals, *coral polyps*, are the primary builders of the hard coral-reef structure. Similar to anemones, the tiny coral polyps attach to hard surfaces, such as volcanic rock or limestone from a previous coral, and then they colonize on the pre-existing framework of limestone or volcanic remains. Therefore, they are found along coastlines and in the open ocean surrounding extinct volcanoes. Large reefs occur only in tropical waters, between about 30 degrees N and 30 degrees S. Most of the largest reefs are found in the tropical Pacific and Indian Oceans, although the world's second largest reef is found off the coast of Belize in the Caribbean Sea. A few small coral reefs occur at shelf breaks along the margins of ocean basins.

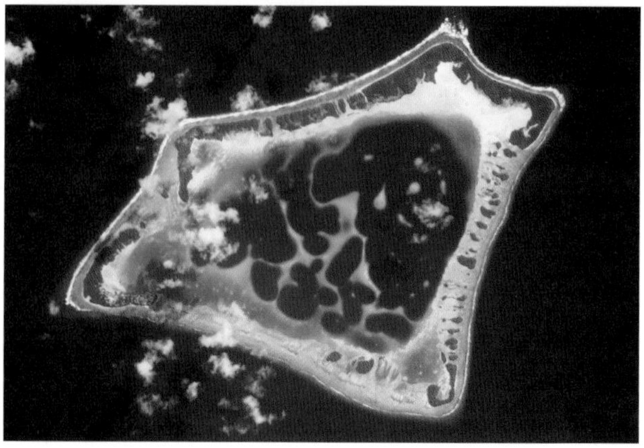

FIGURE 10.13
International Space Station photo of Atafu Atoll in the Pacific Ocean. [NASA]

Coral reefs growing on volcanic islands typically exhibit a characteristic sequence of forms, first described by Charles Darwin (Figure 2.22). While the marine volcano is active, the coral reef grows along the shore as a relatively narrow *fringing reef*, like those bordering the Big Island of Hawaii. As the underlying tectonic plate displaces the volcano from the hot spot that formed it, the volcano becomes inactive and extinct. As the volcano subsides, the coral reef grows upward and forms a *barrier reef*, separated from the main volcanic island by a shallow lagoon. Barrier reefs occur on the older Hawaiian Islands to the northwest of the Big Island, such as off the coast of Oahu. Eventually, as the volcano erodes and sinks deeper into the ocean, the remains of the island are submerged beneath the waves, leaving only a ring of coral reefs surrounding a shallow lagoon where the island once stood. At this stage, the reef is known as an **atoll** (Figure 10.13). Shaped like a ring or horseshoe, atolls vary from 1 to 100 km (0.6 to 60 mi) across.

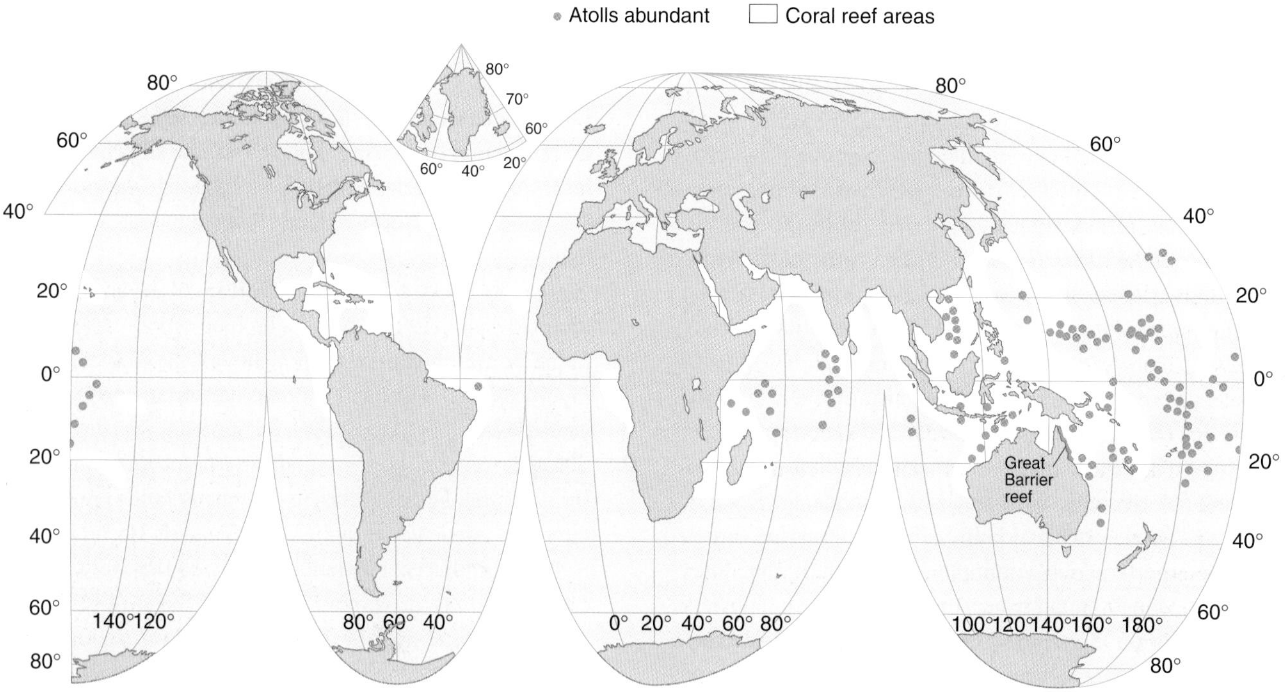

FIGURE 10.14
Location of atolls and coral reefs in the world ocean.

Midway and Kure Islands, at the northwestern end of the Hawaiian Island chain, are examples of atolls. Atolls are most common in the Pacific Ocean (Figure 10.14).

The Great Barrier Reef, with 3000 coral reefs and 600 low-lying islands composed of carbonate sand formed from eroded coral reefs, stretches more than 2000 km (1250 mi) along the Queensland coast of northeastern Australia (Figure 10.15). The world's most extensive reef system, it is the largest structure on Earth made by living organisms, and is visible from space. At its northern end, the reef is nearly continuous only about 50 km (30 mi) offshore. In the south, individual reefs up to 300 km (190 mi) offshore are more common. About 2 million tourists visit the Great Barrier Reef each year.

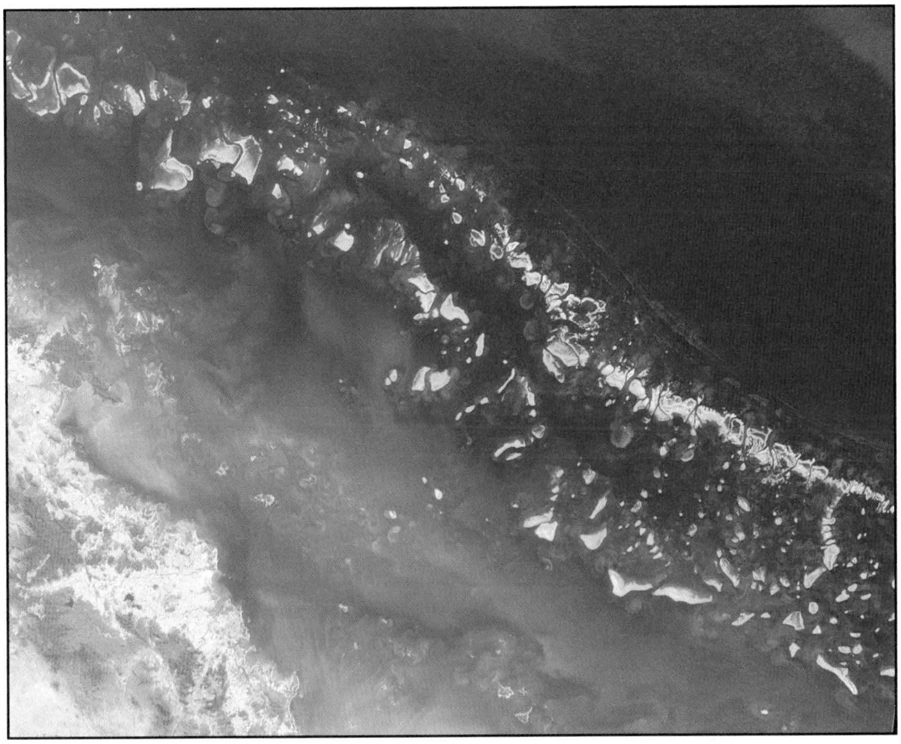

FIGURE 10.15
The Great Barrier Reef extends for 2,000 kilometers along the northeastern coast of Australia. It is not a single reef, but a vast maze of reefs, passages, and coral cays (islands that are part of the reef). [Image courtesy of NASA/GSFC/LaRC/JPL, MISR Team]

Each type of coral has a distinct structure in the reef. Many corals have growth rings, much like trees though in a coral each ring represents an individual organism, and these give hints to past variations in ocean and climate conditions, such as El Niño events in the tropical Pacific Ocean (Chapter 11). Some corals (e.g., brain corals) form robust compact structures while others build delicate, complex branching forms (e.g., sea fans). Coral reefs are among Earth's most productive ecosystems, second only to rainforests in biodiversity. Worldwide, they provide shelter and food for 9 million species of marine life (about a third of known marine species) including fish, invertebrates, mammals, reptiles, and algae. The Great Barrier Reef is home to whales, dolphins, green turtles, fish (1500 species), mollusks (4000 species), and birds (200 species).

The individual coral animals, *coral polyps* (Figure 10.16), capture zooplankton and phytoplankton from the waters flowing over the reef. The majority of their energy, however, is obtained from dinoflagellates, the *zooxanthellae* living within their tissues. It is the photosynthetic zooxanthellae that give corals their bright colors. The polyps get most of their food from the wastes of the zooxanthellae, which in turn, receive shelter and nutrients from the polyps. They share a **symbiotic relationship**, from which both derive benefit. Because zooxanthellae need the sunlight for photosynthesis, corals thrive within the top 200 m (660 ft) of the ocean, which is why fast rising sea level can threaten slow growing corals. If the zooxanthellae do not receive adequate sunlight, they will starve, as will the coral soon after. Many submerged Pacific seamounts are capped by limestone, historic examples of coral growth unable to keep pace with past

FIGURE 10.16
Coral polyps. [NOAA, photo by Brent Deuel]

rapid rise in sea level or sinking platforms. It is hoped that healthy corals will be able to grow upward fast enough to keep their zooxanthellae in the sunlight.

In recent decades, however, the number of coral species threatened by extinction has increased dramatically. An estimated 20% of the world's original coral reef area (285,000 square km or 110,000 square mi) is now dead. According to the Global Coral Reef Monitoring Network's *Status of Coral Reefs of the World: 2008*, an additional 15% of coral reefs may be gone in the next two decades. In February 2011, the World Resources Institute in Washington, DC, published *Reefs at Risk*, an update of a 1998 reef-status report, concluding that about 75% of the world's coral reefs are threatened, with this figure expected to rise to 90% by 2030. Contributing to the decline and loss of these important ecosystems is a combination of stressors, many linked directly or indirectly to climate change (coral bleaching, ocean acidification, sea-level rise, outbreaks of infectious coral diseases) and human activities (overfishing, coastal development). Corals are endangered by sediment runoff from land, oil spills, and other forms of water pollution that reduce sunlight, and smother or poison them. *Eutrophication*, caused by excess nutrients often from agricultural lands and sewage discharge, can stimulate the growth of algae on the surface of coral reefs, also smothering the coral polyps.

Warming of the ocean also threatens corals. Reef-building tropical corals prefer waters having an average annual temperature of 23 °C to 25 °C (73 °F to 77 °F) and corals cannot tolerate prolonged exposure to either low or high water temperatures or large fluctuations in temperature. Even small changes in sea-surface temperatures (SST), associated with large-scale climate variability or climate change, threaten coral reefs. In response to unusually high SST, coral polyps expel their zooxanthellae. Without the colorful zooxanthellae, the corals appear nearly transparent on the white exoskeleton and so the condition is known as **coral bleaching** (Figure 10.17). A rise in SST of 1 to 2 Celsius degrees (2 to 4 Fahrenheit degrees) is sufficient to stress coral, causing temporary bleaching. While corals can recover from bleaching, persistent or severe episodes cause coral polyps, and the reef, to die. Furthermore, higher ocean temperatures and bleaching make corals more vulnerable to infection, adding to reef mortality.

In 1998, elevated SST from a one-two punch of El Niño followed by La Niña caused massive coral bleaching that destroyed 16% of the world's coral reefs. Five years after the event, only 6.4% of the bleached corals had recovered. Then, by August 2005, water temperatures off the Virgin Islands climbed 3 Celsius degrees (5.4 Fahrenheit degrees) above average and remained above average until November. In the eastern Caribbean Sea, bleaching due to thermal stress affected 80% of coral and killed 40%. The decline in populations of the delicate elkhorn corals (*Acropora cervicornis*) and staghorn corals (*Acropora palmata*) was so severe in the Caribbean that the following May the U.S. government listed them as threatened species under the U.S. Endangered Species Act.

FIGURE 10.17
Coral bleaching in Mariana Islands, Guam. [NOAA photo by David Burdick]

An unusually warm episode developed in June 2010 and persisted through the rest of the year. This warm episode affected a much larger area of the western and southern Caribbean, and devastated reefs in the Netherlands Antilles and off the Panama coast. Bleaching was more extensive than observed previously, affecting not only hard corals, but also sponges and sea anemones. NOAA's *Coral Reef Watch (CRW)* reported significant bleaching from thermal stress in the western Gulf of Mexico and Caribbean. Extensive bleaching also occurred in the Indian Ocean and Southeast Asia.

Ocean acidification, due to the uptake of atmospheric carbon dioxide, is predicted to reduce coral fertility by the middle of this century. Researcher Rebecca Albright of the University of Miami, and colleagues, reported this in the 8 November 2010 *Proceedings of the National Academy of Science*. In laboratory simulations, they exposed elkhorn coral to seawater at progressively lower pH. They found that sexual reproduction declined and the supply of young to build coral reefs was cut in half. Furthermore, during the next century, predicted ocean acidification threatens to reduce calcification rates by half, dissolving the limestone reef itself (Chapter 3).

Kent E. Carpenter of the *International Union for Conservation of Nature (IUCN) Species Programme*, and colleagues, rated the risk of extinction for 704 species of reef-building coral worldwide. Reporting in the 28 July 2008 issue of *Science*, they found that 32.4% had an elevated risk because of episodes of thermal stress and the impact of local anthropogenic activities (e.g., coastal pollution and destructive fishing practices). The risk is highest among the corals of the Caribbean Sea and the *Coral Triangle* in the western Pacific (Indo-Malay-Philippine archipelago). Assessment of the extinction threat was based primarily on trends in coral population size and geographical range. Risk categories ranged from least concern (minimum likelihood of extinction) to high risk (critically endangered).

Corals described thus far are *hermatypic*, living in warm, shallow waters on large reefs with zooxanthellae. Solitary corals and small coral reefs also live in cold, deep water, on continental shelf breaks in the ocean (Figure 10.18). Without light, these *ahermatypic corals* depend exclusively on trapping food directly from the water using their stinging cells called nematocysts. Some ahermatypic corals are hard structures whereas others have soft and delicate branching forms. Neither build limestone structures and both provide excellent habitats for deep-water fishes. Unfortunately, trawlers, fishing boats that drag huge nets along the sea floor, are drawn to the marine species found in these areas and their nets destroy the deep-sea corals. Little is known about the distribution and biology of deep-sea corals, and some nations, such as the U.S. and Norway, are beginning to ban harmful fishing practices near them.

FIGURE 10.18
Darkblotched rockfish nestled in the branches of a deep-sea gorgonian soft coral. Courtesy of [NOAA]

Although threats to coral reefs are well documented, practical solutions to increase reef resilience and curb loss are difficult to find. To conserve coral reefs and other underwater marine resources for present and future generations, the U.S. and other nations have designated certain areas of their jurisdictional waters as *marine protected areas*. For more on this topic, refer to this chapter's third Essay.

BENTHIC FEEDING HABITS

Animals living in soft-bottomed habitats—whether in the intertidal zone, shallow coastal areas, or the deep sea—are divided into two categories: infauna and epifauna. **Infauna** inhabit sediment deposits whereas **epifauna** live on the sea floor. Some of these benthic animals are *unselective* in their feeding, moving through or on the mud and simply eating everything. They ingest bottom deposits and digest the organic detritus they contain while the remaining material (mud or very fine sand) is excreted. Other benthic animals are more *selective* in their feeding habits, ingesting food particles from mud and sand, rather than consuming the sediment itself. As benthic animals feed on or in sediment, they disturb the deposits, a process called **bioturbation**. Their feces bind sediments into harder, more durable aggregates that are not readily re-suspended by currents. Close observation of the surface of a water-covered mud flat reveals a variety of feeding structures of infaunal worms and clams extending above the entrances to their burrows.

Soft bottoms also harbor many filter feeders that separate phytoplankton, small zooplankton, and other edible materials from large volumes of water. The distribution of animals in muddy and sandy habitats is controlled primarily by the grain size of the sediment. Filter feeders cannot live in fine-grained mud, because it clogs their feeding apparatus. Hence, they usually inhabit coarser grained sands. Filter feeders are generally rare where deposit feeders are abundant because bioturbation makes the near-bottom water muddy and therefore unsuitable for filter feeding. On the other hand, mud is ideal for unselective feeders. It contains more organic matter brought in from terrestrial sources and raining down from the photic zone plus an abundant supply of decomposing bacteria and other microbes.

Whereas some animals living in the rocky intertidal zone, such as starfish, can absorb dissolved oxygen from the water directly through their body surfaces, benthic animals living in muddy habitats require specialized breathing structures. A respiratory structure may extend outside their burrows or they may pump water through their burrows and past their gills for uptake of dissolved oxygen. Some animals line their burrows with mucus or shell fragments to prevent mud from clogging these structures.

LIFE ON THE DEEP-SEA FLOOR

In the deep ocean, detritus that reaches the sea floor is mostly decomposed and has little nutritional value. For this reason, far fewer organisms can survive in this habitat but it is not the dead zone envisioned by most 19th century scientists. Many unusual creatures live on the deep-sea floor. For example, a starfish-like animal, called a sea lily, lives attached to the bottom by a long, stem-like structure (Figure 10.19). Its arms, covered with a mucous substance, are suspended in the water above the sea floor and trap particles in the near-bottom waters. The particles are moved along grooves to the mouth. Predatory starfish travel along the bottom on long legs that keep their bodies above the sediment surface. Other filter feeders in the deep ocean include sponges, crustaceans, and some kinds of worms and mollusks. Deposit feeders are similar to those living in shallow benthic habitats, although animals in the deep, cold ocean are generally smaller, an adaptation to the scarcity of food. In these dark waters color has no strategic value so that most animals have a drab appearance.

FIGURE 10.19
This delicate sea lily (crinoid), a member of the phylum that includes starfish, can orient toward the current to increase food capture. [Courtesy of NOAA Undersea Research Program]

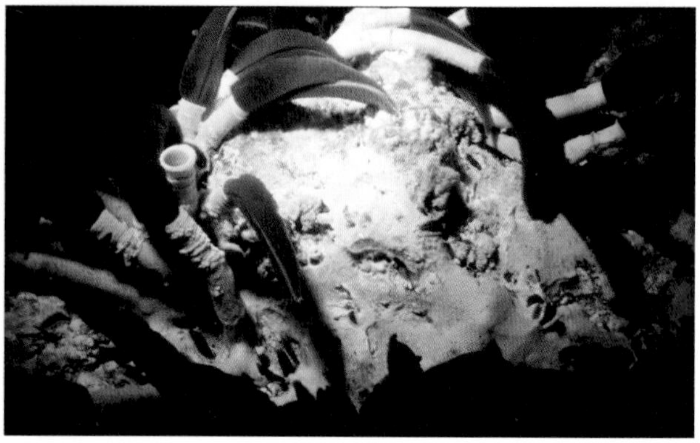

FIGURE 10.20
Tube worms at a hydrothermal vent on the floor of the Pacific Ocean. Discovery of such organisms launched a new avenue of inquiry into our understanding of biological processes operating in the deep ocean (especially chemosynthesis). [OAR/National Undersea Research Program (NURP); College of William & Mary]

Communities of specialized animals have evolved to live on the deep ocean bottom near hydrothermal vents (Chapter 2). Large colonies of tubeworms have a symbiotic relationship with bacteria that live within their bodies (Figure 10.20). The bacteria use hydrogen sulfide (H_2S) as an energy source and make food for the tubeworms through chemosynthesis (Chapter 9). These worms can grow to a length of 3 m (10 ft). Also living in this extreme environment are long legged crabs and specialized mollusks and clams.

Marine Animals

In this section, we briefly describe larger marine animals and their adaptations for life in the sea. These organisms include fishes, mammals, reptiles, and birds.

FISHES

Fishes are the most familiar type of nekton. Unlike mammals, which regulate their body temperature, fish are cold blooded and hence are as warm or cold as the surrounding water. (Some fish that live in water at temperatures slightly below 0 °C contain natural antifreeze.) Fishes range in size from the smallest of all vertebrates, a marine goby only 8 to 10 mm long, to the huge whale shark up to 15 m (50 ft) long. Fish may be herbivorous, carnivorous, or omnivorous. They inhabit all parts of the ocean and possess special adaptations for buoyancy, swimming, and life in the dimly lit twilight zone as well as the greatest depths of the ocean. Many open-ocean pelagic fishes are streamlined for swimming fast as they search for prey and flee from predators. Fishes living among plants or rocks often have body shapes and colors that allow them to blend unnoticed with their background.

There are three major groups of fishes: jawless fishes (Superclass *Agnatha*), cartilaginous fishes (Superclass *Chrondrichthyes*), and bony fishes (Superclass *Osteichthyes*). **Agnathans** or **jawless fishes** have cartilaginous skeletons and long, eel-like bodies, which lack scales or armor. The distinguishing features of this group (in addition to the absence of jaws) include the lack of an identifiable stomach and lack of paired appendages, such as fins. The only surviving members of this ancient group are lampreys and hagfishes. Hagfish are scavengers, which feed on dead and dying fish. They can tie their flexible bodies in knots in order to tear flesh from their prey, escape predators, and remove excess amounts of their excreted slime (a possible defense mechanism) from their body. Lampreys are parasitic fish, which attach to a host organism by their mouths. They use their sharp teeth to cut flesh, complete their feeding, and release the host organism, leaving a potentially fatal wound. As a result, lampreys cause tremendous damage in the commercial fishing industry. Examples of fish that have been adversely affected by the sea lamprey include lake trout and whitefish.

Cartilaginous fishes have skeletons that lack true bones and consist entirely of cartilage. These primitive fishes include sharks, skates, and rays. Their skin feels much like fine sandpaper due to small, embedded tooth-like structures called denticles. Sharks' teeth are large, specialized denticles that occur in rows in the mouth and are continually growing forward, being lost and subsequently replaced from behind.

According to a popular misconception, sharks are ferocious predators who menace swimmers (Figure 10.21). In fact, a few shark species are ferocious but their threat to humans is rare, but real. Shark attacks—no matter how infrequent—are shocking and frightening. Some sharks are selective predators, feeding mostly on preferred types of prey. Other species, such as the tiger shark, are sometimes referred to as "garbage cans of the sea" because they eat almost anything, living or dead. Some types of sharks and rays are bottom feeders, with an upper jaw designed to pick up food from the sea floor. Basking sharks, whale sharks, and many rays are passive filter feeders; that is, they swim continuously, straining plankton from the water passing over their gills. Whale sharks are very docile even though they weigh more than 40,000 kg (about 45 tons) and grow to a length of 15 m (50 ft).

Sharks have some unique limitations in that they do not have a swim bladder for buoyancy, their blood pressure is low, and most cannot pump water over their gills to obtain oxygen. To help compensate for all three of these limitations, most sharks swim ceaselessly their entire life. With swimming motion, their pectoral (side) fins provide lift much like an airplane wing. Movement causes a flow of water over their gills to oxygenate the blood (known as *ram ventilation*), and continuous swimming produces muscle action that helps the heart pump blood. Without active swimming, sharks would sink to the bottom and most would asphyxiate. Some sharks, however, possess spiracles, and others have special muscles that force water over the gills, helping them to obtain oxygen even while stationary.

There are over 800 species of cartilaginous fishes. Their reproductive strategies favor survival of their young. Unlike most bony fishes, cartilaginous fishes have internal fertilization. Although they bear very small numbers of young (whereas bony fishes release thousands of eggs), the chance of reproductive success is increased. Some sharks and rays lay their eggs in leathery cases that harden in a few hours. Beachcombers sometimes find these leathery pouches, called "mermaid's purses," washed up on the beach. The young are hatched fully formed; there is no planktonic larval stage among cartilaginous fishes.

Some species of sharks are *viviparous*; that is, embryonic development occurs entirely within the female shark's body and they bear live young. Hammerhead sharks are a viviparous species. Cartilaginous fishes usually lay their eggs or bear their young in coastal nursery areas where there is less danger of predation by larger fishes.

Cartilaginous fishes are *isosmotic* with their environment; that is, the salt concentration in their body

FIGURE 10.21
A white-tip shark (*Triaenodon obesus*) swimming in waters northwest of the Hawaiian Islands. [NOAA Photo Library, photograph by Dr. Dwayne Meadows, NOAA/NMFS/OPR]

is approximately the same as the surrounding seawater. Hence, there can be an equal amount of water leaving the body as enters it. However, the types of salts are different. They have a very high concentration of organic solutes, mostly urea, in their tissues. The concentration of sodium chloride in their body fluids is only about half that of seawater so that a special rectal gland contains a cell that selectively excretes chloride ions into their urine.

Bony fishes are a large, diverse group that lives in a variety of marine and freshwater environments. They have bony skeletons, scales, a flap covering their gills and, the vast majority of them, a swim bladder, which is an inflatable gas-filled organ that fishes use to adjust their buoyancy.

Salt concentration is much lower, only a quarter to a third, in bony fishes than in seawater. As a result, bony fishes have developed mechanisms to regulate their salt content. Salt molecules cannot cross the membranes of their cells. To prevent dehydration, the kidneys of bony fishes excrete larger ions, such as calcium and magnesium, which helps in retaining water and their gills have special chloride excretory cells that transport the excess salt out of their bodies. Furthermore, bony fishes produce only minute amounts of concentrated urine.

As a consequence of this process, bony fishes are an important source of calcium carbonate in the ocean, accounting for more than one-quarter of the marine carbonate budget. Other sources are marine plankton (i.e., coccolithophores and foraminifera). The calcium carbonate from bony fishes contains more magnesium than that produced by plankton and is more soluble.

Fishes living on or near the ocean bottom are called *demersal fishes* and many of them are commercially important, such as cod, halibut, haddock, and sole. *Flatfishes* are demersal fishes that have a special coloration that camouflages them against the ocean bottom. They include halibut, flounder, and sole. Because these fish spend most of their lives lying on the sea floor, an eye on the lower side would be useless (Figure 10.22). When larval flatfishes hatch as free-swimming zooplankton, they have eyes on either side of their head. As larvae mature, however, one eye moves over the top of its head, so that when a young fish finally takes up life on the bottom, both eyes are looking up on the same side of the head. Also its body thins and is laterally compressed. In addition, they have the ability to quickly change the color and pattern of their skin so that they closely resemble their background, another example of adaptive coloration.

As noted above, many pelagic fishes are fast swimmers. Tuna and mackerel are examples of fishes well adapted for strong swimming over long distances in open waters. For example, bluefin and yellowfin tuna typically cruise at 16 km per hr (10 mph) but are capable of brief bursts of 95 to 130 km per hr (60 to 80 mph). Such high speeds require large amounts of energy and these swift swimmers usually occupy a high trophic level in marine food webs. Their dense muscles make their meat very desirable and they are much in demand at the dinner table.

Smaller pelagic fishes such as herring and sardines are abundant in coastal waters and in upwelling areas where nutrients are plentiful. Some of the world's largest fisheries are based on these species. Many smaller fishes are converted to fishmeal and used as high protein feed for cattle, poultry, and fish farms and also as fertilizer. Some are harvested for human consumption.

Most marine bony fishes spawn; that is, the females release their eggs into the water where they are fertilized after males release sperm. A single female herring can release between 50,000 and 700,000 eggs in a single breeding season. In general, larger and older fishes are the most productive. The eggs hatch, releasing larvae, which exist for varying periods of time as plankton, maturing into stronger swimming juveniles. Only a tiny fraction of the larvae become juvenile fish and far fewer survive to adulthood. The major reasons for this relatively low reproductive success include dispersal of the larvae into unsuitable areas, lack of food while in the larvae stage, and predation by many animals upon the eggs, larvae, and young fishes. Occassionally, low fertilization is a problem for some vertebrates spaced far apart. For example, currents can keep the eggs and sperm apart or they may carry the planktonic larvae away from the protective nursery grounds in estuaries.

Most bony fishes broadcast their eggs and sperm into the water and rely upon chance fertilization. By

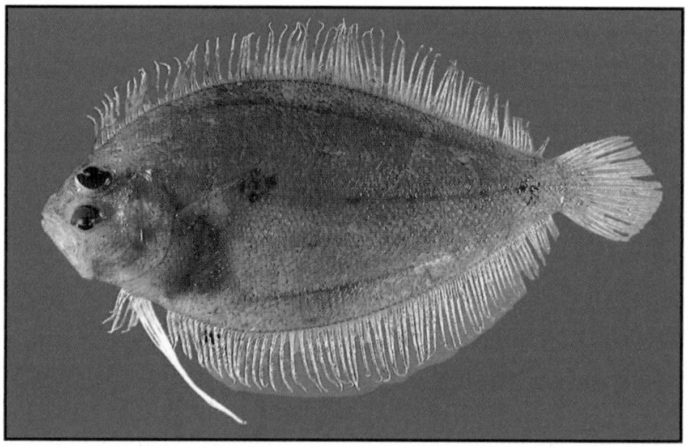

FIGURE 10.22
Note the two eyes on one side of the head of this sash flounder. [SEFSC Pascagoula Laboratory; Collection of Brandi Noble, NOAA/NMFS/SEFSC]

producing thousands of eggs and sperm, they increase the odds of producing a few successful young fishes. Some bony fish have evolved to increase their reproductive success through strategies such as internal fertilization, extended periods of parental care of the young, brooding of the young in the mouth of one of the parents or in a brood pouch (such as in the male seahorse). As mentioned earlier, cartilaginous fish increase their reproductive success through internal fertilization and protection of their young, but they produce only a very small number of offspring.

Many environmental factors influence the reproductive success of bony fishes. Some fishes undertake long migrations. **Anadromous fishes** such as salmon, shad, sturgeon, and striped bass are born in rivers and streams, but spend most of their lives in cold regions of the ocean, returning to the same river to breed when they are sexually mature. Pacific salmon may migrate as far as 1300 km (800 mi) to complete this cycle (Figure 10.23). They reproduce only once, expending considerable energy to reach the headwaters of their particular river. Along the way, they use much of their body mass in this effort or convert it to eggs or sperm. They die after spawning. On the other hand, Atlantic salmon can reproduce many times.

How do spawning fish find their way back to the same stream or river where they were born? In the December 2008 issue of *Scientific American*, Megan McPhee of the University of Montana's Flathead Lake Biological Station provides an answer for salmon, a fish that is particularly adept at locating the specific stream and even the section of shoreline where they were born even after spending several years at sea. Apparently, many environmental cues enable salmon to navigate in the open ocean including solar altitude and length of daylight (Chapter 5), Earth's magnetic field, and seawater salinity and temperature. As spawning time approaches, the salmon orient themselves toward the segment of coastline where the stream of their birth empties into the ocean. When a salmon encounters freshwater, it is mainly its sense of smell that guides it to the correct tributary and upstream to its birthplace. Young salmon are very sensitive to the odors of their surroundings during hatching and as they enter the smolt period to begin their journey to the sea. These odors are imprinted in the salmon's brain and this information is retrieved when as adults they attempt to return to their home stream or river for spawning.

Dams can be formidable obstacles to salmon migration. In the Pacific Northwest, for example, only about 5% of the juvenile salmon survive passage through dams and reservoirs on the Columbia and Snake Rivers.

FIGURE 10.23
A fisheries biologist works with one of the stranded Butte Creek spring run chinook salmon at the net pen before it is transferred to a hatchery truck and a ride up river so it can complete its migratory journey to spawn. [California Department of Fish and Game]

Largely ineffective are fish ladders intended to help the salmon move upstream and other structures that guide them downstream around hydroelectric turbines. These obstacles have been likened to giant food processors for smolt (young salmon) attempting to swim through them. Furthermore, on their downstream passage, smolt are held up in reservoirs where they are exposed to predators, pathogens (disease-producing organisms), and water that is too warm. Atlantic salmon suffer a similar fate. More than 900 dams on New England and European rivers prevent most Atlantic salmon from reaching their freshwater spawning grounds. Consequently, their population has declined to less than 1% of historical levels. Today, almost all "Atlantic" salmon sold at fish markets come from fish farms.

Catadromous fishes are not as well known as anadromous fishes; they breed in the open ocean, but spend their adult lives in fresh water. In North America, the American eel is the only catadromous species (Figure 10.24). Eels from both European and North American rivers travel

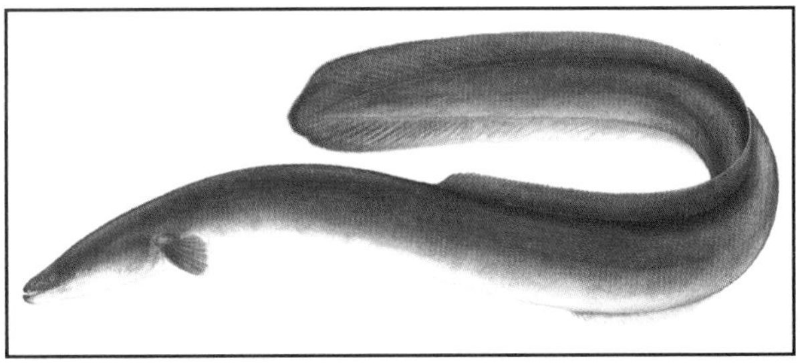

FIGURE 10.24
Drawing of an American Eel. [Original artwork prepared by Ellen Edmonson and Hugh Chrisp as part of the 1927-1940 New York Biological Survey conducted by the Conservation Department.]

thousands of kilometers to congregate in the Sargasso Sea where they spawn. Scientists, however, still do not know where in the Sargasso Sea this takes place. Planktonic eel larvae remain in the Atlantic Ocean for about a year while slowly drifting northward off the U.S. east coast. The eels do not mature into adults while in the ocean, but metamorphose into glass eels as they enter coastal waters, and then either remain in brackish areas or migrate upriver as glass eels. Eels do not become adults until they change into silver eels and begin migrating back to the Sargasso Sea. With their return to the Sargasso Sea, they reproduce once and die. Many questions remain unanswered about how these migratory fishes navigate and identify their home rivers, and how specifics of navigation are transferred from one generation to the next.

Many species of fish, such as herring and mackerel, reduce predation by swimming together as organized groups, keeping a certain distance between one another (Figure 10.25). This **schooling** behavior is particularly advantageous in the open ocean where hiding places are few and far between. Schooling confuses potential predators and functions as an alarm system when would-be intruders approach. Typically, the individuals in a school disperse in all directions when the alarm is sounded and then reassemble after the danger has past. In addition, fish that form schools swim more efficiently over longer distances using less energy than if each individual were swimming alone. This greater efficiency is accomplished by drafting, similar to the strategy employed by racing car drivers. Fish in a given school are of one species, similar in size, and may be a mixture of males and females or a single sex (depending on the species). A hierarchical structure (leaders and followers) is not evident among individuals in the school. Wide-angle vision or sound vibrations enable each fish in the school to sense the location and movement of its neighbors thereby maintaining order.

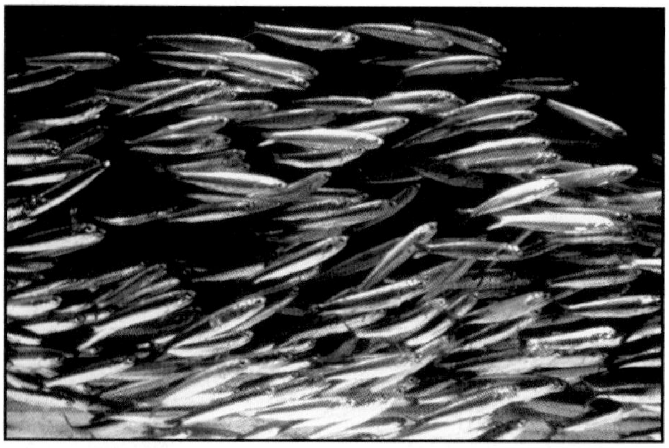

FIGURE 10.25
School of northern anchovies (*Engraulis mordax*). Schooling is a behavioral adaptation that may offer protection from predators. [Courtesy of NOAA National Undersea Research Program]

MARINE MAMMALS

Marine mammals are warm blooded, air breathing animals that bear live young, which they nurse. They include whales, dolphins, seals, walrus, sea lions, and polar bears (a terrestrial mammal that has adapted to a marine habitat). The largest marine mammals, baleen whales, live primarily in the open ocean where they filter zooplankton from the water as they swim slowly with their mouths open. These filter-feeding whales include the largest animal to have lived on Earth, the blue whale. Grey whales are also very large baleen whales. They forage for crustaceans by sucking up large amounts of sediment from the sea floor and expelling it through their baleen, which retains the food. Smaller whales, such as sperm whales, killer whales, porpoises and dolphins are faster swimming, toothed carnivores that actively hunt their prey. Sperm whales feed on squid captured in the mid-depths of the ocean. **Cetaceans** (whales, porpoises, and dolphins) spend their entire lives at sea.

Uniquely adapted to living on sea ice in the far north, polar bears can be considered marine mammals because they spend so much time in the icy waters. They inhabit the coastal areas of Alaska, northern Canada, northern Russia, Greenland, Norway, and the ice covered waters in between. The region's largest land predator, they grow to 2.5 to 3 m (8 to 10 ft) tall and weigh 250 to 800 kg (550 to 1700 lb). Polar bears depend on sea-ice floes as platforms to hunt seals, their principal food source. They also take walruses and small whales (e.g., Beluga whales) but only when they are trapped in small openings within the ice. Their dense fur and blubber enable them to travel comfortably across sea ice and open water for distances as great as 100 km (62 mi). The females make their dens on land or ice floes.

In May 2008, polar bears were added to the "threatened" category of the U.S. endangered species list. Because of their dependence on sea ice, polar bears could face extinction as the already shrinking sea-ice cover in the Arctic and far north thins further in response to higher temperatures (Chapter 12). Shrinking sea ice is evident in western Hudson Bay, near the southern limit of the polar bear's range, where the spring ice break-up occurs earlier, their hunting season is shorter, and less food is available for adult polar bears and their cubs. When the lack of sea ice prevents them from hunting seals, hungry bears may wander into settlements and scavenge for food in garbage

FIGURE 10.26
Large walrus (*Odobenus rosmarus divergens*) on the ice of the Bering Sea, Alaska. [NOAA Photo Library photograph by Captain Budd Christman, NOAA Corps (ret.)]

dumps. Some bears have come to depend on humans for food and are further threatened when residents are forced to kill the potentially dangerous scavengers.

Pinnipeds, named for their distinctive swimming flippers, include seals, sea lions, and walruses; many of them inhabit coastal waters (Figure 10.26). Unlike cetaceans, pinnipeds come ashore to breed, give birth, and rear their young. All pinnipeds are fur bearing and many species have been hunted nearly to extinction, although some populations are recovering due to protective legislation.

Manatees and dugongs are large herbivorous mammals that feed on vegetation in shallow waters along the coast and in the estuaries of Florida and parts of Asia and South America. They are now seriously endangered in many areas because of increasing coastal development and encounters with motorboats whose propellers can severely injure these gentle, slow-moving animals. Loss of sea grass beds, a primary feeding habitat for some manatees and dugongs, is another threat to their survival.

The National Marine Fisheries Service of NOAA, through its Office of Protected Resources, is responsible for enforcing the Marine Mammal Protection Act in the United States. This legislation, enacted in 1972 and re-authorized in 1994, established a moratorium on the hunting of marine mammals in U.S. waters and by U.S. citizens on the high seas.

MARINE REPTILES

The few reptiles that live in the ocean fall into three groups, the best known being sea turtles (Figure 10.27). Marine lizards, such as the iguana of the Galápagos Islands, and sea snakes are examples from the other two groups of marine reptiles.

All sea turtles live in the ocean, but come ashore to lay their eggs. Female sea turtles dig their nests on sandy beaches, lay up to one hundred leathery eggs, and then return to the sea. Newly hatched turtles must fend for themselves and find their way back down the beach to the sea. Along the way, sea birds and other animals prey upon them. If they make it to the sea, predatory fishes reduce their numbers still further. Only a few hatchlings survive to adulthood. Sea turtles have been hunted for their meat as well as their shells, which are used to make combs and jewelry. Plastic litter in the ocean is a serious threat to sea turtles (as well as other marine animals) that frequently mistake drifting plastic bags for jellyfish, a favorite prey. These bags, when swallowed, can block the digestive tract, eventually causing death. Turtles also get caught in illegal drift nets and on long-lines of hooks intended to catch swordfish. We have more on sea turtles in Chapter 14.

Sea snakes are rare, occurring only in the tropical Pacific and Indian Oceans, usually around reefs. Unlike moray eels, which are actually fishes with gills, sea snakes are reptiles and must come to the surface to breathe; they have special valves that close their nostrils while submerged. Some species of sea snakes go ashore to reproduce and lay eggs; others mate and bear live young without leaving the sea. They are among the most venomous animals on Earth. A single snake can produce about 10 to 15 milligrams of venom at a time, but the fatal dose for humans is only one-tenth this amount. Fortunately sea snakes are extremely shy animals and rarely attack humans unless provoked.

Marine iguanas are herbivores, feeding mostly on seaweeds in the intertidal zone. Large males feed offshore, diving to depths of up to ten meters. They occur only in the Galápagos Islands in the equatorial Pacific Ocean.

FIGURE 10.27
The leatherback, an endangered species, is one of the largest sea turtles in the world. It can grow to a length of over 2 m (7 ft). [Photo: Scott R. Benson, NOAA/NMFS Southwest Fisheries Science Center]

Having evolved from land lizards, they need a special adaptation to deal with the large amount of salt water they swallow while feeding. A special gland connected to the nostrils excretes excess salt, which the iguana expels by sneezing. Being cold blooded, iguanas regulate their body temperature by spending much time sunning themselves on the rocky shore.

The Galápagos Islands are the product of hot spot volcanism in the eastern tropical Pacific (Chapter 2). Prevailing trade winds cause equatorial upwelling of cold, nutrient-rich waters that provide sufficient production to support marine iguanas. Apparently, the abundant food supply was a factor in the iguana's adaptation to cold, salty ocean waters. During El Niño episodes, however, upwelling of nutrients is reduced thus lowering production so that populations of iguanas and other inhabitants of the Galápagos Islands are threatened (Chapter 11). Upwards of 10% of the iguana population can die of starvation during a lengthy and severe El Niño episode.

SEA BIRDS

Unlike most birds on land, seabirds are carnivorous predators that occupy a high trophic level in marine food webs. They have relatively high metabolic rates and require energy-rich, fatty foods. Wading birds, such as herons, have long legs and inhabit wetlands and shallow soft-bottom habitats where they find prey in the shallow waters and soft sediment. Some, such as flamingos and spoonbills, have specialized bills that allow them to strain bottom deposits to feed on small plants and animals. Herons have a sharp bill enabling them to snatch small fishes from the shallow water. Other seabirds dive for food; these include sea ducks, cormorants, loons, terns, and pelicans (Figure 10.28). Pelicans are among the largest and heaviest birds on Earth, yet they make spectacular dives as they hunt for fish.

No seabird can stay at sea for its entire life. All of them must come ashore to breed and lay eggs on a solid surface. Also, their body heat is necessary to hatch the eggs. The chicks require feeding by their parents until they are ready to fly and hunt for themselves. However, some seabirds spend very long periods flying over the open ocean. The albatross is a classic example, spending years at a time far from land.

Where fishing grounds are exceptionally productive, seabirds establish huge colonies on rocky shores and cliffs, especially during their breeding season. Some seabirds migrate long distances searching for food. Terns migrate from the Canadian Arctic, where they spend the Northern Hemisphere summer, all the way to Patagonia

FIGURE 10.28
Brown pelican in breeding colors. [NOAA National Marine Fisheries Service, photo by William B. Folsom]

at the southern tip of South America, where they spend the Southern Hemisphere summer.

One type of seabird, however, has lost its ability to fly. Penguins are adapted to live at sea and their streamlined bodies enable them to swim with considerable speed and agility (Figure 10.29). They can dive as deep as 250 m

FIGURE 10.29
Although ungainly on land, the flightless penguin is an expert swimmer. Shown here are Gentoo Penguins (*Pygoscelis papua*), largest of the "bush-tail" penguins. They live on Antarctic islands as well as the Antarctic Peninsula. [NOAA/Vents, Korea Polar Research Institute (KOPRI)]

(820 ft) and remain submerged for up to 20 minutes while searching for fish and small squid. They lay eggs and rear their young on land ice. Nearly all species of penguins (17) live in the Southern Hemisphere, mostly around Antarctica and on the southern extremities of Africa, South America, and Australia. One species lives near the equator on the Galápagos Islands. All these areas are nutrient-rich with cold currents. Some penguin species spend up to 75% of their lives at sea, away from land for months at a time. For birds, penguins have unusually dense bones, an adaptation that counteracts their buoyancy problem.

Conclusions

Most of Earth's living space is in the ocean where habitats are many and diverse, ranging from intertidal zones along the coast to soft sediments and hydrothermal vents on the deep-ocean floor. Organisms exploit all these habitats. But survival in the ocean presents many challenges that differ from those faced by terrestrial organisms. Marine plants and animals have evolved many specialized adaptations involving buoyancy, feeding, reproduction, and protection from predators.

Marine plants and animals are distributed nonuniformly in the ocean; their distribution is controlled primarily by environmental factors such as water temperature, salinity, light, and availability of food. Most marine organisms live near the ocean's surface, edges, or on the bottom. In the open ocean, organisms are most abundant in the surface waters, less so at mid-depths, and are least abundant in the deep ocean.

This chapter completes our survey of marine organisms and how they cope with the physical, chemical, and biological properties of the ocean in the Earth system. In the next chapter, we examine some of the large-scale interactions between the ocean and atmosphere and their impacts on the rest of the Earth system including marine life.

Basic Understandings

- Living space in the ocean not only vastly exceeds that on land but also features very different types of habitats. Organisms live in all parts of the ocean, even extreme environments such as sea ice and hydrothermal vents. Other marine habitats include rocky shores, tide pools, estuaries, coral reefs, polar oceans, deep-sea trenches, and sediments and bedrock on the sea floor.
- Marine life zones are commonly defined in terms of distance from shore and water depth. Open ocean waters constitute the pelagic zone, home to plankton (passive floaters or weak swimmers) and nekton (strong swimmers). The environment of the sea floor at all depths is called the benthic zone, home to organisms that live either on the ocean bottom or within sediment deposits.
- The area along the shore between high- and low-tide lines is the intertidal (or littoral) zone, home to ecosystems such as salt marshes and mangrove swamps. The area seaward from shore to the shelf break at a depth of about 200 m (650 ft), including the intertidal zone, forms the neritic zone—commonly referred to as the coastal zone.
- Oceanic life zones are also defined in terms of nutrient supply and production. Nutrient-poor waters having low primary production are described as oligotrophic. Broad areas of open-ocean within the subtropical gyres are examples of oligotrophic waters. Nutrient-rich waters having high primary production are described as eutrophic. Coastal upwelling zones and many estuaries have eutrophic waters.
- Plankton, members of phytoplankton, zooplankton or microbial communities, evolved adaptations and strategies for life in the pelagic zone. An adaptation is a genetically controlled trait or characteristic that enhances an organism's chances for survival and reproduction in its environment. For example, the shape of some plankton produces a relatively high surface area to volume ratio that provides buoyancy so that they do not sink below the sunlit photic zone.
- Larger, free-swimming pelagic animals, collectively called nekton, include fish of all sizes, squid, sea turtles, and marine mammals. Some migrate over long distances; the distribution of many others is limited by their tolerance for water temperature, salinity, or food availability. Nektonic organisms also have adaptations for buoyancy—the simplest of these are gas bladders.
- Most pelagic fish rely on active swimming to obtain food. Some fish lie on the bottom, waiting for their prey and suddenly lunge to catch it. Others swim all the time, searching for prey over large areas of the ocean.
- The photic zone is where food is most plentiful but it is also the most dangerous for marine

organisms because predators can easily see them. Many types of zooplankton avoid this problem by daily vertical migration. Each day at dusk, they come to the surface zone to feed on phytoplankton. Then, as daylight returns, they return to the relative safety of darker, deep waters.

- Species of smaller fish that are prey for larger ones have adaptations that make them less visible and therefore less likely to be eaten. The most common adaptation is adaptive coloration where the animal's color pattern closely matches its background substrate. Many fish exhibit countershading, that is, their dorsal side is a dark color, making it difficult for predators above to see them against the dark, deep water. Their ventral side is lighter colored, making the fish more difficult for predators to see them from below against the more brightly lit surface waters.
- Some marine animals emit light, a phenomenon known as bioluminescence. The light usually is a product of a chemical reaction that takes place in specialized cells or organs. Marine animals use light emission to attract mates or prey, frighten or confuse predators, or as a disguise.
- The general name for animals that are equipped with features that strain food particles out of large volumes of water is filter feeder. Some planktonic organisms produce large mucus nets, which function much like a spider web to trap small water-borne organisms and other food particles.
- Benthic organisms may live attached to a firm surface, construct burrows or tunnels or simply dig into sediment deposits, or they may move freely on the ocean bottom. Attached animals are filter feeders; mobile animals are active predators; and burrowing organisms ingest sediment, digesting the organic material it contains and excreting the rest.
- A distinction is made among intertidal habitats based on the type of substrate (e.g., rocky, muddy, sandy). The most energetic of these habitats are rocky intertidal shorelines where organisms are frequently subject to strong wave action, especially in winter. Where the distance between low and high tide levels is relatively great, organisms may be exposed to dry air and sunshine for many hours each day. Depending on the season, they may be forced to withstand temperature extremes when they are not covered by seawater.
- Seaweeds are important components of the intertidal zone, providing habitat for many marine organisms. All are algae and are green, brown, or red, depending on the color of the dominant photosynthetic pigment. They are distributed according to their light requirements and their ability to resist drying and freezing during low tide.
- On mud flats and other soft-bottomed habitats, the most important plants are the highly productive sea grasses, also known as submerged aquatic vegetation (SAV). These are angiosperms, flowering and seed bearing vascular plants with true roots. They are limited to water depths where light is sufficient for photosynthesis.
- Salt marshes commonly occur along sheltered shorelines and are ecologically similar to sea grass beds in estuaries. The salt marsh grasses and bushes differ from sea grasses by being salt-tolerant true land plants, pollinating in the air. Salt marshes are flooded at high tide, but the grasses are never completely covered by salt water. They are home to abundant marine life and are refuges for waterfowl and other wildlife.
- In estuaries, plants and animals vary from marine species near the mouth of the estuary to brackish water types where the salinity is lower, to those that prefer a nearly freshwater environment at the head of the estuary.
- In the tropics, mangrove swamps are common along muddy, low-lying coastlines and in estuaries. They consist of tropical tree species that are salt-tolerant and can withstand tidal flooding of their roots and lower stems. Mangroves form dense growths and their aerial roots provide a complex and protective habitat for many organisms such as mangrove oysters.
- Seaward of the intertidal zone, kelp forests grow where waters are cool and nutrient-rich. Kelp includes species of brown algae that grow to an enormous size. They are abundant in coastal upwelling zones off California and the Pacific Northwest. Kelp grows attached to rocky bottoms by root-like structures known as holdfasts and much of the plant is supported at the surface by a bulbous gas filled float below the fronds. Recent research indicates that patches of kelp also occur in tropical latitudes.

- Coral reefs are among the ocean's most spectacular biological features. They grow along coastlines or cap extinct undersea volcanoes. Coral reefs are only thin veneers of living organisms growing on older layers of dead coral (limestone) or volcanic-rock.
- Hermatypic corals live in warm, shallow water and build large reefs. Solitary corals and small coral reefs also live in cold, deep water along continental shelf breaks in some parts of the ocean. These are known as ahermatypic corals. Coral reefs provide shelter for many species of fish, invertebrates, and plants and are among the ocean's most productive habitats.
- Coral reefs growing on volcanic islands typically exhibit a characteristic sequence of forms caused by either sea level rise or sinking of the island: fringing reef, barrier reef, and atoll.
- Animals living on and in soft-bottomed habitats consist of either infauna or epifauna. Infauna live within sediment deposits whereas epifauna live on the sea floor.
- Soft ocean bottoms harbor many filter feeders that separate phytoplankton, small zooplankton, and other edible materials from large volumes of water.
- Many unusual creatures live on the deep-sea floor. Sediment feeders are similar to those living in shallow benthic habitats, although animals in the deep, cold ocean are generally smaller, an adaptation to the scarcity of food. In addition, communities of specialized animals have evolved to live on the deep ocean bottom near hydrothermal vents utilizing chemosynthesis.
- Unlike mammals, which regulate their body temperature, fish are cold blooded. Fish may be herbivorous, carnivorous, or omnivorous. They inhabit all parts of the ocean and possess special adaptations for buoyancy, swimming, and life at the surface, in the dimly lit twilight zone, as well as the greatest depths of the ocean.
- The three major groups of fishes are jawless fishes, cartilaginous fishes, and bony fishes.
- Anadromous fishes are born in freshwater rivers and streams, but spend most of their lives in the ocean, returning to the same river to breed when they are sexually mature. Less well known are catadromous fishes that breed in the open ocean, but spend their adult lives in fresh water.
- Many species of fish reduce predation by swimming together in organized groups called schools. This strategy confuses potential predators and functions as an alarm system when would-be intruders approach.
- Marine mammals are warm blooded, air breathing animals that bear live young, which they nurse. The largest marine mammals, baleen whales, live primarily in the open ocean where they filter zooplankton from the water as they swim slowly with mouths open. Pinnipeds are named for their distinctive swimming flippers and include seals, walruses, and sea lions; many of them live in coastal waters.
- The few reptiles that live in the ocean include sea turtles, marine lizards, and sea snakes.
- Unlike most birds on land, seabirds are carnivorous predators that feed at high trophic levels in marine food webs.
- Penguins are flightless marine birds confined primarily to the Southern Hemisphere.

Enduring Ideas

- Ocean life zones (marine habitats) are defined in terms of location (distance from shore and depth) or nutrient supply and productivity (oligotrophic, eutrophic).
- Adaptations for buoyancy that enable marine organisms to remain within the sunlit photic zone include a shape that gives them a relatively high surface area to volume ratio (e.g., plankton), gelatinous bodies (e.g., jellyfish), gas bladders (e.g., nektonic organisms), and active swimming (e.g., most pelagic fish).
- Adaptations that enable marine organisms to obtain food and avoid predation include vertical migration, adaptive coloration and countershading, bioluminescence (e.g., to attract prey), large sensitive eyes (for vision in the twilight zone), and feeding strategies (e.g., filter feeding).
- Waves, winds, and tidal currents continually disturb the intertidal zone. This dynamic portion of the ocean is alternately exposed to seawater and the atmosphere, accompanied by drastic changes in temperature and salinity. Special adaptations enable animals and plants to survive these stressful conditions.
- Fish are cold blooded animals that may be herbivorous, carnivorous, or omnivorous. Some are cartilaginous and some are bony. Some are anadromous and some are catadromous.

Review

1. Describe the portions of the ocean represented by the pelagic, benthic, and neritic zones.
2. Which areas of the ocean are usually oligotrophic? Which are eutrophic? What is the implication for primary production?
3. Define adaptation.
4. Phytoplankton possesses a special adaptation that significantly slows their sinking motion in the ocean. Describe that adaptation.
5. What is the function of adaptive coloration and countershading for fish?
6. What is bioluminescence and what purpose does it serve?
7. Describe the types of marine organisms found in the benthic zone based on their life strategy.
8. What is the significance of sea grasses for other organisms?
9. Characterize the environmental conditions most favorable for the growth of coral reefs in tropical ocean waters.
10. Are seabirds herbivorous or carnivorous predators? Explain your answer.

Critical Thinking

1. Why is the ocean's neritic zone likely to be eutrophic whereas the open ocean is more likely to be oligotrophic?
2. What causes an algal bloom and what are the potential consequences for marine life in terms of availability of dissolved oxygen?
3. Describe the daily changes in the environmental conditions of the intertidal zone.
4. Why are adaptations for buoyancy essential for the survival of marine animals?
5. Explain how many types of zooplankton avoid predation through vertical migration.
6. How might shrinkage of the Arctic ice cover (because of global warming) impact polar bears?
7. Provide several examples of how marine organisms avoid sinking to the ocean bottom.
8. Birds feed at a relatively high trophic level in marine food webs. What is the significance of this for bioaccumulation of persistent toxins?
9. What is the source of food energy for marine organisms that live in the deep, dark ocean bottom?
10. Summarize the major differences between marine and terrestrial food webs.

ESSAY: Life and Sea Ice

Marine life forms occupy extreme environments upon, within, and beneath the sea-ice cover of the polar regions. In the Arctic Ocean, for example, sea ice provides habitat for a variety of organisms that are members of an ice-specific food web and include bacteria, viruses, unicellular algae, and small invertebrates. These organisms have adapted to dramatic changes in light intensity, temperature, and salinity. In addition, evidence points to an unexpectedly more diverse marine community in the water column and seafloor under ice shelves.

As noted in Chapter 3, dissolved salts lower the freezing point of seawater so, at an average salinity of 35 psu, seawater initially freezes around −1.9 °C. Additionally, salt is excluded as sea ice forms (*brine rejection*). In autumn, as sunlight fades and daylight shortens, the air temperature drops and surface waters cool until ice forms on the ocean surface as porous structures of interlocking ice crystals filled with a salty solution called *brine*. Initially accounting for 10% to 30% of the ice volume, brine occupies tiny channels and pore spaces between and within ice crystals. As winter progresses sea-ice cover solidifies, pore spaces decrease, and the salinity of the brine increases. As long as the temperature remains above −5 °C, the ice continues to be riddled with tiny passages a few micrometers to several centimeters in diameter. But when the temperature drops to lower values, the permeability of the ice becomes minimal and the salinity of the trapped brine can approach values near 250 psu, over 7 times that of seawater.

Sea ice insulates the underlying seawater from the atmosphere so that a considerable temperature difference develops. This *temperature gradient* ranges from below −35 °C at the ice-atmosphere interface to near −2 °C at the ice-seawater interface, which is generally the same temperature as the water. For this reason, most ice organisms concentrate within the lowermost centimeters of the ice. When the sunlight returns in spring, photosynthesis can begin and populations of unicellular ice algae explode into the summer. Often forming chains and filaments that extend many meters into the water, the several hundred species of algae living in the ice are the chief primary producers in the Arctic. Ice algae account for 4% to 26% of total marine primary production in seasonally ice-covered Arctic waters and up to 50% or more in perennially ice-covered waters.

Organisms that dwell within ice have adapted to prevent ice crystals from forming inside their bodies. As mentioned earlier in this chapter, an *adaptation* is a genetically controlled trait or characteristic that enhances an organism's chance for survival and reproduction. For example, some ice organisms accumulate fat-like substances in their bodies that prevent freezing.

Feeding habits is another adaptation, and bacteria in ice specialize in the wastes from ice algae. The algae are eaten directly by crustaceans, rotifers (microscopic invertebrates), and turbellarians (flatworms). Also feeding on the algae are tiny crustaceans (amphipods), which live on the underside of the ice and avoid predation by sheltering in the brine channels. Zooplankton live out their juvenile stages feeding on the ice bottom community. With higher temperatures in the spring, the solid ice cover breaks up into pack ice and ice floes, transporting the organisms thousands of kilometers. When the ice melts, the accumulated organic material sinks into the water column and fuels pelagic and benthic food webs.

In the past decade, scientists have gathered new information regarding life in and under the polar ice. In December 2003, Australian scientists drilled a hole through the 480 m (1575 ft) thick Amery ice shelf in Antarctica, some 100 km (60 mi) from the ocean, lowered a video camera to the seafloor 775 m (2550 ft) below. To their amazement, they identified 24 species of invertebrates (including sponges, mollusks, and sea urchins) in the 2 square m within the camera's range. Scientists previously had assumed that the food supply in the Southern Ocean was insufficient to support life under the ice shelf but it is, in fact, very diverse.

An international scientific team discovered hundreds of new species in the deep waters near Antarctica during three cruises in 2002 and 2005. Onboard the German research icebreaker *Polarstern*, the team scooped up sediment and its inhabitants at 40 sampling sites ranging from 774 to 6348 m (2500 to 20,800 ft) deep. During the Southern Hemisphere summer of 2006-2007, the British Antarctic Survey inaugurated its remotely operated *Isis* vehicle on the deep seafloor surrounding Antarctica. The onboard camera revealed krill (*Euphausia superba*) at a depth of 3000 m (9800 ft). Until this discovery, scientists assumed this important member of the polar ecosystem only lived at much shallower depths.

In the northwest Weddell Sea of the Southern Ocean, enhanced biological productivity was reported in the waters surrounding icebergs during the 2005 austral spring, which is seasonally when the ozone shield is thinnest.

Kenneth L. Smith, Jr., an oceanographer at the Monterey Bay Aquarium Research Institute in Moss Landing, CA, and colleagues discovered a zone of increased productivity extending 3.7 km (2.3 mi) radially from each iceberg. Phytoplankton was about 5 times more abundant near the bergs and populations of organisms dependent on phytoplankton were similarly elevated. Researchers noted high concentrations of chlorophyll, krill, and sea birds. Chemical analysis of water samples within the zone revealed higher than usual levels of nutrients. Another example of how the geosphere, cryosphere, and biosphere are linked within the Earth system, these nutrients are incorporated into the parent glacier as it scraped material off the land then is delivered to the sea as the icebergs melt.

Large warm-blooded animals also live on sea ice. Birds (e.g., penguins in the Antarctic), seals, whales, and polar bears utilize sea ice for migration routes, hunting grounds, and rookeries. Seasonal changes in ice cover and thickness make for a dynamic environment and require animals to have excellent navigation skills.

Polar bears have a special adaptation that enables them to survive the extreme cold of the Arctic climate (Figure 1). Because white polar bears blend in with snow-covered ice, scientists planned a bear census using thermal (infrared) imagery assuming that polar bears would appear as hot spots. However, while scientists could visually observe the polar bears, the bears were invisible to the thermal imagery. Made of hollow tubes for insulation, polar bear fur is so efficient that essentially no body heat is lost to the ambient air, and so there were no heat signals to reach the infrared sensor.

FIGURE 1
Polar bear (*Ursus maritimus*) and cub blend into the background and because of their excellent insulation cannot be detected by airborne infrared sensors. Photograph was taken on the Arctic Ocean ice north of western Russia. [NOAA Photo Library, NOAA Climate Program Office, NABOS 2006 Expedition, photograph by Mike Dunn, NC State Museum of Natural Sciences.]

ESSAY: Coastal Dead Zones

Excessive amounts of nutrients from agricultural and urban runoff create barren underwater landscapes, killing benthic organisms in coastal marine ecosystems. Across large areas of the inner continental shelf, deep waters are so depleted in *dissolved oxygen (DO)* that fish and large marine invertebrates perish. Waters with less than 2 milligrams of DO per liter are *hypoxic*, supporting only *anaerobic bacteria*, which thrive with little or no oxygen while other animals suffer. Those that cannot move elsewhere, such as mollusks, suffocate, and free-swimming animals such as fish, shrimp and crabs are forced to migrate to more favorable areas or die. For this reason, the area of hypoxic waters is known as a *dead zone*.

Every summer since the mid-1970s, the Gulf dead zone (with hypoxic bottom waters) has spread west of the Mississippi River delta and along the Louisiana-Texas Gulf Coast. In fact, there are more than 100 dead-zone sites along the Gulf of Mexico coast (in addition to areas of naturally occurring DO-depleted deep water). In July 2008, the Gulf dead zone encompassed an area the size of Massachusetts, 20,720 square km (7965 square mi), making it the largest in U.S. coastal waters and one of the largest worldwide.

The Gulf dead zone is fed by the discharge of the Mississippi and Atchafalaya Rivers, which contain significant levels of nitrogen. On agricultural fields draining into the rivers, synthetic nitrogen fertilizer has been routinely applied in the spring and summer since the late 1940s. This nutrient input is supplemented by runoff from urban and suburban landscapes (e.g., fertilizer and the effluent of partially treated sewage). As these fertilizers reach the coastal waters, it prompts an *algal bloom*, an explosive growth of algae populations. This high *primary production* on the inner shelf is the principal source of organic matter and oxygen depletion of bottom waters (Figure 1). It turns the water green or brown, shading ocean-bottom sea grasses from essential sunlight. As the sea grasses die, the ecosystem they had sheltered is destroyed, depriving other organisms of habitat for reproduction and growth. As dead algae sink, they add to the accumulating organic matter on the ocean floor, which is decomposed by oxygen-depleting bacteria, and hypoxic conditions develop.

The water column is stratified with a strong vertical density gradient throughout the summer with warm, oxygenated water above and cold, oxygen-depleted water below. The stronger winds in late fall mix the water column, replenishing the dissolved oxygen content of the bottom waters, so life can return.

As shown in Figure 2, the Mississippi River and its tributaries drain a huge area (about 40%) of the coterminous U.S., from the headwaters in Minnesota, west to the Rocky Mountains, and east to the Appalachian Mountains. This drainage basin includes the Corn Belt, which is responsible for much of the runoff in the Gulf dead zone. In the future, the problem is likely to be exacerbated (a larger dead zone) as more corn is used for biofuels, and world food demands increase.

In 2001, G.F. McIsaac, an environmental scientist at the University of Illinois at Urbana-Champaign, reported that the Gulf dead zone could be alleviated significantly if farmers in the Mississippi River drainage basin cut their application rate of nitrogen fertilizer by 12%. McIsaac determined that such a reduction would not necessarily reduce crop yields but would decrease the nitrates reaching the Gulf of Mexico by one-third by the year 2010. Phosphorus, however, also contributes to the problem. While 75% of nitrogen comes from agricultural fertilizer application, 60% of phosphorus does as well. Hence, shrinking the Gulf dead zone requires reducing the input of both nitrogen and phosphorus. Strategies

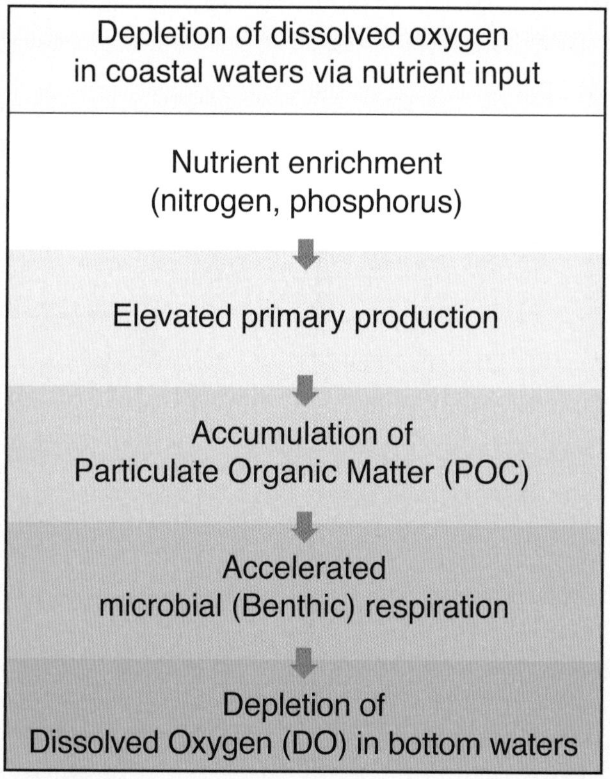

FIGURE 1
Process by which nutrient input leads to depletion of dissolved oxygen in bottom waters, creating the Gulf dead zone.

for accomplishing this include limiting fertilizer use, creating buffers (e.g., wetlands, forests) between agricultural fields and rivers to reduce runoff, and major reductions in sewage runoff from urban areas.

The present status of the nation's dead zones is summarized in the September 2010 report, *Scientific Assessment of Hypoxia in U.S. Coastal Waters*. The report was the product of an interagency working group of the National Science and Technology Council's Committee on Environmental and National Resources. NOAA was the lead agency with contributions by the USGS, EPA, USDA, and the Virginia Institute of Marine Science. The report documents an exponential growth of seasonal dead zones in U.S. coastal waters over the past 50 years (currently a 30-fold increase). Half the 647 waterways reviewed by the working group (including the Gulf of Mexico) experience hypoxic conditions. The report concludes that "overall, management efforts to stem the tide of hypoxia have not made significant headway" principally due to the steady increase in development and population growth in coastal watersheds. The report also emphasizes the "need to understand the complex underlying science of hypoxia and to predict the range of impacts of hypoxia on ecosystems."

The greatest increase in the number of dead zones over the past two decades has occurred along the U.S. Pacific Coast, which has experienced a 6-fold increase. There are now 37 localities that have problems with low DO. Each summer since 2002, large dead zones have developed off the coast of Oregon and Washington, ranking second in size to the Gulf of Mexico dead zone, and the third largest in the world. In this case, upwelling and climate change, not fertilizer, are key factors.

Along the Pacific Northwest coast, upwelling delivers nutrients into the photic zone. As described in Chapter 6, *coastal upwelling* occurs where Ekman transport moves surface waters away from the coast which draws up nutrient-rich waters from below. Episodes of stronger than usual winds, accompanying the recent warming trend, have intensified coastal upwelling and bring more nutrients to the surface waters of coastal Oregon and Washington, allowing for massive algal blooms. The remains of algae sink to the sea floor and, if their decomposition reduces DO faster than it is resupplied, hypoxic conditions develop.

Coastal hypoxia arising from excess nutrient input is a growing problem around the world. One of the largest dead zones developed in the Black Sea near the mouth of the Danube River and covers an area of 40,000 square km (15,400 square mi). Large dead zones have also developed in the northern Adriatic Sea since the 1950s, the Baltic Sea since the 1960s, and the Kattegat since the 1980s. Other locales where large rivers discharge nutrients, creating hypoxic conditions in bottom water on the continental shelf, include the eastern North Sea and the East China Sea.

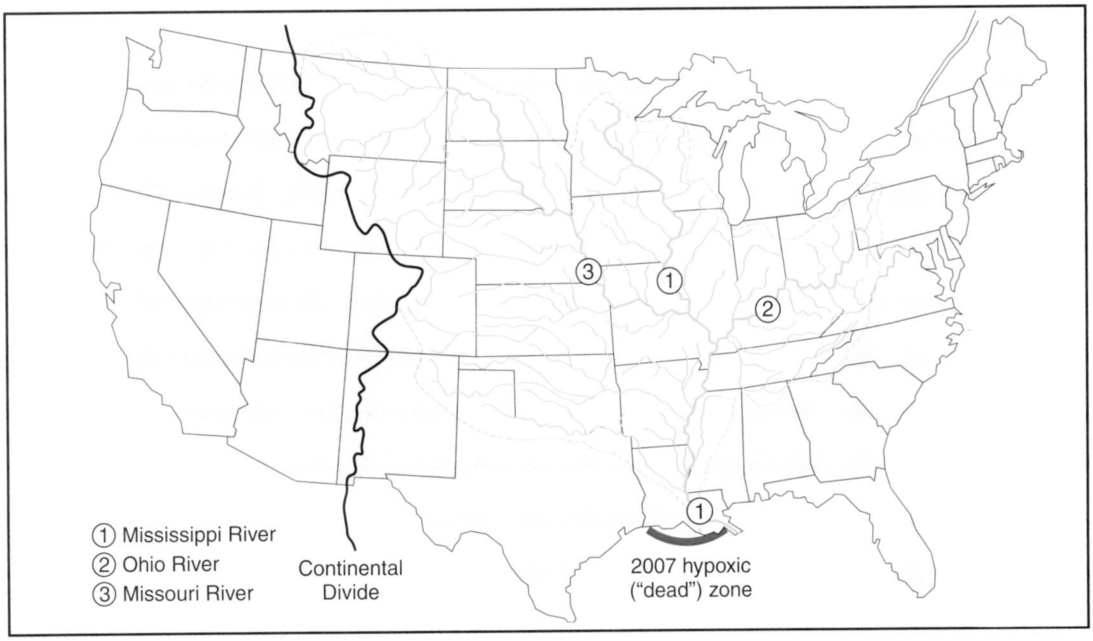

FIGURE 2
In summer, a large volume of bottom water significantly depleted in dissolved oxygen develops in the Gulf of Mexico just offshore of the Louisiana-Texas coast, as represented by the 2007 dead zone (red on the map). Hypoxic conditions eliminate many bottom-dwelling organisms and are caused by excessive amounts of nitrogen and phosphorus fertilizer washed from agricultural land in the Mississippi River drainage basin (shown in blue). [Courtesy of the U.S. Geological Survey (USGS)]

ESSAY: Marine Protected Areas

Presently, over 5000 *marine protected areas (MPAs)* exist worldwide, mostly in coastal areas. Major goals of MPAs are to limit the impact of fishing on biodiversity and prevent the destruction of coral reefs. The number of MPAs is expected to increase in the near future with many established outside national boundaries in the high seas. The *2002 World Summit on Sustainable Development* set internationally agreed targets to create extensive networks of MPAs by 2012.

Internationally, Great Britain created the world's largest marine protected area in April 2010 that encompasses the Chagos Islands, a group of coral reefs located in the Indian Ocean southwest of Sri Lanka and northeast of Madagascar. The 545,000 square km (159,000 square nautical mi) is home to more than 220 coral species which is half of the Indian Ocean's remaining healthy coral species, and more than 1000 species of reef fish. Exceptionally clear waters allow coral to grow at greater depth so that they are less vulnerable to thermal stress bleaching. All potentially disruptive activities such as industrial fishing and deep-sea mining will be prohibited and, except for the U.S. Naval Base on Diego Garcia, the Chagos Islands are uninhabited. According to U.K. Foreign Secretary David Millbrand, the new MPA will "double the global coverage of the world's ocean under protection."

In the United States, a major part of the effort to preserve the natural resources of coastal waters and promote the recovery of overfished and disturbed underwater communities is the *marine sanctuaries* program. At the state level, California has led the way. Point Lobos, CA (near Monterey), was founded in 1960 as the nation's first underwater sanctuary. Since then California has set aside 11 other sites as sanctuaries and has been studying several more potential sites. Twelve years after California, federal legislation moved to protect national marine treasures.

Provisions of the *1972 Marine Protection, Research, and Sanctuaries Act* authorize the President to designate national marine sanctuaries in coastal waters of the continental shelf and in the Great Lakes. Several characteristics qualify a locality as a marine sanctuary, but potential sites must have a special biological, aesthetic, archaeological, cultural, or historical significance. The objective of federal legislation is to preserve and protect those areas by managing the multiple demands placed on them. Most marine sanctuaries are not places of refuge for marine life, although most activities that threaten marine resources are prohibited.

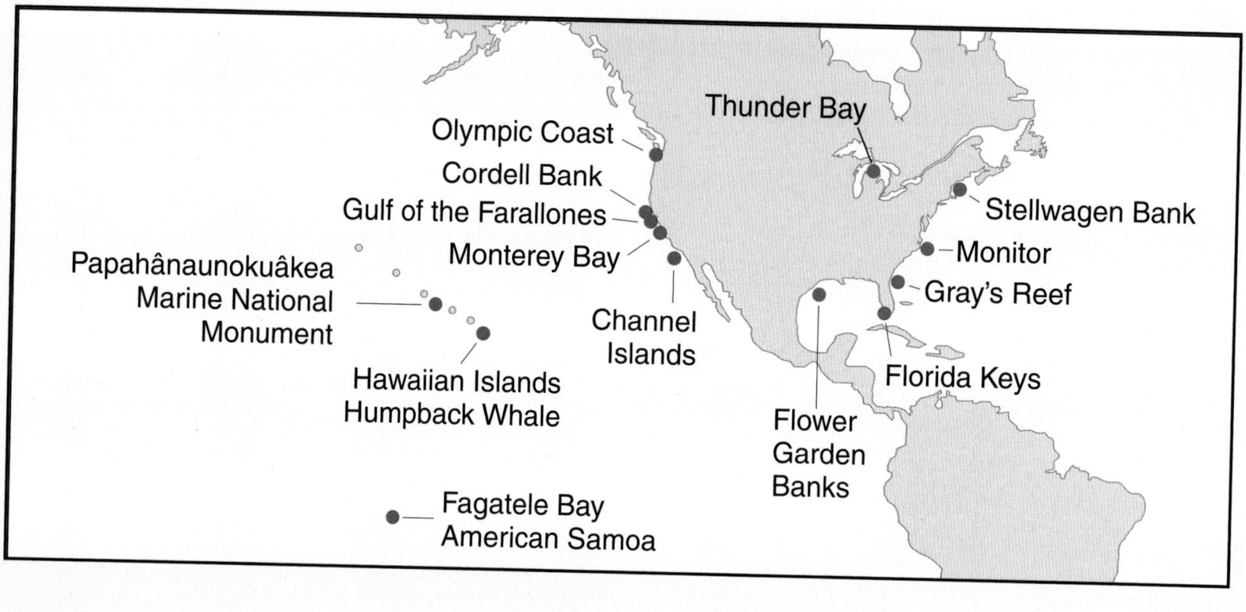

FIGURE 1
Map showing the names and locations of National Marine Sanctuaries in U.S. waters. [NOAA]

Currently, there are 14 designated national marine sanctuaries, one of which is also a national monument, and these include near-shore coral reefs, whale migration corridors, deep-sea canyons, and historical sites (Figure 1), which are administered by *NOAA's National Marine Sanctuary Program*. Established in 1975, three years after the Act was passed and two after the wreck was found, the first national marine sanctuary was established for the Civil War ship, *USS Monitor*, off the North Carolina coast. The first marine sanctuary in the Great Lakes is the resting place of about 160 shipwrecks at the Thunder Bay National Marine Sanctuary and Underwater Preserve in northern Lake Huron near Alpena, MI. The smallest and largest sanctuaries are on islands. Fagatele Bay in American Samoa, home of a tropical coral reef is 0.65 square km (0.19 square nautical mi) whereas Papahānaumokuākea in the waters northwest of Hawaii, the most recently designated marine sanctuary, is 356,894 square km (104,213 square nautical mi).

The Papahānaumokuākea Marine National Monument is also one of the largest marine protected areas in the world (Figure 2). Established by Proclamation by President George W. Bush on 15 June 2006, it includes 70% of all U.S. coral reef habitats, is home to more than 7000 marine species, 25% of which live only there, and is the principal habitat for 1400 Hawaiian monk seals. As a *marine reserve*, any activities, including tourism, that would disturb the ecosystem are prohibited and all commercial and recreational fishing in the region will be phased out by 2012. The "National Monument" designation expedited establishment of the sanctuary because the region was originally designated a national wildlife refuge by President Theodore Roosevelt (1858-1919) in 1909.

Designated a sanctuary in 1980, the Channel Islands National Marine Sanctuary is located 40 km (22 nautical mi) off the coast of Santa Barbara, CA. Waters surrounding the five islands are a rich breeding ground for numerous species of plants and animals, including a kelp forest. Annually, more than 24 species of whales and dolphins visit the sanctuary. The islands also are home to seabird colonies and pinniped rookeries. The Florida Keys National Marine Sanctuary encompasses the waters immediately surrounding most of the 1700 islands of the Florida Keys. Designated in 1990, this sanctuary forms a 355 km (192 nautical mi) arc extending from the southern tip of Key Biscayne (south of Miami) southwest toward, but

FIGURE 2
Pearl and Hermes Atoll, part of the Papahānaumokuākea Marine National Monument. [Courtesy of NOAA]

not including, the Dry Tortugas Islands. This complex of subtropical marine ecosystems has offshore coral reefs, fringing mangroves, and sea grass meadows, supporting more than 6000 species of plants, fishes, and invertebrates. The sanctuary covers 9500 square km (2800 square nautical mi), but only a small portion is closed for fishing.

In November 2000, President Clinton reauthorized the 1972 Marine Protection, Research, and Sanctuaries Act reaffirming the nation's commitment to conserve special areas of the marine environment for the appreciation and enjoyment of present and future generations. However, some conservationists argue that not enough is being done. They point out that at the time of the 1972 legislation the principal threats to the marine environment were oil spills and the plundering of sunken ships. Since then, overfishing has been added to the list of major threats to marine resources (Chapter 14). Nearly half of U.S. marine fisheries are either depleted or overfished and, according to the U.S. Department of Commerce, depletion of the fisheries costs the nation's economy billions of dollars in lost revenue each year. Yet, most marine sanctuaries (not marine reserves) permit fishing and allow recreational boating and mining of some resources, all activities that disrupt marine habitats.

Conservationists argue for more *marine reserves* where all fishing and other activities that threaten marine habitats are prohibited. The goal of marine reserves is to promote recovery of overfished stocks, preservation of seafloor communities such as coral reefs, and restoration of those destroyed by trawling. Studies of existing marine reserves show that compared to the surrounding waters, number of fish, size of fish, and species diversity are greater and fish larger within protected boundaries. These ecological benefits typically develop rapidly and occur regardless the size of the marine reserve. Coral, however, may require 5 to 15 years before rebounding. Prior to the designation of the Papahānaumokuākea Marine National Monument, only about 430 square km (125 square nautical mi) were marine reserves while individual states and local municipalities protect small, no-take areas.

Already beleaguered by dwindling fish stocks (some on the brink of extinction), the fishing industry is adamantly against any new, or expanding, marine reserves that would reduce the area of open fishing grounds. However, research reported over the past decade demonstrates that marine reserves also boost populations of fish, and the size of those fish, in the waters *outside* protected borders. Fish larvae drift some distance into surrounding waters and reseed fish stocks. Some fishers are now beginning to realize the potential benefits of marine reserves.

CHAPTER 11

THE OCEAN, ATMOSPHERE, AND CLIMATE VARIABILITY

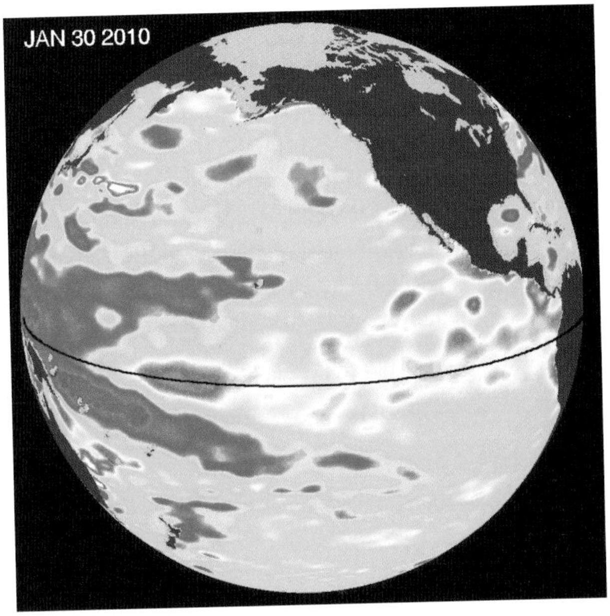

Jason-2 satellite image showing El Niño conditions in the Pacific Ocean during a 10-day period centered on 30 January 2010. [Courtesy of NASA/JPL Ocean Surface Topography Team]

Case in Point

In 1982-83, the weather seemed to go wild in many parts of the world and after it was over the total worldwide impacts included thousands of deaths and an estimated $13 billion in property damage. From mid-November 1982 until late January 1983, excessive rains caused the worst flooding of the century in usually arid Ecuador. Strong winds and torrential rains produced by six tropical cyclones lashed the islands of French Polynesia in a span of only three months. (By comparison, on average one tropical cyclone strikes this region of the eastern South Pacific about every 5 years.) At the other extreme, drought parched eastern Australia, Indonesia, and southern Africa. Huge drought-related wildfires broke out in Australia and Borneo. Australia's drought was that nation's worst in 200 years, causing $2 billion in crop losses and the deaths of millions of sheep and cattle. Meanwhile, persistent drought in sub-Saharan Africa grew worse. Over North America, the winter storm track shifted hundreds of kilometers south of its usual location, bringing episodes of destructive high winds and heavy rains to portions of California. Flooding rains also caused havoc across the southeastern United States. Meanwhile ski resorts in the northern U.S. experienced a snow drought and considerable economic loss as a consequence.

Just prior to these worldwide weather extremes, the ocean circulation off the northwest coast of South America changed drastically with dire implications for marine production. Along the coast of Ecuador and Peru, plankton populations plunged to about 5% of their normal level. The decline in plankton reduced the numbers of anchovy, which feed on plankton, to a record low. Other fishes dependent on plankton, such as jack mackerel, suffered a similar fate. Commercial fisheries off the coast of Ecuador and Peru collapsed. With the decline in fish populations, marine birds (e.g., frigate birds and terns) and marine mammals (e.g., fur seals and sea lions) also experienced major population declines as scarce food supplies caused breeding failures as well as migration or starvation of adults.

The ecological impact of these 1982-83 environmental changes was particularly severe on the remote island of Kiritimati (Christmas Island) at 2 degrees N, 157 degrees W, in the central tropical Pacific. An estimated 17 million seabirds, which feed on fish and squid, normally nest on the island. But scientists arriving on Kiritimati in November 1982 found that nearly all the adult birds had abandoned the island, leaving behind their young to starve. Apparently, sharp declines in food sources forced fish and squid to search for better feeding grounds, and the adult birds followed their prey. By July 1983, the usual ocean circulation returned, plankton populations rebounded, and the adult birds returned to Kiritimati. Nesting began and seabird populations started to recover.

What caused these drastic changes in atmospheric and oceanic conditions? Early on, some scientists attributed the weather extremes to the violent eruption of the Mexican volcano El Chichón in March-April 1982. But it quickly became apparent that the worldwide weather extremes were linked to large-scale ocean/atmosphere circulation changes in the tropical Pacific and soon a new scientific term was added to the public's vocabulary: El Niño. The El Niño of 1982-83 spurred further research on ocean-atmosphere interactions and the deployment of an array of in situ and remote sensing instruments in the tropical Pacific to provide early warning of the development of El Niño. In the late 1990s, when El Niño returned in its full fury, the global community was better prepared thereby lessening the impact.

Driving Question:
How do interactions between the ocean and atmosphere impact worldwide weather and short-term climate variability?

Some weather extremes such as drought and episodes of unusually heavy rains are linked to coupled changes in atmospheric and oceanic circulation. The principal focus of this chapter is short-term climate fluctuations involving interactions between the ocean and atmosphere. One of the most extensively studied of these interactions occurs in the tropical Pacific and is known as El Niño/La Niña. During El Niño, trade winds weaken, upwelling diminishes off the South American coast and along the equatorial Pacific, sea-surface temperatures (SST) rise well above long-term averages over the central and eastern tropical Pacific, and areas of relatively heavy rainfall shift from the western into the central tropical Pacific. La Niña sometimes (but not always) follows El Niño and is a period of exceptionally strong trade winds in the tropical Pacific, vigorous coastal and equatorial upwelling in the Pacific, and lower than usual SST in the central and eastern tropical Pacific. Based on changes in SST in the eastern tropical Pacific, some scientists refer to El Niño as the *warm phase* and La Niña as the *cold phase* of this air/sea interaction.

Broad-scale changes in SST patterns over the tropical Pacific that accompany El Niño and La Niña influence the prevailing circulation of the atmosphere in middle latitudes, especially in winter. Weather extremes that most often accompany La Niña are essentially opposite those that usually occur during El Niño. Although we devote much of this chapter to El Niño and La Niña, we also consider other examples of short-term climate variability stemming from air/sea interactions including the North Atlantic Oscillation, the Arctic Oscillation, and the Pacific Decadal Oscillation. We set the stage for this discussion by first describing controls operating within Earth's climate system.

Earth's Climate System

In Chapter 5, we defined **climate** as weather at a particular location averaged over a specific interval of time (by convention, 30 years). A complete description of climate also includes extremes in weather (e.g., highest and lowest temperatures on record, frequency of drought). In a more general sense, climate is the state of the Earth-atmosphere system (Earth's surface plus atmosphere) that gives a locality its characteristic weather patterns.

Climate varies both spatially and temporally. The globe is a mosaic of many different climate types including, for example, the warm, humid tropics, hot and cold deserts, temperate regions, and polar ice caps. Climate changes over a broad spectrum of temporal scales, from years to decades to centuries to millennia. In this section, we describe the various controls operating in Earth's climate system with emphasis on the role of the ocean.

CLIMATE CONTROLS

Many factors work together in shaping the climate of any locality. Controls of climate consist of (1) latitude, (2) elevation, (3) topography, (4) proximity to large bodies of water, (5) Earth's surface characteristics, (6) net incoming solar radiation, (7) long-term average atmospheric circulation, and (8) prevailing ocean circulation. Over a period of at least several million years, all climate controls are variable. On time scales that extend to hundreds of millions of years, continents have drifted to different latitudes, ocean basins have opened and closed, and mountain ranges have risen and eroded away—all with implications for climate. On shorter time scales (e.g., the range of human existence), for all practical purposes, the first four climate controls on our list are essentially fixed and exert regular and predictable influences on climate.

Seasonal changes in net incoming solar radiation, as well as length of daylight, vary with latitude, and air and sea-surface temperatures respond to those regular variations (Chapter 5). Air temperature decreases with increasing elevation and determines whether precipitation falls in the form of rain or snow. Topography can affect the distribution of clouds and precipitation. For example, the windward slopes of high mountain barriers (facing the oncoming wind) usually are wetter than the leeward slopes (facing downwind). The relatively great thermal inertia of large bodies of water (especially the ocean) moderates the temperature of downwind localities, reducing the temperature contrast between summer and winter and lengthening the growing season. Recall our discussion in Chapter 3 regarding the contrast in maritime versus continental climates. Earth's surface characteristics (e.g., ocean versus land, type of vegetative cover, semi-permanent snow and ice cover) influence the amount of incident solar radiation that is converted to heat and how that heat is used (e.g., raising air temperature, evaporating water).

Atmospheric circulation encompasses the combined influence of all weather systems operating at all spatial and temporal scales ranging from sea breezes to the prevailing winds that encircle the planet (Chapter 5). Although strongly influenced by the other climate controls, atmospheric circulation is considerably less regular and less predictable than the others. This variability is especially evident in weather systems such as thunderstorms and hurricanes that are smaller than the planetary scale. Planetary-scale circulation systems (e.g., the prevailing wind belts, subtropical anticyclones), exert a more systematic influence on climate, determining for example, where precipitation is seasonal and the location of the major subtropical deserts.

ROLE OF THE OCEAN

The ocean is a major player in Earth's climate system operating on temporal scales of days to millennia and spatial scales from local to global. The ocean influences radiational heating and cooling of the planet. Covering about 71% of Earth's surface, the ocean is a primary control of how much solar radiation is absorbed (converted to heat) at the Earth's surface. (Recall from Chapter 5 that the average albedo of the ocean surface is only 8%.) Also, the ocean is the main source of the most important *greenhouse gas* (water vapor) and is a major regulator of the concentration of atmospheric carbon dioxide (CO_2), another greenhouse gas.

On an annual average, the ocean absorbs about 92% of the solar radiation striking its surface; the balance is reflected to space. Most of this absorption takes place within about 100 to 200 m (330 to 650 ft) of the ocean surface with the depth of penetration of sunlight limited by the amount of suspended particles and discoloration caused by dissolved substances. On the other hand, at high latitudes highly reflective multi-year pack ice greatly reduces the amount of solar radiation absorbed by the ocean. The snow-covered surface of sea ice absorbs only about 15% of incident solar radiation and reflects away the rest. At present, multi-year pack ice covers about 7% of the ocean surface with greater coverage in the Arctic Ocean than the Southern Ocean (mostly in Antarctica's Weddell Sea). (The Arctic is an ocean surrounded by continents whereas the Antarctic is a continent roughly centered on the pole and surrounded by ocean. Without an Antarctic continent, much more of the Southern Ocean would be covered by sea ice.)

The atmosphere is relatively transparent to incoming solar radiation but much less transparent to outgoing infrared (heat) radiation. This is the basis of the *greenhouse effect*. Most water vapor, the principal greenhouse gas, enters the atmosphere via evaporation of seawater (Figure 11.1). Carbon dioxide, a lesser greenhouse gas, cycles into and out of the ocean depending on factors including sea-surface temperature, circulation patterns, and biological activity in surface waters (Chapter 9). CO_2 is more soluble in cold water than warm water so that carbon dioxide is absorbed from the atmosphere where surface waters are chilled (at high latitudes) and released to the atmosphere where upwelling brings cool water to the surface and is heated (at low latitudes). Photosynthetic organisms take up carbon dioxide and all organisms release carbon dioxide via cellular respiration. Through the physical and biological pumps, carbon is sequestered in the ocean for varying periods (Chapter 9).

The ocean influences the planetary energy budget not only by affecting the radiational heating and radiational cooling of the entire planet, but also by contributing to the non-radiative latent heat and sensible heat fluxes at the air-sea interface. Recall from Chapter 5 that heat is transferred from Earth's surface to the atmosphere via latent heating (vaporization of water at the surface followed by cloud formation in the atmosphere) and sensible heating (conduction plus convection). On a global average annual basis, about ten times more heat is transferred from the ocean surface to the atmosphere via latent heating than sensible heating.

The ocean and atmosphere are closely coupled. This coupling is most apparent in the ocean's surface waters where temperatures and wind-driven currents respond to variations in atmospheric conditions within hours to days. On the other hand, the deeper basin-scale thermohaline circulation responds more sluggishly to changes in atmospheric conditions, taking decades to centuries or longer to fully adjust. In turn, ocean currents strongly influence climate. Cold surface currents, such as the California Current, are heat sinks; they chill and stabilize the overlying air, thereby increasing the frequency of sea fogs and reducing the likelihood of thunderstorms. Relatively warm surface currents, such as the Gulf Stream, are heat sources; they supply heat and moisture to the overlying air, destabilizing the air, thereby

FIGURE 11.1
The ocean interacts with the atmosphere exchanging heat energy, water, and gases. The ocean is the primary source of atmospheric water vapor, the principal greenhouse gas, through evaporation. The ocean also exchanges carbon dioxide with the atmosphere. [National Park Service]

energizing storm systems. As noted in Chapter 5, ocean surface currents and the thermohaline circulation transport heat from the tropics to higher latitudes.

Although the importance of the ocean in Earth's climate system is apparent, we need to remember that the ocean and atmosphere work together in governing climate. Research findings underscore the ocean-atmosphere climate connection with somewhat surprising observations. At the same latitude, winters are significantly milder in Western Europe than in Eastern North America. Since at least the mid-1800s, scientists have attributed this climate contrast primarily to the moderating influence of the warm Gulf Stream and North Atlantic Current on Western Europe. However, recent research results question this assumption. Some scientists argue that while the Gulf Stream continually replenishes the relatively warm waters of the North Atlantic, it may not be as important as atmospheric circulation in explaining Western Europe's relatively mild winters.

In 2002, Richard Seager of Columbia University's Lamont-Doherty Earth Observatory and David Battisti of the University of Washington in Seattle, WA, reported that the key to Western Europe's mild winters is the pattern exhibited by the prevailing westerlies. Recall from Chapter 5 that aloft the westerlies blow from west to east in a wave-like pattern of ridges and troughs (Figure 5.20). In winter, a cold pool of air (a *trough*) is anchored over eastern North America resulting in cold northwesterly winds on the western flank of the trough, but an intrusion of warm air over the North Atlantic (a *ridge*) is associated with milder southwest winds over Western Europe. That is, in winter, the westerlies tend to blow from the colder northwest over Eastern North America but from the milder southwest over Western Europe. In addition, the relatively great thermal inertia of ocean water means that summer heat persists in the North Atlantic surface waters long after the North American continent has cooled in fall. Hence, southwest winds blowing toward Western Europe cross relatively warm waters and are heated from below. The direction of the prevailing winds over the North Atlantic and Western Europe is primarily responsible for delivering relatively mild air masses in winter. This finding also implies that shifts in the direction of the prevailing winds could alter the winter climate of Western Europe.

Broad scale patterns of sea-surface temperature (SST) strongly influence the location of major features of the atmosphere's planetary scale circulation. When SST patterns change so too do the locations of planetary-scale circulation features. For example, changes in the location of the highest sea-surface temperatures in the tropical Atlantic affect the north-south shifts of the **intertropical convergence zone (ITCZ)**. The ITCZ is a discontinuous belt of showers and thunderstorms paralleling the equator that marks the convergence of the trade winds of the Northern and Southern Hemispheres (Figure 6.7). The ITCZ encircles the globe and shifts north and south with the seasonal excursions of the Sun—more so over land and less so over the ocean—reaching its most northerly location in July and its most southerly location in January. Displacement of the ITCZ over the tropical Atlantic Ocean affects the timing and amount of rainfall along the east coast of South America from Brazil northward into the Caribbean Sea from March to May and in the western part of sub-Saharan Africa in August and September. With north-south shifts in the location of the highest SST, the latitude of the ITCZ changes, and regional rainfall patterns also vary.

A particularly dramatic example of how air-sea interaction governs climate is evident in the Atacama Desert of northern Chile, reputed to be the driest desert on Earth although bordering the Pacific Ocean. For details, refer to this chapter's first Essay.

Tropical Pacific Ocean/Atmosphere

Short-term variations in climate can originate in one region of the world but create weather extremes across the globe. The most apparent of these inter-annual variations in the Earth-atmosphere system are El Niño and La Niña in the tropical Pacific where conditions at the ocean/atmosphere interface vary on a quasi-periodic basis.

HISTORICAL PERSPECTIVE

Initially, El Niño was named by fishermen for the unusually warm south flowing ocean current, and accompanying poor fishing conditions, off the coast of Peru and Ecuador. El Niño (the boy) arrives around Christmas and refers to the Christ child. Typically brief, these warm water episodes lasted a month or two before sea-surface temperatures and the fisheries returned to normal. However, every three to seven years, El Niño would persist for a year or perhaps longer than 18 months. Over vast stretches of the tropical Pacific, there were significant changes in sea-surface temperatures creating major shifts in planetary-scale oceanic and atmospheric circulations, and the collapse of important South American fisheries. Today, the term **El Niño** is reserved for the long lasting ocean/atmosphere anomalies.

The first steps to understanding El Niño came not from South America or the Pacific, but the Indian monsoon failure in 1899-1900, when more than a million lives were lost. In response, Britain appointed Englishman Sir Gilbert Walker (1868-1958) to director general of observatories in India and charged him with developing a method to predict the Indian monsoon. For twenty years, he extensively studied the relationship between monsoon rains and weather conditions around the world and in 1924, discovered the **Southern Oscillation**, a seesaw variation in air pressure across the tropical Indian and Pacific Oceans. The Southern Oscillation is the pattern whereby when air pressure is high in the eastern tropical Pacific, it is low west of the International Dateline over the western tropical Pacific and the Indian Ocean, and monsoon rains are plentiful over India. With the opposite pressure pattern (high pressure west of the International Dateline and low pressure east of the dateline), Indian monsoon rains are lighter than usual.

Today, the *Southern Oscillation Index (SOI)* is based on the difference in air pressure between Darwin, on the north coast of Australia at 12 degrees S, 130 degrees E, and Tahiti, an island in the central south Pacific at 18 degrees S, 149 degrees W. When air pressure is anomalously low at Darwin, it is anomalously high at Tahiti, and when high at Darwin, it is low at Tahiti (Figure 11.2). (There is some question as to the quality of air pressure readings at Tahiti prior to 1935.)

Another four decades passed before the Southern Oscillation was linked to El Niño. In 1966, the Norwegian-American meteorologist Jacob Bjerknes (1897-1975) demonstrated how El Niño and the Southern Oscillation interact by analyzing oceanic/atmospheric observations gathered from the tropical Pacific during the *International Geophysical Year*, 1957-58, which fortuitously coincided with a strong El Niño. Across the tropical Pacific, as the air pressure gradient shifts, the air pressure to the west rising and air pressure to the east falling, the SST also shifts. At the same time, the weakening pressure gradient across the tropical Pacific heralds the slackening of the trade winds, and an El Niño episode begins. This relationship between El Niño and the Southern Oscillation is known by the acronym **ENSO**.

ENSO is a *coupled* phenomenon in that its variability in Earth's climate system cannot be explained as exclusively an oceanic or atmospheric event. *Coupled* refers to more than merely something that occurs in both the ocean and atmosphere; in this case, the phenomenon depends on feedbacks between the ocean and atmosphere. Changes in ocean conditions (primarily SST) can and do drive changes in atmospheric circulation and precipitation patterns. What is unique about ENSO is the strong coupling: changes in the ocean drive changes in the atmosphere which then feedback and further alter the ocean.

Not until the El Niño of 1982-83, one of the two most intense of the 20th century, was the potential

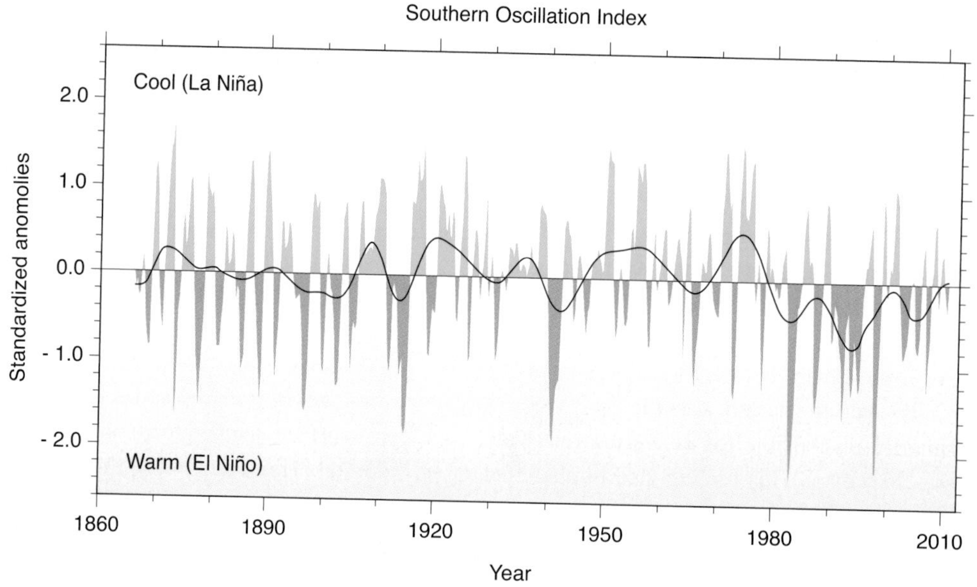

FIGURE 11.2
Variation in the Southern Oscillation Index based on monthly mean sea level pressure anomalies at Darwin, Australia, and Tahiti. Strongly positive values of the index indicate La Niña conditions and strongly negative values of the index indicate El Niño conditions. The thick black line is the 10-year running mean (trend line). [Courtesy of Climate Analysis Section, Climate and Global Dynamics Division, National Center for Atmospheric Research, Boulder, CO]

worldwide impact of ENSO realized, as described in this chapter's Case-in-Point. That event spurred development of numerical models to simulate ENSO as well as the deployment of a network of instrumented buoys and satellites to provide advance warning of a developing El Niño. Also, the last three decades have seen increasing interest in **La Niña** (*the girl*), the name coined in the mid 1980s for an ocean/atmosphere interaction that is essentially opposite El Niño, although typically less intense. The cold La Niña can be thought of as the opposite extreme of the *ENSO cycle*.

El Niño or La Niña are now routinely incorporated into long-range seasonal weather outlooks worldwide because of their importance in year-to-year climate variability. Such outlooks identify areas of expected anomalies in temperature and precipitation, and guide development of regional agricultural and water management strategies. Adoption of these strategies helps lessen the impact of weather extremes on water supply and food production.

NEUTRAL CONDITIONS IN THE TROPICAL PACIFIC

To understand El Niño and La Niña, the long-term average or *neutral* ocean/atmosphere conditions in the tropical Pacific should be examined first. The trade winds impact ocean currents, sea-surface temperatures, and rainfall across the tropical Pacific.

The air pressure gradient directed from high air pressure over the central and eastern tropical Pacific and low air pressure over the western tropical Pacific drives the trade winds. The greater the air pressure gradient from east to west, the stronger the winds blow from east to west. Once in motion, the Coriolis Effect deflects winds to the right in the Northern Hemisphere, creating trade winds from the northeast, and to the left in the Southern Hemisphere, creating trade winds from the southeast.

Prevailing winds blow from the south and southwest along the west coast of South America and Ekman transport drives warm surface waters to the left (westward), away from the coast (Chapter 6). As the warm, nutrient-poor surface waters are pushed offshore, the cold, nutrient-rich waters well up from below the thermocline, which is only 50 to 100 m (165 to 325 ft) deep along the coast. Although this zone of *coastal upwelling* is narrow, typically less than 15 km or 10 mi wide, the abundance of nutrients carried into the photic zone spurs an explosive growth of phytoplankton and supports a diverse ecosystem and highly productive fishery.

Meanwhile, relatively warm surface waters are driven by these same southwest trade winds westward toward Indonesia and northern Australia along the north and south equatorial currents. This wedge of warm water increases the depth of the thermocline to 150 m (490 ft) and raises sea level in the western tropical Pacific. That warm water, piled higher by trans-Pacific trade winds, also expands when heated so that sea level is 60 cm (2 ft) higher in the west than in the east with an 8 Celsius degrees (14.4 Fahrenheit degrees) contrast in SST.

Waters with high sea-surface temperatures heat the overlying air in the western tropical Pacific, further lowering surface air pressure and strengthening convection. Water vapor condenses into towering cumulonimbus (thunderstorm) clouds that produce heavy rainfall (Figure 11.3). Aloft this air flows back eastward

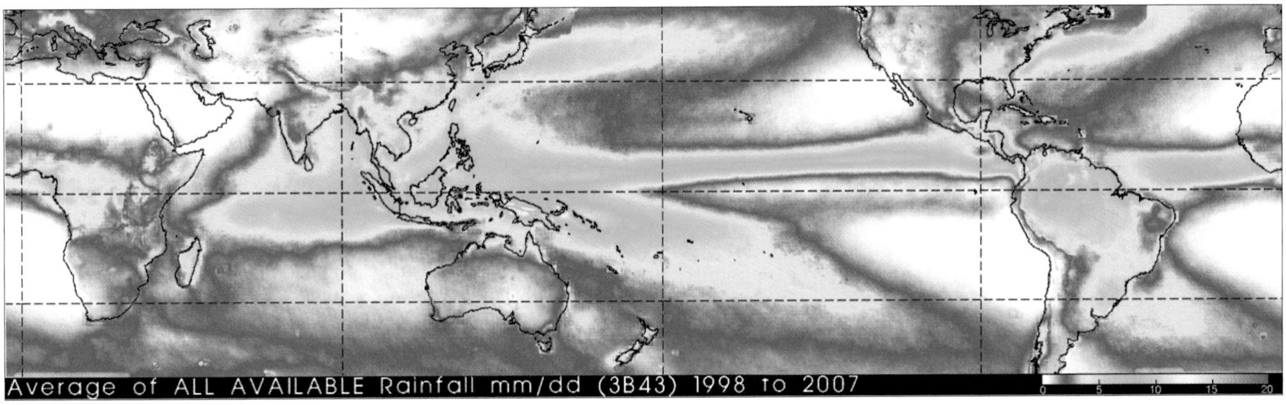

FIGURE 11.3
Benchmark average rainfall in millimeters per day (mm/d) across the tropical Pacific Ocean for the 10-year period 1998 through 2007. The heaviest rainfall is in the western tropical Pacific where sea-surface temperatures and sea levels are highest. These data were obtained from the TRMM Microwave Imager and IR sensors onboard geosynchronous satellites supplemented by conventional rain gauge measurements. [Courtesy of NASA, Tropical Rainfall Measuring Mission (TRMM)]

and sinks over the cooler waters of the eastern tropical Pacific. There, low SST in the central and eastern tropical Pacific chill the overlying air, further raising surface air pressure and suppressing convection. Sinking air is compressed and warmed so that clouds vaporize or fail to develop.

High SST in the western tropical Pacific lower the surface air pressure, whereas low SST in the eastern tropical Pacific raise the surface air pressure. Hence, during neutral conditions the east-west SST gradient reinforces the trade winds by strengthening the east-west air pressure gradient. In flowing over the ocean surface, the trade winds become warmer and more humid. In the western tropical Pacific warm humid air rises, expands, and cools (Figure 11.4). This completes the large convective-type circulation known as the *Walker Circulation*, named for Sir Gilbert Walker.

EL NIÑO, THE WARM PHASE

As part of the Southern Oscillation, air pressure falls over the eastern tropical Pacific and rises over the western tropical Pacific, which can trigger an El Niño episode. As the air pressure gradient weakens across the tropical Pacific, trade winds slacken and, during an intense El Niño, western trade winds may even reverse direction and blow toward the east.

In response to these shifts in atmospheric circulation, surface ocean currents, SST, sea level, thermocline depth, and upwelling all shift (Figure 11.5). The relaxed trade winds cause the ocean surface currents to weaken and, occasionally, reverse direction as well, flowing eastward. Hence, the thick, warm surface water normally in the west drifts slowly eastward, taking several months to reach the Americas where it is deflected by the continents. In the past, it has reached as far north as Canada and as far south as central Chile. In the western tropical Pacific, higher air pressure causes SST to drop, which along with the lack of trade winds, causes sea level to fall and the thermocline to rise. The lower air pressure in the eastern tropical Pacific causes SST to rise, which increases sea level and, without the trade winds to produce upwelling, the thermocline deepens (Figure 11.6).

The impacts of these environmental changes can be severe on marine ecosystems. The reduced upwelling means the cold, nutrient-rich waters along the coast of Ecuador and Peru remain below, causing phytoplankton to decline and fish populations to plummet. Peru's fishing industry was booming in the 1950s and 1960s, and by 1970 was one of the largest in the world, accounting for 20% of the total global catch of anchovies and a third of the nation's foreign income. However, over-fishing and the 1972-73 El Niño decimated the Peruvian fishery from which it has yet to recover.

Warmer surface waters can also stress coral reefs living in shallow tropical waters. Responding to elevated sea-surface temperatures (SST), coral expels zooxanthallae, the symbiotic microscopic algae that supply coral with oxygen and some organic compounds

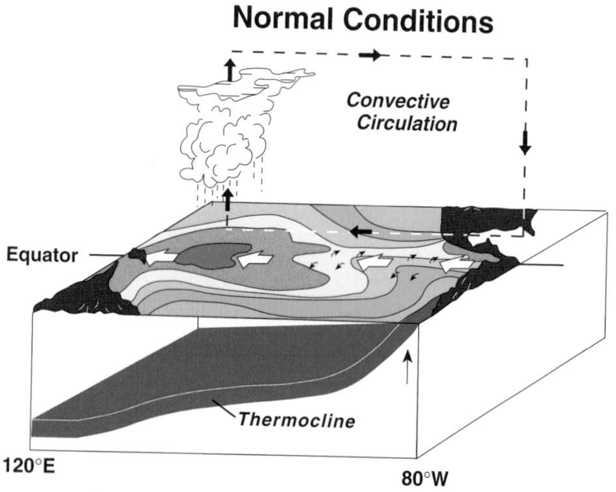

FIGURE 11.4
Schematic block diagram showing ocean/atmosphere conditions in the tropical Pacific during normal or neutral episodes. Red indicates areas of highest sea-surface temperatures. [Courtesy of NOAA, Pacific Marine Environmental Laboratory (PMEL), Tropical Atmosphere Ocean Project]

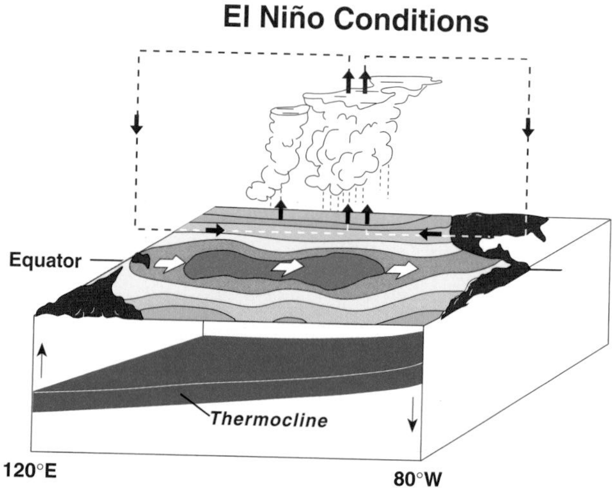

FIGURE 11.5
Schematic block diagram showing ocean/atmosphere conditions in the tropical Pacific during El Niño conditions. Red indicates areas of highest sea-surface temperatures. [Courtesy of NOAA, Pacific Marine Environmental Laboratory (PMEL), Tropical Atmosphere Ocean Project]

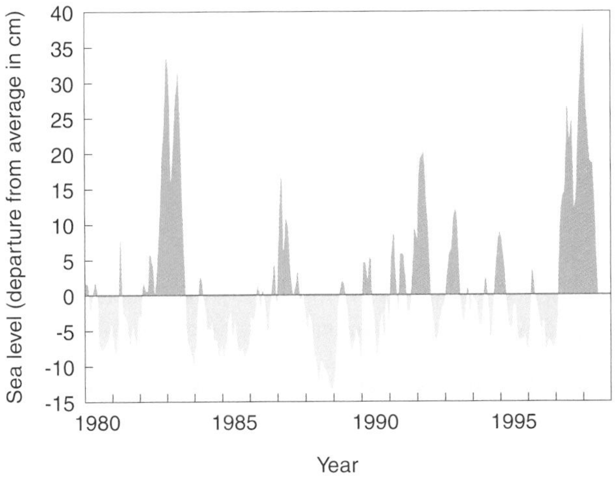

FIGURE 11.6
Sea level record at Galápagos in the eastern tropical Pacific based on tide gauge records and expressed in cm as departure from the long-term average. Relatively high sea levels correspond to El Niño episodes. [Courtesy of NOAA PMEL]

produced through photosynthesis. Recall from Chapter 10 that without zooxanthallae, coral polyps have little pigmentation and appear nearly transparent on the coral's white skeleton, a condition known as **coral bleaching**. Excessive bleaching can kill coral polyps, destroying habitats for a great variety of marine life. In addition, bleaching makes coral more vulnerable to infectious diseases. Coral bleaching was widespread and severe during the 1997-98 El Niño/La Niña episodes, and in 2005 and 2010. Global warming is likely to bring more frequent and severe spikes in SST and more extensive coral bleaching.

In a study of Caribbean coral reefs during the summers of 1983 to 2000, Jennifer Gill and colleagues at the University of East Anglia in the U.K., found that atmospheric aerosols may help prevent coral bleaching. Associated with El Niño is a rise in Caribbean SST that favors coral bleaching. But at least partially offsetting warming is the presence of tiny particles suspended in the atmosphere that scatter and absorb solar radiation thereby lowering SST. Much of this aerosol load is transported by winds aloft blowing from the Sahara to the Caribbean, supplemented by particles from distant volcanic eruptions (Chapter 4). Although El Niño occurred in 1991, the cooling associated with the eruption of Mount Pinatubo in the Philippines that same year appears to have prevented widespread coral bleaching in the Caribbean (Chapter 12). According to Gill and her colleagues, El Niño and atmospheric aerosols account for about 70% of the year-to-year variability in Caribbean coral bleaching.

The decline and loss of coral reef ecosystems due to coral bleaching, ocean acidification (Chapter 3), disease, and human activities (overfishing, coastal development) have serious economic, social, cultural and ecological consequences for people and communities throughout the world. The livelihoods and welfare of an estimated 100 million people living on the coasts of tropical developing nations depend on coral reefs. Coral reefs are valued for subsistence fishing, tourism, and protection of shorelines from storm surges generated by tropical cyclones (Chapter 8). According to Camilo Mora, a coral specialist at Dalhousie University, Halifax, Nova Scotia, coral reefs generate revenue estimated worldwide at $30 billion annually (mostly from fisheries and tourism).

Anomalous weather patterns in the tropics and subtropics develop from the low SST in the western tropical Pacific, high SST in the central and eastern tropical Pacific, and change in the trade winds of El Niño. Normally, abundant rainfall comes from the winds blowing onshore over Indonesia, but during El Niño, prevailing winds are directed offshore and Indonesia is dry. El Niño has also brought droughts to India, eastern Australia, northeastern Brazil, and southern Africa. Meanwhile, warmer surface waters off the west coast of South America spur convection and heavy rainfall along the normally arid coastal plain, causing flash flooding. Wetter conditions tend to occur in southern Brazil, Uruguay, and equatorial East Africa.

Martin Hoerling, a NOAA meteorologist, and colleagues found that central India endured 10 severe droughts during the monsoon rainy season (June, July, and August) in El Niño years for the period 1871 to 2002. When the greatest SST anomalies during El Niño occurred in either central or eastern equatorial Pacific, drought would dominate central India. All 10 droughts occurred during El Niño episodes. However, the converse did not hold; that is, not all El Niño events during the 132-year period were accompanied by drought in central India. In fact, in 13 cases rainy season precipitation was at or slightly greater than the long-term average. In examining the various cases of drought, Hoerling and colleagues found that the location of the greatest SST anomalies during El Niño was either in the central equatorial Pacific or in the eastern equatorial Pacific. Drought occurred in central India when the highest SSTs were in the central equatorial Pacific. Numerical models predicted that these exceptionally warm waters would produce warm, humid air that would rise high in the tropical atmosphere while concurrently losing much of its moisure, move over central India, and subside. The subsiding air would inhibit development of clouds and precipitation and soon a drought would be underway. This is one of many examples that demonstrate the fact that no two El Niño events are the same.

Typically, El Niño also brings dry weather to the Hawaiian Islands. The North Pacific subtropical anticyclone shifts so that the descending air is closer to the Islands, creating a persistent dry weather pattern. Almost all of Hawaii's major droughts during the 20th century coincided with an El Niño event.

The intensity, frequency, and spatial distribution of tropical cyclones (e.g., tropical storms, hurricanes) are also influenced by El Niño. The extensive area of warmer water over the eastern tropical Pacific allows hurricanes to travel farther north and west than usual, altering their intensity and location in the Pacific and Indian Oceans. In the Atlantic Ocean, stronger than usual winds aloft inhibit the development of tropical cyclones, and those few that do develop usually are weaker and short lived (Chapter 8).

El Niño has a ripple effect on the weather of middle latitudes, especially in winter. This link in atmospheric circulation changes in different regions of the globe, often over distances of thousands of kilometers, is known as a **teleconnection**. Caused by latent heat released into the atmosphere during deep convection and the buildup of thunderstorms in the tropical troposphere, teleconnections help control the planetary-scale circulation. During El Niño when high SST in central and eastern tropical Pacific heats and destabilizes the troposphere, wind and weather patterns worldwide shift. A teleconnection is associated with the deep convection that generates towering thunderstorms that help drive atmospheric circulation, governing the course of jet streams, storm tracks, and moisture transport by winds at higher latitudes.

Like large boulders producing eddies in a swiftly flowing stream, these thunderstorm clouds build high into the tropical troposphere and deflect the upper air winds. Moving the boulders will displace the stream of eddies. In the same way, a shift in location of the principal area of convection eastward over the tropical Pacific redirects the atmospheric circulation.

During typical El Niño winters, prevailing storm tracks bring abundant rainfall and cooler than usual conditions to the Gulf Coast states, from Texas to Florida. Over the northern U.S. and Canada, prevailing winds blow from the west, moving cold air masses eastward across the Arctic and northern Canada. Persistence of this circulation pattern prevents cold air masses from invading south, so that mild weather prevails over much of Canada, Alaska, and parts of the northern U.S. West-to-east flow in the westerlies also diminishes the usual spring contrast between warm, humid air masses moving northeastward from the Gulf of Mexico and cold, dry air masses sweeping southeastward from Canada. Consequently, severe thunderstorms and tornadoes may be less frequent than usual in the Ohio and Tennessee River Valleys.

Although some weather extremes almost always accompany El Niño, no two events are the same because El Niño is only one of many factors that influence inter-annual variations in climate. In southern California, for example, heavy winter rains (snows at higher elevations) have occurred during some but not all El Niño events. Record heavy rainfall in Southern California during January 1995 was linked to a southerly shift of the jet stream and storm track over the eastern Pacific. In that case, a change in atmospheric circulation associated with El Niño was the culprit. Whereas the 1982-83 El Niño brought severe drought to eastern Australia, dry conditions in Australia during the 1997-98 El Niño were far less severe.

THE 1997-98 EL NIÑO

With weather extremes and outbreaks of disease from standing water, such as malaria and cholera, claiming an estimated 22,000 lives worldwide and causing $36 billion in economic losses, the 1997-98 El Niño riveled the 1982-83 as the most intense of the 20th century or since (Figure 11.7). This El Niño developed rapidly in early 1997, with trade winds weakening until they reversed direction in the western tropical Pacific. A pool of exceptionally warm surface waters (SST greater than 29 °C or 84 °F) migrated eastward from the western tropical Pacific, and equatorial upwelling ceased that summer in the Northern Hemisphere. By that fall in the eastern tropical Pacific, the SST was at least 5 Celsius degrees (9 Fahrenheit degrees) higher than the long-term average, setting record highs each month from June through September. As 1997 drew to a close, the thermocline flattened within the tropical Pacific, rising 20 to 40 m (65 to 130 ft) in the west and falling more than 90 m (295 ft) in the east. Somewhat lower SST in the west, and much higher than usual SST in the east, weakened the east-west SST gradient and the already weakened trade winds.

The 1997-98 El Niño came to an abrupt end in mid-May 1998. Trade winds strengthened rapidly, resuming upwelling along the equator and off the northwest coast of South America. The SST over the eastern tropical Pacific plummeted in response to upwelling of very cold water, even dropping 8 Celsius degrees (14 Fahrenheit degrees) in only four weeks at one location near 125 degrees W.

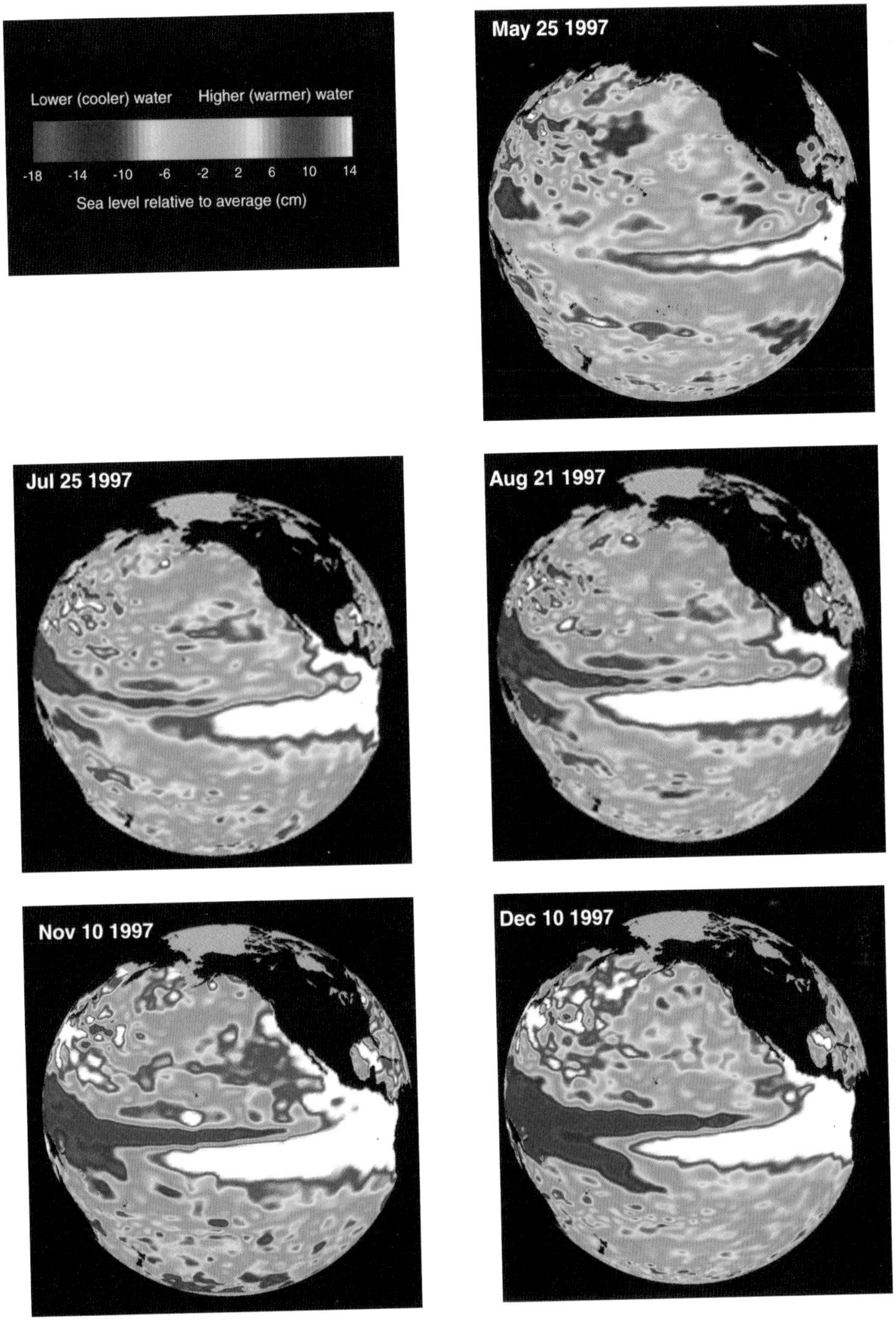

FIGURE 11.7
Evolution of the 1997-98 El Niño as derived from changes in ocean surface height (compared to the long-term average) as measured by altimeter sensors onboard the TOPEX/Poseidon satellite. On the color scale, whites and reds indicate elevated areas (warmer than normal water). In the white areas, the sea surface is 14 to 32 cm (6 to 13 in.) above normal; in the red areas it is about 10 cm (4 in.) above normal. Green indicates normal sea level whereas purple corresponds to areas that are at least 18 cm (7 in.) below normal sea level (colder than normal water). [Courtesy of NASA Goddard Space Flight Center]

LA NIÑA, THE COLD PHASE

A period of unusually strong trade winds and exceptionally vigorous upwelling in the eastern tropical Pacific, **La Niña** is an exaggeration of neutral conditions and the opposite of El Niño conditions, which it can, but does not always, follow (Figure 11.8). Like its warm counterpart, La Niña tends to persist for 12 to 18 months. During La Niña, the air pressure is greater over colder surface waters of the central and eastern tropical Pacific and lower over warmer surface waters of the western tropical Pacific. However, SST anomalies are typically greater during El Niño. SST usually rise 5 to 6 Celsius degrees (9 to 11 Fahrenheit degrees) above average during an intense El Niño but drop only 2 to 3 Celsius degrees (3.6 to 5.4 Fahrenheit degrees) below average during a strong La Niña.

Accompanying La Niña are worldwide weather extremes that are often opposite those observed during El Niño. As with El Niño, the most consistent middle latitude teleconnections appear in winter. In the tropical Pacific, lower than usual SST in the east inhibit rainfall and higher than usual SST in the west enhance rainfall in Indonesia, Malaysia, and northern Australia during the Northern Hemisphere winter and the Philippines during the Northern Hemisphere summer. Elsewhere around the globe, the Indian monsoon rainfall (in summer) tends to be heavier than average (especially in northwest India) and wet conditions prevail over southeastern Africa and northern Brazil (during the Northern Hemisphere winter). Southern Brazil to central Argentina experiences a dry winter. In addition, weak winds aloft during La Niña favor tropical cyclone formation in the Atlantic Basin.

Across middle latitudes of the Northern Hemisphere, westerlies tend to be *meridional*, encircling the globe in great north-south loops during La Niña. They steer cold air masses toward the southeast and warm air masses toward the northeast. Occasionally, a meridional flow pattern becomes so extreme that a broad pool of rotating air separates from the main current and whirls over the same area for weeks or months, like a whirlpool in a swiftly flowing stream. If circling in a clockwise direction when viewed from above, the rotating air is known as a *cutoff anticyclone* (or *cutoff high*). For as long as a cutoff high persists over the same area, the weather remains dry, with an increasing probability of drought. This type of atmospheric circulation pattern was responsible for the severe summer drought that afflicted the central U.S. during the La Niña year of 1988.

In the spring, the likelihood of severe thunderstorms and tornadoes across the central U.S. increases with La Niña's meridional flow pattern, bringing a key ingredient for severe weather development: air masses with contrasting temperature and humidity. In the winter, below average precipitation and mild temperatures form a band from the Southwest, through the central and southern Rockies, and eastward to the Gulf Coast. In the Pacific Northwest, the winter tends to be cool and wet. Lower than usual winter temperatures also occur east of the northern Rockies and north-central states.

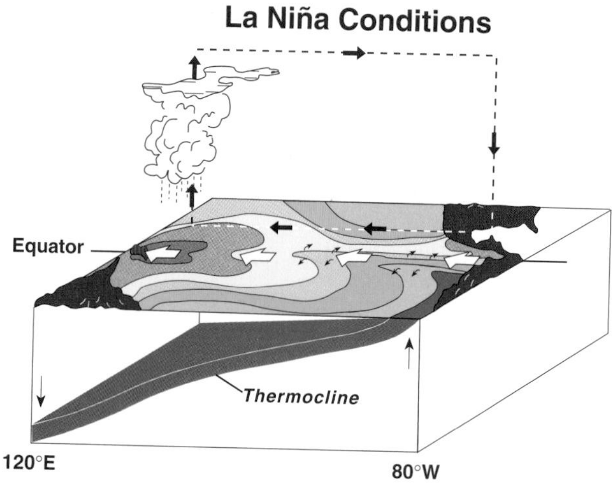

FIGURE 11.8
Schematic block diagram showing ocean/atmosphere conditions in the tropical Pacific during La Niña conditions. Red indicates areas of highest sea-surface temperatures. [Courtesy of NOAA, Pacific Marine Environmental Laboratory (PMEL), Tropical Atmosphere Ocean Project]

PREDICTING AND MONITORING EL NIÑO AND LA NIÑA

Not until after the eastern tropical Pacific had already begun warming in 1997, did official forecasts warn of an impending El Niño. In November, the Climate Prediction Center's winter outlook for December 1997 through February 1998 correctly predicted heavier than usual precipitation across the southern U.S. and anomalous warmth over the northern third of the nation, as well as above average rainfall in Peru, east Africa, and northern Argentina. Unusually dry conditions for Indonesia, northern South America, and southern Africa were also verified. However, the extreme drought expected for northeast Australia did not materialize.

Numerical models simulate the onset, evolution and decay of El Niño and La Niña by approximating oceanic processes that alter sea-surface temperatures, and the atmospheric response, including convection, clouds, and winds. Forecasters rely on two basic types of numerical models to predict El Niño or La Niña: empirical models and dynamical models. An *empirical model* compares current and evolving oceanic and atmospheric conditions with observational data from previous episodes over the prior 40 years. The similarity between past and present conditions is the basis for an empirical model prediction. A *dynamical model* uses mathematical equations to simulate interactions, or couplings in the atmosphere, ocean, and land. These *coupled models* are more sophisticated than empirical models.

Reliable observational data from the tropical Pacific Ocean and atmosphere are essential for detecting a developing El Niño or La Niña, and initializing numerical models. The observed data represent the initial conditions, used as a starting point for predicting future states of the ocean and atmosphere, and verifying the model predictions and results. This is especially important for dynamical models, which depend on reliable data to run their complex coupled equations. Then as El Niño or La Niña unfold, data are continuously assimilated into the models to correct or "nudge" the model.

To improve understanding, detection and prediction of ENSO related variability by producing the observational data needed for models, monitoring systems were deployed in the tropical Pacific as part of the 10-year (1985-1994) international *Tropical Ocean Global Atmosphere (TOGA)* study. By December 1994, the *ENSO Observing System,* was fully operational. The ENSO Observing System consists of an array of moored and drifting instrumented buoys, island and coastal tide gauges, ship-based measurements, and satellites (Figure 11.9).

TAO (Tropical Atmosphere/Ocean), one component of the ENSO Observing System, is an array of moored buoys, small instrumented platforms, in the tropical

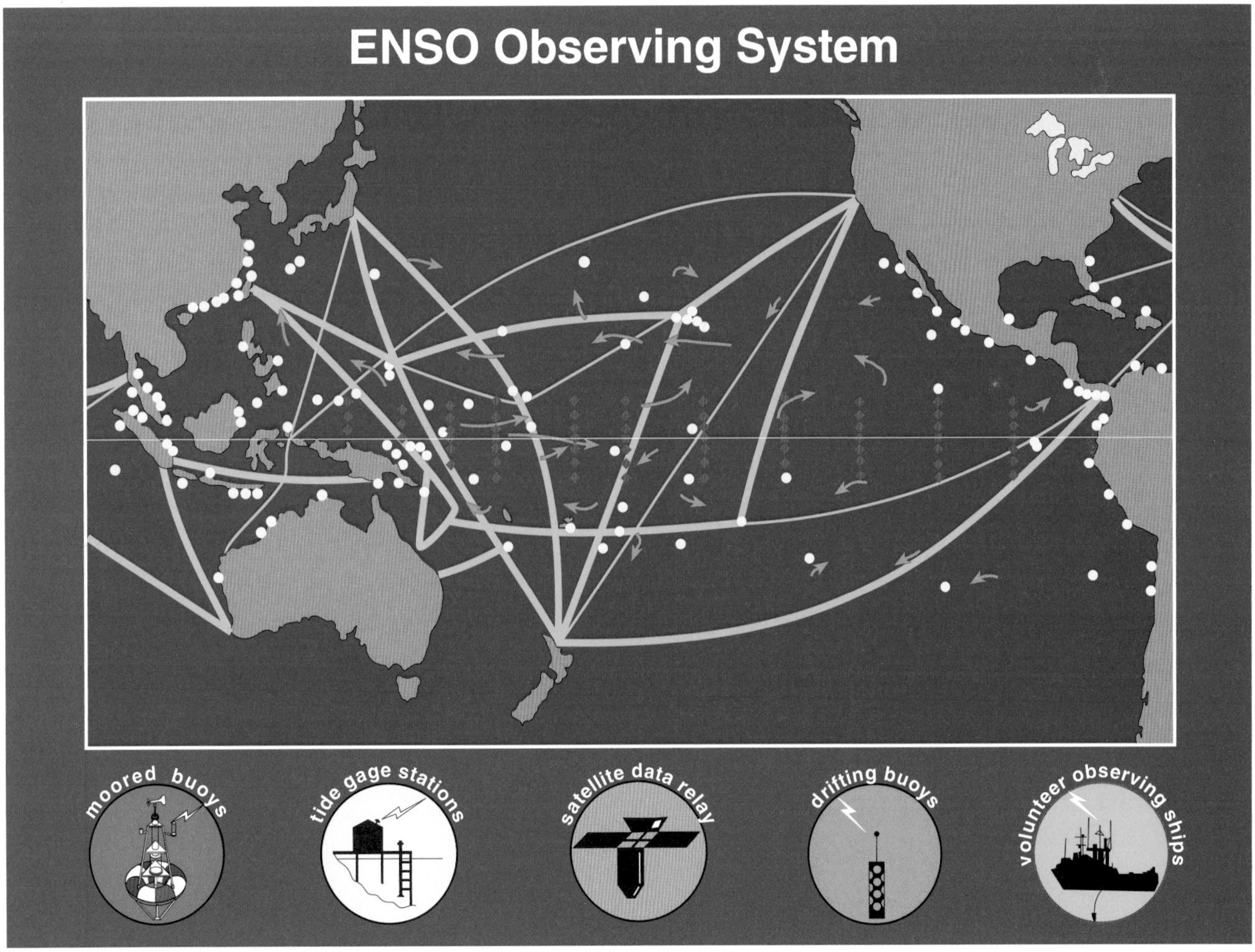

FIGURE 11.9
Components of the ENSO Observing System provide advance warning and monitor the development and decay of El Niño and La Niña events. [Courtesy of OAA, Pacific Marine Environmental Laboratory (PMEL), Seattle, WA]

FIGURE 11.10
An instrumented TAO moored buoy photographed at NOAA's Pacific Marine Environmental Laboratory in Seattle, WA. An array of similar moored buoys gathers oceanic and atmospheric data from the tropical Pacific as part of the ENSO Observing System. [Photo by J.M. Moran]

Pacific Ocean (Figure 11.10). Buoys are strategically placed within 8 degrees N, 8 degrees S, 95 degrees W, and 137 degrees E. As shown in Figure 11.11, approximately 70 deep-sea moorings, renamed TAO/TRITON in 2000, measure atmospheric and oceanic variables, including air temperature, wind, relative humidity, and sea-surface and subsurface temperatures at 10 depths in the upper 500 m (1640 ft). Several newer moorings also have salinity sensors, along with additional meteorological sensors. Five moorings along the equator use Subsurface Acoustic Doppler Current Profilers to measure ocean current velocity. Observational data are transmitted to NOAA's Pacific Marine Environmental Laboratory (PMEL) in Seattle, WA, via a NOAA polar-orbiting satellite, and are available in near real-time on the Internet.

Remote satellite sensing, such as NOAA and NASA satellites that monitor cloud cover and map SST, plays an important role in providing early warning of an evolving El Niño or La Niña. The TOPEX/Poseidon satellite, a joint mission between NASA and the Centre National d'Etudes Spatiales in France, provided images of ocean surface topography (sea level) at 10-day intervals from 1992 until January 2006 after completing 62,000 orbits. In December 2001, NASA and CNES launched *Jason 1*, successor to TOPEX/Poseidon, and in June 2008, *Jason 2* was launched into the same orbit. Radar altimeters onboard these satellites bounce microwaves off the ocean surface to obtain precise distance measurements between satellite and sea surface to produce images of sea-surface height. Elevated topography (hills) indicates warmer water whereas areas of low topography (valleys) indicate colder water (Figures 11.7 and 11.12). Surface ocean currents, based on the geostrophic assumption, can be calculated from these images (Chapter 6). For more on remote sensing of ocean surface elevation, refer to the second Essay in Chapter 7.

The joint U.S.-Japanese *Tropical Rainfall Measuring Mission (TRMM)* launched in November 1997 also detects the onset and evolution of El Niño and La Niña (Figure 11.3). The TRMM satellite uses active radar and passive microwave energy sensors to monitor clouds, precipitation, and radiation between 40 degrees N and 40 degrees S in the Pacific Ocean. Uniquely, the TRMM Microwave Imager (TMI) can "see through" clouds to

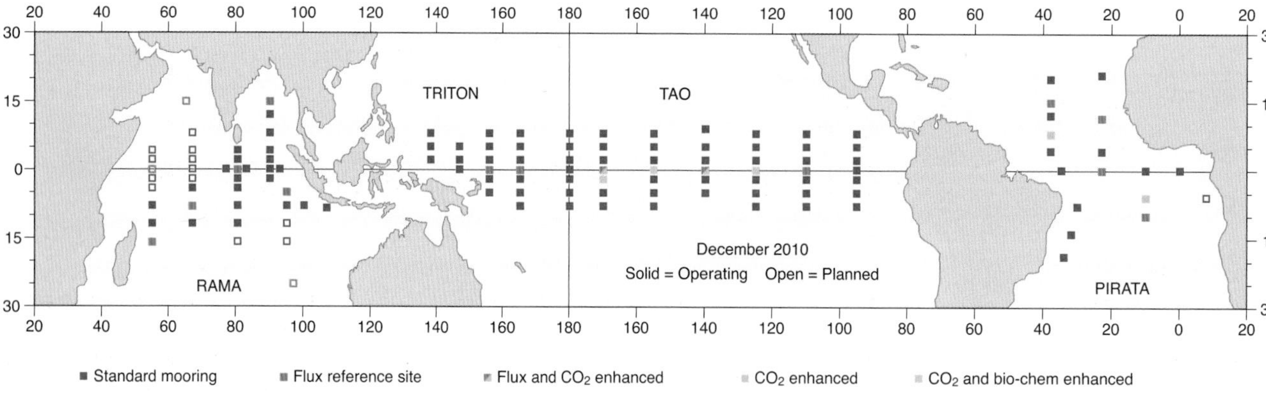

FIGURE 11.11
TAO/TRITON deep-sea mooring locations as of December 2010.

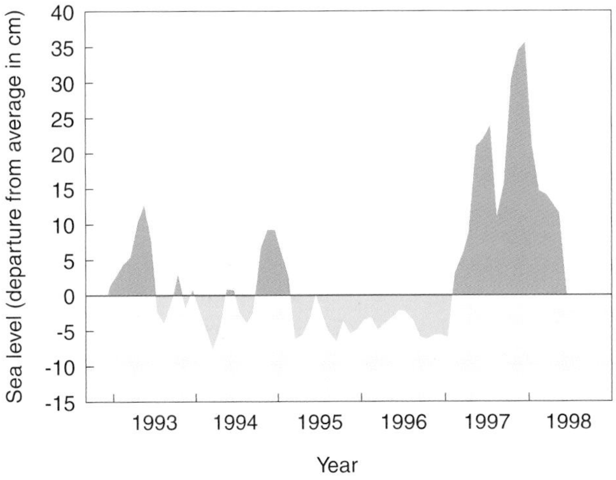

FIGURE 11.12
Sea level record at a location along the equator in the eastern tropical Pacific derived from measurements made by the TOPEX/Poseidon satellite. Sea level is expressed in cm as departure from the long-term average. Relatively high sea levels correspond to El Niño episodes. [Courtesy of NOAA PMEL]

measure SST. The TMI uses microwaves emitted by the sea surface to characterize the IR emission and determine the radiation temperature. However, while microwaves pass through clouds with little attenuation, they are strongly scattered and absorbed by rainfall so TMI can measure SST only during fair weather.

Increasingly accurate predictions of El Niño and La Niña are enabling better handling of the impacts of inter-annual climate variability. Forecasts allow informed strategic planning in agriculture, fisheries, and water resource management. Consider an example: Peru's economy, like that of most developing nations, is very sensitive to climate. As stated earlier, El Niño is bad for fishing in Peru and is often accompanied by destructive flooding. While La Niña benefits fishing, it may bring drought and crop failure. Warning of an impending El Niño or La Niña prior to the start of the growing season allows agricultural interests and government officials to consult on what crops to plant. If the forecast calls for El Niño, rice is favored over cotton because rice thrives during a wet growing season whereas cotton is more drought-tolerant and therefore is more suitable when La Niña is forecasted. Also, in anticipation of the heavy rainfall likely to accompany a full-blown El Niño, water resource managers can direct the gradual draw down of reservoirs to reduce flooding.

FREQUENCY OF EL NIÑO AND LA NIÑA

In September 2003, NOAA scientists provided an index for operational definitions of El Niño and La Niña, based on sea-level air pressure, zonal (east-west) and meridional (north-south) surface winds, surface air temperature, sky cloud cover, and sea-surface temperature measured in the tropical Pacific (Figure 11.13). Sea-surface temperatures are mapped in an area bound by 120 degrees W and 170 degrees W, and 5 degrees N and 5 degrees S, which includes the equatorial cold tongue. By this index, El Niño is characterized by a *positive* SST departure from normal, and La Niña by a *negative* SST departure, greater than or equal to 0.5 Celsius degrees

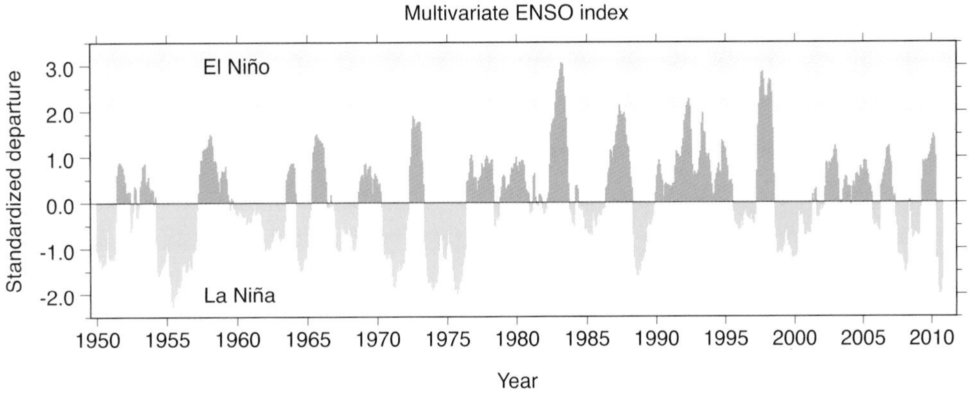

FIGURE 11.13
Variations in the Multivariate ENSO Index showing the sequence of El Niño and La Niña events since 1950. The Index is based on six variables measured in the tropical Pacific: sea-level air pressure, zonal (east-west) component of surface wind, meridional (north-south) component of surface wind, surface air temperature, sky cloud cover, and sea-surface temperature. Sea-surface temperature anomalies (departures from 1971-2000 averages) are measured for the area in the tropical Pacific Ocean between 5 degrees N and 5 degrees S latitude and from 120 degrees W to 170 degrees W longitude. Warm anomalies (greater than about 0.5 Celsius degree) generally indicate El Niño whereas cold anomalies (less than about –0.5 Celsius degree) generally indicate La Niña. [NOAA/ESRL/Physical Science Division-University of Colorado at Boulder/CIRES/CDC]

averaged over three consecutive months, based on the 1971-2000 average.

In February 2009, NOAA's Climate Prediction Center launched its **ENSO Alert System**. An El Niño or La Niña *watch* is issued when conditions in the equatorial Pacific are favorable for their development within three months and an El Niño or La Niña *advisory* when conditions have already developed and are expected to continue.

La Niña, which does not always follow an El Niño is most likely after an especially intense El Niño. An intense El Niño conveys great amounts of heat from the eastern tropical Pacific into higher latitudes, which sets the stage for a following La Niña. Additionally, the unusually cold water just below the warm surface water is poised to well up to the surface as soon as the trade winds strengthen. After the intense 1997-98 El Niño, there was a significant La Niña.

While there are more El Niño events, La Niña events tend to last longer, and the balance of the time neutral or near-neutral conditions prevail. From 1951 to 2000, El Niño conditions prevailed 24% of the time and La Niña occurred 28% of the time (Table 11.1). During the 1980s and 1990s, La Niña was less frequent than El Niño. As of this writing, the most recent El Niño was a major episode from 2009-10, followed by La Niña in July 2010 and was still present in April 2011. At the time of this writing, a transition to neutral conditions was expected by June 2011.

El Niño and La Niña are not just recent phenomena. In fact, documentary, archeological, and geological evidence indicates that ENSO has operated over tens of thousands of years. For more on this topic, refer to this chapter's second Essay.

TABLE 11.1
El Niño and La Niña Events since 1950

El Niño
2009-10, 2006-07, 2004-05, 2002-03, 1997-98, 1994-95, 1991-92, 1986-88, 1982-83, 1977-78, 1976-77, 1972-73, 1969-70, 1968-69, 1965-66, 1963-64, 1957-58, 1951
La Niña
2010-11, 2008, 2007-08, 2000-01, 1998-2000, 1995-96, 1988-89, 1984-85, 1973-76, 1970-72, 1964-65, 1954-57, 1949-51

North Atlantic Oscillation

Research on El Niño has spurred interest in other regular oscillations involving the interaction of the ocean and atmosphere that impact short-term climate variability. These include the North Atlantic Oscillation (NAO), the Arctic Oscillation (AO), and Pacific Decadal Oscillation (PDO). In general, NAO, AO, and PDO affect more restricted geographical areas and operate over longer time periods than either El Niño or La Niña.

Over the North Atlantic, the time-averaged planetary-scale atmospheric circulation features a subpolar low pressure system near Iceland (the *Icelandic low*) and a massive subtropical anticyclone centered near 30 degrees N that stretches from Bermuda to near the Azores (Chapter 5). The **North Atlantic Oscillation (NAO)** refers to a seesaw variation in air pressure between Iceland and the Azores. When air pressure is higher than the long-term average over the Azores, it is lower than the long-term average over Iceland and vice versa. The air pressure gradient between the Bermuda-Azores subtropical anticyclone and the Icelandic low drives winds that steer storms from west to east across the North Atlantic.

The North Atlantic Oscillation influences precipitation and temperatures primarily in winter (December to March) over eastern North America and much of Europe and North Africa. The **NAO Index** is directly proportional to the strength of the North Atlantic air pressure gradient, i.e., the difference in sea level air pressure between the Bermuda-Azores high and the Icelandic low. The NAO Index time series in Figure 11.14 is based on the difference in sea level pressure over Gibraltar and the sea level pressure over southwest Iceland. When the NAO Index is relatively high, stronger than usual winter winds blow across the North Atlantic moving cold air masses over eastern Canada and the U.S. so that winters tend to be colder than usual in that region. But cold air masses modify considerably as they move over the relatively mild ocean surface, warming and becoming more humid so that winters are milder and wetter than usual downstream over Europe. Meanwhile, dry winters prevail in the Mediterranean region. On the other hand, when the NAO Index is low, steering winds over the North Atlantic shift southward so that winters are colder than usual over northern Europe and wet and mild conditions prevail from the Mediterranean eastward into the Middle East. The eastern U.S. and Canada tend to experience relatively mild winters while winters in the southeast U.S. are colder than usual.

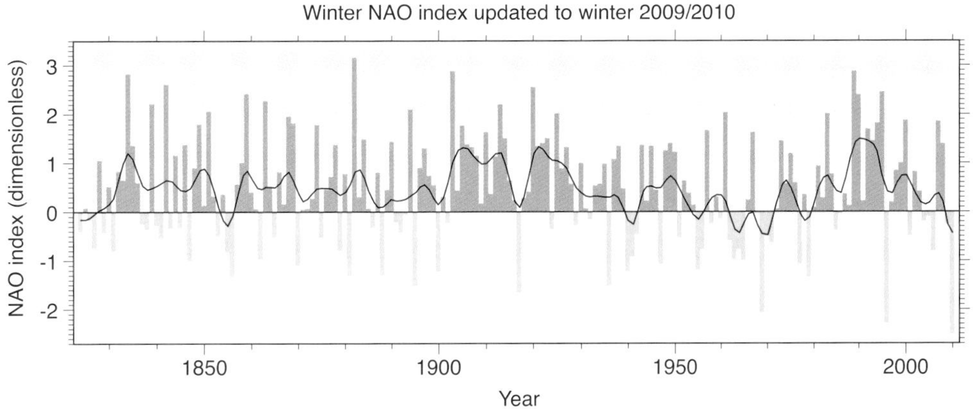

FIGURE 11.14
Record of the North Atlantic Oscillation (NAO) during winter (December to March) through 2009-2010, based on the difference between the normalized sea-level air pressure at Gibraltar and the normalized sea-level air pressure over southwest Iceland. Solid black line is a running mean. [From Tim Osborn, Climate Research Unit, University of East Anglia, Norwich, UK]

The NAO Index varies significantly from one year to the next and from decade to decade and is much less regular than the ENSO cycle. The NAO Index was generally low in the 1950s and trended upward from the 1960s through the early 1990s and then generally downward into the winter of 2010-11.

Changes in winter moisture supply associated with NAO have had varied impacts in Europe and North Africa. During recent decades of relatively high NAO-Index, wetter winters have increased hydroelectric power potential in the Scandinavian nations, lengthened the growing season over northern Eurasia, but also diminished the snow cover for winter recreation. Meanwhile, a moisture deficit has been the problem in the Iberian Peninsula, the watershed of the Tigris and Euphrates Rivers, and the Sahel of North Africa.

Arctic Oscillation

Related to the North Atlantic Oscillation, the **Arctic Oscillation (AO)** is a seesaw variation in air pressure between the North Pole and the margins of the polar region. Associated changes in the horizontal air pressure gradient alter the speed of the band of winds aloft (the *polar vortex*) that blow counterclockwise (viewed from above) around the Arctic. Alternate strengthening and weakening of these polar winds impact winter weather in middle latitudes and contribute to climate variability and changes in ocean circulation.

The Arctic Oscillation shifts between negative and positive phases. During its negative phase, the air pressure gradient is weaker and the polar vortex circulation is not as strong as usual. This allows bitterly cold Arctic air masses to more frequently move out of their source regions in the far north and plunge southeastward into middle latitudes. This brings colder than usual winter weather to most of the U.S., Northern Europe, Russia, China, and Japan. Heavy lake-effect snows around the Great Lakes are more frequent and nor'easters are more likely along the U.S. Eastern Seaboard (Chapter 8). But when the Arctic Oscillation is in its positive phase, the air pressure gradient is greater and winds encircling the Arctic are stronger. These stronger winds act as a dam to impede the southeastward flow of Arctic air. The middle latitude westerlies also strengthen and blow more directly from west to east, flooding much of the U.S. with relatively mild air from off the Pacific Ocean (instead of frigid Arctic air from Canada). Winter temperatures are milder than usual (especially east of the Rocky Mountains) and major snowstorms are less likely. Meanwhile, Alaska, Scotland, and Scandinavia are snowier than usual and California and Spain are drier.

Although during any winter the Arctic Oscillation can shift many times between its negative and positive phases, extended periods occur when either the negative or positive phase dominates the winter season. In the 1960s, the negative phase of the Arctic Oscillation dominated. Since then, the positive phase has been more frequent, a trend that is consistent with observed climate fluctuations in middle latitudes (e.g., less frequent episodes of extreme cold and major snowstorms). Furthermore, during this recent episode of dominantly positive Arctic Oscillation phase, winds have been delivering warmer than usual air and ocean water into the Arctic. As we will see in Chapter 12, this may at least help explain the recent shrinkage of Arctic ice cover.

Pacific Decadal Oscillation

The **Pacific Decadal Oscillation (PDO)** is a long-lived variation in climate over the North Pacific and North America. Sea-surface temperatures fluctuate between the north central Pacific and the west coast of North America. During a PDO *warm phase*, SST are lower than usual over the broad central interior of the North Pacific and above average in a narrow strip along the coasts of Alaska, western Canada, and the Pacific Northwest. In an interesting parallel to what happens off the coast of Ecuador and Chile during El Niño, the layer of relatively warm surface waters off the Pacific Northwest Coast significantly reduces upwelling of nutrient-rich bottom water. Populations of phytoplankton and zooplankton plummet and juvenile salmon migrating to coastal waters from streams and rivers starve (Chapter 10). On the other hand, during a PDO *cold phase*, SST are higher in the North Pacific interior and lower along the coast associated with the return of nutrients, primary production, and salmon.

Key to the climatic impact of PDO is the strength of the subpolar *Aleutian low*, which persists through the winter off the Alaskan coast. During a PDO *warm phase*, the Aleutian low is well developed and its strong counterclockwise winds steer mild and relatively dry air masses into the Pacific Northwest. Winters tend to be mild and dry and water supplies suffer from reduced mountain snow pack. But during a PDO *cold phase*, the Aleutian cyclone is weaker so that cold, moist air masses more frequently invade the Pacific Northwest. Winters are colder and wetter, and the mountain snow pack is thicker. PDO phases tend to last for 20 to 30 years. Cold phases prevailed from 1890 to 1924 and again from 1947 to 1976 whereas warm phases persisted from 1925 to 1946 and 1977 through the 1990s. Recently, the typically decadal cycle has broken down; with a cold phase from 1998-2001, a warm phase from 2002-2005, and a cold phase from 2007-2009 (Figure 11.15).

Conclusions

El Niño and La Niña involve interactions between the tropical Pacific Ocean and atmosphere. These phenomena underscore the importance of the flux of heat energy and moisture between the ocean and atmosphere. Changes in these fluxes during El Niño and La Niña have relatively short-term (one to two year) impacts on ocean circulation and the weather and climate in various parts of the world with implications for marine life, fisheries, and hydrologic budgets on land (e.g., drought, flooding rains). Short-term fluctuations in climate induced by El Niño and La Niña as well as the longer-term NAO, AO, and PDO are superimposed on much longer period variations in climate that have longer lasting impacts on the Earth system. The ocean also plays a central role in these long-term climate changes.

ENSO occurs in the tropics because that is the only place where the ocean can drastically alter its structure on time scales of months to years. At higher latitudes, the ocean structure is more strongly confined because of the greater role of Earth's rotation. Therefore, any large-scale ocean-atmosphere coupled variations at higher latitudes

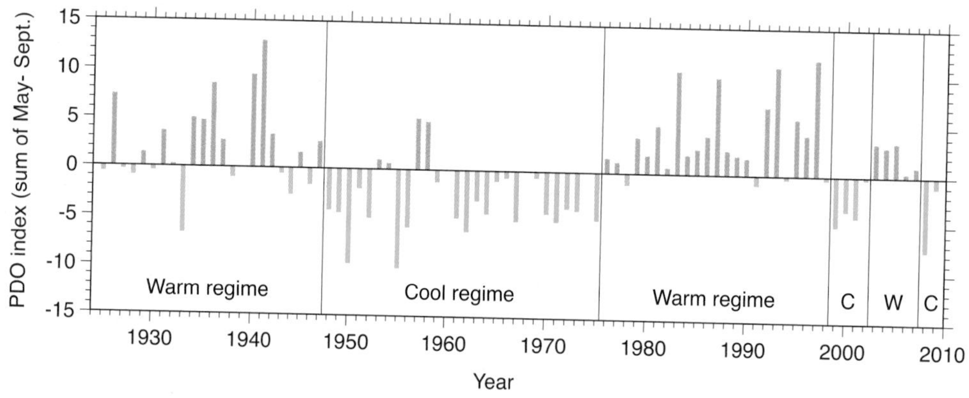

FIGURE 11.15
Variations in the Pacific Decadal Oscillation between warm and cool phases or regimes.

will necessarily have a time scale of many years to decades. Scientists are just beginning to piece together the character of these modes of variation in part because studying longer time-scale variations requires longer detailed records which are beginning to emerge. While the current observational record has captured many ENSO events, scientists have only begun to acquire sufficient data to characterize variability at higher latitudes.

In the next chapter, we examine the nature of climate change, causes of climate change, and possible implications of those changes for the ocean and humankind.

Basic Understandings

- Many factors working together shape the climate of any locality. Controls of climate consist of latitude, elevation, topography, proximity to large bodies of water, Earth's surface characteristics, net incoming solar radiation, long-term average atmospheric circulation, and prevailing ocean circulation.
- Climate responds in periodic fashion to the diurnal and seasonal variations in incoming solar radiation. On time scales of millions of years, all climate controls are variable. On shorter time scales, for all practical purposes, the first four climate controls listed above are essentially fixed and exert regular and predictable influences on climate.
- The ocean is a major player in Earth's climate system operating on time scales of days to millennia and spatial scales from local to global. The ocean influences the radiational heating and cooling of the planet and its circulation has a major influence on the global redistribution of heat energy.
- Ocean currents strongly influence climate. Cold surface currents are heat sinks. They chill and stabilize the overlying air, reducing sea fog frequency and the probability of thunderstorms. Relatively warm surface currents are heat sources. They supply heat and moisture to the overlying air, destabilizing the air, thereby energizing storm systems.
- In 1924, Sir Gilbert Walker discovered the Southern Oscillation, a seesaw variation in air pressure between the western and central tropical Pacific. In the mid-1960s, Jacob Bjerknes found that an El Niño episode begins when the air pressure gradient across the tropical Pacific begins to weaken in response to changes in SST.
- Over the long-term average (neutral conditions), southerly and southeasterly winds along the west coast of South America drive warm surface waters westward (via Ekman transport), away from the coast. Departing warm surface waters are replaced by cold, nutrient-rich water that wells up from below the thermocline (upwelling) fueling high biological production. Upwelling also occurs along the equator, east of the International Dateline, due to convergence of the trade winds of the two hemispheres and Ekman transport.
- During neutral conditions, relatively cool surface waters in the central and eastern tropical Pacific chill the overlying air and suppress convection so that rainfall is light in that region and along the adjacent western coastal plain of South America. Meanwhile, over the western tropical Pacific, relatively warm surface waters heat the overlying air, strengthening convection and giving rise to heavy rainfall. Higher air pressure in the east and lower air pressure in the west result in trade winds blowing from the east piling up a thick layer of warm surface water in the western tropical Pacific.
- At the onset of El Niño, air pressure falls over the eastern tropical Pacific and rises over the western tropical Pacific as part of the Southern Oscillation. The air pressure gradient across the tropical Pacific weakens and the trade winds slacken and may even reverse direction west of the International Dateline. This allows the thick layer of warm surface water piled up in the western tropical Pacific to slosh eastward along the equator into the eastern tropical Pacific, decreasing coastal upwelling.
- As El Niño evolves, over the western tropical Pacific SST decline, sea level falls, and depth of the thermocline decreases. Meanwhile, in the eastern tropical Pacific, SST rise, sea level climbs, and depth of the thermocline increases. Conditions are drier than usual in the western tropical Pacific and wetter than usual in the central tropical Pacific.
- Through teleconnections, El Niño and La Niña have ripple effects on the weather and climate of middle latitudes, especially in winter. During

El Niño, prevailing westerlies tend to blow more directly from west to east (zonal) so that winters are milder than normal in western Canada and across parts of the northern U.S. and wet and cool along the Gulf Coast.

- Although some weather extremes almost always accompany El Niño, no two events are exactly the same because El Niño is only one of many factors that influence inter-annual climate variability.
- La Niña is a period of unusually strong trade winds and vigorous upwelling in the eastern tropical Pacific. During La Niña, SST anomalies are essentially opposite those observed during El Niño. Accompanying La Niña are worldwide weather extremes that often are opposite those observed during El Niño.
- Scientists employ two types of numerical models to predict the evolution of El Niño or La Niña: empirical (or statistical) models and dynamical models. An empirical model bases predictions on records of past ocean/atmosphere conditions. A dynamical model consists of mathematical equations that simulate ocean/atmosphere coupling.
- The accuracy of dynamical models in predicting the evolution of El Niño or La Niña depends not only on how well their constituent equations simulate the coupled ocean/atmosphere system, but also the reliability of observational data used to initialize, correct, and verify the model.
- The ENSO Observing System in the tropical Pacific consists of an array of instrumented moored buoys, current meters, and satellites. Radar (microwave) altimeters onboard Earth-orbiting satellites accurately measure ocean surface elevation (sea level) changes that accompany El Niño or La Niña.
- El Niño occurs about once every 3 to 7 years. La Niña appears more likely to follow a strong El Niño than a weak one.
- Other quasi-regular oscillations involving the ocean and atmosphere that impact climate variability include the North Atlantic Oscillation, Arctic Oscillation, and the Pacific Decadal Oscillation.
- The North Atlantic Oscillation (NAO) refers to a seesaw variation in air pressure between Iceland and the Azores. When air pressure is higher than the long-term average over the Azores, it is lower than the long-term average over Iceland and vice versa. The North Atlantic Oscillation influences precipitation patterns and winter temperatures over eastern North America and much of Europe and North Africa.
- The Arctic Oscillation (AO) is a seesaw variation in air pressure between the North Pole and the margins of the polar region. Associated changes in the horizontal air pressure gradient alter the speed of the band of winds aloft that encircle the Arctic. Strengthening and weakening of these polar winds impact winter weather in middle latitudes and contribute to climate variability and changes in ocean circulation.
- The Pacific Decadal Oscillation (PDO) is a long-lived variation in climate over the North Pacific and North America. Sea surface temperatures fluctuate between the north central Pacific and along the west coast of North America.

Enduring Ideas

- The ocean is a major player in Earth's climate system because of its strong absorption of solar radiation, emission of infrared radiation, great thermal inertia, circulation and global redistribution of heat energy.
- El Niño or La Niña are routinely included in long-range seasonal weather outlooks because of their contribution to year-to-year climate variability.
- Relatively high biological productivity is favored by coastal upwelling in which Ekman transport of surface waters offshore is accompanied by the upward movement of nutrient-rich cold water from below into the photic zone.
- A linkage between changes in atmospheric circulation occurring in widely separated regions of the world is known as a teleconnection. Through teleconnections, El Niño and La Niña have ripple effects on the weather and climate of middle latitudes, especially in winter.
- During La Niña, a period of unusually strong trade winds and vigorous upwelling in the eastern tropical Pacific, SST anomalies are essentially opposite those observed during El Niño. Accompanying La Niña are worldwide weather extremes that often are opposite those observed during El Niño.
- Other quasi-regular oscillations involving the ocean and atmosphere that impact climate variability include the North Atlantic Oscillation, Arctic Oscillation, and the Pacific Decadal Oscillation.

Review

1. Define climate.
2. Identify and describe the principal controls of Earth's climate.
3. Provide three examples of the role played by the ocean in Earth's climate system.
4. How does the extent of sea-ice cover influence the absorption of solar radiation by the ocean?
5. What is the Southern Oscillation? How is the Southern Oscillation Index calculated?
6. During an El Niño event, how does sea-surface temperatures and sea level change over the eastern tropical Pacific?
7. Describe the sea-surface temperature anomalies over the tropical Pacific during a La Niña event.
8. Distinguish between the two principal types of numerical models that forecasters rely upon to predict the evolution of El Niño or La Niña.
9. What is the North Atlantic Oscillation and what is its significance?
10. How do negative and positive phases of the Arctic Oscillation relate to winter weather in the U.S.?

Critical Thinking

1. What is the role of the ocean in Earth's greenhouse effect?
2. In what sense are the ocean and atmosphere coupled during an El Niño event?
3. Explain how changes in patterns of sea-surface temperature can alter the planetary-scale circulation of the atmosphere.
4. What are the early signs of a developing El Niño in the tropical Pacific Ocean?
5. What are the early signs of a developing La Niña in the tropical Pacific Ocean?
6. Why is La Niña more likely to follow a strong El Niño than a weak one?
7. What is meant by a teleconnection and what is its significance in forecasting the impact of ENSO?
8. No two El Niño events are the same. Explain the significance of this statement.
9. How does a horizontal air pressure gradient affect the speed of the trade winds?
10. How does the SST influence the temperature and humidity of the overlying air mass?

ESSAY: Atacama Hyper-arid Coastal Desert

Even though it borders the Pacific Ocean, the Atacama Desert of northern Chile is the driest desert on Earth. It occupies a narrow strip of land extending 1000 km (600 mi) south from the Chile-Peru border to 30 degrees S (Figure 1). Encompassing nearly 105,000 square km (40,600 square mi), the Atacama Desert is mostly barren salt basins, sand, and solidified lava flows; the soil is essentially lifeless (Figure 2). *Hyper-arid conditions* (almost no moisture) prevail in the central depression of the desert (elevation 2300 m or 7500 ft), situated between the Andes Mountains to the east (elevation 5000-6000 m or 16,400-19,700 ft) and the linear Chilean Coast Range to the west (elevation 500-3000 m or 1650-9800 ft). Mean annual rainfall at coastal desert towns is meager, only 28 mm (1.1 in.) at Iquique, 5 mm (0.2 in.) at Arica, and 1.0 mm (0.04 in.) at Antofagasta. Some locations have not recorded rain in over two decades, and there is evidence that some desert localities received no significant rainfall for hundreds of years (e.g., 1570-1971).

The extreme aridity of the Atacama Desert is the product of interactions involving the ocean, atmosphere, and topography (Figure 3). Recall from Chapter 5 that the climate of South America's west coast is dominated by a massive semipermanent subtropical anticyclone (high) centered over the South Pacific Ocean near 30 degrees S, the southern end of the Atacama Desert. Viewed from above, surface winds blow counterclockwise and outward about the center of the South Pacific high. This atmospheric circulation drives the South Pacific gyre, which includes the cold Peru Current (also known as the Humboldt Current) that flows northwest along the west coast of South America from the southern tip of Chile to northern Peru. This broad *eastern boundary current* can extend up to 1000 km (600 mi) offshore.

Cold surface waters transported northward from near Antarctica combined with cold water brought to the surface by coastal upwelling (Chapter 6) explains the exceptionally low SST of the Peru Current, one of the coldest on Earth. This cold surface water chills and stabilizes the overlying air, inhibiting development of thunderstorms and tropical cyclones (Chapter 8). At the same time, the air subsiding over the broad eastern flank of the South Pacific high is warmed by compression. This descending air encounters the *marine air layer*, the shallow, cold, humid layer of air overlying the ocean and influenced by the Peru Current. Because the air descending from above is warmer and considerably drier than the marine air layer, a temperature inversion forms. Within the temperature inversion, the air temperature increases with increasing altitude so the air layer is extremely stable, strongly inhibiting upward air motion and preventing the development of rain-producing clouds.

FIGURE 1
The Atacama Desert of northern Chile.

FIGURE 2
Photo of the Atacama Desert shows hyper-arid conditions. [Courtesy of Rich Cageao, NASA Langley Research Center]

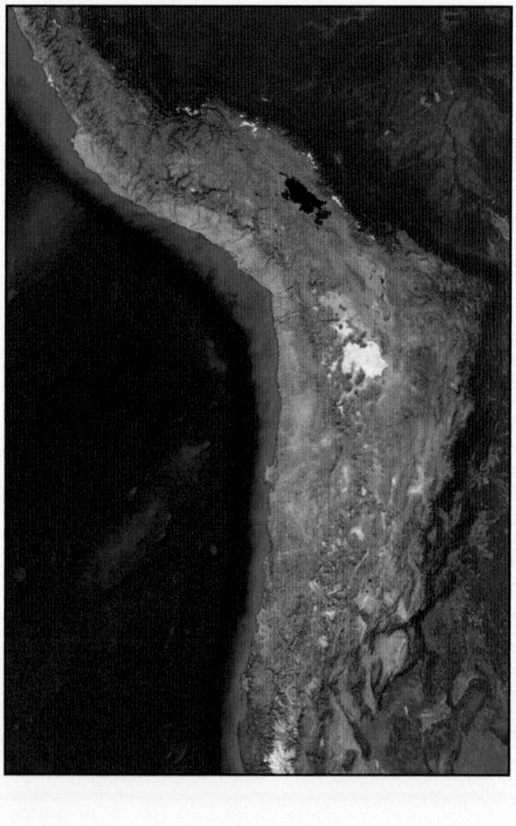

FIGURE 3
Satellite image showing the Atacama Desert of Chile to the southeast of the sharp bend in the coastline. [Courtesy of NASA World Wide open source virtual globe]

Although atmospheric conditions do not favor development of precipitation, low clouds and fog often form within the marine air layer and drift onshore into portions of the Atacama Desert, known locally as Camanchaca. The tiny fog droplets are the only source of moisture for much of the coastal desert, sufficient for survival of lichens, hypolithic algae, and cacti.

The Andes Mountains and Chilean Coastal Range are major topographic features that contribute to the hyper-arid conditions of the Atacama Desert. A mountain range that intercepts the prevailing winds forms a natural barrier, creating a cloudier, wetter climate on one side of the range than on the other side. Air that is forced to ascend the windward slopes (facing the oncoming wind) expands and cools. With sufficient cooling, clouds and precipitation develop. Meanwhile, air descending the leeward slopes (downwind side) is compressed and warms so that clouds vaporize and there is little or no precipitation. Hence, mountain ranges induce contrasting climates: moist climate on the windward slopes and dry climates on the leeward slopes.

Prevailing winds (trade winds) blow from the east, over the Atacama Desert so the desert is on the dry leeward (*rain shadow*) side of the Andes Mountain Range. Similarly, the Andes prevent the warm humid air masses that originate in the Amazon Basin from reaching the western coastal plain. On occasions when winds blow from the west, the desert is on the dry leeward side of the Chilean Coastal Range.

As demonstrated by the Atacama Desert and other coastal deserts (e.g., Baja, CA, Namib Desert of Africa), proximity to the ocean does not necessarily produce a wet climate. Although most water that precipitates on land originates in the ocean, the interaction of subsystems within the Earth system may greatly inhibit precipitation.

ESSAY: El Niño in the Past

Lengthy records of El Niño episodes of the past would provide a useful perspective on more recent events and perhaps help determine whether a connection exists between global warming and the frequency and intensity of El Niño. However, in most areas of the world, reliable instrument-based weather records that signal past El Niño events extend back only to about the mid 19th century. For information on earlier occurrences of El Niño, scientists must rely on *proxy climatic data*, that is, information inferred from documentary (e.g., logs, diaries), geological (e.g., ocean bottom sediment cores, glacial ice cores), or biological (e.g., tree growth rings, tropical corals) indicators of climate. Suppose, for example, that written records from India and Southeast Asia list agricultural losses due to prolonged drought. Meanwhile, a sediment core extracted from a lake bottom contains evidence of heavy rainfall in normally arid coastal Peru during the same time interval. These simultaneous events are consistent with a strong El Niño.

Geological evidence indicates that El Niño episodes were occurring at least as far back as late in the Pleistocene Ice Age (1.8 million to 10,500 years ago). In 1999, based on his analysis of ancient corals from Sulawesi Island, Indonesia, Harvard researcher Daniel Schrag concluded that El Niño was occurring at 3- to 7-year intervals about 124,000 years ago. Schrag used oxygen isotope analysis to distinguish between El Niño and neutral (normal) years. Living corals build their external skeletons of limestone ($CaCO_3$) that incorporates two isotopes of oxygen: O^{16} and O^{18}. The lighter oxygen isotope evaporates more readily than does the heavier oxygen isotope. Hence, during relatively dry weather, evaporation rates are high, and the coral skeleton is enriched in O^{18} relative to O^{16}. On the other hand, during rainy episodes, evaporation rates are low, and the coral skeleton is enriched in O^{16} relative to O^{18}. Corals build their skeletons in annual growth rings that can be dated so that oxygen isotope analysis permits a detailed chronology of general weather conditions that can be tied to El Niño.

David Lea of the University of California, Santa Barbara, and Sandy Tudhope of the University of Edinburgh, Scotland studied coral reef terraces in New Guinea dating back some 130,000 years. In 2001, they reported on their isotopic and chemical analyses of these corals. They determined that El Niño events were more intense over the past 100 years than at any time during the past 130,000 years. Furthermore, they concluded that El Niño was about 50% weaker during the Ice Age and intensified during warm episodes, again spurring speculation that the recent upturn in El Niño intensity is linked to the global warming trend.

The El Niño signal shows up in a 4000-year record of lake sediments extracted from Glacial Lake Hitchcock, which occupied the Connecticut River valley during the waning phase of the last Ice Age. Layers of lake sediments chronicle short-term climate fluctuations spanning the period from about 17,500 to 13,500 years ago (during the final retreat of glacial ice from New England), and apparently resolve both intense and weak El Niño events (having periods of 2.5 to 5 years).

In 2002, Geoffrey Seltzer of Syracuse University and his colleagues reported on their analysis of two 8-m (26-ft) sediment cores extracted from the bottom of Laguna (Lake) Pallcacocha high in the Andes Mountains of southern Ecuador. Rainfall governs the amount of sediment delivered to the lake but rainfall associated with a weak El Niño is not likely to reach the 4200-m (13,800-ft) high lake. Hence, the researchers interpreted anomalously thick layers of silt in the cores as indicating heavy rainfall associated with an intense El Niño. The lake sediment cores span the past 12,000 years and indicate that between 12,000 and 7000 years ago, strong El Niño episodes occurred five or fewer times per century. The frequency of intense El Niño episodes then increased and peaked during the 9th century CE when they occurred about every three years.

Using documentary evidence from explorers and early settlers of the coast of northwest South America, W.H. Quinn and colleagues were able to reconstruct the chronology of El Niño episodes back to the year CE 1525. For example, El Niño conditions prevailed during 1531-32, when Francisco Pizarro (1478-1541) conquered the Incas. Although heavy rains impeded his advance, his horses were well fed by the vegetation made unusually lush by the moist conditions. Records of the maximum discharge of the Nile River at Cairo extend back to CE 622 and are another valuable source of information on past El Niño episodes. Year to year changes in discharge are linked to variations in the summer monsoon rains in the Ethiopian highlands, near the headwaters of the Nile, and decrease during El Niño.

The long-term reconstructed climate record suggests that El Niño has occurred regularly since at least the late Pleistocene—albeit with at least one significant lull from 12,000 to 7000 years ago. However, El Niño intensity has varied over thousands of years (i.e., the tempo is roughly the same but the beat has alternately intensified and weakened). We now appear to be in a period of particularly intense El Niño events.

CHAPTER 12

THE OCEAN AND CLIMATE CHANGE

Sun reflecting off open water and ice floes [Collection of Dr. Pablo Clemente-Colon, Chief Scientist National Ice Center]

Case in Point

In May 2008, when polar bears (*Ursus maritimus*) were added to the "threatened" category of the U.S. endangered species list, the species became the first to be so designated because of climate change. With the warming trend reducing the sea-ice habitat to which the polar bear is adapted, the population is declining over much of its range (Alaska, northern Canada, northern Russia, Greenland, and Norway). In 2011, researchers estimated that the polar bear population in western Hudson Bay had decreased to 900 from 1200 a decade before. Polar bears, large carnivores with low reproductive rates, depend on sea ice as a platform to hunt seals (their principal food source), mate, and travel. If the ice does not persist long enough for them to forage for sufficient food, they starve.

Péter Molnár and colleagues from the University of Alberta, Edmonton, studied the reproductive ecology of the Hudson Bay polar bears and found that their litter size has decreased with declining sea-ice cover. Depending on stores of body fat for survival, pregnant females retreat to a maternity den without food for up to 8 months. As the spring ice break-up occurs earlier, their hunting season shortens and, without their stores of fat, a pregnant bear cannot meet her own needs let alone give birth to and raise cubs. Without sufficient energy, she will either not enter a maternity den or else abort. Using a mathematical model, Molnár and colleagues predicted that if the spring ice break-up in western Hudson Bay were one month earlier than in the 1990s, 40% to 73% of females would not produce a litter. If the break-up occurred two months earlier, well over half and possibly all, female polar bears would fail to reproduce.

The warming trend responsible for the loss of sea-ice habitat is attributed to the buildup of greenhouse gases, primarily caused by human activities (burning of fossil fuels and land clearing). Based on numerical climate models, continuation of the Arctic warming trend will cause a dramatic decrease in the spatial and temporal extent of sea ice. Between 1979 and 2010, the extent of Arctic sea ice at the end of the summer melt season has declined an average of 11.5% per decade. In 2007, the USGS predicted that if greenhouse gas emissions continue unabated, (the "business-as-usual" scenario), two-thirds of the world's 25,000 polar bears could disappear by mid-century due to habitat loss.

Compounding matters, the declining population increases the difficulty of finding a mate. In the 16 December 2010 issue of *Nature*, Brendan Kelly of NOAA's National Marine Mammal Laboratory in Juneau, Alaska, and colleagues, warn that loss of sea-ice habitat also threatens species through interbreeding. As populations and species become more isolated, polar bears are socializing with other bears producing hybrids and reducing diversity. Already, hunters report encounters with grizzly polar bear hybrids. With hybridization, rare species are more likely to go extinct and, as species become mixed, adaptive gene combinations are lost.

Scientists disagree on the rate at which sea-ice habitat will disappear. Sea ice may decrease proportionally to rising temperature or, if a *tipping point* is crossed when the temperature rise passes a critical threshold, sea-ice cover could abruptly and irreversibly disappear seasonally. Modeling outcomes seem to favor the more gradual alternative.

What can be done to slow the loss of polar bear habitat and extinction? In the 16 December 2010 issue of *Nature*, Steven C. Amstrup of the USGS Alaska Science Center, and colleagues, used a general atmospheric circulation model to demonstrate that significantly more sea-ice habitat will persist with successful mitigation of greenhouse gas emissions. If this mitigation keeps the global mean surface temperature to less of an increase than 1.25 Celsius degrees (2.25 Fahrenheit degrees), more polar bears would occupy a larger area. Coupled with traditional wildlife management practices, Amstrup and colleagues concluded that the polar bear population, although lower than at present, would be sustainable through the remainder of this century.

FIGURE 12.1
Polar bears on an ice floe. [Courtesy of NOAA Office of Ocean Exploration]

Driving Question:

How and why does climate change and how does the ocean participate in and respond to climate change?

In this chapter, we examine the instrument-derived and reconstructed climate record. One of our primary goals is to learn more about how climate changes through time. These lessons of the climate past are useful in establishing a perspective on the present climate and how climate might change in the future. Two of the most obvious lessons of the climate past are: (1) climate changes over a broad range of temporal scales, from years to decades to centuries to millennia, and (2) many forcing agents and mechanisms are responsible for driving climate change.

The ocean is an important source of information on the climate record. From analysis of deep-sea sediment cores, scientists reconstruct climate change extending back hundreds of thousands of years, and even millions of years. The ocean's low albedo, great thermal inertia, and role as a sink and source of heat energy and gases (especially water vapor and carbon dioxide) inhibit wild swings in climate. At the same time, the ocean reacts to global climate change. Sea level falls in response to cooling trends that cause glaciers to expand and sea-surface temperatures to fall (contracting sea water). Sea level rises in response to warming trends that melt glaciers (with runoff draining into the ocean) and raise mean ocean temperature (expanding sea water). Such changes in sea level have important implications for the coastal zone including the potential for storm surges, flooding, and erosion.

We begin by examining the climate record for what it tells us about climate behavior. We then identify and describe the many factors that influence the variation of climate through time and finally we summarize what is known about the climate future particularly relating to the world ocean.

The Climate Record

Climate varies from one place to another and with time. Examining the record of the climate past reveals some basic understandings pertaining to the nature of climate variability and climate change and provides a useful perspective on current and future climate. In most places, however, the reliable instrument-based record of past weather and climate is limited to not much more than 130 years or so and almost exclusively for land locations. For times and places where no instrument-derived record of climate exists, scientists infer past climate information from various environmental sensors that substitute for actual weather instruments. These **proxy climate data sources** include historical documents, deep-sea sediment cores, pollen profiles, tree growth rings, and glacial ice cores. Climate reconstructions based on proxy climate data are compared to hindcasts generated by computer models. Agreement between reconstructions and hindcasts helps boost confidence in computer model simulations of future climate.

HISTORICAL DOCUMENTS

Certain historical documents archived in libraries and museums can yield a wealth of information on past climates. Personal diaries, almanacs, old newspapers, and mariner's logbooks yield qualitative and/or quantitative references to weather and climate. Other types of documents refer only indirectly to weather and climate but can be useful nonetheless. Records of grain harvests, quality of wine, or phenological phenomena (e.g., dates of blooming of plants in spring) provide some indication of growing season weather.

As far back as the 17th century, captains of naval and merchant ships took observations of wind force and direction, general weather conditions, and the state of the sea (Chapter 7), and recorded those observations in logbooks. In some cases, observations were made three times daily. Dennis Wheeler, a geographer at the University of Sunderland in the UK, recognized mariner's logbooks as a largely untapped store of weather and climate information at sea. In 2000, Wheeler and colleagues assembled the world's first daily climatological record for 1750-1850 based on logbooks from several European nations. Because navy ships often traveled in fleets, Wheeler and colleagues were able to compare observations among ships and check data for consistency and reliability. In 2003, Wheeler and colleagues completed the *Climatological Database for the World's Oceans (CLIWOC)*; it includes 273,269 observations drawn from 1624 logbooks.

The challenge of assembling and interpreting weather and climate information from mariner's logbooks is far from complete. Logbooks number in the tens of thousands (including those from the US Navy and English East India Company). Also, more than 120,000 mariner's logbooks from the pre-instrument era await analysis.

MARINE SEDIMENTS AND CLIMATE

Cores extracted from sediments that blanket the ocean floor yield a continuous record of sedimentation going back many hundreds of thousands of years and in places, millions of years (Figure 12.2). Much of what we know about the climate fluctuations of the Pleistocene Ice Age is based on analysis of the shell and skeletal remains of microscopic marine organisms that are found in deep-sea sediment cores. Identification of the environmental requirements of these organisms plus **oxygen isotope analysis** of their remains enables scientists to distinguish between cold and mild climatic episodes of the past.

Scientists use a special property of water to reconstruct large-scale climate fluctuations of the Pleistocene Ice Age. A water molecule (H_2O) is composed of one of the two stable isotopes of oxygen, O^{16} or O^{18}. (Isotopes of a single element differ in atomic mass based on the number of neutrons in the nucleus of the atom.) In the Earth system, the lighter isotope (O^{16}) is much more abundant than the heavier isotope (O^{18}); only one O^{18} exists for every thousand or so O^{16}. Nonetheless, small but significant variations occur in the amount of light oxygen compared to heavy oxygen circulating in the global water cycle. These variations have important implications for past fluctuations in glacial ice volume on the planet.

On average, water molecules containing the lighter O^{16} isotope move slightly faster than water molecules containing the heavier O^{18} isotope and evaporate more readily. Water molecules that evaporate from the ocean (or land) are enhanced in light oxygen compared to heavy oxygen. The amount of O^{16} compared to O^{18} is also greater in cloud particles and rain and snow versus liquid water on Earth's surface. Most precipitation either falls back into the ocean directly or drains from land back to the sea, replenishing the ocean's supply of light oxygen and maintaining a relatively constant average ratio of light to heavy oxygen. However, geographical variations in the oxygen isotope ratio of seawater arise because of differences in precipitation amounts and evaporation rates. Seawater has more O^{18} at subtropical latitudes where evaporation exceeds precipitation and less in middle latitudes where rainfall is greater.

FIGURE 12.2
Sediment cores extracted from beneath the ocean floor provide valuable information on the geologic and climatic past. (A) A hollow pipe lined with plastic tubing and coupled to a weight at the top is lowered over the side of a ship. When within about 8 m of the bottom, the corer free-falls into the sediment while a piston suctions sediment into the tube. The coring device is recovered and the sediment core is removed and split lengthwise for analysis. (B) USGS crew on the research vessel *G.K. Gilbert* collect a 20-ft sediment core, using an electric coring system and hydraulic crane. [USGS] (C) Core racks holding a total of 72,000 m of sediment cores at the Deep-Sea Sample Repository of Lamont-Doherty Earth Observatory in Palisades, NY. [Courtesy of the Lamont-Doherty Earth Observatory]

Heavy water molecules condense and precipitate slightly more readily than light water molecules so that moisture plumes moving from the tropics to high latitudes lose heavy oxygen along the way. Hence, snow falling at high latitudes has less O^{18} than rain falling in the tropics. The result is that growing ice sheets, during a glacial climate episode, sequester more and more light oxygen while ocean water ends up with less and less O^{16} proportionately. With a shift to an interglacial climate, ice sheets shrink and melt water rich in O^{16} drains back into the ocean increasing the amount of O^{16} compared to O^{18}. Hence, over time, the proportion of light to heavy oxygen in ocean water decreases with increasing glacial ice cover.

Organic sediments on the ocean floor record fluctuations in the oxygen isotope ratio of seawater. Marine organisms living in the sunlit surface waters, such as foraminifera, build their shell-like tests from calcium carbonate ($CaCO_3$) that is dissolved in seawater. Tests formed during warmer interglacial climatic episodes contain more light oxygen than those formed during colder glacial climatic episodes. When these organisms die, their tests settle to the ocean bottom and mix with other marine sediments (Chapter 4). From specially outfitted deep-sea drilling ships, scientists extract cores from an undisturbed sequence of ocean bottom sediments, with the youngest sediments at the top of the core and the oldest sediments at the bottom. In the laboratory, the core is split open, and tests are extracted and analyzed for their oxygen isotope ratio. Variations in oxygen isotope ratio document changes in the planet's glacial ice volume and hence, past changes in climate.

Oxygen isotope analysis of deep-sea sediment cores indicates that the Pleistocene was punctuated by abrupt changes between numerous glacial and interglacial climatic episodes. Oxygen isotope analysis has also been applied to ice layers within cores extracted from the Greenland and Antarctic ice sheets. These analyses confirm the abrupt change behavior of climate back hundreds of thousands of years.

OTHER PROXY CLIMATIC DATA SOURCES

On land, much has been learned about past climate from study of fossil pollen, tree growth rings, and glacial ice cores. Pollen is a valuable source of information on late Ice Age vegetation and climate, especially over the past 15,000 years. *Pollen* is the tiny dust-like fertilizing component of a seed plant that is dispersed by the wind and accumulates on the bottom of lakes (and in other depositional environments) along with other organic and inorganic sediments. Scientists use a corer to extract a sediment column (core) that chronicles past changes in pollen (and therefore, vegetation). Back at the laboratory, pollen grains are separated from the other sediment in the core, identified as to source vegetation, and counted to determine the dominant vegetation type. Implications for past climate are drawn assuming that (1) the pollen is of local origin and (2) climate largely controls vegetation (and pollen) type. Changes in dominant pollen type in a core signal changes in nearby vegetation, likely in response to climate change.

Tree growth rings can yield a year-to-year record of past climate variations stretching back many thousands of years. Each spring/summer, living trees add a growth-ring whose thickness and density depend on weather conditions during the growing season. A small hollow drill is used to extract a tree's growth-ring record. Although other environmental factors (e.g., soil type, drainage) can be important, tree growth rings are especially sensitive to moisture stress and have been used to reconstruct lengthy chronologies of drought prior to the era of instrument-based records.

Ice cores extracted from glaciers yield a record of past seasonal snowfall preserved as thin layers of ice (Figure 12.3). Cores extracted from the Greenland and Antarctic ice sheets provide detailed decadal-scale climate information going back roughly 420,000 years. Using the oxygen isotope technique (described earlier), scientists can distinguish between cold and mild episodes in the past and through chemical analysis of tiny air bubbles trapped in the ice, determine the chemical composition of

FIGURE 12.3
Desert Research Institute scientist examines a section of the WAIS Divide ice core recovered from a depth of 500 meters. [Photo courtesy of Kendrick Taylor]

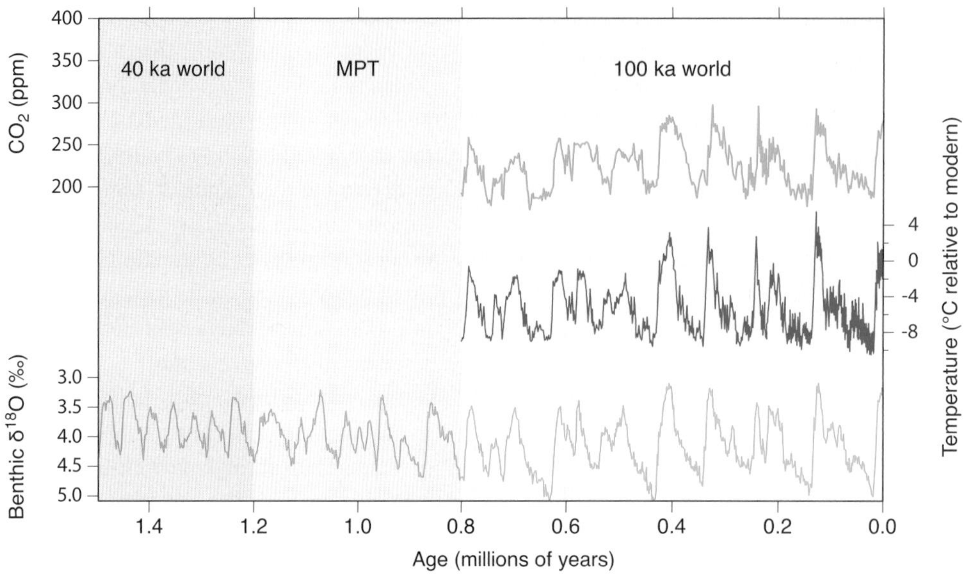

FIGURE 12.4
The record of past temperature and carbon dioxide from the European Project for Ice Coring in Antarctica (EPICA) Dome C core, the oldest such records yet obtained, and the oxygen isotope record from benthic foraminifera, a proxy for global ice volume and therefore global climate conditions. Lower values of oxygen isotope ratios in calcium carbonate of benthic foraminifera correspond to times of lower global ice volume, warmer temperatures in Antarctica, and higher levels of carbon dioxide. The switch from predominantly 41,000-year cycles to 100,000-year cycles in the isotope record at the mid-Pleistocene transition (MPT) took place during the period from about 1.2 million to 800,000 years ago. [Based on EOS, V.91, No. 40, 5 October 2010, page 357]

ancient air. This latter information is important in tracking long-term trends in the atmospheric concentration of greenhouse gases such as carbon dioxide (Figure 12.4). Dust layer thicknesses also allow differentiation between cold and mild episodes as well as an exact counting of annual ice layers, and precise dating, if the dust layers are volcanic in origin.

For more information on climatic inferences drawn from glacial ice cores, refer to this chapter's first Essay.

GEOLOGIC TIME

Throughout most of the 4.5 billion years that constitute **geologic time**, which is divided based on large-scale geological events into eons, eras, periods, and epochs (Figure 12.5), global climate is challenging to reconstruct. Lengthy gaps in proxy climate records, problems dating specific events, and limitations in correlating events in widely separated locations make describing early climate difficult. Furthermore, the movement of tectonic plates complicates climate reconstruction that spans hundreds of millions of years (Chapter 2). Nonetheless, the available evidence supports some general conclusions regarding the climate of geologic time.

About 570 million years ago as the Proterozoic eon neared its end, just before the Phanerozoic eon, the planet experienced extreme climate fluctuations. Along Namibia's Skeleton Coast, in southwest Africa on the Atlantic, are layers of rock that formed in tropical seas amid layers of glacial deposits, indicating abrupt changes in climate between extreme heat and cold. According to the *snowball Earth* hypothesis, during as many as four cold episodes, each lasting 10 million years, the continents were encased in glacial ice and the ocean frozen to a depth of more than 1000 m (3300 ft). At the close of each cold episode, temperatures rose rapidly and, within only a few centuries, all the ice melted.

As noted in the first Essay of Chapter 1, five major mass extinctions occurred over the past 550 million years. Drastic changes in Earth's environment took place during these episodes and more than 75% of plant and animal species died out in a geologically brief interval. Four of the five mass extinctions (at the close of the Ordovician, Devonian, Permian, and Triassic periods) were linked to a combination of chemical and circulation changes in the ocean, coupled with global warming due to an enhanced greenhouse effect.

The Mesozoic era, 251 million to 65.5 million years ago, which would end with the one extinction due to an extraterrestrial cause, was characterized by a generally warm Earth free of large glacial ice sheets. Between the

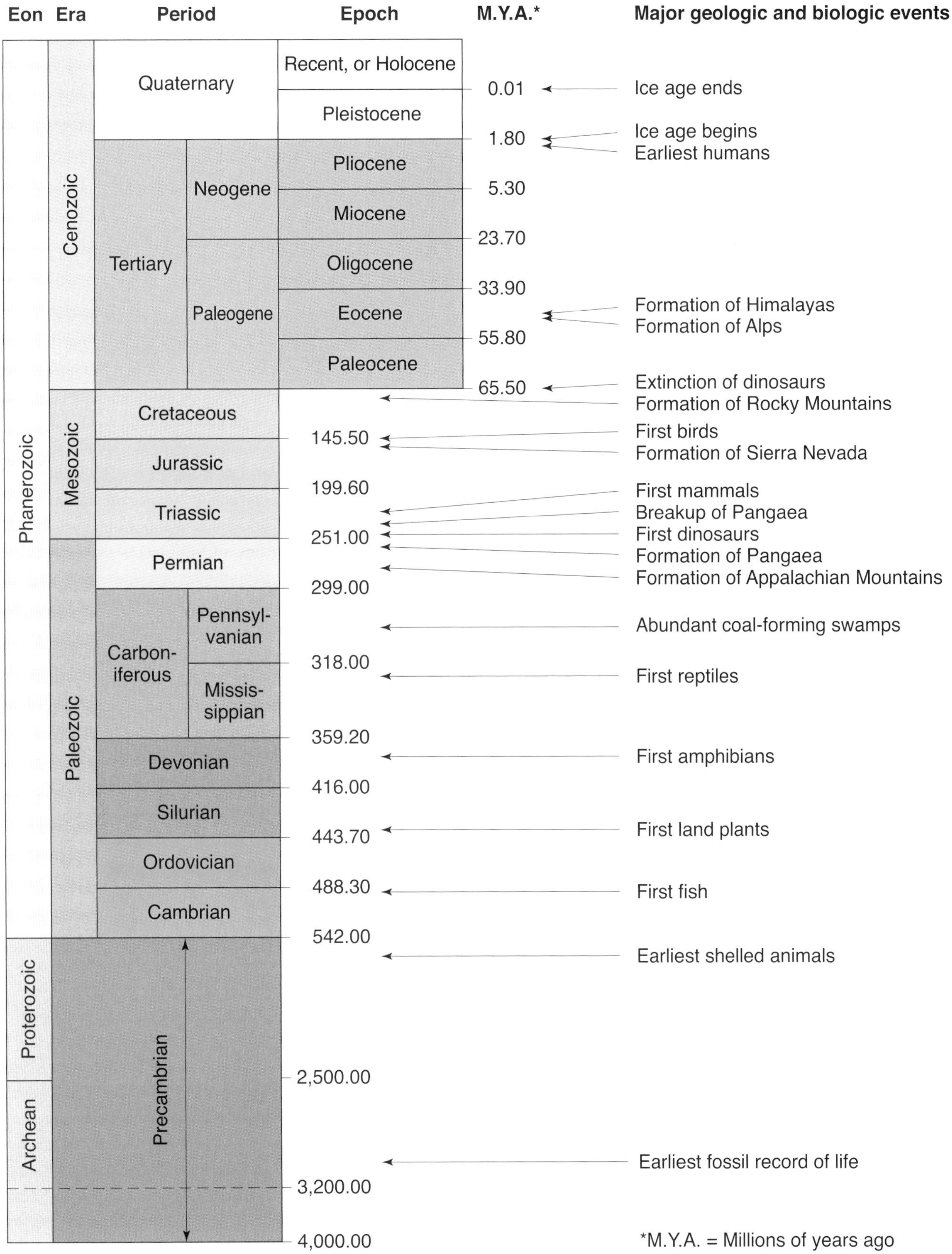

FIGURE 12.5
Geologic time scale.

Triassic and Jurassic periods, the first and second of the Mesozoic era, the global mean temperature rose 3 to 4 Celsius degrees (5.5 to 7 Fahrenheit degrees), contributing to a major extinction and displacement of many species. At peak warming during the Cretaceous period, the last of this era, the global mean temperature was 6 to 8 Celsius degrees (11 to 14 Fahrenheit degrees) higher than now. Subtropical plants and animals lived as far north as 60 degrees N, the present polar region, and dinosaurs roamed what is now the North Slope of Alaska. The fifth mass extinction, 65 million years ago near the close of the Cretaceous, was likely caused by an asteroid impact on Earth's surface, throwing huge quantities of dust into the atmosphere, blocking sunlight and causing cooling that contributed to the demise of the dinosaurs. Without competition, mammal populations exploded and diversified as the atmosphere cleared.

With the warmest episodes of the past 80 million years occurring during the early Paleogene period (from 45 to 65 million years ago), and the glacial climate intervals of the Pleistocene Ice Age beginning1.8 million years ago, drastic climate fluctuations have characterized the present Cenozoic era. About 56 million years ago, near the transition between the Paleocene and Eocene epochs, deep-ocean temperatures rose 6 to 7 Celsius degrees (11 to 13 Fahrenheit degrees), adding to an already warm Earth. This **Paleocene-Eocene Thermal Maximum (PETM)**, which spanned 170 thousand years, is thought to have been caused by massive amounts of methane released from submarine gas hydrate deposits (refer to the second Essay of Chapter 9). The methane escaped from the ocean into the atmosphere and oxidized to carbon dioxide, enhancing the greenhouse effect dramatically. In only a thousand to 10,000 years, the global mean surface air temperature climbed 5 to 10 Celsius degrees (9 to 18 Fahrenheit degrees).

As reported in the 17 March 2011 issue of *Nature*, Philip F. Sexton of Scripps Institution of Oceanography and colleagues examined an ocean drill core dating from the early Paleogene period that provided evidence of six episodes of rapid global warming, though of lesser magnitude and shorter duration than the PETM. These *hyperthermals* (approximately 65.2, 58.2, 53.7, 53.2, 52.5, and 41.8 million years ago) were likely caused by enhancement of the greenhouse effect ultimately from repeated large-scale releases of dissolved organic carbon from the abyssal ocean.

Also revealed by ocean drill cores from the Southern Ocean is a dramatic increase in CO_2 levels during the middle Eocene epoch, after the PETM about 40 million years ago, that enhanced the greenhouse effect and interrupted a cooling trend. The *Middle Eocene Climatic Optimum (MECO)* was among the hottest intervals in Earth history, the SST in the southwest Pacific rising by 3 to 6 Celsius degrees (5.4 to 11 Fahrenheit degrees), as reported by Peter K. Bijl of Utrecht University in the Netherlands, and colleagues, in the 5 November 2010 issue of *Science*.

The Middle Eocene Climatic Optimum may be linked to tectonic processes. After the supercontinent Pangaea broke up, the tectonic plate carrying the Indian subcontinent moved northward until, about 50 million years ago, it slammed into Asia. Prior to this collision, its approach triggered a million years of active volcanism along Asia's southern border, generating 4 million cubic km of basaltic lava from carbonate-rich sediments on the sea floor. This increased atmospheric CO_2 to more than 1000 ppmv, raising temperatures worldwide. When the plates collided, forming the Himalayans, they cut off the supply of carbonate sediments. At the same time, weathering and erosion of rock exposed on the Indian subcontinent pulled CO_2 out of the air, decreasing it to 300 ppmv by 30 million years ago. The consequent weakening of the greenhouse effect was responsible for long-term cooling.

Earth's climate not only became cooler but also drier and more variable, setting the stage for the Pleistocene Ice Age, an epoch of numerous major glacial advances and recessions that began 1.8 million years ago. According to W.F. Ruddiman of the University of Virginia and J.E. Kutzbach of the University of Wisconsin-Madison, mountain building, specifically the rise of the Colorado Plateau, Tibetan Plateau, and Himalayan Mountains, may explain this change in Earth's climate. Prominent mountain ranges influence the geographical distribution of clouds and precipitation, and can alter the planetary-scale circulation. Furthermore, mountain building may alter the global carbon cycle. As noted earlier, weathering of bedrock exposed in mountain ranges sequesters more atmospheric carbon dioxide in sediments thereby weakening the greenhouse effect.

Although mountain building began about 40 million years ago, about half of the total uplift took place between 10 and 5 million years ago. Over the past 10 million years, nearly half of the total Himalayan uplift took place and the Tibetan Plateau and Himalayan Mountains now cover an area of more than 2 million square km (0.8 million square mi) with an average elevation of more than 4500 m (14,700 ft). In the American West, the region from the California Sierras to the Rockies, known as the Colorado Plateau, has an average elevation of 1500 to 2500 m (5000 to 8200 ft).

These plateaus diverted the planetary-scale west winds into a more meridional pattern, increasing the north-south exchange of air masses and altering the climate over a broad region of the globe. Also, seasonal heating and cooling of the plateaus cause low pressure to develop in summer and high pressure in winter, which strengthens the monsoon circulation over southern Asia.

PAST TWO MILLION YEARS

Climate varies over a broad spectrum of time scales so that viewing the climate record of the past two million years in progressively narrower time frames is useful. Such an approach helps to resolve the oscillations of climate into more detailed fluctuations especially over the recent past. Over the past two million years, plate tectonics was not a major factor in climate change. For example, assuming a spreading rate of 10 cm (3.9 in.) per year, in two million years the Atlantic Basin would spread a total distance of 200 km (124 mi), not very significant in a climatic sense. For all practical purposes, over the past two million years mountain ranges, continents, and ocean basins were essentially as they are today.

Compared to the climate that prevailed through most of geologic time, the climate of the last two million years was unusual in favoring the development of huge glacial ice sheets (although evidence also exists of ice ages earlier in geologic time). During much of Earth's history, the average global temperature may have been 10 Celsius degrees (18 Fahrenheit degrees) higher than it was over the past two million years. Cooling that set in about 40 million years ago culminated in the Pleistocene Ice Age that began 1.8 million years ago and ended 10,500 years ago.

During the Pleistocene Ice Age the climate shifted numerous times between glacial climates and interglacial climates. A **glacial climate** favors the thickening and expansion of glaciers whereas an **interglacial climate** favors the thinning and retreat of existing glaciers or no glaciers at all. During major glacial climatic episodes of the Pleistocene, the Laurentide ice sheet developed over central Canada and spread westward to the Rocky Mountains, eastward to the Atlantic Ocean and southward over the northern tier states of the United States (Figure 12.6). At about the same time, mountain glaciers in the Rockies coalesced into the Cordilleran ice sheet, a relatively thin ice sheet covered the Arctic Archipelago, and an ice sheet much smaller than the Laurentide developed over northwest Europe including the British Isles and Scandinavia. The vast quantity of water locked up in these ice sheets caused sea level to drop by 113 to 135 m (370 to 443 ft), exposing portions

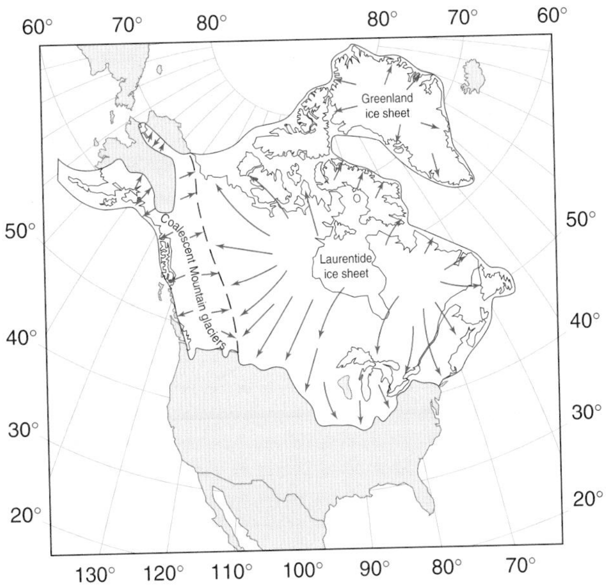

FIGURE 12.6
Extent of glacial ice cover over North America about 18,000 to 20,000 years ago, the time of the last glacial maximum. In places, the ice was as thick as 3 km (1.9 mi).

of the continental shelf, including a land bridge linking Siberia and North America.

The Laurentide and European ice sheets thinned and retreated, and may even have disappeared entirely, during relatively mild interglacial climatic episodes, which typically lasted about 10,000 years. Throughout these interglacials, glacial ice cover persisted over most of Antarctica and Greenland as it still does today.

During glacial climatic episodes, temperatures were lower than they are today but the magnitude of cooling was not the same everywhere. A variety of geologic evidence indicates that during the Pleistocene, temperature fluctuations between major glacial and interglacial climatic episodes typically amounted to as much as 5 Celsius degrees (9 Fahrenheit degrees) in the tropics, 6 to 8 Celsius degrees (11 to 14 Fahrenheit degrees) at middle latitudes, and 10 Celsius degrees (18 Fahrenheit degrees) or more at high latitudes. An increase in the magnitude of a climatic change with increasing latitude is known as **polar amplification**, indicating that polar areas are subject to greater changes in climate.

Oxygen isotope analysis of deep-sea sediment cores shows numerous fluctuations between major glacial and interglacial climatic episodes over the past 600,000 years (Figure 12.7A). Shifting focus to the past 160,000 years, resolution of the climate record improves. The temperature curve in Figure 12.7B is based on analysis of an ice core extracted from the Antarctic ice sheet at

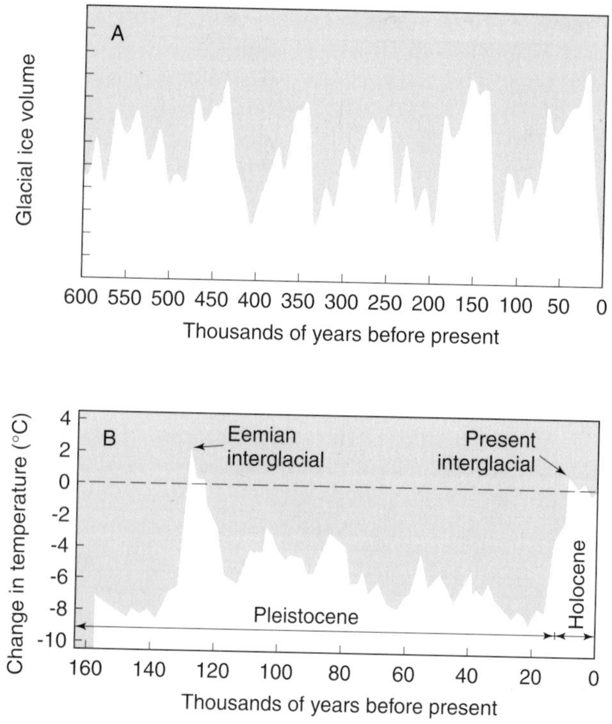

FIGURE 12.7
Reconstructed records of (A) the variation in global glacial ice volume over the past 600,000 years based on analysis of the oxygen isotope ratio of shells in deep-sea sediment cores, and (B) temperature variation over the past 160,000 years derived from oxygen isotope analysis of an ice core extracted from the Antarctic ice sheet at Vostok and expressed as a departure in Celsius degrees from the 1900 global mean temperature. [Compiled by R.S. Bradley and J.A. Eddy from J. Jousel et al., *Nature* 329(1987):403-408 and reported in *EarthQuest* 5, No. 1 (1991).]

Vostok. A relatively mild interglacial episode, referred to as the *Eemian*, began about 127,000 years ago and persisted for about 7000 years. In some localities, temperatures may have been 1 to 2 Celsius degrees (2 to 4 Fahrenheit degrees) higher than during the warmest portion of the present interglacial. The Eemian interglacial was followed by numerous fluctuations between glacial and interglacial episodes. The last major glacial climatic episode began about 27,000 years ago and reached its peak about 18,000 to 20,000 years ago when glacial ice cover over North America was about as extensive as it had ever been (Figure 12.6).

The global mean temperature 20,000 to 18,000 years ago was about 4 to 6 Celsius degrees (7 to 11 Fahrenheit degrees) lower than at present. A general warming trend (and rise in sea level from melting ice) followed the last glacial maximum, punctuated by relatively brief returns to glacial climatic conditions. A notable example is the relatively cold interval from about 11,000 to 10,000 years ago known as the *Younger Dryas* (named for the polar wildflower, *Dryas octopetala*, that reappeared in portions of Europe at the time). The return of colder conditions triggered short-lived re-advances of remnant ice sheets in North America, Scotland, and Scandinavia. Glacial ice finally withdrew from the Great Lakes region about 10,500 years ago ushering in the present interglacial, the **Holocene**.

Events surrounding the *Younger Dryas* point to the role of the ocean's large-scale thermohaline circulation in influencing climate. About 11,000 years ago, glacial ice lobes disrupted drainage patterns, diverting melt water from the Mississippi River to the St. Lawrence River and into the North Atlantic. With the input of fresh water, North Atlantic surface waters became less saline and eventually were not sufficiently dense to sink and form deep water. This weakened the meridional overturning circulation (MOC), which, in turn, diminished the flow of warm water into the central and northern North Atlantic causing a marked cooling of surrounding lands. This was the beginning of the *Younger Dryas*.

At the onset of the Holocene, a warming trend caused glaciers to retreat sufficiently that fresh melt water was diverted from the St. Lawrence back to the Mississippi River. North Atlantic surface waters became saltier and denser, deep water formed, and the meridional overturning circulation strengthened. The climate of the lands surrounding the central and northern North Atlantic warmed, signaling the end of the *Younger Dryas*.

Although the Laurentide ice sheet was melting and would disappear almost entirely by about 5500 years ago, the Holocene has been an epoch of spatially and temporally variable temperature and precipitation. Cores extracted from the Greenland ice sheet and sediment cores taken from the bottom of the North Atlantic reveal that the overall post-glacial warming trend was interrupted by abrupt millennial-scale fluctuations in climate. Post-glacial warming during the Holocene gave way to a cold episode about 8200 years ago; significant cooling also occurred from about 4000 to 2400 years ago. On the other hand, at times during the mid-Holocene (classically known as the *Hypsithermal*) mean annual global temperature was perhaps 1 Celsius degree (2 Fahrenheit degrees) higher than it was in 1900, the warmest in more than 110,000 years, that is, since the Eemian interglacial. A pollen-based climate reconstruction indicates that 6000 years ago, over most of Europe, July mean temperatures were about 2 Celsius degrees (3.6 Fahrenheit degrees) higher than now.

A generalized temperature curve for the past 1000 years, derived mostly from proxy climate data sources,

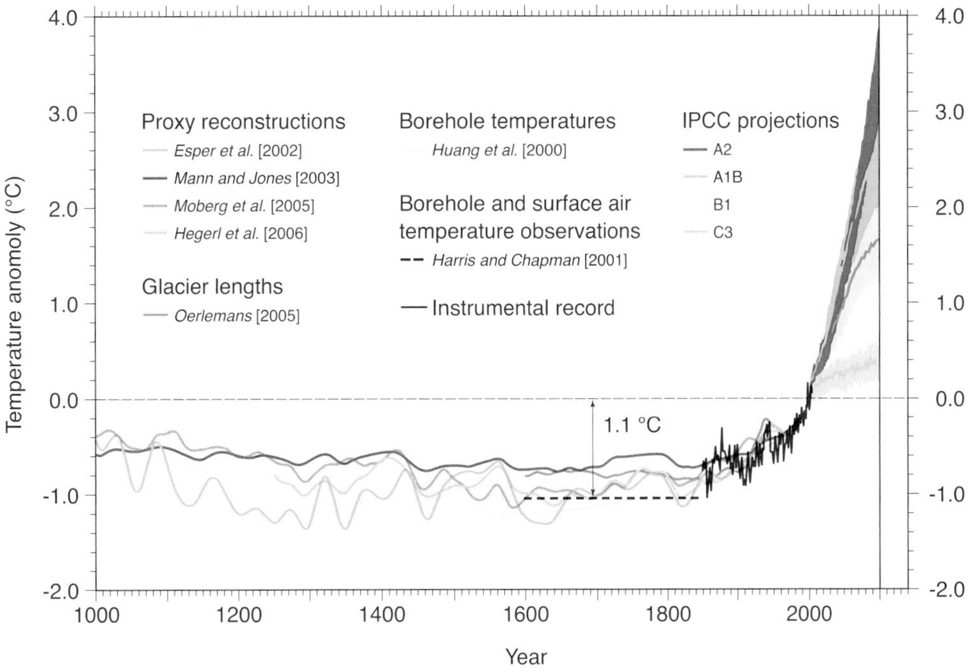

FIGURE 12.8
Records of Northern Hemisphere temperature variation during the last 1000 years. These reconstructions use multiple climate proxy records and an instrumental temperature record (labeled on the image). All series have been smoothed with a Gaussian-weighted filter to remove fluctuations on time scales less than 30 years. All temperatures represent departures from the 1961 to 1990 mean (in Celsius degrees). Various temperature projections through the year 2100 are also displayed. Data from *Climate Change 2007: The Physical Science Basis*, Working Group I Contribution to the Fourth Assessment Report of the IPCC, Cambridge University Press, U.K., 2007]

is shown as Figure 12.8. Notable features of this record are the **Medieval Warm Period** (which has begun about 950 CE) extending to about CE 1250, and the cooling that followed, from about CE 1400 to 1900, a period now known as the **Little Ice Age**. The Medieval Warm Period and the Little Ice Age were not episodes of sustained warming and cooling, respectively. On the contrary, sediment and glacial ice core records plus historical documents point to significant decadal-scale fluctuations in temperature (and precipitation). During much of the Medieval Warm Period, global temperature averaged about 0.5 Celsius degree (0.9 Fahrenheit degree) higher than in 1900. The first Norse settlements were established along the southern coast of Greenland and vineyards thrived in the British Isles. Independent lines of evidence confirm that the Little Ice Age was a relatively cool period in many regions with mean annual global temperatures perhaps 0.5 Celsius degree (0.9 Fahrenheit degree) lower than it was in 1900. Sea-ice cover expanded, mountain glaciers advanced, growing seasons shortened, and erratic harvests caused much hardship for many people, including the end of the Norse settlements in Greenland. Note also the dramatic warming trend underway during the 20th century and first part of the 21st century.

INSTRUMENT-BASED TEMPERATURE TRENDS

Invention of weather instruments, establishment of weather observing networks around the world, plus standardized methods of observation and record-keeping made the climate record much more detailed and dependable. The most reliable temperature records date from the late 1800s with the birth of national weather services including those of the U.S. and Canada, along with the predecessor to today's World Meteorological Organization (WMO). Examination of temperature trends over the past 120 years or so is instructive as to the short-term variability of climate.

Plotted in Figure 12.9 is the 1880 to 2010 instrument-derived record of variations in (1) global (land plus sea-surface) mean annual temperature, (2) global mean sea-surface temperature, and (3) global mean land surface temperature. In all three cases, the temperature is expressed as a departure (in Celsius degrees and Fahrenheit degrees) from the long-term (1901-2000) average. For reasons given in Chapter 3, these temperature series, assembled by NOAA's *National Climatic Data Center (NCDC)*, indicate greater year-to-year variability over land than ocean. The trend in global mean temperature is initially downward and then back upward from 1880 un-

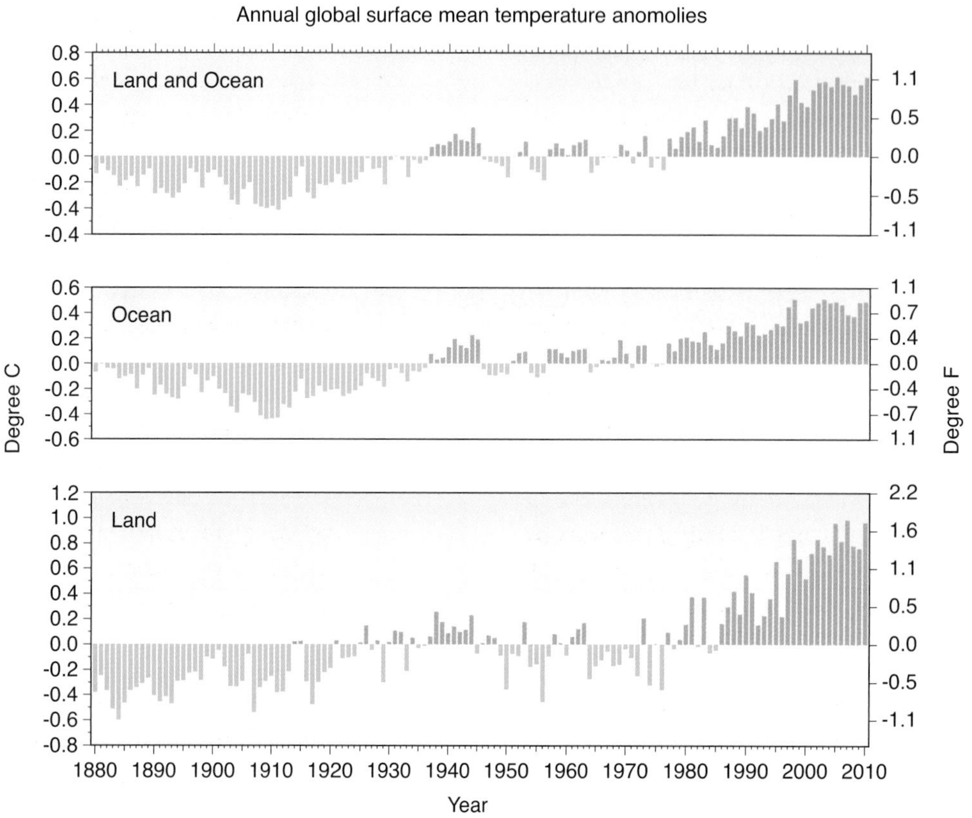

FIGURE 12.9
Instrument-derived trends in mean annual global (land plus ocean), sea-surface, and land temperatures; expressed as departures in Celsius degrees and Fahrenheit degrees from the 125-year period average. [NOAA, National Climatic Data Center]

til about 1940, downward or steady from 1940 to about 1970, and upward again through the 1990s and early 2000s. The overall temperature fluctuation is mostly ±0.5 Celsius degree (±0.9 Fahrenheit degree) about the century average. Note that this temperature record for the globe as a whole is not representative (in direction or magnitude) of all locations worldwide; that is, the trend was amplified or reversed, or both, in specific regions of the world.

In 2011, NOAA reported that 2010 was the 34th consecutive year in which global temperatures were above the 1901-2000 average. Combined global land and ocean annual surface temperatures in 2010 tied with 2005 as the warmest, at 0.62 ± 0.07 Celsius degrees above the 20th century average, since the instrumental record began in 1880. As shown in Figure 12.9, most of the warming of the 20th century took place from 1956 to 2000 (0.65 ± 0.15 Celsius degrees). NASA data shows that climate has warmed by about 0.2 Celisus degrees from the late 1970s to 2010. Climate reconstructions (mostly in the Northern Hemisphere) indicate that the 20th century was the warmest in 1000 years. For example, the ice core record extracted from a glacier at 7163 m (23,500 ft) in the Himalayan Mountains revealed that the 1990s and the last half of the 20th century were the warmest of any equivalent period in a millennium.

Some critics question the integrity of hemispheric or global temperature records. They cite as potential sources of error: (1) improved sophistication and reliability of weather instruments through the period of record, (2) changes in location and exposure of instruments at most long-term weather stations, (3) huge gaps in monitoring networks, especially over the ocean, and (4) the warming influence of urbanization (*urban heat islands*). However, even when careful statistical analysis of all these potential sources of error is taken into account, a clear warming trend over the past century is evident.

A general consensus in the scientific community holds that a global-scale warming trend has prevailed since the end of the Little Ice Age. The simplest scientific explanation for the observed warming trend is the steady build-up of carbon dioxide in the atmosphere (primarily because of combustion of fossil fuels) and the consequent enhancement of the natural greenhouse effect (Chapter 5).

Continuation of the upward trend in the concentration of atmospheric CO_2 (and other greenhouse gases) inevitably will lead to global warming throughout this century and perhaps beyond. We have more to say on this topic later in this chapter.

Lessons of the Climate Past

What does the climate record tell us about the behavior of climate through time? The following lessons of the climatic record are useful in assessing prospects for the climate future and the possible impacts of climate change.

- *Climate is inherently variable over a broad spectrum of time scales ranging from years to decades, to centuries, to millennia.* Variability is an endemic characteristic of climate. The question for the future is not whether the climate will change, but how climate will change, by how much and how fast.
- *Variations in climate are geographically non-uniform in both sign (warming or cooling) and magnitude.* Some areas may experience warming while other areas experience cooling over the same period. Global- and hemispheric-scale trends in climate are not necessarily duplicated at particular locations although the tendency is for the magnitude of temperature change to amplify with increasing latitude (*polar amplification*). Partially for this reason, a large-scale change in climate is not likely to have the same impact everywhere.
- *Climate change may consist of a long-term trend in various elements of climate (e.g., mean temperature or average precipitation) and/or a change in the frequency of extreme weather events (e.g., drought, excessive cold).* Recall from Chapter 5 that climate encompasses mean values plus extremes. A trend toward warmer or cooler, wetter or drier conditions may or may not be accompanied by a change in frequency of weather extremes. On the other hand, a climatic regime featuring relatively little change in mean temperature or precipitation through time could be accompanied by an increase or decrease in frequency of weather extremes.
- *Climate change tends to be more abrupt than gradual.* In the context of the climate record, "abrupt" is a relative term. If the time of transition between climatic episodes is much shorter than the duration of the individual episodes, then the transition is considered to be relatively abrupt. Analysis of cores extracted from the Greenland ice sheet indicate that millennial-scale cold and warm climatic episodes were punctuated by abrupt change over periods as brief as a single decade. The abrupt-change nature of climate would test the resilience of society to respond effectively to climate change.
- *Only a few cyclical variations can be discerned from the long-term climate record.* Regular cycles include diurnal and seasonal changes in incoming solar radiation (the forcing) and temperature (the response). This means simply that days are usually warmer than nights and summers are warmer than winters. Quasi-regular variations in climate include El Niño (occurring about every 3 to 7 years), Holocene millennial-scale fluctuations identified in glacial ice cores, and the major glacial-interglacial climate shifts of the Pleistocene Ice Age unlocked from analysis of deep-sea sediment cores and operating over tens of thousands to hundreds of thousands of years. Study of climate cycles adds to our understanding of the present climate and anthropogenic influences on climate.
- *Climate change impacts society.* History recounts numerous instances when climate change significantly impacted society. Although modern societies may be more capable of dealing with climate change than early peoples, a rapid and significant change in climate would seriously impact all sectors of modern society.

Factors Contributing to Climate Change

There is no single, simple explanation for climate variability and climate change. The complex spectrum of climate variability and change is a response to the interactions of many forcing agents and mechanisms operating both within and external to **Earth's planetary system** (atmosphere-land-ocean system) (Figure 12.10).

One way to organize our thinking on the many possible causes of a change in climate is to match a cause (forcing agent or mechanism) with a specific climate fluctuation (response) based on similar time periods (Figure 12.11). With plate tectonics, for example,

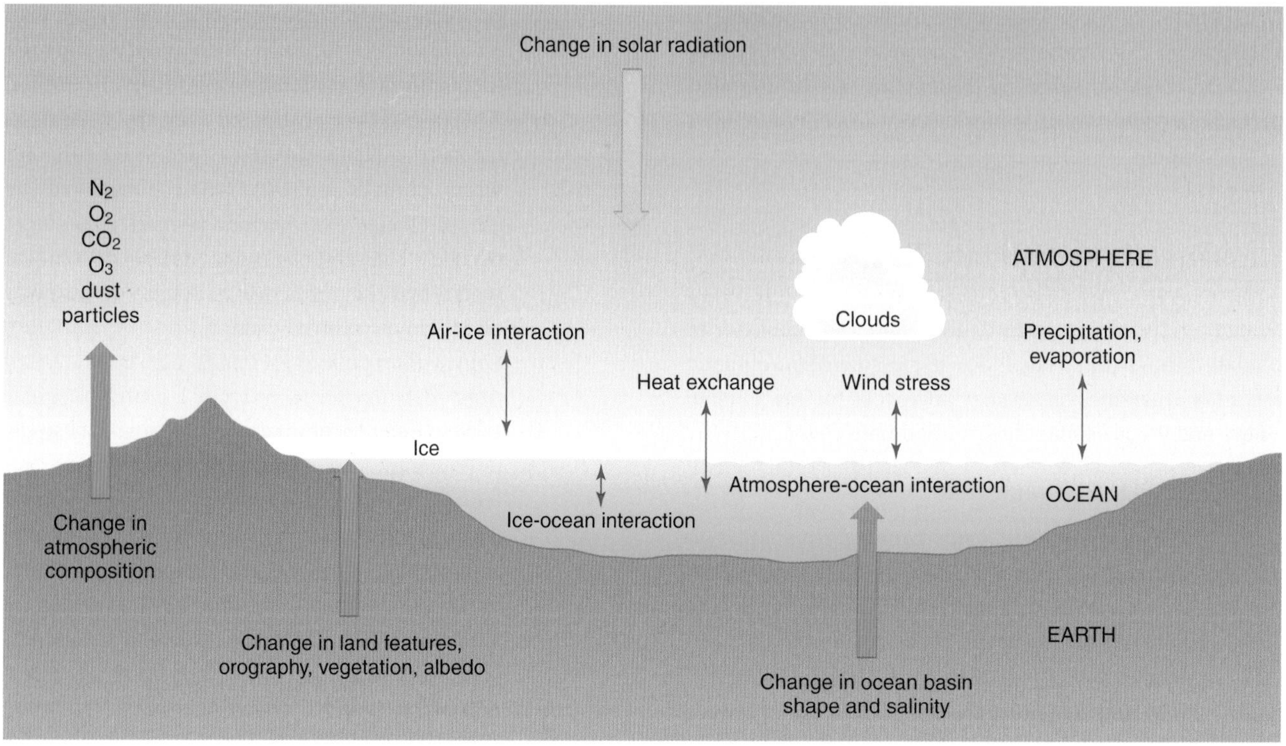

FIGURE 12.10
The complex spectrum of climate variability and climate change is a response to the interactions of many forcing mechanisms that operate both internal and external to the Earth-atmosphere-ocean system. [Modified from W.L. Gates and Y. Mintz, *Understanding Climatic Change: A Program for Action*, 1975, National Academy of Sciences, National Academy Press, Washington, DC]

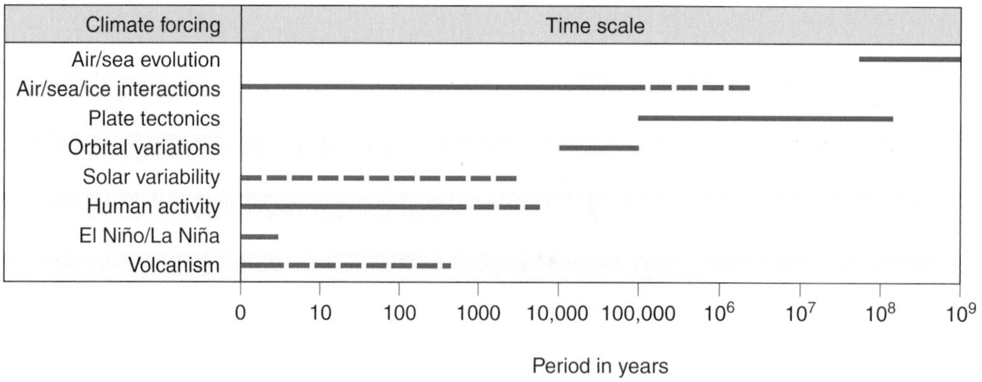

FIGURE 12.11
The potential causes of climate change operate over a broad spectrum of time scales. Dashed lines indicate uncertainty.

atmospheric and oceanic circulation patterns change in response to continental drift, the opening and closing of ocean basins, and mountain building. Hence, plate tectonics explains long-term climate changes that operate over hundreds of millions of years. As well, systematic changes in Earth's orbit about the Sun affect the latitudinal and seasonal distribution of incoming solar radiation, accounting for climate shifts of 10,000 to 100,000 years. Variations in the number of sunspots and energy from the Sun may be associated with climate fluctuations of decades to centuries. Explosive volcanic eruptions, El Niño, or La Niña may account for inter-annual climate variability. But matching a forcing mechanism with a climate response based on a similar periodicity is no guarantee of a real physical relationship.

Another way to think about the possible causes of climate change is through the global energy budget. As noted in Chapter 5, **global radiative equilibrium** means that energy entering Earth's planetary system (absorbed solar radiation) equals the energy leaving the system (infrared radiation emitted to space). Any change in energy input, thus affecting energy output, would shift Earth's planetary system to a new equilibrium and alter the planet's climate. In Chapter 11, we described the influence of El Niño and La Niña on inter-annual climate variability. In this section, we summarize the many other factors that may contribute to climate change over a broad range of time scales.

CLIMATE AND SOLAR VARIABILITY

Fluctuations in the Sun's energy output, sunspots, or regular variations in Earth's orbital parameters are external factors that can alter Earth's climate. Satellite measurements since the 1980s confirmed that the Sun's total energy output at all wavelengths (total *solar irradiance*) is not constant. Numerical global climate models predict that a 1% change in the Sun's energy output could significantly alter the mean temperature of the Earth's planetary system. Such a change is far greater than the variability measured via instruments on space platforms.

Changes in solar energy output are related to the number of sunspots on the *photosphere*, the surface of the Sun. A **sunspot** is an irregularly shaped dark blotch on the face of the Sun, typically thousands of kilometers across, which develops where an intense magnetic field suppresses the flow of gases transporting heat energy from the Sun's interior. A sunspot appears dark because its temperature is about 400 to 1800 Celsius degrees (720 to 3240 Fahrenheit degrees) lower than the temperature of the surrounding photosphere. Typically a sunspot lasts only a few days but the rate of sunspot generation is such that the number of sunspots varies systematically (Figure 12.12). The time between successive sunspot maxima or minima averages about 11 years. Recently, the sunspot number reached a minimum in 1996, a maximum in 2001, and a minimum in late 2008. The next solar maximum is expected in 2013.

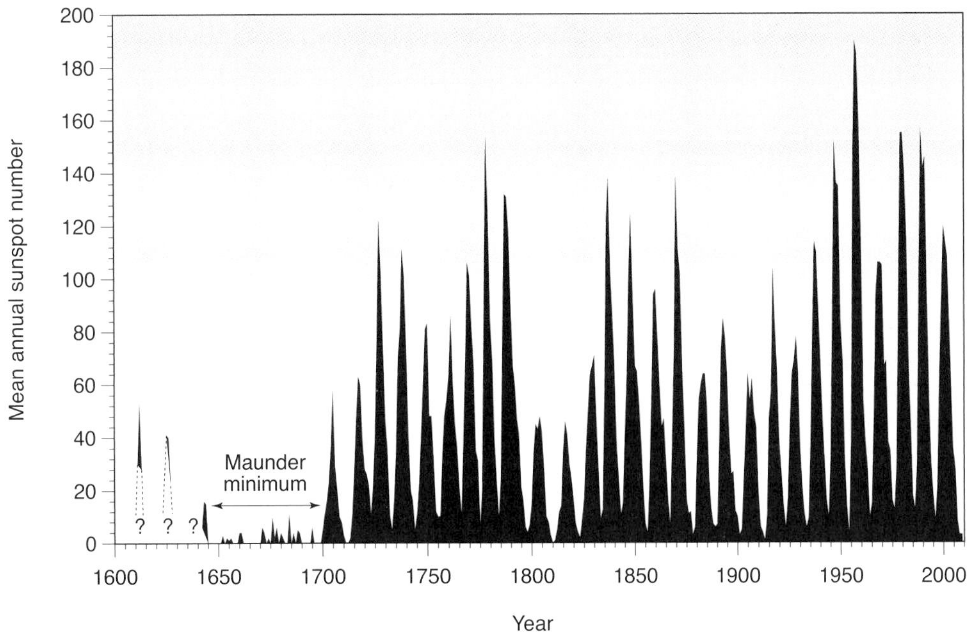

FIGURE 12.12
Variation in mean annual sunspot number since the early 17th century. [National Geophysical Data Center, Solar-Terrestrial Physics Division]

Satellite monitoring reveals that the Sun's energy output varies directly with sunspot number so when the Sun is slightly brighter, it has more sunspots whereas when it is slightly dimmer, there are fewer sunspots. A brighter Sun is associated with more sunspots because of a concurrent increase in bright areas, known as *faculae*, which appear near sunspots on the photosphere. When faculae dominate sunspots, the Sun brightens. More sunspots may contribute to a warmer global climate and fewer sunspots may translate into a colder global climate. However, the total solar energy output varies less than 0.1% through an 11-year sunspot cycle, much less than that expected to cause climate change. Also, most of the variation is in the ultraviolet (UV) portion of the solar spectrum.

In the late 1880s, the German astronomer F.W. Gustav Spörer (1822-1895) and the English solar astronomer E. Walter Maunder (1851-1928) reported that sunspot activity was greatly diminished during the 70-year period from 1645 to 1715, which coincides with a cold episode in Europe. This period of greatly reduced solar activity is now referred to as the **Maunder minimum**. Until American astronomer John A. Eddy (1931-2009) reinvestigated the phenomenon, the scientific community mostly ignored the Maunder minimum. Eddy pointed out that the Maunder minimum plus a prior 90-year period of reduced sunspot number, called the *Spörer minimum* (1460 to 1550), occurred about the same time as relatively cold phases of the Little Ice Age in Western Europe. Furthermore, the Medieval Warm Period was an interval of heightened sunspot activity between about 1100 and 1250.

The connections between sunspots and climate fluctuations are not universally accepted in the scientific community. Cold episodes occurred in Europe just prior to and after the Maunder minimum, and these relatively cool conditions did not persist through the Maunder minimum, nor were they global events. Furthermore, the variation in the solar radiation output during an 11-year sunspot cycle may be too weak to significantly alter Earth's climate. Still, mechanisms operating within the atmosphere could amplify changes in total solar output, which could make the slight brightening and dimming of the Sun an important player in Earth's climate system.

CLIMATE AND EARTH'S ORBIT

In 1842, the French mathematician Joseph Alphonse Adhémar (1797-1862) proposed that regular variations in the shape (eccentricity) of Earth's orbit about the Sun could explain the climate changes responsible for the glacial fluctuations of the Ice Age.

Over twenty years later, James Croll (1821-1890), a self-educated Scottish scientist, attributed the climate changes responsible for the Ice Age to the fluctuations of incoming solar radiation that accompanied regular changes in Earth's orbital parameters (the eccentricity of the orbit, tilt of the rotational axis, and precession of the axis). On this basis, Croll worked out an ice age chronology, arguing that less incoming solar radiation in winter favored greater accumulations of snow and the more extensive snow cover would chill the atmosphere (positive feedback), eventually culminating in glaciation. Croll predicted multiple glaciations, an idea that was later confirmed through field work by the American geologist Thomas C. Chamberlain (1843-1928). However, by the close of the 19th century, Croll's theory fell into disfavor when European geologists discovered a serious discrepancy between his astronomically-based ice age chronology and field evidence.

In the second decade of the 20th century, the Serbian astronomer Milutin Milankovitch (1879-1958) revived Croll's idea, convinced that reduced solar radiation during summer, not winter as Croll had proposed, at northern latitudes was key to initiating glaciation. For more than 25 years, he calculated the latitudinal and seasonal variations in solar radiation striking Earth's surface, which arise from the long-term regular changes in Earth's three orbital parameters. In 1938, he published radiation curves for latitudes ranging from 55 to 65 degrees N, and for the past 600,000 years. While support for Milankovitch's astronomical theory of the Ice Age was considerable in the 1930s and 1940s, especially in Europe, by the mid-1950s most geologists rejected it. It was not until two decades later that independent corroborative evidence from deep-sea sediment cores firmly established the Milankovitch cycles as the pacemaker of major climatic fluctuations of the Pleistocene Epoch.

Milankovitch cycles are regular variations in the eccentricity of Earth's orbit about the Sun, and the tilt and precession of its rotational axis (Figure 12.13). These changes in the Earth-Sun geometry are caused by gravitational influences exerted on Earth by other large planets, the Moon, and the Sun. Combined, Milankovitch cycles drive climate fluctuations operating over tens of thousands to hundreds of thousands of years.

The eccentricity (shape) of Earth's elliptical orbit about the Sun shifts from relatively high (elongated oval) to low (nearly circular) during an irregular cycle every 90,000 to 100,000 years. Variation in orbital eccentricity alters the distance between Earth and Sun at *perihelion*, when Earth is closest to the Sun, and *aphelion*, when Earth is farthest from the Sun. When Earth's orbit over a year is highly elliptical, the Earth intercepts less solar radiation than when the orbit is less elliptical.

The tilt of Earth's spin axis shifts from 22.1 degrees to 24.5 degrees and back over a period of 41,000 years. As the axial tilt increases, winters become colder and summers become warmer in both hemispheres. (Presently, the tilt is 23.5 degrees.) Earth's axial tilt has been decreasing for about 10,000 years and will continue to do so for another 10,000 years.

Over a period of 23,000 years, Earth's spin axis circles like the wobble of a spinning top. This precession cycle changes the dates of perihelion and aphelion, increasing the summer-to-winter seasonal temperature contrast in one hemisphere while decreasing it in the other. At present, perihelion is in early January and aphelion is in early July. In about 10,000 years, those dates will be reversed (perihelion in July and aphelion in January) and the seasonal contrast will be greater in the Northern Hemisphere, with colder winters and warmer summers, but less in the Southern Hemisphere with milder winters and cooler summers. Combined, changes in the eccentricity and the precession cycle significantly influence the length of the individual astronomical seasons.

The greatest significance of the Milankovitch cycles is not the amount of solar energy the Earth's planetary system receives, but where and when it arrives, creating long-term variations at different latitudes throughout the seasons. However, the amplitude of glacial/interglacial climate cycles is not explained by only the relatively small variations in solar radiation associated with the Milankovitch cycles. There are feedback processes within Earth's climate system amplifying the orbital variations. A major player in this regard is the greenhouse gas carbon dioxide. Glacial ice cores from Antarctica, spanning the past 650,000 years, reveal that the concentration of atmospheric CO_2 varies inversely with changes in the volume of glacial ice (Figure 12.4). So the concentration is higher (260 to 280 ppmv) during interglacial episodes, and lower (200 ppmv) during glacial episodes. During interglacial episodes, rising levels of CO_2 adds to the warming caused by orbital variations whereas during glacial episodes, falling levels of CO_2 contributes to the cooling caused by orbital variations.

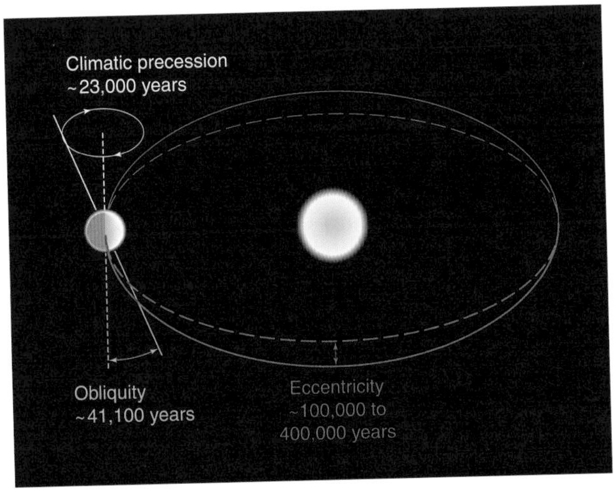

FIGURE 12.13
Milankovitch cycles likely explain the large-scale fluctuations of Earth's glacial ice cover during the Pleistocene. Note that diagrams greatly exaggerate changes in Earth-Sun geometry.

Based on the three orbital cycles, Milankovitch developed a numerical model that calculated incoming solar radiation, and the corresponding surface temperature, by latitude for the 600,000 years prior to 1800. He proposed that glacial climatic episodes began when Earth-Sun geometry favored an extended period of increased solar radiation in winter and decreased solar radiation in summer at 65 degrees N. More intense winter radiation in eastern Canada translated into somewhat higher temperatures, with higher humidity, and therefore more snowfall. If coupled with weaker solar radiation in the summer, some of that winter's snow cover, especially north of 65 degrees N, would survive the summer. A succession of many such cool summers would favor formation of a glacier which, once formed, would reflect away incoming solar radiation, creating a positive feedback loop. This process, according to Milankovitch, was the origin of the Laurentide ice sheet. At other times, Earth-Sun geometry favored enhanced solar radiation in summer at 65 degrees N, triggering higher temperatures, an interglacial climatic episode, and shrinkage of the Laurentide ice sheet.

Meanwhile, Willard F. Libby (1908-1980) of the University of Chicago had developed the radiocarbon method of age-determination for the remains of long dead organisms. By 1951, when geologists applied the new dating technique to organic materials associated with the last major glacial advance, they discovered the astronomical chronology did not match the radiocarbon chronology. (The range of radiocarbon dating was limited to 40,000 years at the time. Now, the range extends to about 50,000 years) However, the radiocarbon chronology was derived mostly from terrestrial sources and some geoscientists believed a more reliable chronology could be obtained from deep-sea sediment cores.

Because the rate of sedimentation in the open ocean is extremely slow, ranging from 1 to 3 mm per century, a relatively short undisturbed core contains information on climatic fluctuations through much of the Pleistocene Epoch with a resolution of 10,000 years. Reconstructing past climate oscillations from analysis of deep-sea sediment cores, using new age-dating techniques (based on reversals in Earth's magnetic field) and oxygen isotope analysis, an international team of scientists analyzed two cores extracted from the bottom of the Indian Ocean that went back 450,000 years. In December 1976, they announced that the major changes in climate occurred at essentially the same frequency as variations in the eccentricity, tilt, and precession of Earth's orbit. They labeled the cycles the "pacemakers of the ice ages."

Today, Milankovitch's astronomical theory is the widely accepted explanation for the major climate fluctuations of the Pleistocene Epoch. Some climate scientists, however, have reservations over the importance of the eccentricity cycle which is much weaker than the other two orbital cycles. While positive feedback (such as albedo or CO_2) may have amplified the climatic response to the eccentricity cycle, it is also conceivable that the 100,000 year climate cycle is due to entirely different mechanisms. The proposed relationship to the eccentricity cycle may be coincidental.

Milankovitch cycles also do not account for the period of climate quiescence prior to the Pleistocene Epoch. While Milankovitch cycles have operated throughout Earth's history, continental-scale glaciations could not be initiated without the appropriate boundary conditions, the arrangement of ocean basins, continents, and mountain ranges, which were achieved with the onset of the Pleistocene Epoch.

CLIMATE AND PLATE TECTONICS

Plate tectonics have operated on the planet, and influenced climate, for 2.5 billion years. Recall from Chapters 1 and 2 that Earth's solid outer skin is divided into a dozen gigantic rigid plates (and many smaller ones) drifting very slowly, between 20 and 200 mm per year, over the face of the planet. As plates move, the continents they carry also move (changing latitude), and ocean basins open and close (altering ocean currents). Tectonic stresses build mountain ranges (changing elevation and atmospheric circulation), and trigger volcanoes and volcanic eruptions (altering the composition of the atmosphere). Compared to the span of human existence, plate movements are so slow, that we often consider topography and the geographical distribution of the ocean and continents as fixed controls of climate. Over the vast expanse of geologic time, however, plate tectonics has been a major player in large-scale climate change.

Changes in the location of continents, *continental drift*, altered the local and regional radiation budget and the response of air temperature. Continental drift explains such seemingly anomalous discoveries as glacial deposits in the Sahara Desert, fossilized tropical plants in Greenland, and coal in Antarctica. These finds reflect climate conditions millions to hundreds of millions of years ago when landmasses were situated at different latitudes. Furthermore, colliding plates cause mountain ranges to rise even as weathering processes erode them away. In a geological time scale, these constantly alter atmospheric circulation and chemistry, prevailing cloud patterns and precipitation.

Opening and closing of ocean basins change the course of heat-transporting ocean currents and alters the thermohaline circulation. About 100 million years ago, the *Tethys Sea* separated Africa from Eurasia. Central America was submerged and warm water currents flowed around the equator connecting the ocean basins. About 40 million years ago, diverging plates pulled what is now Antarctica away from Australia, and, about 10 million years later, the Drake Passage opened between Antarctica and the southern tip of South America. The Drake Passage permitted an ocean current to flow around Antarctica, which eventually became the Southern Ocean as the Antarctic continent situated itself over the South Pole. By blocking transport of heat from the tropics, the Southern Ocean and Antarctic circumpolar current, led to the formation of the Antarctic ice sheets around 17 million years ago.

About 20 million years ago, when Saudi Arabia moved northward against Asia, it sealed off the Tethys Sea and formed the Mediterranean Sea. About 3 million years ago, volcanic eruptions created a narrow isthmus of land, Central America, which blocked the equatorial currents that previously flowed from the Atlantic through the Caribbean and into the Pacific Ocean. For a discussion of how plate tectonics played a role in climate change and major changes in the Mediterranean Sea, refer to this chapter's second Essay.

CLIMATE AND VOLCANOES

The idea that volcanic eruptions influence climate has been around for more than two centuries. Benjamin Franklin (1706-1790) proposed that the severe European winter of 1783-1784 was caused by a volcanic eruption in southern Iceland during the previous summer. The

Laki fissure eruption, in which hot molten lava and gases flowed through fractures in the ground, began in June 1783 and lasted about 8 months. It emitted 14.7 cubic km of lava, along with sulfur dioxide, hydrochloric acid, and hydrofluoric acid. Eruptive materials reached to altitudes as great as 13 km (8 mi). The eruption killed an estimated 10,000 Icelanders, roughly 20% of the population. The principal cause of death was the contamination of food and drinking water by fluoride produced from the hydrofluoric acid that spread across the countryside during the eruption. The Laki eruption likely also contributed to a spike in mortality in Britain and the European continent from weather extremes and famine. (During 1783, other larger volcanic eruptions also occurred in eastern Asia, adding to the aerosol loading of the atmosphere.)

The Laki eruption was one of the two largest fissure eruptions in Iceland over the past 11 centuries. The Eldgjá, also a fissure eruption, lasted six years beginning in 934 CE and emitted twice as much sulfur oxides to the atmosphere as Laki. Evidence of both eruptions appears in glacial ice cores as sulfuric acid fallout and in the stunted growth rings of northern temperate trees.

Richard B. Stothers of NASA's Goddard Institute for Space Studies (GISS) reconstructed the impacts of the Eldgiá and Laki eruptions. In both cases, winds transported sulfurous aerosol clouds of volcanic origin eastward across northern Europe, dimming the Sun and producing red sunrises and sunsets. King Henry of Saxony noted a thick dry fog in 934 after Eldgiá erupted and Benjamin Franklin made a similar observation in France in 1783, when the aerosol veil from Laki persisted over the Northern Hemisphere for more than 5 months. Both eruptions were followed by several cold winters, poor harvests, and famine. Stothers estimated that adverse impacts from the two fissure eruptions persisted for 5 to 8 years after Eldgjá and 2 to 3 years following Laki.

Only explosive volcanic eruptions rich in sulfur dioxide are likely to impact global or hemispheric climate and then only for a few years, at most. The unusually cool summer of 1816 (the *year without a summer* in New England) occurred after the violent eruption of Tambora, an Indonesian volcano, in the spring of 1815. Several relatively cold years followed on the heels of the 1883 eruption of Krakatau, also an Indonesian volcano. Much earlier, the climatic impact of the violent eruption of the Peruvian volcano Huaynaputina in 1600 likely contributed to one of Russia's worst famines.

In explosive eruptions, volcanoes discharge ash particles and sulfur dioxide (SO_2) high into the *stratosphere* (the atmospheric layer above the troposphere). While the larger ash particles settle to Earth's surface in only a few days, the SO_2 within the stratosphere combines with water vapor to form tiny droplets of sulfuric acid (H_2SO_4) and sulfate particles, collectively called **sulfurous aerosols**. The small size of sulfurous aerosols (averaging about 0.1 micrometer in diameter) coupled with the absence of precipitation in the stratosphere, allows sulfurous aerosols to remain suspended in the stratosphere for many months to longer than a year before they cycle back to Earth's surface. Successive volcanic eruptions have produced a sulfurous aerosol veil in the stratosphere at altitudes of about 15 to 25 km (9 to 16 mi). Sulfur dioxide emissions from clusters of volcanic eruptions temporarily thicken the stratospheric aerosol veil and impact climate.

Research following the 1991 eruption of Mount Pinatubo provided new insights on the relationship between sulfurous aerosols of volcanic origin and large-scale climate fluctuations. On 15-16 June 1991, Mount Pinatubo, on Luzon Island in the Philippines, injected 20 megatons of sulfur dioxide into the stratosphere with a violent eruption (Figure 12.14), the most massive stratospheric volcanic aerosol cloud of the 20th century. By altering both incoming and outgoing radiation, sulfurous aerosols affected temperatures in the stratosphere and troposphere, altered atmospheric circulation, and impacted surface air temperatures around the globe.

Sulfurous aerosols absorb both incoming solar and outgoing infrared radiation, which warms the lower stratosphere, especially in the tropics. At the same time, sulfurous aerosols reflect more solar radiation to space. NASA scientists reported that in the months following the Mount Pinatubo eruption, sensors on Earth-orbiting satellites measured a 3.8% increase in solar radiation reflected to space. Additionally, in the presence of chlorine, sulfurous aerosols destroy ozone (O_3), allowing more solar UV radiation to reach Earth's surface. In the two years following the Pinatubo eruption, ozone levels at middle latitudes declined by 5% to 8%. With the stratospheric warming in low latitudes and ozone depletion at high latitudes, the circumpolar vortex strengthened. This change in large-scale circulation created a non-uniform change in surface temperature, as some places cooled while others warmed.

By influencing the flux of radiation and attendant changes in global circulation patterns, the Pinatubo eruption was likely responsible for the cool summer of 1992, with surface temperatures up to 2.0 Celsius degrees (3.6 Fahrenheit degrees) lower than the long-term average over continental areas of the Northern Hemisphere.

FIGURE 12.14
The June 1991 explosive eruption of Mount Pinatubo (on Luzon Island in the Philippines) was rich in sulfur dioxide. The resulting sulfurous aerosol veil in the stratosphere was responsible for cooling at Earth's surface, interrupting the post-1970s global warming trend for a few years. [Courtesy of U.S. Geological Survey]

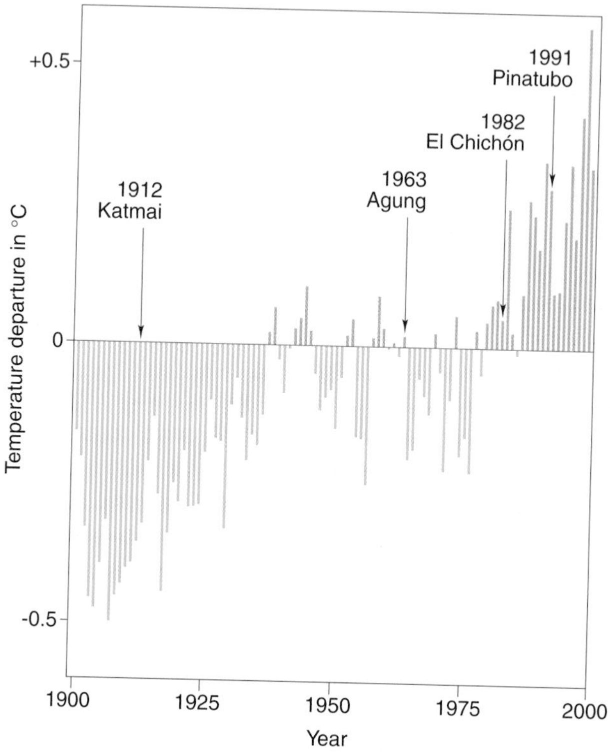

FIGURE 12.15
Large-scale cooling often followed massive volcanic eruptions that emitted sulfur dioxide into the stratosphere.

The Pinatubo eruption also generated the temperature anomalies of the winters of 1991-92 and 1992-93, with higher than average temperatures over most of North America, Europe, and Siberia and lower than average temperatures over Alaska, Greenland, the Middle East, and China.

A violent sulfur-rich volcanic eruption is unlikely to lower the mean hemispheric or global surface temperature by much more than 1.0 Celsius degree (1.8 Fahrenheit degrees), although the magnitude of local and regional temperature change may be greater (Figure 12.15). The 1963 eruption of Agung on the island of Bali lowered the mean temperature of the Northern Hemisphere an estimated 0.3 Celsius degree (0.5 Fahrenheit degree) for a year. The violent eruption of the Mexican volcano El Chichón in March-April 1982 may have produced hemispheric cooling of about 0.2 Celsius degrees (0.4 Fahrenheit degrees). Cooling associated with the Mount Pinatubo eruption temporarily interrupted the post-1970s global warming trend. From 1991 to 1992, the global mean annual temperature dropped 0.4 Celsius degrees (0.7 Fahrenheit degrees).

However, an Australian oceanographer, John A. Church, and his colleagues found that the cooling of the lower troposphere associated with the Mount Pinatubo eruption was great enough to affect sea level. Ocean water contracts when cooled, thus causing sea level to drop. By analyzing tide gauge, temperature, and salinity data from around the world and with the assistance from climate models, Church and colleagues estimated that mean sea level dropped a total of about 5 mm during the 18-month cool episode following the eruption. Then, as sulfurous aerosols settled out of the atmosphere and air temperatures returned to pre-eruption levels, SST increased and sea level rose at a gradual rate of 0.5 mm per year. By comparison, measurements by satellite sensors between 1993 and 2000 indicate that sea level rose 3.2 mm per year.

CLIMATE AND EARTH'S SURFACE PROPERTIES

Earth's surface, which is mostly ocean water, is the principal absorber of solar radiation within the planetary system (Chapter 5). Any change in the physical properties of Earth's water or land surfaces, or in the relative distribution of ocean, land, and ice, may affect Earth's radiation budget and the climate.

Extensive snow cover has a chilling effect on the atmosphere, so any changes in mean regional snow cover contribute to climate variability and climate change. Fresh-fallen snow typically reflects 80% or more of incident solar radiation, substantially reducing the amount of solar heating and lowering the daily maximum air temperature. Snow is also an excellent emitter of infrared radiation, so heat is efficiently radiated to space, especially on nights when the sky is clear, lowering the daily minimum air temperature. Because of this radiational feedback, extensive snow cover tends to be self-sustaining. An unusually extensive winter snow cover favors the persistence of cold weather. On the other hand, less winter snow cover raises average air temperatures.

Whereas changes in regional snow cover might impact climate variability seasonally, changes in Earth's sea ice or glacial ice cover are likely to have long-lasting effects on climate. Worldwide, sea ice (formed from the freezing of seawater) covers an average area of 25 million square km (9.6 million square mi), about the area of the North American continent. Terrestrial ice sheets, ice caps, and mountain glaciers cover a total area of about 15 million square km (5.8 million square mi), roughly 10% of the land area of the planet. Both the ocean or snow-free land strongly absorb incident solar radiation while ice, especially snow-covered ice, reflects a great deal of incident solar radiation. Any change in glacial or sea-ice cover affects climate. As discussed later in this chapter, the Arctic climate is particularly sensitive to change.

Changes in ocean circulation and sea-surface temperatures (SST) also contribute to large-scale climate change. As described in detail in Chapter 11, changes in SST patterns accompanying El Niño and La Niña greatly influence inter-annual climate variability. Ocean circulation includes warm and cold surface currents and the deep-ocean thermohaline circulation that transports heat energy throughout the world. Regular changes in the strength of this circulation may explain millennial-scale (1400- to 1500-year) climate cycles over the past 10,000 years. A strong thermohaline circulation brings a relatively mild climate to Western Europe whereas weakening of the thermohaline circulation triggers cooling. Such climate shifts can be abrupt, underway in a decade or less.

CLIMATE AND HUMAN ACTIVITY

In 2007, the **Intergovernmental Panel on Climate Change (IPCC)** concluded that global warming since the mid-20th century, with an estimated probability greater than 90%, was *very likely* caused primarily by human activities. In a report issued six years earlier, the IPCC described the human role in global warming as *likely*, with an estimated probability higher than 66%. The IPCC, formed in 1988 by the World Meteorological Organization (WMO) and the United Nations Environmental Programme, is charged with evaluating the state of climate science as the basis for policy action.

Many human activities affect climate over broad ranges of spatial and temporal scales. Humans modify the landscape (e.g., urbanization, clear-cutting of forests), altering radiation properties of Earth's surface. Cities are slightly warmer than the surrounding countryside (the *urban heat island effect*). Combustion of fossil fuels (i.e., coal, oil, and natural gas) alters concentrations of certain key gaseous and aerosol components of the atmosphere. Of these human impacts on Earth's climate system, most likely the last has the greatest influence for climate globally.

Many scientists, public policy makers, and the general public are concerned about the steadily rising concentration of infrared-absorbing gases, especially carbon dioxide. Higher levels of these gases enhance the greenhouse effect, contributing to warming on a global scale. The *Synthesis Report of the IPCC Fourth Assessment Report*, issued in November 2007, concluded that: "Warming of the climate system is unequivocal as is now evident from observations of increases in global average air and ocean temperatures, widespread melting of snow and ice, and rising global average sea level." The Report goes on to state that: "Most of the increase in globally-averaged temperatures since the mid-20th century is *very likely* (greater than 90% probability) due to the observed increase in anthropogenic GHG (greenhouse gas) concentrations." This is the *Callendar effect* described in Chapter 5.

Systematic monitoring of atmospheric carbon dioxide levels began in 1957 at NOAA's Mauna Loa Observatory in Hawaii under the direction of Charles D. Keeling (1928-2005) of Scripps Institution of Oceanography. The observatory, situated on the northern slope of Earth's largest volcano 3397 m (11,140 ft) above sea level in the middle of the Pacific Ocean, is sufficiently distant from major sources of air pollution so that carbon dioxide levels are considered representative of at least the Northern Hemisphere. Also since 1957, atmospheric

CO_2 has been monitored at the South Pole station of the U.S. Antarctic Program with a record that closely parallels Mauna Loa. The Mauna Loa record (the *Keeling curve*) shows a sustained increase in average annual atmospheric CO_2 from about 316 ppmv (parts per million by volume) in 1959 to 390 ppmv in 2010 (Figure 12.16). The growth rate of atmospheric CO_2 from 1997 to 2007 was faster than over any other 10-year period since record keeping began at Mauna Loa. Superimposed on this upward trend is the annual CO_2 cycle caused by seasonal changes in photosynthesis, which reaches a minimum in October and a maximum in May, from the Northern Hemisphere vegetation.

The upward trend in atmospheric carbon dioxide was underway long before Keeling's monitoring. Humankind's contribution to the buildup of atmospheric CO_2 may have begun thousands of years ago with the clearing of forests and grasslands for agriculture and settlement. Land clearing contributes via burning of vegetation, decay of wood residue, and reduced photosynthetic removal of CO_2 from the atmosphere. By the middle of the 19th century, the beginning of the Industrial Revolution and growing dependency on coal burning triggered a much more rapid rise. The burning of coal and other fossil fuels produces, among other things, large quantities of CO_2. As we enter the second decade of the twenty-first century, the concentration of atmospheric CO_2 is about 35% higher than it was only two hundred years ago. Fossil fuel combustion accounts for roughly 75% of the increase while deforestation (and other land clearing) is likely responsible for the balance. With continued growth in fossil fuel combustion, the atmospheric carbon dioxide concentration could top 550 ppmv (double the pre-industrial level) by the end of the present century.

As noted in Chapters 3 and 9, the ocean is a major reservoir in the global carbon cycle and plays an important role in governing the amount of carbon dioxide in the atmosphere. While scientists were able to estimate the amount of CO_2 released by human activity since the beginning of the Industrial Revolution, more than half has been missing. The "missing" carbon is distributed between sinks in the ocean and on land. According to IPCC estimates in 2007, the ocean absorbs 56.2% of the carbon dioxide of anthropogenic origin (via photosynthesis, cold surface water absorption, and deep-water sequestration) while terrestrial biomass is a sink for 13.7%. At the air/sea interface, the rising concentration of atmospheric CO_2 drives the net flux of carbon dioxide into the water. Some is transported into the thermohaline circulation at depth where it may be sequestered for 1000 years. However, as the surface ocean continues warming with increasing global temperatures, uptake of carbon dioxide by ocean waters is likely to decline.

Along with carbon dioxide, the concentration of other infrared-absorbing gases associated with human activity is rising. Based on analysis of air bubbles trapped in glacial ice cores, the concentration of methane in the atmosphere is now greater than at any time in the past 400,000 years. Methane is a product of the decay of organic matter in the absence of oxygen (*anaerobic*

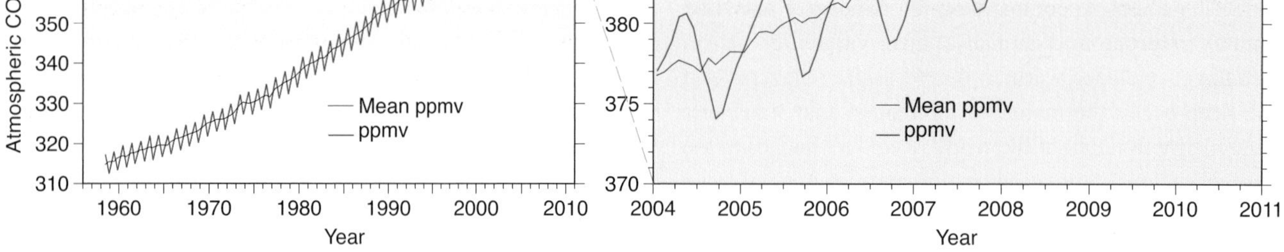

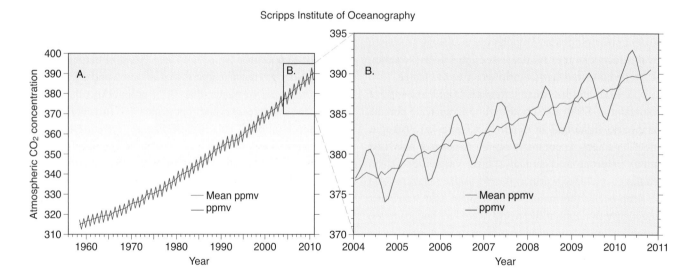

FIGURE 12.16
Trend in atmospheric carbon dioxide concentration since 1957, based on measurements made at the Mauna Loa Observatory in Hawaii. [Source of Mauna Loa data is C.D. Keeling et al., Scripps Institution of Oceanography, University of California, La Jolla, CA.]

decay) and its rise is linked to increasing rice cultivation, cattle, landfills, and/or termites. Industrial air pollution, biomass burning, and chemical fertilizers produce nitrous oxide (N_2O). Like methane, nitrous oxide absorbs infrared radiation very efficiently and at a different wavelength than CO_2, though both concentrations are extremely low (measured in parts per billion).

Aerosols are tiny (nanometer to micrometer in size) solid or liquid particles suspended in the atmosphere that vary in origin, size, shape, and chemical composition. Larger aerosols have short residence times in the atmosphere and settle out quickly, whereas smaller aerosols may remain suspended for many days to weeks. The smaller aerosols can also be transported thousands of kilometers by the wind, impacting climate on a planetary scale. Approximately 90% of anthropogenic aerosols are byproducts of fossil fuel burning in the Northern Hemisphere but they can cause either warming or cooling of the atmosphere.

Sulfur oxides emitted to the atmosphere from power plant smokestacks and boiler vent pipes combine with water vapor in the air to produce tiny droplets of sulfuric acid and sulfate particles, collectively called *sulfurous aerosols*. Unlike greenhouse gases, sulfurous aerosols cause cooling in the atmosphere. They increase the albedo directly by reflecting sunlight back into space and, indirectly, by acting as condensation nuclei that spur cloud development. Greater reflectivity cools the lower atmosphere.

However, sulfurous aerosols in the troposphere have only a short-term impact on climate. Approximately one-fifth of the carbon dioxide released now will still be in the atmosphere after 200 years whereas rain and snow wash sulfurous aerosols from the atmosphere in typically only a few days. Sulfuric acids are also the primary components of acid rain, another serious environmental issue.

Black carbon, which is soot from the incomplete combustion of fossil fuels, strongly absorbs solar radiation and causes warming. For all anthropogenic aerosols, the global net effect is cooling, although the degree to which remains in question.

The Climate Future

What does the climate future hold? Scientists attempt to answer this question by interpreting the output of global climate models, based on projections of future greenhouse gas and aerosol emissions, and studying past and present climates.

GLOBAL CLIMATE MODELS

A **global climate model** simulates Earth's climate system using equations to describe how its components interact and should not be confused with a *numerical weather prediction (NWP) model*, which is used in weather forecasting. NWP models use the current state of the atmosphere, based on near-simultaneous instrument measurements from around the globe, and apply physical laws to make predictions for the subsequent 12, 24, 36, 48 hours or longer. On the other hand, a climate model concentrates on the changes in boundary conditions, such as incoming solar radiation, greenhouse gas concentration, and Earth's surface properties. A climate model predicts how the climate would adjust to new conditions, focusing on broad areas of positive and negative temperature and precipitation *anomalies* (departures from long-term averages) as well as the mean location of important atmospheric circulation features, such as jet streams and principal storm tracks, over much longer time scales.

Global climate models can be used to predict the potential impacts from the rising levels of atmospheric carbon dioxide and other greenhouse gases. Using current boundary conditions, a global climate model can simulate the present climate, then, holding all other conditions constant, the concentration of CO_2 can be elevated and the model run to a new equilibrium state. By comparing the new climate state with the present climate, scientists can deduce the impact of greater CO_2 concentration on patterns of temperature and precipitation.

While largely accurate, global climate models still need considerable refinement. Today's models can not adequately simulate the role of small-scale weather systems such as thunderstorms, or accurately portray local and regional conditions. Currently, limited spatial resolution due to computational speed limits models but, with faster supercomputers, future climate models promise greater resolution.

FEEDBACK

Along with shifting boundary conditions, global climate models also must account for the feedback processes operating in Earth's climate system. In this context, **feedback** refers to a process where a change in one variable interacts with other variables, and alters the original variable. If that interaction enhances the original change, then the feedback is positive. If the change suppresses the original, the feedback is negative. Feedbacks in Earth's climate system can be significant and are thought to be responsible for more than half the global warming expected from human activities.

Consider examples of *positive feedback*. A warming trend in climate is likely to accelerate the rate of melting of snow and ice, producing more bare ground that will absorb more solar radiation, which further raises the air temperature and thus further accelerates melting snow and ice. Alternatively, a cooling trend prolongs snow and ice cover in spring and summer so more solar radiation is reflected to space, instead of absorbed, which causes additional cooling and prolongs snow and ice cover. In both cases, feedback is positive because the initial change in temperature is amplified.

Feedbacks among temperature, cloud cover, and radiation are still not well understood, because the net feedback of clouds is a complex uncertainty. Clouds cause both cooling (by reflecting sunlight to space) and warming (by absorbing and emitting to Earth's surface outgoing infrared radiation). Generally, with an increase in low cloud cover cooling prevails whereas with an increase in high cloud cover, warming prevails. It is thought that a warming trend will increase the rate of evaporation of water from Earth's surface, thereby increasing low cloud cover. A thicker and more extensive low-cloud cover reflects more solar radiation to space and inhibits a further rise in surface temperature. Hence, this dampening of the initial temperature change is an example of *negative feedback*.

Accounting for feedback in Earth's climate system is essential for realistic models of climate to predict the climate future. When the climate is altered, feedback will either amplify (positive feedback) or dampen (negative feedback) that change. While climate forcing agents and mechanisms (e.g., variations in solar energy output, regular fluctuations in Earth-Sun geometry) drive climate change, feedback controls the magnitude of climate change.

Overall, when positive feedback prevails, Earth's climate system is unstable while, when negative feedback predominates, the climate system is stable. In an unstable climate system, the climate regime shifts toward an extreme characterized by excessive cold that would eventually encase the planet in snow and ice ("snowball" or "ice ball" Earth) or toward the other extreme resulting in much higher temperatures, the product of a runaway greenhouse effect. Over the billions of years that constitute geologic time, Earth's climate system has been a mostly stable combination of negative feedbacks compensating for positive feedbacks.

SEARCH FOR CYCLES AND ANALOGS

Another approach to predicting future climate is empirical, seeking to identify factors that contributed to past fluctuations and to extrapolate their influence into the future. With instrument-based records, atmospheric scientists have reconstructed ancient climates in search of cycles and analogs of how regional climates respond to global-scale climate change. Unfortunately, few of the quasi-regular oscillations in the climate record have practical value in forecasting climate over only the next century.

Although climate records yield information on how climate behaves through time, the search for realistic analogs of future global warming has been less successful. Proposed analogs include relatively warm episodes of the mid-Holocene Epoch 12,000 years ago and the Eemian interglacial 127,000 years ago. The change during both of these intervals primarily affected seasonal temperatures and produced only a slight rise in global mean temperature. Furthermore, boundary conditions were drastically different. During the mid-Holocene, sea level was lower, glacial ice sheets were more extensive, and the seasonal and latitudinal receipt of solar radiation (due to different dates of perihelion and aphelion, when Earth is closest and farthest from the Sun) were significantly different. Also, although the level of atmospheric CO_2 trended upward during the mid-Holocene, the rate of increase was only 0.5 ppmv per century whereas now it is more than 60 ppmv per century.

Proposed pre-Pleistocene analogs, such as the *Paleocene-Eocene Thermal Maximum (PETM)*, disregard the lack of ice sheets, and topography and land-ocean distribution differences. Our knowledge of the PETM is based on analysis of deep-sea sediment cores. According to *IPCC Climate Change 2007: The Physical Science Basis Report*, the PETM occurred 56 million years ago, during the Cenozoic Era, near the transition from the Paleocene to Eocene epochs. Lasting only 170,000 years, the PETM was a geologically brief interval of widespread warming associated with a massive buildup of the greenhouse gases carbon dioxide and methane in the atmosphere. Over a period of 1000 to 10,000 years during the PETM, the global mean surface temperature rose 5 to 10 Celsius degrees (9 to 18 Fahrenheit degrees).

A more promising analog for the climate projected for the close of this century in the mid-Pliocene epoch, about 3 million years ago. Based on oxygen isotope analysis of deep-sea sediment cores, and other proxy climatic data, researchers determined that boundary conditions were not much different than today, with continents and ocean currents very similar while atmospheric CO_2 levels were only slightly elevated compared to today. An important difference is the more gradual build-up of atmospheric CO_2 compared to today. The global mean surface temperature was 2 to 3 Celsius degrees higher with much greater warming in polar regions (polar amplification). The prevailing temperature pattern across the tropical Pacific resembled a modern El Niño episode. In addition, mean sea level averaged 25 m (82 ft) higher than now.

ENHANCED GREENHOUSE EFFECT AND GLOBAL WARMING

In the long run, during the next 10,000 to 100,000 years, the Milankovitch Earth-Sun orbital cycles favor an eventual return to Ice Age conditions. Over the next few decades, if all other boundary conditions remain fixed, rising concentrations of atmospheric CO_2 and other greenhouse gases will cause global warming to persist well beyond this century. The magnitude of warming will depend on the continued emissions of greenhouse gases.

In 1979, President Jimmy Carter requested that the National Academy of Sciences (NAS) report on the potential impact of the increasing atmospheric concentration of carbon dioxide. Jule G. Charney (1917-1981) of the Massachusetts Institute of Technology (MIT) led the Academy investigation team that designed the now classic experiment in which numerical models of Earth's climate system doubled the atmospheric concentration of CO_2 while holding all other variables constant.

Addition of CO_2 makes the atmosphere more opaque for outgoing infrared radiation (heat), warming the lower atmosphere and cooling the upper atmosphere. Applying basic radiation laws, Charney found that doubling the atmospheric CO_2 concentration would reduce the net radiative flux (from Earth to space) at the tropopause by a global average of about 4 watts per square meter (W/m^2). How much warmer would Earth's surface become as a consequence of this enhanced greenhouse effect? According to the *Stefan-Boltzmann law*, the radiation emitted by an object is directly proportional to the fourth power of the object's absolute temperature. Following a doubling of atmospheric CO_2, Earth would have to radiate an additional 4 W/m^2 to space, brought about by a global warming of 1.2 Celsius degrees (or 0.3 Celsius degrees per W/m^2), to regain radiative equilibrium. **Climate sensitivity** is defined as the magnitude of increase in global mean surface temperature that will accompany a doubling of the concentration of greenhouse gases (in terms of equivalent CO_2) once the planet has settled into the new radiative equilibrium.

Charney's initial experiment accounted for the effect of a forcing agent (atmospheric CO_2) on global climate but not the influence of feedbacks. As noted above, forcing agents and mechanisms drive climate change while positive and negative feedbacks determine the magnitude. Hence, Charney's "no-feedback" experiment significantly underestimated the potential global warming. With the inclusion of feedbacks, the 1979 Academy study would give a global warming range from 2 to 3.5 Celsius degrees (3.6 to 6.3 Fahrenheit degrees). The most recent IPCC report (AR4) estimates that the magnitude of warming with feedbacks incorporated as 3 Celsius degrees (5.4 Fahrenheit degrees) with an uncertainty range of 2 to 4.5 Celsius degrees (3.6 to 8.1 Fahrenheit degrees). This greater sensitivity depends on the different feedbacks, the three primary ones involving clouds, sea ice, and water vapor, which are both positive and negative, and either amplify or diminish the greenhouse effect.

Once in the atmosphere, the concentration of a greenhouse gas depends on how the rates of emission of the gas into the atmosphere compare to the rates whereby physical, chemical, and biological processes remove it from the atmosphere. These interactions determine the **lifetime of a gas** in the atmosphere, which is the time it takes for the gas to be reduced to 37% of its original amount. Some greenhouse gases have a short lifetime (e,g., 12 years for methane), their concentration declining almost immediately if they are no longer emitted. Other greenhouse gases may linger in the atmosphere for hundreds of years or longer after emissions end. If emissions remain constant or increase with time, the atmospheric concentration also increases regardless of the lifetime of the gas.

For carbon dioxide, more than 50% of the gas emitted is currently removed from the atmosphere within a century, but about 20% of emitted carbon dioxide remains in the atmosphere for many millennia. (The lifetime of carbon dioxide is undefined.). The slow rate of removal of CO_2 from the atmosphere means that even if CO_2 were to stabilize at its current level, the atmospheric concentration of CO_2 would continue to increase. If emissions were substantially reduced, CO_2 would still enter the atmosphere faster than it is cycled out and the concentration would continue to increase, though at a slower rate. Only the unlikely complete elimination of anthropogenic CO_2 would reduce or stabilize the concentration of atmospheric CO_2 at current levels. The lengthy lifetime of CO_2 in the atmosphere implies that continued greenhouse warming is inevitable and we are already committed to such impacts as shrinking glaciers and rising sea level.

Another important consideration in predicting the climate future is the dominating role of the ocean on Earth's climate system. As discussed in Chapter 3, the voluminous ocean has a tremendous capacity for storing heat energy because water has a high specific heat so large quantities of heat energy are required to bring about relatively small changes in ocean temperature. Furthermore, the transport and mixing of heat energy and carbon dioxide throughout the deep ocean operate over thousands of years. Hence, a considerable length of time is required for Earth's climate system to achieve equilibrium under new environmental conditions.

Complicating predictions for the climate future, the global radiation budget is not currently in equilibrium. In 2005, scientists reported that Earth emits 0.85 W/m^2 less energy as infrared radiation to space than it receives from the Sun, creating an imbalance of 0.06% of the total incoming solar radiation at the top of the atmosphere (Chapter 5). This conclusion is supported by satellite sensors and instrumented buoys at sea, which show an increase in ocean temperature over the previous decade. The imbalance represents the delay in the response of Earth's climate (surface temperature) to a forcing (build up of greenhouse gases). This delay, in turn, is the consequence of the considerable thermal inertia in Earth's climate system, which is mostly due to the ocean. On this basis, global climate models predict that as Earth shifts toward a new state of global radiative equilibrium, an as yet unrealized additional warming of 0.6 Celsius degrees (1.1 Fahrenheit degrees) will occur even without a change in atmospheric composition.

As we have seen, since the Little Ice Age ended in the mid 19th century, the global mean temperature has generally trended upward. In recent years, that warming trend has accelerated—probably the consequence of the build up of carbon dioxide and other greenhouse gases in the atmosphere. (As noted in Chapters 3 and 4, more carbon dioxide in the atmosphere also may be leading to *ocean acidification.*) If the warming trend continues, how might this climate change impact the ocean? Higher sea level is one consequence that has worldwide ramifications. In addition, warmer conditions in the Arctic may significantly reduce the sea ice cover with important implications for Earth's climate system. Furthermore, global warming is likely to impact marine life.

Global Warming and Rising Sea Level

Climate changes responsible for the waxing and waning of Earth's glacial ice cover during the Pleistocene Epoch also caused sea level to fall and rise. Geological evidence, such as drowned beaches, submerged river valleys, and submarine canyons, attests to periods when sea level was much lower. Scientists estimate that during the last glacial maximum, about 20,000 to 18,000 years ago, mean sea level was 113 to 135 m (370 to 443 ft) lower than it is today. More than 90% of this drop in sea level was due to a change in the global water cycle brought about by a colder climate. As noted in Chapter 1, the total amount of water in the Earth system is essentially constant. During glacial climatic episodes, glaciers on land thickened and expanded, and the volume of water in the ocean basins decreased (i.e., sea level fell). Conversely, during interglacial climatic episodes, glaciers on land thinned and retreated, and the volume of water in the ocean basins increased (i.e., sea level rose). Furthermore, 7% or 8% of the drop in sea level during the last glacial maximum was due to lower ocean temperature, resulting in contraction of the ocean water and an increase in water density. Recall from Chapter 3 that seawater always contracts with falling temperature and expands with rising temperature.

Waxing and waning of land-based glaciers and ocean temperature fluctuations are two climatic factors that govern **eustasy**, the global variation in sea level brought about by changes in the volume of water occupying ocean basins. Non-climatic contributions to eustasy include tectonic processes, which alter the volume of ocean basins (Chapter 2), reduction in sediment supply to deltas due to dam construction (Chapter 4), and ground subsidence that accompanies extraction of oil and groundwater in coastal areas. Global warming will cause sea level to rise in response to shrinking land-based polar ice sheets and mountain glaciers, coupled with thermal expansion of seawater.

When glacial ice flows from land to sea, the less dense ice floats and sea level rises immediately. An analogous situation is shown in Figure 12.17. The container on the far left is half filled with water. When ice cubes are added to the water, simulating a glacier moving from land to sea, the water level rises nearly to the top of the container (center). But, as the floating ice cubes melt (container on the far right), the water level is unchanged, remaining near the top of the container. Similarly, when glacial ice leaves the land and enters the ocean, it floats, displacing a volume of seawater equal to its own weight and causing a rise in sea level. When the ice melts, the volume of melt water occupies the same volume of sea water that the ice originally displaced. Hence, the melting of ice shelves already floating in the ocean does not alter sea level.

How has *mean sea level (msl)* responded to the present global warming trend? For most of the 20th century, coastal tide gauges were the principal source of data on

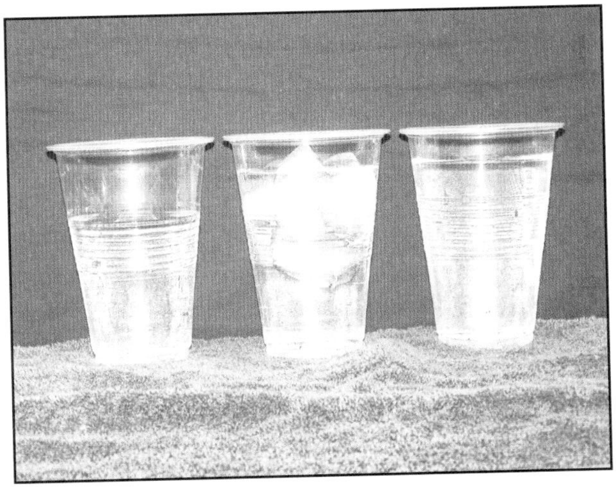

FIGURE 12.17
What happens to mean sea level when glacial ice enters the ocean? In this analogous situation, a glass initially is half filled with water (left). Ice cubes are added to the glass simulating a glacier moving from land to sea and the water level rises to near the top of the glass (center). As the floating ice melts (right), the water level remains unchanged. Hence, glacial ice has an immediate impact on mean sea level but melting of floating ice (e.g., an ice shelf) has no effect on mean sea level. [Modified after Robin E. Bell, "The Unquiet Ice," *Scientific American*, February 2008, page 61.]

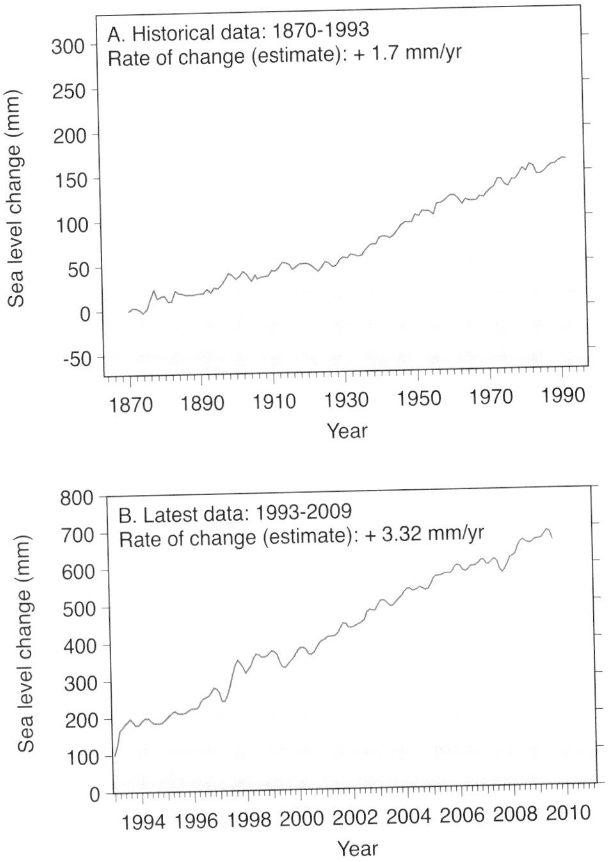

FIGURE 12.18
Upward trend in mean sea level as indicated by (A) coastal tide gauge records, 1870-1993, and (B) measurements by microwave altimeters flown onboard Earth-orbiting satellites from 1993 to 2009. [From CSIRO (A) and CLS/Cnes/Legos (B)]

sea level change. Care must be taken to exclude (or adjust for) tide gauge records that are influenced by geological processes (i.e., tectonic uplift, subsidence, or post-glacial rebound). Based on adjusted tide gauge records, msl from 1870 to 1993 is estimated to have risen at an average rate of 1.7 mm per year (Figure 12.18A). Beginning in 1993, microwave altimeters on Earth-orbiting satellites have provided more precise measurements of the rise in msl (Figure 12.18B). (Refer to the discussion of TOPEX/Poseidon and its successors in the second Essay of Chapter 7.) From 1993 to 2009, msl increased an estimated 3.3 ± 0.4 mm per year on average, showing that sea level rise is accelerating. In total, mean sea level is estimated to have risen about 180 mm (7.1 in.) during the 20th century.

How much of the recent rise in msl was due to melting of glacial ice and how much was due to thermal expansion of warming ocean waters is not known. Most mountain glaciers have been shrinking since the mid 20th century, portions of the Greenland and West Antarctic ice sheets recently have shown signs of accelerating mass loss, and portions of the ocean are warming.

SHRINKING ICE SHEETS

Amplification of the global warming trend at higher latitudes threatens the ice sheets of Antarctica and Greenland. About 90% of the planet's glacial ice blankets Antarctica and its melting could cause a considerable rise in sea level. But how likely is this to happen? Two ice sheets cover most of Antarctica, separated by the Transantarctic Mountains. The larger of the two, the East Antarctic ice sheet is situated on a continent about the size of Australia, accounts for two-thirds of Antarctic ice, averages a little more than 2 km (1.2 mi) thick, and is well above sea level. Complete melting of the East Antarctic ice sheet would raise msl about 60 m (197 ft), though such large scale melting is highly unlikely. The West Antarctic ice sheet sits on a series of islands and the floor of the Southern Ocean, with parts of the ice sheet more than 1.7 km (1 mi) below msl. Complete melting of the West Antarctic ice sheet would raise msl an estimated 5.8 m (19 ft). While geological evidence suggests that the East Antarctic ice sheet has been stable for the past 30 million years and remains stable today, the West Antarctic ice sheet has undergone episodes of rapid disintegration and may have completely melted at least once in the past 600,000 years.

An important source of data on the Antarctic and Greenland ice sheets is remote sensing by satellite. NASA's *Ice, Cloud and Land Elevation Satellite (ICESat)*, launched in January 2003, was the world's first laser-altimeter satellite (Figure 12.19). ICESat monitored variations in the thickness and mass of the Antarctic and Greenland ice sheets, as well as changes in polar sea ice thickness, until October 2009 when the satellite's laser altimeter system failed. NASA continues monitoring polar areas with *Operation Ice Bridge*, a 5-year airborne survey of ice sheets, ice shelves, and sea ice, until the scheduled launch of *ICESat-2* in 2015. Additional information on changes in the mass of the Antarctic and Greenland ice sheets is provided by the NASA/German Aerospace Center's *Gravity Recovery and Climate Experiment (GRACE)*. Launched in March 2002, GRACE consists of twin satellites flying in formation and designed to obtain detailed measurements of Earth's gravity field.

The behavior of ice streams is an important consideration in predicting how the Antarctic and Greenland ice sheets might affect sea level. An **ice stream** is a zone of relatively fast flowing ice within an ice sheet. Just as water on land flows via rivers and streams to the ocean, most of the ice (perhaps 90%) that flows from Antarctica and Greenland to the surrounding ocean occurs via ice streams and *outlet glaciers* (ice streams bounded by mountains). For perspective, the Rutford Ice Stream of West Antarctica is 150 km (93 mi) long, 25 km (16 mi) wide, up to 3 km (2 mi) thick, and moves at an average speed of 1.0 m per day.

Many of Antarctica's ice streams feed **ice shelves**, extensive areas of floating ice attached to land (at the *grounding line*) that fringe about 44% of the coast. Although the disintegration and melting of ice shelves does not affect sea level, they dam (or buttress) the flow of land-based glaciers that slowly feed the ice shelf. Ice shelves are vulnerable to warming seas and when an ice shelf disintegrates, the dam disappears and the glacier surges forward into the ocean.

Although an ice stream flows faster than the surrounding ice, the speed can be highly variable depending on ice temperature and the frictional resistance of the terrain over which the ice flows. Scientists have confirmed the existence of large lakes underlying portions of some ice streams. Lake Vostok, under Russia's Vostok Station in East Antarctica and discovered in 1973 by scientists of the Scott Polar Research Institute, is about the size of Lake Ontario and the largest known *subglacial lake*. Because the ice floats in the water, the lake surface offers essentially no frictional resistance to the moving

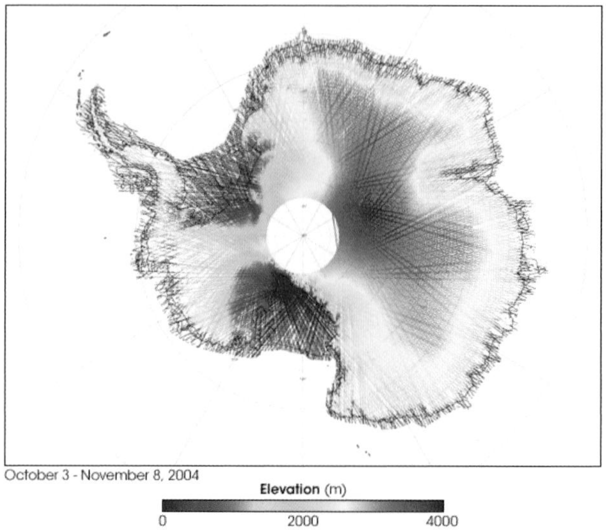

FIGURE 12.19
Topography of Antarctica from measurements by NASA's ICESat mission based on data compiled from 3 October – 8 November 2004. Red indicates the highest elevations (up to 4000 m above msl); yellow, green, and turquoise represent progressively lower elevations with green being 2000 m above msl. Dark blue signifies sea level. [NASA image courtesy of Christopher Shuman, ICESat Deputy Project Scientist, Goddard Space Flight Center]

ice. Additionally, the water warms the ice stream, reducing its viscosity as it accelerates over the subglacial lake. Along the way, however, the ice stream also incorporates sediment, which it deposits as a wedge when it enters the ocean. That wedge acts like a speed bump, slowing the advancing ice.

More than 30 years ago, recognition of the relative instability of the West Antarctic ice sheet prompted speculation that ice streams flowing from the interior to the Ross and Ronne ice shelves might bring about total collapse of the ice sheet in only a few centuries. (The Ross ice sheet, at 180 degrees, is south of New Zealand while the Ronnie ice sheet, at 60 degrees W, is south of the Falkland Islands.) Such a catastrophic event would greatly accelerate the rate of sea level rise. As noted above, complete disintegration of the West Antarctic ice sheet would raise sea level by about 5.8 m (19 ft).

In 2001, the discovery that new snowfall was keeping pace with the loss of ice from bergs breaking off the Ross ice shelf alleviated such concerns. In early 2002, scientists at the California Institute of Technology and the University of California-Santa Cruz reported that, based on satellite measurements of the flow of the Ross ice streams, the West Antarctic ice sheet appeared to be thickening. However, the region of the ice sheet that feeds the Thwaites and Pine Island glaciers is thinning. Pine

Island, stretching between 60 and 120 degrees W, south of both American continents, is the largest ice stream in West Antarctica and is losing ice mass more quickly than any other portion of Antarctica. By 2010, the flow velocity was nearly twice what it was in the mid-1970s. Mass loss is most likely due to the warm ocean currents under the ice shelf. The consequent thinning of the ice shelf and retreat of the grounding line accelerates the flow of the ice stream into the ocean.

Oceanographer Eric Rignot of NASA's Jet Propulsion Laboratory in Pasadena, CA, and colleagues used satellite imagery to measure the thickness of ice fringing the coast of Antarctica. These data along with estimates of ice stream discharge and snowfall enabled them to determine the annual loss of ice from Antarctica. They concluded that ice loss from Antarctica (mostly West Antarctica) was about 75% greater in 2006 than in 1996. Gravity data acquired by GRACE satellites indicate that Antarctica has been losing more than 100 cubic km (24 cubic mi) of ice each year since 2002, with the rate of ice loss accelerating in recent years (Figure 12.20). Nonetheless, the consensus of scientific opinion today is that the West Antarctic ice sheet is unlikely to catastrophically accelerate the current rise in msl.

However, melting (and the rate of rise in msl) could accelerate if the Antarctic ice sheets begin to warm in response to global climate change. Over most of Antarctica, the mean air temperature has fluctuated very little over the past 50 years. The Antarctic Peninsula is the only part of the continent that has shown significant warming, with summer mean temperatures rising more than 2 Celsius degrees (3.6 Fahrenheit degrees) over the past half century. This warming has been accompanied by higher SST around Antarctica and breakup of ice shelves along the Antarctic Peninsula coast (Figure 12.21). A major review of the situation in 2009 found that the majority of glaciers of the Antarctic Peninsula were retreating at accelerating rates.

Unlike the Antarctic ice sheets, which are polar (cold) glaciers, the Greenland ice sheet is a temperate glacier. While the Antarctic ice sheets are well below the pressure-melting point, the temperature of much of the ice on Greenland is near the pressure-melting point. (The melting point of ice decreases with increasing confining pressure or depth within the glacier.) Temperate glaciers more readily generate melt water and, with less frictional resistance, tend to flow faster than polar ice sheets.

NASA research results released in 2000 concluded that while the central interior of the Greenland ice sheet showed no sign of thinning, about 70% of the

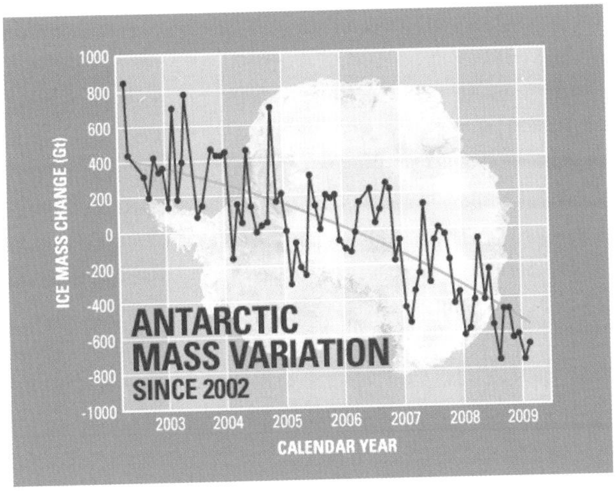

FIGURE 12.20
Based on measurements of changes in Earth's gravity field, Antarctica has been losing ice mass since at least 2002. [NASA Global Climate Change; GRACE mission]

FIGURE 12.21
The northern section of Sector B of the Larsen ice shelf on the eastern side of the Antarctic Peninsula shattered and separated from the continent over a 35-day period beginning on 31 January 2002, sending thousands of icebergs adrift in the Weddell Sea. About 3250 km^2 of shelf (larger area than the state of Rhode Island) about 220 m thick disintegrated. This image was taken by the Moderate Resolution Imaging Spectroradiometer (MODIS) onboard NASA's Terra satellite on 5 March 2002. [Courtesy of NASA]

margin was undergoing substantial thinning. Two research teams—one using a Global Positioning System (GPS) to monitor ice flow and the other relying on an airborne laser altimeter to measure ice thickness—reached the same conclusion based on observations made between 1993 and 1999. The maximum melting rate at the margin was about 1 m per year. An estimated 50 cubic km (12 cubic mi) of Greenland's ice melts each year, enough to raise mean sea level by 0.13 mm annually.

Scientists at the University of Colorado reported that during the summer of 2002, surface melting on the Greenland ice sheet encompassed an area of about 695,000 square km (265,000 square mi)—about 9% greater than observed during any summer since monitoring by satellite began 24 years previously. In addition, melting in the northern and northeastern portion of the ice sheet occurred at elevations as high as 2000 m (6550 ft) where normally temperatures are too low for any melting. The duration and extent of melting at elevations above 2000 m set a new record in 2007. Researchers at NASA's Goddard Space Flight Center monitored snow melt on the ice sheet surface using a satellite microwave sensor and computed a snow melt index by multiplying the area of snowmelt by the duration of melting. During 2007, the melting index above 2000-m elevation was about 2.5 times greater than the annual average from 1988 to 2006. Currently, it appears that winter snowfall is insufficient to offset the summer melt, so that overall the Greenland ice sheet is shrinking (Figure 12.22).

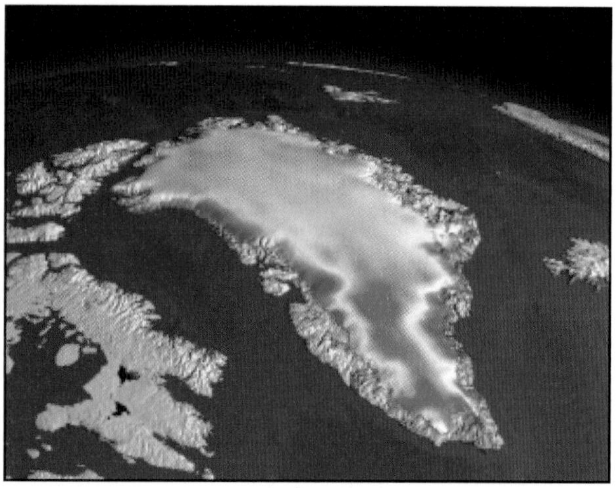

FIGURE 12.22
Measurements of the change in thickness of the Greenland ice sheet between 2003 and 2006 made by sensors onboard NASA's ICESat satellite. Pink and red regions indicate slight thickening while shades of blue and purple signify thinning. [NASA/Goddard Space Flight Center Scientific Visualization Studio]

In 2002, scientists discovered that meltwater lakes that form in summer on the surface of the Greenland ice sheet (known as *supraglacial lakes*) can force open a crevasse below, allowing water to drain catastrophically to the base of the glacier. Researchers from the Woods Hole Oceanographic Institution and the University of Washington documented this phenomenon in July 2006 (Figure 12.23). In only 90 minutes, a supraglacial lake covering 5.7 square km (2.2 square mi) and containing 11.6 billion gallons of water drained through a crevasse, plunging 980 m (3215 ft) to the base of the glacier. Reporting in the 14 November 2008 issue of *Science*, glaciologist Richard Alley of Pennsylvania State University and colleagues proposed that the lubricating effect of this water is not as important as the heat it delivers to the base of the ice sheet, which accelerates glacier flow. Should the global warming trend lengthen the melt season and create

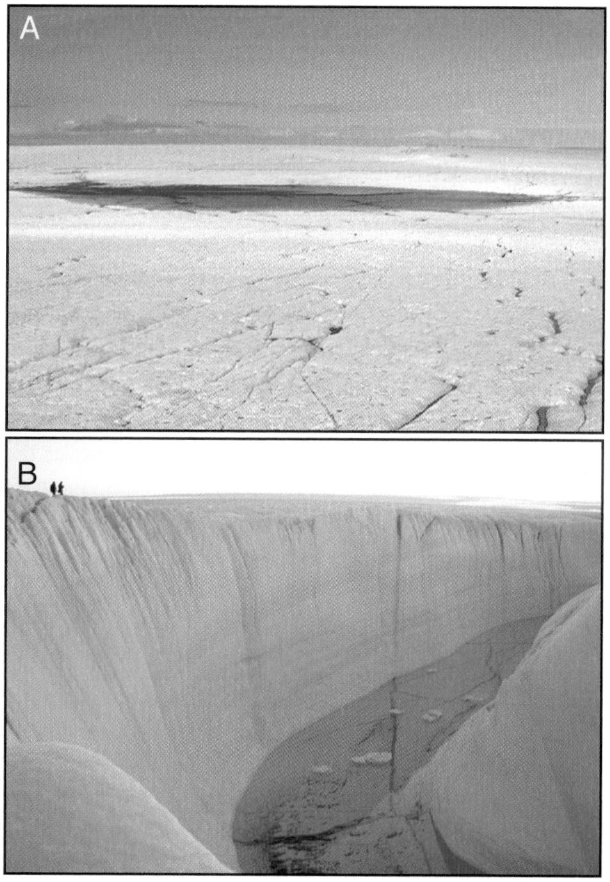

FIGURE 12.23
A meltwater lake on the surface of the Greenland ice sheet, one of thousands that form each summer (A). [Photo by Ian Joughin, University of Washington Polar Science Center] Over many years, a meltwater stream has excavated this large ice canyon in the Greenland ice sheet (B). [Photo by Sarah Das, Woods Hole Oceanographic Institution]

melt-water lakes even further inland on the Greenland ice sheet, the downward cascading melt-water could thaw areas where the glacier is now frozen to the ground. This is likely to significantly accelerate glacier flow and perhaps destabilize the Greenland ice sheet. Complete melting of the Greenland ice sheet would raise mean sea level by an estimated 7.3 m (24 ft).

Alley and colleagues also point to another factor that influences the flow of outlet glaciers to the sea. At the glacier-ocean interface, outlet glaciers move faster when they do not encounter obstacles, such as ice shelves or grounded ice. Ice shelves surround much of Antarctica but grounded ice, bedrock highs, and coastal landforms obstruct the flow of Greenland outlet glaciers. Melting of ice shelves and grounded ice would accelerate the flow of outlet glaciers.

MELTING MOUNTAIN GLACIERS

How are the much smaller glaciers that occupy mountain valleys influencing mean sea level? Roger G. Barry of the *National Snow and Ice Data Center* at the University of Colorado, Boulder, reports that the rate of melting of most of the world's mountain glaciers accelerated after the mid-1900s and especially since the mid-1970s (Figure 12.24). Some mountain glaciers have disappeared entirely. According to the U.S. Geological Survey (USGS), in 2010 only 25 glaciers larger than 10 hectares (25 acres) remain of the 150 glaciers that existed in Montana's Glacier National Park a century ago. Barry estimates that runoff from melting mountain glaciers contributes about 0.4 mm to the annual rise in msl.

Alpine glaciers are also shrinking at accelerating rates. According to Frank Paul at the University of Zurich-Irchel in Zurich, Switzerland, the combined area covered by almost 940 Alpine glaciers decreased about 18% from 1973 to 1999. Paul based his conclusion on analysis of satellite images, aerial photographs, and land surveys. Swiss glaciers are now shrinking at an annual rate six times quicker than they did from 1850 to 1973. At the present rate of shrinkage, Alpine glaciers at elevations below 2000 m (6500 ft) will likely disappear by the year 2070.

With the warming trend, the Rhône glacier in southern Switzerland could disappear by 2100. In 2008, Guillaume Jouvet of the Federal Polytechnic Institute in Lausanne, Switzerland, and colleagues used a glacier-climate model to simulate the retreat of the Rhône glacier since 1874. A 1.0 Celsius degree (1.8 Fahrenheit degree) rise in global mean temperature would reduce the glacier volume an estimated 35% by 2100. The glacier would disappear entirely with a temperature rise of 4 Celsius degrees (7.2 Fahrenheit degrees).

OCEAN WARMING

In spring 2000, NOAA scientists reported that the combined Atlantic, Pacific, and Indian Oceans warmed significantly between 1955 and 1995. The greatest warming, 0.31 Celsius degree (0.56 Fahrenheit degree), occurred in the upper 300 m (985 ft) of the ocean. Meanwhile, water in the upper 3000 m (9850 ft) warmed by an average 0.06 Celsius degree (0.11 Fahrenheit degree). In February 2002, researchers at Scripps Institution of

FIGURE 12.24
The present global warming trend has caused mountain glaciers to shrink with much of the meltwater eventually draining into the ocean. The dramatic recession of Grinnell Glacier in Montana's Glacier National Park is shown in photographs taken in 1938, 1981, 1998, and 2006. [Photos courtesy of Glacier National Park Archives and the USGS]

Oceanography reported that between the 1950s and 1980s temperatures at mid-depths (700-1100 m or 2300-3600 ft) in the Southern Ocean rose 0.17 Celsius degrees. Although these magnitudes of temperature change may sound trivial, recall that water has an unusually high specific heat so that even a very small change in temperature, over such vast volumes of water, involves a tremendous heat input into the ocean (Chapter 3). Sequestering of vast quantities of heat energy in the ocean may help explain why global warming during the 20th century was less than some climate models had predicted, because heating of the ocean partially offset warming in the lower atmosphere. This finding also underscores the importance of the ocean's moderating influence on global climate change.

The additional heat energy caused thermal expansion of seawater, which contributed 25% of the sea level rise since 1960 and increased to 50% from 1993 to 2003. Since 2003, however, warming of the upper ocean has decreased to 30%, thus reducing the contribution of thermal expansion to sea level rise. With the recent accelerated mass loss from the Greenland and West Antarctic ice sheets, their contribution to global sea level rise nearly doubled from 15% recorded between 1993 and 2003. For the period 1993-2009, land ice loss explains about 60% of sea level rise, compensating for the decline in contribution by thermal expansion of sea water. Hence, sea level rise continued at the same rate.

According to the 2007 *IPCC Fourth Assessment Report*, thermal expansion of seawater will be a greater contributor to mean sea level rise than melting land-based glaciers during the 21st century. Climate models predict that global warming will cause a rise in mean sea level in the range of 0.2 to 0.6 m (8 to 24 in.). Thermal expansion of ocean waters would account for more than 60% of the rise, the balance due to melting glaciers. At least over the short-term (the coming few decades), the behavior of ice sheets is the largest unknown regarding the magnitude of sea level rise.

Some scientists view the IPCC estimates as too conservative, underestimating the impact of melting glaciers and ice sheets. For example, W.T. Pfeffer and colleagues at the Institute of Arctic and Alpine Research at the University of Colorado, Boulder, include estimates of glacier melt in their projected msl rise of 0.8 to 2.0 m (2.6 to 6.7 ft) by 2100. Other climate models predict sea level rise to be in the range of 0.3 to 1.8 m (1 to 5.9 ft) by 2100.

IMPLICATIONS

Higher mean sea level would accelerate coastal erosion by wave action, allow seawater to inundate wetlands, estuaries and islands, and make low-lying coastal plains more vulnerable to storm surges. For people currently living on low-lying islands (e.g., Maldives, Tuvalu), abandonment may be their only option as sea level rises. Globally, a 50-cm (20-in.) rise in msl would double the number of people at risk from storm surges from about 45 million, at present, to over 90 million, not counting any additional population growth in the coastal zone.

Rising sea level would disrupt coastal ecosystems, ruin agricultural lands, and threaten historical, cultural, and recreational resources (Figure 12.25). In some coastal areas, higher sea level is likely to exacerbate saltwater intrusion into groundwater. (For more on this problem, refer to this chapter's third Essay.) A 1997 report by the U.S. Office of Science and Technology Policy predicted that a 50-cm (20-in.) rise in msl would result in substantial loss of coastal lands, especially along the U.S. Gulf and southern Atlantic coasts. Particularly vulnerable is south Florida's Everglades where one-third of the land is less than 30 cm (12 in.) above msl.

Along with the compaction of sediments, sea-level rise is likely to mean further loss of coastal Louisiana. In 2009, Harry Roberts of Louisiana State University in Baton Rouge estimated that about 25% of the wetlands of the Mississippi River delta had been claimed by the ocean over the past few centuries without being replenished. Sediments accumulate in reservoirs behind dams upriver, cutting off much of the supply of sediment to the delta. Deprived of new sediment, existing sediments compact, making the delta more vulnerable to erosion as sea level continues to rise. Roberts and Michael Blum of ExxonMobil reported on their use of numerical models to predict how these processes could impact the delta in the decades to come. Currently, at Grand Isle, LA, near the edge of the delta, the land is sinking up to 8 mm per year. With sea level rising about 3 mm per year (and likely to accelerate), by the end of the century as much as 10% of the present land area of Louisiana will be submerged.

While higher temperatures would mean higher sea level, the levels of North America's Great Lakes are likely to fall. Higher summer temperatures combined with less winter ice cover are likely to translate into greater evaporation. And less winter snowfall would reduce spring runoff. Depending on the model used, forecasts call for a drop in mean water level of Lake Michigan of up to 2 m (6.5 ft) by the year 2070. Residents of the western Great Lakes may have previewed the impact of global warming from the late 1990s into the early 2000s when the levels of Lakes Michigan and Huron dropped to near historic record lows. (See Figure 3 in the first Essay of Chapter 7.)

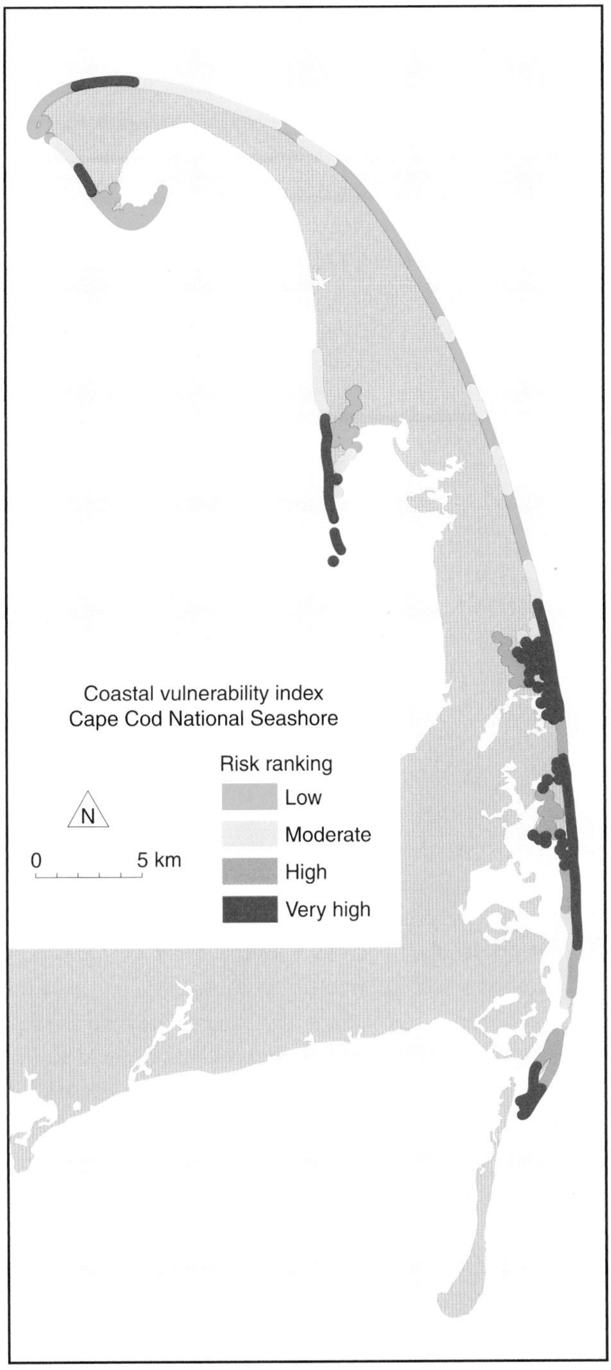

FIGURE 12.25
Map of the U.S. Geological Survey's *Coastal Vulnerability Index (CVI)* for Cape Cod National Seashore, MA, showing the vulnerability of the coast to changes in sea level. The CVI is based on tidal range, wave height, coastal slope, historic shoreline change rates, geomorphology, and historic rates of relative sea level change due to eustatic sea level rise and tectonic uplift or subsidence. [USGS Fact Sheet FS-095-02, September 2002]

Arctic Sea Ice Cover

A potentially major impact of global warming is the shrinking of Arctic sea ice, a trend that has been evident over the past three decades and, if it continues, the Arctic will be seasonally ice-free within 2 to 3 decades. While melting floating sea ice does not raise sea level, it can significantly alter the climate by triggering an **ice-albedo feedback** loop, further accelerating the melting of sea ice and amplify warming (Figure 12.26).

Recent research indicates that episodic sea ice formation at the margins of the Arctic Ocean dates back to about 47.5 million years ago, during the middle Eocene Epoch. In 2009, Catherine Stickley, a scientist with the Norwegian Polar Institute, and colleagues extracted sediments from a submarine ridge in the central Arctic about 250 km (155 mi) from the North Pole. They found fossil diatoms closely related to diatoms living in Arctic sea ice today suggesting that the episodic nature of Arctic ice has changed little, if at all.

Sea ice insulates overlying air from less frigid seawater and reflects much more incident solar radiation than ocean water. Additionally, the albedo of snow-covered sea ice is about 85% whereas ice-free Arctic Ocean water has an average albedo of about 40%. As sea ice cover decreases, the greater area of ice-

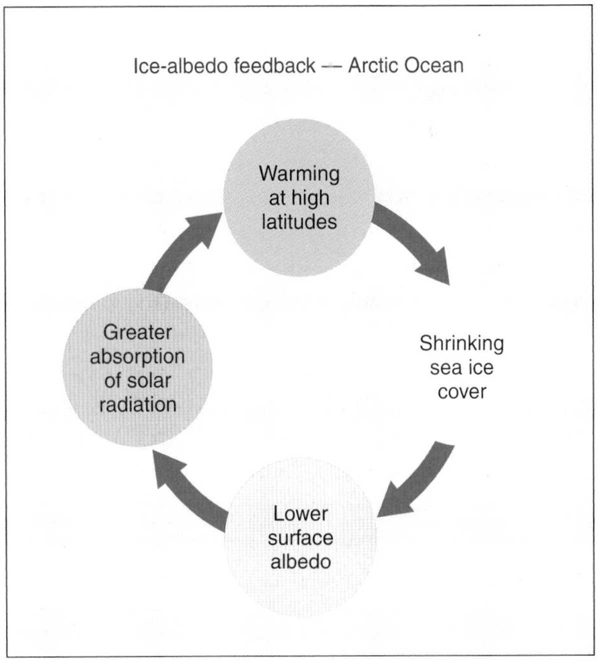

FIGURE 12.26
Positive ice-albedo feedback in the Arctic is likely to accelerate warming of surface waters and shrinkage of sea ice cover.

free ocean waters lowers the albedo and results in the absorption of more solar radiation, creating higher sea-surface temperatures, and more ice melt. In the autumn, the warmer water also slows the formation of ice. This positive feedback could more rapidly reduce Arctic sea ice cover and alter the flux of heat energy and moisture between the Arctic Ocean and atmosphere with serious ramifications for climate.

With less ice cover, there is likely to be more evaporation, increasing the humidity of the overlying air and more cloudiness. As pointed out earlier in this chapter, clouds cause both cooling (by reflecting sunlight to space) and warming (by absorbing and emitting outgoing infrared radiation to Earth's surface). During the long, dark polar winter, the presence of more clouds would have a net warming effect. In summer, the impact of greater cloud cover will depend on the altitude of the clouds. With an increase in low altitude cloud cover, cooling would prevail, whereas warming would likely accompany an increase in high cloud cover.

Since early in the 20th century, a variety of sources have provided information on the extent of Arctic sea ice cover, which varies seasonally and yearly because of variable atmospheric and oceanic conditions. Ship and aircraft observations indicate that the multi-year Arctic ice cover remained essentially constant in all seasons through the first half of the 20th century but, beginning in the 1950s, the summer minimum extent of ice began to shrink while the winter maximum remained nearly constant. By the mid-1970s, surveillance by satellites and submarines, as well as ice-core measurements, also documented a decline in the winter maximum extent of ice.

Based on satellite monitoring of the spectrum of microwave energy emitted by the ice, Norwegian researchers reported that the area covered by multi-year ice in the Arctic decreased 14% between 1978 and 1998. Comparing ice-thickness measurements made by U.S. Navy submarines, which used upward-looking acoustic sounders, from 1958 to 1976 with those made during the *Scientific Ice Expedition* program in 1993, 1996, and 1997, University of Washington scientists discovered that the average ice thickness had decreased from 3.1 m (10 ft) to 1.8 m (6 ft). Thinning at a rate of 15% per decade translates into a total volumetric ice loss of about 40% in three decades.

After 2000, the rate of reduction of Arctic sea ice cover accelerated. From analysis of satellite data, scientists from the National Snow and Ice Data Center (NSIDC) of the Cooperative Institute for Research in Environmental Sciences (CIRES) at the University of Colorado reported that the extent of Arctic sea ice in 2002 was the lowest in the satellite record, likely the lowest since the early 1950s, and perhaps the lowest in several centuries. In September 2002, sea ice covered about 5.3 million square km (2 million square mi) compared to the long-term average of about 6.3 million square km (2.4 million square mi). According to the IPCC, from 1979 through 2006, the extent of Arctic sea ice annual average declined by 2.7 ± 0.6 percent per decade while the end-of-summer (September) declined 7.4 ± 2.4 percent per decade. Between 1981 and 2000, average ice thickness decreased by about 1.13 m (3.7 ft) or 22%.

NSIDC scientists reported that in 2005 the extent of end-of-summer Arctic sea ice cover was the lowest since satellite monitoring of the polar region began in 1979 (Figure 12.27). That record was broken in 2007 when the summer minimum decreased to 4.2 million square km (1.6 million square mi), which was 38% below the long-term average and 23% below the previous record low. This sharp decline in Arctic sea ice cover was attributed to a combination of factors: the persistence of the long-term trend in ice thinning and shrinkage, unusually strong summer winds that conveyed large amounts of ice out of the Arctic basin and created a large area of open water and thin ice in the Arctic Ocean, less than the usual summer cloud cover that allowed more solar radiation to reach the ocean surface, and the ice-albedo feedback that accelerated the warming and melting during the summer.

In September 2008, the extent of Arctic sea ice cover was 4.67 million square km, second to 2007 as the

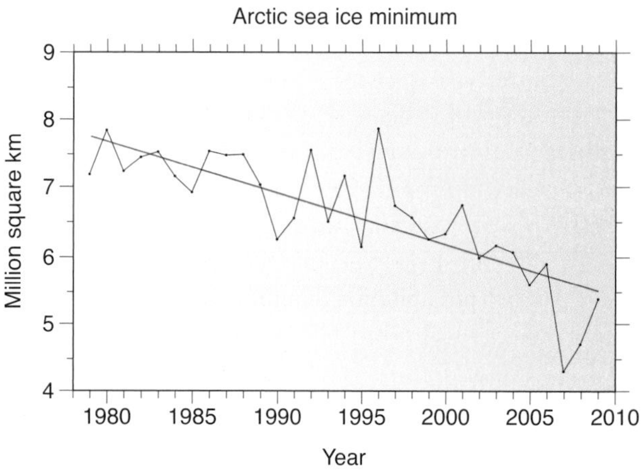

FIGURE 12.27
Arctic sea ice has been shrinking at least since the time satellite monitoring began in 1979, reaching a new end-of-season (September) record low in 2007. [Adapted from NOAA's National Snow and Ice Data Center, Boulder, CO]

lowest since the beginning of the satellite-based record. However, going into the melt season, much of the winter ice was thin (first-year ice) so that the volume of floating ice might have set a new record low. At the end of the 2009 Arctic summer, more ice cover remained than during the prior two record-setting years, but the sea ice extent was third lowest since 1979. Averaged over the month of September 2010, Arctic sea ice extent was only 4.90 million square km (1.89 million square mi) so 2010 replaced 2009 as the third lowest in the satellite-based record.

The shrinkage of Arctic sea ice may be a direct consequence of higher air temperatures or indirectly the result of changes in ocean circulation (i.e., greater input of warmer Atlantic water under the Arctic sea ice). Some scientists argue that these are natural variations in the Arctic climate associated with the *Arctic Oscillation (AO)* and that the ice-cover will return to normal after the AO shifts phase (Chapter 11). However the Arctic sea ice cover has declined to record low levels over the past decade while the phase of the AO has not been consistently positive. Instead, several other sea level air pressure patterns have, perhaps coincidentally, generated winds that exported sea ice out of the Arctic. It is possible the wind more readily transports thinner ice. Furthermore, research shows that less cold water flowing into the Arctic basin remains below the halocline, and has not obviously contributed to sea ice shrinkage.

Mean annual air temperatures in the Arctic region have climbed about 0.5 Celsius degree (0.9 Fahrenheit degree) per decade over the past thirty years. In 2007, mean annual temperatures in the Arctic were more than 3 Celsius degrees (5.4 Fahrenheit degrees) above the long-term (1951-1980) mean. Proxy climate data indicate that present temperatures (especially in winter and spring) may have reached their highest levels in four centuries. In response, mountain glaciers in Alaska are shrinking at historically unprecedented rates, *permafrost* (permanently frozen ground) is beginning to thaw, and freshwater runoff into the ocean has increased. More winter precipitation combined with accelerated flow of groundwater (due to the thawing of permafrost) is likely responsible for a 7% increase in the discharge of six major Eurasian rivers into the Arctic Ocean since the 1930s.

Input of more fresh water from rivers and melting glaciers would impact the ocean thermohaline circulation by reducing the salinity and thus the density of surface ocean waters. Recall from Chapter 6 that relatively dense, cold and salty water sinks at high latitudes of the Atlantic Ocean, so the ocean may become more stratified, with less dense surface waters and cold denser water below. As pointed out earlier in this chapter, changes in strength of the thermohaline circulation could alter the climate of Western Europe. Paradoxically, warming at high latitudes could lead to regional cooling due to a weaker thermohaline circulation regime.

Marine Life

How might climate change impact marine life? As described in Chapters 9 and 10, marine animals and plants are components of ecosystems. An *ecosystem* is a fundamental subdivision of the Earth system in which organisms depend for their survival on other organisms and the physical and chemical constituents of their surroundings. Materials and energy flow from one organism to another via food webs within and between ecosystems. Climate change could alter the physical and chemical conditions in the ocean, perhaps exceeding the tolerance limits of organisms. If these organisms are unable to avoid these stressful conditions (e.g., ocean acidification), they may perish.

Shrinkage of Arctic sea ice may have serious implications for organisms living on, within, or under the ice. To learn more about this extreme environment, in June/July 2005, an international team of scientists explored the Canada Basin (one of the deepest portions of the Arctic Ocean) using a variety of in situ and remote sensing techniques. The expedition, called "The Hidden Ocean,' was funded by NOAA's Office of Ocean Exploration (Figure 12.28).

Global warming could raise the sea-surface temperature sufficiently to threaten coral reefs. As noted in Chapter 10, coral polyps are sensitive to small changes in water temperature and prolonged periods of exposure to excessively warm waters can lead to *coral bleaching* and death. Because of the interdependency of organisms (e.g., for food, habitat), loss of one species can disrupt food webs and the entire ecosystem.

Organisms are particularly vulnerable to environmental change that affects a limiting factor. A *limiting factor* is an essential resource that is in lowest supply compared to what is required by the organism. For example, climate change may indirectly affect the supply of an essential nutrient. Concurrent shifts in atmospheric circulation may suppress upwelling in some locations or reduce the transport of micronutrients to portions of the open ocean. Furthermore, rising sea level is likely

FIGURE 12.28
Mette Nielson and Rolf Gradinger are part of an international team studying the effects of diminishing Arctic ice cover on the marine community living on, within, and below the ice. Here, they are extracting ice cores for later analysis. [Courtesy of The Hidden Ocean/Arctic 2005 Exploration, NOAA-OE]

to alter marine habitats especially in the coastal zone. In some cases—where the coastline has been developed for roads and buildings—flooding may eliminate marine habitats entirely causing the demise of organisms that are dependent on those habitats.

Consider, for example, the fate of salt marshes already threatened by development. Higher temperatures will increase the rate of evaporation from the soil surface thereby elevating the soil salinity. More saline soils will cut biological production and may exceed the tolerance limits of plants. Rising sea level threatens to drown salt marshes. If plants cannot shift inland to higher ground, their fate is sealed and the marsh may be lost. Loss of salt marshes, in turn, will make the coastline more vulnerable to flooding and erosion by storm waves. Furthermore the filtration function of salt marshes will be lost so that higher levels of pollution and nutrients will enter estuaries with runoff.

A key consideration in the potential impact of climate change on marine organisms is the rate of change. Marine ecosystems can more readily adjust to gradual rather than abrupt changes in the ocean environment brought on by global scale climate fluctuations. But as noted earlier in this chapter, the rate of climate change due to the enhanced greenhouse effect could be without precedent in the past 10,000 years.

Conclusions

Climate is inherently variable. In most places, the reliable instrument-based climate record extends back only 130 years or so. For information on prior climate regimes, scientists must rely on a variety of documentary, geological, and biological proxy climate indicators. Although the climate record loses detail, continuity, and reliability with increasing time before present, it is evident that climate changes over a broad spectrum of time scales.

The interaction of many factors is responsible for climate change. Although we can isolate specific climate forcings that are internal or external to the Earth-atmosphere-ocean system, our understanding of how these forcings interact is far from complete. This state of the art limits the ability of scientists to forecast the climate future.

Continued research on climate is needed. Physical laws govern climate change; that is, variations in climate are not arbitrary, random events. As scientists more fully comprehend the laws governing climatic change and especially the role played by the ocean in Earth's climate system, their ability to predict the climate future will improve. Meanwhile, trends in climate must be monitored closely, especially in view of the potential wide ranging impacts of climate change.

Basic Understandings

- For information on climate prior to the reliable instrument-based era, scientists rely on climate inferences drawn from historical documents, and geological/biological proxy climatic evidence such as bedrock, fossils, pollen, tree growth rings, glacial ice cores, and deep-sea sediment cores.
- Much of what is known about the climate fluctuations of the Pleistocene Ice Age is based on analysis of the shell and skeletal remains of microscopic marine organisms extracted from deep-sea sediment cores. Identification of the organism plus oxygen isotope analysis of remains enable scientists to distinguish between cold and mild climatic episodes of the past.
- Plate tectonics complicates reconstruction of climate over periods spanning hundreds of millions of years. In the context of geologic time, topography and the geographical distribution of continents and the ocean basins are variable controls of climate.
- By 40 million years ago, global climate began shifting toward cooler, drier, and more variable conditions. Geoscientists have identified tectonic forces and the building of the Colorado Plateau, Tibetan Plateau, and Himalayan Mountains as possible causes of this major change in climate.
- Cooling culminated in the Pleistocene Ice Age about 1.8 million years ago. During the Pleistocene, the climate shifted numerous times between glacial climatic episodes (favoring expansion of glaciers) and interglacial climatic episodes (favoring shrinkage of glaciers).
- Cooling during the Pleistocene was geographically non-uniform in magnitude with maximum cooling at high latitudes and minimum cooling in the tropics. This is an example of polar amplification.
- Notable post-glacial climatic episodes were the relatively warm mid-Holocene, the Medieval Warm Period, from about CE 950 to 1250, and the Little Ice Age, from about CE 1400 to 1900. In all cases, the temperature change was geographically non-uniform in magnitude.
- The instrument-based record of global mean temperature indicates a gradual warming trend since 1880, interrupted by cooling from about 1940 to 1970. Warming accelerated during the 1990s and into the 2000s.
- Analysis of the climate record reveals many useful lessons regarding the temporal behavior of climate. Climate is inherently variable; climate change is geographically non-uniform in direction and magnitude; climate change may involve changes in mean values of temperature or precipitation as well as changes in the frequency of weather extremes; climate change tends to be more abrupt than gradual; the climate record contains few cycles that are sufficiently reliable to permit their use in forecasting climate over the next century; and climate change impacts society.
- Factors that could alter the global radiative equilibrium and change Earth's climate include fluctuations in solar energy output, Earth-Sun geometry, volcanic activity, changes in Earth's surface properties, and certain human activities.
- Changes in the Sun's total energy output are apparently related to sunspot activity. Solar output varies directly and minutely with sunspot number; that is, a Sun with more sunspots is slightly brighter, whereas a Sun with fewer sunspots is slightly dimmer.
- Milankovitch cycles drive climatic oscillations operating over tens of thousands to hundreds of thousands of years and were likely responsible for the major advances and recessions of the Laurentide ice sheet over North America during the Pleistocene. They consist of regular variations in precession and tilt of Earth's rotational axis and the eccentricity of its orbit about the Sun. These same cycles also show up in deep-sea sediment cores dating from the Pleistocene Ice Age.
- Only violent volcanic eruptions rich in sulfur dioxide gases are likely to impact hemispheric or global-scale climate. Such eruptions are unlikely to lower the mean annual global surface temperature by more than about 1.0 Celsius degree (1.8 Fahrenheit degrees) for more than a year or two.
- Earth's surface, which is mostly ocean water, is the prime absorber of solar radiation so that any change in the physical properties of water or land surfaces or in the relative distribution of ocean and land may impact the global radiation balance and climate.
- Human activity may impact global-scale climate in several ways including elevating the concentration of greenhouse gases (causing

warming) or sulfurous aerosols (causing cooling). The current upward trend in atmospheric carbon dioxide is primarily due to burning of fossil fuels and, to a lesser extent, the clearing of vegetation.

- The ocean, a major reservoir in the global carbon cycle, plays an important role in governing the amount of carbon dioxide in the atmosphere.
- A global climate model simulates Earth's climate system using equations to describe how its components interact. A global climate model is used to predict broad regions of expected positive and negative temperature and precipitation anomalies.
- Global climate models predict that significant global warming will accompany the increases in atmospheric carbon dioxide, possible by the close of this century. Warming will be greater in magnitude than any other prior climate change during the history of civilization with serious implications for all sectors of society.
- Accounting for positive and negative feedback in Earth's climate system is essential for the development of realistic climate models and predictions of the climate future.
- Global warming is likely to cause sea level to rise in response to melting of polar ice sheets and mountain glaciers plus thermal expansion of seawater. Higher sea level would accelerate coastal erosion, inundate wetlands, estuaries and some islands, and make low-lying coastal plains more vulnerable to storm surges. Rising sea level would disrupt coastal ecosystems and could threaten historical, cultural, and recreational resources.
- Shrinkage of the Arctic sea-ice cover raises concerns about a possible ice-albedo feedback mechanism that would accelerate melting of sea-ice and amplify warming in the Arctic. Sea ice insulates the overlying air from warmer sea water and reflects much more incident solar radiation than ocean water. If sea-ice cover shrinks, the greater area of ice-free ocean waters will absorb more solar radiation, sea-surface temperatures will rise, and more ice will melt. This positive feedback could rapidly reduce the sea-ice cover and greatly alter the flux of heat energy and moisture between the ocean and atmosphere with possible ramifications for global climate.
- Global warming could disrupt marine ecosystems by exceeding tolerance limits or destroying habitats.

Enduring Ideas

- Climate-sensitive documentary, geological, and biological data sources provide valuable information on climate change over a broad range of time scales prior to the reliable instrument-based era. Proxy climatic data sources include shell and skeletal remains of microscopic marine organisms obtained from deep-sea sediment cores.
- Plate tectonics, encompassing continental drift, the opening and closing of ocean basins, and mountain building, exerts an important control on climate over periods of hundreds of millions of years.
- Polar amplification refers to the increase in magnitude of temperature change with increasing latitude. This is an example of the geographic non-uniformity of climate change.
- Based on long-term recorded and reconstructed climate records, climate is inherently variable. Climate change tends to be more abrupt than gradual, and climate change is geographically non-uniform in both direction and magnitude.
- Pleistocene climatic fluctuations reconstructed from deep-sea sediment cores have periods that closely match those of the Milankovitch cycles, indicating that regular variations in Earth-Sun geometry (plus feedback) pace the major glacial advances and recessions during the Pleistocene Ice Age.
- Human activity significantly impacts global climate by elevating the atmospheric concentration of greenhouse gases (causing warming) or sulfurous aerosols (causing cooling). According to the IPCC 2007 report, most of the global warming since the mid-20th century is "very likely" due to the observed increase in greenhouse gas concentrations.
- Global climate models predict that significant warming will accompany a doubling of atmospheric levels of CO_2, possibly by the end of this century if the growth of CO_2 emissions is not greatly curtailed.
- Global warming is likely to cause a continued rise in mean sea level, shrinking of Arctic sea-ice cover (involving ice-albedo feedback), and disruption of marine ecosystems (e.g., coral bleaching).

Review

1. What is the basis for subdividing geologic time?
2. Distinguish between glacial climatic episodes and interglacial climatic episodes.
3. What is meant by polar amplification?
4. When did the Medieval Warm Period and the Little Ice Age occur and how did these climatic episodes influence Norse settlements in Greenland?
5. Define global radiative equilibrium and describe its significance for Earth's climate.
6. What is the potential climatic significance of the Maunder Minimum?
7. Describe the Milankovitch cycles and the general time periods over which these cycles drive climate.
8. What human activities have contributed to the significant upward trend in the atmospheric carbon dioxide level over the past three centuries?
9. Does the climate record contain regular cycles and analogs that enable scientists to accurately predict climate change over the next century?
10. How will continued shrinkage of Arctic sea-ice cover likely influence Arctic temperatures?

Critical Thinking

1. What does analysis of deep-sea sediment cores reveal about the major climate fluctuations of the Pleistocene Ice Age?
2. What role does plate tectonics play in global-scale climate change?
3. What are some of the ways to reduce the anthropogenic contribution to global climate change?
4. How might the great thermal inertia of the ocean affect global climate change?
5. What are some of the potential impacts of rising sea level on the coastal zone?
6. Shrinking of the Arctic sea-ice cover is likely to have a positive feedback on any warming trend at high latitudes. Explain why.
7. Does climate change tend to be more abrupt or gradual? Explain your response.
8. Identify some of the limitations of proxy climatic data sources.
9. Why is climate change geographically non-uniform in direction and magnitude?
10. What is the role of the ocean in global climate change?

ESSAY: Climate Rhythms in Glacial Ice Cores

The Pleistocene Epoch, the most recent of Earth's Ice Ages, began about 1.8 million years ago and ended about 10,500 years ago. Since then, conditions in the present Epoch (the Holocene) have been reasonably mild, with relatively minor temperature fluctuations compared to the Pleistocene. In an effort to better understand climate change, scientists are collecting and analyzing climate data from the Pleistocene Ice Age with an eye toward predicting future climate (especially in view of the exploding world population and the influence of climate on energy demand and supplies of food and fresh water). Important sources of data on the climate as well as the chemical composition of air during the Pleistocene Ice Age are ice cores extracted from the polar ice sheets (Figures 1 and 2).

In 1988, Soviet and French scientists reported on their analysis of a 2200-m (7200-ft) ice core extracted at Vostok station on the East Antarctic ice sheet. The ice core spanned 160,000 years. Oxygen isotope analysis yielded a temperature record and chemical analysis of trapped air bubbles revealed trends in the greenhouse gases carbon dioxide and methane. In the mid-1990s, drilling at Vostok recovered a 3100-m (10,170-ft) ice core spanning the past 425,000 years. In 2004, the European Project for Ice Coring in Antarctica (EPICA) extracted an ice core from East Antarctica representing a time interval of 740,000 years. During the summers of 1991-93, two independent scientific teams, one American and the other European, drilled into the thickest portion of the Greenland ice sheet. The two drill sites were located within 30 km (19 mi) of each other, about 650 km (404 mi) north of the Arctic Circle. Both cores were about 3000 m (9840 ft) in length and spanned a time interval of roughly 200,000 years.

Ice cores from both Greenland and Antarctica clearly reveal an approximately 100,000 year Ice Age cycle consisting of cold glacial climatic episodes (e.g., the Wisconsinan stage) sandwiched between mild interglacial climatic episodes (e.g., the Holocene). Perhaps 16 of these long-term cycles operated during the Pleistocene Epoch. As discussed elsewhere in this chapter, evidence from deep-sea sediment cores indicates that regular variations in Earth-Sun geometry (the Milankovitch cycles) drive this approximately 100,000-year glacial/interglacial cycle.

The Greenland and Antarctic ice core records correlate well both in terms of magnitude of temperature change and the timing of events indicating that the 100,000-year Ice Age cycles were globally synchronous. However, comparison of the Greenland and Antarctic ice core data over the most recent Ice Age cycle (i.e., from about 142,000 years ago to 10,500 years ago) reveals marked differences between the Southern and Northern Hemispheres. Whereas the Antarctic record is reasonably smooth and "calm," the Greenland record shows numerous abrupt and drastic flip-flops between glacial and interglacial climatic episodes. Temperatures changed as much as 7 Celsius degrees (12.6 Fahrenheit degrees) over periods of decades or less (in some cases in only 3 years.) These abrupt temperature changes, having two basic periods of 2000 to 3000 years and 7000 to 12,000 years, occurred during the Wisconsinan stage but not during the subsequent Holocene Epoch. The periods of these temperature fluctuations are much shorter than those of the Milankovitch cycles and hence are probably unrelated to changes in Earth-Sun geometry.

The most likely explanation for these short-term abrupt changes in temperature is the weakening and strengthening of the ocean's thermohaline circulation (Chapter 6). For example, as discussed elsewhere in this chapter, this may explain the occurrence of the *Younger Dryas* cold episode 11,000 to 10,000 years ago. The *Younger Dryas* began abruptly when a sudden influx of fresh water discharged by the St. Lawrence River into the North Atlantic prevented the formation of North Atlantic Deep Water (NADW). Temperatures in the North Atlantic and the surrounding lands plunged. The *Younger Dryas* ended just as abruptly as it began when the input of fresh water into the North Atlantic decreased and formation of NADW resumed. The geographic pattern of the *Younger Dryas* climatic impacts (e.g., little in western North America and only a muted response in the Antarctic ice core record) suggests that the *Younger Dryas* was not part of the larger ice age variability driven by the Milankovich cycles. Rather, the *Younger Dryas* was a regional shorter-term climatic fluctuation likely linked to changes in the Atlantic thermohaline circulation.

Changes in the thermohaline circulation that delineate the *Younger Dryas* also occurred at other times. Scientists have interpreted certain layers of lithogenous sediment in cores extracted from the floor of the North Atlantic as materials released during melting of fleets of icebergs. These icebergs surged or slid off glaciated North America and floated out onto the North Atlantic every 2000 to 3000 years as the climate flip-flopped between warm and cold episodes. Melting of the icebergs freshened the North Atlantic surface waters which resulted in the weakening of the thermohaline circulation.

With colder conditions and fewer icebergs, freshening of the surface waters ceased, the water became salty again due to wind-driven evaporation, and the thermohaline circulation strengthened. After two or three of these events, an even greater discharge of icebergs occurred at intervals of 7000 to 12,000 years. The smaller, shorter, more frequent events are called *Dansgaard-Oeschger events* (named for the paleoclimatologists Willi Dansgaard and Hans Oeschger) or "flickers" because of their relatively short period. The Greenland ice core record contains some 23 Dansgaard-Oeschger events between 110,000 and 15,000 years ago. The larger, longer, and less frequent episodes are known as *Heinrich events* (discussed in the first Essay of Chapter 4). Flickers and Heinrich events occurred during both glacial and interglacial regimes and are evident in the temperature record reconstructed from Greenland ice cores.

FIGURE 1
The Greenland Ice Sheet Project (GISP) cores in this photo show the sharp change from clear to silty ice that occurs at a depth of 3040.33 m. The transition is followed by alternating bands of silty and clear ice followed by progressively siltier ice until contact with bedrock at 3053.51 m. [Photo by J. S. Putscher, NOAA/GISP2, University of New Hampshire]

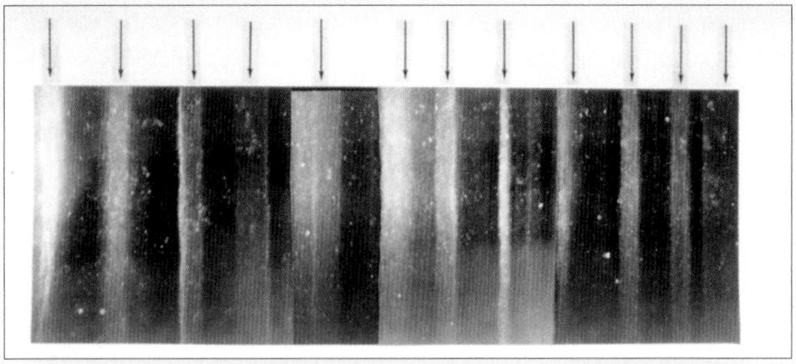

FIGURE 2
A 19-cm long section of GISP2 ice core from 1855 m, showing annual layer structure illuminated from below by a fiber optic source. Section contains 11 annual layers with summer layers (arrows) sandwiched between darker winter layers. [Photo by Anthony Gow, United States Army Corps of Engineers]

ESSAY: The Drying of the Mediterranean Sea

About 5.6 million years ago, late in the Miocene Epoch, the Mediterranean Sea was isolated from the Atlantic Ocean and nearly dried up. Then, with geological abruptness, it refilled again.

Surrounded by Africa, Europe, and Asia, the Mediterranean Sea connects to the Atlantic Ocean via the narrow Strait of Gibraltar (Figure 1). It is the world's largest inland sea with an area of 2,499,350 square km (965,000 square mi), about 3900 km (2400 mi) long and, at its maximum width, about 1600 km (1000 mi). The average water depth is 1500 m (4900 ft), though off the south coast of Greece, it is over three times deeper, at 5150 m (16,900 ft). Within the Strait of Gibraltar, a bathymetric sill separates the West Mediterranean basin from the Atlantic Ocean basin (Figure 2). The sill is never deeper than 284 m (932 ft) at a point where the Strait is about 30 km (18.6 mi) wide, and the Strait is narrowest at 14.3 km (8.9 mi) wide.

The Mediterranean is a remnant of the once vast *Tethys Sea*, but was nearly squeezed shut during the Oligocene Epoch, 34 to 24 million years ago. However, the Mediterranean basin continues to be tectonically active and is slowly shrinking in size as the African plate moves northward, subducting under the Eurasian plate. This subduction is responsible for large-scale bending and uplift of marine sedimentary rocks that formed the Alps, along with volcanic activity at the northern edge of the Mediterranean.

In the 1960s, William Ryan of the Lamont-Doherty Earth Observatory at Columbia University discovered something mysterious about the sea floor of the Mediterranean. While sailing in the Mediterranean on the *R/V Chain* from Woods Hole Oceanographic Institution, Ryan used a 'new' continuous seismic profiler, which could penetrate sea bottom sediments. The acoustic return located a reflecting layer 100 to 200 m (325 to 650 ft) beneath the sea bottom, whimsically labeled the "mysterious layer," or M-layer. Deposited after the deep basin had already formed, the M-layer is ubiquitous across the Mediterranean and has almost the same bathymetry as today.

In 1972, Ryan and Kenneth Hsü, a Swiss geologist, onboard the drill ship *Glomar Challenger* in the western Mediterranean brought up the first cores of the M-layer. The cores resembled marble and were labeled "the pillars of Atlantis." The core material turned out to be anhydrite (calcium sulfate, $CaSO_4$) and stromatolites (mats of the remains of sediment-trapping cyanobacteria) dating back 5.6 million years ago, from the late Miocene Epoch. These two types of

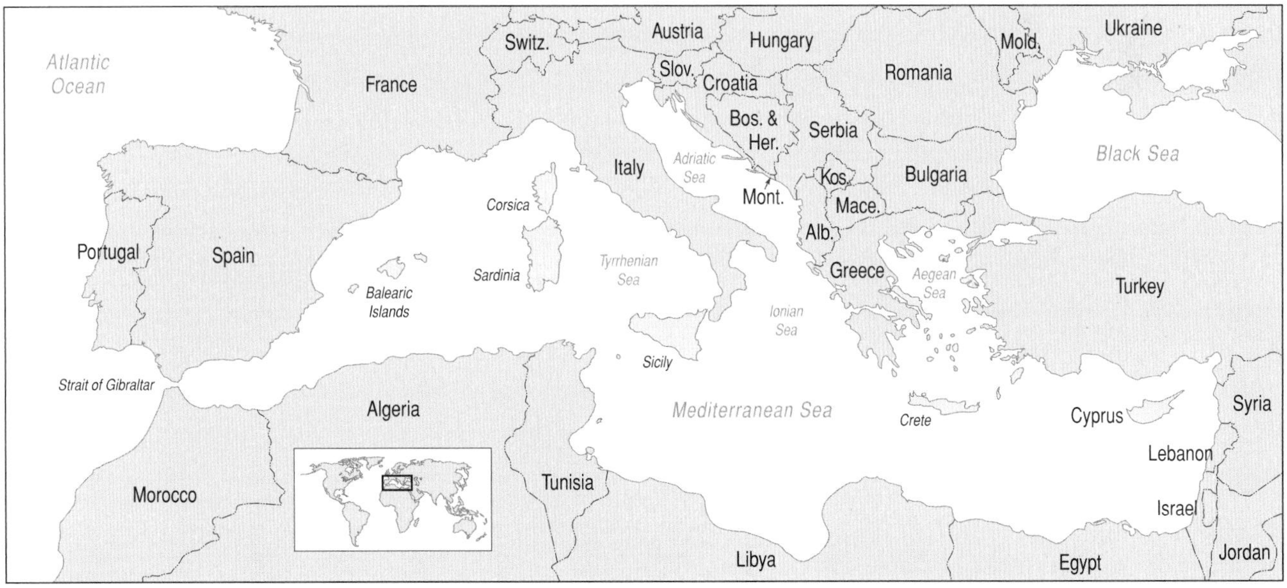

FIGURE 1
The Mediterranean Sea

sediment core material were extremely unusual because anhydrites form only in hot, dry deserts where salty groundwater close to the surface evaporates, causing calcium sulfate to precipitate as a solid. How is it possible for evaporites to form beneath 200 m (650 ft) of marine sediments on the bottom of the Mediterranean Sea? Stromatolites form in broad, intertidal mud flats in the Bahamas and in salty bays in Western Australia, and require light for photosynthesis. What are they doing on the bottom of the Mediterranean, and how could two such different materials form concurrently?

M-layer composition points to one of the most extraordinary events in Earth history, combining plate tectonics, continental-scale glaciation, and sea level fluctuation. About 5.6 million years ago, when the Mediterranean Sea became isolated from the Atlantic Ocean, it almost completely dried up. For hundreds of thousands of years, the basin bottom consisted of deserts, salt marshes, and salty lakes. Then sea level rose and Atlantic water poured over the sill at the Strait of Gibraltar, refilling the Mediterranean basins. Although we know this has happened, no one can yet explain exactly how it occurred.

Several hypotheses seek to explain the drying and subsequent refilling of the Mediterranean. The tectonic hypothesis suggests that the entire basin was uplifted by the subduction of the African plate under the Eurasian plate, which caused similar uplift in the Alps. When the plate dropped down, the Mediterranean Sea refilled, and the cycle was later repeated several times. Another hypothesis attributes the drying to large-scale glaciation, which caused sea level to drop below the sill at Gibraltar, cutting off the inflow of water from the Atlantic Ocean. When the ice sheets melted, sea level rose above the sill and refilled the Mediterranean basin. Other hypotheses propose that the Strait of Gibraltar was squeezed shut, or parts of the Atlantic Ocean floor were deformed and uplifted, forming a gate that alternately closed and opened the Strait of Gibraltar. The most reasonable explanation for the drying of the Mediterranean Sea is likely a combination of the previously mentioned hypotheses. In the late Miocene, tectonic stresses (causing moderate regional uplift) combined with glaciations to drop sea level below the sill at Gibraltar, cutting off the Mediterranean Sea from the Atlantic Ocean.

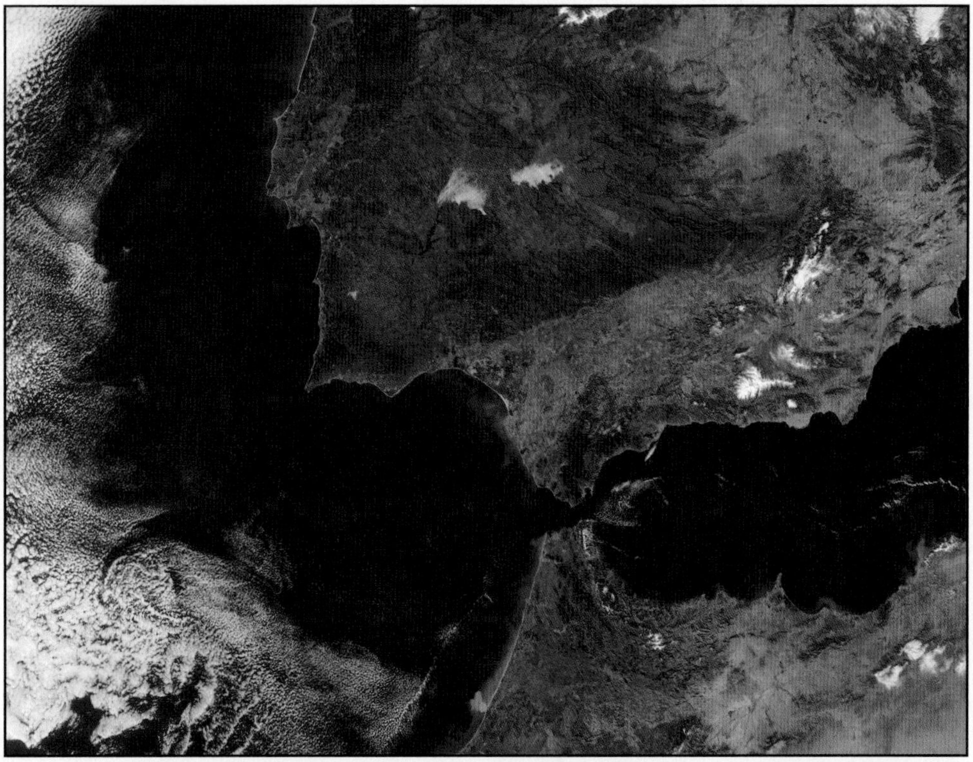

FIGURE 2
This image of the Strait of Gibraltar and the region surrounding it was captured by MODIS on the Aqua satellite on 19 December 2008. [Jeff Schmaltz, MODIS Land Rapid Response Team, NASA GSFC]

Over the subsequent 1000 years, the waters of the Mediterranean evaporated, exposing the bottom of the basins where deserts formed among shallow hypersaline lakes, salt marshes, and salt flats. The rock (geologic) record shows that the region was as hot and arid as it is today and had little inflow from rivers and streams. Such a climate would readily evaporate Mediterranean waters and leave behind the anhydrite deposits.

With the loss of the Sea, the climate of the region probably became cooler and even more arid. In the deeper East Mediterranean basin, a system of salty lakes, similar to North America's Great Salt Lake, was fed by salty overflow from the Black Sea located to the northeast. Such conditions can produce both anhydrites and stromatolites that together accumulate on the bottom of the Mediterranean. Alternating layers of biogenous sediments and evaporites suggest that the Mediterranean began to refill and dry up between 8 and 40 times. Finally about 5.3 million years ago, tectonic forces relaxed, dropping Gibraltar and/or sea level rose, causing Atlantic waters to spill over the Gibraltar sill and refill the Sea.

Hsü calculated that the flow of water over the Gibraltar sill was about 1000 times greater than Niagara Falls today and took a century to fill the Mediterranean basins. However, recent field study supports an alternate view that refilling occurred in a much shorter time span. Key information came from sediment cores and seismic analysis associated with a train tunnel project planned for under the Strait of Gibraltar linking Europe and Africa. Exploration of the sea floor for the proposed tunnel revealed a deep sediment-filled channel with a U-shaped profile measuring 200 km (124 mi) long, 6 to 11 km (4 to 7 mi) wide, and 300 to 650 m (980 to 2100 ft) deep. D. Garcia-Castellanos of the Spanish National Research Council and colleagues concluded that the buried channel was excavated by a megaflood. Writing in the 10 December 2009 issue of *Nature*, the researchers proposed that the flow of Atlantic water into the dry Mediterranean rapidly eroded the sill at Gibralter so that the discharge of water increased exponentially. At its peak, the discharge of the megaflood was about 1000 times greater than that of today's Amazon River. With the water level of the Mediterranean rising more than 10 m (33 ft) per day, the basin could have refilled in as quickly as a few months but no more than two years.

ESSAY: Sea Level Rise and Saltwater Intrusion

Rising sea level threatens to exacerbate the problem of saltwater intrusion in certain low-lying coastal areas. *Saltwater intrusion* is the movement of saltwater into subsurface zones previously occupied by fresh groundwater and is most common along flat coastal plains (e.g., Florida and southeastern Georgia) and islands. Water that completely fills the openings (e.g., pore space, fractures) in sediment and rock is known as *groundwater*. An *aquifer* consists of porous and permeable earth material that is saturated with water and will yield water in usable quantities to a well or spring. In general, groundwater under land is fresh whereas groundwater under the ocean is salty. Fresh groundwater is the single most important source of potable water for humankind.

In coastal areas where fresh groundwater is situated adjacent to salty marine groundwater, excessive withdrawal of fresh groundwater can allow salt water to migrate inland. As saltwater replaces fresh water in an aquifer, coastal wells begin delivering saltwater. One of the most serious consequences of saltwater intrusion is contamination of fresh water sources so that they cannot be used for domestic purposes or irrigating crops.

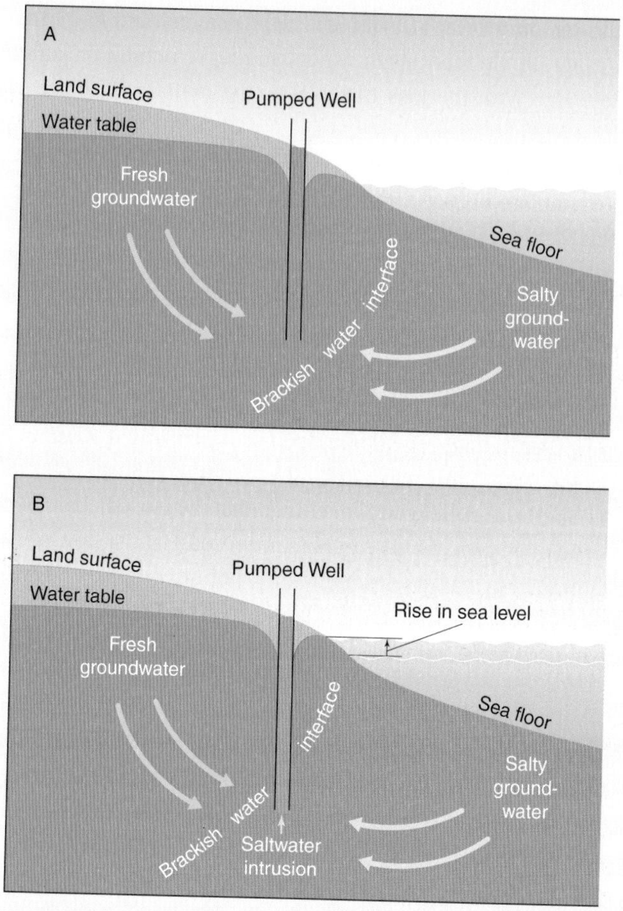

FIGURE 1
Along a low-lying coastal plain, rising sea level causes the interface between salty groundwater and fresh groundwater to migrate inland. This results in saltwater intrusion of wells.

The potential for saltwater intrusion is greatest in freshwater aquifers that are hydraulically connected to seawater (Figure 1). In such a system, a transition zone develops between fresh groundwater and salty groundwater. Saltwater is denser than fresh water so that a higher hydraulic head of fresh water is needed to balance the hydraulic head of saltwater and keep the interface offshore. Because the dry land surface is higher than the ocean bottom, this is usually the case. But when large amounts of water are pumped from a fresh-water coastal aquifer (without replacement), the accompanying change in hydraulic gradient encourages the flow of saltwater toward wells. That is, the interface migrates inland and upward and well water turns salty. Rising sea level increases the potential for saltwater intrusion by increasing the hydraulic head of saltwater.

Islands are particularly vulnerable to saltwater intrusion. On an island subsurface fresh water floats like a lens on the denser underlying saltwater. Excess pumping of wells on an island or rising sea level causes upward movement of saltwater in wells.

What can be done to prevent or alleviate saltwater intrusion? Reducing the rate of groundwater withdrawal and relocating wells so that they are farther apart will help. Artificially recharging the freshwater aquifer (using injection wells) diminishes seawater encroachment. Another strategy is to drill extraction wells to remove seawater from an aquifer before it has a chance to migrate to producing wells. Prior to employing any of these strategies, however, it is important to determine the geologic and hydrologic properties of the aquifer and to install monitoring wells to locate the subsurface salt/fresh water transition zone.

CHAPTER 13

THE FUTURE OF OCEAN SCIENCE

Multibeam sonar systems emit sound waves from directly beneath a ship's hull to produce fan-shaped coverage of the sea floor. These systems measure and record the time elapsed between the emission of the signal from the transducers to the sea floor or object and back again. Multibeam sonars produce a 'swath' of soundings (i.e., depths) to ensure full coverage of an area. [NOAA]

Case in Point

Understanding the ocean, its properties and processes, as well as its role in the Earth system, requires standardized methods of observation, sampling, and analysis. Beginning with the British *Challenger Expedition* of 1872-76, the first phase of ocean exploration depended upon oceanographic ships that serve as platforms for deployment of various types of instruments and later submersible vehicles for in situ measurements of seawater and the ocean floor (Figure 13.1). However, ships are slow and follow specified routes so that vast areas of the ocean remained unexplored. Enter the third phase of ocean exploration in the late 1970s: remote sensing of the ocean by sensors onboard Earth-orbiting satellites. Satellite sensors revolutionized study of the sea because of their capability to rapidly monitor most of the ocean surface

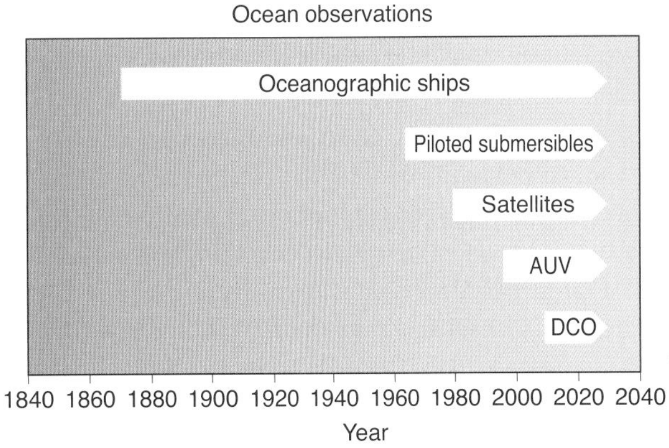

FIGURE 13.1
Principal phases of ocean exploration. Automated Underwater Vehicles (AUV) and Deep-Sea Cabled Observatories (DCO) are the most recent methods for ocean observation.

and measure a wide variety of parameters including, for example, sea surface temperature, sea level, biological production, and wave height. However, satellite sensors are limited in that they can gather data only from surface waters. Now, many ocean scientists see on the horizon the beginnings of a new phase of ocean exploration: deep-sea cabled observatories.

A **deep-sea cabled observatory** is placed on the ocean floor and linked to a mainland facility by fiber-optic and power cables. The cables provide electricity to power sensors and deliver the stream of data obtained by those sensors. A power grid allows for placement of instruments on the sea floor as well as in the water column. A major advantage of a deep-sea observatory is that monitoring is continuous rather than sporadic. (Visits to specific sites of scientific interest by ship may be decades apart.) The real-time aspect of data delivery means much more rapid response time if something interesting or unexpected shows up in the data. The continuous availability of power is an advantage for power-hungry experiments and new instruments such as devices that monitor micro-seismic events or sequence the DNA of marine organisms. Most conventional undersea instruments in use today run on batteries and cannot record or transmit large amounts of data because of very limited power supply. Therefore, data often must be archived for downloading at a later time. Deep-sea observatories will be connected to the Internet making the data stream readily available for scientists, teachers, students, and the public.

Because of high construction, operational, and maintenance costs, only a small number of deep-sea cabled observatories are operational, or even in the planning phase. With funding from the National Science Foundation's Ocean Observatories Initiative (OOI) in 2000, a consortium of U.S. and Canadian scientists, led by the University of Washington, planned a network of cable and sensors to encircle and cross the Juan de Fuca Plate off the Pacific Northwest coast (Figure 13.2). In 2005, they were joined by more than 175 scientists across the U.S. to create a large network of focused experimental sites for the real-time study of seafloor volcanism, hydrothermal vents, earthquakes, seafloor spreading, and subduction zones. Originally named the NEPTUNE (North East Pacific Time-integrated Undersea Networked Experiments) Ocean Observatory, the U.S. component is now known as Regional Scale Nodes and the Canadian as NEPTUNE Canada, focusing on the southern and northern parts of the tectonic plate respectfully.

While NEPTUNE Canada is installed and operational, creating a loop off southern Vancouver Island that explores the continental slope subduction zone, mid-plate abyssal plain and ocean spreading center, Regional Scale Nodes will launch from two sites. The first main cable will begin in Pacific City, OR, and extend 260 km (162 mi) westward to study intense hydrothermal activity and the second will head southeast to a major gas hydrate deposit. With

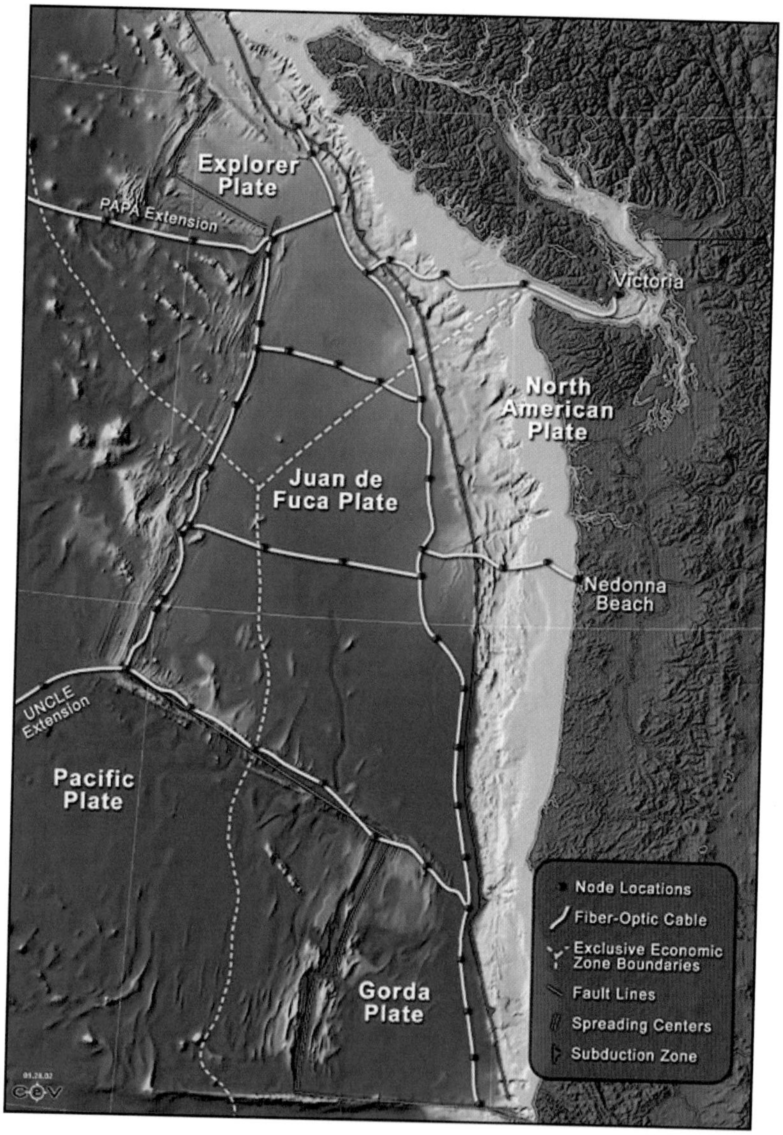

FIGURE 13.2
Regional Scale Nodes and NEPTUNE Canada are collecting data from cable-linked sensors located around and across the Juan de Fuca plate off the Pacific Northwest coast. [Courtesy of the NEPTUNE Project and the University of Washington Center for Environmental Visualization.]

three water column moorings on the cable, each of the sites will be stable and densely populated with sensors. Together, Regional Scale Nodes and NEPTUNE Canada will create an observatory to encompass the entire tectonic plate. The system will observe phenomena throughout the water column, beneath the seafloor, and at the air-sea interface.

In early 2007, south of the Regional Scale Nodes site in Monterey Bay, CA, a 52 km (32 mi) submarine cable equipped with sensor nodes was installed as a part of the testing phase for deep-sea observatories. At a depth of 891 m (2923 ft), the MARS (Monterey Accelerated Research System) Ocean Observatory Testbed features a video camera for monitoring bioluminescent organisms and a robotic microbiology laboratory. Also, sensors on a benthic rover will measure the amount of organic carbon settling on the ocean floor.

Driving Question:
How are advances in technology improving our understanding of the ocean?

The first systematic scientific observations of the ocean were made in the late 19th century from the decks of a converted sailing warship. A hundred years later, sensors onboard Earth-orbiting satellites were routinely providing ocean scientists with a unique perspective of the ocean surface that is impossible to obtain from ships at sea. Prior to remote sensing by satellite, scientists mapped the ocean by combining data from ship-borne observations often separated by thousands of kilometers in distance and decades in time. Today, satellite sensors provide ocean data instantaneously and can monitor the entire surface of the planet in just a few hours. While surface and deep ocean data still must be collected to calibrate and verify remotely sensed observations, using these modern sources of higher resolution data, ocean scientists can observe changes in major ocean currents and sea-surface temperature, map sea surface topography, and detect early signs of a developing El Niño or La Niña.

Throughout history, the ocean has commanded a prominent position in mythology, religion, and literature as a mysterious and threatening place. Today, we continue to regard the ocean with awe and treat it with respect. In this chapter, we examine humankind's efforts to learn more about the ocean, that is, to map the ocean floor, measure the properties of seawater, and monitor marine life as we seek to understand ocean's role in the Earth system.

Humankind's ability to investigate the ocean depends on the availability of appropriate observing systems. The vastness of the ocean continues to challenge the available observational technologies. For one, satellite-borne sensors are limited to observing surface waters. Even though remote sensing by satellite is developing rapidly, the ocean—especially the deep ocean—remains under-sampled so that many of its properties and processes are still poorly understood. Scientists too frequently are limited by their inability to sample and measure the properties of seawater and marine life. As one renowned ocean scientist put it, our scientific models of the ocean often have a "curious, dreamlike quality." This chapter summarizes humankind's efforts to investigate the ocean scientifically—the basic goal of oceanography. In this chapter, we also show where progress is being made, the challenges that remain, and how ocean scientists are seeking to answer questions that have so far gone unanswered. We begin by summarizing the history of exploration of the ocean.

Investigating the Ocean

From the beginning of human existence, people have been fascinated by the ocean and the mystery of what lies beyond the sea-surface horizon. From tentative forays into local waters onboard crude boats, people eventually developed ocean-going vessels capable of circumnavigating the globe.

VOYAGES OF EXPLORATION

The earliest migrations of modern humans (*Homo sapiens*) out of Africa began around 50,000 years ago. Human remains dating from 40,000 years ago were found in Australia. These early migrants apparently knew about fishing and using boats, which they needed to cross the sea separating Australia and Southeast Asia. Further evidence of human migrations via the ocean is the widespread distribution of human populations. Humans had populated all the continents except Antarctica thousands of years before Europeans began their era of ocean exploration in the 15th century. Accounts of voyages and shipwrecks

2000 years ago are found in the Bible and other works of comparable antiquity.

Little is contained in surviving written documents or charts about these ancient ocean explorations for a number of reasons. Many ancient societies had no written language. Most sailors were illiterate and thus unable to document their travels, although oral accounts were passed down from one generation to the next. Perhaps more importantly, knowledge of the sea routes that they explored were trade secrets, too valuable to be shared by the nations or companies that supported the expeditions.

One documented ancient ocean-exploring voyage was the royal trade expedition organized by the Egyptian queen and Pharaoh Hatshepsut around BCE 1450. From northern Egypt, the expedition journeyed through the Red Sea, and southward along the African coast to what is now Mozambique. Scenes from the expedition are displayed at Hatshepsut's funerary temple in southern Egypt near Luxor. The purpose of the expedition was to bring back some of the valuable resources of the region including timber, gold, ivory, and wild animals (e.g., monkeys and baboons). Commercial traffic likely had been conducted for centuries through this part of the Indian Ocean but no written accounts of such voyages have been found. The Egyptians developed sailing ships around BCE 4000, some 2500 years earlier, but probably used them only in the eastern Mediterranean, near the mouth of the Nile River.

A new study, published in *Science News* in 2011, finds evidence for sea voyages between China and Taiwan around BCE 3000. Climate change led to rising sea levels in the Fuzhou Basin of southeastern China beginning about 9000 years ago, as confirmed by sediment cores extracted from that area. The marshy areas needed for traditional rice cultivation diminished, and a maritime culture arose in its place. Island dwellers likely navigated waterways with a type of bamboo raft similar to those used today, although with the addition of sails for longer voyages (Figure 13.3). The study notes that the Fuzhou Basin flooding peaked about 7000 years ago, and remained stable until sea level rapidly declined about 2000 years ago. Genetic, linguistic, and archaelogical evidence suggests that this ancient seafaring may have spurred major colonization of the tropical Pacific Ocean Islands thousands of years later by the Lapita people of modern day Polynesia.

The earliest period of major exploration and colonization of islands in the tropical Pacific Ocean took place from about BCE 1300 to 800 when the Lapita people ventured forth from coastal areas of the western tropical Pacific which they had occupied since BCE 30,000. Using simple seaworthy canoes, the Lapita sailed from Papua New Guinea at least 3200 km (2000 mi) eastward into the tropical south Pacific, settling many islands including New Caledonia, Fiji, Tonga, Samoa, and Vanuatu. Having no navigation devices, the sailors watched for signs of

FIGURE 13.3
Present day bamboo rafts on the Yulong River in China. The maritime culture created between Mainland China and Taiwan about 5000 years ago likely relied on rafts similar to those pictured, but with the addition of sails. [Photo by Qeqertaq/LIcense: Creative Commons Attribution 3.0 Unported]

islands (e.g., volcanic plumes, cloud banks, seabirds, floating tree branches). After a thousand-year hiatus in exploration activity, descendants of the Lapita, known as the Polynesians, sailed eastward into the central and eastern tropical Pacific. The Polynesians reached Hawaii (CE 800-1000), Easter Island (CE 900-1200), French Polynesia (CE 500-1000), and New Zealand (CE 1200-1300).

In sailing eastward, these early explorers encountered the southeast trade winds, a head wind that would have slowed their progress. However, the persistence of the trade winds also meant that if conditions deteriorated, these early sailors could turn around for a swift journey back to where they had come from. Evidence that the Polynesians reached the west coast of South America around CE 1000 suggests that El Niño may have aided their voyage. Recall from Chapter 11 that during an intense El Niño episode, the trade winds not only weaken but may reverse direction blowing from west to east. Such a circumstance would have enabled the Polynesians to sail into the eastern tropical Pacific and eventually encounter the South American continent.

For some time prior to about BCE 1200, the Greek city state of Mycenae was at the center of a great civilization that controlled trade in the Aegean Sea and much of the Mediterranean. For about a thousand years following the decline and fall of Mycenae (at least partially because of prolonged drought), the Phoenicians, from the eastern end of the Mediterranean Sea (now modern Lebanon), dominated sea trade. They established many prosperous colonies along the shores of the Mediterranean, in Sicily, Spain, North Africa, and beyond the Strait of Gibraltar into the Atlantic Ocean and down the Red Sea into the Indian Ocean. Mythic accounts of Phoenician explorations indicate that one expedition reached Cornwall in southwestern England (then called the *Island of Tin*). Another expedition sailed along the West African coast to the mouth of the Niger River.

After the fall of the Phoenician city of Tyre to the Assyrians in BCE 640, the Greeks reestablished their dominance of Mediterranean and Black Sea trade routes. They developed more reliable methods of navigation at sea. To estimate latitude, the Greeks used a calculation based on length of daylight corrected for the time of year. About BCE 450, the Greek Herodotus published an accurate map of the world known to the Greeks. Another Greek astronomer and geographer, Pytheas, sailed out of the Mediterranean, around England and possibly even reached Iceland and the Baltic Sea around BCE 325. He also was the first to determine latitude by measuring the angular distance above the horizon of the North Star (Polaris), a technique that is still used today. About BCE 300, Alexander the Great founded a library at Alexandria, which served as a repository for scrolls from ships and land caravans. For the next 600 years, the library functioned as a maritime studies center, fostering advances in science and celestial navigation by many Greek and Egyptian scholars. About CE 150, the geographer Ptolemy produced an accurate map of the world known to the Romans that influenced geographic thinking for centuries.

The Vikings began sailing from Scandinavia around CE 750 and explored and colonized lands bordering the North Atlantic, including Iceland (CE 810), Greenland (CE 980), and Newfoundland (CE 1000). The relatively mild conditions of the Medieval Warm Period (about CE 950 to 1250) and reduced sea ice cover apparently made their voyages possible. The full extent of Viking colonization of North America during this period is much debated today. Vikings who had settled along the coast of Greenland were unable (or chose not) to adapt to a changing climate that heralded the Little Ice Age (Chapter 12). Their increasing inability to grow European style crops and livestock and to communicate with their homeland by sea (due to more widespread sea ice cover in the North Atlantic) eventually led to the collapse of the Viking settlements in Greenland.

Around CE 1420, during the Ming Dynasty (1368-1644), the Chinese embarked on seven major expeditions involving hundreds of ships technically advanced for the day (Chinese junks), using magnetic compasses. The purpose of these expeditions was to map unknown regions, obtain treasures and exotic animals for the Forbidden City (walled section of Peking containing royal palaces), and spread Chinese influence throughout the region. Some claim that the Chinese circumnavigated the world and may even have reached North America. Economic and political pressures at home terminated these voyages and began the long period of Chinese isolation.

In the 15th century, the Portuguese began the European *Age of Discovery*, an era of ocean exploration and colonization of the Americas, India, Asia, Australia, and New Zealand. In 1420, Prince Henry the Navigator (1394-1460) founded a naval observatory in Portugal, the first school for teaching navigation, astronomy, and cartography. Vasco da Gama (c. 1469-1524) sailed around Africa, reaching India in 1498 to open a lucrative trade in spices. In 1519-1522, ships under the command of Ferdinand Magellan (c.1480-1521) were credited with being the first to sail around the world, although he died in the Philippines in 1521 after reaching the eastern edge of the then known world. His crew completed the voyage back to Spain.

Oceanographic expeditions began with the three Pacific Ocean voyages of Captain James Cook (1728-1779) of the British Royal Navy over the period 1768-1780. These voyages were made onboard, at various times, the HMS *Endeavour*, *Resolution*, *Adventure*, and *Discovery*. Cook relied on the latest navigational tools (precise clocks known as chronometers) to determine longitude accurately. (Refer to the first Essay in Chapter 5.) He mapped the Southern Ocean as well as many other parts of the Pacific, "discovering" Australia, New Zealand, and the Hawaiian Islands. He was the first to circumnavigate Earth at high latitudes and sailed as far south as about 70 degrees S but did not sight Antarctica. The chief objective of Cook's voyages was not science, but to establish a British presence in the South Seas. Nonetheless, some valuable scientific information was acquired. During their exploration of New Zealand, the Great Barrier Reef, Tonga, and the Easter Islands, Cook and his crew collected samples of terrestrial plants and animals, marine life, and the ocean bottom. On his search for a Northwest Passage, Cook prepared charts of the West Coast of North America that remained very useful until World War II. He also "discovered" Hawaii although he was killed on the Big Island during a confrontation with native peoples.

Two early English explorers, Sir John Ross and his nephew Sir James Clark Ross, were interested in conditions in the deep ocean. In 1818, Sir John Ross obtained a bottom sample at a depth greater than 1900 m (6200 ft) off the coast of Greenland. Later, Sir James Clark Ross, discoverer of the Ross Sea and Victoria Land in Antarctica, achieved an ocean depth of almost 4900 m (16,000 ft) in the South Atlantic using a long rope attached to a weight. They discovered abundant marine animals living on the ocean bottom at great depths of water—an extreme environment previously believed to be devoid of all life.

According to the then widely accepted **azoic hypothesis**, deep-ocean waters below 300 fathoms (550 m or 1800 ft) were barren. It was assumed that no species could survive the extreme deep-sea environment characterized by a lack of light (for photosynthesis), high pressure, and low temperature. The hypothesis originated in the mid-1800s with Sir Edward Forbes (1815-1854), at one time considered the father of deep-sea biology. Forbes spent more than a year (1841-42) on the HMS *Beacon* in the eastern Mediterranean Sea. Dredging the sea floor at depths as great as 230 fathoms (425 m or 1390 ft), he found that the number of marine species decreased with increasing depth. By simple extrapolation, Forbes concluded that no life existed below 300 fathoms. However, Forbes was unaware of the role played by *marine snow* in feeding deep-sea organisms (Chapter 4) and the low primary productivity of surface waters of the eastern Mediterranean. Furthermore, dredges in his day were notoriously inefficient at capturing small organisms so that much of the deep-sea fauna was not sampled. Discovery of live animals on the deep-sea floor during several British ocean expeditions debunked the azoic hypothesis and spurred scientific interest in further exploration of the deep ocean basins.

From 1831 to 1836, the HMS *Beagle* undertook a voyage to study the natural science of Galápagos Islands (in the equatorial Pacific off Ecuador) as well as many other locations. Charles Darwin (1809-1882) was onboard as a naturalist. From his observations came his first major published work, *Structure and Distribution of Coral Reefs* (1842). In this treatise, Darwin correctly argued that the form and structure of reefs and atolls develop because they are living organisms growing upward in an effort to remain in the ocean's photic zone as compensation for the sinking sea floor (Chapters 2 and 10). Darwin also wrote on several other marine subjects including barnacle biology and fossils, all of which was overshadowed by his revolutionary work, *On the Origin of Species* (1859).

In 1838-42, the United States launched its Exploring Expedition, a two-pronged voyage in that it was primarily a naval expedition but had more scientific latitude than the British Challenger Expedition of 1872-76 (discussed below). Had it not been for the contentious personality of its leader, Lt. Charles Wilkes, USN, this voyage might have been as famous as those of Cook or the Challenger Expedition. The *Vincennes* was the flagship of a six-vessel fleet. Goals included showing the flag, charting, whale watching, gathering geological specimens, and general scientific observations. Exploration and charting of large parts of the Antarctic coast confirmed that Antarctica was a continent. The most unusual goal of the expedition certainly must have been to test the hypothesis that Earth was hollow with two large entry holes at the North and South Poles. The many specimens and artifacts, scientific and otherwise, collected during the voyage formed the nucleus of the collections of the newly established Smithsonian Institution (1846) in Washington, DC. Wilkes and his scientific staff produced a 19-volume final report.

Matthew Fontaine Maury (1806-73) began his naval career as a midshipman onboard the U.S. sloop-of-war *Falmouth*, serving as navigation officer (Figure 6.2).

After being injured in a stagecoach accident in 1839, Maury was unable to return to sea and was assigned desk duty at home. In 1842, he was appointed Superintendent of the Navy's Depot of Charts and Instruments (later the U.S. Naval Observatory and Hydrographical Office) in Washington, DC. Maury discovered that existing ocean navigation charts were out of date, inaccurate, and supplied little useful information on winds and currents. He decided to revise the charts. Maury and his staff poured over naval logbooks stored at the Depot to glean any useful navigational information. Captains of naval vessels were ordered to supply Maury with navigational, hydrographic, and meteorological data. Later, Maury supplied captains of merchant ships with special logbooks for recording similar observations; the logbooks were mailed to the Depot at the end of a voyage. Out of this work came more accurate navigation charts of winds and currents in the world ocean. The first of Maury's "Wind and Current Charts" was issued in 1847. With these charts, sea captains could take advantage of favorable currents and winds and thereby significantly reduce sailing time on many routes. For his pioneering work on ocean navigation, Maury became known as the "pathfinder of the seas." In 1855, Maury published *The Physical Geography of the Sea*, the first textbook on modern oceanography. With the outbreak of the American Civil War, Maury resigned his position in 1861 and joined the Confederate Navy.

CHALLENGER EXPEDITION (1872-1876)

The Scottish oceanographer Charles Wyville Thomson (1830-1882) led the first voyage dedicated exclusively to marine science. The **Challenger Expedition** started in December 1872 and ended in May 1876 having covered 127,500 km (79,200 mi), about three times the circumference of the planet. Scientists sampled every ocean basin except the Arctic, probed the ocean and the seafloor to depths as great as 9000 m (29,500 ft), and sailed as far south as 61 degrees S before being turned back by sea ice. The Challenger Expedition laid the foundation for modern ocean science.

The Royal Society of London funded and organized the expedition while the Royal Navy provided the vessel, the HMS *Challenger*, a three-masted, square rigged wooden warship, 68.5 m (225 ft) long. The Royal Navy also supplied the captain and crew. In a forerunner of modern oceanographic practice, the captain ran the ship while a board of six chief scientists was in charge of the scientific enterprise. Thomson and his colleague Sir John Murray (1841-1914) are credited with coining the term *oceanography*. On the ship's deck, scientific gear, a laboratory, and facilities for storing samples took the place of all but two of the ship's original 19 guns (Figure 13.4). *Challenger* was equipped with an auxiliary steam engine for maneuvering while taking samples and scientific observations.

The Challenger Expedition tested the hypothesis that no life could exist below an ocean depth of 550 m (1800 ft) because of extreme pressure and no light. The *Challenger* was able to sample to a depth of 8185 m (26,900 ft), making 492 deep soundings, with samplers and nets at 362 stations. The hypothesis was proven false with a total of 4017 previously unknown species of marine animals and plants discovered to all depths.

Scientists onboard the *Challenger* measured air and sea temperature, winds, salinity, density, and currents. In addition, 77 water samples were obtained and stored. Many of the Challenger Expedition's contributions are still used today; an example is the global map of marine sediment distribution (Figure 4.7). Other important

FIGURE 13.4
Dredging and sounding equipment onboard the HMS *Challenger*. The 1872-76 Challenger Expedition laid the foundation for modern ocean science. [Courtesy of NOAA Photo Library]

contributions of the Challenger Expedition include the first systematic map of major ocean currents and water temperatures, a map of ocean bottom features (e.g., mid-ocean ridges), discovery of the *Challenger Deep* in the Mariana Trench in the Pacific Ocean, finding manganese nodules at the bottom of the North Atlantic, and documenting the great diversity of marine plants and animals, especially microscopic plankton. Probably the most important product of the voyage was the 50-volume report of the expedition's scientific findings, the *Challenger Report*, written and published between 1880 and 1895 by Sir John Murray. This report formed the foundation for the science of oceanography and is still highly valued today. In short, the Challenger Expedition marked the beginning of the modern study of the world ocean.

Some of the instruments used to collect samples on the Challenger Expedition differ little from those in use today: towed nets to gather plankton and larger organisms, bottles for collecting water samples, and coring tubes for retrieving seafloor sediment samples. Since the time of *Challenger*, winches for lowering and retrieving instruments have improved greatly and electronic instruments have largely replaced mercury thermometers and other mechanical devices. Navigation, originally done by astronomical observations, is now largely based on the Global Positioning System (described in the first Essay of Chapter 6). Furthermore, satellites and electronic computers have greatly accelerated data collection, relay, and analysis.

FIGURE 13.5
The USS *Albatross* of the United States Fish Commission carried out many expeditions throughout the world from 1887 to 1925 and discovered hundreds of marine species using surface townets and benthic trawls and dredges. [Courtesy of NOAA Photo Library]

Modern Ocean Studies

The Challenger Expedition was a model for modern scientific study of the ocean carried out throughout the 20th century and into the present century. In this section, we examine modern initiatives and innovations aimed at developing a greater understanding of the ocean, its properties, processes, and life forms. We open with an historical perspective of technological innovations that enabled scientists to more extensively probe the sea.

TECHNOLOGICAL INNOVATIONS

The Challenger Expedition inspired two American expeditions that were directed by the naturalist Alexander Agassiz (1835-1910) on the U.S. Coast and Geodetic Survey Ship *Blake* (1877) and later on the survey ship the *Albatross*. The USS *Albatross* sailed throughout the world ocean from 1887 to 1925, collecting hundreds of marine species using specially designed nets and benthic dredges (Figure 13.5). In the period between these two American voyages, the Russians under S.O. Makarov conducted a 3-year cruise onboard the *Vitiaz*, analyzing the physical properties of the North Pacific Ocean waters.

The first major oceanographic expedition of the 20th century was the German Atlantic Ocean Expedition of 1925-27. The Institute of Marine Research in Berlin organized the expedition and the German Navy supplied the ship, the R/V *Meteor* (the first of several German oceanographic research ships to bear that name). The primary mission of the voyage was to gather data on the physical properties of the South Atlantic Ocean using optical, acoustic, and electronic equipment. Probably the most important innovation was the use of an acoustic echo sounder to almost instantaneously measure and record the depth and profile of the sea bottom. These observations drastically changed the long-held view of the ocean floor as monotonously flat to a place of considerable topographic relief—information that would later help support the theory of sea floor spreading and plate tectonics (Chapter 2).

Since its founding in 1902, the Scripps Institution of Oceanography in La Jolla, CA, has spurred the advance of oceanographic research. In 1937, the Institution's schooner, *E.W. Scripps*, began a broad research program on physical, biological, and chemical oceanography in Pacific waters off Southern California. Work at Scripps eventually led to publication of the well-known book on ocean science, *The Oceans*, in 1942.

National defense needs during World War II (1939-1945) and the subsequent Cold War (1945-1990) spurred advances in ocean science and technology by the industrialized nations of the West and the Soviet Bloc.

Many nations supported multi-purpose exploration of specific areas of the ocean using ship-borne observations and backed development of observing instruments for ocean studies. Some of these studies were also aimed at improving long-range weather forecasting and detection of missile-launching submarines. The latter led to a better understanding of sound propagation in the ocean. Enormous amounts of data were collected and analyzed. Eventually these data were declassified and are now available for peacetime scientific research.

Paralleling the increasing use of ships (both military and civilian) for ocean research was the design and development of piloted submersible vehicles that enabled scientists to observe the ocean depths directly. The *bathysphere*, developed in the early 1930s by William Beebe and Otis Barton of the New York Zoological Society, was a hollow steel ball about 1.5 m (5 ft) in diameter, equipped with oxygen tanks, chemicals to absorb carbon dioxide, and portholes and a searchlight for viewing. A cable from a ship lowered and raised the bathysphere into and out of ocean waters. In 1934 just off the Bermuda coast, Beebe and Barton descended in the bathysphere to a depth of 923 m (3028 ft). They reported seeing fish and invertebrates previously unknown.

More maneuverable than the bathysphere for exploration of the deep ocean was the *bathyscaph*, designed by Auguste Piccard (1884-1962) in the late 1930s and tested in 1948. The bathyscaph was not tethered to a ship. It consisted of a thick-walled cabin suspended under a tank (float) containing less-dense-than-water petroleum. Ballast tanks attached to the float kept the bathyscaph afloat while on the surface. Flooding the ballast tanks caused the bathyscaph to dive whereas systematically releasing weights (iron pellets) slowed the descent or allowed the vehicle to surface. In the early 1950s, Piccard and his son Jacques built a new and improved version of the bathyscaph, the *Trieste*, which reached a maximum depth of 3167 m (10,392 ft) in the Mediterranean Sea. Facing high maintenance and operational costs, the Piccards sold the Trieste to the U.S. Navy in 1958. On 23 January 1960, the bathyscaph *Trieste*, piloted by Jacques Piccard and U.S. Navy Lieutenant Don Walsh, dove to a new world record depth of 10,912 m (35,800 ft) very close to the bottom of one of the deepest places in the ocean, the *Challenger Deep* in the Mariana Trench—in the Pacific Ocean about 325 km (200 mi) southwest of Guam.

Alvin, one of the world's first deep-ocean research submersibles, was built in 1964 (Figure 13.6). Today, the 3-passenger (pilot plus two scientists) vehicle is one of only five deep-sea research submersibles operating in the

FIGURE 13.6
On 12 April 2004, the research submersible *Alvin* made its 4000th dive since christening in 1964. The dive took place on the East Pacific Rise off the coast of Mexico to a maximum depth of 2500 m (8200 ft). Operated by the Woods Hole Oceanographic Institution (WHOI), *Alvin* can transport a pilot and two passengers to ocean depths as great as 4500 m (14,765 ft). [WHOI photo by Rod Catanach]

world. Owned by the U.S. Navy, funded by the National Science Foundation (NSF) and NOAA, *Alvin* is operated by the Woods Hole Oceanographic Institution (WHOI) in Falmouth, MA (Figure 13.7). To remain state-of-the-art, *Alvin* is disassembled, re-built, and technologically upgraded about every 3-5 years. Considered the workhorse of submersibles, *Alvin* averages 175 dives per year. As of 2010, *Alvin* had completed a total of more than 4400 dives, each dive lasting 6 to 10 hours and achieving maximum depth of about 4500 m (14,765 ft) so that *Alvin* has had direct access to almost 63% of the world ocean bottom.

Alvin uses its six reversible thrusters to hover, rest on the ocean floor, or maneuver across rugged terrain. It is equipped with three 30-cm (12-in.) viewports, external lights, two robotic arms (Figure 13.8), still and video cameras, and a basket having a maximum capacity of 680 kg (1500 lbs) of seafloor samples. *Alvin* made possible the discovery of many important deep-sea

FIGURE 13.7
Aft view of the Research Vessel *Atlantis* moored at the Woods Hole Oceanographic Institution in Massachusetts. The 83.5-m (274-ft) *Atlantis* is the support ship for the submersible *Alvin.* [Courtesy of NOAA]

features, including the first known hydrothermal vents and associated unique biological communities in the 1970s (Chapters 2 and 9). *Alvin* also located a lost hydrogen bomb in the Mediterranean Sea (1966), the wreck of the RMS *Titanic* (in 1986), and the remains of many other ships. *Alvin* was named for Allyn Vine, a WHOI engineer and geophysicist who helped pioneer deep submergence technology.

In the late 1980s, scientists and engineers at the Woods Hole Oceanographic Institution designed and built *Jason*, a **remotely operated vehicle (ROV)** capable of reaching a depth of 6000 m (19,685 ft). Cables (transmitting power and data) tether a ROV to a surface ship. *Jason* was used to explore and photograph hydrothermal vents along mid-ocean ridges and to survey old shipwrecks among other activities. *Jason* was retired in 2001 and replaced by *Jason 2* the following year. *Jason 2* represents a significant technological upgrade with greater maneuverability and sampling capacity than its predecessor. *Jason 2* has the latest in robotics and is equipped with two mechanical arms that can reach twice

FIGURE 13.8
Alvin's arm and claw grasp a sample of pillow basalt on the ocean floor. [Courtesy of NOAA, WHOI, the Alvin Group, and the 2004 GOA Seamount Exploration Science Party]

as far and lift five times as much material as the single arm on the first *Jason*. *Jason 2* has more power for brighter lighting and is equipped with digital cameras. Designed for routine operation at depths as great as 6500 m (21,325 ft), *Jason 2* can remain on the ocean floor for days at a time. Both *Alvin* and *Jason 2* are components of the National Deep Submergence Laboratory at WHOI, the only facility of its type in the nation.

In recent years, Japan and France have built research submersibles capable of diving to depths as great as 6500 m (21,325 ft). Also, the U.S. Navy has dedicated a nuclear powered submarine, the *NR-1Deep Submergence Craft*, to scientific research. While it can dive to a maximum depth of only 700 m (2300 ft), the submarine can remain underwater for 30 days with a crew of 7 and is well suited for scientific voyages under the Arctic sea ice.

A major impetus for development of ocean instrumentation came from the offshore oil industry, which played a lead role in the design of remotely operated vehicles (Figure 13.9). Since 1970, more than 1000 ROVs have been built primarily for operational needs such as servicing oil pipelines on the ocean bottom, retrieving lost objects, and even investigating wrecked aircraft and ships on the ocean floor.

ROVs and Autonomous Underwater Vehicles (AUV) (discussed below) have greatly increased our knowledge of mid- and deep-water marine life. They are equipped with robotic arms and traps that can capture fragile, gelatinous organisms such as jellyfish and ctenophores that would never survive being caught in traditional net trawls. Such organisms have been

FIGURE 13.9
Launch of the ROV *Innovator*. Sample buckets are attached to the sides. [Photo by Jeremy Potter, NOAA Office of Ocean Exploration]

underrepresented in historical surveys of oceanic fauna but, as noted in Chapters 9 and 10, they play an important role in marine ecosystems. In early 2003, scientists from the Monterey Bay Aquarium Research Institute announced that using a ROV equipped with video cameras, they had discovered an entirely new species of red jellyfish. Nicknamed "Big Red," these animals are up to 1.0 m (3.3 ft) in diameter and live at depths of 650 to 1500 m (2100 to 4900 ft). The fact that scientists overlooked such a large animal until now underscores how little is known about the deep ocean and how modern technology enables scientists to expand our understanding.

Another method for sampling delicate drifting organisms is via a video plankton recorder (VPR). A VPR is essentially a high-powered video microscope mounted on an instrument frame towed behind a ship. The instrument is designed to illuminate and image a small volume of undisturbed water as the frame is pulled through the ocean. Cables relay the video image to the ship where it can be viewed and recorded. Image recognition software is then used to calculate census information for the different taxonomy groups.

REMOTE SENSING

Remote sensing refers to acquisition of data on the properties of some object without the sensor being in direct contact with the object. Remote sensing involves not only Earth-orbiting satellites, but also certain automated observing platforms. For example, arrays of underwater microphones (originally intended for submarine detection) measure water temperatures over great distances within ocean basins based on sound propagation (Chapter 3). Such measurements are unobtainable using ships alone. These observations are the basis for detecting and following the movements of water masses through the ocean, locating submarine volcanic eruptions, and tracking migrating whales.

Some of the most significant advances in the study of the ocean and its basins came in the past four decades with development and application of remote sensing by high altitude aircraft and satellites (Chapter 1). Throughout this book, we use a variety of images processed from data acquired by sensors onboard Earth-orbiting satellites. These space platforms provide a unique global perspective of the ocean, its properties and processes. Initially, satellite sensors were limited to gathering data at only the ocean-atmosphere interface because radiation at most wavelengths cannot penetrate to great depths in the ocean (Chapter 5). Satellite sensors remotely measure such ocean properties as sea-surface temperature (*AVHRR*), sea-level fluctuations and ocean currents (*TOPEX-Poseidon, Jason-1*), surface winds (*QuikSCAT*), sea-ice extent, and ocean color (*SeaWiFS, MODIS*).

Aquarius, a satellite mission planned for launch in mid-2011, is designed to provide the first global salinity maps for ocean surface waters. Developed by NASA and the Space Agency of Argentina, the goal of Aquarius is to gather sea-surface salinity (SSS) measurements over Earth's surface once every 7 days with an accuracy of 0.2 psu. The mission is expected to be operational for at least 3 years. Remote sensing of SSS is a major achievement in itself, but combining global satellite observations of salinity and temperature promises to yield a better understanding of the processes involved in the global water cycle, ENSO, and longer-term changes in ocean circulation. Such data will make possible climate models bridging the ocean-atmosphere-land-ice subsystems for predicting the climate future. According to NASA, Aquarius will provide as much SSS data in only a few months as has been obtained by ships and instrumented buoys over the past 125 years. Prior to profilers and Aquarius, SSS measurements were made mostly in summer, confined to shipping lanes.

Paralleling advances in remote sensing by satellite is development of computers, sufficiently powerful and fast to store and analyze vast quantities of observational data. Today, data collected in the most remote reaches of the ocean can be sent nearly instantaneously to laboratories via Earth-oribiting communication satellites for analysis almost anywhere in the world. Other satellite-based technologies that greatly aid ocean studies include the Global Positioning System (GPS) that enables unmanned instrumented platforms to report their locations accurately and communications satellites that relay data to computers for analysis (Chapter 6).

SCIENTIFIC OCEAN DRILLING

The modern era of sampling marine sediments and the oceanic crust for scientific purposes began in the early 1960s when engineers demonstrated that small vessels and barges equipped with oil-field drilling rigs could be used to sample the deep-ocean bottom. The success of these experiments led to the **Deep Sea Drilling Program (DSDP)**, which operated from 1968 to 1983 with the *Glomar Challenger*. Onboard thrusters, computer-controlled propulsion units responding to an acoustic signal emitted by a beacon on the sea floor, permitted the *Glomar Challenger* to drill in deep waters without anchoring. This technique, known as *dynamical positioning*, kept the drill ship within a fixed distance of the spot where the drill penetrated the sea floor.

In 1983, the *JOIDES Resolution*, a larger and more capable drill ship, continued deep-sea drilling within the **Ocean Drilling Program (ODP)**. ODP operated between 1983 and 2003 with support from 22 nations. Newer drilling ships use GPS, along with the acoustic beacon, to fix the precise location of the drill hole and ship. While drilling, the center of the JOIDES *Resolution* never varies from a 10-m circle.

The **Integrated Ocean Drilling Program (IODP)** succeeded the ODP in 2003. Sponsored initially by the U.S. and Japan, 16 other nations are now involved. The IODP called for a second ship, as well as specialized drilling platforms that can go were drill ships cannot safely or efficiently operate, such as in the Arctic.

By 2006, the *Chikyu* (meaning 'the earth' in Japanese), joined the refurbished *JOIDES Resolution* (Figure 13.10). Built and operated by Japan, and capable of drilling 7000 m (23,000 ft) below the seabed, the *Chikyu* is much larger than earlier drill ships and technologically more advanced. Equipped with specialized equipment that will minimize environmental hazards (e.g., petroleum blowouts), she can drill closer to shore where the chance of encountering oil or natural gas deposits is greater. *Chikyu's* first project was to investigate earthquake mechanisms by boring into the Nankai Trough subduction zone offshore of Honshu, Japan.

Drilling up to 1700 holes annually over the past 40 years, scientists have recovered rock and sediment cores from the ocean floor that, if laid end-to-end, would have a total length of 160 km (100 mi). Many important scientific discoveries have originated from what has become the world's largest internationally supported

FIGURE 13.10
The refurbished *Joides Resolution* on its first voyage into the Pacific Ocean. [Courtesy of National Science Foundation]

Earth science program. Among these is verification of sea-floor spreading from the analysis of rock samples on a ridge extending between South America and Africa (Chapter 2). Scientists have also recovered thick deposits of salt underlying marine sediments in the Gulf of Mexico and the Mediterranean Sea (as described in the second Essay of Chapter 12). These deposits formed when the basins were isolated from the rest of the ocean and their waters almost completely evaporated. Furthermore, deep-sea sediment cores yield a record of climate fluctuations reaching back as far as 190 million years ago.

In the future, scientists will explore the oldest rocks in the ocean basins, those occurring near continental margins (Chapter 2), and gas hydrate deposits, a potential source of natural gas, can be safely sampled (Chapter 9). The deep biosphere contained in marine sediments and oceanic crust (home to perhaps two-thirds of the planet's total microbial population) can be explored beyond the depths of previous studies' drill holes and will add to our understanding of global climate change and earthquake generation.

SHIPS OF OPPORTUNITY

Oceanographic research ships are expensive to operate, ranging from \$10,000 to over \$30,000 per day. Given this cost, specialized research vessels will always be in short supply considering the vastness of the ocean. One way to expand the quantity of ocean measurements is to "piggy-back" scientific instrumentation on ships already operating in the cargo and cruise ship fleets. These *Volunteer Observing Ships (VOS)* can be outfitted with instruments that record atmospheric conditions as well as sea-surface temperature and salinity. Usually, these instrument packages are automated and do not require a scientist or technician onboard.

Some chemical properties of ocean water are measured by VOS including, for example, the CO_2 concentration of surface waters. Measurements of ocean currents were obtained by an *ADCP (Acoustic Doppler Current Profiler)* mounted on the cargo vessel *MV Oleander* which operated a regular route between Port Elizabeth, NJ, and Bermuda. The route took the ship across the Gulf Stream and more than 10 years of data document changes in the flow of this western boundary current. XBTs (expendable bathythermographs) are deployed from a VOS to gather subsurface temperature data.

Use of VOS requires hardy instrumentation which can operate in a mostly autonomous mode. It also requires cooperation and coordination with merchant fleets and the crew who operate their vessels. In general, scientific instruments must be capable of collecting data while the vessel is underway and cannot impede any of the normal operations of the ship. As more types of ocean instrumentation become automated, the opportunity to obtain global ocean measurements from Volunteer Observing Ships should continue to increase.

INTERNATIONAL COOPERATION

The need to study large, remote areas of the ocean prompted groups of nations to cooperate in international scientific studies. The high cost of conducting research at sea, including the operation of expensive research vessels, is a strong incentive for international cooperation. An early example was the scientific initiatives underway during 1957-58, the *International Geophysical Year (IGY)*. Among studies begun during the IGY was regular monitoring of carbon dioxide levels in the atmosphere, an important component of global climate change studies (Chapter 12). The IGY featured the first systematic hydrographic survey of the world ocean and provides the earliest baseline for large-scale changes in ocean temperature and salinity.

In the 1970s, the United Nations Educational, Scientific, and Cultural Organization (UNESCO) sponsored the *International Decade of Ocean Exploration (IDOE)*. IDOE organized the first modern systematic surveys of ocean currents and the chemical composition of seawater. IDOE also initiated field studies that eventually led to our present ability to monitor the evolution of El Niño and La Niña (Chapter 11). Subsequently, other programs were organized to marshal international support for decade-long studies of the ocean, primarily focused on specific processes such as the carbon cycle.

Thousands of scientists from over 60 nations took part in the fourth *International Polar Year (IPY)* from March 2007 to March 2009. Organized by the International Council for Science and the World Meteorological Organization (WMO), the IPY was an interdisciplinary and multilateral scientific program focusing on the Arctic and Antarctic and covered a wide range of topics involving cutting edge physical, biological, and social research. Major themes of IPY included (1) decreasing thickness and extent of glaciers, ice sheets and sea ice cover, (2) changes in snow cover and its duration, (3) degradation of permafrost, (4) implications of these environmental changes for global sea level, freshwater supply, ocean circulation, marine ecosystems, and the people of northern lands.

Emerging Ocean-Sensing Technologies

Sending a well-equipped research vessel to sea for an extended period of time with a complete crew of sailors, technicians, and scientists is costly. While seagoing oceanographers cannot be replaced, new microelectronic instruments, better computers, and improved communications and positioning systems developed in the late 1990s have made possible new types of un-piloted ocean observing systems. These innovations have greatly expanded the range of ocean observations into areas not usually traversed by ships. We consider some of these developments in this section.

AUTONOMOUS INSTRUMENTED PLATFORMS AND VEHICLES

The need to understand the subsurface ocean on all temporal and spatial scales has led to development of new types of research platforms. They are designed to overcome the limitations of oceanographic ships, including their slow speeds, high costs, uncomfortable working conditions, and limited supply versus demand. A variety of such platforms is available and more are under development. Available now are instrumented buoys, Argo floats, Slocum gliders, and Autonomous Underwater Vehicles.

Buoys (Figure 13.11) are small floating un-piloted platforms, typically several meters in diameter that are moored at fixed locations in the ocean (such as the buoys deployed across the equatorial Pacific Ocean as components of the ENSO Observing System, described in Chapter 11). Sensors on these buoys continually observe the lower atmosphere and upper-ocean. Data are then transmitted to polar-orbiting satellites for transmission to computers ashore. These platforms are sources of data at locations considered critical for improving weather and climate prediction. Many buoys have lines suspended below them with sensors attached at various depths to continually measure water temperature, currents, and salinity. A few of these buoys also have cables that connect to sensors placed on the sea floor (discussed later in this chapter). Other smaller *drifting buoys* move with ocean currents and provide information on their speed and direction as well as changes in subsurface ocean properties.

Other autonomous platforms, known as *Argo floats*, are cylindrical devices equipped with sensors that augment satellite-based observations of the ocean surface as well as buoy observations of the upper ocean. As described in detail in Chapter 6, Argo floats obtain profiles of ocean temperature and salinity to depths as great as 2000 m (6600 ft). The array of 3000+ Argo profilers provides near real-time, three-dimensional observations of almost the entire world ocean. These data are essential for numerical models. A drawback of Argo floats is the absence of a propulsion system so that they drift passively with the ocean currents. However, this drift allows determination of currents at depth.

FIGURE 13.11
Two technicians repair an NDBC moored instrumented 3-m discus buoy located off North Carolina. Sensors on the buoy monitor air temperature, wind speed and direction, air pressure, sea temperature at a depth of 0.6 m, wave height, dominant wave period, and mean wave direction. [Courtesy of NOAA, National Data Buoy Center]

A more recent type of float, the *Slocum glider*, covers greater distances (a range of 1500 km or 930 mi) and has a longer sampling life span than the Argo float. Plus it is designed to follow a prescribed trajectory. The Slocum glider is named for New England Sea Captain Joshua Slocum (1844-1909). Slocum was the first person to sail solo around the world, completing the 65,000-km (40,000-mi) voyage in three years and two months (1895-98) in his 11-m (36-ft) long sailboat. The 1.5-m (5-ft) long winged torpedo-shaped glider is highly maneuverable, sinking and rising through the ocean by changing its buoyancy. As it sinks and rises, the glider's wings and rudder control horizontal movements so the vehicle moves independent of ocean currents. At a horizontal speed of about 0.5 m per sec (1 nautical mi per hr), the glider follows a saw-tooth sampling trajectory as it descends to a predetermined depth, measuring temperature, conductivity (salinity), and other water properties at various depths. At regular intervals, the glider surfaces, determines its position by GPS, and relays the stored observational data via satellite.

The Slocum glider employs the same buoyancy changing technique used in the Argo float. Hydraulic oil is pumped between an internal reservoir and an external bladder thereby changing the density of the vehicle by altering its total volume. To ascend, oil is pumped to the bladder that expands and thereby increases the volume and lowers the density of the glider. To sink, oil is pumped back to the internal reservoir, the bladder shrinks, and the overall density of the glider increases.

Until recently, the oil pump that controlled buoyancy was powered by batteries, limiting the range of the glider to the lifetime of the batteries. But scientists at Woods Hole Oceanographic Institution developed a novel engine powered by the temperature difference between surface- and deep-ocean waters that extends the range of the Slocum glider. At the heart of the engine is a tube containing a wax that solidifies and shrinks at about 10 °C (50 °F). The wax melts and expands when the glider is in relatively warm surface waters but solidifies and shrinks when the glider enters the colder deep-ocean waters. Expansion and shrinkage of the wax controls the movement of a device that distributes the oil between the internal reservoir and external bladder. While the wax tube does not replace the glider batteries, it reduces the drain on them.

Gliders have been used to carry instruments that analyze and take samples of toxic algal blooms in the Gulf of Mexico (Chapter 9). Scientists at the Mote Marine Laboratory in Sarasota, FL, developed a winged, robotic-controlled platform that is released at the ocean surface and glides slowly to the bottom taking either observations or water samples. The device provides a relatively inexpensive platform to work within a limited area to obtain background data on the ecosystem, detect and sample a *red tide* should one develop, and bring back water and plankton samples for laboratory analysis.

Autonomous, powered vehicles augment the capabilities of research submersibles and ROVs and move through ocean waters faster than gliders. They are called

FIGURE 13.12
Launch of *ABE*, an autonomous underwater vehicle (AUV) equipped with sensors to measure temperature, conductivity, magnetics, and multibeam bathymetry. [Courtesy of NOAA]

Autonomous Underwater Vehicles (AUVs) because they are un-piloted and do not rely on a cable tethering them to a mother ship, as do ROVs (Figure 13.12). Autonomous vehicles must be carefully programmed with simple artificial intelligence in order to carry out their mission without direct communication links with surface operators. These vehicles have sensors that measure ocean water properties (e.g., temperature, salinity, depth, chlorophyll, light, dissolved oxygen) along trajectories that can be pre-set or controlled while the AUV is underway. The operational range of an AUV is restricted by the limited amount of power that their batteries can provide for propulsion and instrumentation. Periodically, they must return to the mother ship or shore-based facility for recharging. AUVs also have limited space and power for electronic instrument packages.

On the plus side, AUVs can observe the ocean in places and under conditions where research ships and other instrumented platforms cannot. AUVs are especially suited for studies under sea ice (such as the Antarctic ice shelves) and can be equipped with sonar to determine the thickness of the ice-cover. With side-scan sonar, they can efficiently and economically map the ocean floor in considerable detail (Figure 13.13). They can probe coastal waters at the edge of the continental shelf as well as specific locations within submarine canyons. AUVs are also useful in highly energetic ocean systems such as current rings (Chapter 6). In the future, hydrogen-powered fuel cells promise greater range and more power for AUVs.

Autonomous, ultra-light, un-piloted aircraft, once primarily used for military surveillance are also

FIGURE 13.13
Recovery of a side-scan sonar towfish used to map the Pearl Harbor Defense Area. [NOAA photo by J. Smith and C. Kelley]

promising platforms for ocean studies. Groups of such aircraft (called *drones*), can fly for days or weeks over thousands of kilometers of ocean surface and observe the development of tropical cyclones, monitor sea-surface temperature, salinity, and ocean color, or collect atmospheric aerosols. GPS tracks the aircraft and polar-orbiting satellites relay data back to computers ashore. Drones are especially useful in coastal and estuarine areas where satellite sensors do not provide adequate resolution for study of complex environmental processes.

An innovative approach to observing remote areas of the ocean uses small, electronic instrument packages (tags) that are attached to free-swimming marine animals having strong homing instincts such as seals and whales. The tags provide information on the animals' movements as well as the environment in which they live. For more on this unique approach, refer to this chapter's Essay.

OCEAN FLOOR OBSERVATORIES

As noted in this chapter's Case-in-Point, ocean floor observatories are a relatively new but powerful means of studying ocean waters, marine life, and the sea floor. An **ocean floor observatory** is an instrumented facility that can perform experiments, collect data, and communicate observations to data networks and scientists worldwide. The first generation of ocean floor observatories stored data that were retrieved during infrequent visits by surface ships or submersibles. With the next generation of *deep-sea cabled observatories*, scientific inquiry can continue indefinitely as seafloor cables transmit in real-time a continuous stream of observational data to networks on land for widespread distribution. In this section, we describe ocean floor observatories that are precursors to the next generation of deep-sea cabled observatories.

Off the New Jersey coast just north of Atlantic City is an observatory that became operational in 1992. Instruments provide near real-time data for coastal managers, students, teachers, fishers, swimmers, surfers, and beachcombers. The undersea station, known as *LEO-15* for *Long-term Ecosystem Observatory*, is located in 15 m (50 ft) of water with two nodes located at 5 km (3 mi) and 7 km (4.4 mi) offshore on the mid-continental shelf. Observatory nodes monitor ocean currents, water temperature, salinity, light and chlorophyll levels, wave height and period, sediment transport, and phytoplankton blooms. Other data sources include an instrumented tower for atmospheric observations and Doppler radar. A small autonomous underwater vehicle is programmed to gather data from areas surrounding the station. Data obtained by undersea instruments are transmitted via electro-optic cable to the Rutgers Marine Field Station in Tuckerton, NJ.

The *Hawaiian Underwater Geo-Observatory (HUGO)*, situated at the summit of Loihi seamount south of the "Big Island" of Hawaii, was installed and began operating in 1997. Loihi is the youngest volcano in the Hawaiian chain and, like the other volcanoes of the Big Island, is located over a hot spot (Chapter 2). Although its summit is more than 900 m (2950 ft) below sea level, Loihi rises more than 2700 m (8900 ft) above the sea floor—taller than Mount St. Helens was above sea level prior to its explosive eruption in 1980. HUGO is an instrumented titanium platform equipped with seismometers, an underwater microphone, and bottom pressure sensor. These instruments monitored seismic activity, volcanic eruptions, and hydrothermal venting. A 47-km (29-mi) submarine fiber-optic cable linked the observatory to a shore station at Honuapo Bay, Hawaii, and provided scientists with data from the seamount. This was the first deep-water cabled monitoring system. However, the harsh conditions at the seamount disrupted the data flow and HUGO ceased operating after five years. The observatory was recovered by *Jason 2*, a remotely operated vehicle out of Woods Hole Oceanographic Institution but there are no plans for re-deployment.

In 1998, the *Hawaii-2 Observatory* (or *H2O*) was installed on an abandoned AT&T submarine telephone cable that had linked Oahu, Hawaii, to the California coast. Located in 5000 m (16,400 ft) of water near 28 degrees N, 142 degrees W, the Observatory was about 1000 km (620 mi) from the nearest land. A junction box on the cable provided power and communications ports for scientific instruments including a seismometer (for monitoring earthquakes), a deep-water pressure gauge (for detecting tsunamis), and a hydrophone (to listen to underwater sounds). H2O operated until 2003.

Detailed studies of seismic waves generated by earthquakes worldwide require observations that are fairly uniformly spaced but most seismometers are land-based. (The Hawaii-2 Observatory was the first seafloor station in the Global Seismographic Network but as noted above only operated from 1998-2003.) The only seismometers within the ocean basins were located on islands and large portions of the ocean such as the North Pacific have few islands. About 20 deep ocean-floor seismic observatories are being planned to help fill in the gaps. Seismic instruments are installed in holes drilled into the sea floor. At some observatories ships will recover the instrument packages and retrieve their recorded data for later analysis. Others will use acoustic transmission of data to listening stations. Still others will be connected to networks on land by submarine cables providing instantaneous distribution of seismic data. These data are essential for improving our understanding of interactions between oceanic crust and the deeper parts of Earth's geosphere. These data also will aid investigations of some of the longer-term fluctuations in the Earth system such as ocean basin opening and closing (Chapter 2).

Ocean-floor seismic sensors placed in a 1000-m (3300-ft) deep borehole near Japan monitor earthquakes occurring near the Japan Trench where the Pacific Plate is subducting under the Eurasian Plate. When the plates slip past one another, seismic energy is released causing potentially destructive earthquakes, which, along with associated tsunamis (Chapter 7), can cause great loss of life and considerable property damage on the islands of Japan. Plans are underway to have the Japanese drill ship, part of the Integrated Ocean Drilling Project (IODP), install additional seismic stations. One hoped-for outcome of better earthquake surveillance is improved earthquake prediction. Initially these observatories will use replaceable data-storage units to record data but eventually fiber-optic cable will link the observatories to data networks onshore.

COMPUTERS AND NUMERICAL MODELS

Whereas past advances in ocean science depended heavily on sampling platforms and observing instruments, future advances will increasingly depend on access to the world's largest and fastest computers (supercomputers). New high-performance computers will be necessary to assimilate the flood of observational data and to run coupled ocean-atmosphere numerical models. In this way, ocean science is following the lead of atmospheric science where even routine weather predictions require the largest and fastest supercomputers available. NOAA put such a supercomputer into operation early in 2003 in Gaithersburg, MD. In 2010, NOAA announced the construction of a new state-of-the-art supercomputer center in Fairmont, WV, which is expected to be fully operational by the fall of 2011. The NOAA Environmental Security Computing Center (NESCC) provides NOAA with a powerful new tool in numerical modeling and product delivery. NOAA's National Weather Service uses computer output to generate a variety of forecast and other guidance products, including some that deal with ocean conditions (e.g., sea-surface temperature, sea-ice extent, tracks of tropical cyclones). Such computers permit more accurate, longer-range forecasts.

One of the major achievements in the application of coupled air-sea numerical models is the prediction of the onset and evolution of El Niño and La Niña months in advance of the event. As noted in Chapter 11, observational data acquired by an array of moored and drifting buoys in the tropical Pacific along with satellite and tide gauge data are used to initialize these numerical models. Model-based predictions are very helpful for the many government agencies responsible for dealing with the potential impacts of extreme weather associated with El Niño and La Niña.

Challenges in Ocean-Sensing Technologies

The principal benefit of newly available ocean sensing technologies is the promise of a better understanding of how the ocean functions as part of the Earth system and hence, improved prediction capabilities. For example, more complete knowledge of the role of the ocean in Earth's climate system is fundamental to answering questions concerning global climate change (Chapter 12). New observing systems are planned to monitor the global ocean over a broad spectrum of temporal and spatial scales. Such observational data help scientists more accurately model ocean properties and processes, enabling them to predict how Earth's climate system is likely to respond to increased concentrations of greenhouse gases, issue more timely warnings of hazards such as tsunamis, and better manage marine and coastal resources.

The essential requirement for understanding the global environment and its interactions with humans is a systematic record of observations of the Earth system and its sub-systems. Among these sub-systems, the ocean is under sampled both in space and time. Furthermore, monitoring the world ocean requires international cooperation. As noted in this book, much of our knowledge of the ocean is derived from scattered observations made by instrumented buoys, floats, coastal tide gauges, and ship-based measurements. Despite recent advances in ocean observation capabilities, much more observational data are needed and many problems remain to be solved. Some of these problems and possible solutions are considered in this section.

The coastal zone is perhaps an area where substantial advances in ocean-observing technology are most needed, in part because this is where the actual and potential impacts of human population pressures are most severe (Chapter 8). Some of the observing techniques or platforms discussed in this chapter cannot be employed close to the shoreline. Yet this is the most heavily used part of the ocean and one of the most dynamic regions in the Earth system where the hydrosphere, atmosphere, biosphere, and geosphere interact. For example, the coastal zone is where storm surges and tsunamis have their greatest impact.

One emerging technology that may be useful in the coastal zone is remotely controlled ultra-light aircraft capable of flying low and making observations over extended periods. Because the most dramatic alterations of the coastal zone occur during infrequent intense storms, such aircraft in their present form cannot be the only answer. Probably an older but expensive technology will be used in the near future: instrumented towers and shallow-water sea-floor observatories.

Scientists are increasingly relying on high frequency radar to monitor surface currents in the coastal ocean. This specialized radar system, known as *CODAR* (*Coastal Ocean Dynamics Applications Radar*), uses scattered radar waves to measure ocean surface wave speed. The observed wave speed is the sum of the natural wave speed and the underlying current. By knowing the water depth, the natural wave speed can be calculated and thus the underlying current can be determined. CODAR requires stable platforms and is typically used in coastal environments where the radar antenna towers can be positioned on land. An array of CODAR stations along a coastline can determine real-time current information as far as 150 km (93 mi) offshore. In addition to benefiting basic research, maps of surface currents are useful for tracking pollution spills and in search and rescue operations conducted by the Coast Guard.

Better technologies for studying marine life are needed. As we have learned, most of the organisms living in the ocean are single-celled microbes, but many sampling techniques are designed for studying only those organisms that can be caught in nets, from large plankton up to the size of fish (Figure 13.14). One remedy is for scientists to use long tubes that pump water from ocean depths over long periods of time. In this way, they sample and filter large volumes of water to gather sufficient numbers of microbes for study. Many of the new techniques to study marine microbes will probably come from advances in biotechnology. One example is the use of new techniques to study archaea, a little understood but extremely abundant organism that often occupies extreme marine environments (Chapter 9). Archaea cannot be seen using standard microscopes and cannot be grown in laboratories using conventional microbiological methods. However, techniques used to sequence the human genome in the early 21st century may

FIGURE 13.14
Crew members of the NOAA ship *McArthur* inspect a set of bongo nets used to determine the type and concentration of marine plankton. The *McArthur* conducts a range of oceanographic research missions mostly in marine sanctuaries along the U.S. West Coast. [Courtesy of NOAA]

assist in the study of marine life, to identify and assess its abundance. These data will aid in identifying organisms and mechanisms responsible for sequestering carbon in the ocean and in this way add to our understanding of the ocean's role in the global carbon cycle.

New microchips are needed to perform chemical analyses of compounds in seawater, including routine monitoring of essential nutrients, the fertilizers of the sea, as well as toxins released by harmful algal blooms (Chapter 9). This promising technology must overcome one of the oldest problems known to sailors, that is, **bio-fouling**. Organisms rapidly colonize any object placed in ocean water (especially warm ocean water) unless the object is pre-treated by toxic chemicals. First come bacteria and other marine microbes, then diatoms, and eventually mussels and barnacles (as any boat owner knows). Such communities will also coat and impair the functioning of any chemical-sensing chip that is placed in the ocean unless the bio-fouling problem is solved.

Better knowledge of the ocean requires more accurate maps of the ocean floor. Mapping the deep-ocean floor, begun by the Challenger Expedition, is still incomplete. Exploration of Earth's neighbors in space has produced better maps of their surfaces than we have of Earth's ocean bottom. Obtaining such maps of the ocean floor is an expensive and daunting task. Until recently, the best sea-floor maps were made by oceanographic ships equipped with GPS and state-of-the-art bottom-mapping equipment that surveys broad swaths of ocean bottom along the ship's track. NOAA's *National Ocean Survey* ships are equipped to make such maps, but about 125 ship-years would be required to complete the task of mapping the world ocean bottom at a cost of about $1 billion. Because of the high cost, to date only about 10% of the sea floor has been mapped by modern surveying techniques. As noted earlier, an economically attractive alternative for mapping the ocean floor uses autonomous underwater vehicles (AUVs) equipped with side-scan sonar.

The best global map of ocean floor topography is obtained from satellite-borne radar (microwave) altimeters. These instruments measure the shape of the sea surface which is strongly influenced by Earth's gravity field. The mountains of rock which compose the ocean floor topography affect the local gravity field and hence the shape of the sea surface (as discussed in the second Essay of Chapter 7). Combining the limited set of direct depth soundings with global coverage of satellite altimetry measurements has produced an unprecedented map of ocean floor topography. Even this map, however, is limited in fine-scale detail; the shape of the ocean surface tends to "blur" the details of very steep ocean ridges and trenches. A satellite mission called *GRACE (Gravity Recovery and Climate Experiment)*, jointly developed by NASA and the German Aerospace Center, consists of twin orbiting satellites which are capable of measuring fine-scale variations in Earth's gravity. Launched in 2002, the GRACE mission provides even greater resolution of Earth's gravity field and thus the shape of the ocean floor topography. In 2010, the GRACE mission was extended through to the end of its on-orbit life, which is expected to be 2015.

More complete mapping of the sea floor would yield numerous benefits. These maps will permit refinement of plate-tectonic models. More precise depiction of sea-floor roughness would permit better understanding of mixing processes in the deep ocean due to tidal currents flowing over features such as volcanic ridges or seamounts. Commercial applications of deep-ocean maps include determining the optimum sites for laying optical cables for communication networks and managing fishery resources in Exclusive Economic Zones (EEZ) (Chapter 4) and the deep ocean. Tsunami prediction and hazard models would also benefit from more accurate ocean-floor maps.

A technique also exists for obtaining detailed, three-dimensional images of specific areas of the seafloor. This instrument, called *multibeam sonar*, operates from a ship somewhat like a depth sounder, but combines information received from dozens or even hundreds of sonar beams simultaneously. Images obtained in this way are so detailed that in 20 m (66 ft) of water objects as small as a lobster trap can be easily seen. For objects at greater depths, less detail can be resolved, but the images still represent a remarkable step forward in identifying seafloor habitats. For example, fishing grounds can be surveyed so that gear (for example, a scallop dredge) is deployed more carefully to avoid damage to the environment.

Since the early 1990s, a global ocean-observing system has been operating under the auspices of the *Intergovernmental Oceanographic Commission (IOC)* of the United Nations' *Educational, Scientific, and Cultural Organization (UNESCO)*. However, little progress was made toward a global ocean observation system. The IOC is now working with the *Global Earth Observation System of Systems (GEOSS)* to identify ocean observation needs and to integrate it with a global system. Needed, however, is the commitment of nations to sustain an integrated ocean observing system.

Conclusions

Humankind's fascination with the sea no doubt extends back tens of thousands of years. It is known that at least thousands of years ago sailing vessels tapped the energy of the winds and currents enabling humans to explore, discover, settle, and exploit the resources of new lands. With the development of reliable methods of navigation, ocean-going vessels provided access to all continents and commercially important trade routes were established. At first an object of mystery and curiosity, the ocean eventually became the subject of intense scientific scrutiny. Beginning with the Challenger Expedition of the 1870s, understanding of the ocean's properties and processes grew rapidly during the 20th century and into the present century. This new knowledge was partially a product of peacetime application of technologies originally developed for national defense. It was also the result of innovative technologies either adapted from other fields (e.g., deep-sea drilling) or developed specifically for probing the ocean (e.g., *Alvin*, Argo floats, Slocum glider).

Although many challenges need to be overcome and many questions remain to be answered, humankind's understanding of the role of the ocean in the Earth's system is progressing at an accelerating pace. Rock and sediment cores extracted from the ocean floor provide evidence of sea-floor spreading and confirms the hypothesis of plate tectonics, revolutionizing our understanding of Earth's large-scale geological processes. Deep-sea sediment cores unlocked the secrets of the climate fluctuations of the Pleistocene Ice Age and may provide information that is key to understanding future climate. A variety of piloted and un-piloted vehicles provide near real-time data on the physical, chemical, biological, and geological characteristics of the ocean. And sensors onboard Earth-orbiting satellites acquire a flood of valuable data from vast stretches of the ocean.

Basic Understandings

- The earliest humans to migrate from Southeast Asia to Australia around 40,000 years ago apparently knew about fishing and relied on boats for transport. Few written documents or charts refer to ancient ocean explorations because many ancient societies had no written language, most sailors were illiterate, and sea routes were trade secrets.
- The Egyptians developed sailing ships around BCE 4000 and by about BCE 1500 a royal trade expedition traveled from northern Egypt, through the Red Sea, and southward along the African coast to what is now Mozambique. Phoenicians dominated trade in the Mediterranean for about 1000 years following the decline of Mycenae in Greece about BCE 1200. After the fall of the Phoenician city of Tyre in BCE 640, the Greeks reestablished control of Mediterranean and Black Sea trade routes and contributed significantly to the development of navigation techniques.
- Around 3000 BCE, the people of the Fuzhou Basin of southeastern China likely reponded to rising sea levels and the innundation of marsh lands crucial to rice cultivation by creating a maritime culture that eventually included voyages between China and Taiwan. This was a possible precursor to major exploration thousands of years later by the Lapita people of modern day Polynesia.
- The earliest period of major exploration and colonization of islands in the tropical Pacific Ocean took place from about BCE 1300 to 800 when the Lapita people ventured forth from coastal areas of the western tropical Pacific. Using simple seaworthy canoes, the Lapita sailed from Papua New Guinea at least 3200 km (2000 mi) eastward into the tropical south Pacific. After a thousand-year hiatus in exploration activity, descendants of the Lapita, known as the Polynesians, sailed eastward into the central and eastern tropical Pacific.
- The Vikings set out from Scandinavia around CE 750 and later explored and colonized Iceland, Greenland, and Newfoundland.
- In the 15th century, the Portuguese began the European *Age of Discovery*, a time of ocean exploration and colonization of the Americas, India, Asia, Australia, and New Zealand. Vasco da Gama sailed around Africa and reached India; ships commanded by Ferdinand Magellan were the first to circumnavigate the globe; and James Cook mapped the Southern Ocean and many islands in the Pacific Ocean.
- In 1842, Matthew Fontaine Maury was appointed Superintendent of the Navy's Depot of Charts and Instruments in Washington, DC. By thorough scrutiny of logbooks maintained by the captains of both naval and merchant ships, Maury and his staff developed more accurate and useful

navigation charts of winds and currents in the world ocean.

- The Challenger Expedition was the first voyage intended exclusively for marine scientific purposes. Between December 1872 and May 1876, the HMS *Challenger* circumnavigated the globe, sampling all ocean basins except the Arctic, probing the ocean and seafloor to depths of 9000 m (29,500 ft), and laying the foundation for modern ocean science.
- National defense needs during World War II and the subsequent Cold War spurred ocean research and technology development by the industrialized nations of the West and the Soviet Bloc. Paralleling the increasing use of military and civilian ships for ocean research was the development of piloted submersible vehicles that enabled scientists to observe directly the ocean depths. *Alvin*, the workhorse of submersibles, has operated since 1964 and made possible the discovery of many deep-sea features including hydrothermal vents.
- Another impetus for development of ocean instrumentation came from the offshore oil industry, which played a leading role in the design of remotely operated vehicles (ROVs). These un-piloted craft are tethered to a ship and can spend more time on the ocean bottom than a piloted vehicle.
- Earth-orbiting satellites provide a unique global perspective of the ocean. Initially, satellite sensors were limited to monitoring processes operating at the ocean/atmosphere interface, but recently satellite-based methods have been developed to infer the state of the ocean at some depth.
- Sampling marine sediments and oceanic crust in deep waters for scientific purposes began in earnest in the early 1960s, initially using oil-field drilling rigs. The success of these experiments led to the Deep Sea Drilling Program (1968-83), the Ocean Drilling Program (1983-2003), and the Integrated Ocean Drilling Program (2003-). Convincing evidence for sea-floor spreading and plate tectonics is one of the major outcomes of these drilling programs.
- One way to expand the quantity of ocean measurements is to "piggy-back" scientific instrumentation on ships already operating in the cargo and cruise ship fleets. These Volunteer Observing Ships (VOS) can be outfitted with instruments that record atmospheric conditions as well as sea-surface temperature and salinity, dissolved gases, and currents.
- The need to study large, remote areas of the ocean prompted nations to cooperate in international scientific investigations. Examples include the International Geophysical Year (IGY) in 1957, the United Nation's International Decade of Ocean Exploration (IDOE) in the 1970s, and the International Polar Year (IPY) in 2007-2009.
- New un-piloted ocean observing systems include instrumented buoys (both moored and drifting), Argo floats that profile temperature and salinity in the upper 2000 m (6600 ft) of the ocean, the Slocum glider that gathers ocean data along a predetermined trajectory, and Autonomous Underwater Vehicles (AUVs) that are highly maneuverable and can observe the ocean in places where other instrumented platforms cannot.
- An ocean floor observatory is a facility designed for long-term experiments and data gathering (e.g., seismic and volcanic activity).
- Despite recent advances in ocean-sensing capabilities, many technological challenges remain to be solved. Needed are observing methods suited to the dynamic conditions of the coastal zone, better technology for studying marine life, a solution to the problem of bio-fouling, and more accurate maps of the sea floor.
- High quality global maps of ocean floor topography have been obtained from satellite-borne radar (microwave) altimeters. These instruments measure the shape of the sea surface which is strongly influenced by Earth's gravity field. Combining the limited set of direct depth soundings with global coverage of satellite altimetry measurements has produced an unprecedented map of ocean floor topography.

Enduring Ideas

- The *Challenger Expedition* of 1872-76 established the foundation of modern ocean science by being the first voyage intended exclusively for marine scientific investigation.
- National defense priorities and offshore oil exploration helped spur development of new technologies in support of ocean exploration and research including, for example, piloted submersible vehicles, un-piloted remotely operated vehicles (ROVs), and Earth-orbiting satellites.
- Deep-sea drilling to sample seafloor sediments and oceanic crust for scientific purposes gained momentum in the early 1960s, and provided valuable information on plate tectonics, sea-floor spreading, and climate fluctuations of the Pleistocene.
- Among the innovative techniques for obtaining data on the state of the ocean are instrumented buoys (moored and drifting), Argo floats (providing temperature and salinity profiles to depths of 2000 m), and the Slocum glider.
- Traditionally, the properties of seawater and the ocean floor have been investigated from above, that is, via ships, buoys, floats, gliders, and Earth-orbiting satellites. A new approach studies the ocean from below, that is, from observatories situated on the ocean floor and intended for long-term experiments and data gathering.

Review

1. Why does much of the ocean, particularly the deep ocean, remain under-sampled?
2. What is the once widely accepted azoic hypothesis and how was it disproved?
3. Identify the major contributions of Matthew Fontaine Maury.
4. How do remotely operated vehicles (ROVs) and Autonomous Underwater Vehicles (AUVs) aid our understanding of marine life forms and their habitats?
5. What is the goal of the Integrated Ocean Drilling Program (IODP)?
6. Describe the basic principle of the buoyancy-changing technique used in the Argo float and the Slocum glider.
7. What is the advantage of AUVs over ROVs and other research submersibles?
8. What are some of the services provided by undersea observatories?
9. List some of the benefits of a more complete mapping of the ocean floor.
10. Why are remote sensing techniques important in investigating the properties of the ocean?

Critical Thinking

1. In investigating the ocean, what are the advantages of satellite-based remote sensing over voyages of exploration?
2. How did the 1872-76 Challenger Expedition serve as a model for the modern scientific study of the ocean?
3. What are the advantages of a piloted submersible vehicle for exploring the ocean and its inhabitants?
4. Why does bio-fouling pose a serious limitation to modern technologies directed at investigating the ocean?
5. What are some of the advantages of deep-water observatories in acquiring scientific data from the ocean?
6. Speculate on how the mission of near-coastal underwater observatories might differ from that of deep-water observatories.
7. In order to provide adequate advance warning of a harmful algal bloom, ocean scientists monitor what properties of the coastal zone?
8. What natural phenomena in the ocean are regularly monitored by sensors onboard Earth-orbiting satellites?
9. In what way does *in situ* monitoring of ocean properties support monitoring by remote sensing techniques?
10. What are the advantages and disadvantages of ocean data acquisition by satellite versus submersibles?

ESSAY: Monitoring the Ocean using Animal-Borne Instruments

An innovative approach to obtaining observational data from remote areas of the ocean uses small, electronic instruments physically attached, or surgically implanted, onto free-swimming marine animals (Figure 1). These electronic devices, called *tags*, are attached to large, long-lived animals with strong homing instincts, such as tuna and whales. Some tags report data via Earth-orbiting satellites, providing information on the animals' movements as well as conditions of the marine environment through which they travel. In other cases, the data are recovered when the animals return to their home locations. Because these animals tend to follow a regular migration route, they contribute to the global oceanographic database.

Along with monitoring migration, the tags also record other data, such as the depth at which the animal swims, body temperature, ambient temperature, and the light level. With development of smaller solid-state electronic components, the sophistication and communication range of the tags has increased. Currently, seals, whales, tuna, sharks, swordfish, and sea turtles are the principal participants but, with future development of even smaller tags, many other animals including sea birds are likely to be fitted with tags.

Satellite-linked tags enable scientists to study the travel habits of individual animals, sometimes with surprising results. For example, Barbara Block, a marine biologist at Stanford University, and her colleagues surgically implanted tags into two bluefin tuna off the coast of western Ireland. Nearly eight months later one of the fish was located off the coast of Portugal and the other one swimming in waters northeast of Cuba, some 4800 km (3000 mi) apart.

Instrument-bearing large marine animals such as seals, whales, and tuna, were valuable in the world's first *Census of Marine Life*, a 10-year international project that culminated in a comprehensive report in 2010 (Chapter 10). The *Census* was organized to obtain a better understanding of marine biodiversity and the regional distribution and abundance of marine organisms, as well as their relationships to each other and their ocean habitat. An early application of instrumented animals

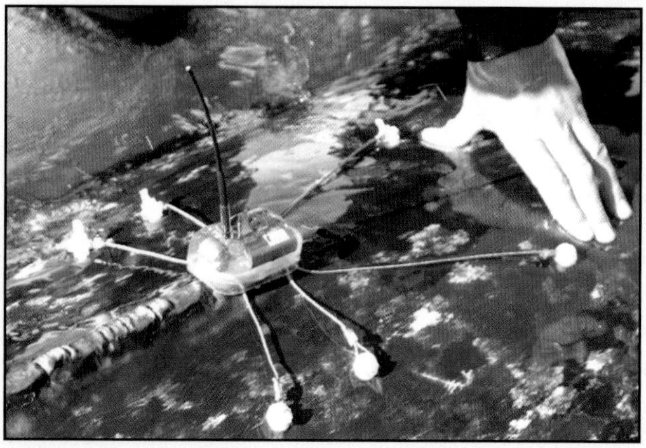

FIGURE 1
Satellite tag on a narwhal. [NOAA Explorer photo courtesy of Kristin Laidre]

in the *Census* involved elephant seals in the northeastern Pacific Ocean, tagged by scientists from the Southwest Center of NOAA's National Marine Fisheries Service in La Jolla, CA. A more recent study was conducted by marine biologist Daniel Costa and colleagues at the University of California, Santa Cruz (Figure 2). Electronic tags glued to the backs of elephant seals recorded the depth of their dives and water temperature while another tag signaled the animal's location.

Elephant seals live in large rookeries on islands offshore from central California, where they breed and raise their pups. Information from the tags revealed that these animals leave their rookeries on regular foraging trips lasting 2 to 9 months; however, males and females follow different migration routes. Males travel along the Alaskan and Aleutian continental shelf and slope whereas females remain in the northeastern Pacific west of California. Scientists found that elephant seals make frequent dives (up to 60 per day) in search of food, and an average dive lasts about 20 minutes,

Figure 2
Scientists attach a tag to the fur of a sedated male elephant seal. [Photo courtesy of Daniel Costa/NASA JPL]

routinely reaching depths of 600 m (2000 ft) but can be as deep as 2000 m (6600 ft). When elephant seals return to the surface, their electronic tags upload data to polar-orbiting satellites for transmission to scientists. Costa reported that the tagged elephant seals from the rookery on Año Nuevo Island near Santa Cruz, CA, produced 75,000 depth profiles in only one season.

The northern elephant seal's cousin, the southern elephant seal, lives in the Southern Ocean. These animals routinely swim thousands of kilometers in the vicinity of Antarctica and off the southern tip of South America. After whales, the male elephant seal goes on the longest feeding migrations of any marine mammal. In 2010, scientists reported the first successful use of tagged southern elephant seals to obtain bathymetry data from around Antarctica. Thick ice cover surrounding Antarctica makes much of the area impassible for ships so that conventional means of sea floor mapping (eco sounding from ships) is not possible. Daniel Costa tagged 57 elephant seals at the U.S. Antarctic Marine Living Resources camp in the South Shetland Islands just north of the Antarctic Peninsula. He followed their movements in the Bellingshausen Sea west of the Antarctic Peninsula from 2005 to 2009. Tagged elephant seals dive under the ice in pursuit of food (reaching the ocean floor about 30% of the time) and their tags, with a pressure sensor calibrated for water depth, relay the information via satellite. These dives provide information on topographic features on the ocean floor which influence ocean circulation and ice formation in an area where ice has been melting rapidly in recent years.

Tags also have been attached to beluga whales (also known as white whales) living in the Arctic and nearby high-latitude waters where ocean data are scarce. In winter, beluga whales often enter ice-covered waters, allowing scientists to study oceanic processes such as sea-ice formation in northern fjords. In one study, oceanographers from the Norwegian Polar Institute tagged two beluga whales, the small instrument package inserted into the blubber so as not to interfere with the animal's activities, from the Storfjorden, Svalbard, Arctic fjord. They were able to determine the whales' location and record salinity and temperature at one-second intervals as the whales ascended from bottom feeding. The data were uploaded via satellite when the whales surfaced for air. One discovery that came out of this experiment, reported in 2003, was the presence of a thin layer of relatively warm North Atlantic water immediately under the ice. Previously, it was assumed that the water beneath the ice was uniformly cold. This discovery may have important implications for understanding recent trends in the extent and thickness of Arctic sea-ice cover (Chapter 12).

In another application, satellite-linked animal tags gathered data via one of Alaska's most endangered marine mammals, the Steller sea lion (*Eumetopias jubatus*). Until the 1960s, the population of Steller sea lions in the North Pacific Ocean was estimated at almost 250,000 but over the following two decades the population plummeted by 80% (Figure 3). Then the southeast Alaska population recovered (at a healthy rate of 3% per year) while populations west of Prince William Sound continued to struggle. In 1997, the Steller sea lion population was divided into an endangered Western Stock (Gulf of Alaska, Bering Sea, and Russia) and a threatened Eastern Stock (California, Oregon, British Columbia, and southeast Alaska).

FIGURE 3
Stellar sea lions. [Courtesy of NOAA]

Based on almost two decades of research on the Steller sea lion mystery, scientists proposed three explanations for the uneven population decline: (1) climate-induced changes in ice cover and ocean currents that altered the nutritional quality of the sea lions' prey; (2) increased mortality perhaps linked to killer whales, entanglement in fishing nets, or intentional shooting by humans; and (3) increased competition with the Bering Sea fishery which greatly expanded (becoming the world's largest commercial fishery) just prior to the sea lions' population decline. In the Bering Sea, annual limits on the catch of groundfish, the sea lions' primary food source, increased from 175,000 metric tons in 1964 to more than 1.5 million metric tons by 1972. Consequently, the food supply for pregnant and nursing female sea lions may have become inadequate resulting in fewer surviving pups.

In August 2010, NOAA's Fisheries Service reported that scientists were most concerned about the fate of the Steller sea lion population in fishery management area 543 in the western Aleutians. Between 2000 and 2008, the number of adults declined 45% and pup production decreased 43% so that the ratio of pups to adult females on rookeries in this management sub-region was the lowest in the western Steller sea lion population.

To investigate the possibility that the western Steller sea lion population decline was due to inadequate food supply, researchers designed a satellite-linked transmitter tag with a battery life of 10 years. The tag was surgically implanted into an adult sea lions' lower abdomen and recorded where the animal foraged for food as well as its body temperature. A sudden drop in temperature suggests a fatal attack by a predator whereas gradual cooling would indicate old age, disease, or starvation. Markus Horning, a pinniped ecologist at Oregon State University, and colleagues followed the movements and temperature variations of 27 Steller sea lions that they tagged and released into the Gulf of Alaska in 2005. In four years, 5 of the animals had died, all but one likely the victims of predators.

More research is needed for us to develop a better understanding of the population dynamics of the Steller sea lion. In this case and in others, electronic tags physically attached or surgically implanted into free-swimming marine animals have proven to be a cost-effective means of gathering valuable data on the marine environment that previously was unavailable.

CHAPTER 14

OCEAN STEWARDSHIP

On NOAA Earth Day 2010, workers aboard the ship *Surveyor II* cut tangled fish and sea stars from fishing gear abandoned in the Puget Sound. Most creatures, particularly mammals and birds, that get caught in these nets do not survive. [Courtesy of NOAA]

Case in Point

In the 17th century, when Europeans first sailed into Chesapeake Bay, they encountered vast oyster reefs that stood above sea level at low tide and, in fact, were a navigation hazard for ships in the Bay. Native Americans and early colonists gathered oysters by hand and readily retrieved them from the shallow Bay bottom. The extensive oyster reefs also provided shelter for many fishes and other species living within the crevices between oyster shells.

Oyster production in Chesapeake Bay peaked in the 1880s and then declined dramatically. Today's production stands at 1% of peak levels. In the early 20th century, sailing vessels, known as skipjacks (Figure 14.1A) were largely replaced by powerboats, and the introduction of powered winches permitted oysters to be harvested from all depths. Not only were the oysters overfished, oyster reefs were destroyed in the process. With the large protective reefs gone, the remaining oysters lived on the muddy Bay bottom where they were exposed to more predators and pathogens. Two oyster diseases, accidentally introduced by non-native oysters, also contributed to their decline. In the early 20th century, commercial oyster production slipped to historically low levels and "oystering" as a traditional way of life for watermen on the Bay was threatened in both Maryland and Virginia.

FIGURE 14.1
(A) Sail powered oyster dredges, known as skipjacks, operated in Chesapeake Bay. Here skipjacks are in port drying their sails. (B) Shells containing oyster spat raised by the Oyster Recovery Partnership are bagged and placed on pallets to grow before being transplanted to a designated oyster reef elsewhere in the Chesapeake Bay. [Photos courtesy of NOAA]

Restoration efforts in recent years have focused on farming (aquaculture) (Figure 14.1B) of native oysters (*Crassostria virginica*) and physically rebuilding oyster reefs that stand above the Bay's muddy bottom. In April 2009, the State of Maryland and Commonwealth of Virginia, together with the U.S. Army Corps of Engineers, issued the *Final Programmatic Environment Impact Statement for Oyster Restoration* in Chesapeake Bay following an extensive two-year study. This statement aligns with the 2009 Executive Order issued by President Barack Obama for a renewed effort to restore and protect Chesapeake Bay and its watershed. Maryland Governor Martin O'Malley announced that new science developed through the joint study will "support both the ecological restoration of our native oyster and the revitilization of our oyster industry with emphasis on new aquaculture opportunities." Oyster farming is extensive in Virginia and beginning to expand in Maryland. Virginia, with its higher salinity waters, has greater spawning and growth rates but also higher incidence of disease.

Hatcheries supply larvae for restoration efforts. Free-swimming oyster larvae preferentially attach to clean oyster shells as oyster spat. Clean oyster shells are available in relatively small quantities from oyster canneries. Shells from fossil reefs are also used but are less suitable as a substrate for juvenile oysters. Mud settling on the shells or growth of benthic organisms inhibits attachment of the larvae.

In 2011, a study led by the Virginia Institute of Marine Science demonstrated that in Chesapeake Bay (and worldwide), protected oyster reefs are showing signs of success. The study called for expansion of these "sanctuary reefs" while taking an ecosystem-based approach to management. Oyster reef restoration had larger Bay ecosystem benefits, including providing habitat for anadromous fish, migratory birds, and some endangered species. Also, oysters are filter feeders that improve water quality.

Driving Question:
How do we balance exploitation of fisheries with efforts to preserve marine species and ecosystems?

In early 2003, two Canadian fisheries scientists released the results of their 10-year study, which showed that commercial fishing had reduced the populations of as many as 90% of the world's largest fish species to such small numbers that their populations may not survive. Even more surprisingly, this population decline happened in only 50 years, the result of intensive, highly industrialized fishing practices. In recent years, there has been considerable debate concerning a 2006 study published in *Science* that warned of a potential collapse in world fisheries by the mid-21st century and striking loss of marine ecosystem biodiversity. Critics of the study questioned the methods use to assess global fish catch and criticized the study for ignoring robust fisheries in different areas of the world, including those that have made strong recoveries from long periods of overfishing. Lead researchers on both sides of the debate have since been collaborating to evaluate the state of global fisheries, including evidence for rebuilding the world's fisheries and supporting ecosystems. An initial joint study was presented in *Science* in 2009.

The rapid decline in the populations of species such as cod, marlin, swordfish, and halibut severely impacts marine ecosystems because these fish occupy the highest trophic levels of marine food webs; most of them are **top predators** in the ecosystem. These large fish are slow growing and take a relatively long time (decades) to reach reproductive maturity. Once they become adults, the older fish tend to be the most fecund (fertile). Hence, removal of the largest fish of a species reduces the reproductive success of the population as a whole.

The United Nations *Food and Agriculture Organization (FAO)* tracks the state of the world fisheries. As of 2008, FAO estimated that about 32% of fish stocks are in the following categories: overexploited (28%), depleted (3%), or recovering from depletion (1%). In comparison, only 10% of fish stocks fell into these three categories in 1974. Slightly over half of fish stocks are fully exploited, a percentage that has remained constant since the mid 1970s. Most fish stocks in the top ten world marine capture fisheries are fully exploited. The largest numbers of depleted stocks are in the Northwest and Northeast Atlantic Ocean, the Mediterranean Sea, and the Black Sea. Although the situation is serious, some fisheries management initiatives have been successful.

In this chapter we examine some examples of adverse human impacts on marine fisheries and ecosystems along with steps that are being taken to manage fisheries and protect threatened marine species. Such efforts are an essential component of **ocean stewardship**, action taken by society to protect the ocean and its living and non-living resources for now and the future.

Stewardship of Ocean Life

Components of the Earth system are highly interdependent so that disturbing one part of the Earth system can impact other parts—perhaps negatively. Thus, stewardship of the ocean and its resources involves responsibly managing all resources to benefit present and future generations. However, despite continuing advances in ocean science and technology (Chapter 13), less than 5% of the ocean bottom has been explored or mapped to the same resolution as the surface of Earth's neighboring planets, Mars and Venus. For this reason, we often lack the basic understandings needed to effectively manage and protect marine living resources and their associated ecosystems.

In the United States, the *National Oceanic and Atmospheric Administration (NOAA)* is the principal government agency charged with stewardship responsibilities for the nation's marine environment and living resources. These responsibilities include maintaining sustainable fisheries in U.S. waters, and protecting and restoring the populations of endangered marine animals, such as sea turtles and whales. NOAA also protects and maintains the viability of the nation's coastal zone. Effective stewardship requires consideration of both the biotic and abiotic components of marine ecosystems to protect living organisms, their habitats, and population sustaining interactions (e.g., predator-prey relationships). Stewardship of ecosystem components also includes managing their exploitation by commercial and recreational fishers for the benefit of people who make their living in these industries or use them for recreation.

One concept that is central to exercising wise stewardship is **sustainability**, defined by the United Nations' *World Commission on Environment and Development* as "developments that meet the needs of the present without compromising the ability of future generations to meet their own needs." Achieving sustainability requires effectively balancing environmental issues with social and economic concerns. Sustainability calls for intergenerational equity in managing resources and the environment; that is, we need to consider the needs of future generations as well as our own. It also includes the concept of maintaining a level of critical ecosystem function and biodiversity.

Fisheries and Sustainable Exploitation

Worldwide, fisheries supply about 16% of all the animal protein consumed by humans* and the fishing industry employs more than 180 million people, over 95% of whom live in developing nations, mainly in Asia. Nearly one-third of the world's fishers and fish farmers live in China, which has been the leading fish exporter since 2002. Global fisheries as a whole contribute between $225 to $240 billion (U.S.) annually to the world economy. In coastal developing nations—especially in the tropics—almost all the protein intake comes from near-shore waters. Fishers in these waters use traditional methods (*artisanal fisheries*) and account for more than half the world's marine fish production. Most of the fish they catch is eaten directly by humans. In developed nations, fishers employ technologies intended to increase the fish catch and a large amount of commercial-fish production is processed into fishmeal to feed pets and livestock, and for use in fertilizers.

In this section, we investigate overfishing, the contributing factors and the implications of overfishing for survival of top predator fish and the viability of associated marine ecosystems. We consider the advantages of a more holistic, ecologically based approach to fisheries management in order to achieve sustainable yields. Following this are discussions of fisheries and habitat restoration and recreational fisheries.

OVERFISHING

According to the authors of an international study published in *Science* in 2006, worldwide an estimated one-third of all exploited fish stocks have collapsed; that is, their numbers have declined to less than 10% of their historical maximum population. **Overfishing** occurs when a fish species is taken at a rate that exceeds the maximum catch that would allow reproduction to replace the population. The two types of overfishing are known as recruitment overfishing and growth overfishing.

With **recruitment overfishing**, adult fish are taken in such great numbers that too few survive to replenish the breeding stock. A 2003 study of global fisheries spanning the previous 50 years showed that within 15 years of when a new fishery opens, about 80% of the largest fish are taken. The largest fish in a population not only have the greatest commercial value, they are also the most valuable ecologically because the largest fish in a species are the most prolific. **Growth overfishing** refers to fish that are taken too small, before the animals have grown to a size that would produce the maximum yield. Growth overfishing can reduce the value of the catch because, if left to grow, individuals would weigh more and bring a higher price in the marketplace. For example, growth overfishing may be the reason that male blue crabs taken from Chesapeake Bay have become fewer in number and smaller in size over the past several decades. (Furthermore, the recent decline in blue crab populations along the Eastern seaboard appears to be linked to changes in the weather and a parasitic infection of a dinoflagellate known as *Hematodinium perezi*.)

For more than 10,000 years, humans have taken food from the sea and for most of this time the ocean's bounty of life was considered limitless. However, even early on there were signs that this view of an inexhaustible food resource from the sea was seriously flawed. In fact, the first known human induced collapse of a marine stock occurred about 3000 years ago along the Peruvian coast. Fishers continued to harvest shellfish even after a natural disaster had greatly reduced shellfish populations.

In the years since the Industrial Revolution, a variety of technological innovations enabled fishers from developed nations to greatly increase the fish catch. These innovations included the steam and internal combustion engines that powered fishing vessels, extending their range from coastal areas to well offshore. These engines also powered heavier fishing gear such as trawls. In addition, refrigeration and factory trawlers made possible the preservation and processing of more fish. In the early 20th century, large, diesel-powered fishing boats permitted fishing on a larger scale in the open ocean and throughout the ocean's depths. In the last half of the 20th century, freezing to preserve freshness, improved fish-locating techniques (e.g., acoustic fish finders), and precision navigational technologies (GPS) aided exploitation of fish stocks worldwide. Today's factory ships can process and store upwards of 1000 metric tons of fish.

Economics is an important factor in overfishing. Governments worldwide spend billions of dollars annually subsidizing their commercial fishing industry. Consequently, many fishery scientists and economists believe that the world fishing capacity is too large. They estimate that the world fishing fleet is about 2.5 times larger than needed to catch the ocean's sustainable fish production. Furthermore, worldwide government subsidies, totaling about $100 billion a year, support an industry that produces about $70 billion worth of fish annually. Each year, subsidies of $15 billion to $30 billion compensate fishers whose income was lost because of government restrictions on fisheries. In addition, trends in price structure can sometimes mask a fishery that is overexploited. It is a classic case of supply and demand. As the supply of a commercially valuable fish declines, its price rises, making the fishery more attractive for fishers. This increase in demand only hastens the decline in population and perhaps ultimately the collapse of the fishery.

Government subsidies and greater fishing efficiency, coupled with the growing demand for fish by the soaring human population drove an increase in total fish catch through the 1950s into the 1980s. The worldwide wild fish catch increased from 19.2 million metric tons (21.2 million tons) in 1950 to 88.6 million metric tons (97.6 million tons) in 1988. (Complicating matters, the People's Republic of China routinely over-reported their fish catch, significantly distorting available statistics on global fish production.) Because the rate of fish production exceeded human population growth rates, marine fisheries were thought to be the solution to the problem of feeding the world's burgeoning human population. At the time, the principal fishery management approach was to expand the areas fished and exploit newly discovered fish populations. Because many marine fishes are fecund, ocean fisheries were assumed to be inexhaustible.

The United Nations *Food and Agriculture Organization (FAO)* reported that after many decades of steady growth, the total annual wild fish catch began to level off in the late 1980s. In 2000, the catch peaked at 87.2 million metric tons (96 million tons) and then declined to 81.7 million metric tons (90 million tons) in 2003. Production levels in 2008 equaled those from 2003. Discovery of new fisheries no longer compensates for the decline in production of existing fisheries. Meanwhile, the total average catch per person fell from 17 kg per year in the late 1980s to 14 kg per year by 2003.

As fishers became more adept at taking fish in larger numbers, stock collapses became more frequent forcing fishery managers to impose quotas on commercially desirable fish, limit fishing time at sea, or close fisheries altogether. The first modern day fishery collapse occurred in 1971-72 involving the Peruvian anchovy fishery. Peru's fishing industry boomed in the 1950s and 1960s and, by 1970, was one of the largest in the world. At that time Peru accounted for about 20% of the total global catch of anchovies. However, overfishing combined with the effects of the 1972-73 El Niño sent the Peruvian fishery into a decline from which it has not recovered.

Overfishing continues to be a serious problem as shown by the collapse of the cod fishery in the North Atlantic off Newfoundland and New England, which was closed in the early 1990s and has only recently reopened (see the Case-in-Point at the beginning of Chapter 10). In the early years of the 21st century, cod fisheries bordering Northern Europe and in the Baltic Sea also showed signs of stress and possible pending collapse. Overfishing has led to the serial depletion of fish populations as each newly discovered stock is fished out and the fishing fleet moves on to the next stock or fishing area. Furthermore, each loss of biodiversity in an ecosystem (such as the collapse of a cod fishery) has been shown to lead to additional losses within the ecosystem and cuts in the odds of recovery. Expansion of production is possible in only about one quarter of the world's fish stocks. According to NOAA's *National Marine Fisheries Service (NMFS)*, more than 60% of the 200 commercially most valuable fish stocks are either overfished or fished to the limit. One consequence is that many marine fishes in fish markets come primarily from fish farms. Wild salmon, for example, are increasingly rare in the market place, replaced by those raised in pens (see the section on mariculture later in this chapter).

Sharks are an example of an over-exploited, top predator fish with a relatively low reproductive capacity. Sharks are slow to mature sexually, some requiring 15 years or longer. Dusky sharks first breed when they are 20 to 25 years old. Sharks bear live young and most have long gestation periods. The spiny dogfish shark, used in British fish and chips, has a 22-month gestation period. Thus, many sharks have annual replacement rates of only 3% to 4%. As their population declines, the slow reproduction rate means that shark species may risk extinction. Even if the fishery is closed, shark populations are slow to recover. For example, in the 1960s, overfishing caused the collapse of the porbeagle fishery in the northwest Atlantic Ocean. Almost three decades passed before these sharks recovered, but when the fishery reopened, it took only about three years for the stock to be overfished again.

Between 1986 and 2000, the numbers of sharks caught in the northwest Atlantic on the long lines used to catch tuna and swordfish diminished significantly. Over the 15-year period, hammerhead shark numbers were down by approximately 89% while white and thresher shark populations declined by about 75%. The thresher shark population may have collapsed; that is, they became extinct for commercial purposes. Some areas of the ocean had no white sharks at all. Only the mako shark did not show a decline. Overall, the rate of shark decline in the northwest Atlantic Ocean indicates that the shark fishery in that area is not sustainable.

Sharks are taken for their meat; some are killed accidentally in collisions with fishing vessels whereas others are caught along with target fish and then dumped overboard (*bycatch*). The most egregious threat to shark populations, however, is catching them exclusively for their fins (Figure 14.2). Demand for pricey Asian shark-fin soups fuels demand for fins that brings a much higher

FIGURE 14.2
NOAA Office of Law Enforcement agent counting seized dried shark fins. [Courtesy of NOAA]

price than shark meat. Crews engaged in "shark finning" typically slice off the fin of a captured shark and then toss the rest of the animal overboard, a tremendous waste of food. In August 1992, the U.S. Coast Guard stopped and boarded the *King Diamond II* out of Honolulu in waters off Acapulco, Mexico. Law-enforcement officials found 29 metric tons (32 tons) of fins onboard but no other shark remains. According to a NOAA official, this discovery indicates that about 30,000 sharks were taken and almost 0.6 million kg (1.3 million lbs) of shark meat was disposed of at sea.

In response to shark finning, the U.S. Shark Finning Prohibition Act has been in effect since March 2002. This law bans U.S. vessels anywhere in the world ocean from possessing shark fins unless the rest of the shark carcass is also onboard. The same regulation applies to foreign vessels fishing in the U.S. Exclusive Economic Zone (EEZ), 370 km (200 nautical mi) or more offshore (Chapter 4). While this law is well intentioned, trade in shark fins continues unregulated throughout most of the world ocean. According to a 2006 study published in *Ecology Letters*, worldwide fishers still take as many as 73 million sharks each year for their fins.

At the 2011 annual meeting of the *American Association for the Advancement of Science (AAAS)*, Villy Christensen of the University of British Columbia and colleagues, reported on the first worldwide analysis of trends in top predator loss. Findings were based on analysis of more than 200 reports of the constituents of local and regional marine ecosystems from the period 1880-2007. A slow, steady loss of top predators prevailed from 1910 to 1970 amounting to a total decline of 10%. But, as fishing fleets became more efficient, beginning in the mid 20th century, the rate of loss of top predators increased dramatically so that today populations are only a third of what they were in 1910.

As increasing fishing pressures cause populations of top predators to decline, other fish at progressively lower trophic levels begin to dominate food webs in marine ecosystems. In terms of biomass, loss of large top predators such as cod, grouper, swordfish, Atlantic halibut, and bluefin tuna is largely offset by a much greater abundance of small, prey fish such as anchovies, menhaden, jellyfish, bearded gobies, and sand lances. Fishers, in turn, must target fish populations at these lower trophic levels. Daniel Pauly of the University of British Columbia coined the phrase "*fishing down the food web*" for this shift in targeted fishes brought about by overfishing of top predators. From an economic perspective, ecosystems that are depleted in top predators and dominated by disproportionate numbers of small fish (along with crustaceans and phytoplankton) are less valuable.

A study of fish catches over the period 1950 to 1994 conducted by the United Nations Food and Agriculture Organization confirmed fishing down the food web. Researchers discovered a gradual shift away from food webs dominated by high trophic level fish to food webs dominated by low trophic level invertebrates and plankton-feeding fish. The FAO found that the rate of fishing down the food web was about 0.1 trophic levels per decade—likely an underestimate because of limited data from tropical fisheries.

Targeting fish at lower trophic levels initially may be accompanied by an increase in total catch. With depletion of top predators, their prey populations (smaller fish) increase in numbers. However, the greater total catch is short-lived as populations of the competitors of the top predators increase. Furthermore, smaller fish and other organisms at lower trophic levels undergo much more rapid population fluctuations than do organisms at the higher trophic levels. This implies a less reliable food supply for the limited number of remaining top predators and a less stable food web; that is, top predators are more vulnerable to environmental changes (including climate change) that impact their prey.

At the 2011 AAAS meeting, fisheries biologist Reg Watson, also of the University of British Columbia, reported that since the 1950s, in response to growing consumer demand for top predator fish, against a background of declining stocks, fishers increased their fishing effort, employing trawlers and helicopter-guided purse seines and other methods aimed at increased efficiency. In 1950, the global harvest of top predators was about 16 million metric tons (17.6 million tons) per year, mostly from fishing along the coast. In the 1980s, intense fishing spread into most of the open ocean and during the 1990s, the global harvest reached 80 million metric tons (88 million tons) although in spite of the considerable effort, the catch was leveling off and declined to about 76 million metric tons (84 million tons) in 2006.

Complicating matters is the possible impact of climate change (Chapter 12) and ocean acidification (Chapter 3) on fisheries. In the tropics, during this century, ocean temperatures are projected to rise about 2 Celsius degrees (3.6 Fahrenheit degrees) and rainfall is expected to be heavier. This will produce fresher, lower density water at the ocean surface, and colder, saltier water at depth—a stable stratification that will inhibit vertical motion and reduce the movement of nutrients into the photic zone. Less food at the base of marine food webs will translate into smaller fish. In addition, ocean acidification will require some fish to utilize energy to clear acid from their bodies. That energy is no longer available for growth and other biogical functions. According to William Cheung of the University of East Anglia, U.K., the combination of climate change plus energy loss due to ocean acidification could reduce fish growth by up to 30% to 40%.

Fisheries scientists estimate that about 8% of the ocean's primary production is now harvested in the global fisheries (Chapter 9). That is, 8% of the total global fixation of carbon by phytoplankton is being removed from the ocean in the form of fish. (This percentage is somewhat misleading because most of the ocean is relatively unproductive.) Fisheries harvest about one quarter of the total fish biomass produced each year. Based on these findings, many scientists believe that the fish catch is already near the limit of the ocean's fish yield. This view is further supported by the apparent decline in global marine fisheries early in the 21st century. Humans are now top predator in many—perhaps most—of the world's major marine ecosystems. Unlike other top predators, however, humans return essentially nothing useful to the ecosystems from which they take their fish.

The life expectancy of a fish is an important consideration in assessing how vulnerable a fish stock is to overfishing. How marine scientists determine the age of a fish is the subject of this chapter's first Essay.

MAXIMUM SUSTAINABLE YIELD

Successful fisheries management requires (1) a solid scientific understanding of the exploited stock, (2) targeted fishing quotas that provide the stock size that exceeds the biomass required to produce the maximum sustainable yield, and (3) supporting regulations (e.g., prohibiting illegal fishing) that are enforceable.

To prevent overfishing of a fish stock, fishery managers typically set catch quotas. The goal is to adhere to the **maximum sustainable yield** of the fish stock; that is, limits are set on fish catches (*total allowable catch*) so that stocks are maintained at a level that will ensure the long-term viability of the target species. Quotas must take into consideration the life span, population growth rate, and reproductive capacity of a particular fish species. A fishery is sustainable if it can be fished indefinitely at reasonable levels while maintaining the ecosystem (function, structure, and diversity) on which the fishery depends, and the integrity of the habitat essential to the fish species. As we have seen, in cases where a fishery has collapsed, the fishery may be closed in order to allow its population to recover.

The maximum sustainable yield is based on a model that simulates the population growth of an exploited fish species. One model consists of a sigmoid (*S*-shaped) growth curve as shown in Figure 14.3A. According to this model, population growth of a fish species is slow at first and then becomes more rapid. Eventually, as the number of fish increases to the point that available food resources begin to become limiting, population growth again slows and eventually levels off at or near the carrying capacity. The **carrying capacity** is the maximum population that can be sustained by the resources of the marine habitat. In the case of fisheries, the carrying capacity is the maximum population that would exist in the absence of commercial fishing.

As shown in Figure 14.3A the population growth rate is relatively low when the population is small or when the population nears the carrying capacity. At some intermediate size, the fish population has the greatest potential for growth and reproduction. For most fish species, reducing the fish population to about half its natural size (i.e., the carrying capacity) makes available more resources per individual and produces the maximum sustainable yield (Figure 14.3B). If the fish population does not drop below this intermediate size, the fish population can be sustained by limiting the catch to no more than the annual population growth rate. Today, the populations of most fish stocks are well below that which would produce the maximum sustainable yield.

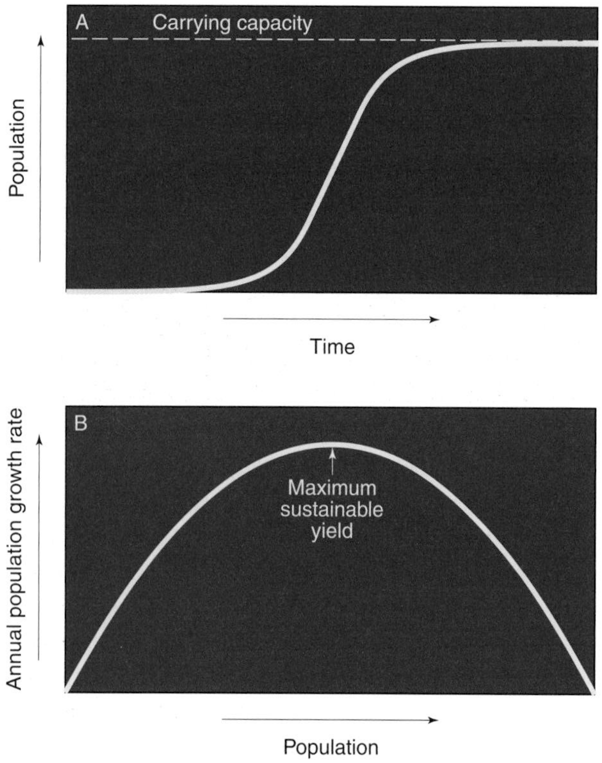

FIGURE 14.3
The maximum sustainable yield of fish is based on a model (A) that simulates the growth of the population of a fish species. The maximum population, the carrying capacity, is determined by the availability of resources in the marine habitat. The annual population growth rate (B) peaks at some intermediate population size; this is the maximum sustainable yield.

In late 2007, economists from the Australian National University and the University of Washington proposed an economics-based approach to the problem of overfishing. Using computer models, they studied four fisheries. As a fish population increases, the cost of catching the fish decreases—costs are lower because less time and fuel are needed to reach the total allowable catch. Hence, reducing the fish catch in the short term (until the population rebounds) translates into greater fishing profits in the long run. The only requirement for this scheme is that fishers who cut back on their catch must be the ones who directly benefit from the reduced cost of fishing. Furthermore, fishers who are not currently fishing cannot be allowed to participate later (after the fishery has recovered).

Also in 2007, J.R. Beddington and colleagues of Imperial College, London, argued that fisheries management strategies are more likely to achieve sustainable fisheries by (1) eliminating subsidies (described earlier), and (2) granting individual fishers a right to a portion of the total catch. In this way, fishers are no longer competing with one another and are more likely to seek to sustain the entire fishery because as the common fish stock increases, each individual's quota will increase. This creates a strong incentive to not overharvest. A 2008 study published in *Science* examined over 11,000 fisheries worldwide and found that only 14% of the 121 fisheries using catch shares had collapsed. Among those without catch share programs, 28% had collapsed.

At the time of this writing, catch shares were being used in 14 U.S. fisheries. In November 2010, NOAA issued a national policy encouraging the use of catch shares as a means to end overfishing and rebuild fisheries. NOAA administrator Jane Lubchenco explained that "catch share programs have proven to be powerful tools to transform fisheries, making them prosperous, stable and sustainable parts of our nation's strategy for healthy and resilient ocean ecosystems."

ECOLOGICALLY SUSTAINABLE YIELD

Fisheries managers are reassessing the traditional approach to sustaining the populations of exploited fish stocks. This reassessment stems from the realization that fish species are components of ecosystems in which they interact with other organisms (e.g., as food sources) and the physical/chemical environment (e.g., for habitat). A decline in the population of a particular species of fish may have a ripple effect on the entire ecosystem, altering its biotic composition and reducing its stability. This is especially the case when fishers decimate the populations of commercially attractive fish such as cod, tuna, swordfish and other species that are at the top of food webs (top predators). This ecosystem perspective argues for a more holistic approach to fisheries management.

In 1993, the 24th Annual Report of the President's Council on Environmental Quality (CEQ) recommended that the President issue a directive establishing a national policy to encourage sustainable development through ecosystem management. More than fifteen years later natural resource managers still struggle with this concept. Managers accept that humans do not manage ecosystems; rather it is human activities that use and impact ecosystems which are managed. Recognizing that, ecosystem management is better referred to as *ecosystem-based management* or an *ecosystem approach to management*. Ecosystem approaches integrate ecological principles, human systems, and goals of sustainability for use in the management decision-making process.

Since the release of the CEQ report, much has been written about the ecosystem approach to management. For marine ecosystems, the concept is arguably best developed for fisheries. While opinions vary regarding what constitutes an ecosystem approach to management, most concerned parties agree that an ecosystem approach to management needs to evolve incrementally from existing approaches. Applications are becoming increasingly sophisticated in response to more intense and diverse public interest, with the use of new knowledge, and by giving explicit consideration to uncertainty. Some key challenges in advancing ecosystem approaches are to (1) improve and create processes for public participation in objective setting and in prioritizing those objectives, (2) develop operational protocols for taking into account food web complexity (recognizing that only a crude level of predictability is likely, at best) and climate change, and (3) determine the functional value of habitat related ecosystem goods (e.g., food) and services (e.g., absorption of atmospheric carbon dioxide) that benefit humanity.

Through an incremental and collaborative transition to an ecosystem approach, managers can expect to achieve healthy ecosystems with increased social and economic value of its resources as well as an informed public by which the benefits of ecosystem management can be maintained.

To be ecologically sound and effective, fisheries management must be grounded in solid scientific understanding. Consider, for example, how a better understanding of Atlantic bluefin tuna may help to head off its extinction. Bluefin tuna is a top predator that is in great demand worldwide as a food fish (Figure 14.4). Since the early 1970s, use of bluefin tuna for sushi (fish with rice and vegetables) and sashimi (sliced raw fish) has soared. Although 75% of the global catch is consumed in Japan, these foods are also popular in American supermarkets, delis, and upscale restaurants. Overfishing, inadequate catch quotas, and illegal fishing are driving Atlantic bluefin toward extinction.

Unlike most of the 20,000 species of fish that are cold-blooded, the bluefin tuna is warm-blooded. An adult bluefin grows to a maximum length of 4 m (13 ft). The largest on record weighed 679 kg (1496 lb) and was caught off the coast of Nova Scotia. The bluefin is sleek and fast, reaching speeds up to 80 km per hr, and has a ravenous appetite, eating almost any animal they encounter.

Bluefin tuna ranges over a huge area of the North Atlantic, as well as portions of the Pacific and Southern Oceans. According to a December 2007 report in *The Washington Post*, two genetically distinct populations of bluefin tuna (*Thunnus thynnus*) live in the North Atlantic Ocean. They swim side-by-side most of the year but

FIGURE 14.4
Bluefin tuna. [Courtesy of NOAA]

separate for breeding; that is, the two populations do not interbreed. The spawning area is the Gulf of Mexico for one population and the Mediterranean Sea for the other population. Mixing of the two populations has resulted in inflated estimates of the western Atlantic bluefin tuna; that is, migrants from the Mediterranean give the erroneous impression of a greater western Atlantic population than actually exists.

In setting catch quotas for Atlantic bluefin tuna, the International Commission for the Conservation of Atlantic Tuna (ICCAT), a regulatory group founded in 1969, erroneously assumed that the two populations foraged for food in different portions of the Atlantic. The ICCAT treated the two populations separately and set strict quotas in the western Atlantic but relatively lenient quotas in the eastern Atlantic where overfishing has become a serious problem. Hence, the Atlantic bluefin tuna is closer to extinction than originally believed. This discovery has prompted conservationists to press NOAA's *National Marine Fisheries Service* for closure of longline fishing in the Gulf of Mexico during the spawning season (April to June) so that the Atlantic bluefin population may begin to recover. Furthermore, interest has developed in breeding bluefin tuna in captivity.

Protection of fisheries requires a thorough understanding of the entire ecosystem, not just the targeted commercial or recreational fish species. This understanding is the basis for achieving an **ecologically sustainable yield**, that is, the yield that a marine ecosystem can sustain without undergoing an undesirable change in state. New management concepts and associated policies must take into account all the species that interact with the targeted species and the habitats they depend on. Even reducing fish catches by 40% may not be enough to protect fish stocks and their ecosystems while allowing both to recover from past overfishing. Thus the goal of sustainable fisheries may be quite different from, and even incompatible with the ecologically sustainable goals of maintaining natural populations of fish and other members of the ecosystem. Furthermore, sustainability will require more extensive monitoring of the marine environment in which these ecosystems function to detect changes in conditions that might significantly alter the system. In short, humans will not only continue to be the dominant predator in marine ecosystems, but will also be responsible for managing, protecting, and restoring them if they are damaged by human activities or climate change.

BYCATCH

Many commercial fishers accidentally catch undersized fish or unwanted species in their nets or hook lines; these are discarded, either dead or dying, because they cannot be sold. Today, about one-third of the annual commercial fish catch worldwide is discarded for this reason. **Bycatch**, referring to fish and other marine animals that are taken in addition to the target species, is a major threat to many endangered species, such as sea turtles, dolphins, and other marine mammals. For example, sea turtles die in large numbers each year after swallowing floating hooks attached to long-lines intended to catch tuna or swordfish, or after they become tangled in shrimp nets. The many commercial fishing techniques are described in Table 14.1.

Ironically, the long-line fishing technique was developed to protect dolphins; it replaced nets that previously were used to catch tuna but inadvertently also killed large numbers of dolphins each year. In the late

TABLE 14.1
Types of Commercial Fishing Gear (modified after Monterey Bay Aquarium)

Gillnetting	Netting that hangs in the ocean from floats. The head of the fish passes through the net but not its body. As the fish tries to escape its gills are caught in the net.
Harpooning	Used to catch large bluefin tuna and swordfish. The fisher stands on a shipboard platform holding an aluminum harpoon, 3 to 4.3 m long, which is attached to a long rope. The fisher uses the harpoon to spear and kill the fish and then it is taken onboard.
Longlining	A central fishing line plus smaller lines are strung with thousands of baited hooks extending distances up to 80 km. May be used to catch fish near the sea surface (e.g., swordfish and tuna) or near the ocean bottom (e.g., cod and halibut).
Pole and Line	Fishers stand on the deck shoulder to shoulder, each holding a long single pole and a short line. Fish are caught one by one and pulled out of the water and onto the ship. Target fish include tuna and mahi-mahi.
Purse Seining	A large net encircles a school of fish. A line enables the crew to close the bottom of the net (like a purse). Target species include tuna, sardines, herring, and mackerel (top image of Figure 14.5).
Traps and Pots	Baited cages (usually attached to a line) are used to trap the catch and keep it alive until the fisher returns after several days. These traps or pots are used to catch lobster, crabs, shrimp, and sometimes bottom-dwelling fish.
Trawling/Dragging	Different types of nets are dragged behind a ship to fish in midwater (for schools of small fish such as anchovies), as shown in the bottom image of Figure 14.5, and along the ocean bottom (for bottom dwelling fish including cod and halibut). In some cases, nets with chain-mesh bottoms are dragged through sediment (e.g., to catch scallops).
Trolling	A boat with long rods pulls fishing lines through the water with different baits and lures.

1950s, fishers in the eastern tropical Pacific Ocean discovered that yellowtail tuna aggregated below schools of dolphins. Since then, the favored method of catching tuna has been to set nets around the schools of dolphin to capture the tuna beneath. For some time, on average more than 350,000 dolphins died each year when they became trapped in the fishing nets. More recently fishers have used special nets that allow dolphins to escape unharmed. Since 1998, about 2000 dolphins were killed each year in these nets, a 99% reduction in mortality. Despite this advance in protecting dolphins, three dolphin species are still listed as depleted (i.e., two species of spotted dolphin and one species of spinner dolphin). NOAA's National Marine Fisheries Service continues to work with other nations fishing in the eastern tropical Pacific to further reduce dolphin mortality.

Trawling for shrimp (also called prawns) is a major and yet little recognized contributor to the bycatch problem worldwide. Nets used to capture shrimp are dragged along the shallow sea bottom where they also catch all kinds of fishes, shellfish, and other animals, including sea turtles. An estimated 150,000 sea turtles are killed each year in shrimp-trawl nets. As noted later in this chapter, however, these turtle deaths can be reduced or avoided by using excluder devices that allow the turtles to escape shrimp nets unharmed.

FIGURE 14.5
Two of the many different fishing types detailed in Table 14.1. The top image is an artist's conception of purse seining operations, in which a large net encircles a school of fish, and a line encloses the net like a purse [NOAA]. The bottom image shows trawling, using a net dragged behind a boat. [Courtesy of U.S. Fish & Wildlife Service/Stockton Fish and Wildlife Office]

Worldwide, shrimp trawling results in a bycatch of 5 to 20 kg (11 to 44 lbs) of unwanted animals and plants for every 1.0 kg (2.2 lbs) of shrimp caught. (In the past, local artisanal fishers would use unwanted fish for their own consumption.) Much shrimp trawling is conducted in shallow tropical waters and the catch is exported to European or North American food markets. The European Union is the world's largest shrimp importer, much of it coming from tropical waters near developing countries. Shrimp fishing is also heavily subsidized by the European Union, which pays nearly half the operating expenses for shrimp trawlers working off the West African (Guinea-Bissau) coast. While representing only about 2% of the global fish catch, trawling for shrimp is responsible for about one-third of the world's total bycatch. Trawl nets also damage the sea bottom (Chapter 10).

Buying farmed shrimp does little good in protecting marine fisheries. Each kilogram of farmed shrimp produced requires feeding them two kilograms of fishmeal processed from wild-fish species. Furthermore, large areas of mangrove swamps are destroyed to construct shrimp ponds, which have a relatively short productive life before disease outbreaks necessitate building new ponds and abandoning the old ones.

The bycatch problem might be partially alleviated by changing or rewriting governmental regulations to relax quotas, permitting accidental catches to be landed legally. Success in this effort may be elusive because a policy that is too relaxed on accidental catches may actually encourage such "accidents." These regulations would apply only within national waters. In open-ocean waters, beyond the boundaries of Exclusive Economic Zones (EEZ), international cooperation would be required. This policy is important for marine mammals and sea turtles.

RESTORING FISHERIES

Few marine fisheries have been restored after experiencing severe depletion. The North Atlantic herring fishery temporarily recovered after fishing ceased during World Wars I and II. Another example is striped bass, also known as rockfish, which recovered from near collapse in the 1980s when Maryland and Virginia declared moratoriums on taking the species in Chesapeake Bay.

To establish sustainable fisheries for other species will require new strategies. Earlier in this chapter, we described the concept of ecologically sustainable yield. The goal of the ecosystem-based approach to fisheries management is to maintain or re-establish marine ecosystems that are home to fisheries. This would require protection of seafloor habitats (e.g., by phasing out destructive trawling practices) and expanding the areas of the world ocean designated as protected areas, called *marine reserves*, where fishing is prohibited. In these so-called *no-take zones*, fishes can grow to maturity and reach their optimal reproductive potential. The largest individual fishes are most desirable because they are the most prolific producers of eggs and sperm. For example, a single female red snapper weighing 12.5 kg (27.5 lb.) produces about the same number of eggs as more than 200 smaller female red snappers each weighing 1.2 kg (2.6 lb.). The resulting increased fish production could be taken in the waters surrounding marine reserves where fishing is permitted. However, existing no-take zones are small and few in number; together they cover only 0.01% of the area of the world ocean.

Within the federal waters portion of the Channel Islands National Marine Sanctuary offshore of Santa Barbara, CA, NOAA established marine reserves (110 square nautical mi) and a marine conservation area (1.7 square nautical mi). Effective 29 July 2007, all extractive activities were prohibited in the marine reserves. Within the Anacapa marine conservation area, recreational fishing for pelagic species and harvesting of lobsters is allowed. Established in 1980, the Sanctuary encompasses the waters surrounding five islands (from mean high tide to 6 nautical miles offshore) in a 1252 square nautical mi area of the Santa Barbara Channel (Figure 14.6). The primary goal of this Sanctuary is to protect the natural and cultural resources within its boundaries.

The Channel Islands National Marine Sanctuary is one of 14 national marine sanctuaries established since 1972 and administered by the NOAA National Ocean Service's *National Marine Sanctuary Program* in U.S. coastal waters and Great Lakes. In a similar vein, New York's Bronx Zoo operates a sanctuary for penguins and

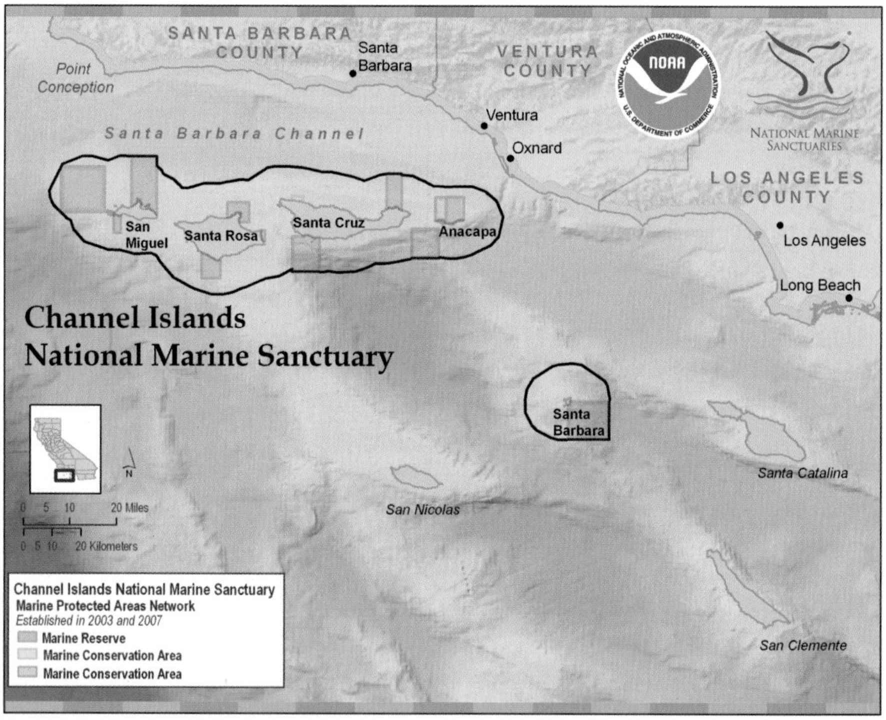

FIGURE 14.6
Map of the Channel Islands National Marine Sanctuary. [Courtesy of NOAA]

other sea birds on two uninhabited rocky islands near the Falkland (or Malvinas) Islands, off Argentina in the South Atlantic. (For more on marine sanctuaries and marine reserves, refer to the third Essay in Chapter 10.)

Other strategies proposed to help re-establish or protect fisheries include reducing the capacity of the global fishing fleet. As noted earlier in this chapter, the world fishing fleet is estimated to be about 2.5 times larger than needed to catch the ocean's sustainable fish production. Related to this strategy is the proposed elimination of government subsidies for unprofitable fishing operations. Furthermore, greater effort is needed to deal with the problems associated with bycatch.

HABITAT DESTRUCTION AND RESTORATION

Habitat destruction has also caused depletion or destruction of major fish stocks. Since steam-powered fishing boats began dragging heavy steel nets over the ocean bottom to catch bottom-feeding fishes (e.g., cod and haddock), benthic ecosystems and habitats have been heavily damaged. For decades the damage was ignored and fishers spoke of "plowing" the sea bottom to improve fish catches, much like farmers cultivate the land. Some fisheries scientists, however, argue that trawling the sea floor is more like clear cutting forests to improve hunting for rabbits or deer. This activity deprives adults of food and young fishes of places to hide from predators. Due to widespread damage to shallow sea bottoms, populations of cod and other long-lived, bottom-feeding fishes have experienced severe declines. Consequently, some commercial fishing operations have shifted to rapidly growing and reproducing open-water organisms, such as squid, which do not require intact benthic ecosystems and habitats for their survival.

Habitat destruction threatens oysters in coastal and estuarine waters. Tomales Bay, along the California coast about 70 km (43 mi) north of San Francisco, was originally home to a small, slow-growing, native oyster. Considered a delicacy, it was essentially destroyed as a commercial fishery in the late 1800s due to local market demand during the Gold Rush days. In an effort to restore this native oyster in the Bay, floats with mesh bags were used to keep the oysters off the bottom so that they could grow faster and avoid predation. Part of the objective is to restore the natural filtering capacity of Tomales Bay's waters. As described in this chapter's Case-in-Point, another example of where habitat destruction resulted in the decline of oyster populations comes from Chesapeake Bay.

The sea grass beds of Chesapeake Bay provide habitat for shallow-water organisms such as blue crabs. Young crabs hide in grass beds to avoid being eaten by predators, including fish, when they are especially vulnerable after shedding their hard shells and before their new, soft shells harden. Trawling to catch blue crabs cuts wide swaths and decimates sea grasses in Chesapeake Bay. Input of excess levels of nutrients from farms, cities, and industry also played a role by promoting growth of algae in the water and on the sea grass blades, depriving the plants of the light they need for photosynthesis. Efforts to restore the sea grass beds include regulation and reduction of anthropogenic nutrient discharges and planting new sea grass beds (Figure 14.7). The goal is

FIGURE 14.7
(A) Volunteers help NOAA scientists prepare seagrass shoots for planting [NOAA]. (B) Other volunteers help restore submerged seagrass near Solomons Island in Chesapeake Bay. [Photo by Mary Hollinger, NODC biologist, NOAA]

to restore about 46,000 hectares (114,000 acres) of the nearly 240,000 hectares (600,000 acres) of submerged aquatic vegetation that originally grew on the Bay bottom providing food for ducks and habitat for many species of small fishes and invertebrates. In 1978, before restoration efforts began, only 16,000 hectares (41,000 acres) of sea grasses remained.

For more on restoration of coastal wetlands, refer to this chapter's second Essay.

RECREATIONAL FISHERIES

Recreational fisheries in the coastal ocean and in lakes are important activities worldwide. They involve large numbers of people who catch fish for sport rather than for profit. This type of fishing is part of the tourism and recreational industry, the world's largest and fastest growing economic activity. In the U.S., more than 12 million people engage in recreational fishing each year and spend about $23 billion; the industry employs about 384,000 people. For some fish species, the value and the amount of the recreational catch equal or exceed that of the commercial fishery. Striped bass, blue fish, and flounder in the Mid-Atlantic region are examples of such prized fishes. More than half the fishes caught by recreational fishers are released alive; unfortunately many die due to the stress of "catch and release."

Important differences exist between recreational and commercial fisheries. Most recreational fishers prefer to fish near home; hence, sport fish populations are most heavily fished near major urban areas. Also recreational fisheries do not damage fish habitat as much as some commercial fishing practices. Furthermore, when the catch of popular recreational fishes declines, fisheries management agencies often resort to fish hatcheries to replenish depleted stocks. The small fishes are released into recreational waters where they grow to a size that will satisfy the sport fishing demand. Lake trout is one example of such a hatchery-raised fish released to many North American lakes; another is salmon on the Pacific coast.

Despite the large numbers of people involved and their associated expenditures, data for recreational fisheries are not as readily available as for commercial fisheries. Based on anecdotal accounts, recreational fisheries have followed some of the same exploitation patterns as commercial fisheries. They expand into new territories as populations of the large-bodied, slow growing but highly prized fishes decline. Like commercial fisheries, the sports fisheries shift to less desirable species and often end up with smaller, fast-growing fish. Collapse of recreational fisheries may be the culmination of a slow population decline over several decades. In time, the expectations of sport fishers change as memories of the preferred species fade.

Protecting Endangered Marine Species

Many large marine animals are protected by government regulations because their populations have been greatly depleted by commercial fishing or other human activities over the past century or so. In 1973, the U.S. *Endangered Species Act* became law providing for the protection of endangered and threatened species (and their habitats). Species determined to be in imminent danger of extinction throughout a significant portion of their range are listed as *endangered*. Species are listed as *threatened* if they are likely to become endangered in the foreseeable future. NOAA's National Marine Fisheries Service (NMFS) has authority for listing (most) marine species as threatened or endangered.

The *Convention on International Trade in Endangered Species of Wild Fauna and Flora (CITES)*, set up by international treaty in 1975, is intended to control international commerce in endangered species and their products. Species listed in Appendix I of CITES are threatened with extinction; trade in these nearly 900 species is prohibited. Species listed in Appendix II of CITES may become threatened and their trade is restricted. At present, there are 175 nations that are parties to CITES, meaning that the Convention has entered into force. In spite of CITES, many marine species including sea turtles, whales, and water birds remain threatened or endangered. *Science* reported that in 2010, CITES faced major challenges in adding several marine species, including three species of hammerhead sharks and the whitetip shark, to Appendix I. Making decisions of trade bans is a constant battle between institutions which profit from international trade and the community of environmentalists and conservation biologists concerned about species loss.

SEA TURTLES

The large graceful marine turtles (called *sea turtles*) inhabit mostly tropical and subtropical ocean waters but are also found occasionally in bays and estuaries where they are most likely to interact with humans. While they resemble their more familiar terrestrial relatives, tortoises and freshwater turtles, sea turtles have special adaptations for the marine environment. Instead of legs, they have flippers for swimming and their shells are lighter and more streamlined to reduce water resistance. All sea turtles, except leatherbacks, have hard upper shells (*carapace*) covered with large scale-like structures (*scutes*) and a hard lower shell (*plastron*). Unlike most other animals, the carapace in turtles incorporates their backbone and ribs. The leatherback sea turtle has boney plates under the leathery skin on its back.

Sea turtles breathe air but can hold their breath for long periods when they dive or rest on the sea bottom. While swimming, they remain near the ocean surface and come up to breathe every few minutes. Some species draw water into their mouth and throat via their nasal passages and oxygen is extracted by the pharynx (similar to a gill). Like all reptiles, sea turtles are cold blooded; that is, their body temperature is the same as the surrounding water. They migrate seasonally, remaining in waters having their desired temperature range and may die if they enter waters that are too cold for their metabolic needs. One species, the leatherback turtle, apparently can create some body heat, permitting them to withstand the cold waters off Canada and Iceland, where they spend the summer. Sea turtles live in the ocean their entire lives except for adult females who come ashore during nesting season to lay eggs in beach sand (Figure 14.8).

Seven species of sea turtles are recognized worldwide. The populations of all of them are listed as either endangered or threatened. Technically, the flatback sea turtle is neither threatened nor endangered—it nests on the remote beaches of northern Australia. But because it resembles the green sea turtle, trade in its products are prohibited by CITES. Many human activities threaten sea-turtle populations: (1) commercial harvesting of adult turtles (e.g., for food, hides, oil) and the poaching of eggs (believed by some to be an aphrodisiac) from nests, (2) development projects (e.g., seawalls) along beaches used by turtles to nest, (3) bycatch in commercial fishing nets, especially shrimp nets, (4) collisions with ships, including recreational boats, (5) ingestion of plastic litter, especially bags, and (6) marine pollution.

Green sea turtles occur throughout the world ocean between about 35 degrees S and 35 degrees N and are considered the most commercially valuable of all sea turtles (Figure 14.9). Hawksbills prefer warmer tropical waters and have a more limited range than the other sea turtles. The shells of hawksbills ("tortoiseshell") are used for jewelry, the principal reason for its endangered status. Kemp's (Atlantic) ridley, the rarest of sea turtles, lives in the Gulf of Mexico and along the U.S. East Coast. The only known nesting beach for this species of sea turtle is Rancho Nuevo, Mexico. The olive (Pacific) ridley is found mostly in the warm waters of the Pacific Ocean but also can be found in the Atlantic and Indian Oceans. It nests along the west coast of Mexico; its status is endangered along the Pacific coast of Mexico and threatened elsewhere in its range. The loggerhead, found throughout the world, has its second greatest population along the U.S. southeast coast. Loggerheads nest on beaches from Texas to New Jersey.

FIGURE 14.8
Federally endangered loggerhead sea turtles lay their eggs on beaches. [Courtesy of U.S. Fish and Wildlife Service photo]

Leatherback turtles are the largest sea turtles (and one of the largest of all marine reptiles); they can reach a length of 2 m (6.6 ft) and weigh up to 590 kg (1300 lb). Leatherbacks have a different lifestyle from other sea turtles, spending most of their lives in deep-ocean waters, between about 50 degrees N and 50 degrees S. In a single year, they can migrate from South America to Nova Scotia. They feed primarily on jellyfish, but can dive to depths of almost 1000 m (3280 ft) in search of other food. These long-lived animals take 10 to 20 years

FIGURE 14.9
A green sea turtle. [Courtesy of NOAA]

to reach sexual maturity. Like all sea turtles, leatherbacks are threatened by poachers who take their eggs from nests on beaches. Their greatest threat, however, comes from long-line fishing for tuna and swordfish. Long-lines are up to 80 km (50 mi) in length and each line has thousands of floating hooks. One scientist calculated that on any given day up to four million hooks are in the ocean for tuna and swordfish. Leatherbacks are caught when they take the bait and swallow the hook or they become entangled in the lines. In the past two decades, almost 95% of Pacific leatherback turtles have disappeared.

The population of all sea turtles declined nearly 90% between 1980 and 2000. Unless this trend is reversed, they may be extinct by 2030. Because they live primarily in open-ocean waters, protecting them will require international action. There are several ways in which the threats to sea turtles by commercial fishing can be reduced. As mentioned above, one strategy is the use of a special trap door in fishing nets that trawl the sea floor for shrimp. These low-cost devices allow turtles to escape the net easily without impeding the shrimp catch. Also, reducing the time a trawl net remains on the bottom lowers the death rate among trapped turtles. Another approach is to declare areas with high seasonal populations of turtles off limits to fishing at these times.

In the United States, NOAA's *Office of Protected Resources* within the National Marine Fisheries Service oversees activities to protect sea turtles. These efforts involve training commercial fishers to use turtle exclusion devices. Because sea turtles nest on land, the U.S. Fish and Wildlife Service co-ordinates conservation efforts on beaches.

WHALES

Hunting whales began early in human history. The first whalers used small boats and hunted close to shore, as some coastal-dwelling native peoples still do primarily around the Arctic Ocean. Around CE 1000, Basque whalers from northern Spain began hunting in the North Atlantic. In the 16th century, these whalers used larger vessels to reach North America where they established whaling stations in Newfoundland. Later, English and Dutch whalers joined them in hunting whales in the Arctic Ocean. These early whalers hunted the North Atlantic right whale, a large (45 metric ton or 50 ton), surface feeding whale. It is called the "right whale" because the animal was so easy to take, that is, it swims slowly, was easy to kill, and floated after death so it could be towed ashore for processing.

Right whales were very common in the Gulf of Alaska nearly 200 years ago. In the 19th century, American whalers harpooned over 23,000 North Pacific right whales. Today, it is estimated that between 30 to 54 right whales reside in the Gulf of Alaska, and only 8 are female.

Whales provide meat for humans (especially in Japan) and for animal feed. Before petroleum became available for oil lamps in the second half of the 19th century, whale oil was used for that purpose. Other uses for whale products included baleen for corset stays, hoop skirts, and umbrellas. Spermaceti, a valuable oily substance from sperm whales, was used to make cosmetics and candles.

When Northern Hemisphere whale stocks declined markedly early in the 20th century, commercial whalers using larger steam-powered vessels moved into the untapped Southern Ocean to exploit the large whale stocks living there. Norwegians dominated commercial whaling worldwide. Initially, whales were taken to factories ashore for processing. After 1924, factory ships were developed to process whales at sea. Increased efficiency was the reason the whale catch rose from 12,000 in 1910 to 40,000 in 1940.

Like fishers, whalers focused on a single species until its numbers were so reduced that they had to hunt another (usually smaller) species. Whales are social animals and some use protected inlets for breeding, making them vulnerable to whalers, who used small boats and hand harpoons to take the whales in these close quarters. In the early 1900s, the huge blue whales were preferred and in 1930-31, 29,000 of these animals were taken. Blue whales were finally protected in 1965 but by then their numbers had dropped to an estimated 6000 in the entire world ocean. After blue whales (135 metric tons or 150 tons) were depleted, whalers took fin whales (45 metric tons or 50 tons) and then the smaller sei whales. Today, Japanese and Norwegian whalers still take the much smaller minke whales (5 metric tons or 6 tons). This is another example of serial depletion and harvesting down the food web.

In 1982, to protect the world's remaining whale stocks, the *International Whaling Commission (IWC)* declared a moratorium on commercial whaling to take effect in 1986. The IWC, founded in 1948, is the decision-making arm of the *International Convention for the Regulation of Whaling (ICRW)*. According to the ICRW charter, the International Whaling Commission was intended "to provide for the proper conservation of whale stocks and this makes possible the orderly development of the whaling industry."

As of this writing, the IWC has 89 member nations (about evenly divided between those that do or do not favor whaling). It is one of the few international regulatory bodies that set policy to protect marine animals from possible extinction. The IWC regulates whaling to conserve stocks, which were greatly depleted during centuries of uncontrolled commercial whaling.

Japan, Norway, and Iceland objected to the 1986 moratorium and continued commercial whaling. These nations claimed that the populations of smaller whales, such as the minke whale (about 8 m in length), are now at healthy levels (numbering in the hundreds of thousands) and can sustain whaling. Another argument is that whales eat many commercially valuable fishes and must be hunted to protect those fish stocks.

Japan later withdrew its objection to the moratorium and opted to take whales for research purposes only, so-called *scientific whaling*, which is permitted under the IWC moratorium. Japan now hunts whales in the North Pacific and Southern Oceans and even intends to take humpback and fin whales in the Southern Ocean Sanctuary. The ICRW sanctions catching whales for scientific study under Article VIII. The original drafters of the document assumed that less than 10 whales would be killed for scientific purposes each year, far fewer than the hundreds of whales that Japan takes annually. Japan defends its continued whaling on cultural, historic and economical grounds. A government agency sponsors Japan's "scientific whaling" program and sells whale meat not used for research and much of it ends up in Japanese restaurants where whale meat is considered a delicacy. Many Japanese view the moratorium, strongly supported by the United States and many other nations, as a form of cultural imperialism. Cultural differences such as this make international conservation policies extremely difficult to negotiate and enforce.

Norway resumed commercial whaling in 1993, defying the IWC's moratorium. In 2005, Norwegian whalers took 639 whales. Most of the meat is eaten locally but small amounts were exported to Iceland before it too resumed whaling. Because no market now exists for whale fat (blubber), it is either burned or dumped at sea. Claiming that whale stocks had recovered sufficiently to permit commercial and scientific whaling, in 2002 Iceland announced plans to resume whaling, targeting minke and fin whales. Iceland intends to study the effect of whales on ecosystems. Many nations and organizations object to whaling, even for research purposes, arguing that it is actually commercial whaling and has little to do with scientific research.

The IWC allows aboriginal people (Inuits) in Alaska, Greenland, and Siberia to conduct small (subsistence) whale hunts for cultural reasons and for food as they had done traditionally. In 2002, their request to take 280 bowhead whales was denied by the IWC. Interestingly, opposition to the Inuit request was led by Japan, Norway and Iceland.

FIGURE 14.10
Two blue whales photographed in the Channel Islands National Marine Sanctuary. [Photo by A. Lombardi, courtesy of NOAA]

Whale watching is an important activity of *ecotourism*, one of the fastest growing sectors of the tourism industry. Iceland's decision to resume whaling was controversial among its citizens at least partially because whale watching is a significant part of that nation's tourism industry, earning Iceland an estimated $8 million per year. Whale watching has become so popular in many places such as Seattle, WA that there is concern that the tour boats are making so much noise that it disturbs the whales, disrupting their normal behavior.

Even after decades of protection, many whale stock populations still have not rebounded to their original numbers. Among endangered whales are the North Atlantic right whale, blue (Figure 14.10), and bowhead whales. A few whales, such as the California gray whale, have recovered to roughly historic levels; their major breeding ground in Baja California is protected by Mexico, which has banned whaling in its coastal waters thus creating the world's largest whale reserve. Some populations of fin and humpback whales have experienced a remarkable recovery (Figure 14.11).

Marine mammals and other animals, including fishes, may be adversely affected by other environmental factors, including increasing noise levels in the ocean. Noise levels in the ocean have been increasing since the beginning of the Industrial Revolution from a variety of activities including ships, petroleum exploration and production, and naval exercises. The ability of mammals to use sound pulses for communication, locating

FIGURE 14.11
Populations of humpback whales have recovered after being listed on the endangered species list. [Courtesy of NOAA]

FIGURE 14.12
While the brown pelican population on the U.S. Atlantic coast has recovered to such an extent that they were removed from the endangered species list, the larger California brown pelican (pictured above) is still endangered. [Courtesy of U.S. Fish and Wildlife Service]

prey, and possibly even to kill prey has been known for many years. Excess noise levels at sea apparently may cause whales to beach themselves onshore where they subsequently die. Sound pulses used in acoustic thermometry to map water temperatures over entire ocean basins have also been challenged as potentially harmful, although these signals are of much lower strength (Chapter 3).

WATER BIRDS

Restoration of some water-bird populations is among the few successes resulting from environmental policies and regulations in the late 20th century. Populations of brown pelicans were greatly depleted by commercial hunters for their feathers to decorate ladies hats and dresses in the late 19th and early 20th centuries. Brown pelicans are large, fish-eating coastal birds. In 1903, President Theodore Roosevelt (1858-1919) designated Florida's Pelican Islands as the nation's first wildlife reserve. In 1918 the Migratory Bird Treaty gave them further protection from hunting.

Commercial fishers often killed pelicans by raiding their nests because they were thought to be voracious predators of commercially valuable fish stocks. As a result, pelican populations declined markedly. After 1940 their numbers were further reduced by the widespread use of DDT (dichlorodiphenyltrichloroethane) and other persistent organic pesticides. These compounds accumulated in the fatty tissues of the fish they ate and in the birds themselves. DDT caused pelicans eggs (and the eggs of other birds) to develop such thin shells that they broke before hatching.

Brown pelican populations began recovering after 1972 when the U.S. Environmental Protection Agency banned the manufacture and use of DDT (Figure 14.12). By 1985, brown pelican populations on the U.S. Atlantic coast had recovered to such an extent that they were removed from the endangered species list. Since then the birds have expanded their range, moving northward as far as Chesapeake Bay. However, populations of brown pelicans living on the Pacific and Gulf coasts as well as in Central and South America remain endangered. DDT is still manufactured and used in some developing countries to control mosquitoes and combat malaria.

The history of North American ospreys during the 20th century was similar to that of the brown pelican. They also are large fish-eating birds, sometimes called fish hawks. They live on rivers, lakes, and estuaries across much of North America in summer and winter in South America. Like brown pelicans, osprey populations were decimated by the egg thinning caused by DDT accumulation in their tissues. The 1972 ban on DDT greatly aided their recovery. Ospreys build their nests in dead trees near the water and also use duck blinds, power poles, or channel markers as bases for their nests. For many years, U.S. Coast Guard personnel destroyed their nests on navigational aids, thus limiting the number of available nesting sites. After the Coast Guard reversed that policy, the number of ospreys rebounded dramatically. Now they are commonly seen and heard near coasts, lakes, and rivers.

Mariculture

Wild fish stocks are declining, increasingly stressed by overfishing and, in some cases, collapsing, such as the once plentiful cod in the North Atlantic. This is taking place as the global human population, currently 6.9 billion, is projected to reach 9.3 billion by 2050. Wild fish caught at sea and beef raised in feed lots cannot meet the growing human demand for protein. In summarizing the situation in the February 2011 issue of *Scientific American*, Sarah Simpson reports that the most sustainable source of protein for humans is **mariculture** (*marine aquaculture*), coastal or offshore farming of fish and shellfish, now accounting for half of all seafood produced worldwide (up from 9% in 1980) (Figure 14.13). Mariculture could supply 62% of the world's total protein by 2050.

For many varieties of fish and shellfish (e.g., salmon, catfish, trout, mussels, oysters, and clams) farmed fish and shellfish dominate retail markets. They are raised within 5 km (3.1 mi) of the shore under state jurisdiction although recent technological advances and soaring demand is beginning to shift fish farming well offshore (within the EEZ) where the practice would be regulated by federal law. Most of these oceanic fish farms are large-scale operations. They have huge floating pens, usually located in protected inlets, holding hundreds of thousands to millions of animals. In many cases, fish farming requires considerable modification of coastal ecosystems. For example, shrimp farming in Mexico and Southeast Asia required clear-cutting of mangrove forests.

The simplest mariculture operations involve filter-feeding mollusks (i.e., mussels, oysters and clams). They are usually grown in mesh bags suspended in seawater from floating platforms. Because these animals feed by filtering plankton from the water, they do not require additional food. As they grow, mollusks must be tended to remove algae that grow on the bags and limit the circulation of oxygenated water. Periodically the bags must be replaced as the animals grow.

Farmed carnivorous fish are fed a diet of fishmeal and fish oils; some of it sinks uneaten to the bottom along with considerable fecal matter. Critics of fish farms compare them to pig farms or cattle feed lots on land in terms of the amount of waste material produced. Decay of these organic deposits can deplete near-bottom waters of dissolved oxygen (DO) and may give rise to harmful algal blooms (HABs). Pens that are anchored well offshore, however, produce less water pollution because relatively strong currents dilute and carry off waste material. In coastal areas, this water pollution problem can be alleviated to some extent by raising seaweed and filter-feeding animals (e.g., mollusks, red sea urchins) nearby the fish pens. These organisms consume organic waste. In some cases, such as European and American salmon, the fish are so densely packed in their enclosure that disease and pathogens (disease-causing organisms) spread rapidly through the population. If infected fish escape their enclosure, they may spread the disease to wild stocks.

A major problem with mariculture is the dependency on small, wild fish, such as anchovies and sardines, as food for farmed fish. Small fish are caught and ground into fishmeal and oil. Currently, mariculture consumes 68% of fishmeal produced worldwide. The increasing demand from fish farms is beginning to stress populations of anchovies and sardines. Fish farming is inefficient in that the biomass of fish in fishmeal is greater than the biomass of the fish intended for market. For example, 0.7 to 0.9 kg (1.6 to 2.0 lb) of anchovies is needed to produce 0.5 kg (1.0 lb) of yellowtail tuna. About 1.1 kg (3.0 lb) of fishmeal is required to grow 0.5 kg (1.0 lb) of salmon. The goal of achieving a break-even ratio is not reached.

A distinct advantage of farmed fish versus top predators caught in the wild involves energy efficiency. Farmed fish expend much less energy avoiding predators, hunting for food, seeking mates, and for reproductive activity. Most of the energy available to farmed fish is for growth so that they can be harvested at a younger age and are less likely to be seriously impacted by bioaccumulation of persistent toxic substances. Furthermore, fish farms eliminate losses due to bycatch, trawling, and dredging.

FIGURE 14.13
Cobia at a Puerto Rican offshore mariculture facility. Mariculture refers to the farming of fish and shellfish in the ocean. [Courtesy of NOAA]

The salmon-farming industry has grown dramatically since the 1960s when it began off Norway. Today, it dominates the world market for salmon, which is supplied by large farms in cold-water inlets along the coast of Norway, Scotland, Atlantic Canada, British Columbia, and Chile. Wild salmon, while highly prized for its flavor, is now rare, but farmed salmon is abundant, inexpensive, and available year-round. The preferred species for these fish farms is Atlantic salmon because it grows faster, is easier to handle, and more animals can be grown in a single pen than other salmon species. The fish are grown in square steel-net pens (about 30 m or 100 ft on a side) suspended in the water from floating platforms.

Farmed salmon must be inoculated against diseases that would otherwise kill them because they live in such crowded conditions. Pesticides are also added to their food to control infestations of fish lice. Toxic chemicals are used to treat the nets to prevent algae growth that would otherwise clog them. Finally the salmon must be fed pigments with their food to make their flesh salmon-pink rather than a less appealing pale gray. In the wild, the salmon's normal color comes from eating pink krill, a variety of large zooplankton. Salmon farms located along the migratory routes of wild salmon can threaten the wild populations as they swim from their freshwater hatching grounds to the ocean. Captive fish escaping when their pens are damaged by storms or sea lions may breed with wild fish, reducing their genetic variability or even replacing them entirely. In contrast, if wild salmon pass too closely to the pens, these salmon may become infested by parasites that normally occur in low concentrations in the wild.

In 2001, researchers discovered sea lice on wild juvenile salmon after they passed salmon farms in British Columbia. Sea lice are tiny crustaceans that feed on fish, creating lesions that disrupt the fish's ability to regulate bodily fluids (an important function in salmon, which start and end their life in fresh water but live most of their time in seawater). In the wild, sea lice occur naturally on adult salmon but not juvenile, which suffer more severely from an infestation. In 2003, Martin Krkošek and colleagues at the University of Alberta, Edmonton, began investigating the salmon sea lice problem. As reported in the 14 December 2007 issue of *Science*, they analyzed 35 years of data on 7 populations of pink salmon in the Broughton Archipelago near Vancouver Island comparing rivers that flow through channels with fish farms and rivers that do not. For most of the study period, mortality differed little among wild salmon but after the lice infestation began in 2001, the mortality of the wild salmon passing close to the pens increased. In fact, it increased so quickly that some populations of wild salmon could be wiped out within only four generations (about 8 years). Based on these findings, Krkošek and colleagues recommended that barriers be erected surrounding the salmon farms to prevent the spread of sea lice and other parasites, but other scientists are not convinced that the evidence supports a major threat to wild salmon.

Marine Exotic Species

Exotic species, sometimes called *alien species*, are animals and plants introduced into ecosystems, usually by humans. Some introductions are intentional (new organisms used for mariculture) whereas others are unintentional (transported in ships' ballast waters). Introduction of exotic species has been taking place for centuries, especially in coastal waters. In most cases the introduced organisms do not survive because they are not well adapted to the physical or chemical conditions of their new environment or they are unable to compete in the new ecosystem. But sometimes an exotic species finds its new environment to be favorable; lacking competitors or diseases, their population expands rapidly. Such introduced species often have no predators to control its numbers. Also they often are outside the range of many of the diseases they normally encounter in their native habitat. Thus, they grow larger than normal, reproduce faster than usual, and take over—perhaps destroying—entire ecosystems. This situation develops in both coastal ocean waters and lakes. (Examples of exotic species in the Great Lakes are lamprey eels and zebra mussels.)

In the late 1950s, an exotic parasite appeared in Chesapeake and Delaware bays and promptly decimated the world's most productive oyster fisheries. A million oysters died in only one year. Oyster production had soared during the 19th century but then plunged because of overfishing. Beginning in the 1920s and up to the arrival of the parasite, the oyster harvest stabilized at a lower but sustainable level. Eventually, scientists identified the parasite as one that lives in the Japanese oyster *Crassostrea gigas*, widely used for aquaculture. The Japanese oyster was introduced to the U.S. east coast to assess its potential. It is also possible that ships returning from duty in Japan and Korea after World War II and the Korean War unintentionally transported the parasite to the U.S. In any event, Chesapeake and Delaware Bay oysters had little or no resistance to the parasite and quickly succumbed. In the 1980s, the parasite spread as far north as Maine and as far south as Florida. Today, the parasite is still present and continues to kill first year spat but fewer adult oysters die apparently because they have developed some resistance to the parasite (especially those living in the lower Chesapeake Bay).

The lionfish may be the first non-native fish to become established in the western North Atlantic and Caribbean Sea (Figure 14.14). Its population primarily consists of *Pterois volitans* with much smaller numbers of the closely related *Pterois miles* (devil firefish). The lionfish occurs naturally throughout the western Pacific while the devil firefish is found in the Indian Ocean and Red Sea. The lionfish, a voracious predator that consumes large amounts of native fish and crustaceans, possesses many characteristics that give it a competitive advantage over most native species. Lionfish have no known predators, resist parasites, can reproduce year round, and are fast growing. Furthermore, they defend against predators using their venomous dorsal, ventral, and anal spines. These venomous spines are hazardous to divers producing a painful sting.

Lionfish are nocturnal, inhabiting crevices in coral reefs during the day (at depths of 10 to 175 m) and moving to deeper water at night in search of food. Researchers report that predation by invasive lionfish is reducing the density of native fish found in coral habitats. Also, they discovered that accumulation of new juvenile fish via settlement of larvae was reduced 79% on reefs with lionfish versus those without.

The first sighting of a lionfish in U.S. waters is attributed to a lobster fisherman off Dania, FL, in October 1985. Subsequently, six specimens were reported as escaping from a south Florida aquarium during Hurricane Andrew in August 1992. (In fact, the aquarium industry is generally assumed to be the means whereby lionfish reached Atlantic and Caribbean waters.) It was not until 2000 that sightings of lionfish became widespread in U.S. waters. With rapid population growth, by 2010, lionfish populations became established along the Atlantic coast of the U.S. from Miami to North Carolina, the Caribbean coast of Central and South America, the Gulf of Mexico, throughout the Greater Antilles, and through the Leeward Islands.

New exotic organisms may also result from subtle exchanges of genetic materials. The tall marsh grass (*Phragmites*) commonly found in the mid-Atlantic salt marshes has become a widespread and invasive species that crowds out other types of marine grasses. For many thousands of years prior to about 1910, the native form of *Phragmites* grew along with other marine grasses. Beginning in the 1970s, the plant became far more intrusive. Apparently some genetic material, perhaps only a small root fragment of a closely related grass, was

FIGURE 14.14
The lionfish is an exotic species that has caused considerable problems in the fishing industry. [Courtesy of NOAA]

brought to North America from Europe or Asia. The genetic material from this exotic variety became dominant and has supplanted the native, less invasive form. Now only plants having genetic material from the Asian form are common.

Exotic marine organisms can also cause health problems among humans. In 1993, a new strain of bacteria, native to Bangladesh and India, appeared in Peru's coastal waters and later spread throughout Central and South America. The bacteria were apparently introduced by ballast water discharged from a ship. Carried by coastal plankton, this strain is now endemic in the region and causes severe cholera-like symptoms when ingested by humans.

Conclusions

For thousands of years humans have looked to the ocean as an important source of food. In fact, the belief was widespread that the sea held an almost limitless bounty of food and inspired some to speculate that marine fisheries could meet the needs of Earth's burgeoning human population well into the future. A variety of technological innovations enabled fishers to become more and more efficient at taking the most desirable top predator fishes and as those fish stocks diminished, they typically moved on to another fishery and repeated the process. However, by 2000, the world fish catch peaked and has been declining ever since. Overfishing forced imposition of quotas and the closure of some fisheries but not before some species were driven to the brink of extinction. Overfishing and its associated problems (e.g., habitat destruction, bycatch) spurred fisheries managers to reassess their management schemes. The move is to shift from the traditional method of managing fisheries for maximum sustainable yield to a more holistic, ecologically based approach that emphasizes the sustainability of the marine ecosystem (structure and function) that is home to the fishery.

Basic Understandings

- Stewardship of the ocean and its resources involves responsibly managing all resources to benefit present and future generations. In the United States, the National Oceanic and Atmospheric Administration (NOAA) is the principal government agency with stewardship responsibilities for the nation's marine environment and living resources.
- Central to exercising wise stewardship is sustainability, defined by the United Nations' *World Commission on Environment and Development* as "developments that meet the needs of the present without compromising the ability of future generations to meet their own needs."
- Overfishing occurs when a fish species is taken at a rate that exceeds the maximum catch that would allow reproduction to replace the population. With recruitment overfishing, adult fish are taken in such large numbers that too few survive to replenish the breeding stock. Growth overfishing takes place when fish are taken too small, before the animals can grow to a size that would produce maximum yield.
- A variety of technological innovations enabled fishers to greatly increase the fish catch. These innovations include new engines to power larger fishing vessels, refrigeration, factory trawlers, fish-locating equipment, and GPS.
- Governments worldwide spend billions of dollars annually subsidizing their commercial fishing industry—often spending more than the fishery is worth.
- As fishers became more adept at taking fish in large numbers, the frequency of stock collapses increased, forcing fishery managers to impose quotas on commercially desirable fish, limit fishing time at sea, or close fisheries altogether.
- Sharks are an example of an over-exploited, top predator with a relatively low reproductive capacity. Sharks are taken for their meat; some are killed accidentally in collisions with fishing boats; others are unintentionally caught along with target fish and then dumped overboard. The most egregious threat to shark populations, however, is catching them just for their fins and disposing the carcasses overboard.
- As increasing fishing pressures cause populations of top predators to decline, organisms at progressively lower trophic levels begin to dominate the food webs of marine ecosystems. Fishers, in turn, must target fish populations at these lower trophic levels. This shift in targeted fishes is described as "fishing down the food web."
- To prevent overfishing of a fish stock, fishery managers typically set catch quotas. The goal is to adhere to the maximum sustainable yield of the fish stock; that is, limits are set on fish catches so that stocks are maintained at a level that will preserve the long-term viability of the target species.
- Protection of fisheries requires a thorough understanding of the functioning of the entire ecosystem, not just the targeted commercial or recreational fish species. This understanding is the basis for achieving an ecologically sustainable yield, that is, the yield that a marine ecosystem can sustain without undergoing an undesirable change in state.
- The ecosystem-based approach to fisheries management requires protection of seafloor habitats (e.g., by phasing out destructive trawling practices) and expanding the areas of the world ocean designated as protected areas (called marine reserves) where fishing is prohibited.
- Many commercial fishers accidentally catch undersized fish or unwanted species; this bycatch is discarded, either dead or dying, because it cannot be sold. Bycatch is a major threat to many endangered species, including sea turtles, dolphins, and other marine mammals.
- Recreational fisheries and commercial fisheries differ in important ways. Most recreational fishers prefer to fish near home; hence, sport fish populations are most heavily fished near major urban areas. Such fisheries do not damage fish habitat as much as some commercial fishing practices. When catches of popular recreational fishes decline, fisheries management agencies often resort to fish hatcheries to replenish depleted populations.

- Around the world, sea-turtle populations are threatened by many human activities including (1) commercial over-harvesting of adult turtles and illegal poaching of eggs from beaches, (2) development along beaches used for nesting, (3) bycatch in commercial fishing nets, especially shrimp nets, (4) collisions with ships, including recreational boats, (5) ingestion of plastic litter, especially plastic bags mistaken for jellyfish, and (6) marine pollution.
- Like fishers, whalers focused on a single species until its numbers were so reduced that they had to hunt another (usually smaller) species. To protect the world's remaining whale stocks, in 1986 the International Whaling Commission (IWC) declared a moratorium on commercial whaling. However, Japan, Norway, and Iceland continued commercial whaling, arguing that the populations of smaller whales (e.g., the minke whale) are now at healthy levels and can sustain whaling, and that whales eat many commercially valuable fishes.
- Restoration of some water-bird populations is among the successes resulting from changed environmental policies and regulations in the late 20th century.
- Mariculture, industrial farming of fish and shellfish in the ocean, is a growing industry worldwide. For many varieties of fish and shellfish (e.g., salmon, catfish, trout, mussels, oysters, and clams), farmed fish and shellfish dominate retail markets.
- Marine exotic species are animals or plants introduced into ecosystems, usually by humans. Some introductions are intentional (new organisms used for mariculture) whereas others are unintentional (brought in with the ballast water of ships). In some cases, an exotic species finds its new environment favorable, lacking competitors or diseases, so that their population explodes. Thus, they grow larger than normal, reproduce faster than usual, and take over, perhaps destroying, the entire ecosystem.

Enduring Ideas

- Central to exercising wise stewardship of the ocean is the concept of sustainability, that is, meeting the needs of current generations without compromising the ability of future generations to meet their needs.
- Overfishing occurs when a fish species is harvested at a rate that exceeds the maximum catch that would allow natural reproduction to replace the population. Promoting overfishing are various technological innovations (e.g., factory trawlers, fish-locating equipment, refrigeration), and government subsidies of commercial fishing.
- Strategies to combat overfishing include imposition of catch quotas on commercially desirable fish, limiting fishing time at sea, or closing fisheries all together.
- As fishing pressure increases, populations of top predators decline, and fishers resort to "fishing down the food web."
- Catch quotas are set for desirable fish at a level corresponding to the maximum sustainable yield and/or the ecologically sustainable yield.
- Bycatch is a major threat to many endangered marine organisms.
- Marine exotic species are animals or plants introduced into an ecosystem usually by humans intentionally or unintentionally. Lacking controls on their population, they may quickly dominate and permanently alter the ecosystem.

Review

1. What are the ocean stewardship responsibilities of the National Oceanic and Atmospheric Administration (NOAA)?
2. What is meant by sustainability?
3. Describe the two types of overfishing.
4. At what general size does a fish population have the greatest potential for growth and reproduction?
5. What is bycatch? Identify some of the ways bycatch occurs.
6. Identify some strategies for establishing sustainable fisheries.
7. How does shrimp trawling threaten sea turtles and what is being done to alleviate this problem?
8. Why are whales in demand by certain nations?
9. Define mariculture and describe how it can damage marine ecosystems.
10. Under what conditions are exotic species likely to infiltrate and possibly dominate a marine ecosystem?

Critical Thinking

1. How do technological innovations contribute to the problem of overfishing?
2. Why are "top predators" particularly subject to overfishing? What are some major implications for the rest of the food web?
3. How would adherence to the maximum sustainable yield help prevent overfishing?
4. What is the principal advantage of achieving ecologically sustainable yield?
5. What measures would help alleviate the bycatch problem?
6. Why is the age of a fish an important consideration in the recovery of a fish population that has been depleted by overfishing?
7. Why is commercial fishing potentially more disruptive of marine ecosystems than recreational fishing?
8. Speculate on why an overfished stock might never recover to its original population level.
9. In what sense are invasive species "exotic"?
10. Briefly outline what is meant by an ecosystem approach to fisheries management.

ESSAY: Determining the Age of a Fish

Knowing how long a fish lives is vitally important to managing fish stocks and protecting them from overfishing by commercial or recreational fishers. As noted elsewhere in this chapter, *overfishing* is the practice of harvesting fish at a rate that causes a dramatic decline in the fish stock, perhaps to the extent that the fishery collapses. In essence, the rate of harvest exceeds the maximum harvest that would allow for replacement of the population through reproduction. Long-lived animals grow slowly, reproduce infrequently, and take a long time to recover from overfishing. Agencies responsible for protecting and regulating fish populations must therefore know the life span of fish that are being harvested.

Orange roughy, a deep-water fish caught near New Zealand, was once presumed to live no more than about 20 to 30 years and catch limits were set accordingly. Within five years, the fishery collapsed due to overfishing. Scientists discovered that the orange roughy actually lives more than 100 years (up to a maximum of about 150 years), is very slow growing, and does not reach sexual maturity until about 30 years of age. It appears likely that recovery of orange roughy stock will take decades.

A long-used indirect method of estimating the age of a fish is to measure the length of large numbers of fish of the same species caught in nets. From these data, scientists estimate age based on models of how fast the particular fish species grows. Another method of age determination is to raise fish in tanks to determine directly the relationship between size and age. Neither of these methods is entirely satisfactory, however.

A more accurate technique is to count the number of annual growth rings in a fish's ear stone, the *otolith*. This calcareous concretion is up to about the size of a nickel. Bone is deposited on the outside of the otolith as the fish grows, thereby forming a banded bone that can be sectioned and studied much like the growth rings of a tree. Some fishes produce annual bands; others take longer to form a single band. By cutting thin sections of an otolith and counting the bands, it is possible to determine a fish's age accurately (Figure 1). Sufficient samples of many fish of the same species provide a reasonable estimate of life span.

Furthermore, chemical analysis of the individual bands in otoliths permits scientists to determine the various environments where the fish has lived during its life cycle (e.g., lakes, estuaries, or open-ocean). Also, oxygen isotope analysis of the otoliths enables scientists to reconstruct the water temperature when the particular band formed. (Oxygen isotope analysis is discussed in Chapter 12.) Finally, these detailed analyses can be used to determine which animals lived together when the bands formed.

Another method of age determination applies to bony fishes. They have scales on their skin that grow by forming rings. The fish's age is determined by analyzing the scales under a microscope and counting the rings. This technique is especially useful in tiny tropical fishes whose otoliths are too small to study conventionally. In general, these smaller, faster-growing tropical fish have shorter life spans than larger, slower-growing fish living in colder waters.

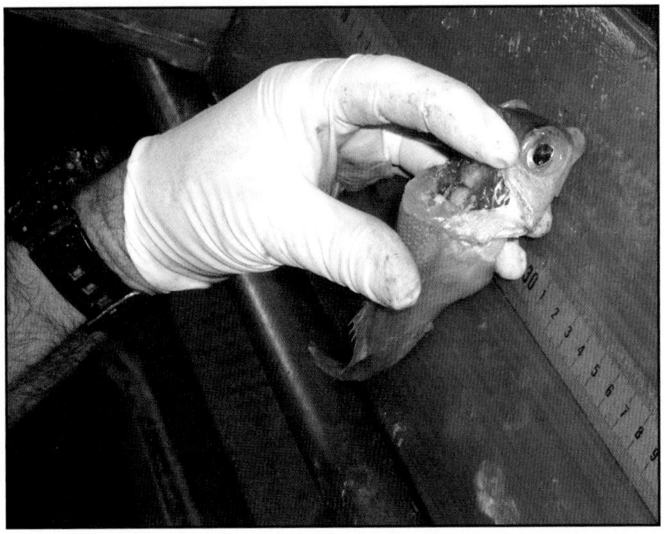

FIGURE 1
Scientist onboard the *Gordon Gunter* removing ear bones from fish to determine its age. [NOAA photo taken by Paul Olson]

ESSAY: Loss and Restoration of Coastal Wetlands

For centuries, people have attempted to reclaim wetlands near cities and in densely populated coastal regions, such as Japan or the Netherlands. Much of Lower Manhattan's shoreline is built on such reclaimed lands, as is New York City's John F. Kennedy International Airport. The Dutch pioneered reclamation techniques, converting shallow sea bottoms to agricultural fields surrounded by dikes; such reclaimed lands are called *polders*. Other former coastal marshes were filled and are now used as landfills for solid waste.

In more recent years, studies have demonstrated the many valuable functions that wetlands perform if maintained in their natural state, including protecting coastal lands against saltwater encroachment and flooding from storm surges and heavy rainfall. Wetlands filter water flowing through as well as acting as sponges, soaking up water at high tide and during floods. Again, the Dutch are leading the way in restoring wetlands by breaking the dikes surrounding certain polders, permitting them to flood again and return to their original condition as salt marshes or shallow sea beds. On smaller scales, individual farmers have removed drains to permit low-lying fields to flood seasonally, thereby encouraging marsh plants to invade and provide suitable habitats. These areas are used by water birds and by fish as nursery grounds during their larval and juvenile stages.

Some marshes were damaged by introduced species that destroyed the native grasses, thereby accelerating erosion as sea level rises. That happened at the Blackwater Natural Wildlife Refuge on Chesapeake Bay's Eastern Shore, which is the largest unbroken stretch of marshland on the Chesapeake. In the 1930s, the *nutria*, a beaver-sized rodent native to South America, was introduced deliberately to provide local trappers with a source of furs. Unlike its cousin the native muskrat, which eats only the top of the marsh grasses, nutria eats the plant roots. Its voracious appetite destroyed the native grasses, leaving large areas of marshland unprotected against erosion by waves and tidal currents (Figure 1). By 2001, the original 9700-hectare (24,000-acre) refuge was losing 60 hectares (150 acres) of salt marsh each year, producing large open-water areas and accelerating erosion of adjoining salt marshes.

FIGURE 1
Research biologist examining nutria damage to marsh at the Blackwater National Wildlife Refuge. [Photo by Mike Haramis, USGS]

In the 1990s, wetland restoration projects began in the Chesapeake including the Chesapeake Marsh Restoration/Nutria Control Project to reclaim marshes in the Blackwater National Wildlife Refuge. This restoration project also helps alleviate another major environmental problem in Chesapeake Bay, that is, the disposal of sediment dredged from shipping channels. A special dredge removes mud from channel bottoms and discharges it over the marshes. The newly deposited sediment layers are sufficiently thin that they do not kill the grasses, which grow over them, thus stabilizing the restored marsh. Each year, many thousands of migrating waterfowl as well as large numbers of bald eagles use the restored marshlands. These areas are also home to endangered fox squirrels. The shallow waters are nursery grounds for crabs, the most important commercial fishery in the Chesapeake. Furthermore, sediment trapping and nutrient retention by the marsh helps improve water quality in the nearby Chesapeake Bay.

According to the Louisiana Coastal Wetlands Conservation and Restoration Task Force, Louisiana has been losing coastal wetlands (i.e., bayous, marshes, and swamps) to the waters of the Gulf of Mexico at an alarming rate of about 65 to 100 square km (25 to 38 square mi) per year for the past several decades (Figure 2). This loss adversely affects fisheries in the Gulf of Mexico and makes the coastal zone more vulnerable to storm surges such as that produced by Hurricane Katrina in August 2005. Since the early 1930s, the state's coastal wetlands have shrunk by an area equivalent to the state of Delaware. According to estimates by the U.S. Geological Survey, an additional 1800 square km (700 square mi) could be lost by mid-century.

As much as 75% of the fish and other marine life in the northern Gulf of Mexico depend on Louisiana's coastal wetlands. The wetlands function as a nursery for commercially important catches of shrimp, crawfish, blue crab, and oysters. It is a food source for larger fish including yellow fin tuna, red snapper, and swordfish. In 2003, about three-quarters of the nation's fish and shellfish catch by weight came from Louisiana's waters. In addition, the wetlands are a stopover for millions of birds migrating between North America and Central/South America. Furthermore, wetlands and associated barrier islands protect the ports, buildings, and other coastal structures from storm surges. Wetlands are particularly important in buffering the levees surrounding New Orleans, much of which is below sea level.

FIGURE 2
Rainey Refuge marsh in Louisiana. Coastal wetlands buffer and absorb storm waves, acting as natural flood protection. [Courtesy of NOAA]

Many factors contribute to the loss of Louisiana's coastal wetlands. Thousands of kilometers of pipelines transporting oil and natural gas through the marshes plus the extensive network of navigation canals allows saltwater to invade the wetlands. Increased salinity of the originally fresh or brackish waters kills wetland grasses, shrubs, and other vegetation that anchor the soil in place. The canals also allow tidal currents to flow further inland, accelerating erosion of wetland soils. Exacerbating the problem was the introduction of the nutria in the 1930s and as was the case in Chesapeake Bay (described earlier), the nutria soon began to devour marsh grasses. The most important factor in the loss of Louisiana's coastal wetlands, however, is the consequence of flood control structures (levees) constructed along the banks of the Mississippi River by the U.S. Army Corps of Engineers in the 1940s. Levees constrict the flow of the river so that water and suspended sediments discharge directly into the Gulf. Deprived of a continuous input of sediment and vegetation-supporting nutrients, the existing sediments compact, wetlands subside, and Gulf waters encroach upon the wetlands. With the expected continued rise in mean sea level due to global climate change, erosion of Louisiana's coastal wetlands may accelerate in the future (Chapter 12).

Plans to reverse the loss of Louisiana's coastal wetlands seek to restore the structure and function of those wetlands while protecting existing wetlands. Although public policy makers were aware of the problem in the 1960s, little was accomplished prior to 1990 when federal legislation provided the state with about \$50 million annually for a variety of wetland restoration projects that, unfortunately, proved to be too small and piecemeal. Federal, state, and local officials proposed a much more ambitious plan to create a "sustainable ecosystem that supports and protects the environment, economy, and culture of southern Louisiana." Known as *Coast 2050*, this proposal carried a price tag of \$14 billion and would take 3 decades to complete. In 2003, the Bush Administration rejected this plan as too costly. The following year, a scaled-down plan, called the *Louisiana Coastal Area (LCA)* study, was developed calling for a budget of \$1.9 billion over 10 years. Among the projects supported by the LCA study was the breaching of some levees along the lower Mississippi River. This partial diversion of the river is intended to increase the supply of sediments to the wetlands. Closing or installing locks on some navigation canals would reduce saltwater encroachment. In addition, dredged sediment would be used to restore wetlands and barrier islands.

However, much more needs to be done. An assessment of the LCA study by the National Academy of Sciences in 2005 concluded that it is "too modest an effort." According to calculations by the U.S. Army Corps of Engineers, the LCA plan would slow the rate of coastal wetland loss by only 20%.

CHAPTER 15

OCEAN PROBLEMS AND POLICY

On 20 April 2010, a sudden explosion and fire occurred on the BP/Transocean Deepwater Horizon oil rig. The accident resulted in the deaths of 11 workers and caused a massive oil spill into the Gulf of Mexico. [Courtesy of U.S. Chemical Safety Board]

Case in Point

On 20 April 2010, an explosion destroyed the Deepwater Horizon oil drilling platform, leased by the energy company BP from Transocean Ltd, in the Gulf of Mexico, 65 km (40 mi) south of the Louisiana coast. The wellhead on the ocean bottom was 1500 m (4900 ft) down, and the just completed well was drilled into the reservoir rock 4300 m (14,100 ft) below. The explosion killed 11 workers and injured 17, sunk the rig, and unleashed a gusher of light sweet crude oil that persisted until the wellhead was finally capped 85 days later in mid July. At least 4.4 million barrels (794 million liters) of oil was released, the largest oil spill in U.S. waters and four times greater than the second largest, produced by the 1989 grounding of the *Exxon Valdez* in Prince William Sound, AK (discussed later in this Chapter).

The spill formed huge oil slicks on the ocean surface that slowly moved toward the Gulf shoreline from Louisiana to Florida, reaching the offshore islands of Alabama and Mississippi by 25 June. At the same time, 3 to 5 km wide and tens of km long underwater plumes of tiny oil droplets fanned out from the gushing wellhead at depths of 1000 to 1500 m (3300 to 4900 ft).

Oil is a complex mixture of numerous hydrocarbons, some of which are toxic and may have long-term effects on health. The greatest concern is *polycyclic aromatic hydrocarbons (PAHs)*, including napthalenes, benzene, toluene, and xylenes. Inhaled or ingested, PAHs can transform into more toxic substances that alter DNA, causing mutations that could reduce fertility or cause cancer in mammals or birds. This threat lessens

TABLE 15.1
Estimated fate of oil spilled into the Gulf of Mexico after a catastrophic explosion destroyed the Deepwater Horizon oil drilling platform in April 2010. Numbers represent thousands of barrels.[a]

Dispersed naturally in small droplets	630	
Chemically dispersed	770	(range: 500-1400)
Formed surface slicks and tar balls, sank to the bottom, or washed to shore	1100	(range: 520-1500)
Evaporated or dissolved	1200	(range: 930-1300)
Removed by emergency operations	1240	(range: 1210-1270)
Recovered from well	820	
Burned	260	
Skimmed	130-190	

[a]From Schrope, M., 2011, "Deep Wounds," *Nature*, vol. 472, 14 April 2011.

with time as liquid oil is emulsified to a mousse (a frothy blend of hydrocarbons and water) or clumps into tar balls. With these changes, the oil loses its volatile more toxic constituents to evaporation. Typically, 20% to 40% of the original hydrocarbons are vaporized after reaching the ocean surface.

Particularly vulnerable are the coastal wetlands that provide habitat and protection for juvenile fish, shrimp, and shellfish, and the offshore islands that are breeding areas for large colonies of birds. Oil that invades Louisiana's fragile wetlands is likely to exacerbate the loss of coastal lands, currently disappearing at 4400 hectares (10,872 acres) per year. Normally grasses anchor the soil, inhibiting erosion, but the oil kills the grasses and the rate of erosion accelerates, especially during storms.

Closure of commercial and recreational fisheries, such as the 118,000 sq km area of the Gulf that NOAA closed on 18 May, the suspicion with which consumers viewed seafood from the area, the closure of recreational beaches, the continued loss of tourism, and the cost of clean-up, have impacted the economy of the region. BP has indicated its intention to reimburse residents for some of their losses.

For over three months, the spill flowed oil like a geyser while the responders struggled to stop it. The initial blowout preventer valves would not close, the 125-tonne containment dome filled up with methane hydrates, and the "top kill" method of pumping heavy fluids into the flow to stop it, failed. In early June, a riser insertion tube was placed in the burst pipe so some of the oil could be collected and burned off, the effectiveness of which was in dispute, but likely drew out less than half the oil. In mid-July, a better fitting containment cap was used and, in mid-September, the relief wells reached the original well 4000 m (13,100 ft) below the ocean floor and, finally pumped cement into the well, effectively killing it.

Government agencies, volunteer organizations, and BP employees attempted to protect the marine and coastal ecosystems by containing, capturing, or neutralizing the oil. Floating booms were used to stop the surface oil from reaching the shores. That oil was then removed with skimmers, burned off with small fires or sprayed with chemical dispersants, as shown in Table 15.1. While long term effects of the chemical dispersants are questionable, they did break the surface slick into tiny droplets of oil, which can disperse more widely and are more readily degraded by microbes. However, the fate of those droplets, either in underwater plumes, as mentioned earlier, or settling in high concentrations on the ocean floor, is still being studied.

There's more about oil spill recovery in marine environments later in this chapter.

Driving Question:

How does national and international ocean policy promote wise stewardship of the world ocean?

This chapter is concerned with the impact of human activities on the ocean and coastal zone, marine ecosystems and habitats, and how society is attempting to minimize that impact, remediate past damage, and head off future problems. Throughout this book, we present many instances of how humans adversely affect the ocean and the coastal zone. Rising sea level, coral bleaching, ocean acidification, and shrinkage of the Arctic sea-ice cover are likely linked to our dependence on fossil fuels and the enhanced greenhouse effect. In Chapter 8, we described how humankind has attempted to stabilize the inherently dynamic coast by constructing seawalls, jetties, breakwaters, and groins often to little or no avail. In fact, we have made some of our coastlines more vulnerable to storm surges, flooding, and erosion by developing barrier islands and removing protective mangrove swamps. The discharge of excessive nutrients into coastal waters has triggered harmful algal blooms (HABs) and created oxygen depleted "dead zones." Our growing demand for fish and shellfish is responsible for overfishing, bycatch, destruction of essential fish habitat, and endangering some marine species.

In attempting to remediate these problems in the ocean and coastal zone, we have adopted *ocean stewardship* (Chapter 14). In this regard we recognize the need for an ecosystems approach to fisheries management and protection of living and non-living marine resources through federal and state sanctuaries programs and fisheries management. To a large measure, ocean stewardship arises from humankind's desire to regulate exploitation of marine resources and space to meet human needs and to achieve *sustainable development*.

While the United States is the principal focus of many of the issues discussed in this book, vast areas of the ocean are in international waters and marine ecosystems adhere to no political boundaries. Hence, in dealing with human impacts on the marine environment, a global perspective and international cooperation are essential. In this chapter, we take a closer look at the human interaction with the ocean and coastal zone with emphasis on remediation, sustainability, and stewardship. We begin with a summary of important milestones in the national and international governance of the ocean.

Milestones in Ocean Governance

On land, humans have long divided space and resources into two categories: private property and public property held in trust for citizens of the nation in which these resources are found. The history of ocean governance, however, differs dramatically from that of land. For most of human history, the open ocean has been beyond the direct management of coastal nations or any other governments, just like outer space. As the human population soared and human exploitation of ocean resources expanded, regulations and institutions to govern the ocean were developed.

FREEDOM OF THE SEAS

In 1609, the Dutch legal scholar Hugo Grotius (1585-1645), a pioneer natural rights theorist, published his influential treatise, *The Freedom of the Seas*. In it, he argued that the ocean was open with free access to all nations; that is, no nation or group of nations has the right to monopolize either its use or access to it. For Grotius, liberty of the sea was key to maintaining communication among peoples and nations. Along with this doctrine of free access, the general consensus was that the sea was limitless and its resources inexhaustible. There was no need to control access if marine resources were essentially unlimited and therefore virtually immune to any possible damage caused by human activities.

Partitioning of the coastal ocean began in the late 1700s, when the new United States of America claimed a 5-km (3-mi) "territorial sea" off its coast. Such a narrow coastal fringe could be guarded and controlled given the range of the military cannons of the day. Authority over these territorial waters was assigned to the individual states. In 1945, the U.S. extended its national claims on coastal resources to the outer edge of the continental shelf. A major reason for this policy change was the expansion of offshore oil and natural gas production on the shallow continental shelf around the Gulf of Mexico and off the California coast. To this day, the U.S. considers ocean waters outside national territorial waters to be international and open to all.

In 1976, the United States extended its jurisdiction over some marine fisheries. The *Magnuson-Stevens Fisheries and Conservation Act* (also known as the *Magnuson Act*) extended the jurisdiction of the nation's fisheries-management seaward 330 km (200 mi) from the shore, waters that were formerly heavily fished by foreign vessels. The Magnuson Act also established fishery councils to manage the fishery resources in each region of the U.S. In 1996, the *Sustainable Fisheries Act* amended the Magnuson Act imposing strict new mandates to halt overfishing, rebuild overfished stocks, reduce bycatch, and protect essential fish habitat.

The first commission to develop a national ocean policy for the United States was the *Commission on Marine Science, Engineering and Resources*, more commonly known as the *Stratton Commission* after its Chair, Julius A. Stratton (1901-94), then also Chair of the Ford Foundation. The Commission worked for two years and presented its final report, *Our Nation and The Sea, A Plan for National Action*, in January 1969. Among other recommendations, it called for creation of the *National Oceanic and Atmospheric Administration (NOAA)* by combining existing government agencies responsible for the ocean and atmosphere. Recognizing that the ocean plays an integral role in the economic, environmental, and security interests of the United States, a second *U.S. Commission on Ocean Policy* was established to make recommendations to the President and Congress for a coordinated and comprehensive national ocean policy. It began work in 2001 and issued its final report at the end of summer 2004. Recommendations called for an overhaul of ocean policy including a shift to ecosystem-based management of marine life (Chapter 14), creation of a National Oceans Council within the Executive Branch of government, and doubling of federal funding for ocean research.

Other nations, including Australia and Canada, have developed ocean policies for their Exclusive Economic Zones (Chapter 4). Such policies are tailored to the interests and priorities of the individual countries. No international ocean policy has yet been developed to deal with open-ocean areas outside the various EEZs.

In 1982, the *United Nations Convention on the Law of the Sea* expanded the narrow territorial waters 20 km (12 nautical mi) from the shoreline and authorized nations to establish *Exclusive Economic Zones (EEZs)*. The U.S. established its EEZ in 1983, extending federal jurisdiction from the seaward limits of the state-controlled territorial seas to 370 km (200 nautical mi) from the coastline. In 1996, EEZs were extended to the edge of the continental shelf when the shelf edge was more than 370 km (200 nautical mi) offshore. Within their individual EEZ, each coastal nation has the same rights and responsibilities that they exercise over their land areas. Migratory fishes, which may cross national boundaries, and range beyond the limits of an EEZ, are regulated under the *Convention on Straddling Stocks*. Cooperation on the conservation of such stocks is the responsibility of those nations where the fish live or where they are taken.

At the beginning of the 21st century, about two-thirds of the ocean was part of the **ocean commons** and beyond the control of any coastal nation. The ocean commons essentially is free to all for unlimited exploitation. Little or no regulation exists over the ocean commons, except for a few specialized treaties such as the 1972 *London Dumping Convention* that bans the discharge of wastes from land by ships or aircraft. Commercial whaling was uncontrolled until banned by the International Whaling Commission in 1986 (Chapter 14). Unregulated commercial fishing has depleted some open-ocean fish populations. An example is the overfishing and eventual collapse of the Canadian cod fishery in the early 1990s (Chapter 10).

As the global human population continues to grow and increasing demands for natural resources exceed supplies on land, ocean resources are likely to be increasingly exploited. In the future, such pressures are likely to drive human governance over more ocean space and resources. Innovative technologies could be used to enforce international regulations. As noted in Chapter 13, various technologies now provide continuous surveillance of the ocean surface and interior. Some experts argue that all ships, planes, and submarines operating in, on, and over the ocean can be tracked, and their activities monitored in various ways. For example, acoustic methods can detect whether fishing boats have nets in the water and even identify the types of nets in use.

Future governance of ocean resource exploitation may apply the *Precautionary Principle*: "When an action causes a threat to human health or the environment, precautionary measures should be taken even if some cause-and-effect relationships are not fully established scientifically." The 1992 United Nations Conference on Environment and Development held in Rio de Janeiro, Brazil, adopted the Precautionary Principle. It was also included in the Rio Declaration's "Agenda 21," which was adopted by 178 nations in June 1992, and later ratified by the United States. The precautionary approach helps to overcome the enormous barrier to action posed by the inevitable scientific uncertainty about cause-effect relationships in complex systems. This is especially important in the ocean where many important sub-systems and processes, and their interactions are still poorly understood.

ANTARCTIC TREATY

Human activity in the Antarctic region includes the slaughter almost to extinction, of millions of seals and whales, and the dumping of substantial amounts of garbage by explorers and scientists. However, in the last half of the 20th century, Antarctica and the Southern Ocean came under the protection of the Antarctic Treaty System. The *Antarctic Treaty System* is a complex of arrangements that regulate relations among nations near Antarctica and those conducting research and exploration there. In 1959, the Antarctic Treaty System was adopted by the dozen nations that engaged in scientific research in the Antarctic region during the *International Geophysical Year* (1957-58). The Treaty has since expanded to 48 nations sponsoring research in Antarctica and/or the Southern Ocean. These countries represent about two-thirds of the world's population. The Antarctic Treaty System ensures that no national claim of territorial rights in Antarctica or the surrounding waters will ever be legally recognized and that Antarctica forever would be used exclusively for peaceful purposes.

The Antarctic Treaty System applies to waters, ice shelves, and islands in the Southern Ocean, that is, south of 60 degrees S. This is the approximate latitude of the *Antarctic convergence* where cold Antarctic waters from the south meet warmer waters from the north (Chapter 6). The Antarctic convergence acts as a biological barrier so that the Southern Ocean is essentially a separate ecosystem. The *Convention on the Conservation of Antarctic Marine Living Resources (CAMLR)* was adopted in 1982 as part of the Antarctic Treaty System. CAMLR was established in response to international concerns that increased catches of krill (a large shrimp-like zooplanktonic organism) in the Southern Ocean could threaten krill populations as well as the large whales, other marine mammals, birds, and fishes that feed primarily on krill and associated high seas fisheries.

Human Impact in the Coastal Zone

As noted in Chapter 8, the majority of the human population lives in the coastal zone where they interact with the ocean (and large lakes). It is in the coastal zone where stewardship over the ocean is most urgent and, in many cases, most challenging. People living in the coastal zone are vulnerable to certain natural hazards (e.g., flooding due to storm surge, tsunamis). They also impact the natural functioning of the coastal zone where the land, ocean, atmosphere, cryosphere, and biosphere interact.

HUMAN POPULATION TRENDS

About 80% of the land on Earth is now inhabited by people or otherwise impacted by human activities such as agriculture, pasturing animals, growing trees, mining, or urban development. Satellite images taken at night of lighted settlements combined with census data (base year 1990) show that half the world's people live within about 30 km (19 mi) of the coastline and nearly 75% within 50 km (31 mi) (Figure 15.1). Population density in Earth's coastal zone is about three times that of the overall Earth average. The most densely populated coastal zones are in Europe and Asia; most of these coastal dwellers live in rural areas and small- to medium-sized cities, rather than large urban areas.

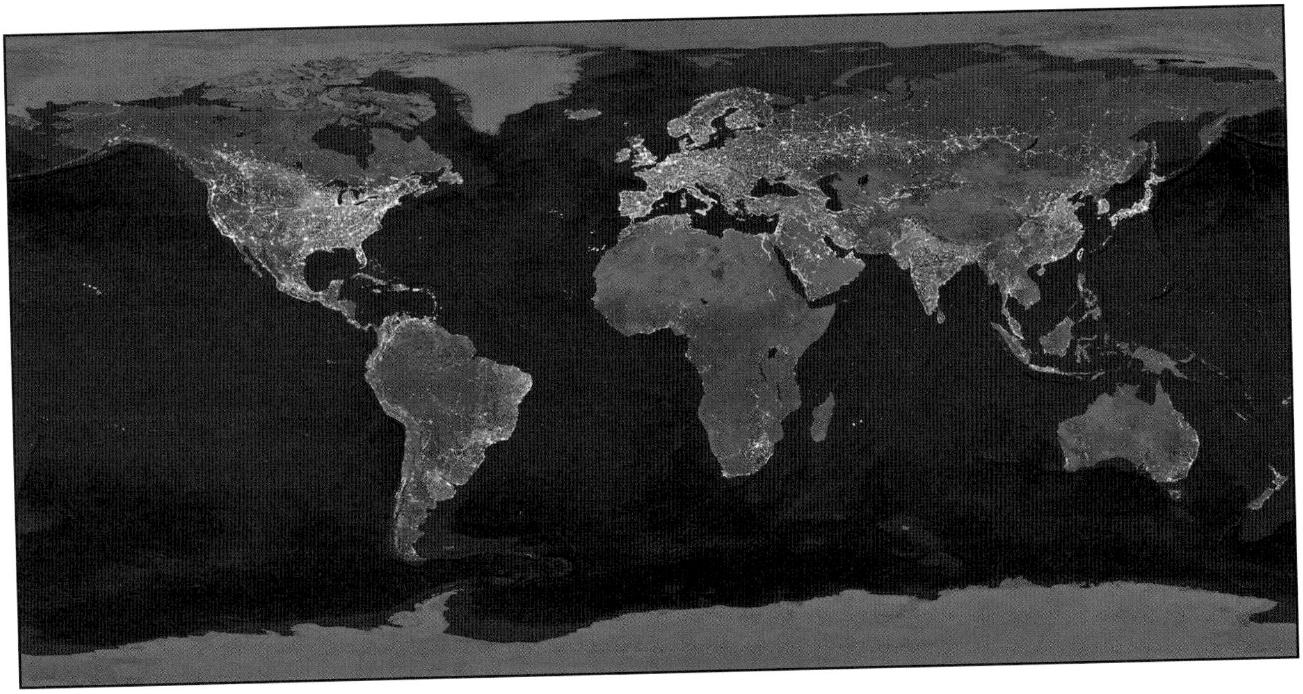

FIGURE 15.1
The image of Earth's lights at night highlights concentration of population and commerce in coastal areas. [Courtesy of NASA].

In earlier chapters, we described how human habitation of the coastal zone exposes people to many natural hazards. The hazard risk is especially great for people living in low-lying coastal plains less than 100 m (330 ft) above sea level. With sea level rising, the principal problems are expected to be in low-lying delta areas, such as the Netherlands in Western Europe, Bangladesh in South Asia, and the Mississippi River delta of North America. In these areas, storm surge flooding can extend inland 100 km (62 mi) or more.

Population growth in the coastal zone during the first half of the 21st century is expected to take place in small- and mid-sized communities in developing countries. Older mega-cities, which are mostly located at shorelines and near large rivers, are expected to grow more slowly. A significant fraction of this coastal-zone population growth will likely come from people moving from inland agricultural communities to small coastal communities, filling in the space between large metropolitan areas, such as the Boston-New York-Washington, DC corridor in the United States or the Tokyo-Osaka corridor in Japan. Consequently, a larger fraction of the world's population will impact the coastal zone and risk exposure to natural hazards.

ENVIRONMENTAL POLLUTION

Increasing human population density in the coastal zone is likely to exacerbate environmental pollution. **Pollution** is defined as an intentional or unintentional disturbance of the environment that adversely affects the wellbeing of organisms (including humans) directly or the natural processes upon which they depend (Figure 15.2). The disturbance might involve the alteration of a biogeochemical cycle as when combustion of fossil fuels alters the global carbon cycle and elevates the concentration of carbon dioxide in the atmosphere (Chapter 12). In other cases, toxic and hazardous materials may be dumped or otherwise introduced into reservoirs where they do not normally occur. As described in the Case-in-Point of Chapter 4, disposal of mercury waste in Minamata Bay, Japan contaminated fish and shellfish that were subsequently consumed by humans who developed debilitating symptoms. In the U.S., all along the Mississippi River, runoff from agricultural lands is transported and eventually discharged into the Gulf of Mexico off the Louisiana coast. This results in excessive primary production off the Mississippi River delta. The subsequent decomposition of this production causes annual oxygen-deficient "dead zones" on the ocean bottom (Chapter 10).

Inevitably, all organisms (including humans) disturb their environment by exploiting and utilizing resources, as well as producing and disposing of waste products. Through the years, humans have been particularly pervasive in their disturbance of the environment so that many areas of the world suffer from air and water pollution. In addition, natural physical forces such as hurricanes, floods, tsunamis, volcanic eruptions, and earthquakes disturb the Earth system. Anyone who has survived a hurricane or destructive earthquake would agree that humankind is unable to prevent such catastrophes. As noted in Chapter 8, even efforts to lessen the impact by structural means (e.g., building sea walls to protect against storm surge) have limitations and often fail in the long term. We can, however, reduce the toll of lives lost and property damaged by planning for such disasters and avoiding habitation of areas that are particularly prone to natural hazards.

On the other hand, we can do something about disturbances for which humans are responsible. For example, we can reduce the input of excess nutrients from agricultural fields into estuaries and other coastal waters, as well as freshwater lakes and systems, and stop the dumping of untreated sewage and industrial waste into rivers and

FIGURE 15.2
Public warning signs like these appear in areas that have been closed due to high levels of toxin in shellfish. [Courtesy of NOAA].

streams that ultimately empty into the ocean. Most long-term efforts to control or prevent harmful algal blooms (HABs) center upon reducing the amount of nutrients in the water that stimulate algal growths, specifically by reducing nitrogen and phosphorus discharges from wastewater treatment plants or runoff from farms, factories, and cities (Chapter 9). In addition, the presence of disease-causing microbes forces the closing of recreational beaches and contaminated shellfish beds. For more on this problem, refer to this chapter's first Essay.

If we are responsible for disturbances that reduce the quality of the coastal environment, then corrective action is needed. Typically, however, not everyone agrees that action is needed or on what specific steps should be taken when there is general agreement that a problem must be dealt with. Some people argue that any disturbance that harms plants, animals, or humans in any way is unacceptable. But how seriously must organisms and/or the ecosystems be impaired before the disturbance is considered unacceptable? Are we concerned about all species or just a select group of species or the ecosystem-species interaction as in ocean fisheries? What are the political and economic ramifications of our choices? There are often no easy answers to these questions.

A better understanding of the problem of pollution in the coastal zone and elsewhere in the ocean is based on the concepts of carrying capacity, assimilative capacity, and limiting factors. Waste disposal in ocean waters is an ongoing problem that can only be exacerbated by human population growth in the coastal zone. The human **carrying capacity** of a region is based on the maximum rate of resource consumption and waste discharge that is sustainable indefinitely without progressively impairing ecosystem productivity and integrity. **Assimilative capacity** is the amount of waste that an ecosystem can assimilate (via decomposition by bacteria, fungi, and invertebrates) without damaging ecosystem functions or building up waste products to levels that may cause unwanted or harmful impacts on living organisms (including humans). The growth and wellbeing of an organism is limited by the essential resource (e.g., nutrients, dissolved oxygen) that is in lowest supply relative to what is required. This most deficient resource is known as the **limiting factor** (Chapter 9).

Carrying capacity and assimilative capacity are especially important considerations for islands, which often are densely populated and lack adequate facilities to treat the wastes of large and seasonally variable populations of tourists. For example, this is a serious problem in many popular beach resorts around the Mediterranean Sea.

COASTAL ZONE MANAGEMENT

The need to balance multiple use and preservation of the coastal zone spurred the federal government to partner with states having coastlines to develop and implement plans for sustainable coastal zone management. The federal government assists coastal states in managing and protecting their coastal resources through the **Coastal Zone Management Program (CZMP)** authorized by the *1972 Coastal Zone Management Act*. Program goals are "to preserve, protect, develop, and where possible, to restore or enhance the resources of the Nation's coastal zone" and to encourage states to develop and implement plans "to achieve wise use of the land and water resources of the coastal zone." At the federal level, NOAA's Office of Ocean and Coastal Resource Management (OCRM), part of the National Ocean Service, administers the CZMP.

The CZMP promotes comprehensive management of coastal resources. This program seeks to protect those resources for future generations while balancing competing economic, cultural, and environmental issues. The federal government supports state efforts at coastal zone management by providing financial assistance (matched by state funds), mediation of conflicts, technical services, and information. Coastal states that show satisfactory progress in implementing their management plan are eligible for special federal grants (matched by state funds) to help states preserve or restore coastal areas, redevelop urban waterfronts and ports, and provide access to public beaches. Actual day-to-day coastal zone management decisions rest with the individual states.

All 34 eligible coastal states and U.S. territories have federally approved coastal zone management plans that apply to much of the more than 160,000 km (100,000 mi) of national coastline. All approved state coastal management plans must address a number of issues in balancing use with preservation. These issues include (1) an inventory and designation of areas of particular concern in the coastal zone, (2) a definition of permitted land and water uses that directly impact coastal waters, (3) identification of how those uses will be controlled, and (4) an outline of broad guidelines to determine priority uses in coastal areas.

Another provision of the 1972 Coastal Zone Management Act established NOAA's *National Estuarine Research Reserve System (NERRS)*. The U.S. Secretary of Commerce can authorize grants to coastal states to

FIGURE 15.3
This National Estuarine Research Reserve at Jug Bay, Chesapeake Bay, MD, is one of 27 sites established for long-term research, education, and stewardship of estuaries. [NOAA Photo Library, NOAA's Office of Ocean and Coastal Resource Management].

acquire, develop, and operate estuarine research reserves (Figure 15.3). Estuarine reserves were established for long-term research, water-quality monitoring, education (including public education), and coastal stewardship. State agencies or universities in partnership with NOAA manage the reserves while NOAA provides funding, national program guidance, and technical assistance. Reserve staff work with local communities and public interest groups on issues such as non-point source pollution, habitat restoration, and invasive species. Since 1972, the NERRS has expanded to include 28 estuarine locations in 22 states along the Atlantic, Gulf, Pacific, and Great Lakes coasts as well as Puerto Rico. In October 2010, a 6757 hectare (16,697 acre) site in the northwest corner of Wisconsin, where the St. Louis River flows into Lake Superior, became the latest addition to the Reserve System, significantly expanding the biogeographic representation of the System. The Lake Superior National Estuarine Research Reserve, the second freshwater estuarine reserve to be established in the Great Lakes (first in the upper Great Lakes), features uplands and submerged lands, riparian and riverine habitats, freshwater marshes, riverine islands, dunes and interdunal wetlands, scrub swamps, forests, and sand beaches.

Marine Oil Spills

Releases of oil, toxic and hazardous waste, and other pollutants continue to threaten fragile wetlands and coastal river and marine ecosystems (Figure 15.4). According to the *Office of Response and Restoration* of NOAA's National Ocean Service, about 200 maritime accidents collectively spill about 7.6 million liters (2 million gallons) of oil each year into the coastal zone. More than 700 hazardous waste sites contaminate the U.S. coastal zone. Rivers and streams carry great amounts of pollutants to the coastal environment, all of which can kill fish and birds, destroy habitats, disrupt marine food webs, and close commercial and recreational fisheries and beaches.

Shipped in huge quantities across the ocean and drilled from beneath the seafloor using offshore drilling platforms, oil is a critical commodity. For many reasons (shipwrecks, collisions, pumping of ships' bilges, well blow-outs, war and terrorism), oil is accidentally and intentionally spilled into the ocean. Even more oil routinely enters from storm sewers, wastewater discharges, and diffused runoff from a variety of sources, as well as an unknown quantity that naturally leaks upwards through the seafloor. While the unnatural discharges adversely affect the ocean and its ecosystems, chemosynthetic ecosystems could not survive without the undersea seeps.

Responders to marine oil spills utilize a number of strategies to mitigate the threat to ecosystems. Floating protective booms enclose oil slicks, which are then skimmed off the sea surface. This method of oil recovery is most effective when winds are light and the sea is calm. In some cases, it is possible to set fire to the oil.

Dispersants are detergent-like chemicals that break oil into tiny droplets which disperse widely and are more readily degraded by microbes. Usually, dispersants are applied to oil slicks floating on the ocean surface but, with the Deepwater Horizon spill described in this chapter's Case-in-Point, large amounts of dispersants were also sprayed on the wellhead near the ocean floor. In

FIGURE 15.4
A heavily oiled Kemp's Ridley turtle recovered near the Deepwater Horizon/BP accident site. [Courtesy of NOAA]

this case, the oil droplets dispersed throughout the water column and may have endangered marine life.

More nutrients and warm climates accelerate the biodegradation of oil. Occasionally, the aerobic microbes that digest the oil droplets deplete the dissolved oxygen (DO) until dead zones develop (Chapter 10). Although large quantities of dispersants were applied to the Deepwater Horizon spill and dissolved oxygen levels declined, no dead zones developed. There is also concern of the toxicity of dispersants but they are believed safer than oil.

As described in this chapter's Case-in-Point, the largest oil spill in U.S. waters was caused by the explosion of the Deepwater Horizon oil drilling platform in the Gulf of Mexico on 20 April 2010. The nation's second largest oil spill occurred in the coastal waters of Alaska in 1989. Because of navigational errors, just after midnight on 24 March 1989, the supertanker *Exxon Valdez* ran aground on Bligh Reef in Prince William Sound, ripping open the tanker's hull. Eventually, nearly 40 million liters (10.6 million gallons) of crude oil spilled into the ocean and onto beaches. The spills occurred in two of the world's most bountiful coastal areas, home to highly diverse and productive ecosystems.

In part due to inadequate planning, technological limitations, and the inability to control and corral the leaking oil during the weeks after the *Exxon Valdez* crashed, strong currents and storm winds transported the spilled oil as a slick through the Sound and along the open North Pacific coast. Some of the oil moved westward, coating beaches along 2400 km (1500 mi.) of the Alaska Peninsula. The Exxon Corporation (now ExxonMobil), the ship's owner, conducted a massive, $2 billion shoreline cleanup. Thousands of people washed beaches with hot and cold water, removed oiled sediment deposits, and spread chemical fertilizers (nutrients) to spur bacterial growth and decomposition of petroleum residue (Figure 15.5). Naturally occurring oil-decomposing bacteria bloomed in the near shore waters after the spill, helping to clean shorelines. Smaller cleanup crews worked during subsequent summers, removing oily sediments and recording the condition of the treated beaches.

While both the Gulf of Mexico and Prince William Sound suffered immensely, there are no natural oil seeps in Alaska. The impact of the Exxon Valdez oil spill was immediate and devastating for many communities of marine organisms, which were covered by oil and thereby exposed to many toxic constituents. Those living attached to the rocky upper and middle parts of the intertidal zone were coated, and, most seaweeds and attached and burrowing animals perished. Weathered oil that sank to the bottom, below low tide level, killed eelgrass beds and the animals living in them. In both cases, seabirds living on the surface were especially vulnerable, in the Sound, an estimated 250,000 died, including common murres and harlequin ducks, and about 250 bald eagles were killed, either directly by the spill or later when they ate oil-coated fish and oil-contaminated carcasses. A year after the Deepwater Horizon spill, the official count holds 8000 birds effected, including 932 pelicans and 3300 laughing gulls, but how to calculate the overall number, taking into account unseen birds along such a popular migratory route, is still being debated.

Marine mammal populations near the spills were also affected. Around the Sound, about 2800 sea otters were killed immediately when their fur coats were oiled, depriving their fur of its insulation property and exposing the otters to low temperatures. Furthermore, oil slicks spread from the spill and contaminated many prime haul out areas used by hundreds of harbor seals, just before pupping season began, killing 300. Large marine mammals, such as humpback whales, apparently were little affected. However, 22 killer whales from a resident pod were missing and presumed dead. An initial 77 marine mammals washed up during the months of the Gulf oil spill, and continuing deaths a year later, in March 2011, were linked to the oil released. Because of ongoing civil and criminal cases involved, the cause has not been released for the 153 dolphin deaths, including 56 newly born or stillborn calves, for just the first three months of 2011.

FIGURE 15.5
Cleanup in Alaska's Prince William Sound area following the 1989 *Exxon Valdez* oil spill. [EXXON VALDEZ Oil Spill Trustee Council, NOAA Photo Library]

Among commercially exploited fishes in Prince William Sound, Pacific herring and salmon were most seriously affected. Billions of salmon died. The herring, which was a $12 million fishery in the late 1980s, recovered quickly to a record high. In 1993, however, it collapsed a second time. Both the poor plankton bloom of 1992, when lack of food made the herring more vulnerable to disease, the oil spill and three sequential years of fishing pressure, have been blamed. For many fish, the cause of death probably will never be known, as they could have sunk, been washed out to sea or eaten by scavengers. Many of the scavengers probably also died from consuming oil-contaminated meat.

A federal-state council of resource agencies implemented a program of natural resource damage assessment. Together with the U.S. Environmental Protection Agency, the Exxon Valdez Oil Spill Trustee Council oversees research and restoration of Prince William Sound and its ecosystems. With the massive cleanup and more than two decades of natural healing, many populations in Prince William Sound have recovered, including the murres, black oystercatchers, bald eagles, cormorants, and salmon.

Some ecosystems may take even longer to recover, or will never return to their original state. Though the sea otter has rebounded in most of Prince William Sound, populations in the heavily oiled intertidal zone remain low. At the time of the spill, two pods of killer whales each lost 40% of its members and, while one pod is recovering slowly, the other is likely to become extinct.

Oil still clings to the underside of rocks on beaches. When 9000 pits were sampled on 91 beaches in 2001 by NOAA chemist Jeffrey Short and colleagues, they estimated 55,000 liters (14,500 gallons) of oil remained, spread over 11 hectares (27 acres). Four years later, Short's team re-sampled 10 of the beaches and found that the oil was decaying at only 0% to 4% per year. On that basis, oil could persist for up to a century.

Some ecologists question whether humans can completely restore a system that has been so extensively disturbed, especially when a significant sum of money is needed for restoration activities. Exxon (now ExxonMobil) settled claims for damages to the marine environment amounting to $1 billion. In addition, the company was fined $5 billion for punitive damages, which was recalculated to $507.5 million and only began payment in 2009. ExxonMobil estimated it spent more than $4.3 billion on compensatory payments, cleanup payments, settlements and fines.

One of the world's largest oil spills was not an accident. In February 1991, retreating Iraqi armies released 4 to 6 million barrels of oil into the shallow Persian Gulf waters, forming a slick that coated nearly 1600 km (1000 mi) of shoreline. While Kuwait's beaches have recovered most quickly, the beaches of Saudi Arabia, including salt marshes, mud flats and mangrove swamps, are still heavily polluted with toxic oil. Covered by sand, buried 0.6 to 2.4 m (2 to 8 ft) below the surface, the oil is even more difficult to clean due to land mines and ordnances.

The *Exxon Valdez* was a single-hulled tanker. Use of double-hulled ships, essentially a ship within a ship, greatly reduces the risk of spills. According to the U.S. Coast Guard, double-hulled ships could eliminate an estimated 95% of all oil spills, but probably not prevent spills in the case of major accidents, such as the one involving the *Exxon Valdez* where both hulls would have been punctured. In 1990, the United States banned single-hulled tankers older than 23 years from operating in its waters and ports after 2010. After the single-hulled *Prestige* split in two and sank offshore of Galicia, Spain in mid-November 2002, the European Union also passed legislation to ban single-hulled tankers older than 23 years from its ports, and seeks to join with the *International Maritime Organization (IMO)* to institute such bans worldwide.

Much of the oil entering the coastal ocean comes from untreated storm-sewer discharges. To deal with these discharges, engineers have developed low cost ways to use natural processes to treat these waters. The initial surge of runoff from a paved road during a heavy rain carries the greatest pollutant loads. Unburned fuel from cars and trucks along with accumulated grease from road surfaces flows into storm drains soon after the rain starts. This water can be diverted into gravel-lined trenches where suspended particles settle into the gravel and the water infiltrates the underlying soil. Bacteria in the trenches could decompose hydrocarbons while metal-rich particles from brake pads and particles from tires are trapped in soils and can be extracted periodically for later disposal in a sanitary landfill. Runoff from later in the rainfall event contains fewer pollutants and flows through drainage channels directly into the ocean. Storm waters flowing through wetlands are also filtered naturally.

Recent technological advances have made possible a more rapid and effective response to oil spills in the ocean. For more on this topic, refer to this chapter's first Essay.

Dams and Marine Habitats

A dam alters the flow of water and sediment in a river or stream and may disrupt coastal river and marine ecosystems. A **dam** is a barrier constructed across a watercourse that impounds water in an upstream reservoir (Figure 15.6). Worldwide about 15% of Earth's renewable freshwater supply is stored in reservoirs behind dams (about 6000 cubic km of water). Almost 3000 of these reservoirs have a combined storage capacity of more than 94 billion liters (25 billion gal), equivalent to all the water in Lakes Michigan and Ontario.

Most dams and associated reservoirs are multipurpose structures. They impound water for irrigation, flood control, recreation (e.g., boating, swimming, fishing), and municipalities (e.g., drinking water supply). They are also used to generate hydroelectric power (supplying about 20% of the global electrical capacity) and regulate water levels for navigation. No reliable statistics exist on the total number of dams worldwide, although one estimate is close to 800,000. In the United States, there are approximately 75,000 dams taller than 2 m (7 ft). Hydrologist Catherine Reidy of Umeå University in Sweden estimates that worldwide about 45,000 dams are greater than 15 m (50 ft) tall. Dams more than 150 m (490 ft) high are considered *major dams*, capable of storing one or more years of average annual flow of a river. The Hoover Dam, constructed on the Colorado River in the 1930s, was the first major dam built. Today, more than 100 major dams operate worldwide.

Major dams are problematic because their large reservoirs dislocate native and indigenous peoples by flooding what had been their homes, farms, and communities. For instance, the world's largest dam, China's Three Gorges Dam, which spans China's largest river, the Yangtze River, has displaced 1.3 million people. Primarily intended for energy generation and flood control, its installed capacity was 18,200 megawatts in 2008, with maximum electrical generating capacity expected to be 22,500 megawatts. The 600 km (373 mi) long reservoir flooded 60,000 hectares (150,000 acres), inundating 160 towns, many villages, 1600 factories, and numerous archeological and cultural sites. Untreated wastewater discharged from upstream cities and factories flow slowly through the reservoir, creating local pollution problems as well as exacerbating the already severe pollution problems in the city of Shanghai, some 1600 km (1000 mi) downstream from the dam.

FIGURE 15.6
Navigational dam located just upstream of Zanesville, OH on the Muskingum River. [Courtesy of NOAA/NWS/Advanced Hydrologic Prediction Service]

Prior to construction of the Three Gorges Dam, the Yangtze was among the largest sediment-transporting rivers in the world (Chapter 4), but now most of that sediment accumulates in the reservoir. Over time, this reduces the reservoir's storage capacity, diminishing hydroelectric capabilities and flood control. Without the former sediment load of the river, the water flow accelerates, increasing erosion along the river banks below the dam. Furthermore, sediment was previously deposited in the Yangtze delta during floods. Without it, the regular renewal of the fertile agricultural land will halt, a situation similar to what happened to the Nile Delta following completion of the Aswan High Dam. (For details, refer to the second Essay of Chapter 4.)

During the 1990s, nearly $40 billion dollars was annually spent on dam construction. In 2003, nearly 1700 dams were under construction, almost 500 in Brazil and more than 700 in India. At present, at least 273 major dams are either planned or under construction on 46 major rivers worldwide, most with funding from the World Bank and the International Monetary Fund. However, the environmental and human cost of dams is becoming more widely recognized and, in many developed nations, emphasis has shifted from dam construction to floodplain management, regulating land use in the floodplain to protect lives and property, and preserve ecosystems. In the U.S., about 25% of all dams are more than 50 years old, many of which have been abandoned or are no longer maintained and pose safety hazards to downstream communities. Over the past decade, more dams have been removed than new ones constructed in the U.S., a trend that is expected to continue, especially for small dams.

During the 20th century, the U.S. government constructed 14 dams on the Columbia River and 13 on the Snake River to produce hydroelectric power, control flooding, and open the region to river-barge traffic. In addition, about 4000 smaller, non-federal dams were built in Oregon and Washington to provide electricity and supply water for factories, farms, and cities. Dams transformed these once cold, swift-flowing rivers and tributary streams into a series of large lakes with warmer, sluggishly flowing waters.

The impact of dams on *anadromous fish* populations in the Pacific Northwest has been especially severe (Chapter 10). Dams block adult salmon when they return from the Pacific Ocean to spawn on upstream gravel bars and sandy river bottoms, and much of the original habitat for spawning has also been eliminated by dams. Salmon populations plummeted from an estimated 16 million to 300,000 and some salmon stocks are now nearly extinct. (Overfishing and climate change have exacerbated the population decline.) *Fish ladders* were built to allow fish to swim past the dams but they are largely ineffectual (Figure 15.7). Furthermore, hydroelectric turbines kill many of the newly hatched salmon as they attempt to swim to the sea. Like a food processor, turbine blades grind up the young fish even as the extreme pressure changes and turbulence from the spinning blades kill an estimated 15% at each dam. Overall about 60% to 70% of juvenile salmon never reach the ocean.

A proposal to protect and restore wild salmon stocks calls for removal or partial breaching of several of the large hydroelectric dams. A portion of the water would flow through a breached dam, permitting fish to swim downstream without encountering the turbines. The problem is that either removal or breaching of dams would reduce the amount of hydropower generated. Having already experienced electrical shortages, residents of the Pacific Northwest are reluctant to lose any more of the generating capacity. Another proposal to remove the lowermost four dams on the Snake River is hotly debated.

In the United States, removal of even small dams on rivers and streams is a relatively new strategy intended to benefit downstream ecosystems, including marine ecosystems, by restoring habitat (Figure 15.8). Hundreds of small dams were removed in the 1990s although few cases have been studied in sufficient detail to demonstrate conclusively the environmental benefits of dam removal. One of the success stories is the breaching of the Edwards Dam on the Kennebec River in Augusta, ME, on 1 July 1999. For the first time in 162 years, the river flowed freely to the ocean. At the time, the Kennebec was the largest river in the nation to have a dam removed

FIGURE 15.7
A new Alaskan Steep Pass fish ladder just prior to installation at a small dam in Byfield, Essex County, MA. The old fish ladder failed to provide anadromous fish with regular access to their spawning grounds in the Parker River. [NOAA Restoration Center, Louise Kane]

and the Edwards Dam is still among the largest dams ever breached.

Built in 1837, the Edwards Dam decimated local fish populations, first by flooding critical habitats and then by preventing anadromous fish from migrating from the ocean to their upstream spawning grounds. Removing the Edwards Dam reopened about 30 km (19 mi) of fish spawning and nursery habitat in the Kennebec River. Environmental improvements in Merrymeeting Bay (at the river's mouth), the largest freshwater tidal complex on the Atlantic coast north of Chesapeake Bay, were reported in the years following removal of the dam. Populations of 10 species of anadromous fish exhibited varying levels of recovery; among them were alewives, American shad, Atlantic salmon, striped bass, and Atlantic sturgeon. Some fish populations will require releases of hatchery-raised young fish to restore their populations to pre-dam levels. Many bird populations also increased (e.g., ospreys, kingfishers, cormorants, and bald eagles) and water quality has improved (i.e., greater water clarity and higher levels of dissolved oxygen). In addition, recovery of the river spurred revitalization of Augusta's waterfront and other riverside communities as tourists, recreational fishers, and boaters returned to the area.

More recently, the Simkins Dam on the Patapsco River in Maryland was breached to reestablish the river's natural sediment flow and by 2010, 32 km (20 mi) of stream habitat were restored for migratory fish such as the American eel and alewife. Eventually 48 km (30 mi) of the river will flow freely, connecting the Patapsco River State Park to Baltimore Harbor and Chesapeake Bay.

FIGURE 15.8
In the summer of 2004, NOAA provided funds to help the town of Henniker, N.H., remove the 18-foot-high West Henniker Dam, which blocked migratory fish and eel passage to upstream spawning and feeding grounds. Removal of the dam restored approximately eight acres of in-stream habitat and opened 15 miles of the Contoocook River that had previously been blocked. Photos show the dam before removal (A); during removal (B) and the Contoocook River after removal of the dam (C). [New Hampshire Department of Environmental Services/NOAA]

Dam removal can cause problems when sediments formerly trapped in the reservoir behind the dam are suddenly released. The influx downstream alters habitats, wiping out organisms, such as insects, algae, and bivalves, and any contaminants in those sediments may be mobilized. The upper Hudson River in New York has several low dams, which formed large, slow flowing pools where PCB-contaminated sediments accumulated over several decades as they were released by upstream manufacturing plants. Most contaminates were sequestered at the bottom of the reservoir behind the dam at Fort Edward until 1973, when the dam's age and poor structural condition prompted its removal. Before it was replaced, major floods in 1974 and 1976 transported the contaminated sediments downstream, depositing them behind the next dam. About the same time, elevated levels of PCBs were detected in fish downstream. Plans call for dredging the contaminants and storing them in an enclosed structure to prevent their further mobilization and pollution of the river.

Restoring Chesapeake Bay

The *Chesapeake Bay Program* became the nation's largest federal-state environmental restoration project in 1983, seeking to halt further degradation of the Chesapeake Bay estuary, restore its water quality, and rebuild its fisheries. The states of Maryland, Virginia, and Pennsylvania, a tri-state legislative body, the District of Columbia, citizen advisory groups, and the Chesapeake Bay Commission are all still involved. The U.S. Environmental Protection Agency (EPA) leads the federal government partnership, which includes ten other departments that have land-holdings in the Chesapeake Bay watershed, and numerous agencies, such as the National Oceanic and Atmospheric Administration (NOAA), with responsibilities and programs in the Bay or its watershed. Total cost for the Bay's restoration is estimated at $20 billion.

Chesapeake Bay is a shallow, moderately stratified (partially mixed) estuary with a slow turnover rate and several tributary estuaries emptying into it (Chapter 8). Draining 14 times as much land surface as its water surface (i.e., a 14:1 land to water ratio), the largest ratio of any coastal bay, Chesapeake Bay drains all or part of New York, Pennsylvania, West Virginia, Delaware, Maryland, West Virginia, and the District of Columbia. All these are part of its **watershed**, the geographical area drained by rivers and their tributaries

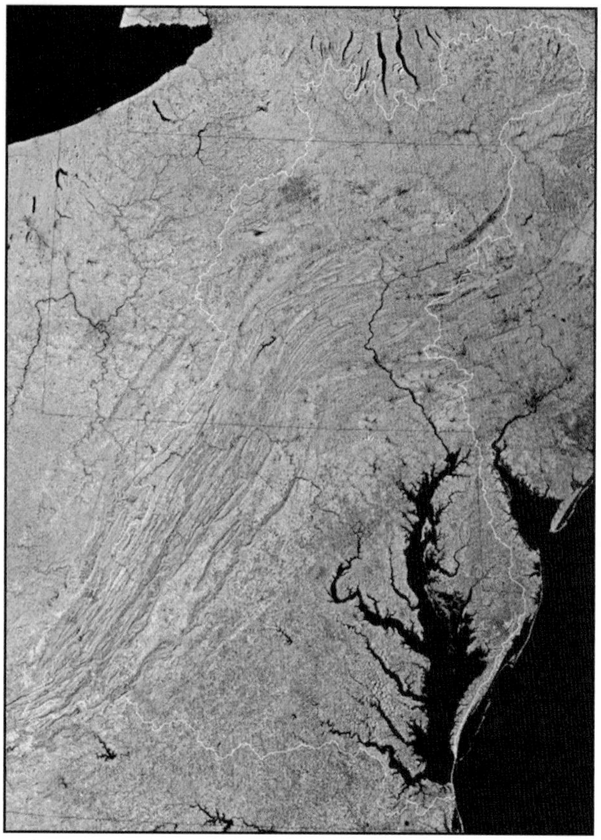

FIGURE 15.9
Since 1983, the Chesapeake Bay estuary has been the focus of the nation's largest federal-state environmental restoration project, the Chesapeake Bay Program. [Courtesy of USGS]

flowing into the Bay (Figure 15.9). The northern end of the Bay is essentially the estuary of the Susquehanna River, flowing from Pennsylvania and New York and contributing about half the river water that reaches the Bay. From the West, the Potomac and James Rivers each bring about a quarter of the total river inflow. An unknown amount of groundwater flows into the Bay, primarily from its eastern side. While sediments enter the Bay directly from the Susquehanna River, those transported by the Potomac and James Rivers are trapped in their lower reaches. Sands from the continental shelf enter the Bay with the tidal currents flowing through its mouth.

Beginning with European settlement early in the 17th century, human activities have greatly modified Chesapeake Bay and the ecosystems within. Much of the original forest that covered its watershed was cleared and converted to agricultural lands, then roads, cities, and suburban developments. Deeply dredged shipping channels lead into port facilities at Norfolk, VA, near the mouth of the Bay and Baltimore, MD, in the north. These modifications accelerated the influx of nutrients (nitrogen and phosphorus), sediments, biocides (pesticides and herbicides), and other pollutants into the Bay.

In essence, human activities converted the Chesapeake Bay watershed from a *closed system* to an *open system*. Prior to settlement, nutrients primarily cycled within the original climax forest with relatively little drainage into the Bay. Clearing the land increased the area of exposed soil, increasing runoff from rain and snowmelt which then accelerates the rate nutrients enter the Bay. Today, the Chesapeake Bay watershed is home to an estimated 16.9 million people, projected to increase to 18.9 million by 2020 and 20.3 million by 2030. About 60% of the 166,000 square km (64,000 square mi) of the Bay's watershed is covered by forests, about 30% by farms and pastures and the remaining 10% by cities, suburbs and highways. The 150 rivers and streams that empty into the Bay carry the additional nutrients into it.

Higher nutrient levels spur the growth of algal populations, which can shadow the sea grasses below and, when these organisms die in mid-summer, their remains settle to the bottom. Their decomposition reduces the dissolved oxygen in the Chesapeake's bottom water (Chapter 10). Also, more sediment increases the water's turbidity, further reducing the sunlight that reaches the sea grasses and other photosynthesizers. These and other habitat changes have a ripple effect on Chesapeake food webs.

Nutrients, as well as toxins, are carried from point and non-point sources within the watershed. A **point source of pollution** is a discernible conduit, such as a pipe, chimney, ditch, channel, sewer, tunnel or vessel, which transports contaminants. A **non-point source of pollution** is a broad area of the modified landscape, such as agricultural fields, lawns or parking lots. In addition, some airborne pollutants (e.g., ozone, nitrogen oxides), primarily from coal fired power plants in the Ohio River Valley, also reach the Bay and settle, or are washed by rain and snow into its waters.

The first U.S. estuary selected for restoration and protection, Chesapeake Bay required immediate attention for its over-enrichment of nutrients, loss of submerged aquatic vegetation (sea grass beds), and pollution by toxic chemicals. The Chesapeake Bay Program focuses on restoring the Bay's living resources, especially shellfish and finfish stocks, and the sea grass habitat. After three decades, accomplishments have been modest, but also encouraging. The size of the watershed, covering so many states, and the slow rate of circulation, have meant that communication must be on an equally large scale.

Regionally, communication among federal and state agencies has improved and plans for fisheries management and habitat restoration, including mapping the expansion of sea grass beds, have been formulated and adopted. Numerical models now describe the amounts and effects of nutrients discharged to the Bay, so goals were set to reduce the influx of nutrients and monitor water quality.

In the early 1970s, the government banned phosphate detergents, which is the primary reason phosphorous concentration in the Bay has decreased. Not dropping as rapidly, the 40% reduction goal for nitrogen was not met by 2000. Although runoff has been reduced, 25% to 30% of the nitrogen entering the Bay and its watershed is airborne, emitted by motor vehicles and coal-fired power plants, which is more difficult to control. On the plus side, the Chesapeake Bay Program has helped to prevent environmental conditions in the Bay from deteriorating as rapidly as they were prior to the 1970s, and significant signs of improvement are appearing.

When the Bay's striped bass (popularly known as rockfish) population nearly collapsed in the mid- and late-1980s, Maryland imposed a moratorium on fishing, which was followed by Virginia, which shares jurisdiction over Chesapeake Bay. With control over the southern half, Virginia kept the moratorium for a much shorter period but the striped bass rebounded and, in 1995, was declared restored and is again a major fishery. This is an example of successful policy action when different states manage a single shared ecosystem.

The recent rebound of the blue crab population demonstrates what can be accomplished when the management plan is science-based. As noted in Chapter 10, restoration of submerged sea grass beds is a key factor in the upturn of Chesapeake Bay blue crab population, but another cooperative effort by Maryland and Virginia to protect female crabs made the greatest difference. Female blue crabs mate only once during their lives, storing sperm internally to fertilize their eggs months later (Figure 15.10). Tens of millions of pregnant female blue crabs migrate up to 240 km (150 mi) south to the high salinity waters at the mouth of Chesapeake Bay and burrow into the mud to hibernate over the winter. The following spring and summer, from May through August, they each release up to 3 million fertilized eggs. While the females remain at the Bay's mouth for the rest of their lives, juveniles migrate north into the Bay for the protection of the sea grasses.

Curtailing the harvest of of the pregnant females, taking no more than 46% of the stock, has alleviated stress on the blue crab population. The spawning season sanctuary has been extended and harvesting now stops early, when the female migration peaks in the fall. In 2010, the last day for harvesting female crabs was 10 November. Winter dredging of hibernating pregnant females, a practice that killed two crabs for every one caught, has also been prohibited, and more lost traps, which contribute to crab mortality, are being retrieved. In only three years, the Chesapeake Bay blue crab population has rebounded from near collapse to healthy and sustainable levels (Figure 15.11).

A 2009 survey found that the female blue crab population had increased 70% while the male population remained level. In spring 2010, the number of females increased 200% compared to 2008, and the total Chesapeake Bay adult blue crab population soared

FIGURE 15.10
Underside of adult female blue crab. Note the red-tipped claws. Both the design of the underbody and color of the claws differentiate a male blue crab from a female blue crab. [Photo by Mary Hollinger, NOAA/NESDIS/NODC (ret.)]

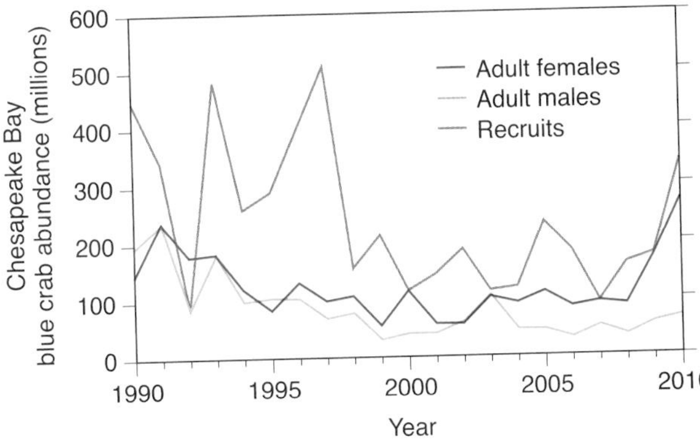

FIGURE 15.11
Rebound of the Chesapeake Bay blue crab population, especially for adult females and recruits (juvenile blue crabs less than 1 year old). [Adapted from *Science*, Vol. 330, 10 December 2010]

from 131 million to 315 million. Fishery management experts anticipate that with an increase in the blue crab's reproduction rate, the harvest season can be extended in the future.

Beyond federal and state regulation, *non-governmental organizations (NGOs)* have helped to improve environmental conditions in Chesapeake Bay. Founded in 1967, the *Chesapeake Bay Foundation (CBF)*, the largest NGO dedicated to restoring the Bay, works with volunteers to prevent pollutants from entering the Bay, as well as to restore the Bay's natural water-filtering systems: forests, wetlands, sea grass beds, and oysters. Along with organizing and informing the public, CBF has educational programs for children and adults to build an environmental ethic in the watershed.

In December 2010, the Chesapeake Bay Foundation issued its most recent State of the Bay report, which contains "some encouraging signs of improvement." The CBF uses the *Bay Health Index*, assigning scores in the three indicator categories of pollution, habitat, and fisheries (Table 15.2). For reference purposes, the Bay Health Index is assigned a theoretical value of 100 for the time when colonial settlers first arrived in the watershed. A score is assigned for each indicator and then the scores of the three categories are averaged to gauge the overall state of the Bay. In 2010, the overall index was 31, up 3 points from that reported in 2007, with a letter grade of D+ (poor).

TABLE 15.2
State of Chesapeake Bay in 2010[a]

		Score	Grade	Change
Pollution	Nitrogen/Phosphorus	16/23	F/D−	−1/+2
	Water Clarity	16	F	+2
	Dissolved Oxygen	19	F	+5
	Toxics	28	D	+1
Habitat	Forested Buffers	58	B+	+2
	Wetlands	42	C+	0
	Underwater Grasses	22	D−	+2
	Resource Lands	31	D+	+1
Fisheries	Rockfish	69	A	−1
	Crabs	50	B+	+15
	Oysters	5	F	+1
	Shad	9	F	0

[a]Source: 2010 State of the Bay, Chesapeake Bay Foundation

From 2008 to 2010, pollution issues improved slightly for water clarity (+2), dissolved oxygen (+5), and toxics (+1) although the overall status remains very low. On slightly better footing initially, habitat restoration has also improved slightly for forested buffers (+2), underwater grasses (+2), and resource lands (+1). Thanks to the blue crabs (+15), fisheries showed the greatest improvement. Already, the rockfish population (−1), with an A overall, had the highest score but oysters (+1) and shad (0), an anadromous fish spawning in fresh water but maturing in the ocean like salmon, continue to struggle desperately.

Waste Disposal in the Ocean

The global human population will continue to grow rapidly during the next century. Population growth will increase demands for fresh water, food, energy, consumer goods, and living space. At the same time, it will be increasingly important to find new and acceptable ways to dispose of waste generated by human activities while maintaining a satisfactory level of environmental quality both on land and in the ocean. The rate of waste production is likely to keep pace with or perhaps outpace population growth. Disposing of waste in an environmentally acceptable manner is likely to be most challenging in the coastal zone where the human population is growing at a much more rapid rate. Sanitary landfills are filling rapidly and high-tech incinerators are expensive to operate. Recycling of discarded materials is among the strategies to reduce the amount of material entering the waste stream. Some cities encourage, whereas others require, recycling of paper, metals, glass, and plastics. Management of toxic and hazardous waste is particularly challenging. Treaties banning the manufacture of some persistent chemicals, such as DDT, PCBs, and CFCs have been most successful in developed countries but have had less adherence in developing countries. Discharge of excess nutrients and sediments eroded from the land will likely continue worldwide.

The ocean is viewed as one possible solution to the problem of waste disposal as terrestrial options are used up. International treaties and national regulations prohibit ocean disposal of most waste. However, the ocean has been used for disposal of a great variety of wastes since humans began navigating and fishing its waters. Shipwrecks, abandoned petroleum production platforms, and other debris from human activities are common on the ocean bottom, especially near land and in shipping lanes. Many nations have directly discharged wastes at sea, including materials dredged from coastal waterways, solids (sewage sludge) from wastewater treatment plants, various industrial and construction wastes, and radioactive materials.

Deliberate discharge of these substances from ships, planes, and platforms is regulated under the *Convention on the Prevention of Marine Pollution by Dumping of Waste and Other Matter*, more commonly known as the *London Dumping Convention*, administered by the United Nations *International Maritime Organization (IMO)*. The treaty was first established in 1972 and strengthened in 1996. Dumping of some materials was banned entirely (e.g., radioactive wastes); others were regulated (e.g., dredge spoils, sewage sludge, mining wastes).

Given the population pressures on land, some people argue that waste disposal in the ocean is an attractive alternative for many waste disposal operations. The ocean is vast and the deep-ocean floor (more than 3000 m or 10,000 ft below the sea surface) covers about half of Earth's surface. Much of this deep-ocean bottom is assumed by many to be a featureless, sediment-covered plain, and seemingly monotonous physically, chemically, and biologically. Unfortunately, our knowledge of this realm is very limited. Thus, it is easy to assume that part of this region could be used for waste disposal without interfering with other uses of the sea. For example, the argument continues, dumping barge-loads of crushed rock from Indonesian gold-mine operations down a submarine canyon off New Guinea would have few, if any, adverse effects on the nearby ocean bottom. This region naturally receives huge amounts of river-borne sediment.

Other people suggest that we should use the deep ocean for disposal of nuclear wastes, or carbon dioxide from industrial chimneys (in liquid or solid form). The ocean is viewed as a potential receptacle for wastes that would be considered undesirable in a land disposal site. But most scientists point out that we do not yet know nearly enough about the chemistry, circulation processes, and marine ecosystems in the deep ocean to be sure that such disposal would be safe and secure.

In addition to scientific concerns, ethical issues are involved in ocean waste disposal. Many individuals and organizations vigorously oppose using the ocean in this way; they consider it immoral to pollute the ocean and are against all forms of waste disposal at sea. This is especially true for disposal of radioactive materials. Disposal of these wastes pose extremely difficult problems for the industries and government agencies that produce them. Radioactive wastes (e.g., from nuclear power plants) emit ionizing radiation that poses serious health hazards. As a general rule, radioactive isotopes must be kept isolated for a period equivalent to 10 half-lives before they can be safely released into the environment. (One *half-life* is the amount of time required for the radioactivity emanating from a particular radioactive substance to decrease by one-half.) For some high-level radioactive isotopes, this means isolation for hundreds of thousands of years. Because human institutions and structures are highly unlikely to survive intact for such long periods, the materials must be safely isolated. Isolation through burial in geologically stable regions is usually the preferred disposal option.

Land disposal sites proposed for radioactive wastes are bitterly opposed because of the risk of escape of radioactive materials to the groundwater or air. The "NIMBY" (not in my back yard) argument is often heard when this issue is debated. Sealing radioactive wastes in special containers and then burying them in the deep-ocean floor is one option that has been discussed. This strategy is hypothesized to keep the radioactive waste out of direct human contact for thousands, perhaps millions, of years. Other schemes would place radioactive-waste containers in subduction zones so that they would be drawn down into Earth's mantle and presumably, out of contact with humans (Chapter 2).

Essentially these disposal plans are resisted because we know so little about the deep ocean. The basic argument is that it is unwise to put the most dangerous waste materials where they may be subjected to processes not fully understood. Furthermore, retrieving such waste from the sea floor might be difficult if not impossible if they had to be relocated. The deep ocean appears to be even more complex, with greater biodiversity, than many ecosystems on land. Some deep-ocean currents are apparently stronger than previously thought, and the ocean's thermohaline circulation cycles over a time scale of about 1000 years, which is far less than the half-life of many radioactive wastes. Furthermore, new fishing techniques permit commercial fishing in deep-ocean waters so that wastes might be accidentally introduced into commercial fish catches, or trawled up.

Deep-Ocean Carbon Storage

As we have learned, levels of atmospheric carbon dioxide have been rising since at least the middle of the 19th century and are likely contributing to global climate change by enhancing Earth's greenhouse effect (Chapters 5 and 12). Burning fossil fuels and deforestation are primarily responsible for the increase in atmospheric CO_2. Schemes to remove carbon dioxide from the atmosphere or to store it directly instead of releasing it to the atmosphere have been proposed.

Replanting tropical forests has been proposed as a low-cost way to remove atmospheric carbon dioxide. Plantation costs are low in many developing countries in the tropics and rain forest vegetation grows rapidly, thus storing carbon for decades to centuries. Unfortunately, such projects do not provide enough carbon storage to be effective over the long term. Growth slows as trees mature, so the carbon uptake also diminishes. Furthermore, forests are either harvested or burned due to natural causes (e.g., lightning) every 30 to 50 years. As populations grow and more agricultural land is needed, forests will probably shrink, rather than expand.

Another potential storage place for carbon dioxide is in rock formations below the sea floor. This type of repository has been used at some offshore natural-gas production facilities where carbon dioxide is separated from methane. Methane, the chief component of natural gas, is piped ashore and the carbon dioxide is injected into bedrock fractures and pore spaces below the facility (rather than being vented to the atmosphere). Using this technique, Norway's state-controlled oil company has disposed of about 1 million tons of carbon dioxide per year since 1996 at its natural gas operation in the North Sea. Other options for sequestration of carbon dioxide include injection into aquifers, deep coal seams, or depleted oil reservoir rock.

Ocean iron fertilization (OIF) is proposed as a means of boosting the ocean's uptake of carbon dioxide from the atmosphere (Figure 15.12). Recall from Chapter 9 that broad areas of the subtropical open-ocean receive abundant sunlight and have relatively high concentrations of nitrogen and phosphorus compounds, all essential ingredients for photosynthesis by planktonic algae. Yet, primary production is meager in these high-nutrient low-chlorophyll (HNLC) areas. In the 1990s, ocean scientists demonstrated that these waters are unproductive because they lack essential micronutrients, such as iron (refer to the first Essay in Chapter 9). Clouds of iron-rich dust particles blown from the land and settling into the ocean's surface waters trigger infrequent pulses of short-lived algal blooms. As of this writing, ocean scientists have conducted at least a dozen OIF experiments since 1993, releasing a soluble form of iron into these waters to artificially stimulate longer-term algal blooms. Thereby, carbon dioxide is taken up from the atmosphere that would not otherwise be used in photosynthesis. When the algal cells die, some sink through the surface layer and are eaten or decomposed in the deep ocean or are deposited on the ocean floor via the physical and biological pumps (Chapter 9).

Researchers continue to investigate the feasibility of OIF and any possible adverse effects it might have on marine ecosystems. Field experiments conducted in the Southern Ocean and in Pacific equatorial waters showed temporary increases in phytoplankton populations and primary production following the addition of iron. But as yet no proof exists that such a method could be used economically on a large scale to bring about long-term removal of carbon from ocean surface waters and the overlying atmosphere. In fact, it appears unlikely that ocean iron fertilization would sequester atmospheric carbon in the deep ocean for lengthy periods of the order of centuries to millennia. Most areas of the ocean where OIF would work (HNLC regions) are locales of upwelling (e.g., tropical Pacific Ocean, Southern Ocean). Models

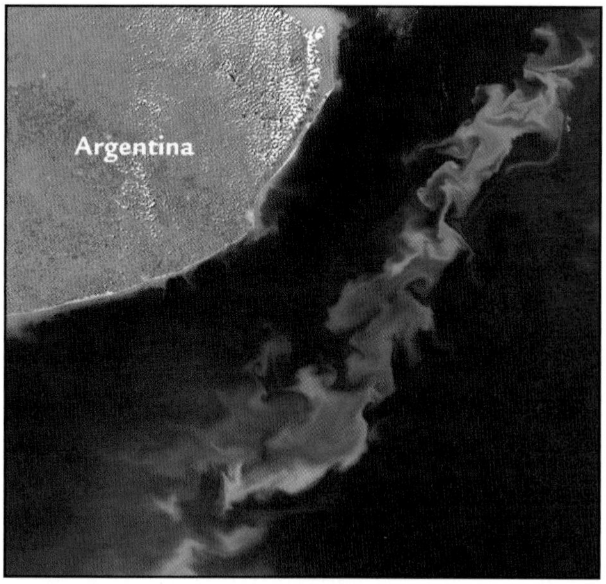

FIGURE 15.12
A large algal bloom is indicated by the swirls of green and blue in the Atlantic Ocean off the coast of Argentina. Via photosynthesis, this multitude of phytoplankton takes up carbon dioxide from the ocean and atmosphere. Adding iron to the ocean would stimulate more blooms and removal of more carbon dioxide. Scientists are trying to determine how effective ocean iron fertilization would be in offsetting Earth's enhanced greenhouse effect and global warming. [Image courtesy of NASA]

suggest that most carbon pumped to ocean depths soon would be upwelled again and re-vented to the atmosphere over a time span of only years to decades.

Furthermore, open-ocean and deep-ocean ecosystems are too poorly understood to accurately predict possible consequences of such iron releases. More field experiments are planned in other areas of the ocean and over greater temporal and spatial scales to identify possible adverse effects. For example, too much phytoplankton production might deplete dissolved oxygen as the organic matter subsequently decomposes.

Many scientists and conservationists argue that a more practical, economical, and less ecologically risky strategy would be to focus on reducing anthropogenic carbon dioxide emissions in the first place. Much of the technology to accomplish this already exists and is being used in several industrialized nations of Europe. In North America, however, many people do not strongly support this approach and the measures it would require, such as increasing the fuel efficiency of motor vehicles and reducing reliance on fossil fuels for generating electricity.

Obstacles to Ocean Policy Making

We have considered examples of policies designed to protect the ocean and its resources in this and the previous chapter. They include local, regional, national and international policies, forged by local, state and national governments, and implemented through international treaties and organizations. However, the agreements required to forge an acceptable environmental policy at any level of government are usually complex and fraught with obstacles and controversy. A number of problems arise repeatedly. Issues are very often divisive and opposing sides hold rigid views. For example, fishers rarely believe the warnings of scientists and government agencies that fish stocks are declining. They fish with more determination, further hastening the decline, until draconian measures, such as a complete moratorium on fishing, are necessary.

Often, as in the effort to restore and protect Chesapeake Bay, multiple levels of government are involved. As noted earlier, the Bay Program involves agencies in three states, the District of Columbia, and many agencies of the federal government. Political differences sometimes get in the way of forging agreements on the policies needed to achieve common goals. Various governments and organizations must respond to competing interest groups and all parties must be invited to negotiations even when they have different agendas. Diverse sectors of society may blame each other for the damage done while failing to make progress toward remediation. The agricultural industry in rural areas surrounding Chesapeake Bay feels that it is unfairly blamed for nutrient pollution in the Bay and argues that the cities and industries should bear more of the responsibility and cost of cleanup because of their sewage and other waste discharges.

At the international level, the problems are even more complex. Industrialized nations possessing advanced technology and many resources to combat pollution dominate the Northern Hemisphere. These nations have small population growth rates (if any) and an educated, affluent society that can push governments to take action to combat environmental problems. In many cases the environmental damage was done a century or more ago and is now forgotten or at least partially repaired. Developing nations are trying to improve their economies and reduce poverty by establishing industries that will create jobs. Unfortunately, the rush to industrialize and the huge population pressures in many developing countries have resulted in enormous pollution problems and damage to ecosystems. For example, the use of dynamite or cyanide to kill coral reef fish for easier harvesting causes severe harm to the reef ecosystem. The practice is banned in much of the world, but is still common in some developing countries. DDT is still used in some areas as a cheap and effective means of combating the mosquito that carries malaria (Chapter 14).

The attitude of leaders in many developing countries is that economic development must be the top priority if they are to alleviate widespread poverty. Developed countries blame them for many pollution and environmental problems. In return the developing countries point to the damage done in the past by industrialized countries (e.g. whaling, overfishing, strip mining, clear-cut logging, pollution) and demand financial and technological assistance if they are to take action on their own environmental problems. This conflict frequently frustrates or greatly delays attempts to develop international policies to protect the ocean and marine life.

Conclusions

Humans long assumed that the vast ocean holds a bounty of virtually limitless resources that are there for the taking by almost anyone. They have also assumed that the ocean's ability to assimilate waste was limitless so that the ocean was considered an appropriate repository for a wide variety of wastes ranging from sewage sludge to construction debris. In more recent times, people have begun to reevaluate these assumptions as they become more aware of issues such as overfishing, accelerated shoreline erosion, pollution, harmful algal blooms, oxygen-deficient "dead zones," and exotic species invasions.

Guided by an Earth system perspective and the fundamental understanding that marine life depends on the orderly functioning of ecosystems, scientists and public policy makers have joined ranks to rectify past practices that disrupt marine ecosystems and to formulate new policies that promise to safeguard the ecological integrity of the ocean and coastal zone for future generations. This stewardship effort is becoming more urgent as the human population density in the coastal zone continues to increase. Individual nations have taken more control over human activities in their EEZs (e.g., closing stressed fisheries) and the international community is beginning to take steps to further regulate human activities in the ocean commons. In order for ocean policy to be appropriate and effective, however, more resources need to be directed toward developing a more complete understanding of the properties and processes of the ocean.

Basic Understandings

- The history of ocean governance differs dramatically from that of the land. For most of human history, the open ocean has been beyond the direct management of coastal nations or any other governments. As the human population soared and exploitation of ocean resources expanded, regulations and institutions to govern human activities in the ocean have developed.
- In 1609, the Dutch legal scholar Hugo Grotius argued that the ocean was open with free access to all nations. For him, liberty of the sea was key to maintaining communication among peoples and nations. Along with this doctrine of free access, the general consensus was that the sea was limitless and its resources inexhaustible. There was no need to control access if marine resources were essentially unlimited and therefore virtually immune to any possible damage caused by human activities.
- Partitioning of the coastal ocean began in the late 1700s, when the new United States claimed a 5-km (3-mi) "territorial sea" off its coasts. In 1945, the U.S. extended its claims on coastal resources to the outer edge of the continental shelf—principally because of the expansion of offshore oil and natural gas production on the shallow continental shelf.
- The U.S. established its Exclusive Economic Zone (EEZ) in 1983, extending federal jurisdiction from the seaward limits of the state-controlled territorial seas (5 km or 3 nautical mi) to 370 km (200 nautical mi) from the coastline. In 1996, EEZs were extended to the edge of the continental shelf in the cases when the shelf edge was farther offshore than 370 km (200 nautical mi).
- Future governance of ocean resource exploitation may be guided by the Precautionary Principle: "When an action causes a threat to human health or the environment, precautionary measures should be taken even if some cause-and-effect relationships are not fully established scientifically."
- In 1959, Antarctica and the Southern Ocean came under the protection of the Antarctic Treaty System ensuring no national claim of territorial rights in Antarctica or the surrounding waters will ever be legally recognized and that Antarctica would be forever used exclusively for scientific and other peaceful purposes.
- Increasing human population density in the coastal zone is likely to further exacerbate the problem of environmental pollution. Pollution is defined as an intentional or unintentional disturbance of the environment that adversely affects the wellbeing of an organism (including humans) directly or the natural processes (e.g., ecosystem) upon which it depends.
- A better understanding of the pollution problem in the coastal zone and elsewhere in the ocean is based on application of the concepts of carrying capacity, assimilative capacity, and limiting factors.
- The human carrying capacity of a region is based on the maximum rate of resource consumption and waste discharge that is sustainable indefinitely without progressively impairing ecosystem productivity and integrity.
- Assimilative capacity is the amount of waste that an ecosystem can decompose without damaging ecosystem functions or building up waste products to levels that may cause unwanted or harmful impacts on living organisms (including humans).

- The growth and wellbeing of an organism is limited by the essential resource that is in lowest supply relative to what is required; this most deficient resource is known as the limiting factor.
- The need to balance multiple use and preservation of the coastal zone spurred the federal government to partner with states having coastlines to develop and implement plans for sustainable coastal zone management.
- For many reasons (shipwreck, collisions, pumping of ships' bilges, war, and terrorism), oil is accidentally or deliberately spilled into the ocean. Some oil enters the ocean routinely from storm sewer discharges and an unknown quantity of oil seeps into the ocean naturally, leaking through the sea floor into the overlying water.
- Dams have caused major changes in river and ocean ecosystems. The effects of dams on *anadromous fish* populations have been especially severe to the point that some salmon stocks are now nearly extinct. Fish ladders were built to allow fish to swim past the dams but they did not work very well. Furthermore, hydroelectric turbines kill many of the newly hatched salmon as they attempt to swim to the sea. The most radical proposal to protect and restore wild salmon stocks calls for removal or partial breaching of some large dams.
- Since 1983, Chesapeake Bay has been the focus of the nation's largest federal-state environmental restoration project, the Chesapeake Bay Program. The Bay Program seeks to stop further degradation of the Bay, restore its water quality, and rebuild its fisheries.
- Within the Chesapeake Bay watershed, numerous point and non-point sources discharge wastes into the Bay waters, primarily through rivers that empty into the Bay. A point source of pollution is a discernible conduit, such as pipes or channels which transport contaminants. A non-point source of pollution is a broad area of the landscape, such as agricultural fields or parking lots.
- The vast ocean is considered by some as a possible solution to the growing problem of waste disposal as terrestrial options are used up. But most scientists argue that we do not yet know nearly enough about the chemistry, circulation processes, and marine ecosystems in the deep ocean to be sure that such disposal would be safe.
- To reduce global warming due to a build-up of atmospheric carbon dioxide (a greenhouse gas), many schemes have been proposed to remove carbon dioxide from the atmosphere or to store it directly without releasing it to the atmosphere. One possibility is to store CO_2 in rock formations below the sea floor. This technique has been used at some offshore natural-gas production facilities where carbon dioxide is separated from methane.
- Fertilization of ocean surface waters (e.g., iron as a micronutrient) has been proposed as a means of increasing the ocean's uptake of carbon dioxide from the atmosphere and letting nature deliver it to the ocean bottom via the biological and physical "pumps".
- It appears unlikely that ocean iron fertilization (OIF) would sequester atmospheric carbon in the deep ocean for lengthy periods of the order of centuries to millennia. Most high-nutrient low-chlorophyll (HNLC) areas of the ocean where OIF would work are locales of upwelling (e.g., tropical Pacific Ocean, Southern Ocean). Models suggest that most carbon pumped to ocean depths soon would be upwelled again and re-vented to the atmosphere over a time span of only years to decades.
- Agreements required to forge an acceptable national and/or international ocean environmental policy at any level of government are usually complex and fraught with obstacles and controversy.

Enduring Ideas

- Early in the 17th century, Dutch legal scholar Hugo Grotius proposed that the ocean was open with free access to all nations. At the time, the seas were considered limitless and their resources inexhaustible.
- Jurisdiction over the coastal zone and its resources began in the late 1700s, and by 1983, the U.S. had designated its Exclusive Economic Zone (EEZ) as extending 370 km (200 nautical mi) from the coastline.
- In 1959, the Antarctic Treaty System ensured that no national claim of territorial rights in Antarctica or the surrounding ocean waters would ever be legally recognized.
- The concepts of carrying capacity, assimilative capacity, and limiting factor aid our understanding of the potential impacts of human activity on the coastal zone.
- The response to a major oil spill in the ocean depends on a wide variety of environmental factors.
- Dams are multipurpose structures that have both positive and negative impacts on the orderly functioning of the environment.
- Restoration of the Chesapeake Bay ecosystem is a challenging project that must address the complex of ecological relationships if it is to be successful.

Review

1. How are the resources of Antarctica and the Southern Ocean protected?
2. How far does the U.S. Exclusive Economic Zone (EEZ) extend seaward from the shore?
3. What is the ocean commons?
4. Define limiting factor.
5. List the potential purposes of a dam constructed across a river.
6. How might a dam influence marine ecosystems?
7. What are the goals of the Chesapeake Bay Program?
8. Distinguish between a point source and a non-point source of pollution.
9. What does the London Dumping Commission regulate?
10. What are some ocean-based options for removal and storage of carbon dioxide?

Critical Thinking

1. How might human activities be regulated in the ocean commons?
2. How does human population growth in the coastal zone impact the quality of the marine environment?
3. What are some of the disadvantages of dams constructed across large rivers that empty into the ocean?
4. What can be done to restore a beach that was extensively eroded following construction of a dam across a river?
5. Explain why water quality specialists use a drainage-basin (or watershed) approach when inventorying pollution sources for Chesapeake Bay.
6. Why is a federal-state partnership needed for restoration of the Chesapeake Bay estuary?
7. Why are physical and chemical properties important considerations in regulating the type of waste materials that are dumped into the ocean?
8. What is the purpose of the Precautionary Principle?
9. Why is Antarctica and the surrounding Southern Ocean given special status for exploitation of marine resources?
10. What fundamental assumption underlies the concept of an ocean commons?

ESSAY: Pathogens in the Coastal Zone

Along with rapid development and human population growth in the coastal zone has come more frequent contamination of beaches and shellfish beds by disease-causing micro-organisms (*pathogens*) derived primarily from animal and human fecal waste. The public health threat is forcing officials to close more beaches and shellfish beds for longer periods. The National Resources Defense Council reported that in 2009, coastal states ordered a total of over 18,000 days of closure and pollution advisories were issued for about one-third of all beaches regularly monitored by public health officials.

People are exposed to waterborne microbes by consuming contaminated shellfish or by engaging in recreational activities at the beach (e.g., swimming, surfing, waterskiing). People who eat raw or inadequately cooked shellfish may contract gastroenteritis (an unpleasant malady with vomiting and diarrhea) or other more serious illnesses. At the beach, people run the risk that waterborne microbes will enter the body through the mouth, eyes, or an open wound. Potential illnesses include gastroenteritis and infections of the eyes, ear, respiratory system, and skin. In addition, more serious diseases such as hepatitis and Guillain-Barré syndrome have been reported. In addition to the health hazards, contaminated beaches have implications for the local economy: tourism dollars decrease and real estate values decline.

What is the connection between development of the coastal zone and contamination of beaches and shellfish beds? Prior to development, a typical coastal zone was covered by soils, forests, and wetlands—surfaces through which water gradually seeps. Through this infiltration process, water is cleansed of most pathogens. With development, however, forests were clear-cut, wetlands were drained and filled, and much of the coastal zone was covered over by impervious asphalt and concrete with buildings, roads, and parking lots (Figure 1). Now, instead of seeping into the ground, most rainwater (and snowmelt) runs off and is usually conveyed by drain pipes or ditches to the ocean (or large lake). Along the way, the storm water runoff picks up animal feces and associated microbes. Added to this input may be human waste from sewage overflow or leaking septic tanks depending upon the local sewage treatment system.

FIGURE 1
The image shows the extent of impervious surfaces in and around Washington, DC and Baltimore, MD. Red represents high concentration of impervious surfaces. Blue represents moderate concentration and green represents low concentration of impervious surfaces. The map of impervious surfaces was derived via data from both Landsat and Space Imaging's high-resolution IKONOS satellite. [Courtesy of NASA]

Two separate networks of sewer pipes are under the streets of most U.S. cities (Figure 2). In this *separated sewer system*, the larger-diameter storm-sewer pipe collects runoff and channels it to the nearest waterway (e.g., ocean, lake, river) without any waste treatment. The smaller-diameter sanitary-sewer pipe conveys wastewater from dwellings and commercial establishments to a sewage treatment plant, where it is treated (including removal of pathogens and other contaminants) and then discharged into a waterway. In some cities, a single large-diameter pipe transports both runoff and wastewater to a sewage treatment plant; this is called a *combined sewer system*. During rainy episodes, runoff combines with wastewater and both are treated and then discharged. If the rainfall is excessive, however, the plant's capacity may be exceeded and the overflow (containing raw sewage) bypasses treatment and is discharged directly into the ocean or other waterway.

Many coastal areas are not served by a centralized wastewater treatment plant (either separated or combined systems). In that case, most residents rely on a septic tank system in which wastewater is cleansed by gravitational settling of solids, decomposition of organic materials, and seeping through soil and sediment. However, in some locales, the substrate is highly permeable (e.g., sands, fractured limestone) and very little filtration is achieved. In that case, appropriate fill must be hauled in and mounded or a holding tank is installed and periodically pumped out.

The greater runoff in a developed coastal zone also translates into more rapid erosion of stream banks, drainage ditches, and construction sites. The suspended sediments not only turns the runoff turbid, they also bind with the fecal bacteria and viruses, thereby protecting the pathogens from potentially lethal ultraviolet radiation.

What can be done about this problem of pathogens in the coastal zone? Writing in the June 2006 issue of *Scientific American*, Michael A. Mallin, an aquatic ecologist at the University of North Carolina, Wilmington, calls for "smart-growth" strategies in the coastal zone. He urges developers to minimize clear-cutting of forests, draining of wetlands, use of impervious pavement, and runoff of sediments from construction sites. In addition, Mallin calls for more vegetated areas. These strategies promise to reduce runoff and increase infiltration that will remove contaminants.

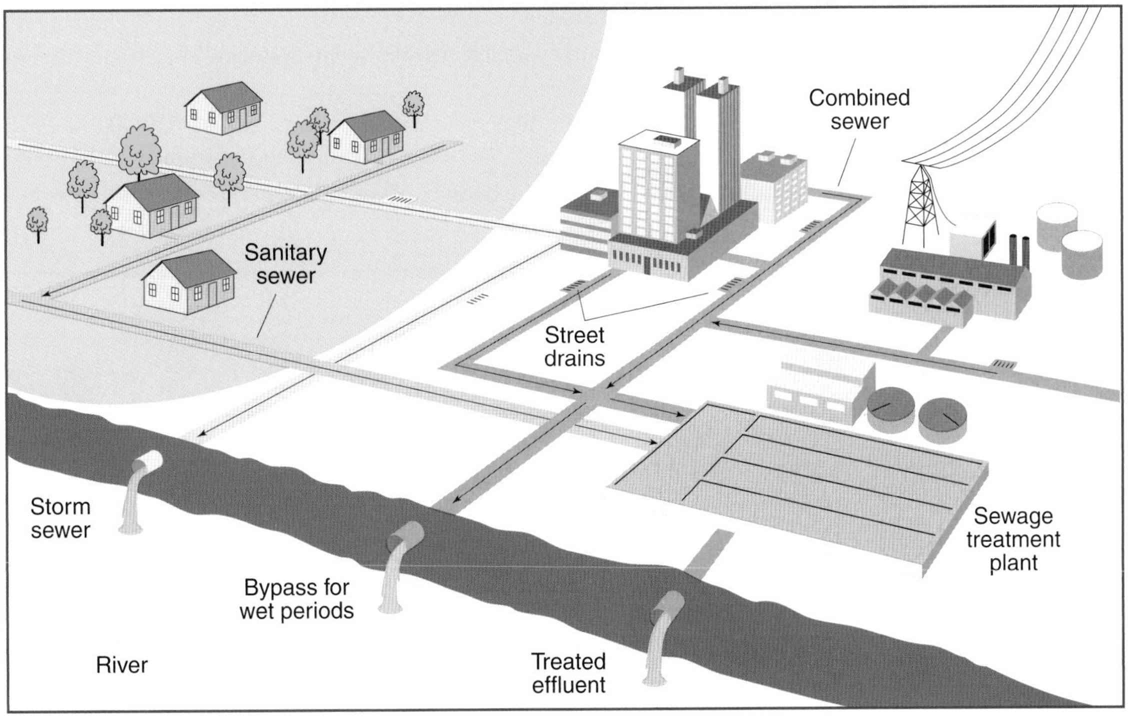

FIGURE 2
Separated sewer system beneath the streets of most U.S. cities.

ESSAY: Predicting Oil-Spill Trajectories

Accidents involving spilled oil are all too frequent occurrences in coastal regions. Many factors (e.g., winds, surface currents, tides, air and water temperatures, and salinity) control the movement of spilled oil. Furthermore, the type and amount of spilled oil, and local shoreline and bottom features also influence movements of an oil slick. An effective response to an oil spill requires the input of scientists representing many different specialties plus information on the chemical composition of the spilled oil, ocean currents, and weather. All these data are needed for mathematical models that predict movements (known as *trajectory analysis*) of the oil. When combined with biological resource information, trajectory analyses can be used to identify those areas most vulnerable to damage from the oil so that equipment to contain the oil spill can be deployed to where it will be most effective. Sensitive areas requiring protection include marine sanctuaries and other unique habitats, especially those that are home to endangered species. Protection is key because oil not only causes immediate contamination but also has long-term effects on coastal ecosystems.

NOAA's *Coastal Change Analysis Program* uses remote sensing data, primarily aerial photography, to identify and classify sensitive habitats of bottom-dwelling organisms, such as sea grasses. Oil spills can affect sea grasses in many ways. Toxic materials in the spilled oil mix with sediments and enter the water and ultimately living organisms. Oil can also block sunlight from reaching the plants. Heavy oils sink and mix with sediment and can coat sea grasses. When these grasses are damaged or killed, organisms that depend upon them as a food source or for habitat are also adversely affected. This has a ripple effect on the local ecosystem and ultimately can damage economically valuable fish and shellfish, thus impacting local economies for many years.

Major sectors of Maine's economy, especially tourism and fishing, require clean coastal waters, aesthetically appealing coastlines, and functioning coastal ecosystems. Thus, the state requires timely and appropriate responses to oil spills. To demonstrate the importance of accurate oil-spill trajectory analysis and prompt response to mediate the effects of oil spills, we examine two cases separated by nearly 25 years.

In 1972, the tanker *Tamano* spilled 380,000 liters (100,000 gal) of oil into Casco Bay, near Portland in southern Maine, then the largest oil spill in the state's history. Plans for dealing with spills were antiquated and slow to be implemented. During the several days required to mobilize response teams, the oil spread along about 75 km (47 mi) of coastline, including beaches on 18 small islands, damaging commercially valuable fish and shellfish stocks. The cleanup that followed took 11 years and cost nearly $4 billion.

On 27 September 1996, the tanker *Julie N.* struck a bridge entering Casco Bay, spilling 680,000 liters (180,000 gal) of light fuel oil and much heavier bunker-C, a mix of oils. The light fuel oil evaporated quickly but left behind toxic components. The less toxic but heavier bunker-C sank to the bottom, covering vegetation and wildlife with a thick, sticky coating. The response to this incident was rapid and effective. The damaged tanker quickly docked and was immediately surrounded by floating barriers, called booms, to contain floating oil (Figure 1). Other booms were deployed quickly to prevent oil from reaching vulnerable biological resources or economically valuable locations. In addition, special clean-up vessels skimmed and collected oil from the water surface.

NOAA officials note that the spill occurred during the spring ebb tides, with south to southwest winds that initially kept the oil within the inner harbor. However, the following night stronger winds pushed the oil through the protection booms at the upriver side of the inner harbor. Timely response combined with successful predictions of oil movements under those circumstances meant that damage to the extent of that which had occurred 25 years earlier in the same area was not repeated.

FIGURE 1
Oil tanker *Julie N* surrounded by booms. [Photo by S. Mierzykowski, USFWS]

APPENDIX I

CONVERSION FACTORS

	Multiply	*By*	*To obtain*
LENGTH	inches (in.)	2.54	centimeters (cm)
	centimeters	0.3937	inches
	centimeters	10000	micrometers (μm)
	feet (ft)	0.3048	meters (m)
	feet	0.167	fathoms (fm)
	meters	3.281	feet
	meters	0.547	fathoms
	statute miles (mi)	1.6093	kilometers (km)
	statute miles	0.869	nautical miles
	kilometers	0.6214	statute miles
	kilometers	0.54	nautical miles
	nautical miles	1.852	kilometers
	kilometers	3281	feet
	feet	0.0003048	kilometers
	fathoms	1.8288	meters
	fathoms	6	feet
SPEED	miles per hour (mph)	1.6093	kilometers per hour (kph)
	miles per hour	0.869	knots (kts)
	miles per hour	0.447	meters per second (m/s)
	knots	1.151	miles per hour
	knots	0.514	meters per second
	knots	1.852	kilometers per hour
	kilometers per hour	0.6214	miles per hour
	kilometers per hour	0.540	knots
	meters per second	1.944	knots
	meters per second	2.237	miles per hour
WEIGHTS AND MASS	ounces (oz)	28.35	grams (g)
	grams	0.0353	ounces
	pounds (lb)	0.4536	kilograms (kg)
	kilograms	2.205	pounds
	tons	0.9072	metric tons

	metric tons	1.102	tons
	metric tons	1000	kilograms
LIQUID MEASURE	fluid ounces (fl oz)	0.0296	liters (L)
	gallons (gal)	3.785	liters
	liters	0.2642	gallons
	liters	33.814	fluid ounces
AREA	acres (A)	0.4047	hectares (ha)
	square yards (yd^2)	0.8361	square meters (m^2)
	square miles (mi^2)	2.590	square kilometers (km^2)
	hectares	0.010	square kilometers
	hectares	2.471	acres
	hectares	10000	square meters
	square kilometers	1000000	square meters
PRESSURE/ FORCE	pounds force (lb)	4.448	newtons (N)
	newtons	0.2248	pounds
	millimeters of mercury at 0 °C	133.32	pascals (Pa; N per m^2)
	pounds per square inch (psi)	6.895	kilopascals (kPa; 1000 pascals)
	pascals	0.0075	millimeters of mercury at 0 °C
	kilopascals	0.1450	pounds per square inch
	bars	1000	millibars (mb)
	bars	100000	pascals
	bars	0.9869	atmospheres (atm)
ENERGY	joules (J)	0.2389	calories (cal)
	kilocalories (kcal)	1000	calories
	joules	1.0	watt-seconds (W-sec)
	kilojoules (kJ)	1000	joules
	calories	0.00397	Btu (British thermal units)
	calories	4.186	joules
	Btu	252	calories
POWER	joules per second	1.0	watts (W)
	kilowatts (kW)	1000	watts
	megawatts (MW)	1000	kilowatts
	kilocalories per minute	69.80	watts
	watts	0.00134	horsepower (hp)
	kilowatts	56.87	Btu per minute
	Btu per minute	0.0235	horsepower

APPENDIX II

OCEAN TIMELINE

BCE 4000	Egyptians developed shipbuilding and ocean piloting capabilities, as well as trading on the Nile.
BCE 3800	First maps showing water (river charts).
BCE 3000	Evidence for seafaring between China and Taiwan that may have spurred colonization of the tropical Pacific Ocean islands by the Polynesians.
BCE 2000-500	The Polynesians voyaged across the Pacific Ocean and settled all the major islands, first inhabiting Tonga, Samoa, ca. 1000 BCE. Micronesian stick chart used.
BCE 1200-850	Phoenicians explored entire Mediterranean Sea and sailed into the Atlantic to West Africa and to Cornwall, England for trading.
BCE 900	Greeks first use the term *okeanos*, a mythical god, to describe the great river that flowed in a circle around the Earth and the root of the present word ocean.
BCE 800	First graphic aids to marine navigation.
BCE 600	The Greek Pythagoreans assumed a spherical Earth.
	Thales of Miletus, a Greek natural philosopher, developed the origin of science as we define it today and he is alleged to have written a book on navigation.
BCE 450	The Greek Herodotus compiled a map of the known world centered on the Mediterranean region.
BCE 325	The Greek Pytheas explored the coasts of England, Norway, and perhaps Iceland. He developed a means of determining latitude from the angular distance of the North Star and proposed a connection between the phases of the moon and the tides.
	The Greek philosopher, Aristotle, published *Meteorologica,* which described the geography of the Greek world, and *Historia Animalium* the first known treatise on marine biology—a catalog of marine organisms.
BCE 300	Library founded at Alexandria by Alexander the Great, which served as the repository for scrolls from ships and land caravans; for the next 600 years, it was a maritime studies center, fostering advances in science, celestial navigation by a variety of Greek and Egyptian scholars.
BCE 276-192	The Greek Eratosthenes, a scholar and librarian at Alexandria, calculated the circumference of a spherical Earth with remarkable accuracy using trigonometry noting the specific angle of sunlight

	that occurred at Alexandria and Syene (now called Aswan), Egypt. He also invented latitude and longitude lines based on landmarks.
BCE 127	Hipparchus of Nicaea, a Greek mathematician, developed a 360 degree system and arranged latitude and longitude in regular grid by degrees.
BCE 54 - CE 30	The Roman Seneca devised the hydrologic cycle to show that, despite the inflow of river water, the level of the ocean remained stable because of evaporation.
1st century CE	Chinese invent first compass.
CE 150	The Greek-Egyptian scientist, Claudius Ptolemy, added minutes and seconds to the latitude-longitude system and compiled a map of the Roman World, but erred in estimating Earth's circumference, which remained uncorrected for hundreds of years.
CE 415	The Alexandrian library with an estimated 700,000 scrolls destroyed and the last librarian, Hypatia, murdered by religious mob; Earth considered flat again.
CE 673--735	The English monk Bede published *De Temporum Ratione,* which discussed the lunar control of the tides and recognized monthly tidal variations and the effect of wind drag on tidal height.
CE 780	Viking raids begin.
CE 800-1000	Hawaii colonized by Polynesians
CE 950-1250	Medieval Warm Period
CE 982	Eric the Red, a Norse chieftain, completed the first transatlantic voyage, reaching present-day Baffin Island, Canada.
CE 995	Leif Ericson, son of Eric the Red, established the settlement of Vinland in what is now Newfoundland, Canada.
CE 1000	Norwegian colonies established in North America
1050	The astrolabe, a navigation instrument used to measure the height of celestial bodies above the horizon, first arrived in Europe from the East.
1400-1900	The Little Ice Age.
1405	Admiral Cheng Ho, the commander of the Treasure Fleet of the Dragon Throne began his voyages for Emperor Chu Ti, ultimately ruling the South Pacific and Indian Ocean until 1433.
1420	Prince Henry the Navigator, of Portugal, founded a naval observatory and the first school for teaching navigation, astronomy, and cartography.
1452-1519	Leonardo da Vinci, the famous Italian scientist, observed, recorded and interpreted characteristics of currents and waves and noted that fossils in Italian mountains implied that sea level had been higher in the ancient past.

1460	Prince Henry the Navigator died.
1492	Christopher Columbus rediscovered North America, sailing to the islands of the West Indies.
1497–1499	Vasco da Gama, a Portuguese navigator, journeyed to India by sea by rounding the Cape of Good Hope with four ships, thereby establishing the first all water trade route between Europe and India.
1500	The Portuguese navigator and explorer, Pedro Alvares Cabral, traveled westward in search of a route to India, but reached and explored Brazil.
1513	Juan Ponce de Leon described the swift and powerful Florida Current.
1513-1518	Vasco de Nunez de Balboa crossed the Isthmus of Panama and sailed in the Pacific Ocean.
1515	Peter Martyr proposed an origin for the Gulf Stream.
1519-1522	Ferdinand Magellan embarked on the first circumnavigation of the globe; the crew led by Sebastian del Cano completed the voyage. On the voyage, the Pacific Ocean was named.
1569	Geradus Mercator, the Flemish mapmaker, constructed a map projection of the world that was adapted to navigational charts.
1605	Bishop Resen of Copenhagen drew the first map of the Gulf Stream.
1609	Hugo Grotius, a Dutch legal scholar, published *Mare Liberum,* the foundation for all modern law of the sea.
1674	The British scientist Robert Boyle investigated the relation among temperature, salinity, and pressure of seawater with depth and reported his findings in *Observations and Experiments on the Saltiness of the Sea.*
1687	Sir Isaac Newton published *Principia Mathematica*, which includes an explanation of the operation of gravity and that tides are the result of gravitational attraction of the moon and sun.
1714	Gabriel D. Fahrenheit invents a temperature scale.
1725	Count Luigi Marsigli from Bologna, Italy compiled *Histoire Physique de la Mer*, the first book pertaining entirely to oceanography, examining the geological formation of ocean basins along with the marine life that lived in the sea.
1733	English clockmaker John Harrison invented the first of his five accurate and durable nautical chronometers for determining longitude at sea.
1735	George Hadley proposed an explanation for the trade wind regime that involved a consideration of the Earth's rotation.
1740	Leonhard Euler, noted Swiss mathematician, developed an approximation of the three-body problem to calculate the magnitude of the lunar and solar attractive forces that generate ocean tides.
1742	Anders Celsius invents a temperature scale.

1758	Carolus Linnaeus publishes 10th edition of *Systema Naturae*, formalizing biological nomenclature.
1760	John Harrison invented the Number Four chronometer in his quest to solve the longitude problem.
	Joseph Black proposed the concept of specific heat.
1768-1771, 1772-1775, 1776-1780	Captain James Cook commanded three major ocean voyages gathering extensive data on geography, geology, biota, currents, tides, and water temperatures of all of the principal oceans.
1769 or 1770	The American scientist, Benjamin Franklin, published the first chart of the Gulf Stream (by Folger), which was used by ships to cross the North Atlantic Ocean faster.
1779	James Cook died in Hawaii.
1802	Nathaniel Bowditch of Massachusetts published the *New American Practical Navigator,* a navigational resource that continues to be revised and published to this day.
1805	Admiral Sir Francis Beaufort of the Royal Navy developed his 12-category wind force scale to aid mariners estimate wind speed; this scale was later modified.
1807	President Thomas Jefferson mandated coastal charting of the entire United States and established the U.S. Coast and Geodetic Survey, now known as the National Geodetic Survey.
1811	British chemist Sir Humphry Davy made the first gas hydrates in his laboratory.
1817-1818	Sir John Ross ventured into the Arctic Ocean to explore Baffin Island, where he sounded the bottom successfully (took the first deep-water and sediment samples) and recovered starfish and mud worms from a depth of 1.8 km.
1820	Alexander Marcet, a London physician, noted that the proportion of the chemical ingredients in seawater is unvarying in all ocean basins.
1831-1836	The epic five-year journey of Charles Darwin aboard the British research ship HMS *Beagle* led to a theory of atoll formation and later the theory of evolution by natural selection.
1835	Gaspard-Gustav de Coriolis, a French mathematician, published the first papers on an object's horizontal motion across the surface of the rotating Earth.
1836	William Henry Harvey, an Irish botanist, devised a taxonomy of seaweeds.
1838-1842	Departure of the United States Exploring Expedition led by Lieutenant Charles Wilkes on the flagship *Vincennes*. Six vessels mapping coastal areas, collected specimens.
1839-1843	Sir James Clark Ross led an expedition to Antarctica, recovering samples of deep-sea benthos down to a depth of 4.9 km.
1841, 1854	Sir Edward Forbes published *The History of British Star-Fishes* (1841) and then his *Distribution of Marine Life* (1854), in which he argued that sea life cannot exist below a depth of about 600 m (the so called azoic hypothesis).

1847	Hans Christian Oersted observed plankton.
1851	The first telegraph cable was laid across the Straits of Dover, stimulating new technologies to protect, raise and lower cables to the sea floor.
1853	The first international conference convened in Brussels by Lt. Matthew Fontaine Maury, USN for the purpose of establishing a uniform means of meteorological observations at sea.
1855	Matthew Fontaine Maury, USN compiled and standardized the wind and current data recorded in U.S. ship logs and summarized his findings in *The Physical Geography of the Sea*, credited as the first textbook of modern oceanography.
1859	Darwin's *Origin of Species* published.
1865	Fr. Pietro Angelo Secchi, a papal scientific advisor, perfected the Secchi disc for determination of the transparency of Mediterranean Sea water to sunlight.
1868-1870	Charles Wyville Thomson, aboard HMS *Lightning* and HMS *Porcupine*, made the first series of deep-sea temperature measurements and collected marine organisms from great depths, disproving Forbes' azoic hypothesis.
1870s	William Ferrel, an American meteorologist, drew attention to the deflecting effect of the Earth's rotation on ocean currents (the Coriolis Effect).
1871	The U.S. Fish Commission was established with a modern laboratory at Woods Hole, MA.
1872-1876	Lead by Charles Wyville Thomson, *HMS Challenger* conducted worldwide scientific expeditions, collecting data and specimens that were later analyzed in the more than 50 volumes of the *Challenger Reports.*
1873	Charles Wyville Thomson published a general oceanography book called the *Depths of the Sea.*
1877-1880	Alexander Agassiz, an American naturalist, founded the first U.S. marine station, the Anderson School of Natural History, on Penikese Island, Buzzards Bay, MA.
1880	William Dittmar determined major salts in seawater.
1884-1901	USS *Albatross* was designed and constructed specifically to conduct scientific research at sea and undertook numerous oceanographic cruises.
1880s	Development of the civil time zones and the ultimate specification of the Greenwich Mean Time (GMT) as the world-wide reference.
1888	The Marine Biological Laboratory was established at Woods Hole, MA.
1890	Alfred Thayer Mahan completed *The Influence of Sea Power upon History.*
1891	Sir John Murray and Alphonse Renard classified marine sediments.
1893	Discovery of Maunder sunspot minimum by astronomer E. Walter Maunder.

1893–1896	The Norwegian Fridtjof Nansen, while aboard the *Fram,* studied the circulation pattern of the Arctic Ocean and confirmed that a northern continent did not exist.
1900	On 8 September, a hurricane storm surge struck Galveston, TX. More than 8000 people died, mostly by drowning, in the most deadly natural disaster in U.S. history.
1902	Danish scientists with government backing established the International Council for the Exploration of the Sea (ICES) to investigate oceanographic conditions that affect North Atlantic fisheries. Council representatives were from Great Britain, Germany, Sweden, Norway, Denmark, Holland, and the Soviet Union.
1903	The Friday Harbor Oceanographic Laboratory was established at the University of Washington in Seattle.
	The Laboratory that became the Scripps Institution of Biological Research, and later the Scripps Institution of Oceanography, was founded in San Diego, CA.
1905	Marconi perfected wireless telegraphy, which became important for communication on the ocean.
	The Swedish oceanographer, V.W. Ekman, published his classic work on the cause of the spiral nature of wind driven ocean currents near the surface.
1906	The first Sonar type listening device was invented by Lewis Nixon in order to detect icebergs.
	Prince Albert I of Monaco established the Muscée Océanographique.
1907	Bertram Boltwood calculated age of Earth by radioactive decay.
1911	Roald Amundsen was first at South Pole.
1912	Following the sinking of the *Titanic*, the International Ice Patrol was formed to monitor icebergs.
	German meteorologist Alfred Wegener proposed his theory of continental drift.
1912	Scripps Institution allied with the University of California.
1918	Vilhelm Bjerknes and colleagues at the Bergen (Norway) school formulated theories of air masses, atmospheric fronts, and midlatitude storm systems.
1921	International Hydrographic Bureau founded.
1924	Discovery of the Southern Oscillation by Sir Gilbert Walker.
1925-1927	A German expedition aboard the research vessel *Meteor* studied the physical oceanography of the Atlantic Ocean, using an echo sounder extensively for the first time.
1929	British geologist Arthur Holmes proposed that convection in Earth's mantle was the driving force for continental drift. This explanation was revived in the 1960s by Harry Hess and Robert S. Dietz.
1930	The Woods Hole Oceanographic Institution was established on the southwestern shore of Cape Cod, MA.

1931	*Atlantis* launched.
1932	The International Whaling Commission was organized to collect data on whale species and to enforce voluntary regulations on whaling.
1937	*E.W. Scripps* launched.
1938	Serbian astronomer Milutin Milankovitch began work on his astronomical theory of the Ice Age.
1942	Harald Sverdrup, Richard Fleming, and Martin Johnson published the first modern oceanographic reference text, *The Oceans*, which is still consulted today.
1943	Jacques Cousteau and Emile Gagnan invented the scuba regulator and tank combination, the "aqualung" for diving underwater.
1946	On 1 April, a tsunami destroyed most of the waterfront of Hilo, Hawaii causing 159 deaths.
1949	Maurice Ewing formed the Lamont (later changed to Lamont Doherty) Geological Observatory at Columbia University in New York.
1957	Systematic monitoring of atmospheric carbon dioxide levels began at the Mauna Loa Observatory, Hawaii, under the direction of Charles D. Keeling.
1957-1958	The International Geophysical Year (IGY) was organized as an international effort to coordinate geophysical investigations of the Earth, including the ocean.
1958	The first U.S. nuclear submarine, USS *Nautilus*, also made the first submerged transit of the Arctic ice pack, passing the geographic North Pole.
1959	Antarctic Treaty System adopted by 12 nations.
1959-1965	The International Indian Ocean Expedition was established under United Nations auspices to intensively investigate Indian Ocean oceanography.
1960	The bathyscaphe *Trieste* carrying Jacques Piccard and Don Walsh reached the bottom of the deepest (Mariana) trench (10,912 m, or 35,800 ft).
1960s	Jacob Bjerknes developed concept of teleconnections involving El Niño events and the Southern Oscillation.
1962	Rachel Carson's book *Silent Spring* initiated the U.S. environmental movement.
1963	F.J. Vine and D.H. Matthews of Cambridge University interpreted sea floor magnetic field anomaly patterns in support of sea floor spreading.
1964	*Alvin*, the world's first deep-ocean research submersible was launched.
	On 28 March, a tsunami struck the Seward, Alaska area, claiming more than 100 lives.
1965	Discovery of a natural deposit of methane hydrates under permafrost in Siberia.

1966	The U.S. Congress adopted the Sea Grant College and Programs Act to provide nonmilitary funding for marine science education and research.
1968	*Glomar Challenger* returns first deep-sea cores, indicating the age of the Earth's crust. The cores support theories of plate tectonics.
	The U.S. National Science Foundation organized the Deep Sea Drilling project (DSDP) to core through the sediments and crust of the ocean bottom.
1969	Santa Barbara, CA, oil well blowout captured national attention.
	Stratton Commission developed the first national ocean policy for the U.S.
1970	The U.S. government created the National Oceanic and Atmospheric Administration (NOAA) to oversee and coordinate government activities related to oceanography and meteorology.
	Completion of the Aswan High Dam on the Nile River.
1970s	The United Nations initiated the International Decade of Ocean Exploration (IDOE) to improve scientific knowledge of the ocean.
	Saffir-Simpson Hurricane Intensity Scale rated hurricanes from 1 to 5 corresponding to increasing intensity. It was revised in 2010 and is now known as the Saffir Simpson Hurricane Wind Scale.
1970	John Tuzo Wilson, a Canadian geologist, proposed a cyclic model of tectonic revolution.
1971	Jason Morgan of Princeton University proposed hotspot volcanism.
1972	The Geochemical Ocean Sections Study (GEOSECS) was organized to study seawater chemistry, ocean circulation and mixing, and the biogeochemical recycling of chemical substances.
	Marine Mammal Protection Act established a moratorium on the hunting of marine mammals in U.S. waters and by U.S. citizens on the high seas (reauthorized in 1994).
	Provisions of the 1972 Marine Protection, Research and Sanctuaries Act authorized the President to designate national marine sanctuaries in coastal waters of the continental shelf and in the Great Lakes (reauthorized in 2000).
1972	London Dumping Convention banned the discharge of waste from land into the ocean by ships or aircraft.
	Coastal Zone Management Act passed into law whereby the federal government assists coastal states in managing and protecting coastal resources.
1973	U.S. Endangered Species Act became law.
1974	Project FAMOUS (French-American Mid-Ocean Undersea Study) mapped and sampled the Mid-Atlantic Ridge, a zone of seafloor spreading.
1974	F.S. Rowland and M.J. Molina first warned of the threat of CFCs to the stratospheric ozone shield.

1975	Sinking, on 10 November, of the ore carrier Edmund Fitzgerald on Lake Superior with the loss of her 29-member crew.
1976	Scientists discover indications of Milankovitch cycles in deep-sea sediment cores.
	Magnuson-Stevens Fisheries and Conservation Act established fishery councils to manage the fishery resources in each region of the nation.
1977	The submersible *Alvin* finds hydrothermal vents in the Galápagos rift.
1978	*Seasat-A,* the first satellite dedicated to the remote sensing of the ocean, was launched.
	Coastal Zone Color Scanner (CZCS) flown aboard NASA's Nimbus-7 satellite.
1979	NOAA's SLOSH numerical model developed to predict the location and height of a storm surge.
1981	The Burgess Shale declared a UNESCO World Heritage Site.
1982	The UN Convention on the Law of the Sea established, which granted jurisdiction over an exclusive economic zone (EEZ) to 151 coastal nations.
1983	The Chesapeake Bay Program became the nation's largest federal-state environmental restoration project; the nation's first estuary selected for restoration and protection.
1985	The scientific research vessel *JOIDES Resolution* replaced *Glomar Challenger* in Deep Sea Drilling Project.
	Tropical Ocean-Global Atmosphere (TOGA) commenced a 10-year experiment in the equatorial Pacific, with the deployment of the TOGA Tropical Atmosphere Ocean (TAO) Array.
	R. D. Ballard located wreck of *Titanic.*
1986	Moratorium on commercial whaling issued by the International Whaling Commission (IWC).
1987	Joint Global Ocean Flux Study (JGOFSS) founded to assess processes controlling carbon fluxes.
1988	The Intergovernmental Panel on Climate Change (IPCC) formed by the World Meteorological Organization and the UN Environmental Programme to evaluate the state of climate science as the basis for policy action.
	Heinrich layers first described by German researcher Hartmut Heinrich. Sediment released to the floor of the North Atlantic during the melting of fleets of ice bergs.
1989	On 24 March, the supertanker *Exxon Valdez* ran aground on a reef in Prince William Sound, Alaska, producing the 2nd largest oil spill in U.S. history.
1990	The *JOIDES Resolution* retrieved a sediment sample estimated to be 170 million years old.
1991	JOI researchers bore to a depth of 2 km (1.24 mi) beneath the seafloor near the Galápagos Islands.

	The NOAA National Weather Service Forecast Office in Miami, FL, began to issue outlooks for rip currents.
1992	The U.S.-French *TOPEX/Poseidon* satellite launched to monitor global ocean topography.
	The Tropical Ocean Global Atmosphere Coupled Ocean-Atmosphere Response Experiment (TOGA COARE) was conducted in western Pacific.
1993	Beginning of a series of ocean iron fertilization experiments.
1994	The Law of the Sea Treaty entered into force.
1995	*Keiko*, a small remotely controlled Japanese submersible, sets a new depth record, reaching 10,978 m (36,008 ft) in the Challenger Deep.
1998	UN declared the "Year of the Oceans" to increase awareness of the importance of the ocean.
	Galileo spacecraft found possible evidence of an ocean on Jupiter's moon Europa.
	Scientists from 30 nations began deploying Argo profiling floats in the world ocean.
2000	Beginning of 10-year international project, *Census of Marine Life*.
2004	On 26 December, tsunami originated off the west coast of Sumatra, Indonesia; death toll is estimated at 227,900.
	U.S. Commission on Ocean Policy made recommendations for a new national ocean policy.
2005	On 28 August, Hurricane Katrina struck the Gulf coast of Louisiana and Mississippi. At least 1500 people lost their lives in the most destructive hurricane in U.S. history.
2007-09	Fourth International Polar Year (IPY).
2007	Modern Japanese drill ship, *Chikyu*, became fully operational.
	Argo array is completed with over 3000 active profiling floats deployed in the world ocean.
2008	The U.S.-French Jason-2 satellite launched to continue monitoring global ocean topography.
2009	IceBridge, a six-year NASA mission, begins the largest airborne survey of polar ice.
2010	An explosion on 20 April at the Deepwater Horizon drilling platform, leased by BP, releases almost 5 million barrels of oil into the Gulf of Mexico before it is capped in mid-July.
	The Census of Marine Life completes a decade of ocean life research.
	The National Ocean Council is established by Executive Order.
2011	On 11 March, a tsunami originates off the east coast of Japan with a death toll of 15,093 in Japan and 9,093 persons missing.

GLOSSARY

A

absorption—The process whereby incident radiant energy is retained by a substance. The absorbed radiation is then converted to another form of energy (e.g., **heat**).

abyssal plain—The flat surface of the seafloor, usually at the base of a **continental rise** and formed by the deposition of **sediments** that obscure the preexisting topography.

abyssal storm—Unusually strong bottom current (of the order of 1.0 knot) that scours the ocean floor generating moving clouds of suspended **sediment**.

acid—A hydrogen-containing compound that releases positively charged hydrogen ions (H^{+1}) when dissolved in water. Strong acids more readily release hydrogen ions than weak acids.

acid rain—Precipitation with a **pH** of less than 5.6.

acoustic thermometry—A method of determining ocean **temperature** by measuring the speed of sound propagating through the ocean.

adaptation—A genetically controlled trait or characteristic that enhances an organism's chance for survival and reproduction in its environment.

adaptive coloration—Camouflage whereby an organism's color pattern closely matches its background substrate.

aerosols—Minute solid and liquid particles which are suspended in the **atmosphere**.

Agnathans—Also known as **jawless fishes**, members of this major fish group have cartilaginous skeletons and long, eel-like bodies, which lack scales or armor. The distinguishing features of this group (in addition to the absence of jaws) include the lack of an identifiable stomach and lack of paired appendages, such as fins. Agnathans are the most primitive of the three major fish groups.

air mass—A large widespread volume of air, the thermal, moisture, and stability properties of which are characteristic of its source region and are modified as the air mass moves away from its source.

air pressure—The force exerted per unit area by the atmosphere as a consequence of gravitational attraction upon the molecules in a column of air lying directly above a specified location.

air pressure gradient—A change in **air pressure** from one place to another.

albedo—The ratio of the amount of **electromagnetic radiation** reflected by a body to the amount incident on it; commonly expressed as a percentage. Usually, albedo refers to radiation in the visible range or to the full spectrum of solar radiation.

alkaline substance—Also known as a base, this type of substance releases negatively charged hydroxyl ions (OH^{-1}) when dissolved in water and may be weak or strong.

anadromous fishes—Fishes which travel from salt water to fresh water or up rivers to spawn. They include salmon, shad, sturgeon and striped bass.

Arctic Oscillation (AO)—A seesaw variation in **air pressure** between the North Pole and the margins of the polar region. Changes in the horizontal **air pressure gradient** alter the speed of horizontal winds in the polar vortex.

artificial beach nourishment—A process whereby sand dredged from offshore (or an inland source) is moved to badly eroded **beaches** with an aim of re-establishing the natural balance between **sediment** input and output on the **beach**.

assimilative capacity—The amount of waste that an **ecosystem** can assimilate (via decomposition by bacteria, fungi, and invertebrates) without damaging **ecosystem** functions or building up waste products to levels that may cause unwanted or harmful impacts on living organisms (including humans).

asthenosphere—A region of the upper mantle between 200 and 400 km deep which exhibits plastic-like behavior and readily deforms in response to stress; underlies the **lithosphere**.

astronomical tides—The periodic rise and fall of sea level resulting from the gravitational interaction and motions of the Moon, Sun, and Earth system.

atmosphere—A relatively thin envelope of gases and suspended particles surrounding the Earth and held there by gravity.

atmospheric window—**Wavelength** bands in which atmospheric constituents absorb very little or no **electromagnetic radiation**; allows the escape of Earth's heat to space.

atoll—A series of **coral reefs** surrounding a **lagoon** that remain after a volcanic island sinks beneath the waves or erodes away.

Autonomous Underwater Vehicles (AUVs)—Un-piloted, remotely controlled powered vehicles that are not tethered to a mother ship. They carry sensors that measure ocean water properties (e.g., **temperature**, **salinity**, dissolved oxygen).

azoic hypothesis—The erroneous belief that deep-ocean waters (below 300 fathoms or 550 m) lacked sufficient dissolved oxygen to support marine life.

B

barrier island—An elongated, narrow accumulation of sand oriented parallel to a shoreline, but separated from the mainland by a **lagoon**, **estuary**, or bay.

beach—A deposit of unconsolidated **sediment** (usually sand and gravel), extending landward from low tide to a change in topography or where permanent vegetation begins.

beach sediment budget—The sum of all **sediment** outputs and inputs on a **beach**; determines whether the **beach** is growing or shrinking.

benthic zone—One of the two basic subdivisions of the marine biome which includes the seafloor and bottom dwelling organisms.

benthos—A collective term for marine organisms which live on or near the seafloor, or **benthic zone**.

berm—A platform of sand, flat-topped and sloping steeply seaward, formed near the mean high-water mark.

bioaccumulation—The process by which persistent materials (that resist chemical, physical, or biological breakdown) gradually become increasingly concentrated in living tissue as one organism consumes another within a **food web**.

bio-fouling—The undesirable accumulation of organisms on any submerged surface.

biogenous sediment—Marine **sediment** formed from the excretions, secretions, and remains (e.g., shells) of organisms.

biogeochemical cycle—Pathways along which solid, liquid, and gaseous materials flow among the various reservoirs of **Earth's planetary system**.

biological pump—The process whereby carbon cycles through the ocean as organic matter decomposes.

bioluminescence—Production of light by living organisms. Light is generally the product of a chemical reaction that takes place in specialized cells or organs.

biosphere—Consists of all living organisms on Earth; composed of **ecosystems**.

bioturbation—The churning and stirring of **sediment** deposits by benthic animals in the course of feeding or movement.

black smokers—Substances that precipitate from **hydrothermal vent** waters appear as dark clouds and accumulate in the form of conical chimneys. They are distinguished from **white smokers** by their darker appearance and the types of minerals precipitated.

bony fishes—Members of this large, diverse group live in a variety of marine and freshwater environments. They have bony skeletons, scales, a flap covering their gills and, the vast majority of them, a swim bladder.

Bowen ratio—For any moist surface, the ratio of **heat** energy used for **sensible heating** (conduction and convection) to the heat energy used for **latent heating** (**evaporation** of water or **sublimation** of snow or ice).

breaker—The collapsing of a wave as it becomes steeper and less stable upon entering shoaling water.

breakwater—A long, narrow offshore structure, usually constructed of large blocks of rock or concrete, oriented parallel to the shoreline and intended to provide calm waters for docking boats and to protect **beaches** from **erosion**.

brine rejection—Salts are excluded from the ice structure as **seawater** freezes and the remaining unfrozen water becomes saltier and therefore freezes at still lower **temperatures**.

bubble injection—A mechanism of gas transfer at the air/sea interface whereby breaking waves introduce a foam composed of small bubbles below the surface greatly enhancing exchange rates.

buffer—A substance that causes chemical equilibrium.

bycatch—Fish and other marine animals that are caught in addition to the target species; they are often disposed of.

C

calcareous oozes—Deep-sea **pelagic deposits** containing at least 30% calcareous (calcium carbonate) skeletal remains by weight.

Callendar effect—Theory that global **climate** change can be brought about by enhancement of the natural **greenhouse effect** by increased levels of atmospheric CO_2 from anthropogenic sources, principally the burning of fossil fuels. The theory is named for the British engineer Guy Stewart Callendar who investigated the link between global warming and fossil fuel combustion beginning in the late 1930s.

calving—The breaking away of a mass of ice from the leading edge of a **glacier** forming icebergs upon entering the ocean.

capillary waves—Small ocean **waves** with a **wavelength** of less than 1.7 cm (0.7 in.).

carbonate compensation depth (CCD)—The depth of the ocean below which material composed of calcium carbonate ($CaCO_3$) dissolves and does not accumulate.

carrying capacity—The maximum population of a species that can be sustained by the resources of the habitat.

cartilaginous fishes—These generally primitive fishes lack true bones and their skeletons consist of cartilage. Examples include sharks, skates, and rays.

catadromous fishes—Fish that breed in the open ocean, but spend their adult lives in fresh water. An example is the American eel.

celerity—The rate at which a surface **wave** progresses outward in still water from the point where the water was disturbed.

cellular respiration—The process whereby food is broken down liberating energy for maintenance, growth and reproduction, while releasing carbon dioxide, water and heat energy to the environment.

cetaceans—Group of marine mammals (including whales, porpoises, and dolphins) that spend their entire lives at sea.

Challenger Expedition—The first voyage dedicated exclusively to marine exploration. Conducted from December 1872 to May 1876, scientists sampled every ocean basin except the Arctic.

chemosynthesis—The process whereby marine organisms in the absence of sunlight derive energy from substances such as hydrogen sulfide (H_2S) or methane.

climate—The **weather** of some locality averaged over some specific interval of time (e.g., 30 years) plus extremes in **weather** observed during the same period or during the entire period of record.

climate sensitivity—Refers to the equilibrium change in global mean annual surface **temperature** caused by an increment in downward infrared radiative flux that would result from sustained doubling of atmospheric CO_2 concentration compared to its preindustrial level.

coast—A strip of land of indefinite width that extends from the low-tide line inland to the first major change in landform features; transitional between land and ocean.

coastal downwelling—Downward motion of warm surface waters along the coast caused by **Ekman transport** onshore.

coastal upwelling—Upward motion of cold, nutrient-rich deep water along the coast caused by **Ekman transport** offshore.

Coastal Zone Management Program (CZMP)—A program authorized in 1972 whereby the federal government assists coastal states in managing and protecting their coastal resources.

coastline— The farthest inland extent of storm waves, in some cases marked by sand dunes or wave-cut cliffs.

coccolithophorids—Single-cell photosynthesizing organisms covered with tiny calcium carbonate plates. They are unusual in that they appear to thrive in nutrient-poor waters, forming large blooms that turn surface waters greenish blue as viewed from space.

cold-core rings—Eddies having a core of relatively cold water that break off from an ocean surface current; viewed from above in the Northern Hemisphere, they rotate in a counterclockwise direction.

compensation depth—The ocean depth below which no **net primary production** occurs, usually where the light level diminishes to about 1% of what it is at the surface.

condensation—The process whereby water changes phase from vapor to liquid.

conservation of angular momentum—For a body rotating in a plane about a point, a decrease in the radius of the orbit is accompanied by an increase in the angular velocity.

constructive wave interference—The result when the crests of sets of **waves** coincide to form a **wave** of greater height.

consumer—An organism that is unable to manufacture its food from nonliving materials but is dependent on the energy stored in other living things; also known as a **heterotroph.**

continental climate—Characterizes a middle or high latitude locale well inland from the moderating influence of the ocean or large lake; characterized by a significant difference between mean summer and winter **temperatures**, contrast with **maritime climate**.

continental crust—The outermost part of the **lithosphere** overlying the mantle and forming the solid surface of Earth's landmasses; mostly granite in composition.

continental drift—The slow movement of landmasses as part of **tectonic plates**. The continents of today were once a single landmass (Pangaea) that broke apart with the various fragments moving over the surface of the planet.

continental rise—The gently sloping area of the seafloor beyond the base of the **continental shelf**.

continental shelf—The submerged zone between the **shoreline** and **continental slope** where the seafloor slopes seaward at a gradient of less than one degree.

continental slope—The relatively steep downward sloping area from the shelf-slope break to the more gently sloping **continental rise** or directly into an **ocean trench**.

convergent plate boundary—The boundary between two tectonic plates that are moving toward one another; responsible for **subduction**.

copepods—A diverse group of tiny crustaceans covered with an exoskeleton made of chitin.

coral atoll—A ring-shaped island surrounding a **seawater lagoon**.

coral bleaching—The whitening of coral colonies due to the loss of symbiotic zooxanthellae from the tissues of coral polyps; may be triggered by a rise in **seawater temperature**.

coral reef—A calcareous reef in relatively shallow, tropical seas composed of a thin veneer of living coral growing on older layers of dead coral or volcanic rock.

Coriolis Effect—An apparent force relative to Earth's surface due to Earth's rotation that causes deflection of moving objects to the right in the Northern Hemisphere and to the left in the Southern Hemisphere.

cosmogenous sediment—Particles entering the Earth system from outer space, often originating from meteorite fragments.

cotidal line—A line plotted on a chart connecting points at which high water (high tide) occurs simultaneously. Lines show the time lapse, in lunar-hour intervals, between the Moon's passage over a reference meridian (usually the prime meridian) and the succeeding high water at a specified location.

countershading—Protective coloration found in fish whereby their dorsal (back) side is a dark color making it difficult to see them from above and their ventral (underbelly) side is a light color making it difficult to see them from below.

cryosphere—The frozen portion of the **hydrosphere**, encompassing glacial ice, icebergs, sea ice, and the ice in permafrost.

D

dam—A barrier constructed across a watercourse that impounds water in an upstream reservoir.

decomposers—**Consumers**, usually microscopic, that feed on dead organic matter, either on the ocean bottom or in the water column, thus aiding in recycling nutrients.

deep layer—Dark, cold nearly isothermal ocean water below the **pycnocline**; accounts for most of the ocean's mass.

deep-sea cabled observatory—Ocean observing system placed on the ocean floor and linked to a mainland facility by fiber-optic and power cables. The cables provide electricity to power sensors and deliver the stream of data obtained by those sensors.

Deep Sea Drilling Program (DSDP)—Using the drill ship *Glomar Challenger*, this project sampled marine sediments and oceanic crust for scientific purposes from 1968 to 1983.

deep-water wave—A **wave** on the surface of a body of water whose depth is more than one-half the **wavelength**.

delta—A landform at the mouth of a river produced by the sudden divergence and dissipation of a stream's velocity and the resulting deposition of the river's suspended **sediment** load in the shape of the Greek letter *delta* when viewed from above.

density—Mass per unit volume.

deposition—The process whereby water changes directly from vapor to solid (ice crystals) without first becoming liquid.

destructive wave interference—When the troughs of sets of **waves** coincide with the crests of another set of **waves** producing **waves** of reduced height.

diatoms—A diverse group of minute shell-covered phytoplanktonic marine organisms having silica exoskeletons.

dinoflagellates—Any of a class of single-cell marine organisms with two flagella (thread-like tails) that can be **phytoplankton** or **zooplankton**.

distillation—Purification of water through phase changes (e.g., **evaporation** followed by **condensation**). When water vaporizes, all suspended and dissolved substances such as sea salts are left behind.

diurnal inequality—The difference in heights between the two successive high waters (high tides) or the two successive low waters (low tides) of a tidal day.

diurnal tide—An **astronomical tide** with only one high water (high tide) and one low water (low tide) occurring each lunar (tidal) day.

divergent plate boundary—The zone between tectonic plates that are pulling apart with **magma** and new crust moving in to fill the gap; most often occurring at a mid oceanic ridge.

doldrums—An east-west equatorial belt of light and variable surface winds where the trade winds of the two hemispheres converge.

E

Earth-atmosphere system— The interaction of processes operating at the Earth's surface with those of the overlying **atmosphere**.

Earth's planetary system—Encompasses the **at mosphere**, ocean, and land, and interactions between these systems.

eastern boundary currents—Generic term for the relatively slow, broad, and shallow flow of ocean water (current) that runs along the eastern flank of a subtropical ocean gyre; the Canary Current is an example.

ebb tides—Tidal currents flowing seaward with falling sea levels.

ecological efficiency—The fraction of the total energy input of a system that is transformed into work or some other usable form of energy.

ecologically sustainable yield—The maximum catch that a marine **ecosystem** can sustain without undergoing an undesirable change in state.

ecosystem—Community of plants and animals that interact with one another, together with the physical conditions and chemical constituents in a specific geographical area.

Ekman spiral—In response to a steady wind blowing over the ocean surface, water at increasing depths moves in directions more and more to the right (in the Northern Hemisphere) until at about 100 m depth the water is moving in a direction opposite to that of the wind; the spiral is in the opposite direction in the Southern Hemisphere.

Ekman transport—The net transport of water due to the **Ekman spiral**; 90 degrees to the right of the surface wind in the Northern Hemisphere and 90 degrees to the left of the surface wind in the Southern Hemisphere.

El Niño—Anomalous warming of ocean surface waters in the eastern tropical Pacific; accompanied by suppression of upwelling off the coasts of Ecuador and northern Peru and along the equator east of the international dateline. Typically lasting for 12 to 18 months and occurring every 3 to 7 years, El Niño is accompanied by changes in oceanic and atmospheric circulation plus **weather** extremes in various parts of the world.

electromagnetic radiation—Energy in the form of **waves** that have both electrical and magnetic properties; these waves can travel through gases, liquids, and solids and require no physical medium. All objects emit all forms of electromagnetic radiation, although each object emits its peak radiation at a certain **wavelength** within the **electromagnetic spectrum**. Forms of electromagnetic radiation include gamma rays, x-rays, ultraviolet, visible light, infrared, microwaves, and radio waves.

electromagnetic spectrum—The various forms of **electromagnetic radiation** arranged and distinguished by type by their **wavelength** (or frequency).

ENSO—Contraction for **El Niño/Southern Oscillation**. The term for the coupled ocean-atmosphere interactions in the tropical Pacific Ocean characterized by episodes of anomalously high sea surface **temperatures** in the equatorial and tropical eastern Pacific. It is associated with large-scale swings in surface air pressure between the western and eastern tropical Pacific and is the most prominent source of inter-annual variability in **weather** and **climate** around the world.

ENSO Alert System—Under this system, launched by NOAA's Climate Prediction Center in 2009, an **El Niño** or **La Niña** watch is issued when conditions in the equatorial Pacific are favorable for their development within three months and an **El Niño** or **La Niña** advisory is issued when conditions have already developed and are expected to continue.

epifauna—Organisms that live on the surface rather than within marine **sediments** on the ocean floor.

equatorial upwelling—Upward circulation of cold, nutrient-rich bottom water toward the ocean surface; the consequence of convergence of the trade winds of the North and Souh Hemispheres plus **Ekman transport**.

erosion—Removal and transport of **sediments** by running water, **glaciers**, wind, or gravity.

estuary—The highly productive portion of a river affected by ocean tides; the semi-enclosed region in the vicinity of a river's mouth, in which the freshwater of the river mixes with the saltwater of the ocean.

euphausiid—Larger members of the **zooplankton** community, including krill.

eustasy—Condition of world-wide sea level and its fluctuations which may be caused by changes in the global water cycle (e.g., the waxing and waning of Earth's glacial ice sheets).

evaporation—The process whereby water changes phase from a liquid to a vapor.

evaporative cooling—The cooling a surface, such as Earth's, experiences as water evaporates, absorbing and transferring heat to the atmosphere via water vapor. **Evaporation** of water requires **latent heat of vaporization**.

exclusive economic zone (EEZ)—The jurisdictional area initially established in 1983 for marine resources; the inner boundary of that zone is coterminous with the seaward boundary of coastal nations and extends seaward 370 km (200 nautical miles).

exotic species—Animals and plants introduced into an **ecosystem** usually by humans intentionally or unintentionally; also known as alien or invasive species.

extratropical cyclone—Any synoptic-scale storm system that is not a **tropical cyclone**, usually referring only to the migratory low pressure systems of middle and high latitudes.

eye—The roughly circular area of comparatively light winds and fair **weather** found at the center of an intense **tropical cyclone** (i.e., a **hurricane**).

eyewall—The organized ring of intense thunderstorms surrounding the **eye** of a **tropical cyclone**, typically a **hurricane**.

F

feedback—A process where a change in one variable interacts with other variables, and alters the original variable. If that interaction enhances the original change, then the feedback is positive. If the change suppresses the original, the feedback is negative.

fecal pellets—Organic excrement found especially in marine **sediment** deposits and made up of the undigested organic matter secreted by animals.

fetch—The distance the wind blows over a continuous water surface.

filter feeders—Marine organisms that use tiny hairs (called *cilia*), mucous-covered surfaces or other strategies to capture food particles suspended in the water.

fjord—A narrow inlet or arm of the sea bordered by steep cliffs; formed when the post-glacial rise in sea level flooded a glacially eroded river valley.

flood tides—Tidal currents directed toward land, causing water levels to rise in harbors and rivers.

food chain—A sequence of feeding relationships among organisms whereby energy and biomass is transferred from one trophic (feeding) level to the next higher trophic level (i.e., from **autotrophs** to **heterotrophs**).

food web—A complex of feeding relationships consisting of **food chains** linked together in an **ecosystem**.

G

geologic time—A standard division of time on Earth into eons, eras, periods, and epochs based on occurrence of large-scale geological events; spans millions to billions of years in the past.

geosphere—The solid portion of planet Earth consisting of rocks, minerals, and **sediments**.

geostationary satellite—A satellite which revolves around Earth at the same rate and in the same direction as the planet rotates so that it always remains over the same point on the equator and monitors the same field of view. The satellite orbits Earth at an altitude of about 36,000 km (22,300 mi).

geostrophic flow—Horizontal movement of surface water parallel to ocean height contours and arising from a balance between the **pressure gradient force** and the **Coriolis Effect**.

glacial climate—A **climate** which favors the thickening and expansion of **glaciers**.

glacier—A mass of ice that flows internally under the influence of gravity.

global climate model—A simulation of Earth's **climate** system usually consisting of a series of mathematical equations and used to predict climatic anomalies.

global radiative equilibrium—The balance between net incoming solar radiation and infrared radiation emitted to space by the **Earth-atmosphere system**.

global water cycle—The ceaseless movement of water among its various reservoirs on a planetary scale.

greenhouse effect—Heating of Earth's surface and lower **atmosphere** as a consequence of differences in atmospheric transparency to **electromagnetic radiation**. The **atmosphere** is nearly transparent to incoming solar radiation, but much less transparent to outgoing infrared radiation. Terrestrial infrared radiation is absorbed and radiated principally by water vapor and, to a lesser extent, by carbon dioxide and other trace gases, thereby slowing the loss of heat to space by the **Earth's planetary system** and significantly elevating the average **temperature** of Earth's surface.

greenhouse gases—Those atmospheric gases that absorb and emit terrestrial infrared radiation and contribute to the **greenhouse effect** in the **Earth-atmosphere system**. The principal greenhouse gas is water vapor; others are carbon dioxide, ozone, methane, and nitrous oxide.

groin—A low artificial structure often composed of rock rubble that is built perpendicular to the **shoreline** to trap **littoral drift**; similar to a **jetty** but smaller.

gross primary production—The total amount of carbon fixed into organic matter through **photosynthesis** in a given unit of time; usually expressed in units of grams of carbon per square m per day or year.

growth overfishing—Type of **overfishing** in which fish are taken too small, before the animals have grown to a size that would produce the maximum yield.

guyot—A flat-topped **seamount** of volcanic origin rising more than 1 km above the seafloor.

gyre—Refers to the nearly circular motion of surface ocean currents in each of the major ocean basins centered under **subtropical anticyclones**. Viewed from above the subtropical gyres rotate clockwise in the Northern Hemisphere and counterclockwise in the Southern Hemisphere.

H

HNLC regions—Portions of the ocean featuring *H*igh *N*utrients and *L*ow *C*hlorophyll. Biological production is low even though surface waters have relatively high concentrations of nutrients (nitrogen and phosphorus compounds).

harmful algal blooms (HABs)—A proliferation of a single species of **phytoplankton** that can cause serious harm to the environment, demise of marine organisms, economic losses, and human health problems, including illness and death; also known as **red tides**.

halocline—A layer of water characterized by a relatively large change in **salinity** with increasing depth.

heat—The name applied to a form of energy transferred between systems in response to a difference in **temperature**. Heat is always transferred from a warmer system to a colder system.

high biomass bloom—An excessive proliferation of **phytoplankton** (due to input of excessive levels of nutrients) that upon decomposition greatly depletes the dissolved oxygen supply of **seawater.**

Holocene—The time interval since the end of the Pleistocene Ice Age (about 10,500 years ago); the current **interglacial climate**.

horse latitudes—A nautical term describing the latitude belts over the ocean at approximately 30 to 35 degrees N and S where winds are usually light or the air is calm and the **weather** is hot and dry. These zones coincide with the location of the **subtropical anticyclones**.

hot spot—A long-lived source of **magma** caused by rising plumes of hot material originating deep in the mantle. As a tectonic plate moves over a hot spot or as a hot spot moves, **magma** may break through the crust and form a new volcano.

hurricane—An intense **tropical cyclone** originating over tropical ocean waters, usually in late summer or early fall, with a sustained wind speed of 119 km per hr (74 mph) or higher.

hydrogen bonding—An attractive force whereby a positively charged (hydrogen) pole of a water molecule attracts the negatively charged (oxygen) pole of another water molecule.

hydrogenous sediment—Particles which are chemically precipitated from **seawater**.

hydrosphere—One of the major interacting subsystems of the **Earth's planetary system** which includes water in all three phases (ice, liquid, and vapor) that continually cycles from one reservoir to another within that system (i.e., the global water cycle).

hydrothermal vent—A site where ocean water enters fractures in newly formed oceanic crust, is heated, and then surfaces again through fractures; usually located near oceanic ridge systems.

I

ice-albedo feedback—Warming that causes a reduction in Arctic sea-ice cover is enhanced by the lower **albedo** of the open ocean waters; an example of positive **feedback**.

ice stream—A zone of relatively rapidly flowing ice within a glacial ice sheet.

ice shelves—Extensive areas of floating ice attached to land (at the grounding line) that fringe the coast. In Antarctica, ice shelves fringe about 44% of the coast.

igneous rock—A crystalline or glassy rock formed from the cooling and solidification of **magma** (or *lava*).

infauna—Marine organisms that burrow into and live within marine **sediments** on the ocean floor.

Integrated Ocean Drilling Project (IODP)—A program of ocean floor exploration underway in October 2003 with the goal of using two drill ships plus specialized drilling platforms to retrieve more cores from deeper holes in the ocean floor.

interglacial climate—A **climate** which favors the thinning and retreat of existing **glaciers** or no **glaciers** at all.

Intergovernmental Panel on Climate Change (IPCC)—Established in 1988 by the World Meteorological Organization (WMO) and the United Nations Environmental Programme (UNEP). The IPCC is charged with evaluating the state of **climate** science as the basis for policy action and serving the interests of scientists, public policymakers, and through them the public at large.

internal energy—A measure of the molecular activity of a system or the summation of total energies of all molecules in a specific mass.

internal tides—**Waves** generated by tides well below the sea surface and occurring at tidal frequencies.

internal waves—**Waves** that form and propagate well below the sea surface along the boundary between layers of water that differ in density.

intertidal zone—The shore area between high and low tides.

intertropical convergence zone (ITCZ)—A narrow, discontinuous belt of convective clouds and thunderstorms paralleling the equator and marking the convergence of the trade winds of the two hemispheres; this zone shifts north and south seasonally.

island arc—A curved chain of volcanic islands that parallels a deep-sea trench where oceanic **lithosphere** is subducted causing volcanism and producing volcanic islands.

J

jawless fishes—Also referred to as **Agnathans**, members of this major fish group have cartilaginous skeletons and long, eel-like bodies, which lack scales or armor. The distinguishing features of this group (in addition to the absence of jaws) include the lack of an identifiable stomach and lack of paired appendages such as fins.

jetty—A **breakwater** oriented perpendicular to the **shoreline** and extending seaward up to a kilometer or more; intended to protect a harbor or tidal inlet from filling by **littoral drift**.

K

kelp—Includes various species of brown algae which grow to enormous size, found in cool waters worldwide, especially in **coastal upwelling** zones.

krill—Shrimp-like crustaceans that are the major food source for whales and other organisms in the Southern Ocean.

L

La Niña— An episode of strong trade winds and unusually low sea surface **temperatures** in the central and eastern tropical Pacific. Essentially the opposite of **El Niño**, La Niña is accompanied by **weather** extremes in various parts of the world.

lagoon—A partially enclosed shallow stretch of **seawater** separating a barrier island from the mainland.

latent heat—The **heat** energy that is used to change the phase of water but not the **temperature** of the water; hence the term *latent*, meaning hidden.

latent heat of condensation—**Heat** released to the environment during the change in phase of water from vapor to liquid.

latent heat of deposition—**Heat** released to the environment during the change in phase of water from vapor to solid (ice).

latent heat of fusion—**Heat** released to the environment when water changes phase from liquid to solid.

latent heat of sublimation—**Heat** absorbed from the environment when ice or snow vaporizes.

latent heat of vaporization—**Heat** absorbed from the environment when water changes phase from liquid to vapor.

latent heating—The transfer of **heat** energy from one place to another as a consequence of the phase changes of water.

law of the minimum—The growth and well being of an organism is limited by the essential resource that is in lowest supply relative to what is required by the organism.

law of universal gravitation—The force of attraction between two objects is directly proportional to the product of their masses and inversely proportional to the square of the distance between them.

lifetime of a gas—The time it takes for a gas to be reduced to 37% of the original amount emitted.

limiting factor—The most deficient of the essential resources an organism requires for growth and well being.

lithification—The process whereby **sediment** is converted to **sedimentary rock** involving compaction and/or cementation; part of the **rock cycle**.

lithogenous sediment—Marine **sediment** formed mostly by the **weathering** and **erosion** of pre-existing rock.

lithosphere—The outer, rigid part of the Earth, consisting of the upper part of mantle, oceanic crust and continental crust.

Little Ice Age—A relatively cool interval during the **Holocene** interglacial, from about CE 1400 to 1900, when average **temperatures** were lower in many areas, and alpine **glaciers** advanced down mountain valleys. The Little Ice Age followed the **Medieval Warm Period**.

littoral drift—**Sediment** transport by a **longshore current** along a coast, either nourishing or cutting back **beaches**.

longshore current—The component of water motion parallel to the shore and in the **surf** zone; the consequence of waves breaking at an angle to the shore.

M

magma—Hot molten rock material formed deep in the crust or upper mantle which wells up and migrates along rock fissures; called *lava* when it flows onto Earth's surface.

manganese nodules—Irregularly shaped, sooty black or brown nodules on the seafloor which contain a high concentration of manganese and iron.

mangrove—Any of various coastal or aquatic salt-tolerant trees that form large colonies in swamps or shallow water.

mangrove swamp—A marshy coastal wetland area with a dense growth of **mangroves** and other tropical plant species which tolerate salt water flooding.

mariculture—Industrial farming of fish and shellfish in the ocean.

marine snow—A continuous flow of white particles through the ocean depths, composed of the remains of dead organisms from the upper layer of the ocean, fecal pellets, and various forms of non-living (inorganic) matter.

maritime climate—Characterizes a middle or high latitude locale downwind from the moderating influence of the ocean or large lake; characterized by minimal contrast between mean summer and winter **temperatures**; contrast with **continental climate**.

Maunder minimum—The period of greatly diminished **sunspot** activity between CE 1645 and 1715 identified by and named for E. Walter Maunder.

maximum sustainable yield—The maximum catch of a fish species that will ensure the long-term viability of its population.

Medieval Warm Period—A relatively mild **climatic** episode during the **Holocene** from about CE 950 to 1250.

meridional overturning circulation (MOC)—This large scale ocean overturning transports **heat** energy, salt, and dissolved gases (e.g., the greenhouse gas carbon dioxide) over great distances and to great depths in the world ocean and plays an important role in Earth's **climate** system. In the North Atlantic, for example, a warm surface ocean current flows north and eastward from the Florida Strait. At high latitudes, the surface waters cool, sink, and flow southward as cold bottom water.

mesoscale systems—**Weather** phenomena that are so small that they influence atmospheric conditions over only a portion of a city or county; includes thunderstorms and sea breezes. These systems have dimensions of 1 to 100 km (1 to 60 mi) and last from hours to a day or so.

metamorphic rock—A crystalline rock derived from other rocks that were subjected to high **temperatures**, high pressures, and chemically active fluids.

microbial loop—A micro-**food web** that works within (or along side) the classical **food web**. The smallest organisms, heterotrophic bacteria, use dissolved inorganic material directly as carbon and energy sources.

microscale system—**Weather** phenomena representing the smallest spatial subdivision of atmospheric circulation, such as a weak tornado. These systems have dimensions of 1 m to 1 km (3 ft to less than a mile) and last from seconds to an hour or so.

Milankovitch cycles—Systematic variations in the precession and tilt of the Earth's rotational axis and the eccentricity of its orbit about the Sun; affects the seasonal and latitudinal distribution of incoming solar radiation and influences **climate** fluctuations operating over tens of thousands to hundreds of thousands of years.

mixed layer—The surface layer of the ocean that is mixed by the action of waves and tides so that the waters are nearly isothermal and isohaline; underlain by a **pycnocline**.

mixed tide—An **astronomical tide** with two high waters (high tides) and two low waters (low tides) occurring during a tidal day and having a marked **diurnal inequality**.

mixotrophs—Marine organisms having characteristics of both plants (**autotrophs**) and animals (**heterotrophs**).

model—An approximate representation or simulation of a real system.

mud flat—A nearly level area of fine silt near the shore and an intertidal habitat for submerged aquatic vegetation and benthic animals.

N

NAO Index—A measure of the strength of the horizontal **air pressure gradient** between Iceland and the Azores in the North Atlantic; influences the **climate** of eastern North America and much of Europe and North Africa; varies from year to year and decade to decade.

neap tides—**Astronomical tides** that have the least monthly **tidal range** occurring at the first and third quarter phases of the Moon.

nekton—Pelagic animals that are free-swimmers such as fish, adult squid, turtles, and marine mammals.

neritic deposits—River-borne **lithogenous sediment** which settles along the continental margin.

neritic zone—Also called the **coastal zone**, the area seaward from shore to the **continental shelf** break at a depth of about 200 m (650 ft); includes the **intertidal zone**.

net primary production—The amount of organic matter produced by living organisms within a given volume or area in a given time, minus that which is consumed through **cellular respiration** by the organisms.

net production—The amount of organic matter produced during **photosynthesis** that exceeds the amount consumed in the process of **cellular respiration**.

neutrally stable system—A system that, following a disturbance, does not return to its initial state and is easily mixed.

new production—**Primary production** based on nutrients brought into the **ecosystem** by processes such as **upwelling** or winter mixing.

Newton's first law of motion—An object in constant straight-line motion or a rest remains that way unless acted upon by an unbalanced force.

non-point source of pollution—A broad area of the landscape that yields contaminants to the air or waterways.

nor'easter—Common contraction for northeaster, an intense **extratropical cyclone** that tracks along the East Coast of North America and is named for the direction from which its most destructive winds blow.

North Atlantic Oscillation (NAO)—A seesaw variation in **air pressure** between Iceland (the Icelandic low) and the Azores (the Bermuda-Azores subtropical high); influences the **climate** of eastern North America and much of Europe and North Africa over periods up to decades.

Northeast Snowfall Impact Scale (NESIS)—A rating system for snowstorms in the northeast U.S. based upon the amount of snowfall, the area affected by the snowstorm and the population in the path of the storm. Snowstorms are assigned to one of five categories: 1 (notable), 2 (significant), 3 (major), 4 (crippling), or 5 (extreme).

O

ocean acidification—With the continuing upward trend in the concentration of atmospheric carbon dioxide, concern is focused on the potential effects on the ocean's buffering ability and the viability of certain marine organisms. CO_2 that is absorbed by the ocean participates in chemical reactions that increase the acidity (lowers the **pH**) of ocean waters.

ocean basin—A large topographic depression occupied by ocean water.

ocean commons—The vast portion of the ocean that is beyond the control of any coastal nation and essentially free to all for unlimited exploitation.

Ocean Drilling Program (ODP)—Project that obtained samples of oceanic crust and **sediment** using the drill ship JOIDES *Resolution*; operated from 1983 to 2003.

ocean floor observatory—A long-term facility on the seafloor designed to collect data, perform experiments, and communicate information to scientists onshore.

ocean stewardship—Action taken by society to protect ocean resources for now and the future.

ocean trench—A long narrow deep depression in the ocean floor caused by subduction of tectonic plates.

oceanic crust—The outermost part of the **lithosphere** overlying the mantle and lying beneath the ocean; mostly composed of the fine-grained ferromagnesian igneous rock known as basalt.

open ocean convection—The sinking of relatively dense surface-ocean waters and the ascent of less dense ocean waters.

overfishing—A fish species is taken at a rate that exceeds the maximum catch that would allow reproduction to replace the population.

oxygen isotope analysis—A technique used to identify **climate** fluctuations of the past by examining the ratio of light and heavy isotopes of oxygen (O^{16} and O^{18}) found, for example, in shells extracted from deep-sea **sediment** cores.

P

pH—The acidity of water (or any other liquid) is expressed in terms of the hydrogen ion concentration.

pH scale—The **pH** increases from 0 to 14 as the hydrogen ion concentration decreases. Pure water has a **pH** of 7, which is considered neutral; a **pH** above 7 is increasingly alkaline whereas a **pH** below 7 is increasingly acidic. The **pH** scale is logarithmic; that is, each unit increment corresponds to a tenfold change in acidity.

Pacific Decadal Oscillation (PDO)—A long-lived variation in **climate** over the North Pacific Ocean and North America in which sea surface **temperatures** fluctuate between the north central Pacific and the west coast of North America; linked to changes in strength of the Aleutian low.

Paleocene-Eocene Thermal Maximum (PETM)—A period about 56 million years ago, near the transition between the Paleocene and Eocene epochs, where deep-ocean **temperatures** rose 6 to 7 Celsius degrees (11 to 13 Fahrenheit degrees), adding to an already warm Earth. The PETM, which spanned 170 thousand years, is thought to have been caused by massive amounts of methane released from submarine gas hydrate deposits

partial tides—Harmonic components that comprise the **astronomical tide** at any point. The periods of partial tides are derived from the tidal forces of the Moon and Sun.

partially mixed estuary—A partially isolated body of water where fresh water from rivers and streams mixes with seawater to the extent that stratification is weak and the **salinity** typically varies by less than 10 parts per thousand from bottom to top. Mixing is greater than in a **salt-wedge estuary** but not as great as in a **well-mixed estuary**.

particulate organic carbon (POC)—The carbon contained in organic particles that sink out of the ocean's surface layer and is then consumed by **zooplankton** or decomposed by bacteria and converted back to dissolved inorganic carbon.

pelagic deposit—Fine-grained **sediments** that accumulate over time on the deep-ocean floor.

pelagic zone—The open-ocean environment, divided into the **neritic zone** (seaward to a depth of 200 m) and the oceanic zone (depth greater than 200 m).

photic zone—The upper sunlit layer of the ocean where **photosynthesis** takes place.

photosynthesis—The process whereby **autotrophs** use light energy from the Sun to combine carbon dioxide from the **atmosphere** with water to produce sugar, a form of carbohydrate that contains a relatively large amount of energy.

physical pump—The physical process whereby carbon dioxide sinks deeply in the cold ocean water at high latitudes and is sequestered in the deep ocean for varying lengths of time.

phytoplankton—Microscopic unicellular algae and photosynthetic bacteria living in the ocean and responsible for considerable biological production.

pinnipeds—Marine mammals which have distinctive swimming flippers; species include seals, walruses and sea lions.

placer deposits—Relatively dense and resistant metals and gemstones left behind as a lag concentrate after the mechanical sorting action of river or ocean currents remove the less dense sand grains.

planetary albedo—The fraction (or percent) of incident solar radiation that is scattered and reflected back to space by the **Earth's planetary system**; satellite sensors indicate a planetary **albedo** of about 30%.

planetary-scale systems—**Weather** phenomena operating at the largest spatial scale of atmospheric circulation; includes the global wind belts (i.e., trade winds, westerlies, polar easterlies) and semipermanent pressure systems (e.g., **subtropical anticyclones**).

plankton—Plant and animal life found floating or drifting in the ocean and used as food by nearly all marine animals.

plate tectonics—The process by which the massive plates of the **lithosphere** are slowly driven across the face of the globe by huge convection currents in Earth's mantle.

point source of pollution—A discernible localized source, e.g. a conduit such as a pipe or chimney that transports contaminants to a waterway or the **atmosphere**.

polar amplification—An increase in the magnitude of a **climate** change with increasing latitude.

polar-orbiting satellite—Satellite in a relatively low-altitude orbit that passes near the north and south geographical poles. Earth rotates through the plane of the satellite's orbit, which is at an altitude of about 800 to 1000 km (500 to 620 mi).

poleward heat transport—Meridional flow of **latent heat** and **sensible heat** from tropical to middle and high latitudes in response to latitudinal imbalances in radiational heating and cooling; brought about by air mass exchange, storms, and ocean circulation.

pollution—Intentional or unintentional disturbance of the environment that adversely affects the wellbeing of organisms (including humans) directly or the natural processes upon which they depend.

precipitation—Water in frozen or unfrozen forms (rain, snow, drizzle, ice pellets, hail) that falls from clouds and reaches Earth's surface.

pressure—Force per unit area.

pressure gradient force—A three-dimensional force operating in a fluid that accelerates the fluid from regions of high **pressure** and toward regions of low **pressure**. See **air pressure gradient**.

primary production—The amount of organic matter synthesized by simple organisms from inorganic substances.

principle of constant proportions—The major constituents of **seawater** occur in the same relative concentrations throughout the ocean system.

producers—Simple marine organisms also called **autotrophs** that manufacture the food they need from the physical environment.

profiling float—A free-moving instrument package that measures vertical profiles of ocean water **temperature** and **salinity** plus current velocity.

progressive wave—Wind-driven waves that move through a body of water.

proxy climate data sources—Various environmental sensors from which scientists infer past **climate** information. These sources, acting as substitutes for actual **weather** instruments, include historical documents, deep-sea sediment cores, pollen profiles, tree growth rings, and glacial ice cores.

pycnocline—Layer of ocean water in which **density** increases rapidly with depth (due to vertical changes in **temperature** and/or **salinity**); in low and middle latitudes situated between the mixed layer and deep layer.

R

recreational fisheries—Fishing for sport rather than commercial purposes.

recruitment overfishing—Type of **overfishing** where adult fish are taken in such great numbers that too few survive to replenish the breeding stock.

red tide—A discoloration of surface ocean waters usually in the coastal zone caused by a high concentration of microscopic organisms (e.g., dinoflagellates); also known as a **harmful algal bloom (HAB)**.

reflection—Process whereby a portion of the radiation striking the interface between two different media (e.g., **atmosphere** and ocean) is redirected such that the angle of reflection equals the angle of incidence.

refraction—The bending of a **wave** in response to changing **wave** speed.

regenerated production—**Primary production** using recycled nutrients.

remote sensing—Acquisition of data on the properties of some object without the sensor being in direct contact with the object.

remotely operated vehicle (ROV)—A data-gathering submersible tethered to a ship by cables that transmit power and data.

residence time—The average length of time for a substance in a reservoir to be replaced completely.

resonance—A buildup of amplitude in a physical system when the frequency of an applied force is close to the natural frequency of the system.

rings—Large turbulent rotating warm-core and cold-core eddies that break off from the relatively swift **western boundary currents** (e.g., the Gulf Stream).

rip current—Whereas the shoreward transport of water occurs over broad areas of the surf zone, the return seaward flow of water tends to concentrate in narrow widely spaced belts often corresponding to depressions in the seafloor or breaks in sand bars. The narrow seaward flow of water must balance the broad shoreward flow of water so that the offshore flow occurs as a relatively swift surface or near-surface current.

rock cycle—A sequence of events/processes involving the formation, alteration, destruction, and reformation of **igneous**, **sedimentary**, and **metamorphic rocks** as a result of such processes as **erosion**, transportation, **deposition**, **lithification**, metamorphism, melting, and crystallization.

rogue wave—Unusually high wave that occurs in the open ocean as well as some coastal areas; may be the product of **constructive wave interference**.

S

Saffir-Simpson Hurricane Wind Scale—This scale identifies **hurricanes** in five categories (1 to 5) based on ranges of wind speed; a category five hurricane has the highest range of wind speeds and is most intense. The scale also describes damage potential at each category.

salinity—A measure of the quantity of dissolved salt in **seawater.**

salt marsh—Coastal wetland consisting of salt-tolerant grasses regularly covered with **seawater**.

salt-wedge estuary—An **estuary** where river inflow is swift and tidal currents are weak; the denser high-**salinity seawater** forms a distinct layer beneath the low-**salinity** river water.

sand spit—A finger-like ridge of sand or gravel that projects from the shore into the ocean.

scattering—The process by which small particles suspended in a medium such as air or water diffuse a portion of the incident radiation in all directions.

schooling—The characteristic behavior of many fish species to swim together in organized groups to avoid predation.

sea—The state of the surface of the ocean with regard to waves or swells.

sea wave—An oscillation on the ocean surface that propagates along the interface between the **atmosphere** and the ocean.

seamounts—A structure of volcanic origin rising more than 1 km above the seafloor.

seawall—A concrete embankment intended to protect **beaches**, roads, buildings, and shoreline cliffs from **erosion** by storm waves.

seawater—The water of the ocean, distinguished from fresh water by its higher **salinity**.

secondary production—The organic material produced in the growth of consumers (**heterotrophs**).

sediment—Particles of organic or inorganic origin that are transported from their place of origin and deposited by wind, water, or ice; typically in unconsolidated form.

sedimentary rock—Consolidated rock composed of any one or a combination of compacted and cemented fragments of rock and mineral grains, partially decomposed remains of dead plants and animals (e.g. shells, skeletons), and minerals precipitated from solution.

seiche—A rhythmic oscillation of water in an enclosed basin or partially enclosed coastal inlet; a type of **standing wave**.

semi-diurnal tide—An **astronomical tide** with two high waters (high tides) and two low waters (low tides) occurring during a tidal (lunar) day and having a small diurnal inequality.

sensible heating—Transport of **heat** energy from one location or object to another through conduction, convection, or both, which brings about **temperature** changes.

shallow-water wave—**Wave** in water shallower than the **wave-base** (half the **wavelength**).

shore—Land exposed at low tide up to the coastline.

shoreline—The boundary line between a water body and land, usually taken at mean high tide.

significant wave height—The average height of the tallest one-third of the waves observed in a patch of ocean.

siliceous ooze—**Pelagic deposit** made from the shells of silica-secreting organisms.

SLOSH (Sea, Lake, and Overland Surges from Hurricanes)—A numerical **model** which accurately predicts the location and height of a **storm surge**.

SOFAR channel—*SO*und *F*ixing *A*nd *R*anging; a zone at an ocean depth of about 1000 m (3300 ft) where the speed of sound reaches a minimum value.

solar altitude—The angle of the Sun above the horizon; varies from 0 degree (horizon) to 90 degrees (zenith).

sorting—The dynamic process by which sedimentary particles are separated by size; well-sorted **sediment** has a narrow range of particle sizes whereas poorly-sorted **sediment** has a broad range of particle sizes.

Southern Oscillation—Opposing swings of surface **air pressure** between the western and central tropical Pacific Ocean; associated with intense **El Niño** events.

specific heat—The amount of **heat** required to raise the **temperature** of 1 gram of a substance by 1 Celsius degree.

spring bloom—A dramatic increase in **phytoplankton** populations in the ocean which occurs as a consequence of more sunlight and abundant nutrients.

spring tide—An **astronomical tide** occurring twice each month at or near the times of new Moon and full Moon when the gravitational pull of the Sun reinforces that of the Moon, and having an unusually large or increased **tidal range**.

stable system— A system which tends to persist in its current state without changing and, following a disturbance, tends to return to its original state or condition.

standing wave—This is in contrast to the wind-driven **waves** that are **progressive waves** in that they move through a body of water. With wind-driven **waves**, crests and troughs travel along the water surface but with standing **waves**, crests alternate vertically with troughs but at fixed locations; a **seiche** is an example of a standing wave.

storm surge—An abnormal local rise in sea level accompanying a **tropical cyclone** or other intense storm system and whose height is the difference between the observed level of the sea surface and the level that would have occurred in the absence of the storm.

sub-polar gyres—Roughly circular surface current systems, smaller than their subtropical counterparts, occur at high latitudes of the Northern Hemisphere; the Alaska gyre in the far North Pacific and the gyre south of Greenland in the far North Atlantic. The counterclockwise surface winds in the Aleutian and Icelandic sub-polar low pressure systems drive the ocean currents in the sub-polar gyres.

subduction zone—A long, narrow zone at a **convergent plate boundary** where an oceanic plate descends beneath another plate, either oceanic or continental.

sublimation—The process by which water changes directly from a solid to a vapor without first becoming liquid.

submarine canyon—A steep-sided canyon below sea level in the continental shelf and slope.

submarine fan—A cone-shaped sedimentary deposit that accumulates on the continental slope and rise.

subtropical anticyclone—A massive semi-permanent high pressure system that occurs in subtropical latitudes of both hemispheres; a major control of **climate**.

subtropical gyres—Large-scale roughly circular surface current systems in the ocean basins. Depending on location, gyres are either subtropical or sub-polar and are the dominant type of flow within the ocean's mixed layer. The trade winds, on the equatorward flank of a **subtropical anticyclone**, and the westerlies, on the poleward flank of a **subtropical anticyclone**, drive the subtropical gyres, centered near 30 degrees latitude in the North and South Atlantic, the North and South Pacific, and the Indian Ocean.

sulfurous aerosols—The tiny droplets of sulfuric acid (H_2SO_4) and sulfate particles which form in the stratosphere when sulfur dioxide combines with moisture.

sunspot—A dark blotch on the face of the Sun, typically thousands of kilometers across that develops where an intense magnetic field suppresses the flow of gases transporting heat from the Sun's interior.

surf—A nearly continuous train of waves breaking along a shore.

surface tension—The attraction between molecules at or near the surface of a liquid.

swell—Large long-period ocean waves that radiate away from the region where they were generated by strong storm winds.

symbiotic relationship—A mutually beneficial association between organisms.

synoptic-scale systems—**Weather** phenomena operating at the continental or oceanic spatial scale, including migrating **tropical cyclones** and **extratropical cyclones**. These systems have dimensions of 100 to 10,000 km (60 to 6000 mi) and last from days to a week or so.

system—An interacting set of components that behave in an orderly way according to the laws of physics, chemistry, geology, and biology.

T

tektite—Black fragments of glass formed from rock which has been liquefied when a meteor strikes the Earth.

teleconnection—A linkage between **weather** changes occurring in widely separated regions of the globe.

temperature—A measure of the average kinetic energy of the individual atoms or molecules composing a substance.

tephra—**Sediments** derived from explosive volcanism.

terminal velocity—Constant downward-directed motion of a particle within a fluid due to a balance between gravity (directed downward) and fluid resistance (directed upward).

thermal inertia—Resistance to a change in **temperature**.

thermocline—A layer of water in which the **temperature** decreases rapidly with increasing depth (e.g., between the warmer mixed layer and the colder, deep layer in a thermally stratified ocean).

thermohaline circulation—Subsurface movement of water masses caused by **density** contrasts arising from differences in **temperature** and **salinity**.

tidal bore—In some coastal areas where the tidal range is relatively large and the flood tide enters a narrow bay or channel, a wall of turbulent water, usually less than a meter in height, forms and moves upstream in a river or shallow **estuary**.

tidal currents—Alternating horizontal movements of water accompanying the rise and fall of **astronomical tides**.

tidal day—The Moon-based day based on the interval of time between two successive passes of the Moon over a meridian (approximately 24 hours, 50 minutes).

tidal period—The elapsed time between successive high tides or successive low tides.

tidal range—The difference in height between consecutive high water (high tide) and low water (low tide).

tide pool—A volume of water left behind in a rock basin or other intertidal depression by an ebbing tide.

tombola—A **sand spit** linking an offshore island to the mainland or another island.

top predators—Organisms that occupy a high trophic level in a marine **food web**.

transform plate boundary—The region where adjacent tectonic plates slide laterally past one another.

transitional wave—A **wave** entering water having a depth of between one-twentieth and one-half of the **wavelength**.

transpiration—The process whereby water that is taken up from the soil by plant roots eventually escapes as vapor through the tiny pores (stomates) on the surface of green leaves.

trophic level—The feeding position of an organism within a **food chain** or **food web**.

tropical cyclone—Generic term for a non-frontal synoptic-scale cyclone originating over warm tropical or subtropical ocean waters with cyclonic surface wind circulation (e.g., **hurricane**, **tropical storm**).

tropical depression—A **tropical cyclone** in which the sustained surface wind is at least 37 km per hr (23 mph) but less than 63 km per hr (39 mph); an early stage in the development of a **hurricane**.

tropical disturbance—A discrete system of organized convection in the tropics or subtropics with a detectable center of low air pressure; the initial stage in the development of a **hurricane**.

tropical storm—A **tropical cyclone** having a sustained surface wind speed of 63 to 118 km per hr (39-73 mph).

troposphere—The lowest thermal layer of the **atmosphere**; where the **atmosphere** interfaces with the ocean, **cryosphere**, **lithosphere**, and **biosphere** and where most **weather** takes place.

tsunami—A rapidly propagating **shallow-water wave** that develops when a submarine earthquake, landslide or volcanic eruption disturbs deep ocean water; known to build to tremendous wave height in shoaling coastal areas.

turbidites—Sedimentary deposits produced by **turbidity currents**.

turbidity current—A **sediment**-water mixture denser than normal **seawater** that flows down slope to the deep-seafloor.

twilight zone—The intermediate zone of the ocean, below the **photic zone**, but above the greater depths where there is less biologic activity. Also known as the disphotic zone.

U

unstable system—A system which following a disturbance tends not to return to its original state or condition.

V

vertical migration—The daily migration of **zooplankton** from deep waters to the surface zone to feed on **phytoplankton**.

W

warm-core rings—Eddies having a core of relatively warm water that break off from an ocean surface current; viewed from above in the Northern Hemisphere, they rotate in a clockwise direction.

washover—Material deposited by high waves on the back side of a barrier island

water mass—A large, homogenous volume of ocean water featuring a characteristic range of **temperature** and **salinity**.

watershed—The geographical area drained by a river and all its tributaries.

wave—A regular oscillation that occurs in a solid, liquid, or gaseous medium as energy is transmitted through that medium.

wave-base—A depth of about one-half **wavelength**, where the diameter of the orbits of water particles in **waves** is essentially zero; the depth below which water is not affected by surface **waves**.

wave crest—The highest point reached by an oscillating water surface.

wave frequency—The number of **waves** passing a fixed point over an interval of time.

wave height—A measurement of the vertical distance between **wave crest** and **wave trough**.

wave period—A measurement of the time needed for two successive **wavelengths** to pass a fixed point.

wave trough—The lowest point in an oscillating water surface.

wavelength—The distance between successive **wave crests** (or equivalently, the distance between successive **wave troughs**).

weather—The state of the **atmosphere** at some place and time, described in terms of such variables as **temperature,** precipitation, cloud cover, and wind speed.

weathering—The physical disintegration, chemical decomposition, or solution of exposed rock which takes place where the **lithosphere** (mainly the crust) interfaces with the other Earth subsystems.

well-mixed estuary—An **estuary** where strong tidal currents dominate the inflow from rivers and thoroughly mix the fresh water and saltwater.

western boundary current—Generic term for a relatively strong and narrow flow of ocean water (current) that runs along the western edge of a major ocean basin; the Gulf Stream is an example.

wetland—Low-lying flat areas that are covered by water or have soils that are saturated with water for at least part of the year.

white smokers—Substances that precipitate from **hydrothermal vent** waters and accumulate in the form of conical chimneys. They are distinguished from **black smokers** by their lighter appearance and the types of minerals precipitated.

wind-waves—Sea waves that are the produced as the kinetic energy of the wind is transferred to surface waters.

Wilson cycles—Cycles of ocean basin spreading and closing operating over hundreds of millions of years.

Z

zooplankton—Single-celled and multi-cellular animals that drift passively with ocean currents.

INDEX

C

D

E

F

G

H

I

J

K

N

O

P

Q

R

S

T

U

V

W

Y

Z